P9-CDJ-441

NOT TO BE
TAKEN FROM
THE
LIBRARY!

Comprehensive Natural Products Chemistry

Comprehensive Natural Products Chemistry

Editors-in-Chief
Sir Derek Barton†
Texas A&M University, USA
Koji Nakanishi
Columbia University, USA

Executive Editor
Otto Meth-Cohn
University of Sunderland, UK

Volume 2
ISOPRENOIDS INCLUDING CAROTENOIDS AND STEROIDS

Volume Editor
David E. Cane
Brown University, USA

1999
ELSEVIER
AMSTERDAM – LAUSANNE – NEW YORK – OXFORD – SHANNON – SINGAPORE – TOKYO

Elsevier Science Ltd., The Boulevard, Langford Lane, Kidlington, Oxford, OX5 1GB, UK

First edition 1999

Library of Congress Cataloging-in-Publication Data
Comprehensive natural products chemistry / editors-in-chief, Sir Derek Barton, Koji Nakanishi ; executive editor, Otto Meth-Cohn. -- 1st ed.
p. cm.
Includes index.
Contents: v. 2. Isoprenoids including carotenoids and steroids / volume editor David E. Cane
1. Natural products. I. Barton, Derek, Sir, 1918-1998. II. Nakanishi, Koji, 1925- . III. Meth-Cohn, Otto.
QD415.C63 1999
547.7--dc21 98-15249

British Library Cataloguing in Publication Data
Comprehensive natural products chemistry
1. Organic compounds
I. Barton, Sir Derek, 1918-1998 II. Nakanishi Koji III. Meth-Cohn Otto
572.5

ISBN 0-08-042709-X (set : alk. paper)
ISBN 0-08-043154-2 (Volume 2 : alk. paper)

∞™ The paper used in this publication meets the minimum requirements of the American National Standard for Information Sciences—Permanence of Paper for Printed Library Materials, ANSI Z39.48–1984.

Typeset by BPC Digital Data Ltd., Glasgow, UK.
Printed and bound in Great Britain by BPC Wheatons Ltd., Exeter, UK.

Contents

Introduction

For many decades, Natural Products Chemistry has been the principal driving force for progress in Organic Chemistry.

In the past, the determination of structure was arduous and difficult. As soon as computing became easy, the application of X-ray crystallography to structural determination quickly surpassed all other methods. Supplemented by the equally remarkable progress made more recently by Nuclear Magnetic Resonance techniques, determination of structure has become a routine exercise. This is even true for enzymes and other molecules of a similar size. Not to be forgotten remains the progress in mass spectrometry which permits another approach to structure and, in particular, to the precise determination of molecular weight.

There have not been such revolutionary changes in the partial or total synthesis of Natural Products. This still requires effort, imagination and time. But remarkable syntheses have been accomplished and great progress has been made in stereoselective synthesis. However, the one hundred percent yield problem is only solved in certain steps in certain industrial processes. Thus there remains a great divide between the reactions carried out in living organisms and those that synthetic chemists attain in the laboratory. Of course Nature edits the accuracy of DNA, RNA, and protein synthesis in a way that does not apply to a multi-step Organic Synthesis.

Organic Synthesis has already a significant component that uses enzymes to carry out specific reactions. This applies particularly to lipases and to oxidation enzymes. We have therefore, given serious attention to enzymatic reactions.

No longer standing in the wings, but already on-stage, are the wonderful tools of Molecular Biology. It is now clear that multi-step syntheses can be carried out in one vessel using multiple cloned enzymes. Thus, Molecular Biology and Organic Synthesis will come together to make economically important Natural Products.

From these preliminary comments it is clear that Natural Products Chemistry continues to evolve in different directions interacting with physical methods, Biochemistry, and Molecular Biology all at the same time.

This new Comprehensive Series has been conceived with the common theme of "How does Nature make all these molecules of life?" The principal idea was to organize the multitude of facts in terms of Biosynthesis rather than structure. The work is not intended to be a comprehensive listing of natural products, nor is it intended that there should be any detail about biological activity. These kinds of information can be found elsewhere.

The work has been planned for eight volumes with one more volume for Indexes. As far as we are aware, a broad treatment of the whole of Natural Products Chemistry has never been attempted before. We trust that our efforts will be useful and informative to all scientific disciplines where Natural Products play a role.

D. H. R. Barton† K. Nakanishi O. Meth-Cohn

Preface

It is surprising indeed that this work is the first attempt to produce a "comprehensive" overview of Natural Products beyond the student text level. However, the awe-inspiring breadth of the topic, which in many respects is still only developing, is such as to make the job daunting to anyone in the field. Fools rush in where angels fear to tread and the particular fool in this case was myself, a lifelong enthusiast and reader of the subject but with no research base whatever in the field!

Having been involved in several of the *Comprehensive* works produced by Pergamon Press, this omission intrigued me and over a period of gestation I put together a rough outline of how such a work could be written and presented it to Pergamon. To my delight they agreed that the project was worthwhile and in short measure Derek Barton was approached and took on the challenge of fleshing out this framework with alacrity. He also brought his long-standing friend and outstanding contributor to the field, Koji Nakanishi, into the team. With Derek's knowledge of the whole field, the subject was broken down into eight volumes and an outstanding team of internationally recognised Volume Editors was appointed.

We used Derek's 80th birthday as a target for finalising the work. Sadly he died just a few months before reaching this milestone. This work therefore is dedicated to the memory of Sir Derek Barton, Natural Products being the area which he loved best of all.

OTTO METH-COHN
Executive Editor

SIR DEREK BARTON

Sir Derek Barton, who was Distinguished Professor of Chemistry at Texas A&M University and holder of the Dow Chair of Chemical Invention died on March 16, 1998 in College Station, Texas of heart failure. He was 79 years old and had been Chairman of the Executive Board of Editors for Tetrahedron Publications since 1979.

Barton was considered to be one of the greatest organic chemists of the twentieth century whose work continues to have a major influence on contemporary science and will continue to do so for future generations of chemists.

Derek Harold Richard Barton was born on September 8, 1918 in Gravesend, Kent, UK and graduated from Imperial College, London with the degrees of B.Sc. (1940) and Ph.D. (1942). He carried out work on military intelligence during World War II and after a brief period in industry, joined the faculty at Imperial College. It was an early indication of the breadth and depth of his chemical knowledge that his lectureship was in physical chemistry. This research led him into the mechanism of elimination reactions and to the concept of molecular rotation difference to correlate the configurations of steroid isomers. During a sabbatical leave at Harvard in 1949–1950 he published a paper on the "Conformation of the Steroid Nucleus" (*Experientia*, 1950, **6**, 316) which was to bring him the Nobel Prize in Chemistry in 1969, shared with the Norwegian chemist, Odd Hassel. This key paper (only four pages long) altered the way in which chemists thought about the shape and reactivity of molecules, since it showed how the reactivity of functional groups in steroids depends on their axial or equatorial positions in a given conformation. Returning to the UK he held Chairs of Chemistry at Birkbeck College and Glasgow University before returning to Imperial College in 1957, where he developed a remarkable synthesis of the steroid hormone, aldosterone, by a photochemical reaction known as the Barton Reaction (nitrite photolysis). In 1978 he retired from Imperial College and became Director of the Natural Products Institute (CNRS) at Gif-sur-Yvette in France where he studied the invention of new chemical reactions, especially the chemistry of radicals, which opened up a whole new area of organic synthesis involving Gif chemistry. In 1986 he moved to a third career at Texas A&M University as Distinguished Professor of Chemistry and continued to work on novel reactions involving radical chemistry and the oxidation of hydrocarbons, which has become of great industrial importance. In a research career spanning more than five decades, Barton's contributions to organic chemistry included major discoveries which have profoundly altered our way of thinking about chemical structure and reactivity. His chemistry has provided models for the biochemical synthesis of natural products including alkaloids, antibiotics, carbohydrates, and DNA. Most recently his discoveries led to models for enzymes which oxidize hydrocarbons, including methane monooxygenase.

The following are selected highlights from his published work:

The 1950 paper which launched Conformational Analysis was recognized by the Nobel Prize Committee as the key contribution whereby the third dimension was added to chemistry. This work alone transformed our thinking about the connection between stereochemistry and reactivity, and was later adapted from small molecules to macromolecules e.g., DNA, and to inorganic complexes.

Barton's breadth and influence is illustrated in "Biogenetic Aspects of Phenol Oxidation" (*Festschr. Arthur Stoll*, 1957, 117). This theoretical work led to many later experiments on alkaloid biosynthesis and to a set of rules for *ortho-para*-phenolic oxidative coupling which allowed the predication of new natural product systems before they were actually discovered and to the correction of several erroneous structures.

In 1960, his paper on the remarkably short synthesis of the steroid hormone aldosterone (*J. Am. Chem. Soc.*, 1960, **82**, 2641) disclosed the first of many inventions of new reactions—in this case nitrite photolysis—to achieve short, high yielding processes, many of which have been patented and are used worldwide in the pharmaceutical industry.

Moving to 1975, by which time some 500 papers had been published, yet another "Barton reaction" was born—"The Deoxygenation of Secondary Alcohols" (*J. Chem. Soc. Perkin Trans. 1*, 1975, 1574), which has been very widely applied due to its tolerance of quite hostile and complex local environments in carbohydrate and nucleoside chemistry. This reaction is the chemical counterpart to ribonucleotide→ deoxyribonucleotide reductase in biochemistry and, until the arrival of the Barton reaction, was virtually impossible to achieve.

In 1985, "Invention of a New Radical Chain Reaction" involved the generation of carbon radicals from carboxylic acids (*Tetrahedron*, 1985, **41**, 3901). The method is of great synthetic utility and has been used many times by others in the burgeoning area of radicals in organic synthesis.

These recent advances in synthetic methodology were remarkable since his chemistry had virtually no precedent in the work of others. The radical methodology was especially timely in light of the significant recent increase in applications for fine chemical syntheses, and Barton gave the organic community an entrée into what will prove to be one of the most important methods of the twenty-first century. He often said how proud he was, at age 71, to receive the ACS Award for Creativity in Organic Synthesis for work published in the preceding five years.

Much of Barton's more recent work is summarized in the articles "The Invention of Chemical Reactions—The Last 5 Years" (*Tetrahedron*, 1992, **48**, 2529) and "Recent Developments in Gif Chemistry" (*Pure Appl. Chem.*, 1997, **69**, 1941).

Working 12 hours a day, Barton's stamina and creativity remained undiminished to the day of his death. The author of more than 1000 papers in chemical journals, Barton also held many successful patents. In addition to the Nobel Prize he received many honors and awards including the Davy, Copley, and Royal medals of the Royal Society of London, and the Roger Adams and Priestley Medals of the American Chemical Society. He held honorary degrees from 34 universities. He was a Fellow of the Royal Societies of London and Edinburgh, Foreign Associate of the National Academy of Sciences (USA), and Foreign Member of the Russian and Chinese Academies of Sciences. He was knighted by Queen Elizabeth in 1972, received the Légion d'Honneur (Chevalier 1972; Officier 1985) from France, and the Order of the Rising Sun from the Emperor of Japan. In his long career, Sir Derek trained over 300 students and postdoctoral fellows, many of whom now hold major positions throughout the world and include some of today's most distinguished organic chemists.

For those of us who were fortunate to know Sir Derek personally there is no doubt that his genius and work ethic were unique. He gave generously of his time to students and colleagues wherever he traveled and engendered such great respect and loyalty in his students and co-workers, that major symposia accompanied his birthdays every five years beginning with the 60th, ending this year with two celebrations just before his 80th birthday.

With the death of Sir Derek Barton, the world of science has lost a major figure, who together with Sir Robert Robinson and Robert B. Woodward, the cofounders of *Tetrahedron*, changed the face of organic chemistry in the twentieth century.

Professor Barton is survived by his wife, Judy, and by a son, William from his first marriage, and three grandchildren.

A. I. SCOTT
Texas A&M University

Reprinted from *Tetrahedron*, 1998, **54**, 8847
Photograph courtesy of Library and Information Centre, Royal Society of Chemistry

Contributors to Volume 2

Dr. I. Abe
University of Shizuoka, School of Pharmaceutical Sciences, 52-1 Yata, Shizuoka 422-8002, Japan

Professor D. Arigoni
Laboratorium für Organische Chemie, Eidgenössiche Technische Hochschule, Universitätstrasse 16, CH-8092 Zürich, Switzerland

Dr. G. A. Armstrong
Institute for Plant Sciences/Plant Genetics, Eidgenössiche Technische Hochschule, Universitätstrasse 2, CH-8092 Zürich, Switzerland

Dr. M. H. Beale
IACR Long Ashton Research Station, Department of Agricultural Sciences, University of Bristol, Long Ashton, Bristol, BS18 9AF, UK

Mr. D. A. Bochar
Department of Biochemistry, Purdue University, 1153 Biochemistry Building, West Lafayette, IN 47907-1153, USA

Dr. B. R. Boettcher
Metabolic and Cardiovascular Diseases, Novartis Pharmaceuticals Corporation, 556 Morris Avenue, Summit, NJ 07901, USA

Professor D. E. Cane
Department of Chemistry, Box H, Brown University, Providence, RI 02912-9108, USA

Professor R. Croteau
Institute of Biological Chemistry, Washington State University, Pullman, WA 99164-6340, USA

Mr. J. A. Friesen
Department of Biochemistry, Purdue University, 1153 Biochemistry Building, West Lafayette, IN 47907-1153, USA

Professor M. H. Gelb
Department of Chemistry and Biochemistry, University of Washington, Box 351700, Seattle, WA 98195-1700, USA

Dr. G. Guan
Uniformed Services University of the Health Sciences, Department of Biochemistry, 4301 Jones Bridge Road, Bethesda, MD 20814, USA

Dr. T. M. Hohn
National Center for Agricultural Utilization Research, USDA/ARS, 1815 North University Street, Peoria, IL 61604, USA

Dr. G.-F. Jang
Department of Chemistry and Biochemistry, University of Washington, Box 351700, Seattle, WA 98195-1700, USA

Mr. T. Koyama
Department of Biochemistry and Engineering, Faculty of Engineering, Tohoku University, Aramaki, Aoba-ku, Sendai 980-77, Japan

Professor J. MacMillan
IACR Long Ashton Research Station, Department of Agricultural Sciences, University of Bristol, Long Ashton, Bristol, BS18 9AF, UK

Dr. P. McGeady
Department of Chemistry and Biochemistry, University of Washington, Box 351700, Seattle, WA 98195-1700, USA

Professor K. Ogura
Institute for Chemical Reaction Science, Tohoku University, 2-1-1 Katahira, Aoba-ku, Sendai 980, Japan

Professor K. Poralla
Eberhard-Karls-Universität Tübingen, Mikrobiologie/Biotechnologie im Biologisches Institut, Auf der Morgenstelle 28, D-72076 Tübingen, Germany

Professor G. D. Prestwich
Department of Medicinal Chemistry, University of Utah, 308 Skaggs Hall, Salt Lake City, UT 84112, USA

Professor V. W. Rodwell
Department of Biochemistry, Purdue University, 1153 Biochemistry Building, West Lafayette, IN 47907-1153, USA

Professor M. Rohmer
Institut Le Bel, Université Louis Pasteur, 4, rue Blaise Pascal, F-67070 Strasbourg, France

Dr. M. K. Schwarz
Affymax Research Institute, 4001 Miranda Avenue, Palo Alto, CA 94304, USA

Dr. I. Shechter
Uniformed Services University of the Health Sciences, Department of Biochemistry, 4301 Jones Bridge Road, Bethesda, MD 20814, USA

Dr. C. V. Stauffacher
Department of Biological Sciences, Purdue University, 1153 Biochemistry Building, West Lafayette, IN 47907-1392, USA

Dr. M. Wise
Institute of Biological Chemistry, Washington State University, Pullman, WA 99164-6340, USA

Dr. K. Yokoyama
Department of Chemistry and Biochemistry, University of Washington, Box 351700, Seattle, WA 98195-1700, USA

Abbreviations

The most commonly used abbreviations in *Comprehensive Natural Products Chemistry* are listed below. Please note that in some instances these may differ from those used in other branches of chemistry

A	adenine
ABA	abscisic acid
Ac	acetyl
ACAC	acetylacetonate
ACTH	adrenocorticotropic hormone
ADP	adenosine 5'-diphosphate
AIBN	2,2'-azobisisobutyronitrile
Ala	alanine
AMP	adenosine 5'-monophosphate
APS	adenosine 5'-phosphosulfate
Ar	aryl
Arg	arginine
ATP	adenosine 5'-triphosphate
B	nucleoside base (adenine, cylosine, guanine, thymine or uracil)
9-BBN	9-borabicyclo[3.3.1]nonane
BOC	*t*-butoxycarbonyl (or carbo-*t*-butoxy)
BSA	*N*,*O*-bis(trimethylsilyl)acetamide
BSTFA	*N*,*O*-bis(trimethylsilyl)trifluoroacetamide
Bu	butyl
Bu^n	*n*-butyl
Bu^i	isobutyl
Bu^s	*s*-butyl
Bu^t	*t*-butyl
Bz	benzoyl
CAN	ceric ammonium nitrate
CD	cyclodextrin
CDP	cytidine 5'-diphosphate
CMP	cytidine 5'-monophosphate
CoA	coenzyme A
COD	cyclooctadiene
COT	cyclooctatetraene
Cp	η^5-cyclopentadiene
Cp*	pentamethylcyclopentadiene
12-Crown-4	1,4,7,10-tetraoxacyclododecane
15-Crown-5	1,4,7,10,13-pentaoxacyclopentadecane
18-Crown-6	1,4,7,10,13,16-hexaoxacyclooctadecane
CSA	camphorsulfonic acid
CSI	chlorosulfonyl isocyanate
CTP	cytidine 5'-triphosphate
cyclic AMP	adenosine 3',5'-cyclic monophosphoric acid
CySH	cysteine
DABCO	1,4-diazabicyclo[2.2.2]octane
DBA	dibenz[*a*,*h*]anthracene
DBN	1,5-diazabicyclo[4.3.0]non-5-ene

DBU	1,8-diazabicyclo[5.4.0]undec-7-ene
DCC	dicyclohexylcarbodiimide
DEAC	diethylaluminum chloride
DEAD	diethyl azodicarboxylate
DET	diethyl tartrate (+ or -)
DHET	dihydroergotoxine
DIBAH	diisobutylaluminum hydride
Diglyme	diethylene glycol dimethyl ether (or bis(2-methoxyethyl)ether)
DiHPhe	2,5-dihydroxyphenylalanine
Dimsyl Na	sodium methylsulfinylmethide
DIOP	2,3-*O*-isopropylidene-2,3-dihydroxy-1,4-bis(diphenylphosphino)butane
dipt	diisopropyl tartrate (+ or -)
DMA	dimethylacetamide
DMAD	dimethyl acetylenedicarboxylate
DMAP	4-dimethylaminopyridine
DME	1,2-dimethoxyethane (glyme)
DMF	dimethylformamide
DMF-DMA	dimethylformamide dimethyl acetal
DMI	1,3-dimethyl-2-imidazalidinone
DMSO	dimethyl sulfoxide
DMTSF	dimethyl(methylthio)sulfonium fluoroborate
DNA	deoxyribonucleic acid
DOCA	deoxycorticosterone acetate
EADC	ethylaluminum dichloride
EDTA	ethylenediaminetetraacetic acid
EEDQ	*N*-ethoxycarbonyl-2-ethoxy-1,2-dihydroquinoline
Et	ethyl
EVK	ethyl vinyl ketone
FAD	flavin adenine dinucleotide
Fl	flavin
FMN	flavin mononucleotide
G	guanine
GABA	4-aminobutyric acid
GDP	guanosine 5'-diphosphate
GLDH	glutamate dehydrogenase
gln	glutamine
Glu	glutamic acid
Gly	glycine
GMP	guanosine 5'-monophosphate
GOD	glucose oxidase
G-6-P	glucose-6-phosphate
GTP	guanosine 5'-triphosphate
Hb	hemoglobin
His	histidine
HMPA	hexamethylphosphoramide (or hexamethylphosphorous triamide)
Ile	isoleucine
INAH	isonicotinic acid hydrazide
IpcBH	isopinocampheylborane
Ipc_2BH	diisopinocampheylborane
KAPA	potassium 3-aminopropylamide
K-Slectride	potassium tri-*s*-butylborohydride

LAH	lithium aluminum hydride
LAP	leucine aminopeptidase
LDA	lithium diisopropylamide
LDH	lactic dehydrogenase
Leu	leucine
LICA	lithium isopropylcyclohexylamide
L-Selectride	lithium tri-*s*-butylborohydride
LTA	lead tetraacetate
Lys	lysine
MCPBA	*m*-chloroperoxybenzoic acid
Me	methyl
MEM	methoxyethoxymethyl
MEM-Cl	ß-methoxyethoxymethyl chloride
Met	methionine
MMA	methyl methacrylate
MMC	methyl magnesium carbonate
MOM	methoxymethyl
Ms	mesyl (or methanesulfonyl)
MSA	methanesulfonic acid
MsCl	methanesulfonyl chloride
MVK	methyl vinyl ketone
NAAD	nicotinic acid adenine dinucleotide
NAD	nicotinamide adenine dinucleotide
NADH	nicotinamide adenine dinucleotide phosphate, reduced
NBS	*N*-bromosuccinimider
NMO	*N*-methylmorpholine *N*-oxide monohydrate
NMP	*N*-methylpyrrolidone
PCBA	*p*-chlorobenzoic acid
PCBC	*p*-chlorobenzyl chloride
PCBN	*p*-chlorobenzonitrile
PCBTF	*p*-chlorobenzotrifluoride
PCC	pyridinium chlorochromate
PDC	pyridinium dichromate
PG	prostaglandin
Ph	phenyl
Phe	phenylalanine
Phth	phthaloyl
PPA	polyphosphoric acid
PPE	polyphosphate ester (or ethyl *m*-phosphate)
Pr	propyl
Pr^i	isopropyl
Pro	proline
Py	pyridine
RNA	ribonucleic acid
Rnase	ribonuclease
Ser	serine
Sia_2BH	disiamylborane
TAS	tris(diethylamino)sulfonium
TBAF	tetra-*n*-butylammonium fluoroborate
TBDMS	*t*-butyldimethylsilyl
TBDMS-Cl	*t*-butyldimethylsilyl chloride
TBDPS	*t*-butyldiphenylsilyl
TCNE	tetracyanoethene

TES	triethylsilyl
TFA	trifluoracetic acid
TFAA	trifluoroacetic anhydride
THF	tetrahydrofuran
THF	tetrahydrofolic acid
THP	tetrahydropyran (or tetrahydropyranyl)
Thr	threonine
TMEDA	*N,N,N',N'*,tetramethylethylenediamine[1,2-bis(dimethylamino)ethane]
TMS	trimethylsilyl
TMS-Cl	trimethylsilyl chloride
TMS-CN	trimethylsilyl cyanide
Tol	toluene
TosMIC	tosylmethyl isocyanide
TPP	tetraphenylporphyrin
Tr	trityl (or triphenylmethyl)
Trp	tryptophan
Ts	tosyl (or *p*-toluenesulfonyl)
TTFA	thallium trifluoroacetate
TTN	thallium(III) nitrate
Tyr	tyrosine
Tyr-OMe	tyrosine methyl ester
U	uridine
UDP	uridine 5'-diphosphate
UMP	uridine 5'-monophosphate

Contents of All Volumes

An Historical Perspective of Natural Products Chemistry

KOJI NAKANISHI

Columbia University, New York, USA

To give an account of the rich history of natural products chemistry in a short essay is a daunting task. This brief outline begins with a description of ancient folk medicine and continues with an outline of some of the major conceptual and experimental advances that have been made from the early nineteenth century through to about 1960, the start of the modern era of natural products chemistry. Achievements of living chemists are noted only minimally, usually in the context of related topics within the text. More recent developments are reviewed within the individual chapters of the present volumes, written by experts in each field. The subheadings follow, in part, the sequence of topics presented in Volumes 1–8.

1. ETHNOBOTANY AND "NATURAL PRODUCTS CHEMISTRY"

Except for minerals and synthetic materials our surroundings consist entirely of organic natural products, either of prebiotic organic origins or from microbial, plant, or animal sources. These materials include polyketides, terpenoids, amino acids, proteins, carbohydrates, lipids, nucleic acid bases, RNA and DNA, etc. Natural products chemistry can be thought of as originating from mankind's curiosity about odor, taste, color, and cures for diseases. Folk interest in treatments for pain, for food-poisoning and other maladies, and in hallucinogens appears to go back to the dawn of humanity

For centuries China has led the world in the use of natural products for healing. One of the earliest health science anthologies in China is the Nei Ching, whose authorship is attributed to the legendary Yellow Emperor (thirtieth century BC), although it is said that the dates were backdated from the third century by compilers. Excavation of a Han Dynasty (206 BC–AD 220) tomb in Hunan Province in 1974 unearthed decayed books, written on silk, bamboo, and wood, which filled a critical gap between the dawn of medicine up to the classic Nei Ching; Book 5 of these excavated documents lists 151 medical materials of plant origin. Generally regarded as the oldest compilation of Chinese herbs is Shen Nung Pen Ts'ao Ching (Catalog of Herbs by Shen Nung), which is believed to have been revised during the Han Dynasty; it lists 365 materials. Numerous revisions and enlargements of Pen Ts'ao were undertaken by physicians in subsequent dynasties, the ultimate being the Pen Ts'ao Kang Mu (General Catalog of Herbs) written by Li Shih-Chen over a period of 27 years during the Ming Dynasty (1573–1620), which records 1898 herbal drugs and 8160 prescriptions. This was circulated in Japan around 1620 and translated, and has made a huge impact on subsequent herbal studies in Japan; however, it has not been translated into English. The number of medicinal herbs used in 1979 in China numbered 5267. One of the most famous of the Chinese folk herbs is the ginseng root *Panax ginseng*, used for health maintenance and treatment of various diseases. The active principles were thought to be the saponins called ginsenosides but this is now doubtful; the effects could well be synergistic between saponins, flavonoids, etc. Another popular folk drug, the extract of the Ginkgo tree, *Ginkgo biloba* L., the only surviving species of the Paleozoic era (250 million years ago) family which became extinct during the last few million years, is mentioned in the Chinese Materia Medica to have an effect in improving memory and sharpening mental alertness. The main constituents responsible for this are now understood to be ginkgolides and flavonoids, but again not much else is known. Clarifying the active constituents and mode of (synergistic) bioactivity of Chinese herbs is a challenging task that has yet to be fully addressed.

The Assyrians left 660 clay tablets describing 1000 medicinal plants used around 1900–400 BC, but the best insight into ancient pharmacy is provided by the two scripts left by the ancient Egyptians, who were masters of human anatomy and surgery because of their extensive mummification practices. The Edwin Smith Surgical Papyrus purchased by Smith in 1862 in Luxor (now in the New York Academy of Sciences collection), is one of the most important medicinal documents of the ancient Nile Valley, and describes the healer's involvement in surgery, prescription, and healing practices using plants, animals, and minerals. The Ebers Papyrus, also purchased by Edwin Smith in 1862, and then acquired by Egyptologist George Ebers in 1872, describes 800 remedies using plants, animals, minerals, and magic. Indian medicine also has a long history, possibly dating back to the second millennium BC. The Indian materia medica consisted mainly of vegetable drugs prepared from plants but also used animals, bones, and minerals such as sulfur, arsenic, lead, copper sulfate, and gold. Ancient Greece inherited much from Egypt, India, and China, and underwent a gradual transition from magic to science. Pythagoras (580–500 BC) influenced the medical thinkers of his time, including Aristotle (384–322 BC), who in turn affected the medical practices of another influential Greek physician Galen (129–216). The Iranian physician Avicenna (980–1037) is noted for his contributions to Aristotelian philosophy and medicine, while the German-Swiss physician and alchemist Paracelsus (1493–1541) was an early champion who established the role of chemistry in medicine.

The rainforests in Central and South America and Africa are known to be particularly abundant in various organisms of interest to our lives because of their rich biodiversity, intense competition, and the necessity for self-defense. However, since folk-treatments are transmitted verbally to the next generation via shamans who naturally have a tendency to keep their plant and animal sources confidential, the recipes tend to get lost, particularly with destruction of rainforests and the encroachment of "civilization." Studies on folk medicine, hallucinogens, and shamanism of the Central and South American Indians conducted by Richard Schultes (Harvard Botanical Museum, emeritus) have led to renewed activity by ethnobotanists, recording the knowledge of shamans, assembling herbaria, and transmitting the record of learning to the village.

Extracts of toxic plants and animals have been used throughout the world for thousands of years for hunting and murder. These include the various arrow poisons used all over the world. *Strychnos* and *Chondrodendron* (containing strychnine, etc.) were used in South America and called "curare," *Strophanthus* (strophantidine, etc.) was used in Africa, the latex of the upas tree *Antiaris toxicaria* (cardiac glycosides) was used in Java, while *Aconitum napellus*, which appears in Greek mythology (aconitine) was used in medieval Europe and Hokkaido (by the Ainus). The Colombian arrow poison is from frogs (batrachotoxins; 200 toxins have been isolated from frogs by B. Witkop and J. Daly at NIH). Extracts of *Hyoscyamus niger* and *Atropa belladonna* contain the toxic tropane alkaloids, for example hyoscyamine, belladonnine, and atropine. The belladonna berry juice (atropine) which dilates the eye pupils was used during the Renaissance by ladies to produce doe-like eyes (belladona means beautiful woman). The Efik people in Calabar, southeastern Nigeria, used extracts of the calabar bean known as esere (physostigmine) for unmasking witches. The ancient Egyptians and Chinese knew of the toxic effect of the puffer fish, fugu, which contains the neurotoxin tetrodotoxin (Y. Hirata, K. Tsuda, R. B. Woodward).

When rye is infected by the fungus *Claviceps purpurea*, the toxin ergotamine and a number of ergot alkaloids are produced. These cause ergotism or the "devil's curse," "St. Anthony's fire," which leads to convulsions, miscarriages, loss of arms and legs, dry gangrene, and death. Epidemics of ergotism occurred in medieval times in villages throughout Europe, killing tens of thousands of people and livestock; Julius Caesar's legions were destroyed by ergotism during a campaign in Gaul, while in AD 994 an estimated 50,000 people died in an epidemic in France. As recently as 1926, a total of 11,000 cases of ergotism were reported in a region close to the Urals. It has been suggested that the witch hysteria that occurred in Salem, Massachusetts, might have been due to a mild outbreak of ergotism. Lysergic acid diethylamide (LSD) was first prepared by A. Hofmann, Sandoz Laboratories, Basel, in 1943 during efforts to improve the physiological effects of the ergot alkaloids when he accidentally inhaled it. "On Friday afternoon, April 16, 1943," he wrote, "I was seized by a sensation of restlessness... ." He went home from the laboratory and "perceived an uninterrupted stream of fantastic dreams ... ." (*Helvetica Chimica Acta*).

Numerous psychedelic plants have been used since ancient times, producing visions, mystical fantasies (cats and tigers also seem to have fantasies?, see nepetalactone below), sensations of flying, glorious feelings in warriors before battle, etc. The ethnobotanists Wasson and Schultes identified "ololiqui," an important Aztec concoction, as the seeds of the morning glory *Rivea corymbosa* and gave the seeds to Hofmann who found that they contained lysergic acid amides similar to but less potent than LSD. Iboga, a powerful hallucinogen from the root of the African shrub *Tabernanthe iboga*, is used by the Bwiti cult in Central Africa who chew the roots to obtain relief from fatigue and hunger; it contains the alkaloid ibogamine. The powerful hallucinogen used for thousands of years by the American Indians, the peyote cactus, contains mescaline and other alkaloids. The Indian hemp plant, *Cannabis sativa*, has been used for making rope since 3000 BC, but when it is used for its pleasure-giving effects it is called cannabis and has been known in central Asia, China, India, and the Near East since ancient times. Marijuana, hashish (named after the Persian founder of the Assassins of the eleventh century, Hasan-e Sabbah), charas, ghanja, bhang, kef, and dagga are names given to various preparations of the hemp plant. The constituent responsible for the mind-altering effect is 1-tetrahydrocannabinol (also referred to as 9-THC) contained in 1%. R. Mechoulam (1930–, Hebrew University) has been the principal worker in the cannabinoids, including structure determination and synthesis of 9-THC (1964 to present); the Israeli police have also made a contribution by providing Mechoulam with a constant supply of marijuana. Opium (morphine) is another ancient drug used for a variety of pain-relievers and it is documented that the Sumerians used poppy as early as 4000 BC; the narcotic effect is present only in seeds before they are fully formed. The irritating secretion of the blister beetles, for example *Mylabris* and the European species *Lytta vesicatoria*, commonly called Spanish fly, was used medically as a topical skin irritant to remove warts but was also a major ingredient in so-called love potions (constituent is cantharidin, stereospecific synthesis in 1951, G. Stork, 1921–; prep. scale high-pressure Diels–Alder synthesis in 1985, W. G. Dauben, 1919–1996).

Plants have been used for centuries for the treatment of heart problems, the most important being the foxgloves *Digitalis purpurea* and *D. lanata* (digitalin, diginin) and *Strophanthus gratus* (ouabain). The bark of cinchona *Cinchona officinalis* (called quina-quina by the Indians) has been used widely among the Indians in the Andes against malaria, which is still one of the major infectious diseases; its most important alkaloid is quinine. The British protected themselves against malaria during the occupation of India through gin and tonic (quinine!). The stimulant coca, used by the Incas around the tenth century, was introduced into Europe by the conquistadors; coca beans are also commonly chewed in West Africa. Wine making was already practiced in the Middle East 6000–8000 years ago; Moors made date wines, the Japanese rice wine, the Vikings honey mead, the Incas maize chicha. It is said that the Babylonians made beer using yeast 5000–6000 years ago. As shown above in parentheses, alkaloids are the major constituents of the herbal plants and extracts used for centuries, but it was not until the early nineteenth century that the active principles were isolated in pure form, for example morphine (1816), strychnine (1817), atropine (1819), quinine (1820), and colchicine (1820). It was a century later that the structures of these compounds were finally elucidated.

2. DAWN OF ORGANIC CHEMISTRY, EARLY STRUCTURAL STUDIES, MODERN METHODOLOGY

The term "organic compound" to define compounds made by and isolated from living organisms was coined in 1807 by the Swedish chemist Jons Jacob Berzelius (1779–1848), a founder of today's chemistry, who developed the modern system of symbols and formulas in chemistry, made a remarkably accurate table of atomic weights and analyzed many chemicals. At that time it was considered that organic compounds could not be synthesized from inorganic materials *in vitro*. However, Friedrich Wöhler (1800–1882), a medical doctor from Heidelberg who was starting his chemical career at a technical school in Berlin, attempted in 1828 to make "ammonium cyanate," which had been assigned a wrong structure, by heating the two inorganic salts potassium cyanate and ammonium sulfate; this led to the unexpected isolation of white crystals which were identical to the urea from urine, a typical organic compound. This well-known incident marked the beginning of organic chemistry. With the preparation of acetic acid from inorganic material in 1845 by Hermann Kolbe (1818–1884) at Leipzig, the myth surrounding organic compounds, in which they were associated with some vitalism was brought to an end and organic chemistry became the chemistry of carbon compounds. The same Kolbe was involved in the development of aspirin, one of the earliest and most important success stories in natural products chemistry. Salicylic acid from the leaf of the wintergreen plant had long been used as a pain reliever, especially in treating arthritis and gout. The inexpensive synthesis of salicylic acid from sodium phenolate and carbon dioxide by Kolbe in 1859 led to the industrial production in 1893 by the Bayer Company of acetylsalicylic acid "aspirin," still one of the most popular drugs. Aspirin is less acidic than salicylic acid and therefore causes less irritation in the mouth, throat, and stomach. The remarkable mechanism of the anti-inflammatory effect of aspirin was clarified in 1974 by John Vane (1927–) who showed that it inhibits the biosynthesis of prostaglandins by irreversibly acetylating a serine residue in prostaglandin synthase. Vane shared the 1982 Nobel Prize with Bergström and Samuelsson who determined the structure of prostaglandins (see below).

In the early days, natural products chemistry was focused on isolating the more readily available plant and animal constituents and determining their structures. The course of structure determination in the 1940s was a complex, indirect process, combining evidence from many types of experiments. The first effort was to crystallize the unknown compound or make derivatives such as esters or 2,4-dinitrophenylhydrazones, and to repeat recrystallization until the highest and sharp melting point was reached, since prior to the advent of isolation and purification methods now taken for granted, there was no simple criterion for purity. The only chromatography was through special grade alumina (first used by M. Tswett in 1906, then reintroduced by R. Willstätter). Molecular weight estimation by the Rast method which depended on melting point depression of a sample/camphor mixture, coupled with Pregl elemental microanalysis (see below) gave the molecular formula. Functionalities such as hydroxyl, amino, and carbonyl groups were recognized on the basis of specific derivatization and crystallization, followed by redetermination of molecular formula; the change in molecular composition led to identification of the functionality. Thus, sterically hindered carbonyls, for example the 11-keto group of cortisone, or tertiary hydroxyls, were very difficult to pinpoint, and often had to depend on more searching experiments. Therefore, an entire paper describing the recognition of a single hydroxyl group in a complex natural product would occasionally appear in the literature. An oxygen function suggested from the molecular formula but left unaccounted for would usually be assigned to an ether.

Determination of C-methyl groups depended on Kuhn–Roth oxidation which is performed by drastic oxidation with chromic acid/sulfuric acid, reduction of excess oxidant with hydrazine, neutralization with alkali, addition of phosphoric acid, distillation of the acetic acid originating from the C-methyls, and finally its titration with alkali. However, the results were only approximate, since *gem*-dimethyl groups only yield one equivalent of acetic acid, while primary, secondary, and tertiary methyl groups all give different yields of acetic acid. The skeletal structure of polycyclic compounds were frequently deduced on the basis of dehydrogenation reactions. It is therefore not surprising that the original steroid skeleton put forth by Wieland and Windaus in 1928, which depended a great deal on the production of chrysene upon Pd/C dehydrogenation, had to be revised in 1932 after several discrepancies were found (they received the Nobel prizes in 1927 and 1928 for this "extraordinarily difficult structure determination," see below).

In the following are listed some of the Nobel prizes awarded for the development of methodologies which have contributed critically to the progress in isolation protocols and structure determination. The year in which each prize was awarded is preceded by "Np."

Fritz Pregl, 1869–1930, Graz University, Np 1923. Invention of carbon and hydrogen microanalysis. Improvement of Kuhlmann's microbalance enabled weighing at an accuracy of 1 μg over a 20 g range, and refinement of carbon and hydrogen analytical methods made it possible to perform analysis with 3–4 mg of sample. His microbalance and the monograph *Quantitative Organic Microanalysis* (1916) profoundly influenced subsequent developments in practically all fields of chemistry and medicine.

The Svedberg, 1884–1971, Uppsala, Np 1926. Uppsala was a center for quantitative work on colloids for which the prize was awarded. His extensive study on ultracentrifugation, the first paper of which was published in the year of the award, evolved from a spring visit in 1922 to the University of Wisconsin. The ultracentrifuge together with the electrophoresis technique developed by his student Tiselius, have profoundly influenced subsequent progress in molecular biology and biochemistry.

Arne Tiselius, 1902–1971, Ph.D. Uppsala (T. Svedberg), Uppsala, Np 1948. Assisted by a grant from the Rockefeller Foundation, Tiselius was able to use his early electrophoresis instrument to show four bands in horse blood serum, alpha, beta and gamma globulins in addition to albumin; the first paper published in 1937 brought immediate positive responses.

Archer Martin, 1910–, Ph.D. Cambridge; Medical Research Council, Mill Hill, and Richard Synge, 1914–1994, Ph.D. Cambridge; Rowett Research Institute, Food Research Institute, Np 1952. They developed chromatography using two immiscible phases, gas–liquid, liquid–liquid, and paper chromatography, all of which have profoundly influenced all phases of chemistry.

Frederick Sanger, 1918–, Ph.D. Cambridge (A. Neuberger), Medical Research Council, Cambridge, Np 1958 and 1980. His confrontation with challenging structural problems in proteins and nucleic acids led to the development of two general analytical methods, 1,2,4-fluorodinitrobenzene (DNP) for tagging free amino groups (1945) in connection with insulin sequencing studies, and the dideoxynucleotide method for sequencing DNA (1977) in connection with recombinant DNA. For the latter he received his second Np in chemistry in 1980, which was shared with Paul Berg (1926–, Stanford University) and Walter Gilbert (1932–, Harvard University) for their contributions, respectively, in recombinant DNA and chemical sequencing of DNA. The studies of insulin involved usage of DNP for tagging disulfide bonds as cysteic acid residues (1949), and paper chromatography introduced by Martin and Synge 1944. That it was the first elucidation of any protein structure lowered the barrier for future structure studies of proteins.

Stanford Moore, 1913–1982, Ph.D. Wisconsin (K. P. Link), Rockefeller, Np 1972; and William Stein, 1911–1980, Ph.D. Columbia (E. G. Miller); Rockefeller, Np 1972. Moore and Stein cooperatively developed methods for the rapid quantification of protein hydrolysates by combining partition chromatography, ninhydrin coloration, and drop-counting fraction collector, i.e., the basis for commercial amino acid analyzers, and applied them to analysis of the ribonuclease structure.

Bruce Merrifield, 1921–, Ph.D. UCLA (M. Dunn), Rockefeller, Np 1984. The concept of solid-phase peptide synthesis using porous beads, chromatographic columns, and sequential elongation of peptides and other chains revolutionized the synthesis of biopolymers.

High-performance liquid chromatography (HPLC), introduced around the mid-1960s and now coupled on-line to many analytical instruments, for example UV, FTIR, and MS, is an indispensable daily tool found in all natural products chemistry laboratories.

3. STRUCTURES OF ORGANIC COMPOUNDS, NINETEENTH CENTURY

The discoveries made from 1848 to 1874 by Pasteur, Kekulé, van't Hoff, Le Bel, and others led to a revolution in structural organic chemistry. Louis Pasteur (1822–1895) was puzzled about why the potassium salt of tartaric acid (deposited on wine casks during fermentation) was dextrorotatory while the sodium ammonium salt of racemic acid (also deposited on wine casks) was optically inactive although both tartaric acid and "racemic" acid had identical chemical compositions. In 1848, the 25 year old Pasteur examined the racemic acid salt under the microscope and found two kinds of crystals exhibiting a left- and right-hand relation. Upon separation of the left-handed and right-handed crystals, he found that they rotated the plane of polarized light in opposite directions. He had thus performed his famous resolution of a racemic mixture, and had demonstrated the phenomenon of chirality. Pasteur went on to show that the racemic acid formed two kinds of salts with optically active bases such as quinine; this was the first demonstration of diastereomeric resolution. From this work Pasteur concluded that tartaric acid must have an element of asymmetry within the molecule itself. However, a three-dimensional understanding of the enantiomeric pair was only solved 25 years later (see below). Pasteur's own interest shifted to microbiology where he made the crucial discovery of the involvement of "germs" or microorganisms in various processes and proved that yeast induces alcoholic fermentation, while other microorganisms lead to diseases; he thus saved the wine industries of France, originated the process known as "pasteurization," and later developed vaccines for rabies. He was a genius who made many fundamental discoveries in chemistry and in microbiology.

The structures of organic compounds were still totally mysterious. Although Wöhler had synthesized urea, an isomer of ammonium cyanate, in 1828, the structural difference between these isomers was not known. In 1858 August Kekulé (1829–1896; studied with André Dumas and C. A. Wurtz in Paris, taught at Ghent, Heidelberg, and Bonn) published his famous paper in Liebig's *Annalen der Chemie* on the structure of carbon, in which he proposed that carbon atoms could form C–C bonds with hydrogen and other atoms linked to them; his dream on the top deck of a London bus led him to this concept. It was Butlerov who introduced the term "structure theory" in 1861. Further, in 1865 Kekulé conceived the cyclo-hexa-1:3:5-triene structure for benzene (C_6H_6) from a dream of a snake biting its own tail. In 1874, two young chemists, van't Hoff (1852–1911, Np 1901) in Utrecht, and Le Bel (1847–1930) in Paris, who had met in 1874 as students of C. A. Wurtz, published the revolutionary three-dimensional (3D) structure of the tetrahedral carbon Cabcd to explain the enantiomeric behavior of Pasteur's salts. The model was welcomed by J. Wislicenus (1835–1902, Zürich, Würzburg, Leipzig) who in 1863 had demonstrated the enantiomeric nature of the two lactic acids found by Scheele in sour milk (1780) and by Berzelius in muscle tissue (1807). This model, however, was criticized by Hermann Kolbe (1818–1884, Leipzig) as an "ingenious but in reality trivial and senseless natural philosophy." After 10 years of heated controversy, the idea of tetrahedral carbon was fully accepted, Kolbe had died and Wislicenus succeeded him in Leipzig.

Emil Fischer (1852–1919, Np 1902) was the next to make a critical contribution to stereochemistry. From the work of van't Hoff and Le Bel he reasoned that glucose should have 16 stereoisomers. Fischer's doctorate work on hydrazines under Baeyer (1835–1917, Np 1905) at Strasbourg had led to studies of osazones which culminated in the brilliant establishment, including configurations, of the Fischer sugar tree starting from D-(+)-glyceraldehyde all the way up to the aldohexoses, allose, altrose, glucose, mannose, gulose, idose, galactose, and talose (from 1884 to 1890). Unfortunately Fischer suffered from the toxic effects of phenylhydrazine for 12 years. The arbitrarily but luckily chosen absolute configuration of D-(+)-glyceraldehyde was shown to be correct sixty years later in 1951 (Johannes-Martin Bijvoet, 1892–1980). Fischer's brilliant correlation of the sugars comprising the Fischer sugar tree was performed using the Kiliani (1855–1945)–Fischer method via cyanohydrin intermediates for elongating sugars. Fischer also made remarkable contributions to the chemistry of amino acids and to nucleic acid bases (see below).

4. STRUCTURES OF ORGANIC COMPOUNDS, TWENTIETH CENTURY

The early concept of covalent bonds was provided with a sound theoretical basis by Linus Pauling (1901–1994, Np 1954), one of the greatest intellects of the twentieth century. Pauling's totally interdisciplinary research interests, including proteins and DNA is responsible for our present understanding of molecular structures. His books *Introduction to Quantum Mechanics* (with graduate student E. B. Wilson, 1935) and *The Nature of the Chemical Bond* (1939) have had a profound effect on our understanding of all of chemistry.

The actual 3D shapes of organic molecules which were still unclear in the late 1940s were then brilliantly clarified by Odd Hassel (1897–1981, Oslo University, Np 1969) and Derek Barton (1918–1998, Np 1969). Hassel, an X-ray crystallographer and physical chemist, demonstrated by electron diffraction that cyclohexane adopted the chair form in the gas phase and that it had two kinds of bonds, "standing (axial)" and "reclining (equatorial)" (1943). Because of the German occupation of Norway in 1940, instead of publishing the result in German journals, he published it in a Norwegian journal which was not abstracted in English until 1945. During his 1949 stay at Harvard, Barton attended a seminar by Louis Fieser on steric effects in steroids and showed Fieser that interpretations could be simplified if the shapes ("conformations") of cyclohexane rings were taken into consideration; Barton made these comments because he was familiar with Hassel's study on *cis*- and *trans*-decalins. Following Fieser's suggestion Barton published these ideas in a four-page *Experientia* paper (1950). This led to the joint Nobel prize with Hassel (1969), and established the concept of conformational analysis, which has exerted a profound effect in every field involving organic molecules.

Using conformational analysis, Barton determined the structures of many key terpenoids such as ß-amyrin, lanosterol, cycloartenone, and cycloartenol (Birkbeck College). At Glasgow University (from 1955) he collaborated in a number of cases with Monteath Robertson (1900–1989) and established many challenging structures: limonin, glauconic acid, byssochlamic acid, and nonadrides. Barton was also associated with the Research Institute for Medicine and Chemistry (RIMAC), Cambridge, USA founded by the Schering company, where with J. M. Beaton, he produced 60 g of aldosterone at a time when the world supply of this important hormone was in mg quantities. Aldosterone synthesis ("a good problem") was achieved in 1961 by Beaton ("a good experimentalist") through a nitrite photolysis, which came to be known as the Barton reaction ("a good idea") (quotes from his 1991 autobiography published by the American Chemical Society). From Glasgow, Barton went on to Imperial College, and a year before retirement, in 1977 he moved to France to direct the research at ICSN at Gif-sur-Yvette where he explored the oxidation reaction selectivity for unactivated C–H. After retiring from ICSN he made a further move to Texas A&M University in 1986, and continued his energetic activities, including chairman of the *Tetrahedron* publications. He felt weak during work one evening and died soon after, on March 16, 1998. He was fond of the phrase "gap jumping" by which he meant seeking generalizations between facts that do not seem to be related: "In the conformational analysis story, one had to jump the gap between steroids and chemical physics" (from his autobiography). According to Barton, the three most important qualities for a scientist are "intelligence, motivation, and honesty." His routine at Texas A&M was to wake around 4 a.m., read the literature, go to the office at 7 a.m. and stay there until 7 p.m.; when asked in 1997 whether this was still the routine, his response was that he wanted to wake up earlier because sleep was a waste of time—a remark which characterized this active scientist approaching 80!

Robert B. Woodward (1917–1979, Np 1965), who died prematurely, is regarded by many as the preeminent organic chemist of the twentieth century. He made landmark achievements in spectroscopy, synthesis, structure determination, biogenesis, as well as in theory. His solo papers published in 1941–1942 on empirical rules for estimating the absorption maxima of enones and dienes made the general organic chemical community realize that UV could be used for structural studies, thus launching the beginning of the spectroscopic revolution which soon brought on the applications of IR, NMR, MS, etc. He determined the structures of the following compounds: penicillin in 1945 (through joint UK–USA collaboration, see Hodgkin), strychnine in 1948, patulin in 1949, terramycin, aureomycin, and ferrocene (with G. Wilkinson, Np 1973—shared with E. O. Fischer for sandwich compounds) in 1952, cevine in 1954 (with Barton Np 1966, Jeger and Prelog, Np 1975), magnamycin in 1956, gliotoxin in 1958, oleandomycin in 1960, streptonigrin in 1963, and tetrodotoxin in 1964. He synthesized patulin in 1950, cortisone and cholesterol in 1951, lanosterol, lysergic acid (with Eli Lilly), and strychnine in 1954, reserpine in 1956, chlorophyll in 1960, a tetracycline (with Pfizer) in 1962, cephalosporin in 1965, and vitamin B_{12} in 1972 (with A. Eschenmoser, 1925–, ETH Zürich). He derived biogenetic schemes for steroids in 1953 (with K. Bloch, see below), and for macrolides in 1956, while the Woodward–Hoffmann orbital symmetry rules in 1965 brought order to a large class of seemingly random cyclization reactions.

Another central figure in stereochemistry is Vladimir Prelog (1906–1998, Np 1975), who succeeded Leopold Ruzicka at the ETH Zürich, and continued to build this institution into one of the most active and lively research and discussion centers in the world. The core group of intellectual leaders consisted of P. Plattner (1904–1975), O. Jeger, A. Eschenmoser, J. Dunitz, D. Arigoni, and A. Dreiding (from Zürich University). After completing extensive research on alkaloids, Prelog determined the structures of nonactin, boromycin, ferrioxamins, and rifamycins. His seminal studies in the synthesis and properties of 8–12 membered rings led him into unexplored areas of stereochemisty and chirality. Together with Robert Cahn (1899–1981, London Chemical Society) and Christopher Ingold (1893–1970, University College, London; pioneering mechanistic interpretation of organic reactions), he developed the Cahn–Ingold–Prelog (CIP) sequence rules for the unambiguous specification of stereoisomers. Prelog was an excellent story teller, always had jokes to tell, and was respected and loved by all who knew him.

4.1 Polyketides and Fatty Acids

Arthur Birch (1915–1995) from Sydney University, Ph.D. with Robert Robinson (Oxford University), then professor at Manchester University and Australian National University, was one of the earliest chemists to perform biosynthetic studies using radiolabels; starting with polyketides he studied the biosynthesis of a variety of natural products such as the C_6–C_3–C_6 backbone of plant phenolics, polyene macrolides, terpenoids, and alkaloids. He is especially known for the Birch reduction of aromatic rings, metal–ammonia reductions leading to 19-norsteroid hormones and other important products (1942–) which were of industrial importance. Feodor Lynen (1911–1979, Np 1964) performed studies on the intermediary metabolism of the living cell that led him to the demonstration of the first step in a chain of reactions resulting in the biosynthesis of sterols and fatty acids.

Prostaglandins, a family of 20-carbon, lipid-derived acids discovered in seminal fluids and accessory genital glands of man and sheep by von Euler (1934), have attracted great interest because of their extremely diverse biological activities. They were isolated and their structures elucidated from 1963 by S. Bergström (1916–, Np 1982) and B. Samuelsson (1934–, Np 1982) at the Karolinska Institute, Stockholm. Many syntheses of the natural prostaglandins and their nonnatural analogues have been published.

Tetsuo Nozoe (1902–1996) who studied at Tohoku University, Sendai, with Riko Majima (1874–1962, see below) went to Taiwan where he stayed until 1948 before returning to Tohoku University. At National Taiwan University he isolated hinokitiol from the essential oil of *taiwanhinoki*. Remembering the resonance concept put forward by Pauling just before World War II, he arrived at the seven-membered nonbenzenoid aromatic structure for hinokitiol in 1941, the first of the troponoids. This highly original work remained unknown to the rest of the world until 1951. In the meantime, during 1945–1948, nonbenzenoid aromatic structures had been assigned to stipitatic acid (isolated by H. Raistrick) by Michael J. S. Dewar (1918–) and to the thujaplicins by Holger Erdtman (1902–1989); the term tropolones was coined by Dewar in 1945. Nozoe continued to work on and discuss troponoids, up to the night before his death, without knowing that he had cancer. He was a remarkably focused and warm scientist, working unremittingly. Erdtman (Royal Institute of Technology, Stockholm) was the central figure in Swedish natural products chemistry who, with his wife Aulin Erdtman (dynamic president of the Swedish Chemistry Society), worked in the area of plant phenolics.

As mentioned in the following and in the concluding sections, classical biosynthetic studies using radioactive isotopes for determining the distribution of isotopes has now largely been replaced by the use of various stable isotopes coupled with NMR and MS. The main effort has now shifted to the identification and cloning of genes, or where possible the gene clusters, involved in the biosynthesis of the natural product. In the case of polyketides (acyclic, cyclic, and aromatic), the focus is on the polyketide synthases.

4.2 Isoprenoids, Steroids, and Carotenoids

During his time as an assistant to Kekulé at Bonn, Otto Wallach (1847–1931, Np 1910) had to familiarize himself with the essential oils from plants; many of the components of these oils were compounds for which no structure was known. In 1891 he clarified the relations between 12 different monoterpenes related to pinene. This was summarized together with other terpene chemistry in book form in 1909, and led him to propose the "isoprene rule." These achievements laid the foundation for the future development of terpenoid chemistry and brought order from chaos.

The next period up to around 1950 saw phenomenal advances in natural products chemistry centered on isoprenoids. Many of the best natural products chemists in Europe, including Wieland, Windaus, Karrer, Kuhn, Butenandt, and Ruzicka contributed to this breathtaking pace. Heinrich Wieland (1877–1957) worked on the bile acid structure, which had been studied over a period of 100 years and considered to be one of the most difficult to attack; he received the Nobel Prize in 1927 for these studies. His friend Adolph Windaus (1876–1959) worked on the structure of cholesterol for which he also received the Nobel Prize in 1928. Unfortunately, there were chemical discrepancies in the proposed steroidal skeletal structure, which had a five-membered ring B attached to C-7 and C-9. J. D. Bernal, Mineralogical Museums, Cambridge University, who was examining the X-ray patterns of ergosterol (1932) noted that the dimensions were inconsistent with the Wieland–Windaus formula. A reinterpretation of the production of chrysene from sterols by Pd/C dehydrogenation reported by Diels (see below) in 1927 eventually led Rosenheim and King and Wieland and Dane to deduce the correct structure in 1932. Wieland also worked on the structures of morphine/strychnine alkaloids, phalloidin/amanitin cyclopeptides of toxic mushroom *Amanita phalloides*, and pteridines, the important fluorescent pigments of butterfly wings. Windaus determined the structure of ergosterol and continued structural studies of its irradiation product which exhibited antirachitic activity "vitamin D." The mechanistically complex photochemistry of ergosterol leading to the vitamin D group has been investigated in detail by Egbert Havinga (1927–1988, Leiden University), a leading photochemist and excellent tennis player.

Paul Karrer (1889–1971, Np 1937), established the foundations of carotenoid chemistry through structural determinations of lycopene, carotene, vitamin A, etc. and the synthesis of squalene, carotenoids, and others. George Wald (1906–1997, Np 1967) showed that vitamin A was the key compound in vision during his stay in Karrer's laboratory. Vitamin K (K from "Koagulation"), discovered by Henrik Dam (1895–1976, Polytechnic Institute, Copenhagen, Np 1943) and structurally studied by Edward Doisy (1893–1986, St. Louis University, Np 1943), was also synthesized by Karrer. In addition, Karrer synthesized riboflavin (vitamin B_2) and determined the structure and role of nicotinamide adenine dinucleotide phosphate ($NADP^+$) with Otto Warburg. The research on carotenoids and vitamins of Karrer who was at Zürich University overlapped with that of Richard Kuhn (1900–1967, Np 1938) at the ETH Zürich, and the two were frequently rivals. Richard Kuhn, one of the pioneers in using UV-vis spectroscopy for structural studies, introduced the concept of "atropisomerism" in diphenyls, and studied the spectra of a series of diphenyl polyenes. He determined the structures of many natural carotenoids, proved the structure of riboflavin-5-phosphate (flavin-adenine-dinucleotide-5-phosphate) and showed that the combination of NAD-5-phosphate with the carrier protein yielded the yellow oxidation enzyme, thus providing an understanding of the role of a prosthetic group. He also determined the structures of vitamin B complexes, i.e., pyridoxine, *p*-aminobenzoic acid, pantothenic acid. After World War II he went on to structural studies of nitrogen-containing oligosaccharides in human milk that provide immunity for infants, and brain gangliosides. Carotenoid studies in Switzerland were later taken up by Otto Isler (1910–1993), a Ruzicka student at Hoffmann-La Roche, and Conrad Hans Eugster (1921–), a Karrer student at Zürich University.

Adolf Butenandt (1903–1998, Np 1939) initiated and essentially completed isolation and structural studies of the human sex hormones, the insect molting hormone (ecdysone), and the first pheromone, bombykol. With help from industry he was able to obtain large supplies of urine from pregnant women for estrone, sow ovaries for progesterone, and 4,000 gallons of male urine for androsterone (50 mg, crystals). He isolated and determined the structures of two female sex hormones, estrone and progesterone, and the male hormone androsterone all during the period 1934–1939 (!) and was awarded the Nobel prize in 1939. Keen intuition and use of UV data and Pregl's microanalysis all played important roles. He was appointed to a professorship in Danzig at the age of 30. With Peter Karlson he isolated from 500 kg of silkworm larvae 25 mg of α-ecdysone, the prohormone of insect and crustacean molting hormone, and determined its structure as a polyhydroxysteroid (1965); 20-hydroxylation gives the insect and crustacean molting hormone or ß-ecdysone (20-hydroxyecdysteroid). He was also the first to isolate an insect pheromone, bombykol, from female silkworm moths (with E. Hecker). As president of the Max Planck Foundation, he strongly influenced the postwar rebuilding of German science.

The successor to Kuhn, who left ETH Zürich for Heidelberg, was Leopold Ruzicka (1887–1967, Np 1939) who established a close relationship with the Swiss pharmaceutical industry. His synthesis of the 17- and 15-membered macrocyclic ketones, civetone and muscone (the constituents of musk) showed that contrary to Baeyer's prediction, large alicyclic rings could be strainless. He reintroduced and refined the isoprene rule proposed by Wallach (1887) and determined the basic structures of many sesqui-, di-, and triterpenes, as well as the structure of lanosterol, the key intermediate in cholesterol biosynthesis. The "biogenetic isoprene rule" of the ETH group, Albert Eschenmoser, Leopold Ruzicka, Oskar Jeger, and Duilio Arigoni, contributed to a concept of terpenoid cyclization (1955), which was consistent with the mechanistic considerations put forward by Stork as early as 1950. Besides making the ETH group into a center of natural products chemistry, Ruzicka bought many seventeenth century Dutch paintings with royalties accumulated during the war from his Swiss and American patents, and donated them to the Zürich Kunsthaus.

Studies in the isolation, structures, and activities of the antiarthritic hormone, cortisone and related compounds from the adrenal cortex were performed in the mid- to late 1940s during World War II by Edward Kendall (1886–1972, Mayo Clinic, Rochester, Np 1950), Tadeus Reichstein (1897–1996, Basel University, Np 1950), Philip Hench (1896–1965, Mayo Clinic, Rochester, Np 1950), Oskar Wintersteiner (1898–1971, Columbia University, Squibb) and others initiated interest as an adjunct to military medicine as well as to supplement the meager supply from beef adrenal glands by synthesis. Lewis Sarett (1917–, Merck & Co., later president) and co-workers completed the cortisone synthesis in 28 steps, one of the first two totally stereocontrolled syntheses of a natural product; the other was cantharidin (Stork 1951) (see above). The multistep cortisone synthesis was put on the production line by Max Tishler (1906–1989, Merck & Co., later president) who made contributions to the synthesis of a number of drugs, including riboflavin. Besides working on steroid reactions/synthesis and antimalarial agents, Louis F. Fieser (1899–1977) and Mary Fieser (1909–1997) of Harvard University made huge contributions to the chemical community through their outstanding books *Natural Products related to Phenanthrene* (1949), *Steroids* (1959), *Advanced Organic Chemistry* (1961), and *Topics in Organic Chemistry* (1963), as well as their textbooks and an important series of books on Organic Reagents. Carl Djerassi (1923–, Stanford University), a prolific chemist, industrialist, and more recently a novelist, started to work at the Syntex laboratories in Mexico City where he directed the work leading to the first oral contraceptive ("the pill") for women.

Takashi Kubota (1909–, Osaka City University), with Teruo Matsuura (1924–, Kyoto University), determined the structure of the furanoid sesquiterpene, ipomeamarone, from the black rotted portion of spoiled sweet potatoes; this research constitutes the first characterization of a phytoallexin, defense substances produced by plants in response to attack by fungi or physical damage. Damaging a plant and characterizing the defense substances produced may lead to new bioactive compounds. The mechanism of induced biosynthesis of phytoallexins, which is not fully understood, is an interesting biological mechanistic topic that deserves further investigation. Another center of high activity in terpenoids and nucleic acids was headed by Frantisek Sorm (1913–1980, Institute of Organic and Biochemistry, Prague), who determined the structures of many sesquiterpenoids and other natural products; he was not only active scientifically but also was a central figure who helped to guide the careers of many Czech chemists.

The key compound in terpenoid biosynthesis is mevalonic acid (MVA) derived from acetyl-CoA, which was discovered fortuitously in 1957 by the Merck team in Rahway, NJ headed by Karl Folkers (1906–1998). They soon realized and proved that this C_6 acid was the precursor of the C_5 isoprenoid unit isopentenyl diphosphate (IPP) that ultimately leads to the biosynthesis of cholesterol. In 1952 Konrad Bloch (1912–, Harvard, Np 1964) with R. B. Woodward published a paper suggesting a mechanism of the cyclization of squalene to lanosterol and the subsequent steps to cholesterol, which turned out to be essentially correct. This biosynthetic path from MVA to cholesterol was experimentally clarified in stereochemical detail by John Cornforth (1917–, Np 1975) and George Popják. In 1932, Harold Urey (1893–1981, Np 1934) of Columbia University discovered heavy hydrogen. Urey showed, contrary to common expectation, that isotope separation could be achieved with deuterium in the form of deuterium oxide by fractional electrolysis of water. Urey's separation of the stable isotope deuterium led to the isotopic tracer methodology that revolutionized the protocols for elucidating biosynthetic processes and reaction mechanisms, as exemplified beautifully by the cholesterol studies. Using MVA labeled chirally with isotopes, including chiral methyl, i.e., -CHDT, Cornforth and Popják clarified the key steps in the intricate biosynthetic conversion of mevalonate to cholesterol in stereochemical detail. The chiral methyl group was also prepared independently by Duilio Arigoni (1928– , ETH, Zürich). Cornforth has had great difficulty in hearing and speech since childhood but has been helped expertly by his chemist wife Rita; he is an excellent tennis and chess player, and is renowned for his speed in composing occasional witty limericks.

Although MVA has long been assumed to be the only natural precursor for IPP, a non-MVA pathway in which IPP is formed via the glyceraldehyde phosphate-pyruvate pathway has been discovered (1995–1996) in the ancient bacteriohopanoids by Michel Rohmer, who started working on them with Guy Ourisson (1926–, University of Strasbourg, terpenoid studies, including prebiotic), and by Duilio Arigoni in the ginkgolides, which are present in the ancient *Ginkgo biloba* tree. It is possible that many other terpenoids are biosynthesized via the non-MVA route. In classical biosynthetic experiments, ^{14}C-labeled acetic acid was incorporated into the microbial or plant product, and location or distribution of the ^{14}C label was deduced by oxidation or degradation to specific fragments including acetic acid; therefore, it was not possible or extremely difficult to map the distribution of all radioactive carbons. The progress in ^{13}C NMR made it possible to incorporate ^{13}C-labeled acetic acid and locate all labeled carbons. This led to the discovery of the nonmevalonate pathway leading to the IPP units. Similarly, NMR and MS have made it possible to use the stable isotopes, e.g., ^{18}O, ^{2}H, ^{15}N, etc., in biosynthetic studies. The current trend of biosynthesis has now shifted to genomic approaches for cloning the genes of various enzyme synthases involved in the biosynthesis.

4.3 Carbohydrates and Cellulose

The most important advance in carbohydrate structures following those made by Emil Fischer was the change from acyclic to the current cyclic structure introduced by Walter Haworth (1883–1937). He noticed the presence of α- and ß-anomers, and determined the structures of important disaccharides including cellobiose, maltose, and lactose. He also determined the basic structural aspects of starch, cellulose, inulin, and other polysaccharides, and accomplished the structure determination and synthesis of vitamin C, a sample of which he had received from Albert von Szent-Györgyi (1893–1986, Np 1937). This first synthesis of a vitamin was significant since it showed that a vitamin could be synthesized in the same way as any other organic compound. There was strong belief among leading scientists in the 1910s that cellulose, starch, protein, and rubber were colloidal aggregates of small molecules. However, Hermann Staudinger (1881–1965, Np 1953) who succeeded R. Willstätter and H. Wieland at the ETH Zürich and Freiburg, respectively, showed through viscosity measurements and various molecular weight measurements that macromolecules do exist, and developed the principles of macromolecular chemistry.

In more modern times, Raymond Lemieux (1920–, Universities of Ottawa and Alberta) has been a leader in carbohydrate research. He introduced the concept of *endo*- and *exo*-anomeric effects, accomplished the challenging synthesis of sucrose (1953), pioneered in the use of NMR coupling constants in configuration studies, and most importantly, starting with syntheses of oligosaccharides responsible for human blood group determinants, he prepared antibodies and clarified fundamental aspects of the binding of oligosaccharides by lectins and antibodies. The periodate–potassium permanganate cleavage of double bonds at room temperature (1955) is called the Lemieux reaction.

4.4 Amino Acids, Peptides, Porphyrins, and Alkaloids

It is fortunate that we have China's record and practice of herbal medicine over the centuries, which is providing us with an indispensable source of knowledge. China is rapidly catching up in terms of infrastructure and equipment in organic and bioorganic chemistry, and work on isolation, structure determination, and synthesis stemming from these valuable sources has picked up momentum. However, as mentioned above, clarification of the active principles and mode of action of these plant extracts will be quite a challenge since in many cases synergistic action is expected. Wang Yu (1910–1997) who headed the well-equipped Shanghai Institute of Organic Chemistry surprised the world with the total synthesis of bovine insulin performed by his group in 1965; the human insulin was synthesized around the same time by P. G. Katsoyannis, A. Tometsko, and C. Zaut of the Brookhaven National Laboratory (1966).

One of the giants in natural products chemistry during the first half of this century was Robert Robinson (1886–1975, Np 1947) at Oxford University. His synthesis of tropinone, a bicyclic amino ketone related to cocaine, from succindialdehyde, methylamine, and acetone dicarboxylic acid under Mannich reaction conditions was the first biomimetic synthesis (1917). It reduced Willstätter's 1903 13-step synthesis starting with suberone into a single step. This achievement demonstrated Robinson's analytical prowess. He was able to dissect complex molecular structures into simple biosynthetic building blocks, which allowed him to propose the biogenesis of all types of alkaloids and other natural products. His laboratory at Oxford, where he developed the well-known Robinson annulation reaction (1937) in connection with his work on the synthesis of steroids became a world center for natural products study. Robinson was a pioneer in the so-called electronic theory of organic reactions, and introduced the use of curly arrows to show the movements of electrons. His analytical power is exemplified in the completion of the structure determination of strychnine in 1946. Barton clarified the biosynthetic route to the morphine alkaloids, which he saw as an extension of his biomimetic synthesis of usnic acid through a one-electron oxidation; this was later extended to a general phenolate coupling scheme. Morphine total synthesis was brilliantly achieved by Marshall Gates (1915–, University of Rochester) in 1952.

The yield of the Robinson tropinone synthesis was low but Clemens Schöpf (b.1899) , Ph.D. Munich (Wieland), Universität Darmstadt, improved it to 90% by carrying out the reaction in buffer; he also worked on the stereochemistry of morphine and determined the structure of the steroidal alkaloid salamandarine (1961), the toxin secreted from glands behind the eyes of the salamander.

Roger Adams (1889–1971, University of Illinois), was the central figure in organic chemistry in the USA and is credited with contributing to the rapid development of its chemistry in the late 1930s and 1940s, including training of graduate students for both academe and industry. After earning a Ph.D. in 1912 at Harvard University he did postdoctoral studies with Otto Diels (see below) and Richard Willstätter (see below) in 1913; he once said that around those years in Germany he could cover all *Journal of the American Chemical Society* papers published in a year in a single night. His important work include determination of the structures of tetrahydrocannabinol in marijuana, the toxic gossypol in cottonseed oil, chaulmoogric acid used in treatment of leprosy, and the Senecio alkaloids with Nelson Leonard (1916–, University of Illinois, now at Caltech). He also contributed to many fundamental organic reactions and syntheses. The famous Adams platinum catalyst is not only important for reducing double bonds in industry and in the laboratory, but was central for determining the number of double bonds in a structure. He was also one of the founders of the *Organic Synthesis* (started in 1921) and the *Organic Reactions* series. Nelson Leonard switched interests to bioorganic chemistry and biochemistry, where he has worked with nucleic acid bases and nucleotides, coenzymes, dimensional probes, and fluorescent modifications such as ethenoguanine.

The complicated structures of the medieval plant poisons aconitine (from *Aconitum*) and delphinine (from *Delphinium*) were finally characterized in 1959–1960 by Karel Wiesner (1919–1986, University of New Brunswick), Leo Marion (1899–1979, National Research Council, Ottawa), George Büchi (1921–, mycotoxins, aflatoxin/DNA adduct, synthesis of terpenoids and nitrogen-containing bioactive compounds, photochemistry), and Maria Przybylska (1923–, X-ray).

The complex chlorophyll structure was elucidated by Richard Willstätter (1872–1942, Np 1915). Although he could not join Baeyer's group at Munich because the latter had ceased taking students, a close relation developed between the two. During his chlorophyll studies, Willstätter reintroduced the important technique of column chromatography published in Russian by Michael Tswett (1906). Willstätter further demonstrated that magnesium was an integral part of chlorophyll, clarified the relation between chlorophyll and the blood pigment hemin, and found the wide distribution of carotenoids in tomato, egg yolk, and bovine corpus luteum. Willstätter also synthesized cyclooctatetraene and showed its properties to be wholly unlike benzene but close to those of acyclic polyenes (around 1913). He succeeded Baeyer at Munich in 1915, synthesized the anesthetic cocaine, retired early in protest of anti-Semitism, but remained active until the Hitler era, and in 1938 emigrated to Switzerland.

The hemin structure was determined by another German chemist of the same era, Hans Fischer (1881–1945, Np 1930), who succeeded Windaus at Innsbruck and at Munich. He worked on the structure of hemin from the blood pigment hemoglobin, and completed its synthesis in 1929. He continued Willstätter's structural studies of chlorophyll, and further synthesized bilirubin in 1944. Destruction of his institute at Technische Hochschule München, during World War II led him to take his life in March 1945. The biosynthesis of hemin was elucidated largely by David Shemin (1911–1991).

In the mid 1930s the Department of Biochemistry at Columbia Medical School, which had accepted many refugees from the Third Reich, including Erwin Chargaff, Rudolf Schoenheimer, and others on the faculty, and Konrad Bloch (see above) and David Shemin as graduate students, was a great center of research activity. In 1940, Shemin ingested 66 g of ^{15}N-labeled glycine over a period of 66 hours in order to determine the half-life of erythrocytes. David Rittenberg's analysis of the heme moiety with his home-made mass spectrometer showed all four pyrrole nitrogens came from glycine. Using ^{14}C (that had just become available) as a second isotope (see next paragraph), doubly labeled glycine $^{15}NH_2{}^{14}CH_2COOH$ and other precursors, Shemin showed that glycine and succinic acid condensed to yield δ-aminolevulinate, thus elegantly demonstrating the novel biosynthesis of the porphyrin ring (around 1950). At this time, Bloch was working on the other side of the bench.

Melvin Calvin (1911–1997, Np 1961) at University of California, Berkeley, elucidated the complex photosynthetic pathway in which plants reduce carbon dioxide to carbohydrates. The critical $^{14}CO_2$ had just been made available at Berkeley Lawrence Radiation Laboratory as a result of the pioneering research of Martin Kamen (1913–), while paper chromatography also played crucial roles. Kamen produced ^{14}C with Sam Ruben (1940), used ^{18}O to show that oxygen in photosynthesis comes from water and not from carbon dioxide, participated in the *Manhattan* project, testified before the House UnAmerican Activities Committee (1947), won compensatory damages from the US Department of State, and helped build the University of California, La Jolla (1957). The entire structure of the photosynthetic reaction center (>10 000 atoms) from the purple bacterium *Rhodopseudomonas viridis* has been established by X-ray crystallography in the landmark studies performed by Johann Deisenhofer (1943–), Robert Huber (1937–), and Hartmut Michel (1948–) in 1989; this was the first membrane protein structure determined by X-ray, for which they shared the 1988 Nobel prize. The information gained from the full structure of this first membrane protein has been especially rewarding.

The studies on vitamin B_{12}, the structure of which was established by crystallographic studies performed by Dorothy Hodgkin (1910–1994, Np 1964), are fascinating. Hodgkin also determined the structure of penicillin (in a joint effort between UK and US scientists during World War II) and insulin. The formidable total synthesis of vitamin B_{12} was completed in 1972 through collaborative efforts between Woodward and Eschenmoser, involving 100 postdoctoral fellows and extending over 10 years. The biosynthesis of fascinating complexity is almost completely solved through studies performed by Alan Battersby (1925–, Cambridge University), Duilio Arigoni, and Ian Scott (1928–, Texas A&M University) and collaborators where advanced NMR techniques and synthesis of labeled precursors is elegantly combined with cloning of enzymes controlling each biosynthetic step. This work provides a beautiful demonstration of the power of the combination of bioorganic chemistry, spectroscopy and molecular biology, a future direction which will become increasingly important for the creation of new "unnatural" natural products.

4.5 Enzymes and Proteins

In the early days of natural products chemistry, enzymes and viruses were very poorly understood. Thus, the 1926 paper by James Sumner (1887–1955) at Cornell University on crystalline urease was received with ignorance or skepticism, especially by Willstätter who believed that enzymes were small molecules and not proteins. John Northrop (1891–1987) and co-workers at the Rockefeller Institute went on to crystallize pepsin, trypsin, chymotrypsin, ribonuclease, deoyribonuclease, carboxypeptidase, and other enzymes between 1930 and 1935. Despite this, for many years biochemists did not recognize the significance of these findings, and considered enzymes as being low molecular weight compounds adsorbed onto proteins or colloids. Using Northrop's method for crystalline enzyme preparations, Wendell Stanley (1904–1971) at Princeton obtained tobacco mosaic virus as needles from one ton of tobacco leaves (1935). Sumner, Northrop, and Stanley shared the 1946 Nobel prize in chemistry. All these studies opened a new era for biochemistry.

Meanwhile, Linus Pauling, who in mid-1930 became interested in the magnetic properties of hemoglobin, investigated the configurations of proteins and the effects of hydrogen bonds. In 1949 he showed that sickle cell anemia was due to a mutation of a single amino acid in the hemoglobin molecule, the first correlation of a change in molecular structure with a genetic disease. Starting in 1951 he and colleagues published a series of papers describing the alpha helix structure of proteins; a paper published in the early 1950s with R. B. Corey on the structure of DNA played an important role in leading Francis Crick and James Watson to the double helix structure (Np 1962).

A further important achievement in the peptide field was that of Vincent Du Vigneaud (1901–1978, Np 1955), Cornell Medical School, who isolated and determined the structure of oxytocin, a posterior pituitary gland hormone, for which a structure involving a disulfide bond was proposed. He synthesized oxytocin in 1953, thereby completing the first synthesis of a natural peptide hormone.

Progress in isolation, purification, crystallization methods, computers, and instrumentation, including cyclotrons, have made X-ray crystallography the major tool in structural. Numerous structures including those of ligand/receptor complexes are being published at an extremely rapid rate. Some of the past major achievements in protein structures are the following. Max Perutz (1914, Np 1962) and John Kendrew (1914–1997, Np 1962), both at the Laboratory of Molecular Biology, Cambridge University, determined the structures of hemoglobin and myoglobin, respectively. William Lipscomb (1919–, Np 1976), Harvard University, who has trained many of the world's leaders in protein X-ray crystallography has been involved in the structure determination of many enzymes including carboxypeptidase A (1967); in 1965 he determined the structure of the anticancer bisindole alkaloid, vinblastine. Folding of proteins, an important but still enigmatic phenomenon, is attracting increasing attention. Christian Anfinsen (1916–1995, Np 1972), NIH, one of the pioneers in this area, showed that the amino acid residues in ribonuclease interact in an energetically most favorable manner to produce the unique 3D structure of the protein.

4.6 Nucleic Acid Bases, RNA, and DNA

The "Fischer indole synthesis" was first performed in 1886 by Emil Fischer. During the period 1881–1914, he determined the structures of and synthesized uric acid, caffeine, theobromine, xanthine, guanine, hypoxanthine, adenine, guanine, and made theophylline-D-glucoside phosphoric acid, the first synthetic nucleotide. In 1903, he made 5,5-diethylbarbituric acid or Barbital, Dorminal, Veronal, etc. (sedative), and in 1912, phenobarbital or Barbipil, Luminal, Phenobal, etc. (sedative). Many of his syntheses formed the basis of German industrial production of purine bases. In 1912 he showed that tannins are gallates of sugars such as maltose and glucose. Starting in 1899, he synthesized many of the 13 α-amino acids known at that time, including the L- and D-forms, which were separated through fractional crystallization of their salts with optically active bases. He also developed a method for synthesizing fragments of proteins, namely peptides, and made an 18-amino acid peptide. He lost his two sons in World War I, lost his wealth due to postwar inflation, believed he had terminal cancer (a misdiagnosis), and killed himself in July 1919. Fischer was a skilled experimentalist, so that even today, many of the reactions performed by him and his students are so delicately controlled that they are not easy to reproduce. As a result of his suffering by inhaling diethylmercury, and of the poisonous effect of phenylhydrazine, he was one of the first to design fume hoods. He was a superb teacher and was also influential in establishing the Kaiser Wilhelm Institute, which later became the Max Planck Institute. The number and quality of his accomplishments and contributions are hard to believe; he was truly a genius.

Alexander Todd (1907–1997, Np 1957) made critical contributions to the basic chemistry and synthesis of nucleotides. His early experience consisted of an extremely fruitful stay at Oxford in the Robinson group, where he completed the syntheses of many representative anthocyanins, and then at Edinburgh where he worked on the synthesis of vitamin B_1. He also prepared the hexacarboxylate of vitamin B_{12} (1954), which was used by D. Hodgkin's group for their X-ray elucidation of this vitamin (1956). M. Wiewiorowski (1918–), Institute for Bioorganic Chemistry, in Poznan, has headed a famous group in nucleic acid chemistry, and his colleagues are now distributed worldwide.

4.7 Antibiotics, Pigments, and Marine Natural Products

The concept of one microorganism killing another was introduced by Pasteur who coined the term antibiosis in 1877, but it was much later that this concept was realized in the form of an actual antibiotic. The bacteriologist Alexander Fleming (1881–1955, University of London, Np 1945) noticed that an airborne mold, a *Penicillium* strain, contaminated cultures of *Staphylococci* left on the open bench and formed a transparent circle around its colony due to lysis of *Staphylococci*. He published these results in 1929. The discovery did not attract much interest but the work was continued by Fleming until it was taken up further at Oxford University by pathologist Howard Florey (1898–1968, Np 1945) and biochemist Ernst Chain (1906–1979, Np 1945). The bioactivities of purified "penicillin," the first antibiotic, attracted serious interest in the early 1940s in the midst of World War II. A UK/USA team was formed during the war between academe and industry with Oxford University, Harvard University, ICI, Glaxo, Burroughs Wellcome, Merck, Shell, Squibb, and Pfizer as members. This project resulted in the large scale production of penicillin and determination of its structure (finally by X-ray, D. Hodgkin). John Sheehan (1915–1992) at MIT synthesized 6-aminopenicillanic acid in 1959, which opened the route for the synthesis of a number of analogues. Besides being the first antibiotic to be discovered, penicillin is also the first member of a large number of important antibiotics containing the ß-lactam ring, for example cephalosporins, carbapenems, monobactams, and nocardicins. The strained ß-lactam ring of these antibiotics inactivates the transpeptidase by acylating its serine residue at the active site, thus preventing the enzyme from forming the link between the pentaglycine chain and the D-Ala-D-Ala peptide, the essential link in bacterial cell walls. The overuse of ß-lactam antibiotics, which has given rise to the disturbing appearance of microbial resistant strains, is leading to active research in the design of synthetic ß-lactam analogues to counteract these strains. The complex nature of the important penicillin biosynthesis is being elucidated through efforts combining genetic engineering, expression of biosynthetic genes as well as feeding of synthetic precursors, etc. by Jack Baldwin (1938–, Oxford University), José Luengo (Universidad de León, Spain) and many other groups from industry and academe.

Shortly after the penicillin discovery, Selman Waksman (1888–1973, Rutgers University, Np 1952) discovered streptomycin, the second antibiotic and the first active against the dreaded disease tuberculosis. The discovery and development of new antibiotics continued throughout the world at pharmaceutical companies in Europe, Japan, and the USA from soil and various odd sources: cephalosporin from sewage in Sardinia, cyclosporin from Wisconsin and Norway soil which was carried back to Switzerland, avermectin from the soil near a golf course in Shizuoka Prefecture. People involved in antibiotic discovery used to collect soil samples from various sources during their trips but this has now become severely restricted to protect a country's right to its soil. M. M. Shemyakin (1908–1970, Institute of Chemistry of Natural Products, Moscow) was a grand master of Russian natural products who worked on antibiotics, especially of the tetracycline class; he also worked on cyclic antibiotics composed of alternating sequences of amides and esters and coined the term depsipeptide for these in 1953. He died in 1970 of a sudden heart attack in the midst of the 7th IUPAC Natural Products

Symposium held in Riga, Latvia, which he had organized. The Institute he headed was renamed the Shemyakin Institute.

Indigo, an important vat dye known in ancient Asia, Egypt, Greece, Rome, Britain, and Peru, is probably the oldest known coloring material of plant origin, Indigofera and Isatis. The structure was determined in 1883 and a commercially feasible synthesis was performed in 1883 by Adolf von Baeyer (see above, 1835–1917, Np 1905), who founded the German Chemical Society in 1867 following the precedent of the Chemistry Society of London. In 1872 Baeyer was appointed a professor at Strasbourg where E. Fischer was his student, and in 1875 he succeeded J. Liebig in Munich. Tyrian (or Phoenician) purple, the dibromo derivative of indigo which is obtained from the purple snail Murex bundaris, was used as a royal emblem in connection with religious ceremonies because of its rarity; because of the availability of other cheaper dyes with similar color, it has no commercial value today. K. Venkataraman (1901–1981, University of Bombay then National Chemical Laboratory) who worked with R. Robinson on the synthesis of chromones in his early career, continued to study natural and synthetic coloring matters, including synthetic anthraquinone vat dyes, natural quinonoid pigments, etc. T. R. Seshadri (1900–1975) is another Indian natural products chemist who worked mainly in natural pigments, dyes, drugs, insecticides, and especially in polyphenols. He also studied with Robinson, and with Pregl at Graz, and taught at Delhi University. Seshadri and Venkataraman had a huge impact on Indian chemistry. After a 40 year involvement, Toshio Goto (1929–1990) finally succeeded in solving the mysterious identity of commelinin, the deep-blue flower petal pigment of the Commelina communis isolated by Kozo Hayashi (1958) and protocyanin, isolated from the blue cornflower Centaurea cyanus by E. Bayer (1957). His group elucidated the remarkable structure in its entirety which consisted of six unstable anthocyanins, six flavones and two metals, the molecular weight approaching 10 000; complex stacking and hydrogen bonds were also involved. Thus the pigmentation of petals turned out to be far more complex than the theories put forth by Willstätter (1913) and Robinson (1931). Goto suffered a fatal heart attack while inspecting the first X-ray structure of commelinin; commelinin represents a pinnacle of current natural products isolation and structure determination in terms of subtlety in isolation and complexity of structure.

The study of marine natural products is understandably far behind that of compounds of terrestrial origin due to the difficulty in collection and identification of marine organisms. However, it is an area which has great potentialities for new discoveries from every conceivable source. One pioneer in modern marine chemistry is Paul Scheuer (1915–, University of Hawaii) who started his work with quinones of marine origin and has since characterized a very large number of bioactive compounds from mollusks and other sources. Luigi Minale (1936–1997, Napoli) started a strong group working on marine natural products, concentrating mainly on complex saponins. He was a leading natural products chemist who died prematurely. A. Gonzalez Gonzalez (1917–) who headed the Organic Natural Products Institute at the University of La Laguna, Tenerife, was the first to isolate and study polyhalogenated sesquiterpenoids from marine sources. His group has also carried out extensive studies on terrestrial terpenoids from the Canary Islands and South America. Carotenoids are widely distributed in nature and are of importance as food coloring material and as antioxidants (the detailed mechanisms of which still have to be worked out); new carotenoids continue to be discovered from marine sources, for example by the group of Synnove Liaaen-Jensen, Norwegian Institute of Technology). Yoshimasa Hirata (1915–), who started research at Nagoya University, is a champion in the isolation of nontrivial natural products. He characterized the bioluminescent luciferin from the marine ostracod *Cypridina hilgendorfii* in 1966 (with his students, Toshio Goto, Yoshito Kishi, and Osamu Shimomura); tetrodotoxin from the fugu fish in 1964 (with Goto and Kishi and co-workers), the structure of which was announced simultaneously by the group of Kyosuke Tsuda (1907–, tetrodotoxin, matrine) and Woodward; and the very complex palytoxin, $C_{129}H_{223}N_3O_{54}$ in 1981–1987 (with Daisuke Uemura and Kishi). Richard E. Moore, University of Hawaii, also announced the structure of palytoxin independently. Jon Clardy (1943–, Cornell University) has determined the X-ray structures of many unique marine natural products, including brevetoxin B (1981), the first of the group of toxins with contiguous *trans*-fused ether rings constituting a stiff ladder-like skeleton. Maitotoxin, $C_{164}H_{256}O_{68}S_2Na_2$, MW 3422, produced by the dinoflagellate *Gambierdiscus toxicus* is the largest and most toxic of the nonbiopolymeric toxins known; it has 32 alicyclic 6- to 8-membered ethereal rings and acyclic chains. Its isolation (1994) and complete structure determination was accomplished jointly by the groups of Takeshi Yasumoto (Tohoku University), Kazuo Tachibana and Michio Murata (Tokyo University) in 1996. Kishi, Harvard University, also deduced the full structure in 1996.

The well-known excitatory agent for the cat family contained in the volatile oil of catnip, *Nepeta cataria*, is the monoterpene nepetalactone, isolated by S. M. McElvain (1943) and structure determined by Jerrold Meinwald (1954); cats, tigers, and lions start purring and roll on their backs in response to this lactone. Takeo Sakan (1912–1993) investigated the series of monoterpenes neomatatabiols, etc.

from Actinidia, some of which are male lacewing attractants. As little as 1 fg of neomatatabiol attracts lacewings.

The first insect pheromone to be isolated and characterized was bombykol, the sex attractant for the male silkworm, *Bombyx mori* (by Butenandt and co-workers, see above). Numerous pheromones have been isolated, characterized, synthesized, and are playing central roles in insect control and in chemical ecology. The group at Cornell University have long been active in this field: Tom Eisner (1929–, behavior), Jerrold Meinwald (1927–, chemistry), Wendell Roeloff (1939–, electrophysiology, chemistry). Since the available sample is usually minuscule, full structure determination of a pheromone often requires total synthesis; Kenji Mori (1938–, Tokyo University) has been particularly active in this field. Progress in the techniques for handling volatile compounds, including collection, isolation, GC/MS, etc., has started to disclose the extreme complexity of chemical ecology which plays an important role in the lives of all living organisms. In this context, natural products chemistry will be play an increasingly important role in our grasp of the significance of biodiversity.

5. SYNTHESIS

Synthesis has been mentioned often in the preceding sections of this essay. In the following, synthetic methods of more general nature are described. The Grignard reaction of Victor Grignard (1871–1935, Np 1912) and then the Diels–Alder reaction by Otto Diels (1876–1954, Np 1950) and Kurt Alder (1902–1956, Np 1950) are extremely versatile reactions. The Diels–Alder reaction can account for the biosynthesis of several natural products with complex structures, and now an enzyme, a Diels–Alderase involved in biosynthesis has been isolated by Akitami Ichihara, Hokkaido University (1997).

The hydroboration reactions of Herbert Brown (1912–, Purdue University, Np 1979) and the Wittig reactions of Georg Wittig (1897–1987, Np 1979) are extremely versatile synthetic reactions. William S. Johnson (1913–1995, University of Wisconsin, Stanford University) developed efficient methods for the cyclization of acyclic polyolefinic compounds for the synthesis of corticoid and other steroids, while Gilbert Stork (1921–, Columbia University) introduced enamine alkylation, regiospecific enolate formation from enones and their kinetic trapping (called "three component coupling" in some cases), and radical cyclization in regio- and stereospecific constructions. Elias J. Corey (1928–, Harvard University, Np 1990) introduced the concept of retrosynthetic analysis and developed many key synthetic reactions and reagents during his synthesis of bioactive compounds, including prostaglandins and gingkolides. A recent development is the ever-expanding supramolecular chemistry stemming from 1967 studies on crown ethers by Charles Pedersen (1904–1989), 1968 studies on cryptates by Jean-Marie Lehn (1939–), and 1973 studies on host–guest chemistry by Donald Cram (1919–); they shared the chemistry Nobel prize in 1987.

6. NATURAL PRODUCTS STUDIES IN JAPAN

Since the background of natural products study in Japan is quite different from that in other countries, a brief history is given here. Natural products is one of the strongest areas of chemical research in Japan with probably the world's largest number of chemists pursuing structural studies; these are joined by a healthy number of synthetic and bioorganic chemists. An important Symposium on Natural Products was held in 1957 in Nagoya as a joint event between the faculties of science, pharmacy, and agriculture. This was the beginning of a series of annual symposia held in various cities, which has grown into a three-day event with about 50 talks and numerous papers; practically all achievements in this area are presented at this symposium. Japan adopted the early twentieth century German or European academic system where continuity of research can be assured through a permanent staff in addition to the professor, a system which is suited for natural products research which involves isolation and assay, as well as structure determination, all steps requiring delicate skills and much expertise.

The history of Japanese chemistry is short because the country was closed to the outside world up to 1868. This is when the Tokugawa shogunate which had ruled Japan for 264 years was overthrown and the Meiji era (1868–1912) began. Two of the first Japanese organic chemists sent abroad were Shokei Shibata and Nagayoshi Nagai, who joined the laboratory of A. W. von Hoffmann in Berlin. Upon return to Japan, Shibata (Chinese herbs) started a line of distinguished chemists, Keita and Yuji Shibata (flavones) and Shoji Shibata (1915–, lichens, fungal bisanthraquinonoid pigments, ginsenosides); Nagai returned to Tokyo Science University in 1884, studied ephedrine, and left a big mark in the embryonic era of organic chemistry. Modern natural products chemistry really began when three extraordinary organic chemists returned from Europe in the 1910s and started teaching and research at their respective faculties:

Riko Majima, 1874–1962, C. D. Harries (Kiel University); R. Willstätter (Zürich): Faculty of Science, Tohoku University; studied urushiol, the catecholic mixture of poison ivy irritant.

Yasuhiko Asahina, 1881–1975, R. Willstätter: Faculty of pharmacy, Tokyo University; lichens and Chinese herb.

Umetaro Suzuki, 1874–1943, E. Fischer: Faculty of agriculture, Tokyo University; vitamin B_1(thiamine).

Because these three pioneers started research in three different faculties (i.e., science, pharmacy, and agriculture), and because little interfaculty personnel exchange occurred in subsequent years, natural products chemistry in Japan was pursued independently within these three academic domains; the situation has changed now. The three pioneers started lines of first-class successors, but the establishment of a strong infrastructure takes many years, and it was only after the mid-1960s that the general level of science became comparable to that in the rest of the world; the 3rd IUPAC Symposium on the Chemistry of Natural Products, presided over by Munio Kotake (1894–1976, bufotoxins, see below), held in 1964 in Kyoto, was a clear turning point in Japan's role in this area.

Some of the outstanding Japanese chemists not already quoted are the following. Shibasaburo Kitazato (1852–1931), worked with Robert Koch (Np 1905, tuberculosis) and von Behring, antitoxins of diphtheria and tetanus which opened the new field of serology, isolation of microorganism causing dysentery, founder of Kitazato Institute; Chika Kuroda (1884–1968), first female Ph.D., structure of the complex carthamin, important dye in safflower (1930) which was revised in 1979 by Obara *et al.*, although the absolute configuration is still unknown (1998); Munio Kotake (1894–1976), bufotoxins, tryptophan metabolites, nupharidine; Harusada Suginome (1892–1972), aconite alkaloids; Teijiro Yabuta (1888–1977), kojic acid, gibberrelins; Eiji Ochiai (1898–1974), aconite alkaloids; Toshio Hoshino (1899–1979), abrine and other alkaloids; Yusuke Sumiki (1901–1974), gibberrelins; Sankichi Takei (1896–1982), rotenone; Shiro Akabori (1900–1992), peptides, C-terminal hydrazinolysis of amino acid ; Hamao Umezawa (1914–1986), kanamycin, bleomycin, numerous antibiotics; Shojiro Uyeo (1909–1988), lycorine; Tsunematsu Takemoto (1913–1989), inokosterone, kainic acid, domoic acid, quisqualic acid; Tomihide Shimizu (1889–1958), bile acids; Kenichi Takeda (1907–1991), Chinese herbs, sesquiterpenes; Yoshio Ban (1921–1994), alkaloid synthesis; Wataru Nagata (1922–1993), stereocontrolled hydrocyanation.

7. CURRENT AND FUTURE TRENDS IN NATURAL PRODUCTS CHEMISTRY

Spectroscopy and X-ray crystallography has totally changed the process of structure determination, which used to generate the excitement of solving a mystery. The first introduction of spectroscopy to the general organic community was Woodward's 1942–1943 empirical rules for estimating the UV maxima of dienes, trienes, and enones, which were extended by Fieser (1959). However, Butenandt had used UV for correctly determining the structures of the sex hormones as early as the early 1930s, while Karrer and Kuhn also used UV very early in their structural studies of the carotenoids. The Beckman DU instruments were an important factor which made UV spectroscopy a common tool for organic chemists and biochemists. With the availability of commercial instruments in 1950, IR spectroscopy became the next physical tool, making the 1950 Colthup IR correlation chart and the 1954 Bellamy monograph indispensable. The IR fingerprint region was analyzed in detail in attempts to gain as much structural information as possible from the molecular stretching and bending vibrations. Introduction of NMR spectroscopy into organic chemistry, first for protons and then for carbons, has totally changed the picture of structure determination, so that now IR is used much less frequently; however, in biopolymer studies, the techniques of difference FTIR and resonance Raman spectroscopy are indispensable.

The dramatic and rapid advancements in mass spectrometry are now drastically changing the protocol of biomacromolecular structural studies performed in biochemistry and molecular biology. Herbert Hauptman (mathematician, 1917–, Medical Foundation, Buffalo, Np 1985) and Jerome Karle (1918–, US Naval Research Laboratory, Washington, DC, Np 1985) developed direct methods for the determination of crystal structures devoid of disproportionately heavy atoms. The direct method together with modern computers revolutionized the X-ray analysis of molecular structures, which has become routine for crystalline compounds, large as well as small. Fred McLafferty (1923–, Cornell University) and Klaus Biemann (1926–, MIT) have made important contributions in the development of organic and bioorganic mass spectrometry. The development of cyclotron-based facilities for crystallographic biology studies has led to further dramatic advances enabling some protein structures to be determined in a single day, while cryoscopic electron micrography developed in 1975 by Richard Henderson and Nigel Unwin has also become a powerful tool for 3D structural determinations of membrane proteins such as bacteriorhodopsin (25 kd) and the nicotinic acetylcholine receptor (270 kd).

Circular dichroism (c.d.), which was used by French scientists Jean B. Biot (1774–1862) and Aimé Cotton during the nineteenth century "deteriorated" into monochromatic measurements at 589 nm after R.W. Bunsen (1811–1899, Heidelberg) introduced the Bunsen burner into the laboratory which readily emitted a 589 nm light characteristic of sodium. The 589 nm $[\alpha]_D$ values, remote from most chromophoric maxima, simply represent the summation of the low-intensity readings of the decreasing end of multiple Cotton effects. It is therefore very difficult or impossible to deduce structural information from $[\alpha]_D$ readings. Chiroptical spectroscopy was reintroduced to organic chemistry in the 1950s by C. Djerassi at Wayne State University (and later at Stanford University) as optical rotatory dispersion (ORD) and by L. Velluz and M. Legrand at Roussel-Uclaf as c.d. Günther Snatzke (1928–1992, Bonn then Ruhr University Bochum) was a major force in developing the theory and application of organic chiroptical spectroscopy. He investigated the chiroptical properties of a wide variety of natural products, including constituents of indigenous plants collected throughout the world, and established semiempirical sector rules for absolute configurational studies. He also established close collaborations with scientists of the former Eastern bloc countries and had a major impact in increasing the interest in c.d. there.

Chiroptical spectroscopy, nevertheless, remains one of the most underutilized physical measurements. Most organic chemists regard c.d. (more popular than ORD because interpretation is usually less ambiguous) simply as a tool for assigning absolute configurations, and since there are only two possibilities in absolute configurations, c.d. is apparently regarded as not as crucial compared to other spectroscopic methods. Moreover, many of the c.d. correlations with absolute configuration are empirical. For such reasons, chiroptical spectroscopy, with its immense potentialities, is grossly underused. However, c.d. curves can now be calculated nonempirically. Moreover, through-space coupling between the electric transition moments of two or more chromophores gives rise to intense Cotton effects split into opposite signs, exciton-coupled c.d.; fluorescence-detected c.d. further enhances the sensitivity by 50- to 100-fold. This leads to a highly versatile nonempirical microscale solution method for determining absolute configurations, etc.

With the rapid advances in spectroscopy and isolation techniques, most structure determinations in natural products chemistry have become quite routine, shifting the trend gradually towards activity-monitored isolation and structural studies of biologically active principles available only in microgram or submicrogram quantities. This in turn has made it possible for organic chemists to direct their attention towards clarifying the mechanistic and structural aspects of the ligand/biopolymeric receptor interactions on a more well-defined molecular structural basis. Until the 1990s, it was inconceivable and impossible to perform such studies.

Why does sugar taste sweet? This is an extremely challenging problem which at present cannot be answered even with major multidisciplinary efforts. Structural characterization of sweet compounds and elucidation of the amino acid sequences in the receptors are only the starting point. We are confronted with a long list of problems such as cloning of the receptors to produce them in sufficient quantities to investigate the physical fit between the active factor (sugar) and receptor by biophysical methods, and the time-resolved change in this physical contact and subsequent activation of G-protein and enzymes. This would then be followed by neurophysiological and ultimately physiological and psychological studies of sensation. How do the hundreds of taste receptors differ in their structures and their physical contact with molecules, and how do we differentiate the various taste sensations? The same applies to vision and to olfactory processes. What are the functions of the numerous glutamate receptor subtypes in our brain? We are at the starting point of a new field which is filled with exciting possibilities.

Familiarity with molecular biology is becoming essential for natural products chemists to plan research directed towards an understanding of natural products biosynthesis, mechanisms of bioactivity triggered by ligand–receptor interactions, etc. Numerous genes encoding enzymes have been cloned and expressed by the cDNA and/or genomic DNA-polymerase chain reaction protocols. This then leads to the possible production of new molecules by gene shuffling and recombinant biosynthetic techniques. Monoclonal catalytic antibodies using haptens possessing a structure similar to a high-energy intermediate of a proposed reaction are also contributing to the elucidation of biochemical mechanisms and the design of efficient syntheses. The technique of photoaffinity labeling, brilliantly invented by Frank Westheimer (1912–, Harvard University), assisted especially by advances in mass spectrometry, will clearly be playing an increasingly important role in studies of ligand–receptor interactions including enzyme–substrate reactions. The combined and sophisticated use of various spectroscopic means, including difference spectroscopy and fast time-resolved spectroscopy, will also become increasingly central in future studies of ligand–receptor studies.

Organic chemists, especially those involved in structural studies have the techniques, imagination, and knowledge to use these approaches. But it is difficult for organic chemists to identify an exciting and worthwhile topic. In contrast, the biochemists, biologists, and medical doctors are daily facing exciting life-related phenomena, frequently without realizing that the phenomena could be understood or at least clarified on a chemical basis. Broad individual expertise and knowledge coupled with multidisciplinary research collaboration thus becomes essential to investigate many of the more important future targets successfully. This approach may be termed "dynamic," as opposed to a "static" approach, exemplified by isolation and structure determination of a single natural product. Fortunately for scientists, nature is extremely complex and hence all the more challenging. Natural products chemistry will be playing an absolutely indispensable role for the future. Conservation of the alarming number of disappearing species, utilization of biodiversity, and understanding of the intricacies of biodiversity are further difficult, but urgent, problems confronting us.

That natural medicines are attracting renewed attention is encouraging from both practical and scientific viewpoints; their efficacy has often been proven over the centuries. However, to understand the mode of action of folk herbs and related products from nature is even more complex than mechanistic clarification of a single bioactive factor. This is because unfractionated or partly fractionated extracts are used, often containing mixtures of materials, and in many cases synergism is most likely playing an important role. Clarification of the active constituents and their modes of action will be difficult. This is nevertheless a worthwhile subject for serious investigations.

Dedicated to Sir Derek Barton whose amazing insight helped tremendously in the planning of this series, but who passed away just before its completion. It is a pity that he was unable to write this introduction as originally envisaged, since he would have had a masterful overview of the content he wanted, based on his vast experience. I have tried to fulfill his task, but this introduction cannot do justice to his original intention.

ACKNOWLEDGMENT

I am grateful to current research group members for letting me take quite a time off in order to undertake this difficult writing assignment with hardly any preparation. I am grateful to Drs. Nina Berova, Reimar Bruening, Jerrold Meinwald, Yoko Naya, and Tetsuo Shiba for their many suggestions.

8. BIBLIOGRAPHY

"A 100 Year History of Japanese Chemistry," Chemical Society of Japan, Tokyo Kagaku Dojin, 1978.
K. Bloch, *FASEB J.*, 1996, **10**, 802.
"Britannica Online," 1994–1998.
Bull. Oriental Healing Arts Inst. USA, 1980, **5**(7).
L. F. Fieser and M. Fieser, "Advanced Organic Chemistry," Reinhold, New York, 1961.
L. F. Fieser and M. Fieser, "Natural Products Related to Phenanthrene," Reinhold, New York, 1949.
M. Goodman and F. Morehouse, "Organic Molecules in Action," Gordon & Breach, New York, 1973.
L. K. James (ed.), "Nobel Laureates in Chemistry," American Chemical Society and Chemistry Heritage Foundation, 1994.
J. Mann, "Murder, Magic and Medicine," Oxford University Press, New York, 1992.
R. M. Roberts, "Serendipity, Accidental Discoveries in Science," Wiley, New York, 1989.
D. S. Tarbell and T. Tarbell, "The History of Organic Chemistry in the United States, 1875–1955," Folio, Nashville, TN, 1986.

2.01
Isoprenoid Biosynthesis: Overview

DAVID E. CANE
Brown University, Providence, RI, USA

2.01.1 INTRODUCTION

Isoprenoid natural products are found in essentially all forms of life.[1] Although the function of the vast majority of the tens of thousands of known isoprenoids is still largely unknown, the demonstrated functions include mediation of cell-wall and glycoprotein biosynthesis (dolichol diphosphates), electron transport and redox chemistry (ubiquinones), photooxidative protection and photosynthetic light harvesting (carotenoids), contribution to lipid membrane structure (cholesterol in eukaryotes, archaebacterial lipids), modification of proteins involved in signal transduction (prenylated proteins), modification of *t*-RNA (N^6-isopentenyladenine) intercellular signaling and developmental control (estrogens, gibberellins), interspecies defense (microbe–microbe, plant–microbe, plant–insect) as antibiotics and phytoalexins (trichothecin, capsidiol), to interspecies attack (e.g., fungal–plant) (trichothecanes and gibberellins), among a myriad of activities. Numerous other isoprenoids exhibit potent, medicinally useful, physiological activities whose relationship to their native function is still obscure, for example the widely used diterpene gingkolides (cardiovascular agent) and taxoids (antitumor compound) (Figure 1).

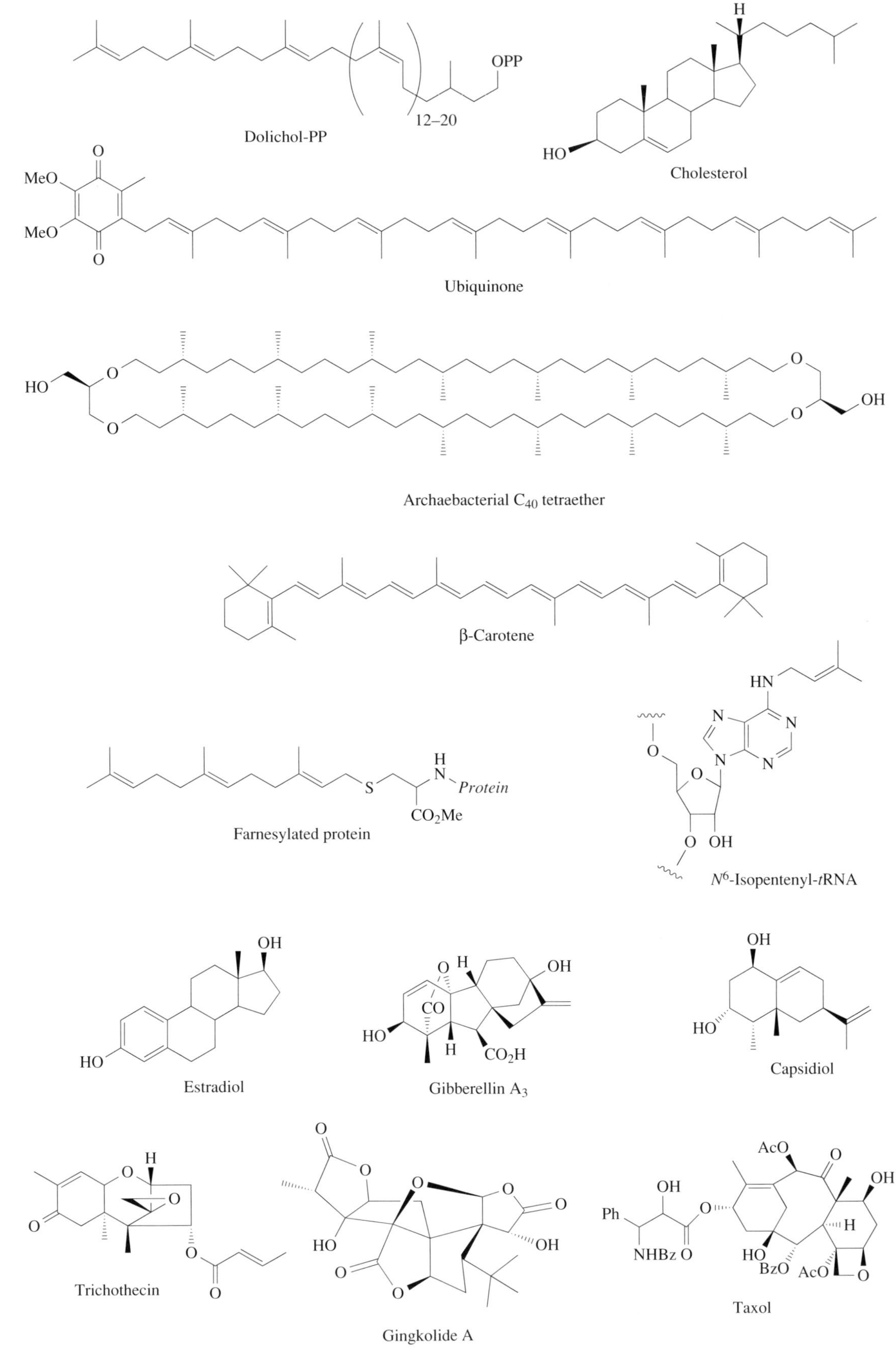

Dolichol-PP

Cholesterol

Ubiquinone

Archaebacterial C_{40} tetraether

β-Carotene

Farnesylated protein

N^6-Isopentenyl-*t*RNA

Estradiol

Gibberellin A_3

Capsidiol

Trichothecin

Gingkolide A

Taxol

Figure 1 Structures of physiologically important isoprenoids.

The isolation and structure determination of naturally occurring terpenes played a central role in the development of organic structural theory and studies of the chemistry of these complex substances contributed significantly to our understanding of reactivity and mechanism. From the very beginning, chemists have speculated about the origins of this diverse family of natural products. The early proposals of Wallach[2] and Ruzicka[3,4] ultimately found mechanistic expression in the proposal of the Biogenetic Isoprene Rule by Ruzicka in 1953,[5] marking the beginning of the modern era of isoprenoid biosynthetic study. Contemporaneous with Ruzicka's proposal were investigations of the biosynthesis of cholesterol being carried out in the laboratories of both Bloch[6] and Lynen[7]. These studies led to the isolation and identification of isopentenyl diphosphate (IPP, (**1**)). In fact, this simple five-carbon, branched-chain pyrophosphate ester, which has become known as the "biological isoprene unit," is the fundamental building block of all isoprenoid natural products.

2.01.2 BIOSYNTHESIS OF ISOPENTENYL DIPHOSPHATE

2.01.2.1 Mevalonic Acid

Early investigations of the formation of sterols in both yeast and vertebrates established that in these organisms isopentenyl diphosphate is derived from (3*R*)-3,5-dihydroxy-3-methylvaleric acid, known more commonly as mevalonic acid (**2**)[8,9] (Scheme 1). Following ATP-dependent formation of 5-pyrophosphomevalonate (**3**), coupled dehydration and decarboxylation of (**3**), presumably through the intermediacy of 3-phospho-5-pyrophosphomevalonate (**4**), gives rise to IPP (**1**). Mevalonate itself has been found to be formed by successive condensations of acetyl-CoA to generate first acetoacetyl-CoA (**5**) and then 3-hydroxy-3-methylglutaryl-CoA (HMG-CoA, (**6**)) (see Chapter 2.02). The latter metabolite undergoes reduction to mevalonic acid by a process requiring two equivalents of NADPH. Indeed, HMG-CoA reductase is a major regulatory enzyme of isoprenoid metabolism in a wide range of organisms and has served as a critical target in the control of cholesterol metabolism in humans, most notably by the fungal metabolite mevinolin and structurally related HMG-CoA reductase inhibitors. The enzymology of all three steps of mevalonate formation has been intensively studied using proteins from a wide range of sources. The formation of acetoacetyl-CoA (**5**) is an enzyme-catalyzed Claisen condensation, involving an intermediate acetyl-cysteine thioester as the electrophilic component, with generation of the nucleophilic partner requiring deprotonation of the α-carbon of the nucleophilic acetyl-CoA. The succeeding formation of HMG-CoA (**6**) is mechanistically analogous, involving nucleophilic attack of the enolate of an acetyl-*S*-enzyme derived from acetyl-CoA on the β-ketone moiety of acetoacetyl-CoA.

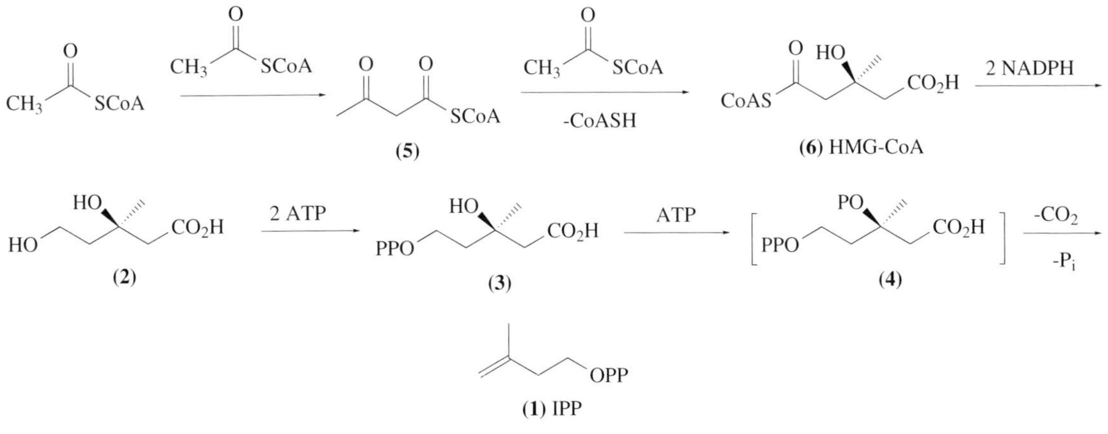

Scheme 1

2.01.2.2 The Deoxyxylulose Phosphate Pathway

Since the 1950s, extensive investigations of isoprenoid biosynthesis in a wide range of species, from yeast and fungi, through lower and higher plants, to vertebrates and higher mammals, have demonstrated the central role of mevalonic acid in the formation of every structural class of isoprenoids, from linear to cyclized mono-, sesqui-, and diterpenes, through sterols and polycyclic triterpenes, as well as carotenoids and prenylated proteins. Based on the successful incorporation of labeled mevalonic acid into isoprenoids of every structural class from a wide range of organisms, mevalonic acid became firmly accepted as the universal precursor of all isoprenoids. It was therefore particularly startling when two independent research groups each made the discovery that there is a second biochemical pathway to IPP, and that this newly discovered metabolic pathway involves neither acetate nor mevalonate (see Chapters 2.03 and 2.14). Moreover, far from being a metabolic rarity, this nonclassical route to IPP occurs widely in nature, being the primary or sole source of isoprenoids in prokaryotes, and existing side-by-side with the mevalonate pathway in higher plants. Archaebacteria, on the other hand, appear to utilize exclusively the mevalonate route. In the first step of the novel pathway to IPP, pyruvate (**7**) and glyceraldehyde-3-phosphate (**8**) have been shown to undergo a thiamine pyrophosphate-dependent coupling reaction to give a deoxypentulose, 1-deoxy-D-xylulose-5-phosphate (**9**) (Scheme 2). Strikingly, this same metabolite, either as the free alcohol or as the phosphate ester, has also been shown to be a key intermediate in the formation of two vitamins, pyridoxol phosphate (vitamin B_6)[10] and thiamine (vitamin B_1).[11] Although the role of 1-deoxyxylulose phosphate as a precursor of isoprenoids in bacteria and in the plastids of higher plants has been unambiguously established by a variety of labeling studies, the biochemical steps linking (**9**) and IPP (**1**) are still uncertain. Evidence has been presented implicating 2-*C*-methylerythrose, presumably as its 4-phosphate ester (**10**), as a rearranged intermediate which then undergoes reduction to the corresponding erythritol derivative (**11**). Formation of IPP from (**11**) would require an additional phosphorylation and two two-electron reduction steps. Experiments under way in several laboratories are likely to lead to the isolation of the remaining intermediates within a relatively brief period and result in the identification of the responsible enzymes and the corresponding structural genes as well.

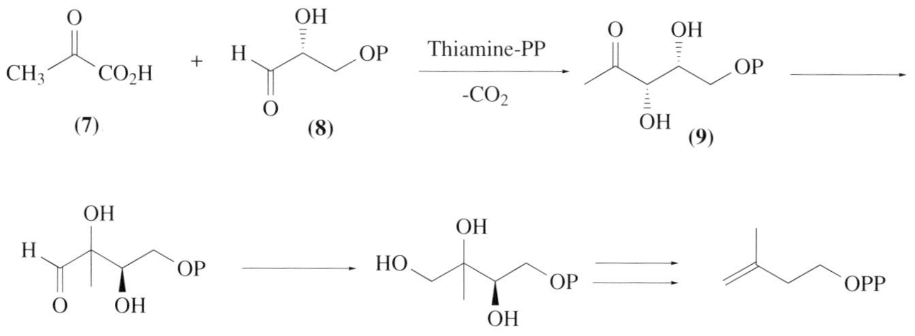

(**7**) + (**8**) —Thiamine-PP, $-CO_2$→ (**9**) → (**10**) → (**11**) → (**1**) IPP

Scheme 2

2.01.3 ISOPENTENYL DIPHOSPHATE METABOLISM AND THE BIOCHEMISTRY OF CARBOCATIONS

Up to the stage of isopentenyl diphosphate formation, either by the mevalonate or the deoxyxylulose phosphate pathway, all the biochemical reactions involved in the early stages of isoprenoid biosynthesis are more or less straightforward analogues of well-known reactions of central metabolism, with the major C—C bond-forming transformations being nucleophilic aldol, Claisen, and transketolase reactions, supplemented by modifications of functionality and oxidation level involving phosphoryl transfer, thioester hydrolysis, and nicotinamide-dependent reduction. Beginning with the isomerization of isopentenyl diphosphate to dimethylallyl diphosphate, however, the further stages of isoprenoid biosynthesis are dominated by the chemistry of carbocations, themselves generated by the attack of electrophilic species on carbon–carbon double bonds. This striking mechanistic and biochemical dichotomy was pointed out in 1968 by Cornforth in a classic review of olefin alkylation in biosynthesis.[12] Although a number of the specific suggestions that were put forward in the latter review on the relative timing of bond-making and bond-breaking events have had to be modified in light of subsequent results, the fundamental insights which were offered by

Cornforth on the nature of these enzyme-catalyzed carbocation–π reactions were remarkably prescient and remain a thought-provoking guide to these types of biochemical reactions.

The electrophilic reactions of isoprenoid biosynthesis can be divided into three major reaction classes according to the manner in which the initial carbocationic intermediate is generated: (1) by protonation of a double bond, (2) by protonation of an epoxide, or (3) by ionization of an allylic diphosphate ester. Each of these classes can be further subdivided according to the subsequent reactions of the initially generated cationic species.

2.01.3.1 Generation of Carbocations: Protonation of a Double Bond

2.01.3.1.1 Protonation–deprotonation

The conceptually most simple electrophilic reaction in isoprenoid biosynthesis is the prototropic rearrangement of isopentenyl diphosphate (IPP, (**1**)) to its isomer, dimethylallyl diphosphate (DMAPP, (**2**)) (Scheme 3) (see Chapter 2.04.2). This conversion, catalyzed by IPP isomerase, is initiated by the addition of a proton to C-4 of the IPP double bond to generate an intermediate carbocationic species (**13**) that then undergoes deprotonation at C-2 to generate DMAPP (**12**). Ingenious isotopic labeling experiments have demonstrated that this electrophilic allylic addition–elimination process takes place by attack of the proton on the *si* face of the C-3,4 double bond of IPP, followed by loss of the H-2*re* proton of (**13**), corresponding to net *anti* stereochemistry in the $S_{E'}$ reaction. Both active site labeling and site-directed mutagenesis experiments have implicated Glu-207 as an essential active site residue, plausibly acting as the Lewis acid that protonates the C-3,4 double bond of IPP.

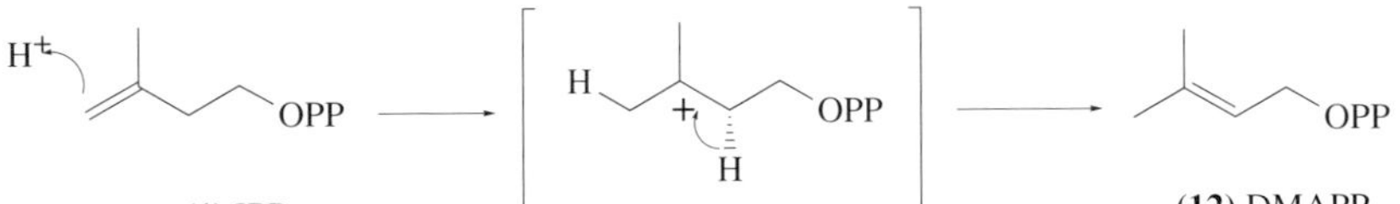

H^+ OPP (**1**) IPP → [H OPP H (**13**)] → OPP (**12**) DMAPP

Scheme 3

2.01.3.1.2 Protonation–olefin alkylation

Many types of isoprenoid cyclization reactions involve protonation of a double bond followed by capture of the resulting carbocation by a neighboring double bond. For example, in a key step in carotene biosynthesis, lycopene cyclase catalyzes protonation of the terminal trisubstituted double bond of lycopene (**14**) and electrophilic attack of the resulting cation on the neighboring C-5,6 double bond, followed by regiospecific deprotonation to generate a substituted cyclohexene. Cyclization of both ends of lycopene gives rise to β-carotene (**15**) (Scheme 4(a)) (see Chapter 2.13.5.2.4). Analogous cyclizations of cationic intermediates generated by double bond protonation and cyclization followed by electrophilic attack on one or more remaining double bonds can result in a cascade of cyclization reactions leading to the formation of polycyclic products. Thus (*E*,*E*,*E*)-geranylgeranyl diphosphate (GGPP, (**16**)) is cyclized to the bicyclic diterpene *ent*-copalyl diphosphate (**17**) by a reaction sequence initiated by protonation of the terminal double bond, followed by a successive pair of electrophilic cyclizations and deprotonation of a methyl group to yield (**17**) (Scheme 4(b)) (see Chapter 2.08). Similarly, hopene cyclase (hopene synthase) catalyzes protonation of the terminal double bond of the acyclic triterpene squalene (**18**) to initiate a cascade of cyclizations that result in the formation of the pentacyclic hydrocarbon hopene (**19**) containing nine contiguous stereogenic centers (Scheme 4(c)) (see Chapter 2.12.4). In some cases, protonation of a double bond results in further cyclization of an enzyme-bound intermediate. Thus protonation of germacrene A (**20**), which is an intermediate of numerous sesquiterpene cyclizations, initiates further cyclization to eudesmane or rearranged eremophilene products such as *epi*-aristolochene (**21**) (Scheme 4(d)) (see Chapters 2.06.3.10 to 2.06.3.12). Each of these cyclizations involves net *anti* additions, in which the electrophile, be it a proton or another carbocation, adds to the face of a double bond opposite that to which the neighboring double bond or external nucleophile becomes attached. In none of the examples cited has the amino acid residue that serves as the Lewis acid in these cyclizations been definitively identified, although X-ray crystallographic results obtained on *epi*-aristolochene synthase suggest a possible role for protonated aspartate-444.

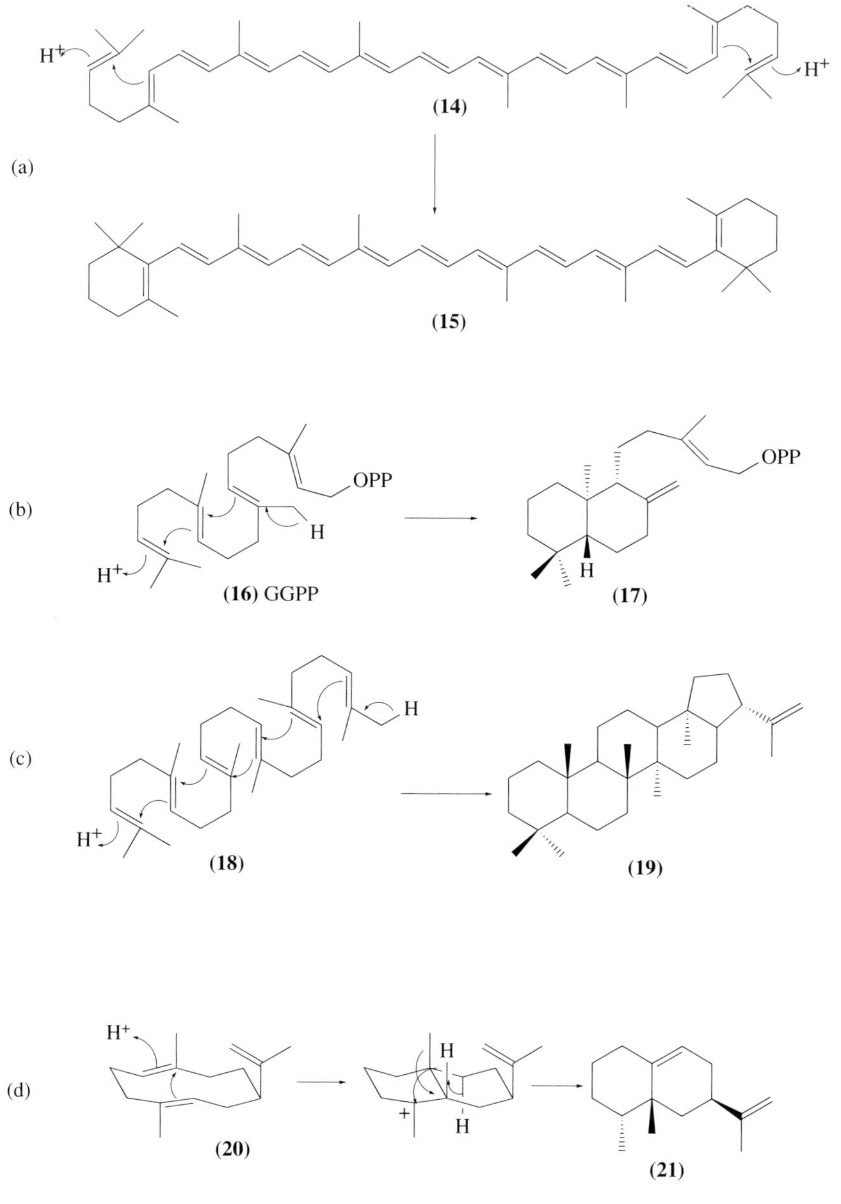

(a) H+ (14) H+ (15)

(b) H+ (16) GGPP OPP H (17) OPP H

(c) H+ (18) H (19)

(d) H+ (20) H + H (21)

Scheme 4

2.01.3.2 Generation of Carbocations: Protonation of an Epoxide

Squalene oxide (2,3-oxidosqualene, (**22**)) serves as the precursor of a large number of hydroxyl-bearing sterols and triterpenes (see Chapters 2.10 and 2.11). The formation of these tetracyclic and pentacyclic triterpenes is typified by the reaction catalyzed by lanosterol synthase in which squalene oxide undergoes acid-catalyzed ring-opening of the epoxide moiety, with capture of the resultant cation by the neighboring double bond as the first step in a cascade of cyclizations and rearrangements leading to the formation of lanosterol (**23**) by way of a cationic protosterol intermediate (**24**) (Scheme 5). The elucidation of the mechanism of formation of lanosterol, based on the brilliant proposal originally put forward by Woodward and Bloch,[13] remains one of the great triumphs of modern organic chemistry. Extensive site-directed mutagenesis experiments have strongly suggested that Asp-456 of yeast lanosterol synthase is the source of the proton that initiates the cyclization.[14] Corey *et al.* have also reported that the corresponding 6-normethyl and 6-chloromethyl analogues of 2,3-oxidosqualene ((**25**) and (**26**)) undergo cyclization with a diminished V_{max}/K_m and have argued from these results that formation of ring A is concerted with epoxide ring-opening.[15] This conclusion is not required by the available data, however, since substrate binding may play a role in organizing the cyclase into its catalytically active conformation and anomalous binding of the substrate analogues might account for the observed changes in the steady-state kinetic parameters.

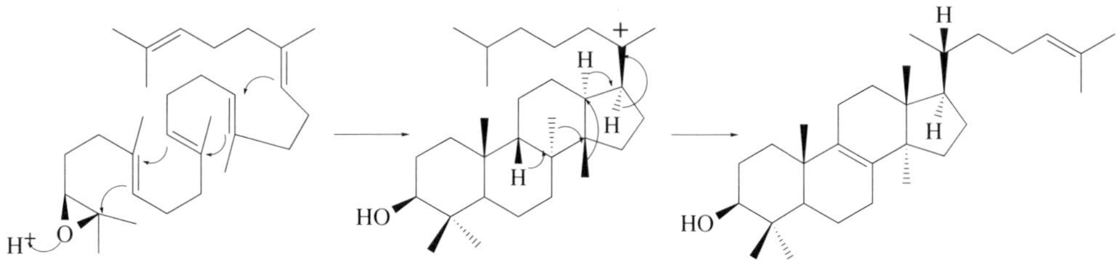

(**22**) (**24**) (**23**)

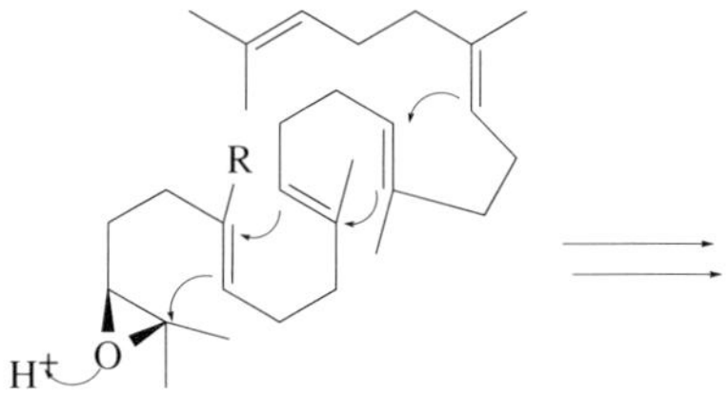

(**25**) R = H
(**26**) R = CH_2Cl

Scheme 5

2.01.3.3 Generation of Carbocations: Ionization of an Allylic Diphosphate

A vast number of isoprenoid biosynthetic reactions are initiated by the ionization of an allylic diphosphate ester. The substrates for these transformations include the acyclic isoprenoids dimethylallyl diphosphate (DMAPP, (**12**)), geranyl diphosphate (GPP, (**27**)), farnesyl diphosphate (FPP, (**28**)), and geranylgeranyl diphosphate (GGPP, (**16**)), as well as bicyclic intermediates such as ent-copalyl diphosphate (**17**) (see Chapters 2.04, 2.05, 2.06, 2.07, 2.08, and 2.09). The enzymes that mediate these reactions share a common requirement for a divalent metal cofactor, most often Mg^{2+}, with Mn^{2+} and, in the case of the protein prenyltransferases, Zn^{2+}, also being used (see Chapter 2.13).

2.01.3.3.1 Isoprenoid diphosphate chain elongation

Isoprenoid chain elongation, catalyzed by the family of enzymes known as prenyl transferases, has been intensively studied at the mechanistic, biochemical, protein structural, and genetic level. The fundamental chain-building reaction involves ionization of an allylic diphosphate substrate and electrophilic attack of the resulting allylic cation on the terminal methylene double bond of the cosubstrate isopentenyl diphosphate (Scheme 6) (see Chapter 2.04.3). Deprotonation of the resultant carbocation gives an allylic diphosphate product that has been extended by a single isoprenoid unit. Depending on the particular prenyl transferase, the product itself may be released from the enzyme surface, or may serve as the substrate for one or more additional condensations with isopentenyl diphosphate, leading to the formation of linear isoprenoids with 2, 3, 4, or more isoprene units, up to the million or more that are found in rubber. Isoprenoid chain length is tightly controlled by the relevant enzyme. Crystallographic investigations have established the key role played by clusters of aspartates in chelating the divalent Mg^{2+} cofactors that in turn both bind and activate the pyrophosphate groups of the substrates.

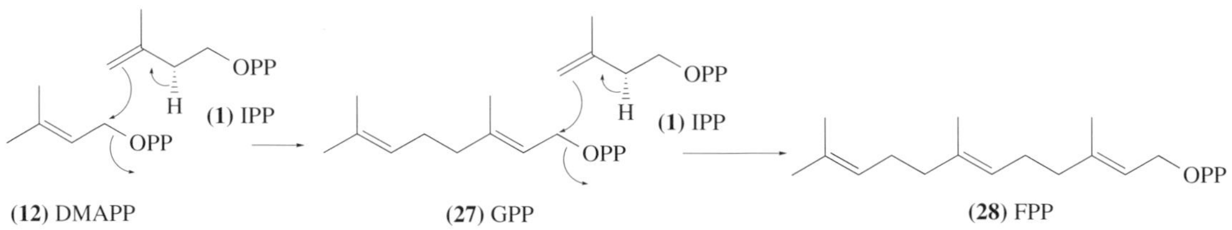

Scheme 6

2.01.3.3.2 Isoprenoid diphosphate isomerization and cyclization

The majority of the tens of thousands of known isoprenoid metabolites are cyclic monoterpenes, sesquiterpenes, and diterpenes generated by cyclization of the acyclic substrates geranyl, farnesyl, and geranylgeranyl diphosphate, respectively (see Chapters 2.05, 2.06, and 2.08). The theoretical basis for the mechanistic understanding of isoprenoid cyclizations was first articulated by Ruzicka,[3,4] who proposed that ionization of an activated derivative of the commonly occurring terpenoid alcohol farnesol, subsequently shown to be the corresponding diphosphate ester, FPP (**28**), will give rise to an allylic cation that can cyclize by electrophilic attack on the central or distal double bonds, as illustrated in Scheme 7, leading eventually to the formation of any given sesquiterpene carbon skeleton (see Chapter 2.06). Recognition that there is a stringent geometric barrier to direct formation of six-membered rings from the *trans*-allylic diphosphate precursor eventually led to the discovery that this barrier can be overcome by initial isomerization of *trans,trans*-farnesyl diphosphate (**28**) to its tertiary allylic isomer, nerolidyl diphosphate (NPP, (**29**)), which has the necessary conformational flexibility to adopt a conformation capable of cyclization (Scheme 8). Although the isomerization and cyclization events have structurally distinct consequences, the two processes are in fact mechanistically closely related, involving ionization of an allylic diphosphate substrate to give a transoid or cisoid allylic cation–pyrophosphate ion pair, which can be quenched either by recapture of the paired pyrophosphate counterion or by backside attack by the neighboring double bond, respectively. Completely analogous mechanistic schemes can account for cyclization of geranyl diphosphate (**27**) to monoterpenes by way of the tertiary allylic isomer, linalyl diphosphate (LPP, (**30**)) (Scheme 9(a)) (see Chapter 2.05). Similarly, cyclization of diterpene substrates such as *ent*-copalyl diphosphate (**17**) involves ionization of the allylic diphosphate and backside capture of the allylic cation at the tertiary allylic position in a process corresponding to a net *anti* $S_{N'}$ reaction (Scheme 9(b)) (see Chapter 2.08). All of these cyclizations are in effect the *intramolecular* analogue of the *intermolecular* prenyl transferase-mediated reactions. It is therefore of considerable interest that not only do the isoprenoid cyclization and chain elongation reactions share considerable

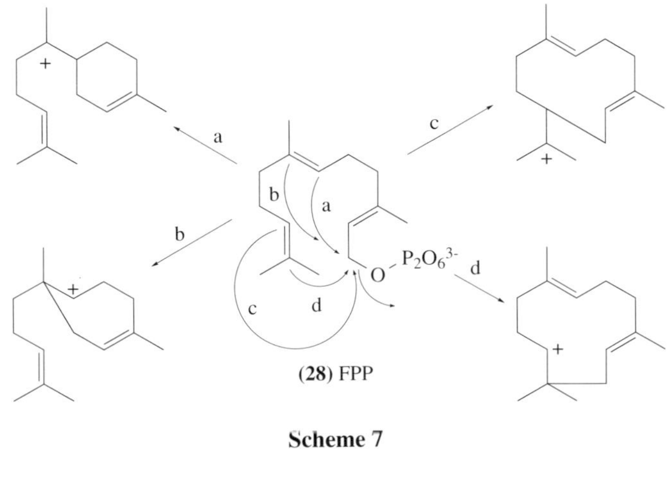

Scheme 7

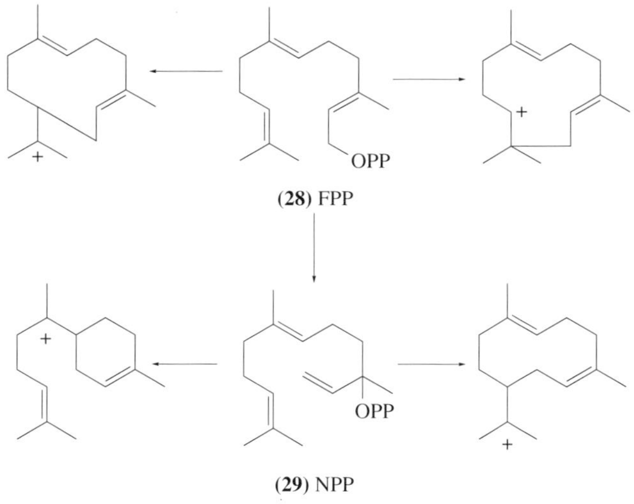

(28) FPP

(29) NPP

Scheme 8

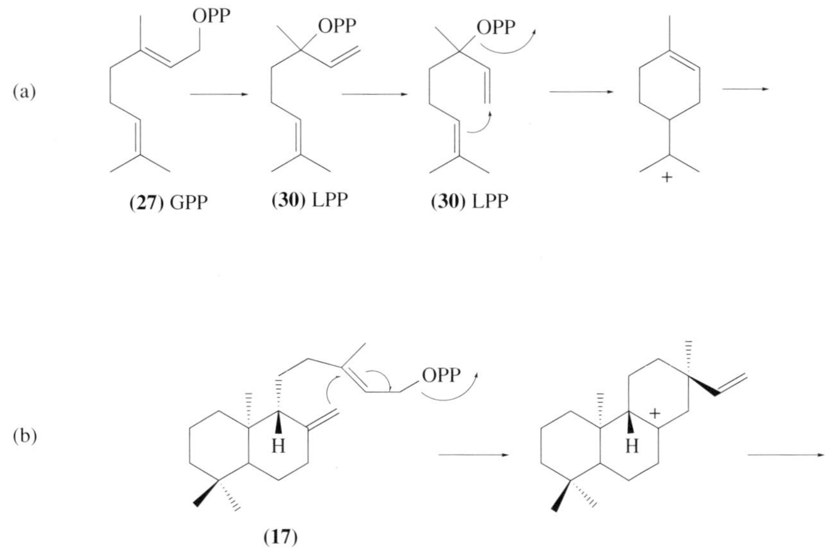

(a)

(27) GPP (30) LPP (30) LPP

(b)

(17)

Scheme 9

mechanistic and biochemical similarity, but crystallographic data suggest that the two classes of proteins responsible for these transformations exhibit striking similarities in overall topology and active site organization, in spite of insignificant similarity at the amino acid sequence level, suggesting an evolutionary conservation of function and form between the two groups of proteins. Among the most salient conserved features is the presence of the aspartate-rich sequence that is responsible for binding the required divalent cation as well as the presence of a rich array of aromatic amino acids within the active site. The latter observation is particularly intriguing in light of experimental and theoretical investigations which suggest that aromatic rings, besides their favorable interaction with hydrophobic ground-state substrates, are particularly well suited to the stabilization of positively charged intermediates as a consequence of strong quadrupole–carbocation interactions.

2.01.3.3.3 Isoprenoid diphosphate chain coupling

The formation of the squalene (**18**), the universal precursor of steroids and triterpenes, involves the reductive head-to-head coupling of two equivalents of farnesyl diphosphate (**28**) (Scheme 10(a)) (see Chapter 2.09). In fact, the actual coupling involves two discrete steps, each initiated by the ionization of an activated diphosphate ester. In the first stage of the coupling process, ionization of farnesyl diphosphate is followed by electrophilic attack on the 2,3-bond of the paired FPP cosubstrate, with subsequent deprotonation to give the cyclopropane ring of presqualene diphosphate (**31**). In the second step, the cyclopropylcarbinyl diphosphate undergoes an isomerization and rearrangement, with an unusual reductive quenching of the resulting carbocationic intermediates by NADPH to yield squalene. A closely related sequence of reactions, involving two equivalents of geranylgeranyl diphosphate (**16**), but terminated by deprotonation rather than reduction, is responsible for the formation of the carotenoid precursor phytoene (**32**) (Scheme 10(b)) (see Chapter 2.12).

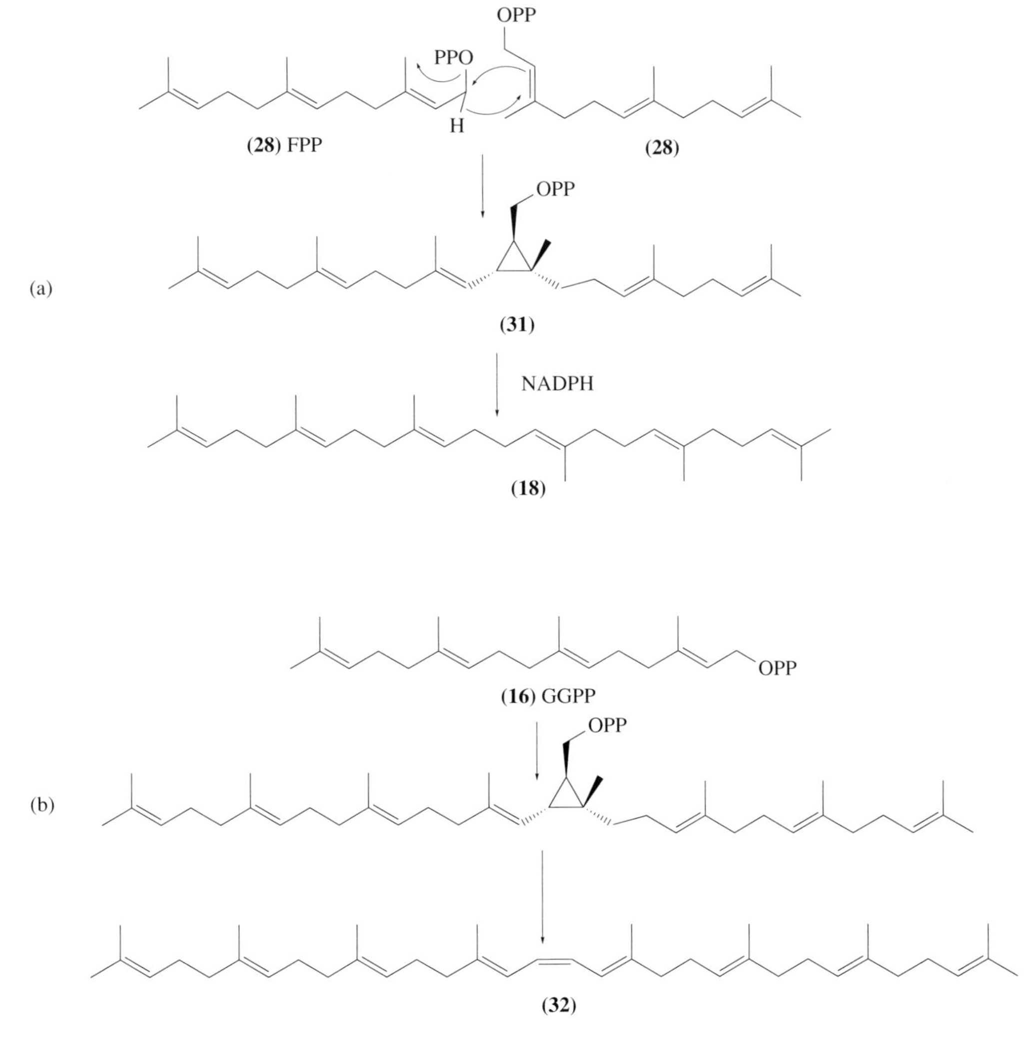

Scheme 10

2.01.3.3.4 *Isoprenoid diphosphate alkylation of heteroatoms*

There are numerous examples of the alkylation of phenolic oxygens by dimethylallyl diphosphate to give allyl ethers, yet few of these processes have been studied at the enzyme level. The reaction of allylic diphosphate esters with heteroatoms that has drawn the most attention is the prenylation of proteins by attachment of farnesyl or geranylgeranyl moieties to cysteine thiols found close to the C-terminus of target proteins (see Chapter 2.13). These reactions, which depend on a Zn^{2+} cofactor, have been shown to be closely related mechanistically to isoprenoid chain elongation reactions, involving initial formation of an allylic cation that is captured by the nucleophilic thiol residue with net inversion of configuration.

2.01.3.4 Reactions of Carbocations

Once cyclizations have been initiated, the resultant carbocations undergo a variety of transformations, many of them well precedented in laboratory organic reactions. Each cation can further cyclize by electrophilic attack, with formation of rings of a range of sizes, from three-membered rings upwards. In these cyclizations, Markovnikov's rules do not seem to be relevant, with examples of electrophilic attack on either the more-substituted or the less-substituted end of the target double bond. Apparent formation of the nominally less-stable cation is poorly understood, but may be controlled by precise folding of the cyclization substrate as well as selective stabilization of reactive intermediates by the protein itself. In this regard, calculations by Jenson and Jorgensen have shown that direct formation of the C-ring of a protosterol catalyzed by lanosterol synthase (Scheme 5), with concomitant generation of a secondary cation, involves no net increase in energy compared with that of the bicyclic intermediate.[16] In spite of considerable speculation, there is no direct evidence for the formation of a covalently bound intermediate in any terpenoid cyclization.

Enzymatically generated carbocations have also been shown to undergo a variety of well-precedented carboskeletal rearrangements, involving methyl migrations, ring expansions, and ring contractions, as well as numerous types of hydride shifts. Although both model studies and experimental investigations of triterpene cyclizations have demonstrated *anti* additions to double bonds[17] (Scheme 11(a)), numerous examples have been documented of net *syn* addition to double bonds (Scheme 11(b)). Similarly, the successive Wagner–Meerwein rearrangements and hydride shifts characteristic of many terpenoid cyclization intermediates have been found to take place with both net *syn* and *anti* stereochemistry (see Schemes 4(d) and 5).

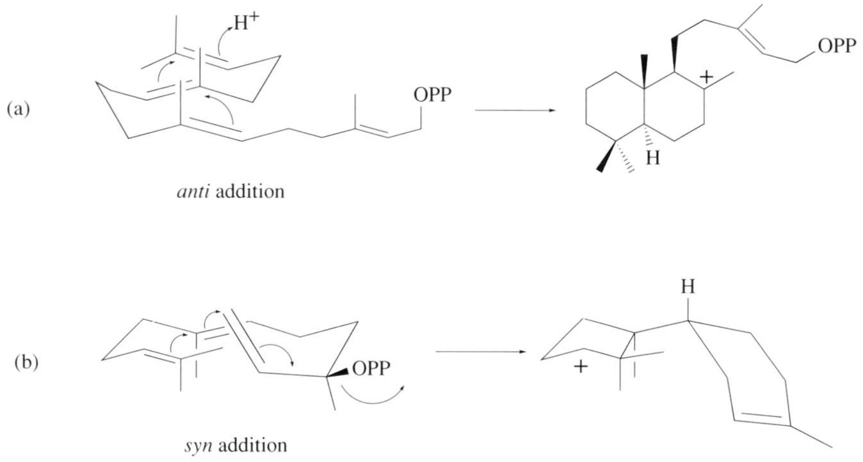

(a) H^+ OPP OPP H

anti addition

(b) OPP H +

syn addition

Scheme 11

2.01.4 MULTIDISCIPLINARY APPROACHES TO THE STUDY OF ISOPRENOID BIOSYNTHESIS

The study of isoprenoid biosynthesis has benefited enormously from powerful experimental advances in spectroscopy, enzymology, and molecular genetics. The early years of biosynthetic study were dominated by experiments involving intact organism, tissue, or cell preparations to which substrates labeled with radioactive isotopes such as ^3H and ^{14}C were administered. In the field of cholesterol biosynthesis, these studies were soon followed by isolation and partial purification of enzymes catalyzing individual steps in the formation of many of the key intermediates in the sterol biosynthetic pathway. In all cases, following isolation and rigorous purification of the derived labeled metabolites, often separated from the precursors by numerous intervening metabolic steps, determination of the site or sites of isotopic labeling served to confirm proposed precursor–product relationships and to identify the origin of individual carbon and hydrogen atoms in the products being analyzed. Rigorous determination of the sites of labeling usually required labor-intensive and experimentally demanding chemical degradations. In the early 1970s the study of natural products biosynthesis took a major leap forward when the widespread introduction of ^{13}C nuclear magnetic resonance provided a means for the direct determination of the distribution of ^{13}C labels. Using multiple isotopic labeling techniques, the ^{13}C NMR method was soon extended to allow the indirect detection of a variety of other isotopes, including ^{2}H, ^{15}N, and ^{18}O, while the use of ^{13}C–^{13}C double labeling provided a direct means to monitor the conservation, formation, and consumption of individual bonds. The contemporaneous introduction of ^{2}H NMR further extended the power of these methods and resulted in a substantial increase in the number of reported isoprenoid biosynthetic studies.

By the early 1980s, a growing number of laboratories had begun to investigate isoprenoid biosynthesis at the cell-free level, first with relatively crude extracts and ultimately using homogeneous enzyme preparations. Early experiments in many cases simply extended the basic methodology of the classical whole-cell precursor–product experiments, with labeled "substrates" replacing "precursors," "incubations" replacing "feedings," and products being analyzed to determine the distribution of isotopic labels. In fact, the move to enzyme systems was no small advance, as now individual enzyme transformations could be studied in place of entire metabolic pathways, and old problems of poor precursor uptake or metabolic degradation of labeled products were simply avoided. Nevertheless, new challenges have arisen, since many isoprenoid enzymatic reactions are themselves multistep chemical transformations that take place at a single active site. With the availability of purified native and, in more recent years, recombinant enzymes, the full arsenal of modern mechanistic enzymology has been brought to bear on these complex, mechanistically intriguing reactions. In 1995 and 1997, the first crystal structures of proteins mediating isoprenoid biosynthesis appeared,[18–21] opening up a new avenue for investigation of these transformations.

Finally, as in all areas of biological science, the study of isoprenoid biosynthesis has benefited enormously from the revolution in molecular biology. Since 1990, numerous isoprenoid biosynthetic genes have been cloned and expressed (see Chapter 2.07). Many older problems which had once appeared intractable have now been revisited and solved. Thus structural genes of almost all the enzymes responsible for the conversion of acetyl-CoA to lanosterol have been identified from a variety of microbial, plant, and mammalian sources and many of the recombinant enzymes are now available in substantial quantities. Several crystal structures have already been reported and more can be expected. Site-directed mutagenesis has already seen extensive use in probing the function of active site residues and in exploring the molecular basis of substrate recognition and control of product formation. In some cases, genetic studies have outstripped the existing enzymology. This is particularly true of studies of carotenoid biosynthesis in which many of the structural genes from a variety of sources have been characterized, but few of the reactions have been studied in detail at the enzyme level (see Chapter 2.12).

The chapters that follow in this volume give a comprehensive and authoritative account of the most important advances since 1980 in all the currently active areas of isoprenoid biosynthesis, from the origins of the biological isoprene unit, isopentenyl diphosphate, through the rich field of terpenoid cyclizations, to carotenoid biosynthesis and protein prenylation. As should be evident, no single approach or group of approaches has been applied to all these fields of study and many questions remain unanswered. The regulation and control of most isoprenoid pathways is still understood at only a primitive level and little is known about the intracellular localization of most terpenoid biosynthetic enzymes. Almost totally unexplored is the molecular basis for the evolution of protein function across diverse species. Each of the authoritative accounts found here will no doubt serve not only as a narrative of what has already been accomplished but as an inspiration and guide to future research in this fascinating and scientifically challenging area.

2.01.5 REFERENCES

1. J. S. Glasby (ed.), "Encyclopedia of Terpenoids," Wiley, Chichester, 1982.
2. O. Wallach, "Terpene und Campher," 2nd edn., Veit, Leipzig, 1914.
3. L. Ruzicka, *Proc. Chem. Soc.*, 1959, 341.
4. A. Eschenmoser, *Chimia*, 1990, **44**, 1.
5. L. Ruzicka, *Experientia*, 1953, **9**, 357.
6. S. Chaykin, J. Law, A. H. Phillips, T. T. Chen, and K. Bloch, *Proc. Natl. Acad. Sci. USA*, 1958, **44**, 998.
7. F. Lynen, H. Eggerer, U. Henning, and I. Kessel, *Angew. Chem.*, 1958, **70**, 738.
8. S. L. Spurgeon and J. W. Porter, in "Biosynthesis of Isoprenoid Compounds," eds. J. W. Porter and S. I. Spurgeon, Wiley, New York, 1981, vol. 1, p. 1.
9. N. Qureshi and J. W. Porter, in "Biosynthesis of Isoprenoid Compounds," eds. J. W. Porter and S. I. Spurgeon, Wiley, New York, 1981, vol. 1, p. 47.
10. K. Himmeldirk, I. A. Kennedy, R. E. Hill, B. G. Sayer, and I. D. Spenser, *J. Chem. Soc., Chem. Commun.*, 1996, 1187.
11. T. P. Begley, *Nat. Prod. Rep.*, 1996, **13**, 177.

12. J. W. Cornforth, *Angew. Chem., Int. Ed. Engl.*, 1968, **7**, 903.
13. R. B. Woodward and K. Bloch, *J. Am. Chem. Soc.*, 1953, **75**, 2023.
14. E. J. Corey, H. Cheng, C. H. Baker, S. P. T. Matsuda, D. Li, and X. Song, *J. Am. Chem. Soc.*, 1997, **119**, 1289.
15. E. J. Corey, H. Cheng, C. H. Baker, S. P. T. Matsuda, D. Li, and X. Song, *J. Am. Chem. Soc.*, 1997, **119**, 1277.
16. C. Jenson and W. L. Jorgensen, *J. Am. Chem. Soc.*, 1997, **119**, 10 846.
17. A. Eschenmoser, L. Ruzicka, O. Jeger, and D. Arigoni, *Helv. Chim. Acta*, 1955, **38**, 1890.
18. L. C. Tarshis, M. Yan, C. D. Poulter, and J. C. Sacchettini, *Biochemistry*, 1994, **33**, 10 871.
19. K. U. Wendt, K. Poralla, and G. E. Schulz, *Science*, 1997, **277**, 1811.
20. C. M. Starks, K. Back, J. Chappell, and J. P. Noel, *Science*, 1997, **277**, 1815.
21. C. A. Lesburg, G. Zhai, D. E. Cane, and D. W. Christianson, *Science*, 1997, **277**, 1820.

2.02
Biosynthesis of Mevalonic Acid from Acetyl-CoA

DANIEL A. BOCHAR, JON A. FRIESEN,
CYNTHIA V. STAUFFACHER, and VICTOR W. RODWELL
Purdue University, West Lafayette, IN, USA

2.02.1 INTRODUCTION

2.02.1.1 Acetyl-CoA to Mevalonate

Eukaryotic organisms employ 3 mol of acetyl coenzyme A (acetyl-CoA) and 2 mol of reduced nicotinamide adenine dinucleotide phosphate (NADPH) to form 1 mol of the isoprenoid precursor (*R*)-mevalonate. The overall reaction is thus as shown in Equation (1) (CoA-SH denotes CoA in which the thiol group is protonated).

$$\text{3 acetyl-CoA} + \text{2 NADPH} + \text{2 H}^+ \rightleftharpoons (R)\text{-mevalonate} + \text{2 NADP}^+ + \text{3 CoA-SH} + \text{H}_2\text{O} \quad (1)$$

As might be expected, this complex biosynthetic process involves not one, but a series of three enzyme-catalyzed reactions (Scheme 1). We therefore first briefly introduce the three enzymes that catalyze these reactions.

$$\text{acetyl-CoA} + \text{acetyl-CoA} \rightleftharpoons \text{acetoacetyl-CoA} + \text{CoA-SH}$$

$$\text{acetoacetyl-CoA} + \text{acetyl-CoA} \rightleftharpoons (S)\text{-3-hydroxy-3-methylglutaryl-CoA} + \text{CoA-SH} + \text{H}_2\text{O}$$

$$(S)\text{-3-hydroxy-3-methylglutaryl-CoA} + \text{2 NADPH} + \text{2 H}^+ \rightleftharpoons (R)\text{-mevalonate} + \text{2 NADP}^+ + \text{CoA-SH}$$

$$\text{SUM:} \quad \text{3 acetyl-CoA} + \text{2 NADPH} + \text{2 H}^+ \rightleftharpoons (R)\text{-mevalonate} + \text{2 NADP}^+ + \text{3 CoA-SH} + \text{H}_2\text{O}$$

Scheme 1

2.02.1.2 Acetoacetyl-CoA Synthase

IUB name: acetyl-CoA:acetyl-CoA *C*-acetyltransferase, EC 2.3.1.9
Common names: acetoacetyl-CoA thiolase; acetoacetyl-CoA synthase

As indicated by its International Union of Biochemistry (IUB) name, Equation (2) involves transfer of an acetyl group from one molecule of acetyl-CoA to the methyl carbon of a second acetyl-CoA. While the equilibrium strongly favors the thiolysis of acetoacetyl-CoA, to call this enzyme acetoacetyl-CoA thiolase downplays the biosynthetic function of Equation (2). We therefore term this enzyme acetoacetyl-CoA synthase to emphasize the biosynthetic role of this Claisen condensation in isoprenoid biosynthesis.

$$\text{acetyl-CoA} + \text{acetyl-CoA} \rightleftharpoons \text{acetoacetyl-CoA} + \text{CoA-SH} \quad (2)$$

2.02.1.3 HMG-CoA Synthase

IUB name: (*S*)-3-hydroxy-3-methylglutaryl-CoA acetoacetyl-CoA-lyase (CoA-acetylating), EC 4.1.3.5

Common name: hydroxymethylglutaryl-CoA synthase; HMG-CoA synthase.

The IUB name fails to emphasize the biosynthetic character of Equation (3), the second Claisen condensation of mevalonate biosynthesis. We therefore term the enzyme that catalyzes Equation (3) (*S*)-3-hydroxy-3-methylglutaryl-CoA synthase, or, more simply, HMG-CoA synthase.

$$\text{acetoacetyl-CoA} + \text{acetyl-CoA} + H_2O \rightleftharpoons (S)\text{-3-hydroxy-3-methylglutaryl-CoA} + \text{CoA-SH} \quad (3)$$

2.02.1.4 HMG-CoA Reductase

IUB name: (*R*)-mevalonate : $NADP^+$ oxidoreductase (CoA acylating), EC 1.1.1.34

Common names: (*S*)-3-hydroxy-3-methylglutaryl-CoA reductase; HMG-CoA reductase

Equation (4) is the first reaction unique to isoprenoid biosynthesis. The preferred name for the enzyme catalyst, HMG-CoA reductase, indicates that the reaction favors mevalonate biosynthesis. In mammals, HMG-CoA is a branch point compound whose carbons serve both for the biosynthesis of isoprenoids and, following cleavage by HMG-CoA lyase (EC 4.2.1.18), for synthesis of the ketone bodies (acetoacetate, β-hydroxybutyrate, and acetone). Located where two metabolic pathways diverge, the branch point enzyme HMG-CoA reductase constitutes the major focus for control of carbon flow from acetyl-CoA to isoprenoids.

$$(S)\text{-HMG-CoA} + 2\,\text{NADPH} + 2\,H^+ \rightleftharpoons (R)\text{-mevalonate} + 2\,NADP^+ + \text{CoA-SH} \quad (4)$$

2.02.2 ACETOACETYL-CoA SYNTHASE

2.02.2.1 The Acetoacetyl-CoA Synthase Reaction is a Classical Claisen Condensation

The nucleophilic substitution reaction catalyzed by acetoacetyl-CoA synthase is a classical Claisen condensation similar to that which forms ethyl acetoacetate from ethyl acetate (Figure 1). In this head-to-tail condensation, one acetyl-CoA molecule serves as an electrophile at C-1 and the other as the equivalent of a C-2 carbanion. The reaction proceeds in three steps which involve an active site cysteinyl acyl–enzyme intermediate (Figure 2).

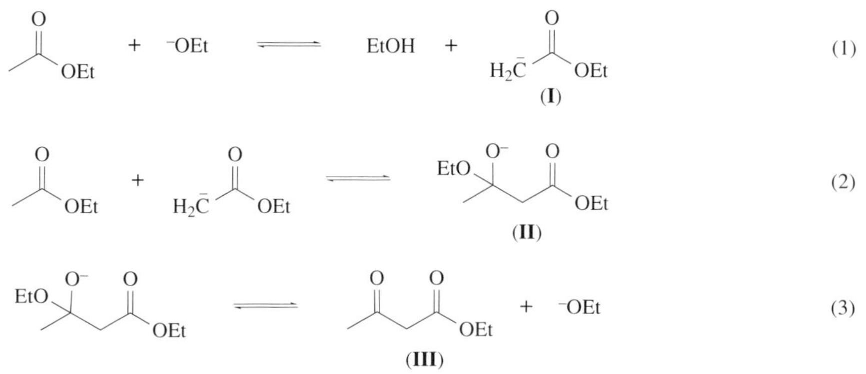

$$CH_3C(=O)OEt + {}^-OEt \rightleftharpoons EtOH + {}^-H_2C{-}C(=O)OEt\ (\mathbf{I}) \quad (1)$$

$$CH_3C(=O)OEt + {}^-H_2C{-}C(=O)OEt \rightleftharpoons CH_3C(O^-)(OEt){-}CH_2C(=O)OEt\ (\mathbf{II}) \quad (2)$$

$$CH_3C(O^-)(OEt){-}CH_2C(=O)OEt \rightleftharpoons CH_3C(=O)CH_2C(=O)OEt\ (\mathbf{III}) + {}^-OEt \quad (3)$$

Figure 1 Mechanism of the chemical Claisen condensation that forms ethyl acetoacetate from ethyl acetate. Stage 1: ethoxide abstracts a proton from the α-carbon of ethyl acetate, forming carbanion (**I**). Stage 2: the powerful nucleophile carbanion I attacks the carbonyl carbon of a second molecule of ethyl acetate, forming (**II**). Stage 3: displacement of the ethoxide anion yields ethyl acetoacetate (**III**).

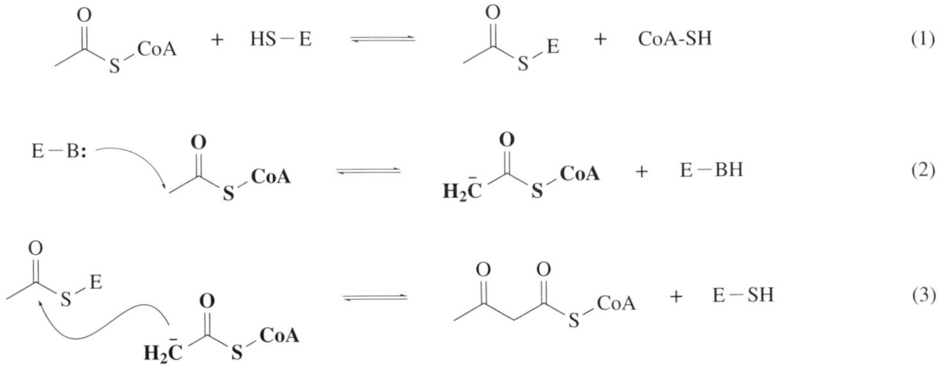

$$CH_3C(=O)S\text{-}CoA \;+\; HS\text{-}E \;\rightleftharpoons\; CH_3C(=O)S\text{-}E \;+\; CoA\text{-}SH \qquad (1)$$

$$E\text{-}B{:} \;+\; \mathbf{CH_3C(=O)S\text{-}CoA} \;\rightleftharpoons\; \mathbf{H_2\bar{C}\text{-}C(=O)S\text{-}CoA} \;+\; E\text{-}BH \qquad (2)$$

$$CH_3C(=O)S\text{-}E \;+\; \mathbf{H_2\bar{C}\text{-}C(=O)S\text{-}CoA} \;\rightleftharpoons\; CH_3C(=O)CH_2C(=O)S\text{-}CoA \;+\; E\text{-}SH \qquad (3)$$

Figure 2 The reaction catalyzed by acetoacetyl-CoA synthase. Stage 1: condensation of one molecule of acetyl-CoA with the enzyme, represented as HS—E, releases CoA-SH and forms an enzyme-bound thioester. Stage 2: a basic group on the enzyme (E—B:) facilitates loss of the α-proton of a second molecule of acetyl-CoA (bold), forming a C-2 anion. Stage 3: nucleophilic attack by the C-2 anion on the carbonyl carbon of the enzyme-bound thioester then forms acetoacetyl-CoA, releasing the enzyme (E—SH).

2.02.2.2 Cells Elaborate Multiple Enzymes that Catalyze Acetoacetyl-CoA Biosynthesis

Mammals, yeast, and bacteria contain multiple enzymes that catalyze the reversible synthesis and cleavage of acetoacetyl-CoA. Based on substrate specificity, we distinguish two broad classes of enzymes that catalyze Equation (2). 3-Ketoacyl-CoA thiolases (EC 2.3.1.16) exhibit broad specificity for substrate chain length, reside in mitochondria and peroxisomes, function in fatty acid oxidation, and play no role in isoprenoid biosynthesis. They thus will not be considered further. By contrast, acetoacetyl-CoA synthases (EC 2.3.1.9) are highly substrate-specific enzymes that function in the biosynthesis of acetoacetyl-CoA. Among acetoacetyl-CoA synthases, we further distinguish those that form acetoacetyl-CoA destined for ketone body synthesis from those that form acetoacetyl-CoA that ultimately forms mevalonate. Acetoacetyl-CoA synthases of animal subcellular organelles function in ketone body formation. By contrast, the cytosolic synthases of eukaryotes function in the biosynthesis of acetoacetyl-CoA destined for mevalonate and isoprenoids. Both classes of enzymes nevertheless exhibit many similarities in mechanism of catalysis, active site residues, primary and quaternary structure, and kinetic parameters.[1–3]

2.02.2.3 Several Cloned cDNAs Encode Cytosolic Acetoacetyl-CoA Synthases

cDNAs that encode cytosolic acetoacetyl-CoA synthases have been cloned from yeast, plants, and humans. A yeast cDNA encodes a 398-residue acetoacetyl-CoA synthase polypeptide.[4] A cloned cDNA for a biosynthetic acetoacetyl-CoA synthase from the radish plant encodes a slightly larger, 406-residue polypeptide that, when expressed in yeast, is located exclusively in the cytosol. Expression of this cDNA in radish cotyledons appears to be light stimulated.[5] A cDNA that encodes human acetoacetyl-CoA synthase has been cloned,[6] and the gene has been localized to a specific chromosome.[7]

2.02.2.4 A Bacterial Acetoacetyl-CoA Synthase Functions in Poly(β-hydroxybutyrate) Synthesis

The acetoacetyl-CoA synthases of certain bacteria function in a second biosynthetic process that, like isoprenoid biogenesis, results in the formation of a biopolymer. The biopolymer poly(β-hydroxybutyrate) serves as an energy reserve in certain bacteria. Acetoacetyl-CoA destined for poly(β-hydroxybutyrate) synthesis is reduced to β-hydroxybutyrate by an NADPH-dependent acetoacetyl-CoA reductase (EC 1.1.1.36), then polymerized by poly(β-hydroxybutyrate) synthase. The acetoacetyl-CoA synthase gene of *Zoogloea ramigera*, a bacterium that performs an important role in waste water treatment, encodes a 391-residue polypeptide.[8] The enzyme, a homotetramer, has been purified and extensively characterized.[8–12]

2.02.2.5 Catalysis Involves a Histidine and an Acylated Cysteinyl Intermediate

As noted by Gilbert, the α proton of a thioester is not as acidic as is generally supposed.[13] Possibly for this reason, catalysis by acetoacetyl-CoA synthase, a reaction that favors thiolysis over acetoacetate synthesis, involves an acyl enzyme intermediate[3,13] (Figure 2). Several residues of the acetoacetyl-CoA synthase of *Z. ramigera* that function in catalysis have been identified.[14] This enzyme, while it functions in the biosynthesis of poly(β-hydroxybutyrate), fulfills a biosynthetic role analogous to that of acetoacetyl-CoA synthases that function in isoprenoid biosynthesis.

Catalytic roles of two active site residues and a possible role for a third have been implicated by chemical modification and site-directed mutagenesis. Involvement of a cysteinyl and a histidyl residue was initially implicated by the loss of activity that followed treatment with iodoacetamide or diethylpyrocarbonate. Involvement of an active site cysteine was confirmed by isolation of an acetyl-S(Cys89) intermediate.[9–12] Mutagenesis was subsequently used to identify Cys378 as the active site base that abstracts the α proton from acetyl-CoA, generating the nucleophilic C-2 carbanion of acetyl-CoA (Figure 2). (Mutant enzymes are designated in the form X123Y or X123Z, in which residue X at position 123 has been replaced by residue Y or Z, and where X, Y, and Z represent the single letter designations for specific aminoacyl residues. Multiply-mutated enzymes are designated X123Y/Z345X/Y456Z, etc.) Mutation of Cys378 to glycine yielded mutant enzyme C378G, which while it formed the acetyl-enzyme intermediate, was essentially inactive.[11] By contrast, when Cys378 was replaced by serine, an alternative active-site base, mutant enzyme C378S retained detectable activity.[14] Cognate cysteinyl residues probably will be shown to perform analogous functions in the acetoacetyl-CoA synthases that function in isoprenoid biogenesis.

2.02.2.6 Crystal Structure of a β-Ketothiolase

No crystal structure is presently available for any cytosolic acetoacetyl-CoA synthase. The 0.28 nm resolution structure of the homodimeric, 417-residue peroxisomal β-ketothiolase of *Saccharomyces cerevisiae* (Figure 3) should, however, provide insights into the structure of cytosolic acetoacetyl-CoA synthases. Both active sites are on the same face of this dimeric β-ketothiolase. The floor of the active site, a shallow pocket lined by highly conserved residues, contains the conserved cysteines and the histidine implicated as functional in catalysis.[15] Sequence similarities and conserved key catalytic residues imply similar reaction mechanisms. The overall topology may also resemble that of other forms of the enzyme, except in the substrate-binding pockets of synthases that specifically bind acetoacetyl-CoA rather than long-chain β-ketoacyl-CoA thioesters.

2.02.3 HMG-CoA SYNTHASE

2.02.3.1 Chemistry of the HMG-CoA Synthase Reaction

The formation of HMG-CoA from acetyl-CoA and acetoacetyl-CoA is catalyzed by the enzyme HMG-CoA synthase. The stoichiometry of the reaction was first determined for the enzyme from yeast. The thioester group of acetoacetyl-CoA becomes that of HMG-CoA, whereas the coenzyme A of acetyl-CoA is released as free coenzyme A. Carbons 5 and 6 of HMG-CoA derive from carbons 2 and 1 of acetyl-CoA, and carbons 1–3 and the methyl group of HMG-CoA arise from carbons 1–4 of acetoacetyl-CoA[16] (Figure 4).

2.02.3.2 Substrate Specificity

HMG-CoA synthase is specific for the acyl moiety of both substrates. Neither propionyl-CoA nor butyryl-CoA can replace acetyl-CoA, nor can 3-ketohexanoyl-CoA replace acetoacetyl-CoA. The enzyme is not, however, absolutely specific for the coenzyme A moiety of its substrates. Acetyl-3-dephospho-CoA, acetylpantetheine, and acetylglutathione can all replace acetyl-CoA.[18–22]

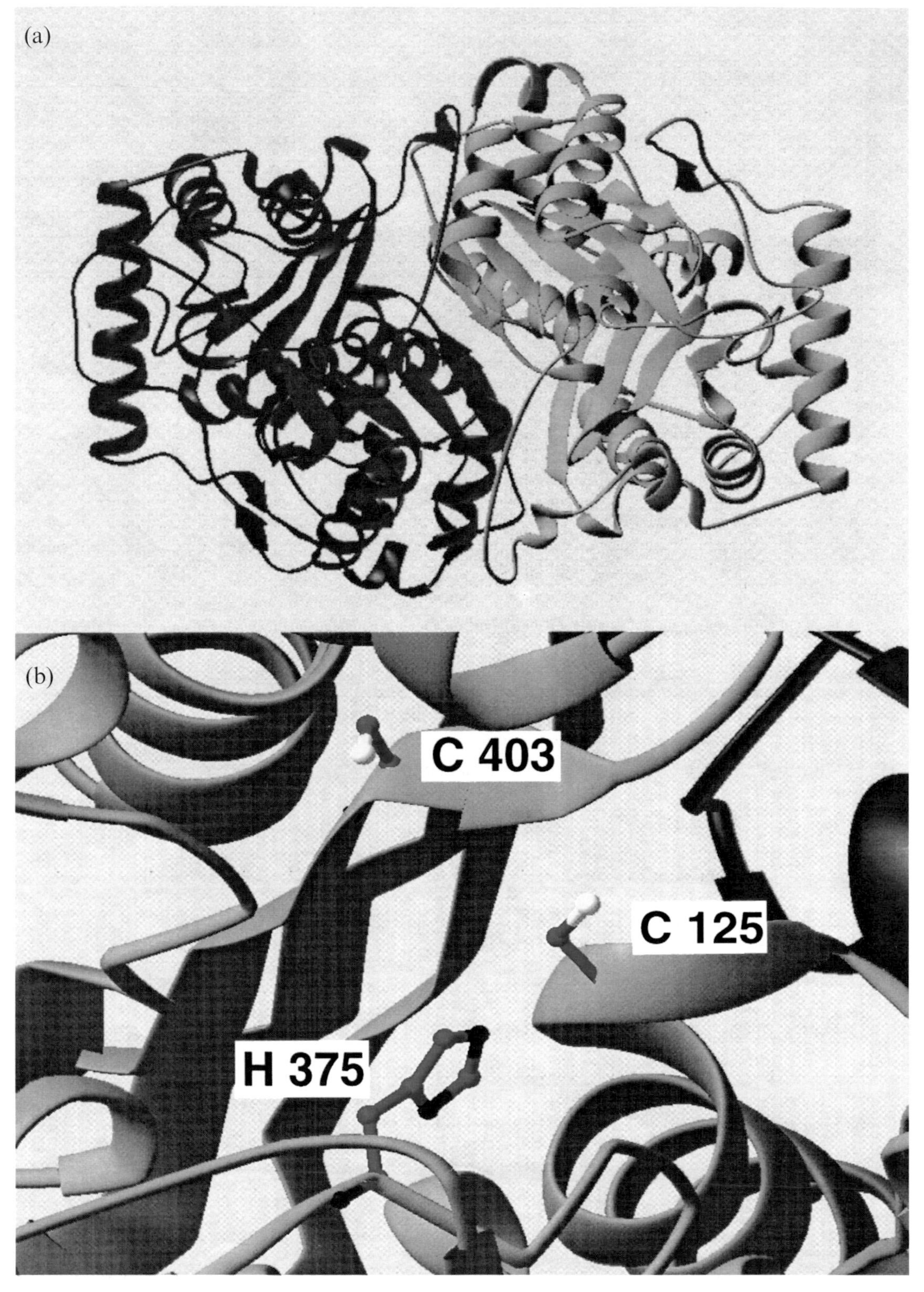

Figure 3 Crystal structure of β-kethothiolase. (a) View along the dimer axis toward the two active sites. The amino and carboxy termini are on the side furthest from the viewer. (b) The active site of subunit 1. Active site residues Cys125, His375, and Cys403, identified both structurally and by reference to the reaction mechanism of the *Z. ramigera* enzyme, form the floor of the active site. No charged polar lysine, arginine, glutamate, or aspartate residues are within 0.1 nm of the active site. (Drawn employing coordinates deposited in the Brookhaven Protein Data Base, accession code 1PXT, by Mathieu *et al.*[15].)

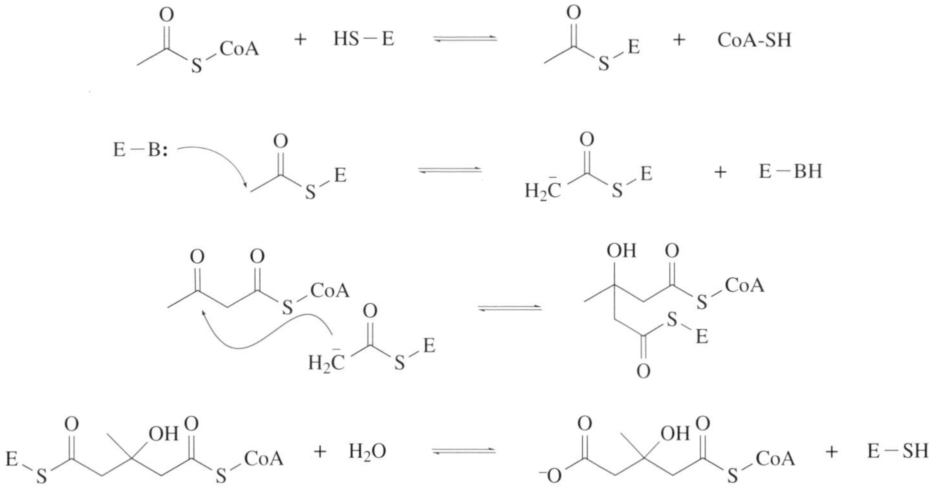

Figure 4 The reaction catalyzed by HMG-CoA synthase. This Claisen ester condensation can be viewed as a three-step process[17] whose chemistry is analogous to that for the reaction catalyzed by acetoacetyl-CoA synthase (see Figure 2).

2.02.3.3 Isozymes Perform Distinct Physiologic Roles

Isoenzymes, or isozymes, are distinct, often readily separable forms of an enzyme elaborated by the same organism. Isozymes catalyze the same chemical reaction, but typically differ with respect to their primary structure, intracellular location, and physiological role. Multimeric isozymes may be the products of different genes or differential gene splicing events. Animals elaborate HMG-CoA synthase isozymes that, based on their intracellular location, are termed the mitochondrial and cytosolic isozymes. These are believed to be the products of different genes rather than being transcribed through differential splicing of a single gene product.[23,24] The N-termini of mitochondrial, but not cytosolic, HMG-CoA synthases contain aminoacyl residues that serve as a leader peptide for import into mitochondria. The kinetic properties of isozymes can also differ. For example, Mg^{2+} stimulates the cytosolic HMG-CoA synthase[25] but inhibits the activity of the mitochondrial enzyme.[26]

2.02.3.4 Cytosolic HMG-CoA Synthases Function in Isoprenoid Biogenesis

Mammalian mitochondrial HMG-CoA synthase performs no known role in isoprenoid biogenesis. Intramitochondrial HMG-CoA serves instead as a precursor of the ketone bodies acetoacetate, β-hydroxybutyrate, and acetone. The mitochondrial isozyme is therefore also termed the ketogenic isozyme. The cytosolic or cholesterogenic isozyme provides the HMG-CoA destined for isoprenoid biogenesis, a process that occurs in the cytosol. Linkage analysis revealed that the mouse genes that encode the cytosolic and mitochondrial HMG-CoA synthases are located on different chromosomes.[27] Despite their differences, both enzymes can form HMG-CoA for incorporation into isoprenoids. A mutant strain of Chinese hamster ovary cells lacking detectable cytosolic HMG-CoA synthase activity are auxotrophic for mevalonate. Following transfection with a cDNA that encoded the mitochondrial enzyme, mevalonate auxotrophy was abolished and the ability to synthesize sterols from acetate was restored.[28]

2.02.3.5 Cloned cDNAs Encode Cytosolic HMG-CoA Synthases

Cloned cDNAs that encode cytosolic HMG-CoA synthases include those of the human,[29,30] Chinese hamster,[31] rat,[23] bird[32] the yeast *Schizosaccharomyces pombe*,[33] and *Arabidopsis thaliana*,[34] and also two from the cockroach.[35,36] The inferred subunit sizes of the encoded enzymes are 50 and 51 kDa (cockroach), 51 kDa (*A. thaliana*), 53 kDa (hamster), 57 kDa (rat and human), and 58 kDa (avian). The degree of sequence identity is higher between cytoplasmic HMG-CoA synthases of different organisms than between the cytoplasmic and mitochondrial isozymes of a given organism. The amino acid sequence of human cytoplasmic HMG-CoA synthase is 94% identical to that of the hamster, 93% identical to that of the rat, 83% identical to that of the chicken, 75% identical to that of the cockroach, and 42% identical to that of *A. thaliana*, but only 66% identical to that of the human mitochondrial isozyme.

2.02.3.6 Catalysis Involves an Active Site Cysteine

One cysteine of cytosolic HMG-CoA synthase, Cys129 of the avian enzyme, is conserved across species and is critical for catalytic activity. No other cysteine appears to function in catalysis.[37] This active site cysteine participates in the formation of a covalent acyl-S-enzyme intermediate during catalysis. Serine cannot substitute for cysteine. Mutation to serine of Cys129 of the cytoplasmic avian liver HMG-CoA synthase yielded a recombinant mutant enzyme that retained the ability to form noncovalent complexes with acetyl-CoA, but was essentially inactive.[38] Cytoplasmic human HMG-CoA synthase mutant enzymes C129S and C129A were also inactive.[29] Mutagenic and kinetic analysis has also implicated His264 of avian HMG-CoA synthase as an active site residue whose postulated role is to anchor the acetoacetyl-CoA via interaction of the imidazole ring with the carbonyl oxygen of the thioester group of acetoacetyl-CoA.[39]

2.02.3.7 Inhibitors of HMG-CoA Synthase Activity

3-Propionyl coenzyme A irreversibly inhibits HMG-CoA synthase, and chloropropionyl-CoA alkylates the active site cysteine.[40,41] Compounds developed and tested as inhibitors of cytosolic HMG-CoA synthases include derivatives of 2-oxetanones,[42] the triyne carbonate L-660,631,[43] Lifibrol,[44] and β-lactones such as L-659,699.[45,46]

2.02.3.8 Formation of HMG-CoA May Involve a Dual-function Enzyme

Evidence is emerging that some life forms can convert acetyl-CoA to HMG-CoA without prior formation or participation of free acetoacetyl-CoA. A purified enzyme system from radish seedlings catalyzed the conversion of acetyl-CoA to HMG-CoA without the addition of acetoacetyl-CoA, and no formation of free acetoacetyl-CoA could be detected in a preparation from mint leaves.[47,48] Data implicate a radical mechanism that would facilitate the otherwise energetically unfavorable dual Claisen condensation. Consistent with this model of a dual-functional enzyme, a search for intermediates of the isoprenoid pathway in plant cells detected all intermediates except free acetoacetyl-CoA.[49] Catalysis of this overall reaction by a single enzyme would allow direct substrate channeling in plants, an apparent advantage considering the many enzymes that compete for acetyl-CoA.

2.02.4 HMG-CoA REDUCTASE

2.02.4.1 The HMG-CoA Reductase Reaction

In eukaryotes and the archaea, or archaebacteria, (*R*)-mevalonate is the precursor of all isoprenoids. Particularly in higher plants, these isoprenoids comprise a formidable array of over 20 000 diverse products. Formation of (*R*)-mevalonate involves the reductive deacylation of the thioester moiety of (*S*)-HMG-CoA to a primary alcohol. The reaction requires 2 mol of NADPH, and is catalyzed by the biosynthetic enzyme HMG-CoA reductase, EC 1.1.1.34 (Figure 5).

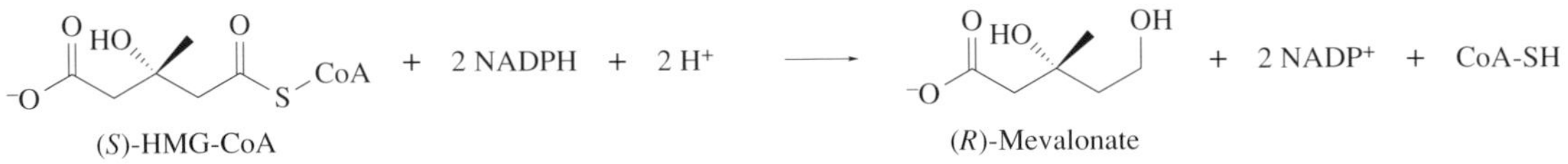

$$(S)\text{-HMG-CoA} + 2\ \text{NADPH} + 2\ \text{H}^+ \longrightarrow (R)\text{-Mevalonate} + 2\ \text{NADP}^+ + \text{CoA-SH}$$

Figure 5 The reaction catalyzed by HMG-CoA reductase. Reductive deacylation of HMG-CoA requires 2 mol of NADPH per mol of HMG-CoA, and results in the conversion of the thioester group of HMG-CoA to the primary alcohol group of mevalonate.

2.02.4.2 Biosynthetic HMG-CoA Reductases

Characterized HMG-CoA reductases include those of higher animals, plants, yeast, and the archaea. Investigation of these HMG-CoA reductases, particularly those of rat liver and of yeast, has spawned a voluminous literature. Numerous reviews and their associated references document the literature concerning eukaryotic HMG-CoA reductases.[25,47,50–64] HMG-CoA reductase is not, however, unique to eukaryotes. The enzyme also plays a central role in isoprenoid biosynthesis by the archaea. Cell membrane phospholipids of the archaea, unlike those of either true bacteria or eukaryotes, contain major quantities of mevalonate-derived isoprenoids rather than fatty acids that are linked to glycerol by an ether bond rather than by an ester bond. These isoprenoids arise from mevalonate formed by HMG-CoA reductase.[65] Genes that encode the HMG-CoA reductases of a halophile[66–68] and of a thermophile[69] have been cloned and their encoded proteins have been purified and characterized. In contrast to archaebacteria, no biosynthetic HMG-CoA reductase has ever been detected in a true bacterium, and the complete genome sequences of *Bacillus subtilis*,[70] *Escherichia coli*,[71] *Helicobacter pylori*,[72] *Hemophilus influenzae*,[73] *Mycoplasma pneumoniae*,[74] a *Synechocystis* species,[75] and *Mycoplasma genitalium*[76] appear to lack a cognate of any known HMG-CoA reductase. The putative attribution of sequence MG085 of *M. Genitalium* as HMG-CoA reductase[76] probably is incorrect, since the translated sequence lacks the DAMG, ENVIG, and GTVGG signature sequences (Figure 6) and the catalytic glutamate, aspartate, and histidine (see below) are not readily apparent. A subsequently completed genome sequence suggests, however, that certain bacteria contain genes that might encode a biosynthetic HMG-CoA reductase. The Lyme disease spirochaete *Borrelia burgdorderi*[77] contains an apparent isoprenoid biogenesis operon that includes genes that may encode HMG-CoA reductase, HMG-CoA synthase, mevalonate kinase, phosphomevalonate kinase, and mevalonate pyrophosphaste decarboxylase, and genes that might encode HMG-CoA reductase are also present in *Streptococcus pneumoniae* and *Streptococcus pyogenes*.[78]

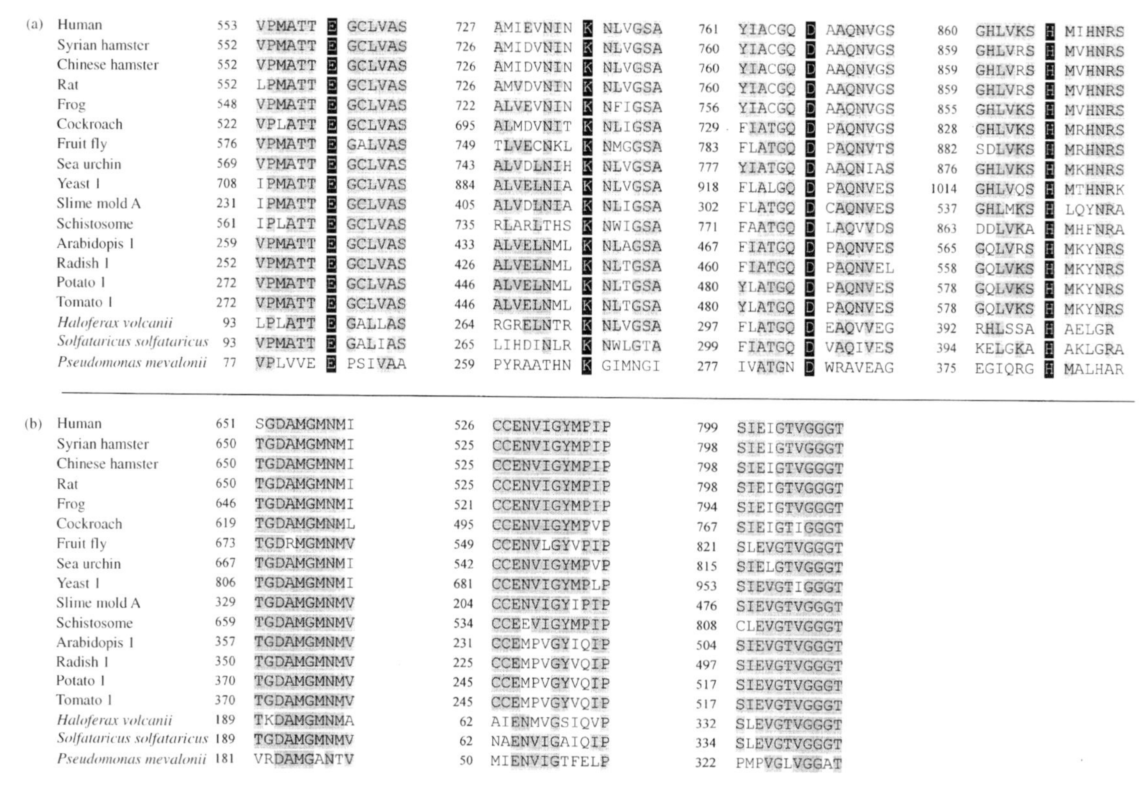

Figure 6 Selected conserved regions of primary structure in representative HMG-CoA reductases. Numbers are for the N-terminal residue shown for each sequence. (a) Highlighted in black are the active site glutamate, lysine, aspartate, and histidine. (b) Shown are selected sequences characteristic of all known HMG-CoA reductases.

2.02.4.3 *Pseudomonas mevalonii* Elaborates a Biodegradative HMG-CoA Reductase

While mevalonate serves as a precursor of isoprenoids in eukaryotes and archaea, a comparable role for mevalonate in bacteria remains to be established. A biodegradative, NAD-dependent HMG-CoA reductase from the bacterium *Pseudomonas mevalonii* has, however, been extensively characterized. *P. mevalonii* was originally isolated from soil based on its ability to grow on mevalonate as its sole source of carbon.[79] In this organism, HMG-CoA reductase does not serve to produce mevalonate for isoprenoid biogenesis, but to convert (*R*)-mevalonate to (*S*)-HMG-CoA (Figure 7). Subsequent cleavage of HMG-CoA by HMG-CoA lyase (EC4.1.3.4) then forms acetyl-CoA and acetoacetate that enter known pathways of two-carbon metabolism.[80] Despite its different physiologic role and specificity for the biodegradative coenzyme NAD rather than for the biosynthetic coenzyme NADPH,[81] *P. mevalonii* HMG-CoA reductase shares many important features with its biosynthetic relatives, and has provided valuable mechanistic insights.

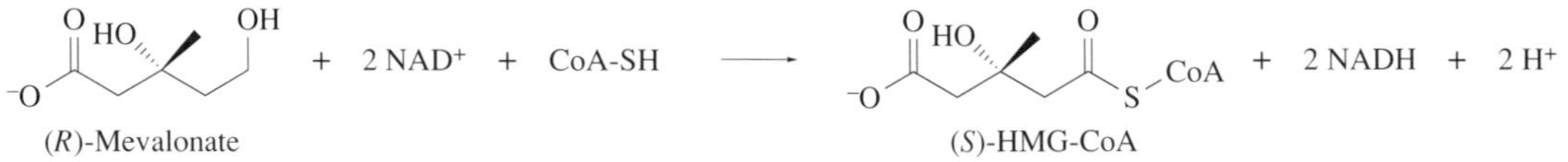

(*R*)-Mevalonate + 2 NAD^+ + CoA-SH → (*S*)-HMG-CoA + 2 NADH + 2 H^+

Figure 7 The reaction catalyzed by the biodegradative HMG-CoA reductase of *P. mevalonii*. Two moles of NAD are reduced during the oxidative acylation of the primary alcohol group of 1 mol of mevalonate to the thioester group of HMG-CoA.

2.02.4.4 Domain Structure of HMG-CoA Reductases

The biosynthetic HMG-CoA reductases of the archaea and the biodegradative HMG-CoA reductase of *P. mevalonii* are soluble, cytosolic enzymes. By contrast, the HMG-CoA reductases of most eukaryotes possess distinct catalytic and membrane anchor domains. Hydrophobicity plots predict that plant and animal HMG-CoA reductases consist of a membrane anchor domain joined to the catalytic domain by a short linker region. From one-eighth to one-half of the N-terminal portion of the polypeptide forms the N-terminal membrane anchor domain. The C-terminal catalytic domain contains all known residues that function in catalysis or regulation of catalytic activity.

While primary structures of the catalytic domain are highly conserved in all eukaryotic HMG-CoA reductases, the membrane anchor domain and the neighboring linker region exhibit considerable sequence diversity. Anchor domains can consist of from as few as two to as many as eight membrane-spanning helices (Figure 8).[82] The smaller number of inferred helices typify the plant HMG-CoA reductases and the larger number the enzymes of yeast and higher animals. A short portion of the N-terminus of the anchor domain is responsible for targeting the enzyme to the endoplasmic reticulum.[82] That the anchor domain plays no essential role in catalysis may be inferred from its absence in the HMG-CoA reductases of *P. mevalonii*[73] and the archaea[73,76] and from the observation that the catalytic domains of the human and Syrian hamster enzymes are active when expressed lacking the membrane anchor domain.[83,84] It is not known why the anchor domains of eukaryotes vary so greatly in size, particularly since as few as two membrane-spanning regions suffice to anchor the enzyme. For example, when cDNAs that encoded two full-length radish HMG-CoA reductases were expressed in a yeast strain auxotrophic for mevalonate, their two membrane-spanning sequences sufficed for direct targeting of nascent radish HMG-CoA reductase into yeast membranes.[85]

2.02.4.5 HMG-CoA Reductase and Cholesterol Homeostasis

As described in subsequent chapters, the mevalonate formed by HMG-CoA reductase serves as the precursor of a vast array of isoprenoid products. Despite this diversity and the central role of HMG-CoA reductase in the biosynthesis of natural products by plants, the scientific literature has until comparatively recently been heavily biased toward the HMG-CoA reductases of animals, and particularly the HMG-CoA reductase of human subjects. This emphasis reflects the central role of HMG-CoA reductase in cholesterol homeostasis and human health. Hypercholesterolemia, an elevated level of cholesterol circulating as plasma lipoproteins, has long been known to constitute a major risk factor for coronary artery disease. Cholesterol arises from two sources: dietary intake

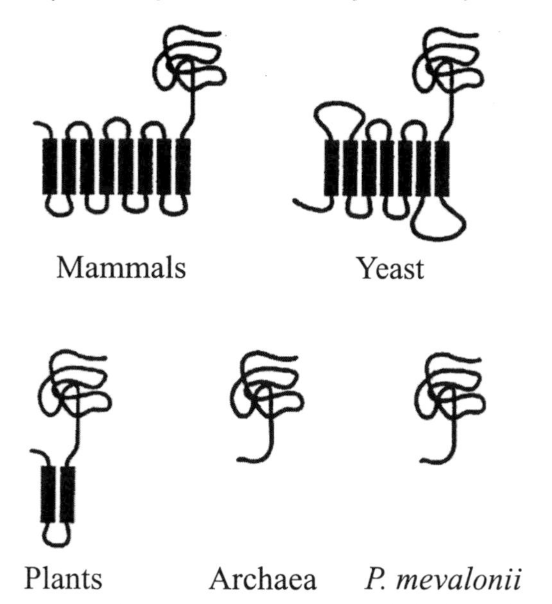

Figure 8 Schematic representation of the domain structures of HMG-CoA reductases from different life forms. The inferred structures of the N-terminal membrane anchor domain of the mammalian and yeast enzymes contain eight and seven transmembrane helices, respectively. The HMG-CoA reductases of plants contain only two transmembrane helices, whereas those of the archaea and *P. mevalonii* lack a membrane anchor domain.

and biosynthesis from acetyl-CoA, primarily in human liver and intestinal tissue. For biosynthesis of cholesterol in both organs, the reaction catalyzed by HMG-CoA reductase is rate limiting under most circumstances of physiological importance in human subjects.[53,61]

To achieve cholesterol homeostasis, animal cells must balance *de novo* biosynthesis of cholesterol with dietary cholesterol intake. Maintaining cholesterol homeostasis involves complex processes that regulate the quantity of low-density lipoprotein (LDL) receptors present in cell membranes, the rate of formation of HMG-CoA, and the rate of its subsequent reductive deacylation to mevalonate. Mammalian cells not exposed to LDL exhibit high levels of HMG-CoA synthase and HMG-CoA reductase. When LDL is present, the activities of both enzymes decline rapidly and dramatically, resulting in a reduced rate of mevalonate biosynthesis. A reduced rate of mevalonate formation complements preferential shunting of mevalonate into nonsterol pathways, in part due to the higher substrate affinities of enzymes of nonsterol biosynthetic pathways.

2.02.4.6 Drugs that Control Cholesterol Biosynthesis Target HMG-CoA Reductase

High circulating levels of LDL cholesterol are causally related to increased risk of coronary heart disease, and lowering LDL cholesterol levels can significantly reduce the incidence of coronary heart disease. As the catalyst for the rate-limiting reaction of cholesterogenesis in human subjects, HMG-CoA reductase thus constitutes the focus of most medical attempts at drug intervention aimed at lowering plasma levels of LDL cholesterol by reducing the rate of cholesterol biosynthesis. Of the numerous cholesterol-lowering drugs that have been investigated (nicotinic acid, clofibrate, cholestyramine, plant sterols, tripanarol, D-thyroxine, and estrogenic hormones), each has undesirable side effects in some or all individuals.

The discovery of mevastatin initiated the current era in drug therapy directed against HMG-CoA reductase. A portion of mevastatin and of the structurally related family of statin drugs (e.g. lovastatin and simvastatin) resembles that of HMG-CoA (Figure 9), suggesting a rational basis for its ability to inhibit HMG-CoA reductase competitively with HMG-CoA with an inhibition constant in the nanomolar range.[86] The voluminous recent medical literature that concerns the amelioration of hypercholesterolemia in human subjects, characterization of existing inhibitors, announcements of new inhibitors, and clinical studies of side effects and relative efficacies lies beyond the scope of this review. For a lucid introduction to this subject, interested readers are referred to the review by Endo, "The Discovery and Development of HMG-CoA Reductase Inhibitors."[86]

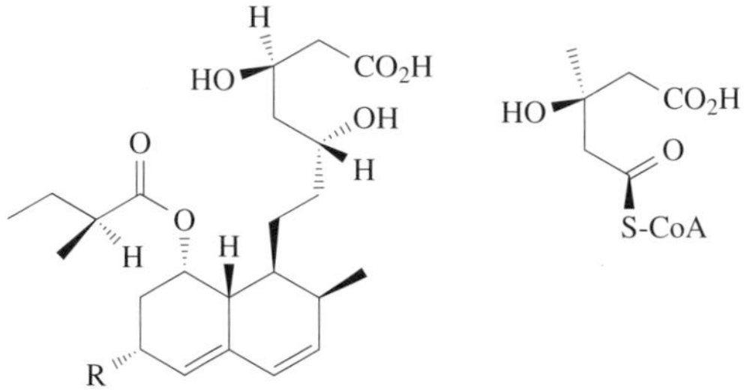

Figure 9 Structures of mevastatin and lovastatin. Shown on the left is the acid form of mevastatin (compactin) and lovastatin (mevinolin), and on the right of HMG-CoA. The substituent R at C-3 of the hexahydronaphthalene ring is H in mevastatin and Me in lovastatin. Like mevalonate, but unlike HMG-CoA, mevastatin and lovastatin can form inner esters, or lactones, by elimination of water from the carboxylic acid group on C-1 and the OH group on C-5 of both structures. The water-soluble, open-chain forms are the more pharmacologically potent forms of these drugs.[86]

2.02.4.7 Plants Elaborate Multiple HMG-CoA Reductase Isozymes

Plant HMG-CoA reductases have been the subject of several recent reviews.[46,49,87,88] The body of knowledge of plant HMG-CoA reductases, while not inconsiderable, is nevertheless less extensive than for the vertebrate and other forms of the enzyme. Particularly in the areas of structure, catalysis, and regulation of activity, we thus shall frequently cite observations and conclusions drawn from nonplant forms of the enzyme, indicating where possible, implications for the HMG-CoA reductases of plants. Plants and yeast,[88] unlike mammals, elaborate a rich diversity of HMG-CoA reductase isozymes. For example, there are at least two HMG-CoA reductase genes in *A. thaliana*,[89,90] radish,[85] and wheat,[91] three in the rubber tree[92] and potato,[93] and four in tomato.[95] The high degree of sequence homology between plant HMG-CoA reductase isozymes is consistent with their having arisen as a consequence of gene duplication and subsequent sequence divergence.[95]

2.02.4.8 Plant HMG-CoA Reductase Isozymes Fulfill Discrete Physiologic Functions

Isozymes of plant HMG-CoA reductases are membrane-associated proteins present in plastids, mitochondria, or the cytoplasmic face of the endoplasmic reticulum. Different isozymes fulfill distinct physiologic functions, either in primary metabolism such as fruit ripening[96] or in response to environmental challenges.[97] Metabolic channeling of mevalonate, a consequence of multiple genes, differential subcellular localization of gene products, and the varied responses of isozymes to physiologic events and environmental challenges, plays a central role in isoprenoid metabolism by plants. HMG-CoA reductase isozymes of higher plants generate distinct metabolic pools of mevalonate destined for the biosynthesis of specific isoprenoid products. Localization in specific subcellular organelles can facilitate independent regulation of parallel pathways that produce different products. For example, carotenoids, ubiquinone, and sterols are synthesized in the chloroplasts, mitochondria, and cytoplasm, respectively.[98]

2.02.4.9 Stimuli Elicit Differential Effects on Plant HMG-CoA Reductase Isozymes

HMG-CoA reductase isozymes are differentially responsive to light. The activity of plastid HMG-CoA reductase rose, whereas that of the microsomal isozyme fell in pea seedlings transiently irradiated with red light,[99] and expression of the *HMG1* gene of *A. thaliana* was differentially responsive to light of different wavelengths.[100] HMG-CoA reductase isozymes are also differentially expressed during plant development. HMG-CoA reductase activity and mRNA levels were high in early stages of tomato fruit development, but low in ripening fruit.[101] Many molecules that plants employ to counteract physical injury or invasion by pathogens are isoprenoids or isoprenoid derivatives. Following injury or exposure to pathogens or pathogen-derived elicitor compounds, these insults initiate induction or repression of genes that encode certain HMG-CoA reductase isozymes, which can result in large changes in HMG-CoA reductase activity.[93,97,102]

2.02.4.10 Yeast also Synthesizes HMG-CoA Reductase Isozymes

The Hmg1p and Hmg2p isozymes of yeast consist of a highly conserved C-terminal catalytic domain linked to a hydrophobic N-terminal domain. The amino acids of the catalytic domains encoded by the genes *hmgr*-1 and *hmgr*-2 are 93% identical to one another and 65% identical to those present in the catalytic domain of human HMG-CoA reductase. By contrast, the hydrophobic domains encoded by *hmgr*-1 and *hmgr*-2 are only about 50% identical to one another, and exhibit no sequence similarity to the anchor domains of other HMG-CoA reductases.[88,103] A recent review[55] discusses the signals that regulate yeast HMG-CoA reductase and the differential responses of each yeast isozyme to a given signal. For example, the two yeast isozymes respond differentially to changing oxygen concentration. The molecular sensor that triggers these changes is the intracellular level of heme, whose synthesis requires oxygen. At low partial pressures of oxygen, Hmg2p predominates. By contrast, Hmg1p predominates during growth under aerobic conditions.[55]

2.02.4.11 Animals do not Appear to Elaborate HMG-CoA Reductase Isozymes

Mevalonate serves as the precursor of diverse isoprenoids in all eukaryotes. In animals these include, in addition to cholesterol, ubiquinone, dolichols, and the prenyl groups of prenylated proteins and certain transfer RNAs. In view of this product diversity, it is somewhat surprising that animals appear to possess only a single HMG-CoA reductase, anchored so as to orient its catalytic domain toward the cytosolic face of the endoplasmic reticulum.

2.02.4.12 Purified HMG-CoA Reductases

Several HMG-CoA reductases that lack a membrane anchor domain have been purified to homogeneity. These include the HMG-CoA reductases of the noneukaryotes *P. mevalonii*, *Haloferax volcanii*, and *Sulfolobus solfataricus*. The apolar character of the anchor domain of the HMG-CoA reductases of eukaryotes constitutes, however, a formidable obstacle to their purification. The intact holoproteins can be solubilized using detergents, but only partial purification of the detergent-solubilized rat liver enzyme has been achieved.[104] Most reports of purified eukaryotic HMG-CoA reductases concern a catalytically active fragment derived from the C-terminal portion of the enzyme. This fragment, or catalytic domain, is liberated spontaneously during most enzyme preparations. When liver microsomes are frozen and thawed in the absence of the protease inhibitor leupeptin, the catalytic domain of HMG-CoA reductase is released with high efficiency.[105] The rupture of lysosomes that contaminate crude microsomal preparations releases proteases that liberate a soluble, catalytically active fragment of approximate subunit mass 50 000–56 000 kDa, or about half that of the intact mammalian protein. As reviewed by Kleinsek *et al.*,[106] this catalytically active fragment has been purified and studied in many laboratories. As described below, purification now generally employs expression in *E. coli* of truncated eukaryotic HMG-CoA reductase cDNAs that encode only the C-terminal catalytic domain.[83,84]

Sequence databases contain over four dozen complete and partial sequences of HMG-CoA reductase genes. These provide an invaluable guide to detection of conserved regions of structure and potentially important aminoacyl residues (see Figure 6). Of these sequenced genes, a smaller number have been expressed in *E. coli*, and the enzymic and regulatory properties of the encoded enzymes examined. Examples include the genes that encode (i) the catalytic domain of the human,[83] Syrian hamster,[84] and *A. thaliana*[107] enzymes; (ii) the *H. volcanii*[66,68] and *S. solfataricus* enzymes;[69] and (iii) the biodegradative enzyme from the soil bacterium *P. mevalonii*[80] (Table 1). As the only

HMG-CoA reductase whose three-dimensional structure has been solved,[113] *P. mevalonii* HMG-CoA reductase has provided numerous mechanistic insights into biosynthetic HMG-CoA reductases.

Table 1 HMG-CoA reductase genes that have been expressed in *E. coli.*

HMG-CoA reductase gene	*Genbank accession number*	*Ref.*
Homo sapiens (human)	M11058	76
Mesocricetus auratus (Syrian hamster)	M12705	77
Cricetulus griseus (Chinese hamster)	X00494	108
	L00165	
Blatella germanica (cockroach)	X70034	109
Arabidopsis thaliana isoform 1	X15032	107
Raphanus sativus 1 (radish)	M21329	110
Ustilago maydis (maize fungal pathogen)	L19262	111
Haloferax volcanii (archaeon)	L19349	66
Sulfolobus solfataricus (archaeon)	M22002	69
Schistosoma mansoni (schistosome)	M22255	112
Pseudomonas mevalonii (bacterium)[a]	M83531	73

[a] A biodegradative enzyme that uses NAD^+ rather than $NADP^+$.

2.02.4.13 Crystallographic Structure of *Pseudomonas mevalonii* HMG-CoA Reductase

The 0.3 nm crystallographic structure of *P. mevalonii* HMG-CoA reductase[113] reveals a tightly bound dimer which brings together residues of importance to substrate binding and catalysis at the subunit interface. Each monomer contains a large and small domain which bind one substrate apiece, and which pack together to form an elongated ellipsoidal molecule (Figure 10). The large domain, which contains residues 4–109 and 216–376, has an unusual structure comprised of a central α-helix surrounded by a triangular set of walls of β-sheets and α-helices. This domain makes the major dimer contacts, with a highly intertwined N-terminus and a four-helix bundle which forms at the center of the molecule. The large domain is the primary binding site for HMG-CoA, and contains three residues implicated in catalysis, Glu83, Lys267, and Asp283. The small domain, which contains residues 110–215, also has an α/β-fold consisting of an interdigitated four-strand antiparallel β-sheet with helices packed against one side. In the dimer, the small domain packs against a long helix in one wall of the large domain, forming a shallow pocket at the dimer interface that is the active site. The small domain serves primarily to bind NAD(H). Unlike enzymes whose NAD-binding domains are located towards the N-terminus (e.g., lactate dehydrogenase) or the C-terminus (e.g., alcohol dehydrogenase), the NAD-binding domain of *P. mevalonii* HMG-CoA reductase is centrally located. The C-terminal portion of the monomer (residues 377–428), which is disordered in the structure of the unliganded enzyme, comprises a mobile domain or flap that closes over the active site during catalysis.[114] This flap domain contains the conserved histidine, His381, and Arg387, the residue which occupies the position that corresponds to the phosphorylation target serine responsible for regulation of HMG-CoA reductases of higher eukaryotes.

2.02.4.14 Ternary Complex Structures Reveal a Substrate-induced Closure of the Flap Domain

The structure of binary and non-productive ternary complexes of *P. mevalonii* HMG-CoA reductase have now been solved.[113,114] The HMG-CoA binds to the large domain in a curled conformation which extends the scissile thioester bond into the intersubunit active site. NAD binds to the small domain in an extended configuration with the nicotinamide ring roughly parallel to the thioester carbonyl plane to facilitate hydride transfer. In the ternary complex, the flap domain can be seen closed over the active site (Figure 11). This substrate-induced closure appears to trap the HMG-CoA and intermediates, but permits the exchange of NAD(H) essential for this two-stage reaction.

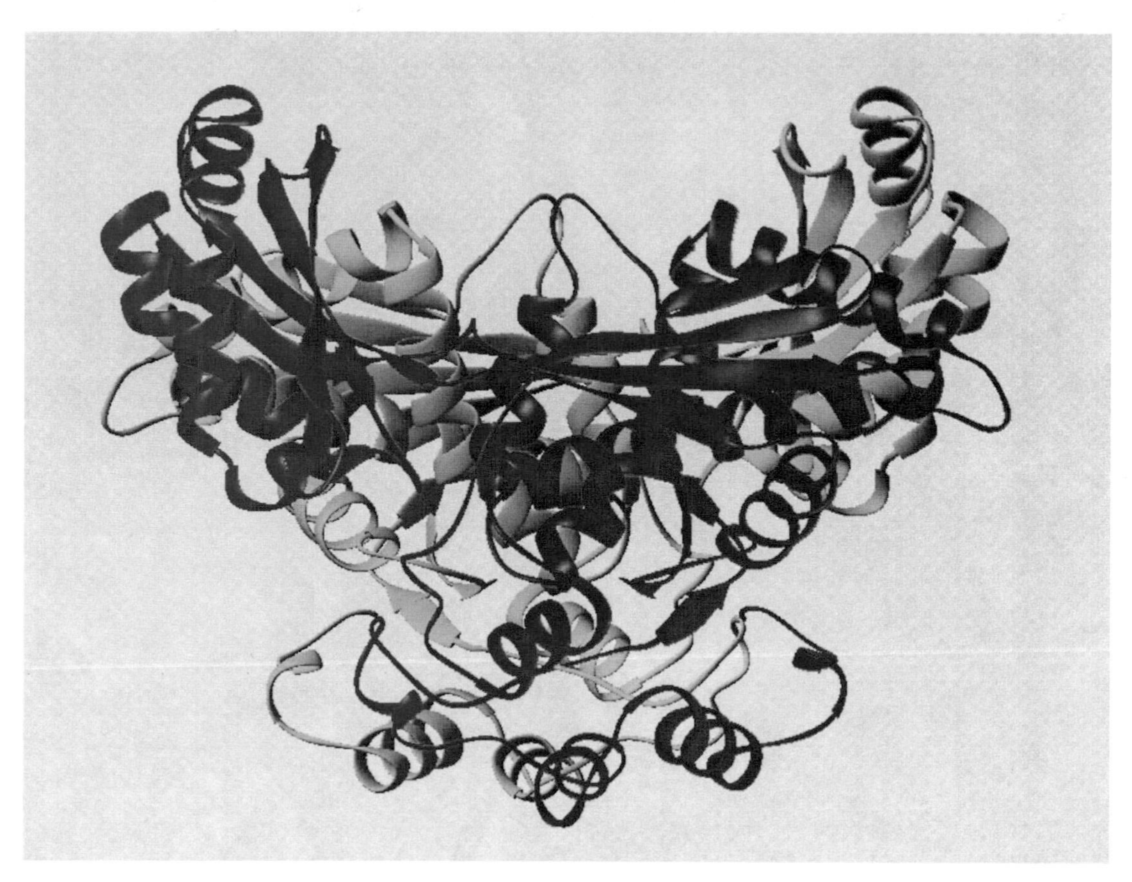

Figure 10 The *P. mevalonii* HMG-CoA reductase dimer. A side view of the dimer of HMG-CoA reductase is shown with one monomer subunit in red and the second in gray. In the red monomer in the foreground, the small domain is to the left and the large domain to the right. The two active sites in the dimer are located under the "wings" of the structure at the interface between subunits. The C-terminal flap domain, which has no defined position in the unliganded structure, would extend from the C-terminus of the long central helix of the large domain near the active site. The tightly intertwined N-terminal domains at the bottom of the figure are where it may be inferred that the membrane anchor of a eukaryotic HMG-CoA reductase should attach.[113]

2.02.4.15 The Small Domain of *Pseudomonas mevalonii* HMG-CoA Reductase has an Unusual NAD-binding Fold

The dinucleotide binding fold of the small domain of *P. mevalonii* HMG-CoA reductase differs from a classical Rossmann fold[115] and from other non-Rossmann dinucleotide binding folds.[116–118] The α/β-fold of the small domain consists of four antiparallel rather than the six parallel β-strands of a classic dinucleotide binding fold (Figure 12). The critical helix, which interacts at its amino terminus with the adenine phosphate, also lies in a different part of the fold, and the configuration of the NAD seen in binary complex structures is different relative to the β-sheet direction. Although the folds of the small domain and of a classical dinucleotide binding fold differ, certain features are common to both. The adenine phosphate interacts with the critical helix in much the same fashion. In addition, the position of Asp146 of *P. mevalonii* HMG-CoA reductase (Figure 13) suggests that it performs a function analogous to that of the invariant aspartate of a Rossmann NAD-binding fold, hydrogen bonding to the 2′-hydroxyl of the adenine ribose of NAD, discriminating against NADP(H).

2.02.4.16 Determinants of Coenzyme Specificity of the *Pseudomonas mevalonii* Enzyme

Consistent with its catabolic function in *P. mevalonii*, and unlike the biosynthetic HMG-CoA reductases for which coenzyme specificity has been established, the HMG-CoA reductase of *P. mevalonii* uses NAD(H) as an oxidoreductant. The initial binary complex of the enzyme with NAD

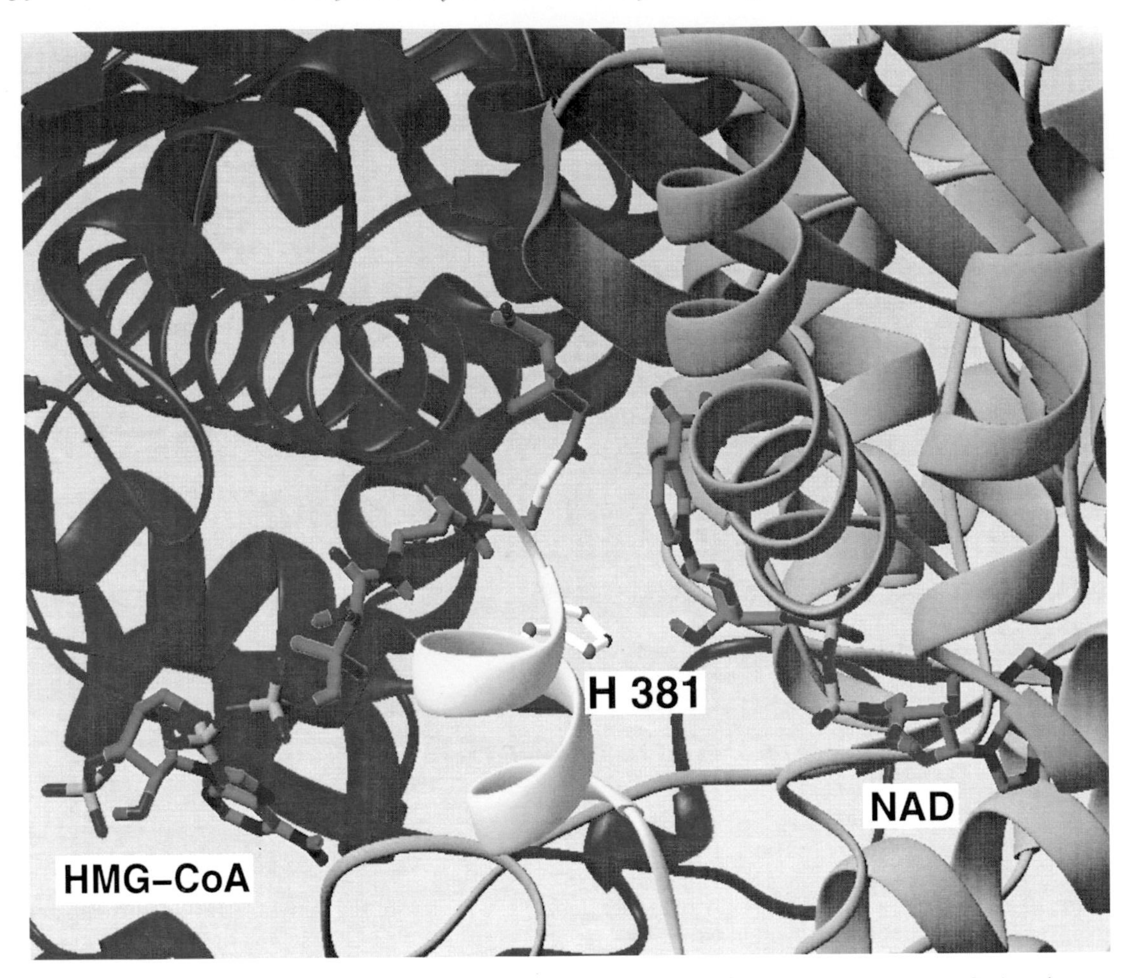

Figure 11 The active site of the HMG-CoA–NAD ternary complex of *P. mevalonii* HMG-CoA reductase. Shown is a close-up view of the ternary complex active site from a view 90° away from that shown in Figure 10. The structure shown in yellow is a portion of the flap domain, which has closed over the active site in the presence of the substrates (drawn as stick models in which C = green, N = blue, O = red, P = gray, and S = yellow). HMG-CoA binds to the large subunit of the red monomer, extending the scissile thioester bond into the active site pocket. NAD binds to the small subunit of the gray monomer with the nicotinamide ring parallel to the thioester carbonyl plane. The first helix of the flap domain (yellow) closes over the active site, trapping the substrate and bringing His381 into position for catalysis.[114]

showed that Asp146 was within hydrogen-bonding distance of the 2′-hydroxyl of the adenine ribose of NAD, suggesting that removal of bulk, charge, or hydrogen-bonding capability at position 146 of the *P. mevalonii* enzyme might enhance its ability to use NADP(H). The primary role of Asp146 in discriminating against NADP was supported by the 1200-fold improvement in NADP specificity, expressed as the ratio of k_{cat}/K_m for NADP to k_{cat}/K_m for NAD, when Asp146 was replaced by alanine, and by the 6700-fold increase in NADP specificity when Asp146 was replaced by glycine. Further improvements in NADP specificity resulted when the adjacent residues Thr192 and Leu148 were mutated to lysine or arginine to provide charge stabilization of the 2′-phosphate on the adenine ribose of NADP, which would occupy roughly the same position as the carbonyl oxygens of Asp146 (Figure 13). For the most effective mutant enzyme, D146A/L148A, NADP specificity had increased 83 000-fold relative to the wild-type enzyme.[119]

2.02.4.17 Catalysis Involves Two Reductive Stages

In the biosynthetic reaction catalyzed by HMG-CoA reductase, 2 mol of NADPH serve as the reductant during the deacylation of HMG-CoA to the isoprenoid precursor mevalonate (Equation

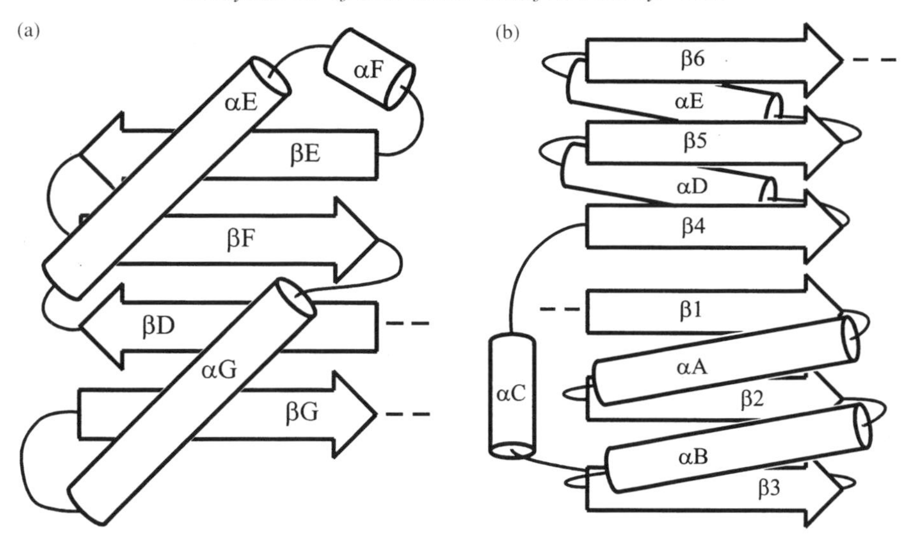

Figure 12 Comparison of secondary structural elements of the nucleotide-binding domains of (a) *P. mevalonii* HMG-CoA reductase[113] and of (b) a classical Rossmann dinucleotide-binding domain.[115] Arrows represent β-strands and cylinders α-helices.

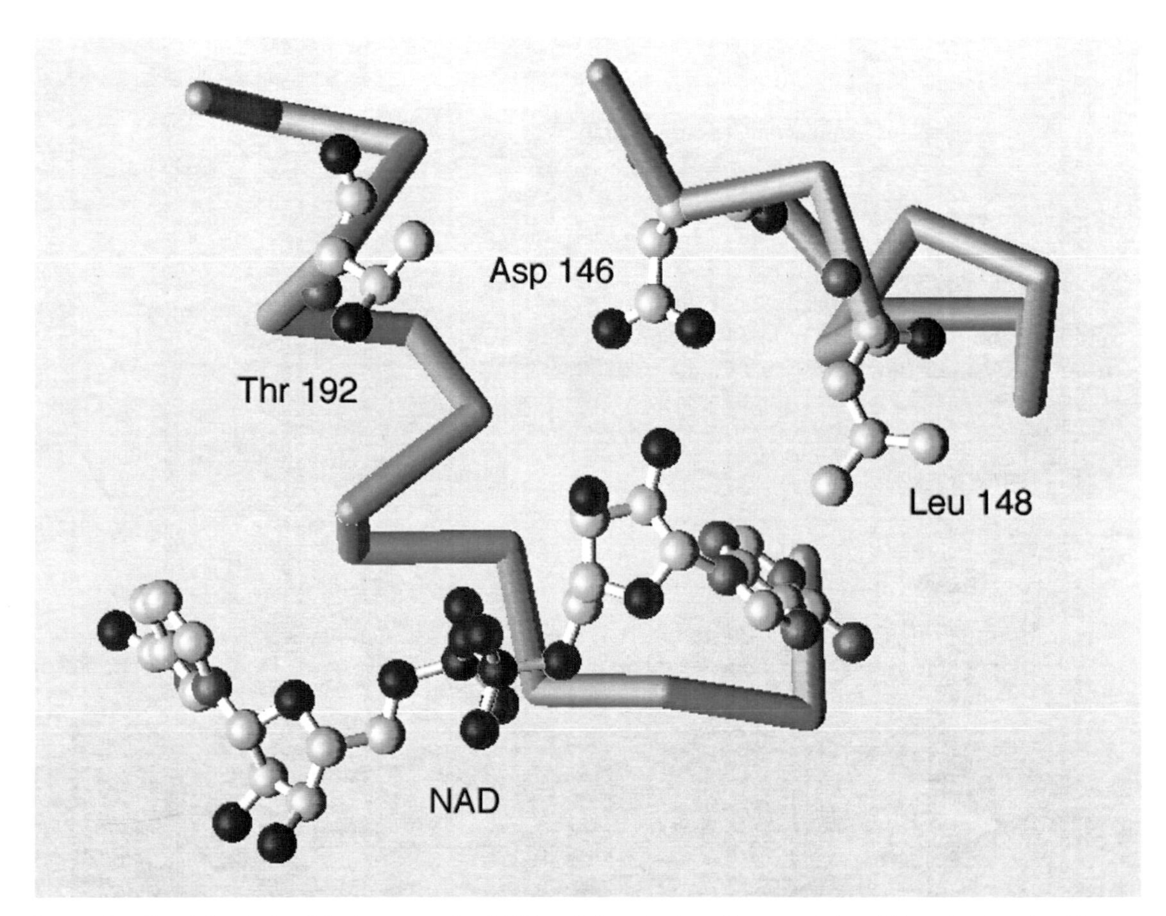

Figure 13 Spatial relationship between the 2′-OH of the adenine ribose and Asp146. α-Carbon backbone segments (residues 145–155 and 182–195) of the small domain of the enzyme are colored blue-green. The NAD and selected side chains are drawn as ball-and-stick models (C = yellow, N = blue, O = red, and P = magenta). The 2′-OH makes hydrogen bond contacts with both ω-carbonyl oxygens of Asp146. Also shown are residues Leu148 and Thr192 that were mutated to lysine or arginine to stabilize the charge on the NADP(H) phosphate.

(4)). While freely reversible *in vitro*, in eukaryotes and archaea the HMG-CoA reductase reaction proceeds unidirectionally toward mevalonate biosynthesis.

By analogy with Equation (5), the biodegradative HMG-CoA reductase of *P. mevalonii* catalyzes the acylation of mevalonate to HMG-CoA, but also utilizes NAD as an oxidant. Following cleavage by HMG-CoA lyase, the carbon atoms of mevalonate then enter pathways of two-carbon metabolism.[79,120]

$$(R)\text{-mevalonate} + \text{CoA-SH} + 2\ \text{NADP}^+ \longrightarrow (S)\text{-HMG-CoA} + 2\ \text{NADPH} + 2\ \text{H}^+ \quad (5)$$

While no intermediates are released or have been isolated during either the biosynthetic or biodegradative reactions, postulated intermediates include mevaldehyde and mevaldyl-CoA[114] (Figure 14). The overall reaction catalzyed by HMG-CoA reductase therefore appears to proceed via two successive reductive stages and to involve the intermediate formation of enzyme-bound mevaldehyde and a mevaldehyde derivative (Equations (6) and (7)).

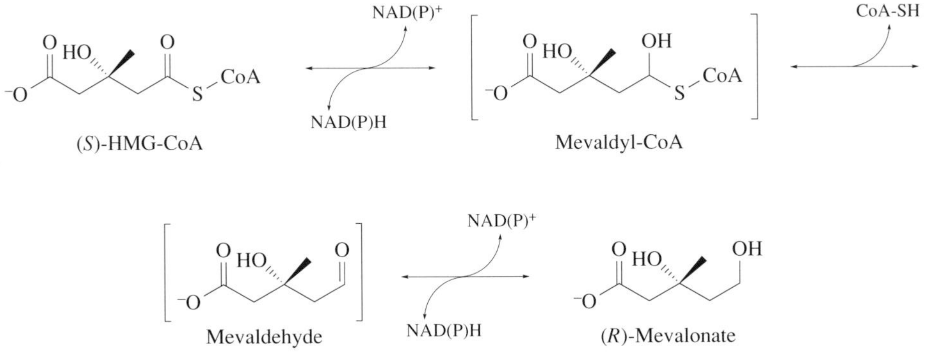

NAD(P)$^+$ CoA-SH
(*S*)-HMG-CoA NAD(P)H Mevaldyl-CoA
NAD(P)$^+$
Mevaldehyde NAD(P)H (*R*)-Mevalonate

Figure 14 Postulated enzyme-bound intermediates in the HMG-CoA reductase reaction. Mevaldyl-CoA and mevaldehyde are in brackets. Although both are substrates (see Equations (8)–(11)) neither has been isolated as an intermediate during catalysis by any HMG-CoA reductase.

First reductive stage $$\text{HMG-CoA} + \text{NADPH} + \text{H}^+ \longrightarrow [\text{mevaldyl-CoA}] + \text{NADP}^+ \quad (6)$$

Second reductive stage $$[\text{mevaldehyde}] + \text{NADPH} + \text{H}^+ \longrightarrow \text{mevalonate} + \text{NADP}^+ \quad (7)$$

Kinetic analysis of product inhibition implies that binding of HMG-CoA and NADPH may be either ordered or random. For both reductive stages, hydride transfer is from the A side of NADPH. Hydride transfer in the first reductive stage forms an enzyme-bound thiohemiacetal, mevaldyl-CoA. Rearrangement to mevaldehyde and coenzyme A, which remain enzyme bound, is followed by hydride transfer in the second reductive stage, forming mevalonate. Mevalonate, and subsequently coenzyme A and NADP, then dissociate from the enzyme.[122–124]

2.02.4.18 Analysis of Additional Reactions Facilitates the Study of Catalysis

In addition to their ability to interconvert HMG-CoA and mevalonate, HMG-CoA reductases also catalyze the oxidation of free mevaldehyde or free mevaldyl-CoA to HMG-CoA (Equations (8) and (10)) and their reduction to mevalonate (Equations (9) and (11)).

$$\text{mevaldehyde} + \text{CoA-SH} + \text{NADP}^+ \longrightarrow \text{HMG-CoA} + \text{NADPH} + \text{H}^+ \quad (8)$$

$$\text{mevaldehyde} + \text{NADPH} + \text{H}^+ \longrightarrow \text{mevalonate} + \text{NADP}^+ \quad (9)$$

$$\text{mevaldyl-CoA} + \text{NADP}^+ \longrightarrow \text{HMG-CoA} + \text{NADPH} + \text{H}^+ \quad (10)$$

$$\text{mevaldyl-CoA} + \text{NADPH} + \text{H}^+ \longrightarrow \text{mevalonate} + \text{NADP}^+ + \text{CoA-SH} \quad (11)$$

Equation (8), the oxidative acylation of free mevaldehyde to HMG-CoA appears to model the reverse of the first reductive stage. Both for Equations (8) and for glyceraldehyde 3-phosphate dehydrogenase (EC 1.2.1.12), a general acid–base facilitates hydride transfer during interconversion of an aldehyde and a high-energy derivative of a carboxylic acid. Both reactions also involve attack on the carbonyl carbon of an aldehyde by a thioanion, that of an active site cysteine (glyceraldehyde-3-phosphate dehydrogenase) or of coenzyme A (HMG-CoA reductase). Equation (9), which appears to model the second reductive stage of the overall reaction, shares features with the reaction catalyzed by alcohol dehydrogenase (EC 1.1.1.1). In both instances a general acid–base facilitates hydride transfer for reduction of an aldehyde to an alcohol.

2.02.4.19 General Acids and Bases Facilitate Catalysis

The first indication that general acids–bases participate in catalysis was provided by investigation of the pH dependence of kinetic parameters for catalysis of Equations (8) and (9) by yeast HMG-CoA reductase.[125] The investigators concluded that a histidine and an acidic residue participate in catalysis of Equations (8) and (9), and by inference also in Equation (4). The deduced amino acid sequences of HMG-CoA reductases subsequently suggested conserved residues that might fulfill these proposed catalytic roles. These conserved residues were then targeted for study by site-directed mutagenesis. Kinetic analysis of point mutant enzymes implicated a histidine,[68,126,127] and not one but two acidic residues, a glutamate[128] and an aspartate,[129] as functional in catalysis. We hereafter refer to these residues as the active site histidine, glutamate, and aspartate. Table 2 lists the locations of the general acids–bases that have been implicated by mutation and subsequent kinetic analysis as functional in catalysis.

Table 2 Identified active site residues of selected HMG-CoA reductases.

HMG-CoA reductase of	*Numbering of active site residues*
Syrian hamster	His865, Glu558, Asp766
P. mevalonii	His381, Glu83, Asp283
H. volcanii	His398

2.02.4.20 Cysteines Play No Role in Catalysis

HMG-CoA reductase has long been known to be inhibited by oxygen and by organic reagents that modify cysteinyl residues.[130] The mammalian enzyme in particular is extremely sensitive to oxidative inactivation by micromolar concentrations of glutathione disulfide even in the presence of millimolar concentrations of glutathione.[131] That cysteines play no known role in catalysis by HMG-CoA reductase may be inferred from examination of *P. mevalonii* HMG-CoA reductase. This bacterial enzyme contains only two cysteines. Neither cysteine is conserved across genera, and their replacement by alanine is accompanied by retention of full catalytic activity.[132] Cys296, the cysteine of the *P. mevalonii* enzyme whose derivatization results in loss of activity,[130] is now known to be located at the subunit interface.[113] The loss of activity that accompanies derivatization of Cys296 thus appears to be a consequence of an altered dimer interface, which disrupts the active site. Similar considerations may account for loss of activity consequent to derivatization of cysteines of other forms of the enzyme.[132]

2.02.4.21 A Histidine is Essential for Catalysis of the Overall Reaction

Diethylpyrocarbonate, a reagent that derivatizes histidyl residues, inactivates *P. mevalonii* and Syrian hamster HMG-CoA reductases, and subsequent treatment with hydroxylamine restores their catalytic activity.[126,127] Sequence alignments of all known HMG-CoA reductases revealed that His381 of the *P. mevalonii* enzyme was the only conserved histidine.[126] Analysis of *P. mevalonii* mutant enzyme H381Q revealed that His381 was essential for catalysis of the overall catabolic reaction, the oxidative acylation of mevalonate to HMG-CoA (see Figure 7).[126] This histidine is not visible in the initial crystal structure of the unliganded enzyme, which has disorder in the flap domain beyond residue 378.[113] The cognate residues of His381 His865 of the Syrian hamster enzyme[127] and His398 of the *H. volcanii* enzyme,[68] also are essential for catalysis of Equation (4).

Despite the ability of hamster mutant enzyme H865Q to catalyze the reduction of mevaldehyde to mevalonate (Equation (9)) in the absence of coenzyme A at 80% of the wild-type rate, this mutant enzyme exhibited less than 0.2% wild-type activity for the overall reaction (Equation (4)). These observations would appear inconsistent with the initially proposed functions of the conserved histidine,[119] accepting a proton from the OH of the thiohemiacetal group of mevaldyl-CoA during its oxidation to HMG-CoA, and donating a proton to mevaldehyde during its reduction to mevalonate.

2.02.4.22 Histidine Appears to Protonate the Coenzyme A Thioanion Released Subsequent to the First Reductive Stage of the Overall Reaction

Coenzyme A has long been known to stimulate catalysis of the reduction of exogenous mevaldehyde to mevalonate (Equation (9)).[121,122,125] By contrast, coenzyme A severely inhibited catalysis of this reaction by hamster mutant enzyme H865Q. Unlike coenzyme A, desthio-CoA, which lacks only the sulfur atom of coenzyme A, stimulated catalysis by both the wild-type and mutant enzyme. Together with the observation that inhibition by coenzyme A decreased at low pH, these data imply that the inhibitory species is the coenzyme A thioanion, CoA-S$^-$.[133] The active site histidine thus appears to be the general acid that protonates the CoA-S$^-$ thioanion released during catalysis of the overall reaction (Equation (4)), and the general base that deprotonates coenzyme A during the reverse of this reaction. An inability to protonate the CoA-S$^-$ thioanion would allow it to attack mevaldehyde, blocking the course of the overall reaction.[133] These inferences are consistent with the inability of mutant enzyme H865Q to catalyze mevaldehyde reduction, but not the overall reaction.

2.02.4.23 A Glutamyl Residue Participates in Catalysis of the Second Reductive Stage

A role in catalysis has been proposed for Glu83 of the *P. mevalonii* enzyme[128] and for its cognate residue, Glu558 of the hamster enzyme.[129] An early crystallographic analysis of the binary complex of HMG-CoA with *P. mevalonii* HMG-CoA reductase[113] revealed that the γ-carboxyl of Glu83 was within hydrogen-bonding distance of the carbonyl oxygen of HMG-CoA. As anticipated, substitution of glutamine for Glu558, the cognate glutamate of the hamster enzyme, severely impaired catalysis. Mutant enzyme E558Q catalyzed the overall reaction (Equation (4)) at less than 0.3% the rate catalyzed by the wild-type enzyme. The pH profile for the oxidative acylation of mevaldehyde (Equation (8)) catalyzed by enzyme E558Q suggested that Glu558 may serve as a general base during catalysis of this reaction.[133]

2.02.4.24 Aspartate Appears to Act as a General Acid–Base in Both Reductive Stages

Mutagenesis of conserved Asp766 of the hamster enzyme suggested that it functions in both reductive stages of the overall reaction. Despite K_m values and an optimal pH for activity similar to that of the wild-type enzyme, hamster mutant enzyme D766N catalyzed the overall reaction at less than 0.2% the wild-type rate, and catalyzed no other reaction at a detectable rate. This suggested that Asp766 may be the general acid that protonates the carbonyl oxygen of HMG-CoA and of the mevaldehyde, facilitating hydride transfer.[133] However, the binary complex structure of HMG-CoA with HMG-CoA reductase[113] revealed that the β-carboxyl of the cognate aspartate, Asp283, was greater than 0.6 nm from the carbonyl oxygen of HMG-CoA, too far to readily fulfill this postulated function.

2.02.4.25 Proposed Mechanism of Catalysis Based on Mutagenesis and Structural Evidence

The roles of the residues identified by mutagenesis as important for catalysis have recently been clarified by examination of the ternary complex structures of *P. mevalonii* HMG-CoA reductase.[114] The acidic residues Asp283 and Glu83, which are contributed by two different monomers, lie close to each other in the shallow active site pocket (Figure 15). A lysine residue, Lys267, is positioned next to the carbonyl oxygen of HMG-CoA and is tightly hydrogen bonded to Asp283, which in turn contacts Glu83 to create a three-residue link that appears to serve as the general acid–base in the reaction. Substrate binding induces closure of the flap domain (see Figure 11), which positions the catalytic histidine, His381, so that it can donate a proton to the CoA-S$^-$ leaving group. NADH bound to the small domain of the opposite monomer is situated for an B-side transfer of hydride, which is facilitated by polarization of the carbonyl and proton transfer. Site-directed mutagenesis of the active site lysine identified in these structures is consistent with its postulated role, as mutant enzyme K267A catalyzes the reaction shown in Figure 7 with a specific activity less than 0.06% that of the wild-type enzyme.[114] The revised mechanism based on these coordinated structural and biochemical studies is closely related to previous proposals, but uses different residues to accomplish the catalytic steps shown in Figure 16.

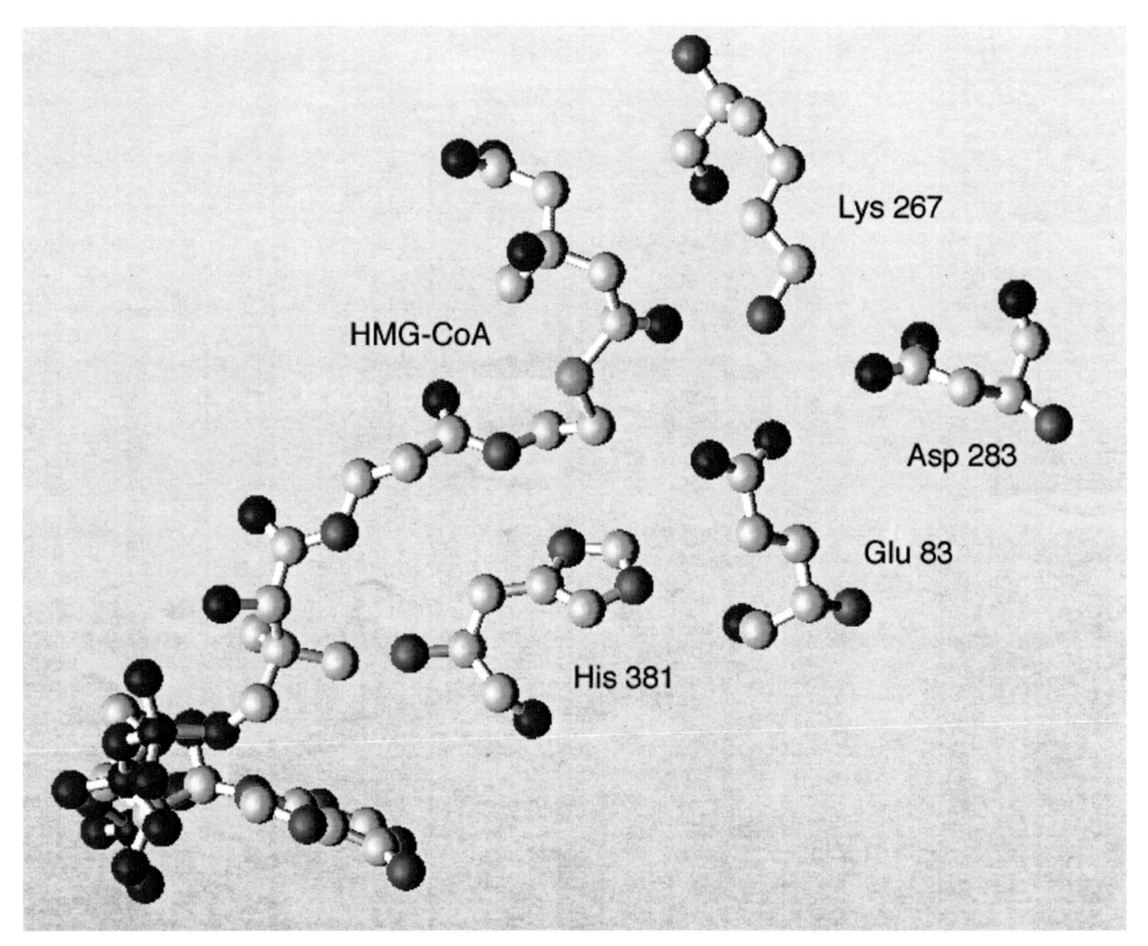

Figure 15 Configuration of active site residues in the ternary complex of *P. mevalonii* HMG-CoA reductase. HMG-CoA and active site residues are shown as ball-and-stick models (C = yellow, N = blue, O = red, P = magenta, and S = green). Lys267 is centrally located in the active site, hydrogen bonded to the thioester carbonyl oxygen of HMG-CoA on one side and to Asp283 on the other. Glu83 in turn is hydrogen bonded to Asp283, and is in close contact with Lys267. These three residues form the active site proton donor triad. His381, which is part of the flap domain, is positioned to protonate the CoA-S$^-$ leaving group (S shown in green) created by cleavage of the thioester bond of HMG-CoA.

2.02.4.26 The Minimal Functional Unit of HMG-CoA Reductase is a Dimer that Incorporates Residues from Different Subunits

Determination of the molecular mass of HMG-CoA reductase in its physiologic state is complicated by its immobilization in subcellular membranes and organelles. Edwards *et al.*[135] approached this problem by studying the kinetics of radiation inactivation of the HMG-CoA reductase activity of rat liver microsomes and intact hepatocytes. These investigators concluded that the minimal functional size of the mammalian HMG-CoA reductase is a dimer. Supportive evidence for a dimer as the minimal functional unit was subsequently provided by the simultaneous expression in *E. coli* of two different, catalytically inactive, forms of the catalytic domain of Syrian hamster HMG-CoA

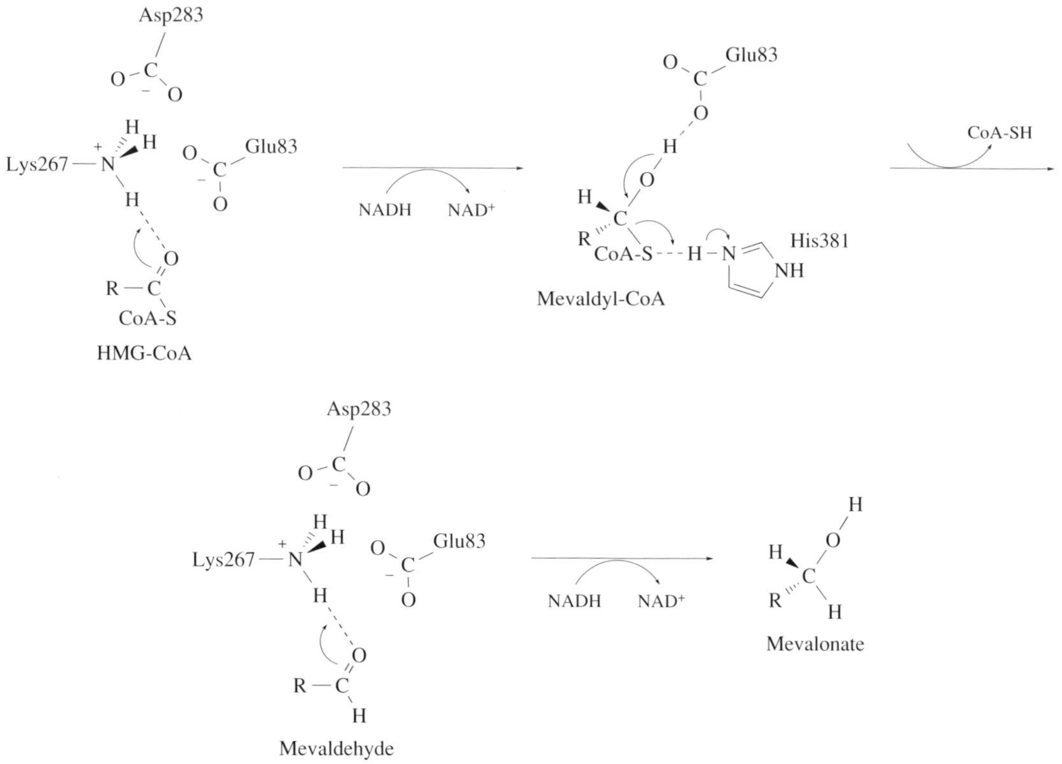

Figure 16 Proposed mechanism for catalysis by HMG-CoA reductase. The model incorporates evidence from crystallographic and mutagenesis studies. Stage 1: protonation of HMG-CoA by Lys267 of the Lys–Asp–Glu triad facilitates hydride transfer from NADH, forming enzyme-bound mevaldyl-CoA. Stage 2: His381 donates a proton to the coenzyme A thioanion formed as a consequence of the abstraction by Glu83 of a proton from bound mevaldyl-CoA. Stage 3: protonation of bound mevaldehyde by Lys267 of the Lys–Asp–Glu triad facilitates a second hydride transfer, forming mevalonate. Reprotonation of the histidine and the Lys–Asp–Glu triad could then occur from the solvent subsequent to flap opening and product release. The reductant shown, NADH, is for the *P. mevalonii* enzyme. For all other forms of the enzyme for which nucleotide specificity has been determined, the reductant is NADPH. (Redrawn by permission of Tabernero *et al.*[114])

reductase.[129] The simultaneous expression in a single cell of genes encoding inactive mutant enzymes E558Q and D766N yielded an enzyme preparation that had 25% wild-type activity for catalysis of the overall reaction (Equation (4)). This was interpreted as resulting from the presence of equal quantities of inactive homodimers and partially active heterodimers (Figure 17). It was furthermore concluded that the active site resides at the dimer interface and that it recruits active site residues from both subunits. Simultaneous x-ray crystallographic solution of the structure of *P. mevalonii* HMG-CoA reductase revealed that it exists in the crystal as a trimer of dimers[113] with an obligate dimer as the minimal catalytic unit. This structure revealed that the active site was indeed present at the dimer interface, and that the active site recruits the active site aspartyl and glutamyl residues from different subunits.

2.02.4.27 Multiple Controls Regulate HMG-CoA Reductase

Regulation of carbon flux from HMG-CoA to mevalonate is achieved via multiple levels of control of HMG-CoA reductase. While most firmly established in animals, emerging evidence suggests a comparable complexity of controls in plants and yeast. Both transcription and translation of the mammalian enzyme are subject to feedback regulation by sterols and nonsterol products derived from mevalonate. The enzyme is also rapidly degraded in the presence of excess sterols and nonsterol products. Altering the rates of synthesis and degradation can vary the steady-state levels of HMG-CoA reductase over 200-fold. In higher eukaryotes, the activity of the pre-existing enzyme is also regulated post-translationally by phosphorylation and dephosphorylation. These multiple controls make HMG-CoA reductase one of the most highly regulated enzymes in nature.[54]

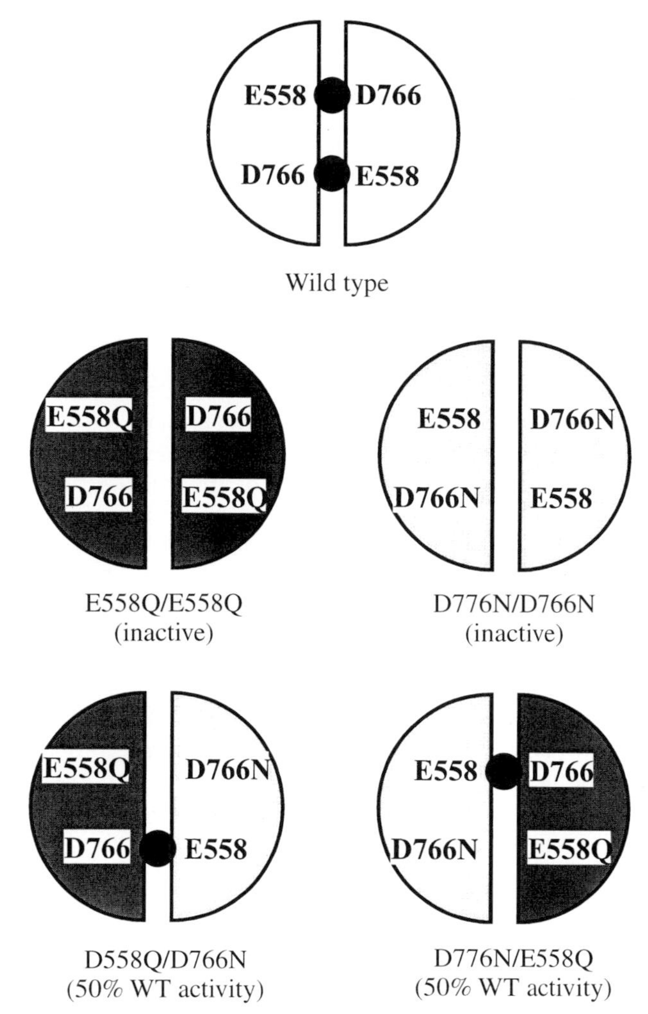

Figure 17 Representation of the location of acidic residues of the intersubunit active site of hamster HMG-CoA reductase. Represented are wild-type enzyme, inactive homodimers of mutant enzymes E558Q and D766N, and active heterodimers composed of one E558Q subunit and one D766N subunit. The symbol ● represents a functional active site.[129]

2.02.4.28 Regulation of Transcription

Multiple factors affect the expression of HMG-CoA reductase genes. In mammals, most observations concern regulation by sterols. Cellular sterol levels regulate the transcription of genes that encode the LDL receptor, HMG-CoA synthase, and HMG-CoA reductase. Mutation of the promoters for these genes has identified a putative common *cis*-acting sterol regulatory element, SRE-1, that confers sterol regulation when incorporated into the promoter for herpes simplex virus thymidine kinase.[54] Additional mutations of the HMG-CoA reductase promoter suggested that its sterol regulatory site may be distinct from SRE-1, and that it differs from, and is more complex than, the promoters for the LDL receptor or HMG-CoA synthase.[135] The 43 kDa protein Red 25, which does not bind to the sterol regulatory region of the HMG-CoA synthase or LDL receptor promoters, binds to the HMG-CoA reductase promoter at a site that overlaps a portion of SRE-1.[136] While the binding specificity of Red 25 corresponds to the sequence required for sterol regulation, a physiologic role for Red 25 remains to be established.[137]

A review has summarized the control mechanisms that regulate transcriptional control of cholesterol homeostasis.[137] Partial amino acid sequences of a protein that binds to the SRE-1s of the HMG-CoA synthase and the LDL receptor were used to clone genes encoding two sterol regulatory element binding proteins, SREBP-1 and SREBP-2. SREBPs are members of the basic helix–loop–helix leucine zipper transcription factor family, and are anchored to the endoplasmic reticulum. When sterol levels fall, proteolysis releases an approximately 68 kDa N-terminal domain that enters the nucleus and activates transcription. Although the HMG-CoA reductase promoter appears different from that of HMG-CoA synthase or the LDL receptor, genetic evidence suggests that the HMG-CoA reductase promoter responds to SREBP-2. Mutant cell lines whose SREBP-2 lacks the membrane anchor domain overproduce HMG-CoA reductase, HMG-CoA synthase, and LDL receptor.[137] Subsequently, it was shown that SREBP-1 bound to the HMG-CoA reductase promoter, and that mutations in the HMG-CoA reductase promoter that abolished sterol regulation also abolish SREBP-1 binding. The SREBP-1 binding site, however, did not correspond to the previously identified SRE-1. Additionally, transient expression of mature, soluble SREBP-1 and SREBP-2 activates the reductase promoter.[138]

2.02.4.29 Regulation of Translation

HMG-CoA reductase is regulated at the translational level both by sterols and mevalonate-derived nonsterols. Regulation in yeast involves feedback regulation by mevalonate-derived products formed prior to squalene. These lower the rate of translation of the mRNA that encodes yeast isozyme Hmg1p. By contrast, regulation of Hmg2p appears to be modulated by signals derived from intermediates formed late in the pathway.[55] The effect of nonsterols on regulation in animals can be seen by treating cells with inhibitors of HMG-CoA reductase activity plus exogenous sterols. In cells treated with inhibitors of HMG-CoA reductase at concentrations that block the formation of mevalonate for nonsterol products, levels of HMG-CoA reductase protein rise, even in the presence of exogenous sterols. When mevalonate is added, levels of HMG-CoA reductase protein then fall precipitously. This decrease results from an approximately five-fold increase in the rate of enzyme degradation and an equivalent decrease in translation, without affecting the total HMG-CoA reductase mRNA levels.[139,140] The decrease in translational efficiency appears to be at the level of initiation.[141] Inhibitors of squalene synthase and squalene cyclase were used to infer that a product lying between mevalonate and squalene increased protein degradation, but did not affect translation, whereas a product between squalene and lanosterol decreased translation of HMG-CoA reductase mRNA.[142] Regulation of the rate of translation of HMG-CoA reductase mRNA and of degradation of the enzyme are dictated by the need of a cell for nonsterol isoprenoids. These independent controls bypass sterol-mediated regulation at the transcriptional level, and permit synthesis of mevalonate for nonsterols even when sterols are abundant.[54] A less extensively characterized mode of post-transcriptional regulation involves control of the degradation of HMG-CoA reductase mRNA mediated by sterols through the 3′-untranslated region of the HMG-CoA reductase gene.[143,144]

2.02.4.30 Regulation of Protein Degradation

Addition of mevalonate or sterols to the medium of cultured mammalian cells is accompanied by accelerated degradation of HMG-CoA reductase. Despite earlier suggestions that phosphorylation signals degradation, the basal rate of degradation of HMG-CoA reductase in isolated rat hepatocytes is unaffected by its phosphorylation state.[145] Two classes of cellular signals have been implicated in accelerated degradation of HMG-CoA reductase: sterol-triggered degradation that appears also to require a mevalonate-derived, nonsterol product, and degradation triggered by a nonsterol, mevalonate-derived metabolite. One nonsterol isoprenoid implicated in regulation is farnesyl pyrophosphate, a compound whose analogues increase degradation and decrease translation without affecting mRNA levels.[146] Farnesol, which arises by dephosphorylation of farnesyl pyrophosphate catalyzed by allyl pyrophosphatase, has recently been implicated as playing a role in the regulated degradation of HMG-CoA reductase.[147] Sterols and nonsterol isoprenoid metabolites target degradation of HMG-CoA reductase via different pathways. However, since both pathways are inhibited by peptide inhibitors of neutral cysteine protease activity, they probably share the same terminal proteolytic step.[148] Proteolysis requires regions of the N-terminal membrane anchor domain of the enzyme. Deletion of two or more membrane-spanning regions of the anchor domain abolished accelerated degradation of the enzyme.[54,149]

Degradation of HMG-CoA reductase in partially purified extracts of hamster kidney cells that had been pretreated with mevalonate or sterols revealed that the protease responsible for accelerated degradation resides in the endoplasmic reticulum. Accelerated degradation occurred in the absence of exogenous ATP or cytosolic components, and hence does not appear to proceed via the ubiquitin pathway. As for *in vivo* degradation, degradation *in vitro* was inhibited by peptide inhibitors of neutral cysteine proteases. A role for the proteasome, a multicatalytic protease, is suggested since degradation was inhibited by lactacystin, an inhibitor of proteasome activity.[150]

Yeast isozymes Hmg1p and Hmg2p differ with respect to their rates of degradation. High levels of presqualene intermediates do not affect degradation of isozyme Hmg1p, but are associated with high rates of degradation of isozyme Hmg2p.[55]

2.02.4.31 Phosphorylation Attenuates Catalytic Activity

As for several key regulated enzymes of lipid and carbohydrate metabolism, the activity of the HMG-CoA reductases of higher eukaryotes is regulated by reversible phosphorylation.[151,152] It was noted in the early 1970s that incubation of washed rat liver microsomes with ATP, Mg^{2+}, and a liver cytosol fraction attenuated HMG-CoA reductase activity, and that subsequent treatment with a second cytosolic fraction restored activity.[153–155] These observations led to the identification and subsequent purification of a protein kinase, HMG-CoA reductase kinase,[156] and a protein phosphatase, HMG-CoA reductase phosphatase.[157] Since HMG-CoA reductase kinase also phosphorylates acetyl-CoA carboxylase and hormone-sensitive lipase[158] and its activity is activated by 5′-AMP, it now is customarily referred to as AMP-activated protein kinase (AMPK). AMPK responds to the intracellular AMP:ATP ratio and appears to regulate mammalian fatty acid and cholesterol biosynthesis by attenuating the activities of HMG-CoA reductase and acetyl-CoA carboxylase in response to lowered energy charge.[159]

Phosphorylation-mediated regulation of HMG-CoA reductase activity involves a single target serine. This serine, Ser872 of the human enzyme[160] and Ser871 of the rat and hamster enzymes,[161,162] is the only serine or threonine of the catalytic domain phosphorylated by AMPK. While it plays no role in catalysis or substrate recognition,[162] the target serine is located only six residues from the catalytic histidine, a histidine-to-serine spacing that is conserved in the HMG-CoA reductases of all higher eukaryotes[163,164] (Figure 18). The activity of yeast HMG-CoA reductase is unaffected by rat liver AMPK. This can now be understood by inspection of the sequence of amino acids in the neighborhood of the putative active site histidine. HMG-CoA reductases that lack an appropriately located target serine include those of yeast, *Schistosoma mansoni*, *Dictyostelium discoideum*, *H. volcanii*, *S. solfataricus*, *Methanococcus jannaschii*, and *P. mevalonii*.

The activity of purified HMG-CoA reductase isozyme 1 from *A. thaliana* is also regulated by phosphorylation *in vitro*.[107] HMG-CoA reductase kinase activity is present in plant tissues, although since its activity is unaffected by 5′-AMP,[165,167] to term this protein kinase AMPK would seem inappropriate. The location of the target serine of *A. thaliana*, Ser577 corresponds to that of the target serine of the mammalian enzymes (Figure 18). Despite the availability of purified HMG-CoA reductases from the rubber tree,[168] radish,[169] and potato,[170] little or no phosphorylation data are available for the enzyme from other plants. Phosphorylation-mediated regulation of the activity of most or all plant HMG-CoA reductases nevertheless seems highly likely. In addition to the analogy with *A. thaliana*, plant genes encode an apparent target serine located in an appropriate primary structural spatial relationship to the putative catalytic histidine and surrounded by an apparent AMPK recognition motif (Figure 18).

2.02.4.32 Dephosphorylation Restores Catalytic Activity

Following attenuation of activity by phosphorylation, complete restoration of catalytic activity accompanies protein phosphatase-catalyzed dephosphorylation of both animal and *A. thaliana* HMG-CoA reductase. For vertebrate HMG-CoA reductase, *in vitro* dephosphorylation is catalyzed by protein phosphatase 1 (PP1), protein phosphatase 2A (PP2A), and protein phosphatase 2B (PP2B).[171] *In vivo*, however, dephosphorylation is thought to be catalyzed predominantly by PP2A.[172] The phosphorylation state of the enzyme is therefore under hormonal control through PP2A as well as being subject to the level of intracellular AMP.

2.02.4.33 Proposed Mechanism by which Phosphorylation Attenuates HMG-CoA Reductase Activity

Attenuation of catalytic activity accompanies phosphorylation of Ser872 of human HMG-CoA reductase[160] and of Ser871 of the rat[161] and hamster[162] enzymes. By analogy, and from the ability of

Mesocricetus auratus (Syrian hamster)	H	M	V	H	N	R	S	K	I	N	L
Mus musculus (mouse)	H	M	V	H	N	R	S	K	I	N	L
Cricetulus grisieus (Chinese hamster)	H	M	V	H	N	R	S	K	I	N	L
Homo sapiens (human)	H	M	I	H	N	R	S	K	I	N	L
Rattus norvegicus (rat)	H	M	V	H	N	R	S	K	I	N	L
Xenopus laevis (frog)	H	M	V	H	N	R	S	K	I	N	L
Strongylocentrus purpuratus (sea urchin)	H	M	K	H	N	R	S	A	L	N	I
Blatella germanica (cockroach)	H	M	R	H	N	R	S	S	V	S	T
Drosophila melanogaster (fruit fly)	H	M	R	H	N	R	S	S	I	A	V
Arabidopsis thaliana 1	H	M	K	Y	N	R	S	S	R	D	I
Arabidopsis thaliana 2	H	M	K	Y	N	R	S	S	R	D	I
Raphanus sativus 1 (radish)	H	M	K	Y	N	R	S	S	R	D	I
Raphanus sativus 2	H	M	K	Y	N	R	S	S	R	D	I
Solanum tuberosum 1	H	M	K	Y	N	R	S	I	K	D	I
Solanum tuberosum 2 (potato)	H	M	K	Y	N	R	S	T	K	A	S
Solanum tuberosum 3	H	M	K	Y	N	R	S	C	K	D	V
Lycopersicon esculentum 1 (tomato)	H	M	K	Y	N	R	S	I	K	D	I
Lycopersicon esculentum 2	H	M	K	Y	N	R	S	T	K	D	V
Lycopersicon esculentum 3	H	M	K	Y	N	R	S	S	K	D	V
Nicotiana sylvestris (tobacco)	H	M	K	Y	N	R	S	T	K	D	V
Hevea brasiliensis 1 (rubber tree)	H	M	K	Y	N	R	S	S	K	D	M
Hevea brasiliensis 2	H	M	K	Y	N	R	S	S	K	D	V
Hevea brasiliensis 3	H	M	K	Y	N	R	S	A	K	D	V
Camptotheca acuminata (magnolia)	H	M	K	Y	N	R	S	N	K	D	V
Catharanthus roseus (periwinkle)	H	M	K	Y	N	R	S	S	K	D	I
Consensus sequence for AMPK		ϕ	X	β,X	β,X	X	S	X	X	X	ϕ

Figure 18 Spatial relationship of the catalytic histidine and the target serine in the sequences of selected eukaryotic HMG-CoA reductases. The catalytic histidine and the target serine are shown in black squares. Also shown is the consensus motif for AMPK,[165] where ϕ is a hydrophobic amino acid (Met, Ile, Leu, or Val), β is a basic amino acid (Arg, Lys, or His), and X is any amino acid. Sequences were obtained from Genbank, and alignments were produced by the Pileup program of the Wisconsin package.[166]

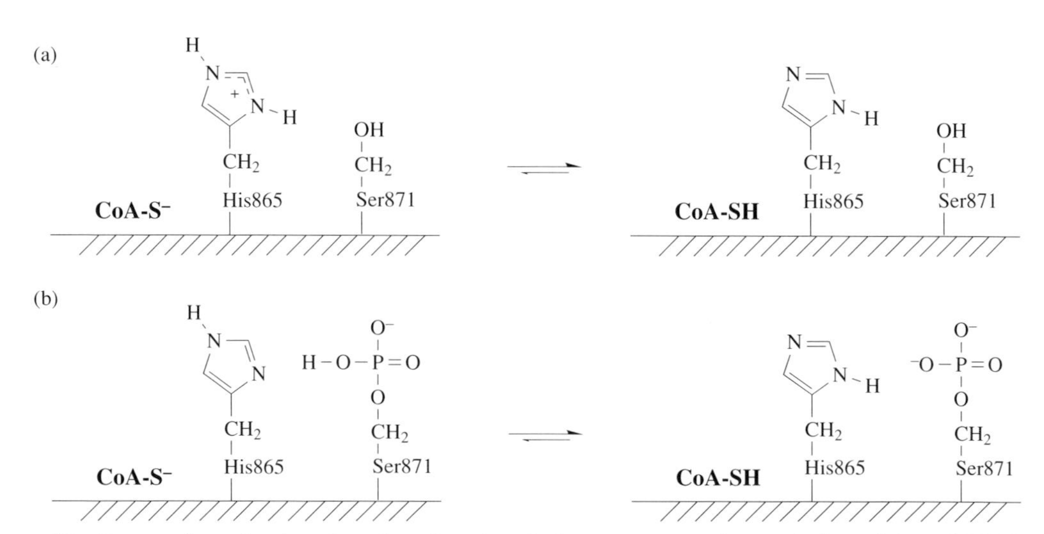

Figure 19 Proposed mechanism for why phosphorylation attenuates the catalytic activity of HMG-CoA reductase. Residue numbers are for the Syrian hamster enzyme. (a) The imizalolium proton of the active site histidine (His865) of the unphosphorylated enzyme can protonate the potentially inhibitory coenzyme A thioanion released after the first reductive stage of the overall reaction. This permits the second reductive stage to proceed, completing the overall reaction. (b) Phosphorylation of the target serine (Ser871) introduces adjacent negative charge that neutralizes the histidinium charge. The equilibrium now favors the coenzyme A thioanion, which can react with mevaldehyde, reversing stage 1 and removing substrate essential for stage 2 of the overall reaction.[165]

phosphorylation to attenuate the activity of the HMG-CoA reductases of many higher eukaryotes,[173] phosphorylation would appear to derivatize a serine located six residues from the inferred catalytic histidine, a spacing that is conserved in the HMG-CoA reductases of all higher eukaryotes (Figure 18). A novel mechanism has been proposed for why phosphorylation attenuates catalytic activity. The phosphate of the phosphoserine could interact directly with the catalytic histidine, negating its ability to protonate the inhibitory coenzyme A thioanion, blocking the ability of the thioanion to depart from the active site (Figure 19).[164] Alternatively, phosphorylation could interfere with closure of the flap domain.[114] In the HMG-CoA reductase of *P. mevalonii*, which lacks both a target serine and a protein kinase recognition sequence,[163] the arginine residue which corresponds to the serine is separated from the histidine by two turns of the α-helix.[113] This enzyme has, however, been modified to introduce both a target serine six residues from the catalytic histidine and a kinase recognition sequence. Consistent with this proposed mechanism, the activity of this mutant enzyme is regulated by phosphorylation–dephosphorylation.[119] Solution of the structures of the phosphorylated and unphosphorylated forms of this mutant enzyme should determine the precise mechanism whereby phosphorylation attenuates HMG-CoA reductase activity.

2.02.5 REFERENCES

1. J. D. Bergstrom and J. Edmond, *Methods Enzymol.*, 1985, **110**, 3.
2. B. Middleton, *Methods Enzymol.*, 1975, **35B**, 128.
3. U. Gehring and F. Lynen, in "The Enzymes," 3rd edn., ed. P. Boyer, Academic Press, New York, 1970, p. 391.
4. L. Hiser, M. E. Basson, and J. Rine, *J. Biol. Chem.*, 1994, **269**, 31 383.
5. , *K. U. Vollack and T. J. Bach, Plant Physiol.*, 1996, **111**, 1097.
6. X. Q. Song, T. Fukao, S. Yamaguchi, S. Miyazawa, T. Hashimoto, and T. Orii, *Biochem. Biophys. Res. Commun.*, 1994, **201**, 478.
7. M. Masuno, T. Fukao, X. Q. Song, S. Yamaguchi, T. Orii, N. Kondo, K. Imaizumi, and Y. Kuroki, *Genomics*, 1996, **36**, 217.
8. O. P. Peoples, S. Masamune, C. T. Walsh, and A. J. Sinskey, *J. Biol. Chem.*, 1987, **262**, 97.
9. J. T. Davis, R. N. Moore, B. Imperiali, A. J. Pratt, K. Kobayashi, S. Masamune, A. J. Sinskey, C. T. Walsh, T. Fukui, and K. Tomita, *J. Biol. Chem.*, 1987, **262**, 82.
10. J. T. Davis, H. H. Chen, R. Moore, Y. Nishitani, S. Masamune, A. J. Sinskey, and C. T. Walsh, *J. Biol. Chem.*, 1987, **262**, 90.
11. M. A. Palmer, E. Differding, R. Gamboni, S. F. Williams, O. P. Peoples, C. T. Walsh, A. J. Sinskey, and S. Masamune, *J. Biol. Chem.*, 1991, **266**, 8369.
12. S. Thompson, F. Mayerl, O. P. Peoples, S. Masamune, A. J. Sinskey, and C. T. Walsh, *Biochemistry*, 1989, **28**(14), 5735.
13. H. F. Gilbert, *Biochemistry*, 1981, **20**, 5643.
14. S. F. Williams, M. A. Palmer, O. P. Peoples, C. T. Walsh, A. J. Sinskey, and S. Masamune, *J. Biol. Chem.*, 1992, **267**, 16041.
15. M. Mathieu, J. P. Zeelen, R. A. Pauptit, R. Erdmann, W. H. Kunau, and R. K. Wierenga, *Structure*, 1994, **2**, 797.
16. H. Rudney and J. J. Ferguson, Jr., *J. Biol. Chem.*, 1959, **234**, 1076.
17. H. M. Miziorko and M. D. Lane, *J. Biol. Chem.*, 1977, **252**, 1414.
18. B. Middleton and P. K. Tubbs, *Methods Enzymol.*, 1975, **35**, 173.
19. H. M. Miziorko, *Methods Enzymol.*, 1985, **110**, 19.
20. W. D. Reed and M. D. Lane, *Methods Enzymol.*, 1975, **35**, 155.
21. H. Rudney, P. R. Stewart, P. W. Majerus, and P. R. Vagelos, *J. Biol. Chem.*, 1966, **241**, 1226.
22. P. R. Stewart and H. Rudney, *J. Biol. Chem.*, 1966, **241**, 1212.
23. J. Ayté, G. Gil-Gómez, and F. G. Hegardt, *Nucleic Acids Res.*, 1990, **18**, 3642.
24. J. Ayté, G. Gil-Gómez, D. Haro, P. F. Marrero, and F. G. Hegardt, *Proc. Natl. Acad. Sci. USA*, 1990, **87**, 3874.
25. K. D. Clinkenbeard, T. Sugiyama, and M. D. Lane, *Methods Enzymol.*, 1975, **35**, 160.
26. W. D. Reed, K. D. Clinkenbeard, and M. D. Lane, *J. Biol. Chem.*, 1975, **250**, 3117.
27. C. L. Welch, Y. R. Xia, I. Shechter, R. Farese, M. Mehrabian, S. Mehdizadeh, C. H. Warden, and A. J. Lusis, *J. Lipid Res.*, 1996, **37**, 1406.
28. J. A. Ortiz, G. Gil-Gómez, R. P. Casaroli-Marano, S. Vilaro, F. G. Hegardt, and D. Haro, *J. Biol. Chem.*, 1994, **269**, 28 523.
29. L. L. Rokosz, D. A. Boulton, E. A. Butkiewicz, G. Sanyal, M. A. Cueto, P. A. Lachance, and J. D. Hermes, *Arch. Biochem. Biophys.*, 1994, **312**, 1.
30. A. P. Russ, V. Ruzicka, W. Maerz, H. Appelhans, and W. Gross, *Biochim. Biophys. Acta*, 1992, **1132**, 329.
31. G. Gil, J. L. Goldstein, C. A. Slaughter, and M. S. Brown, *J. Biol. Chem.*, 1986, **261**, 3710.
32. P. A. Kattar-Cooley, H. H. Wang, L. M. Mende-Mueller, and H. M. Miziorko, *Arch. Biochem. Biophys.*, 1990, **283**, 523.
33. S. Katayama, N. Adachi, K. Takao, T. Nagakgawa, H. Matsuda, and M. Kawamukai, *Yeast*, 1995, **11**, 1533.
34. F. Montamat, M. Guilloton, F. Karst, and S. Deltrot, *Gene*, 1995, **167**, 197.
35. C. Buesa, J. Martínez-Gonzalez, N. Casals, D. Haro, M. D. Piulachs, X. Bellés, and F. G. Hegardt, *J. Biol. Chem.*, 1994, **269**, 11 707.
36. J. Martínez-González, C. Buesa, M. D. Piulachs, X. Bellés, and F. G. Hegardt, *Eur. J. Biochem.*, 1993, **217**, 691.

37. I. Misra, H. A. Charlier, Jr., and H. M. Miziorko, *Biochem. Biophys. Acta*, 1995, **1247**, 253.
38. I. Misra, C. Narasimhan, and H. M. Miziorko, *J. Biol. Chem.*, 1993, **268**, 12 129.
39. I. Misra and H. M. Miziorko, *Biochemistry*, 1996, **35**, 9610.
40. H. M. Miziorko and C. E. Behnke, *Biochemistry*, 1985, **24**, 3174.
41. H. M. Miziorko and C. E. Behnke, *J. Biol. Chem.*, 1985, **260**, 13 513.
42. H. Hashizume, H. Ito, T. Morikawa, N. Kanaya, H. Nagashima, H. Usui, H. Tomoda, T. Sunazuka, H. Kumagai, and S. Omura, *Chem. Pharm. Bull. (Tokyo)*, 1994, **42**, 2097.
43. M. D. Greenspan, J. B. Yudkovitz, J. S. Chen, D. P. Hanf, M. N. Chang, P. Y. Chiang, J. C. Chabala, and A. W. Alberts, *Biochem. Biophys. Res. Commun.*, 1989, **163**, 548.
44. M. I. Schimerlik and W. W. Cleland, *Biochemistry*, 1977, **16**, 576.
45. M. D. Greenspan, H. G. Bull, J. B. Yudkovitz, D. P. Hanf, and A. W. Alberts, *Biochem. J.*, 1993, **289**, 889.
46. M. D. Greenspan, J. B. Yudkovitz, L. Chia-Yee, J. S. Chen, A. W. Alberts, V. M. Hunt, M. N. Chang, S. S. Yang, K. L. Thompson, Y. C. Chiang, J. C. Chabala, R. L. Monaghan, and R. L. Schwartz, *Proc. Natl. Acad. Sci. USA*, 1987, **84**, 7488.
47. T. J. Bach, T. Weber, and A. Motel, *Recent Adv. Phytochem.*, 1990, **24**, 1.
48. T. Weber and T. J. Bach, *Biochim. Biophys. Acta*, 1994, **1121**, 85.
49. D. McCaskill and R. Croteau, *Anal. Biochem.*, 1993, **215**, 142.
50. T. J. Bach, *Lipids*, 1995, **30**, 191.
51. W. E. Brown and V. W. Rodwell, in "Dehydrogenases Requiring Nicotinamide Coenzymes," ed. J. Jeffery, Birkhauser, Basel, 1980, p. 232.
52. R. E. Dugan, in "Biosynthesis of Isoprenoid Compounds," eds. J. W. Porter and S. L. Spurgeon, Wiley, New York, 1981, vol. 1, p. 95.
53. D. M. Gibson and R. A. Parker, in "The Enzymes, XVIIIB," eds. P. D. Boyer and E. G. Krebs, Academic Press, New York, 1987, p. 179.
54. J. Goldstein and M. Brown, *Nature*, 1990, **343**, 425.
55. R. Hampton, D. Dimster-Denk, and J. Rine, *Trends Biochem. Sci.*, 1996, **21**, 140.
56. D. J. McNamara and V. W. Rodwell, in "Biochemical Regulatory Mechanisms in Eukaryotic Cells," eds. E. Kun and S. Grisolia, Wiley, New York, 1972, p. 205.
57. S. Miller, R. Parker, and D. Gibson, *Adv. Enzyme Regul.*, 1989, **28**, 65.
58. M. Monfar, C. Caelles, L. Balcells, A. Rerrer, F. G. Hegardt, and A. Boronat, *Recent Adv. Phytochem.*, 1990, **24**, 83.
59. D. J. Monger, *Methods Enzymol.*, 1985, **110**, 51.
60. V. W. Rodwell, D. J. McNamara, and D. J. Shapiro, *Adv. Enzymol. Relat. Areas Mol. Biol.*, 1973, **38**, 373.
61. V. W. Rodwell, J. L. Nordstrom, and J. J. Mitschelen, *Adv. Lipid Res.*, 1976, **14**, 1.
62. D. W. Russell, *Methods Enzymol.*, 1985, **110**, 26.
63. R. Sato and T. Takano, *Cell Struct. Funct.*, 1995, **20**, 421.
64. A. Sipat, *Methods Enzymol.*, 1985, **110**, 40.
65. M. De Rosa, A. Gambacorta, and A. Gliozzi, *Microbiol. Rev.*, 1986, **50**, 70.
66. K. M. Bischoff and V. W. Rodwell, *J. Bacteriol.*, 1996, **178**, 19.
67. W. L. Lam and W. F. Doolittle, *J. Biol. Chem.*, 1992, **267**, 5829.
68. K. M. Bischoff and V. W. Rodwell, *Protein Sci.*, 1997, **6**, 156.
69. D. A. Bochar, J. R. Brown, W. F. Doolittle, H.-P. Klenk, W. Lam, M. E. Schenk, C. V. Stauffacher, and V. W. Rodwell, *J. Bacteriol.*, 1997, **179**, 3632.
70. GenBank Accession Number AL009126.
71. GenBank Accession Number U00096.
72. GenBank Accession Number AE000511.
73. GenBank Accession Number L42023.
74. GenBank Accession Number U00089.
75. GenBank Accession Number AB001339.
76. C. M. Fraser, J. D. Gocayne, O. White, M. D. Adams, R. A. Clayton, R. D. Fleischmann, C. J. Bult, A. R. Kerlavage, G. Sutton, J. M. Kelley, J. L. Fritchman, J. F. Weidman, K. V. Small, M. Sandusky, J. Fuhrmann, D. Nguyen, T. R. Utterback, D. M. Saudek, C. A. Phillips, J. M. Merrick, J. F. Tomb, B. A. Doughtery, K. F. Bott, P. C. Hu, T. S. Lucier, S. N. Peterson, H. O. Smith, C. A. Hutchison, III, and J. C. Venter, *Science*, 1995, **270**, 397.
77. GenBank Accession Number AE000783.
78. Incomplete sequences from the TIGR Microbial Database, Institute for Genomic Research, Rockville, MD. http://www.tigr.org/tdb/mdb/mdb.html
79. J. F. Gill, Jr., M. J. Beach, and V. W. Rodwell, *J. Biol. Chem.*, 1985, **260**, 9393.
80. M. Beach and V. W. Rodwell, *J. Bacteriol.*, 1989, **171**, 2994.
81. J. A. Friesen, C. M. Lawrence, C. V. Stauffacher, and V. W. Rodwell, *Biochemistry*, 1996, **35**, 11 945.
82. E. H. Olender and R. D. Simoni, *J. Biol. Chem.*, 1992, **267**, 4223.
83. R. J. Mayer, C. Debouck, and B. W. Metcalf, *Arch. Biochem. Biophys.*, 1988, **267**, 110.
84. K. Frimpong, B. G. Darnay, and V. W. Rodwell, *Protein Expression Purif.*, 1993, **4**, 337.
85. K.-U. Vollack, B. Dittrich, A. Ferrer, A. Boronat, and T. J. Bach, *J. Plant Physiol.*, 1994, **143**, 479.
86. A. Endo, *J. Lipid Res.*, 1992, **33**, 1569.
87. T. J. Bach, A. Boronat, C. Caelles, A. Ferrer, T. Weber, and A. Wettstien, *Lipids*, 1991, **26**, 637.
88. M. E. Basson, M. Thorsness, and J. Rine, *Proc. Natl. Acad. Sci. USA*, 1986, **83**, 5563.
89. C. Caelles, A. Ferrer, L. Balcells, F. G. Hegardt, and A. Boronat, *Plant. Mol. Biol.*, 1989, **13**, 627.
90. M. Enjuto, L. Balcells, N. Campos, C. Caelles, M. Arro, and A. Boronat, *Proc. Natl. Acad. Sci. USA.*, 1994, **91**, 927.
91. K. Aoyagai, A. Beyou, K. Moon, L. Fang, and T. Ulrich, *Plant Physiol.*, 1993, **102**, 623.
92. M. Chye, C. Tan, and N. Chua, *Plant Mol. Biol.*, 1992, **19**, 473.
93. Z. Yang, H. Parks, G. H. Lacy, and C. L. Cramer, *Plant Cell*, 1991, **3**, 397.
94. H. Park, C. J. Denbow, and C. L. Cramer, *Plant Mol. Biol.*, 1992, **20**, 327.
95. B. Stermer, G. Bianchini, and K. Korth, *J. Lipid Res.*, 1994, **35**, 1133.
96. N. D. Daraselia, S. Tarchevskaya, and J. O. Narita, *Plant Physiol.*, 1996, **112**, 727.

97. D. Choi, B. L. Ward, and R. M. Bostock, *Plant Cell*, 1992, **4**, 1333.
98. T. W. Goodwin and E. I. Mercer, "Introduction to Plant Biochemistry," 2nd edn., Pergamon, Oxford, 1983, p. 400.
99. R. J. Wong, D. K. McCormack, and D. W. Russell, *Arch. Biochem. Biophys.*, 1982, **216**, 631.
100. R. M. Learned, *Plant Physiol.*, 1996, **110**, 645.
101. J. O. Narita and W. Gruissem, *Plant Cell*, 1989, **1**, 181.
102. H. Suzuki, K. Oba, and I. Uritani, *Physiol. Plant Pathol.*, 1975, **7**, 265.
103. M. E. Basson, M. Thorsness, J. Finer-Moore, R. M. Stroud, and J. Rine, *Mol. Cell. Biol.*, 1988, **8**, 3797.
104. P. J. Kennelly, K. G. Brandt, and V. W. Rodwell, *Biochemistry*, 1983, **22**, 2784.
105. G. C. Ness, S. C. Way, and P. S. Wickham, *Biochem. Biophys. Res. Commun.*, 1981, **102**, 81.
106. D. A. Kleinsek, R. E. Dugan, T. A. Baker, and J. W. Porter, *Methods Enzymol.*, 1981, **71C**, 462.
107. S. Dale, M. Arro, B. Becerra, N. G. Morrice, A. Boronat, D. G. Hardie, and A. Ferrer, *Eur. J. Biochem.*, 1995, **233**, 506.
108. Y. P. Ching, S. P. Davies, and D. G. Hardie, *Eur. J. Biochem.*, 1996, **237**, 800.
109. J. Martínez-González, C. Buesa, M. D. Piulachs, X. Bellés, and F. G. Hegardt, *Eur. J. Biochem.*, 1993, **213**, 233.
110. A. Ferrer, C. Aparicio, N. Nogues, A. Wettstein, T. J. Bach, and A. Borona, *FEBS Lett.*, 1990, **266**, 67.
111. R. Croxen, M. W. Goosey, J. P. Keon, and J. A. Hargreaves, *Microbiology*, 1994, **140**, 2363.
112. A. Rajkovic, J. N. Simonsen, R. E. Davis, and F. M. Rottman, *Proc. Natl. Acad. Sci. USA*, 1989, **86**, 8217.
113. C. M. Lawrence, V. W. Rodwell, and C. V. Stauffacher, *Science*, 1995, **268**, 1758.
114. L. Tabernero, D. A. Bochar, V. W. Rodwell, and C. V. Stauffacher, *Science*, 1998, in press.
115. M. Rossmann, A. Liljas, C. I. Branden, and L. Banaszak, *Enzymes*, 1975, **11**, 61.
116. R. Chen, A. Greer, and A. M. Dean, *Proc. Natl. Acad. Sci. USA*, 1995, **92**, 11 666.
117. P. Rowland, A. K. Basak, S. Gover, H. R. Levy, and M. J. Adams, *Structure*, 1994, **2**, 1073.
118. J. Kim and J. Wu, *Proc. Natl. Acad. Sci. USA*, 1988, **85**, 6677.
119. J. A. Friesen and V. W. Rodwell, *Biochemistry*, 1997, **36**, 2173.
120. V. W. Rodwell, D. H. Anderson, M. J. Beach, J. F. Gill, Jr., T. C. Jordan-Starck, D. S. Scher, and Y. Wang, in "*Pseudomonas*: Biotransformations, Pathogenesis, and Evolving Technology," eds. S. Silver, A. M. Chakrabarty, B. Iglewski, and S. Kaplan, *American Society of Microbiology*, Washington, DC, 1990, p. 151.
121. J. Rétey, E. von Stetten, U. Coy, and F. Lynen, *Eur. J. Biochem.*, 1970, **15**, 72.
122. N. Qureshi, R. E. Dugan, W. Cleland, and J. Porter, *Biochemistry*, 1976, **15**, 4191.
123. N. Qureshi and J. W. Porter, in "Biosynthesis of Isoprenoid Compounds," eds. J. W. Porter and S. L. Spurgeon, Wiley, New York, 1981, vol. 1, p. 47.
124. D. G. Sherban, P. J. Kennelly, K. G. Brandt, and V. W. Rodwell, *J. Biol. Chem.*, 1985, **260**, 12 579.
125. D. Veloso, W. Cleland, and J. Porter, *Biochemistry*, 1981, **20**, 887.
126. B. G. Darnay, Y. Wang, and V. W. Rodwell, *J. Biol. Chem.*, 1992, **267**, 15 064.
127. B. G. Darnay and V. W. Rodwell, *J. Biol. Chem.*, 1993, **268**, 8429.
128. Y. Wang, B. G. Darnay, and V. W. Rodwell, *J. Biol. Chem.*, 1990, **265**, 21 634.
129. K. Frimpong and V. W. Rodwell, *J. Biol. Chem.*, 1994, **269**, 1217.
130. T. C. Jordan-Starck and V. W. Rodwell, *J. Biol. Chem.*, 1989, **264**, 17 913.
131. R. E. Cappel and H. F. Gilbert, *J. Biol. Chem.*, 1993, **268**, 342.
132. T. C. Jordan-Starck and V. W. Rodwell, *J. Biol. Chem.*, 1989, **264**, 17 919.
133. K. Frimpong and V. W. Rodwell, *J. Biol. Chem.*, 1994, **269**, 11 478.
134. P. A. Edwards, E. S. Kempner, S. F. Lan, and S. K. Erickson, *J. Biol. Chem.*, 1985, **260**, 10 278.
135. T. F. Osborne, *J. Biol. Chem.*, 1991, **266**, 13 947.
136. T. F. Osborne, M. Bennett, and K. Rhee, *J. Biol. Chem.*, 1992, **267**, 18 973.
137. T. F. Osborne, *Crit. Rev. Eukyotic Gene Expression*, 1995, **5**, 317.
138. S. M. Vallett, H. B. Sanchez, J. M. Rosenfeld, and T. F. Osborne, *J. Biol. Chem.*, 1996, **271**, 12 247.
139. N. Nakanishi, J. L. Goldstein, and M. S. Brown, *J. Biol. Chem.*, 1988, **263**, 8929.
140. M. S. Straka and S. R. Panini, *Arch. Biochem. Biophys.*, 1995, **317**, 235.
141. D. M. Peffley and A. K. Gayen, *Somatic Cell. Mol. Genet.*, 1995, **21**, 189.
142. D. M. Peffley and A. K. Gayen, *Arch. Biochem. Biophys.*, 1997, **337**, 251.
143. J. W. Choi and D. M. Peffley, *Biochem. J.*, 1995, **307**, 233.
144. J. W. Choi, E. M. Lundquist, and D. M. Peffley, *Biochem. J.*, 1993, **296**, 859.
145. V. A. Zammit and A. M. Caldwell, *Biochem. J.*, 1992, **284**, 901.
146. D. L. Bradfute and R. D. Simoni, *J. Biol. Chem.*, 1994, **269**, 6645.
147. T. E. Meigs, D. S. Roseman, and R. D. Simoni, *J. Biol. Chem.*, 1996, **271**, 7916.
148. J. Roitelman and R. D. Simoni, *J. Biol. Chem.*, 1992, **267**, 25 264.
149. H. Jingami, M. S. Brown, J. L. Goldstein, R. G. Anderson, and K. L. Luskey, *J. Cell Biol.*, 1987, **104**, 1693.
150. T. P. McGee, H. H. Cheng, H. Kumagai, S. Omura, and R. D. Simoni, *J. Biol. Chem.*, 1996, **271**, 25 630.
151. P. J. Kennelly and V. W. Rodwell, *J. Lipid Res.*, 1985, **26**, 903.
152. V. A. Zammit and R. A. Easom, *Biochim. Biophys. Acta*, 1987, **927**, 223.
153. Z. Beg, D. Allmann, and D. Gibson, *Biochem. Biophys. Res. Commun.*, 1973, **54**, 1362.
154. Z. Beg, J. Stonik, and H. Brewer, Jr., *Proc. Natl. Acad. Sci. USA*, 1978, **75**, 3678.
155. J. L. Nordstrom, V. W. Rodwell, and J. J. Mitschelen, *J. Biol. Chem.*, 1977, **252**, 8924.
156. J. Harwood, Jr., K. G. Brandt, and V. W. Rodwell, *J. Biol. Chem.*, 1984, **259**, 2810.
157. W. E. Brown and V. W. Rodwell, *Biochim. Biophys. Acta*, 1983, **751**, 218.
158. A. J. Garton, D. G. Campbell, D. Carling, D. G. Hardie, R. J. Colbran, and S. J. Yeaman, *Eur. J. Biochem.*, 1989, **179**, 249.
159. D. Carling, P. Clarke, V. Zammitt, and D. Hardie, *Eur. J. Biochem.*, 1989, **186**, 129.
160. R. Sato, J. Goldstein, and M. Brown, *Proc. Natl. Acad. Sci. USA*, 1993, **90**, 9261.
161. P. Clarke and D. Hardie, *EMBO J.*, 1990, **9**, 2439.
162. R. V. Omkumar, B. G. Darnay, and V. W. Rodwell, *J. Biol. Chem.*, 1994, **269**, 6810.
163. J. A. Friesen and V. W. Rodwell, *Biochemistry*, 1997, **36**, 1157.
164. R. V. Omkumar and V. W. Rodwell, *J. Biol. Chem.*, 1994, **269**, 16 862.

165. S. Dale, W. Wilson, A. Edelman, and D. Hardie, *FEBS Lett.*, 1995, **361**, 191.
166. J. Devereux, P. Haeberli, and O. Smithies, *Nucleic Acids. Res.*, 1984, **12**, 387.
167. K. Ball, J. Barker, N. Halford, and D. Hardie, *FEBS Lett.*, 1995, **377**, 189.
168. R. Wititsuwannakul, D. Wititsuwannakul, and P. Suwanmanee, *Phytochemistry* (*Oxf.*), 1990, **29**, 1401.
169. T. J. Bach, D. H. Rogers, and H. Rudney, *Eur. J. Biochem.*, 1986, **154**, 103.
170. K. Kondo and K. Oba, *J. Biochem.* (*Tokyo*), 1986, **100**, 967.
171. D. Hardie, *Biochim. Biophys. Acta*, 1991, **1123**, 231.
172. V. A. Zammit and A. M. Caldwell, *Biochem. J.*, 1990, **269**, 373.
173. C. F. Hunter and V. W. Rodwell, *J. Lipid Res.*, 1980, **21**, 399.

2.03
A Mevalonate-independent Route to Isopentenyl Diphosphate

MICHEL ROHMER
Université Louis Pasteur, Strasbourg, France

2.03.1 INTRODUCTION

2.03.1.1 Isoprenoids

Isoprenoids belong to a huge family of natural products.[1] Many of the ~22 000 known compounds are described in other chapters of this volume. They include, for instance, essential metabolites such as sterols acting as membrane stabilizers or as precursors of steroid hormones, carotenoids of photosynthesizing organisms, acyclic prenyl chains found in chlorophylls (phytol), in ubiquinone, menaquinone, or plastoquinone from electron transfer chains, in prenylated proteins, or in dolichols and polyprenols required for the biosynthesis of polysaccharides or protein glycosylation, as well as a multitude of "secondary" metabolites of less obvious fundamental role in cell biology. All isoprenoids share a common feature: they are formally derived from the branched C_5 skeleton of isoprene. The formation of isopentenyl diphosphate (IPP), i.e., of the biological equivalent of isoprene, via the acetate–mevalonate pathway, has been largely documented and is described in detail in Chapter 2.02 of this volume. However, the main steps will be briefly described here as they are required for the discussion of the topic presented in this chapter.

2.03.1.2 The Acetate–Mevalonate Pathway—the Role of IPP

The pioneering work of Bloch, Lynen, Cornforth, and many others allowed an understanding of how living cells synthesize their isoprenoids from acetate, activated as acetyl CoA (**1**) (Figure 1).[2–4] Two Claisen-type condensations yield successively acetoacetyl CoA (**2**) and hydroxymethylglutaryl CoA (**3**). An NADPH-dependent reductase catalyzes the first supposed committed step of isoprenoid biosynthesis, the reduction of hydroxymethylglutaryl CoA to mevalonate (**4**), which was accepted until the mid-1990s as the universal precursor of isoprenoids. Three successive phosphorylations followed by a decarboxylation yield finally the branched five-carbon atom skeleton of IPP (**5**).

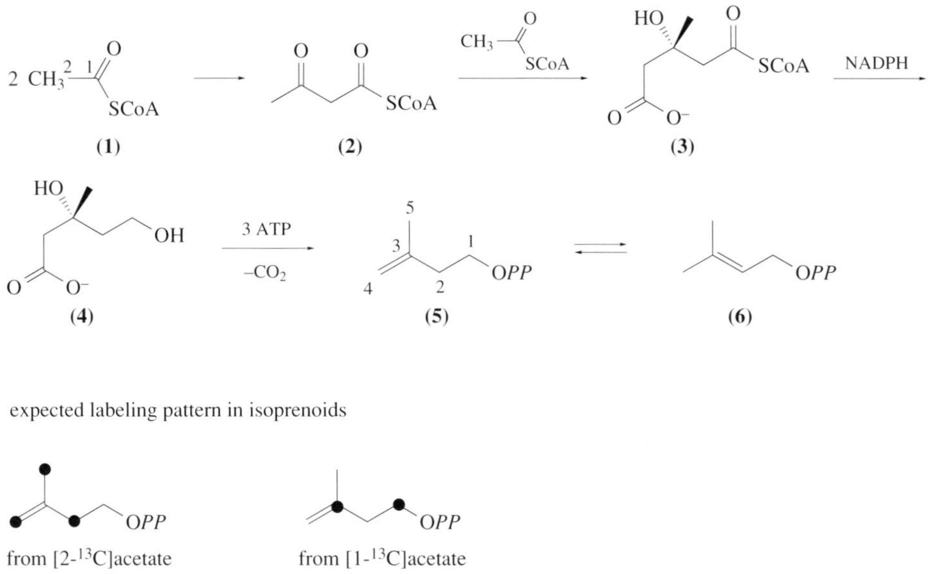

Figure 1 The acetate–mevalonate pathway for isoprenoid biosynthesis.

IPP can be considered as the universal building block for all isoprenoids, even in the bacteria and the eukaryotes that have been shown to possess the mevalonate-independent route and that will be discussed in this chapter. The role of IPP as isoprenoid precursor in eukaryotes has been extensively discussed. We will therefore only focus on the data available from prokaryotic organisms. IPP was always successfully incorporated into bacterial isoprenoids. This is in accordance with its universal role, shared with its isomer dimethylallyl diphosphate (**6**), as isoprenoid precursor. Carbon-14-labeled IPP could be incorporated into ubiquinone, menaquinone, carotenoids, and polyprenols by broken cell preparations from *Rhodospirillum rubrum*,[5] into ubiquinone and polyprenol derivatives by freeze-dried *Escherichia coli* cells upon rehydration,[6,7] and into polyprenyl diphosphates by the polyprenyl diphosphate synthases from several *Micrococcus* strains,[8–12] *Bacillus subtilis*,[13] *Lactobacillus plantarum*,[14] and *Salmonella newington*.[15] In the presence of ATP, NADPH, and dimethylallyl diphosphate, ^{14}C-labeled IPP was also incorporated by cell-free systems from *Zymomonas mobilis* into all isoprenoid series of this bacterium: geranyl and farnesyl diphosphates, squalene resulting from the condensation of two farnesyl diphosphate moieties, as well as diploptene resulting from squalene cyclization.[16] All these data point out the universal role of IPP as isoprenoid precursor.

This is most important, as the list of the above-mentioned bacteria contains species possessing the mevalonate pathway such as *L. plantarum*, as well as species such as *E. coli* and *Z. mobilis* that have been shown to possess a mevalonate-independent route.[17]

2.03.2 BACTERIAL TRITERPENOIDS OF THE HOPANE SERIES: THE TOOLS FOR THE DISCOVERY OF THE MEVALONATE-INDEPENDENT PATHWAY

2.03.2.1 Hopanoid Structures

Hopanoids (Figure 2) represent a family of pentacyclic triterpenes.[18–21] They can be divided into two subgroups depending on the presence or absence of an oxygenated function at carbon atom C-3. Those bearing such a function at this position are mainly found in scattered taxa of higher plants and in a few lichens. They are derived from enzymatic cyclization of (3*S*)-squalene epoxide, much like almost all eukaryotic triterpenoids. Those without an oxygen atom at C-3 were quite often found in ferns and mosses, in a few fungi and several lichens (where they are most probably synthesized by the fungal symbiont), as well as in many bacteria. These 3-deoxyhopanoids were shown to be derived in bacteria from the direct cyclization of squalene. In all hopanoid-producing bacteria, the most simple C_{30} compounds, diploptene (**7**) and diplopterol (**8**), are usually present, at least in trace amounts.[22] The major compounds are always the C_{35} bacteriohopane derivatives, such as bacteriohopanetetrol (**9**) or aminobacteriohopanetriol (**10**) which represent the most common moieties among composite bacterial hopanoids. They are characterized by a carbon–carbon bond between the triterpenic skeleton and the additional C_5 polyhydroxylated side chain derived from a D-pentose and linked via its C-5 carbon atom to the hopane isopropyl group.[22,23] According to the stereochemistry, the polyhydroxylated side chain should be derived in most bacteria from a D-ribose derivative[24,25] or in the *Acetobacter* and *Nostoc* species, eventually from a D-arabinose derivative.[26,27] Bacterial hopanoids are characterized by a huge structural diversity.[21] Variations may involve the triterpenic moiety with additional methyl groups at C-2α, C-2β, or C-3β, additional double bonds at C-6 and/or C-11 and different stereochemistries at C-22. In addition, side chains may differ by the number of hydroxy groups, the presence of a hydroxy or amino group at the terminal C-35 position, an additional methyl group at C-31, or by the presence of polar moieties derived from sugar derivatives (**11**, **12**) or amino acids (**13**) and linked to the C-35 functional group.[21] An extraordinary nucleoside analogue, adenosylhopane (**14**), has even been found in several bacteria such as *Rhodopseudomonas acidophila*,[21] *Nitrosomonas* sp., or *Bradyrhizobium japonicum* (Seemann, Bravo, Poralla, and Rohmer, unpublished results).

2.03.2.2 Distribution and Role of Bacterial Hopanoids

About 300 strains[22] (and Rohmer and co-workers, unpublished results) have been investigated for their hopanoid content, mainly by the author's group. They were found in many species with a phylogenetic relationship, e.g., in most cyanobacteria (but not in all of them), in many obligate methanotrophs, in all *Acetobacter* species with a very peculiar pattern of unsaturated and methylated bacteriohopanetetrols, but also in many scattered taxa of Gram-positive and Gram-negative bacteria. It is rather difficult to draw any definitive phylogenetic conclusions concerning the distribution of this triterpene series, as the species of prokaryotes that can be grown in the laboratory represent only a very minor fraction of the existing bacterial biodiversity.[28] The known distribution therefore reflects the availability of strains in collections and their ease of isolation and cultivation, rather than their significance in natural environments.

Hopanoids share with sterols common structural features which are required for biological membrane stabilizers: amphiphilic compounds with a polar head (hydroxy group or polyfunctionalized polar side chain) and a hydrophobic polycyclic skeleton, planar structure owing to the *trans* ring junctions, and finally similar dimensions with a length corresponding to half the cross-section of the phospholipid bilayers. These similar structural features suggested similar physiological roles.[29] This was verified in several biological systems as well as on artificial membrane models. Hopanoids modulate membrane permeability and fluidity in bacteria in the same way, at least qualitatively, as sterols do in eukaryotic membranes.[18,20] Other roles cannot be excluded and should be further investigated, e.g., for adenosylhopane (**14**), a nucleoside analogue, or for *N*-tryptophanyl aminobacteriohopanetriol (**13**) with its hydrophobic amino acid moiety (Figure 2).

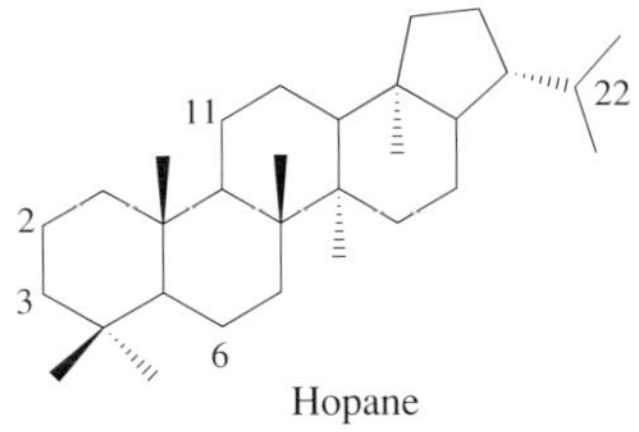

Hopane

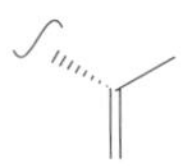

(7)

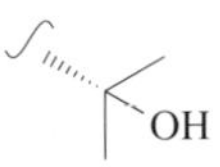

(8)

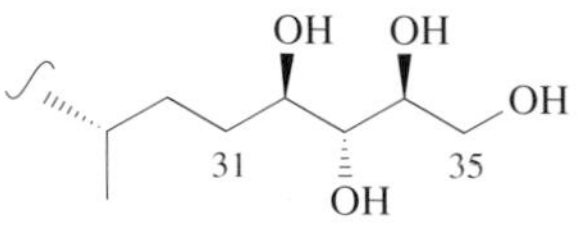

(9)

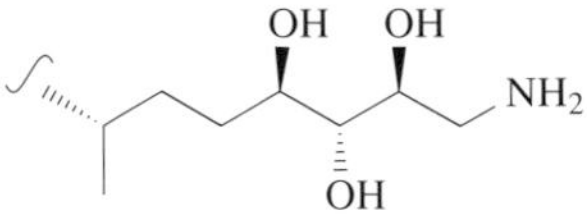

(10)

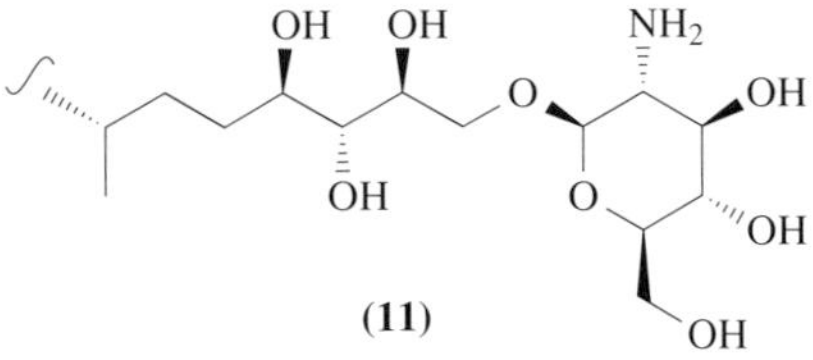

(11)

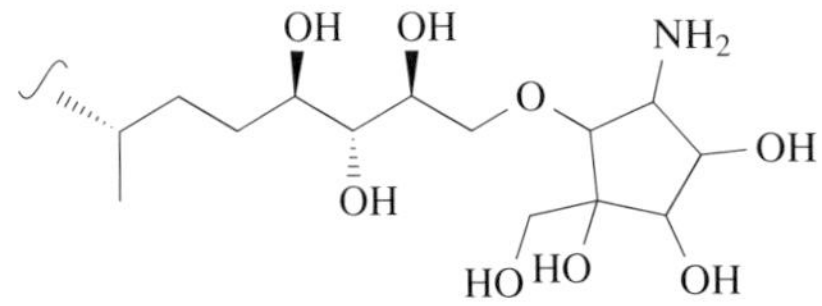

(12)

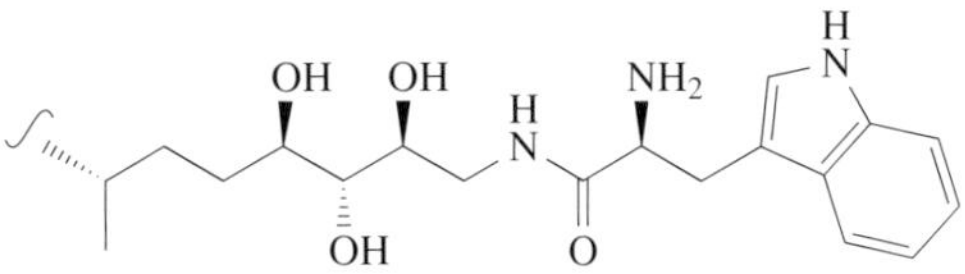

(13)

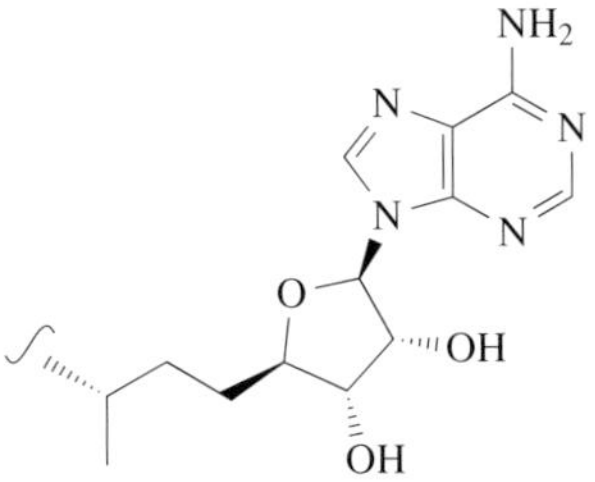

(14)

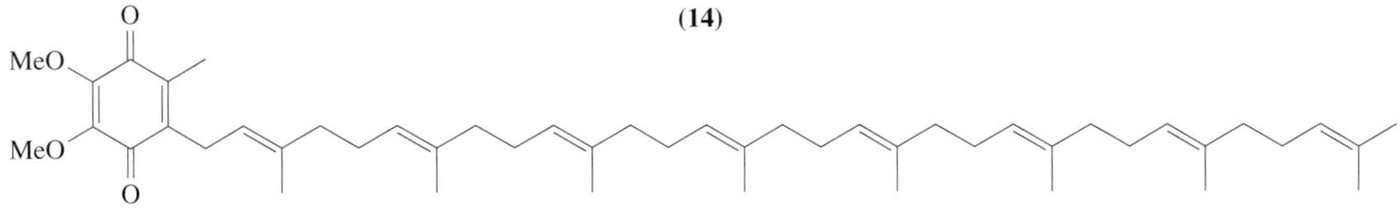

(15)

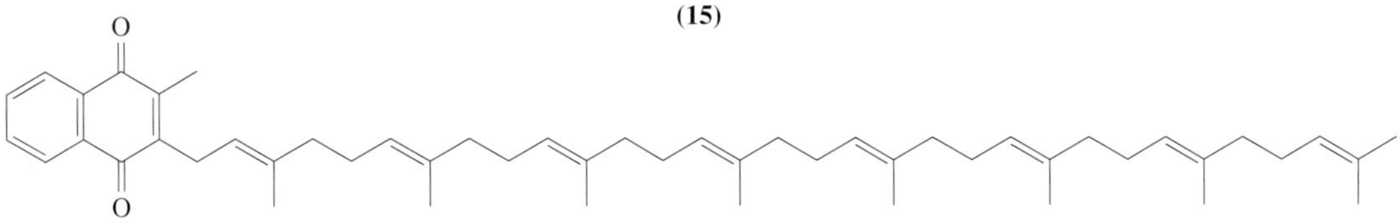

(16)

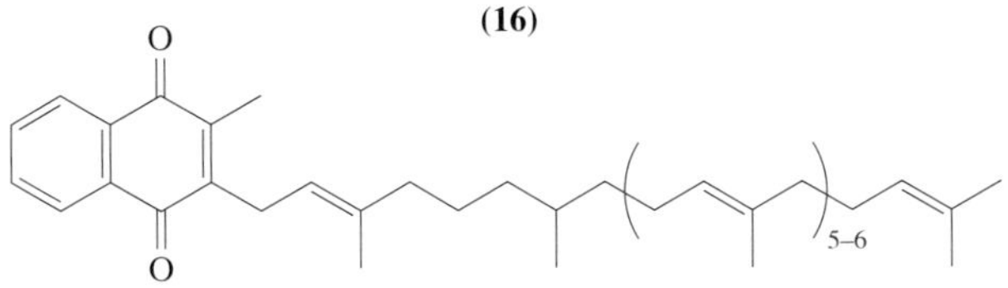

(17)

Figure 2 Bacterial isoprenoids. Triterpenoids of the hopane series (**7**)–(**14**), ubiquinone Q-8 (**15**), menaquinone MK-8 (**16**), dihydromenaquinones MK-8/9(II-H_2) (**17**).

2.03.2.3 Incorporation of ^{13}C-Labeled Acetate into Bacterial Hopanoids

Hopanoid concentrations are usually rather high in bacterial cells, usually between 1 and 5 mg g^{-1} (dry weight) with a maximum of up to 30 mg g^{-1} for *Z. mobilis*.[22] This is much more than the concentrations of all other bacterial isoprenoids such as the probably ubiquitous bactoprenol diphosphate involved as sugar carrier in the biosynthesis of cell wall polysaccharides or the ubiquinones and menaquinones with their prenyl side chains. Hopanoids are chemically stable (much more than carotenoids which are also fairly often found in bacteria) and easily isolated. This makes them marvelous tools for biosynthetic studies using incorporation of ^{13}C-labeled precursors and detection of the resulting labeling patterns by NMR spectroscopy.

The polyhydroxylated side chain of the hopanoids is a unique feature in natural product chemistry. The original goal was therefore the elucidation of the formation of the C_{35} bacteriohopane skeleton. The first incubations were performed with [1-^{13}C]- or [2-^{13}C]acetate in order to determine the origin of this additional side chain.[23] Most important were the labeling conditions. The bacteria were grown on a synthetic mineral medium containing one single carbon and energy source. Nearly all further labeling experiments, including those made later with algae and higher plants, were carried out under such culture conditions, avoiding on the one hand isotopic dilution by other unlabeled carbon sources and on the other competition for the utilization of different carbon sources, as is often observed using complex culture media. This methodology assured that all metabolites were synthesized from the labeled precursor and that only a few, usually clearly identified, catabolic and anabolic pathways were expected to be utilized. Interpretation of the labeling patterns and the isotopic enrichments in such culture conditions was therefore much easier than in the case of cells grown on complex media of indefinite composition, and the possible origins and structures of the precursors could be deduced. This was the major difference between these experiments and most of those previously performed on the biosynthesis of bacterial terpenoids. The bacteriohopanepolyol side chains were thus shown to be derived from a D-pentose, but the most striking and fully unexpected observation concerned the isoprenic moiety (Figure 3, patterns (a) and (b)).[23]

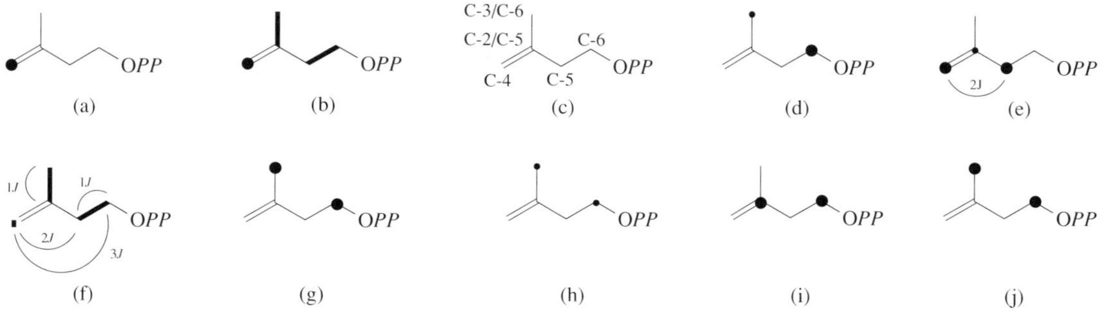

Figure 3 Labeling patterns of isoprenic units (represented by the carbon skeleton of IPP): (a) in bacteria possessing the nonmevalonate pathway after feeding of [1-^{13}C]acetate; (b) or after feeding of [2-^{13}C]acetate; (c) in *Z. mobilis* after feeding of ^{13}C-labeled glucose isotopomers (origin of the carbon atoms); (d) in *M. fujisawaense* after feeding of [6-^{13}C]glucose; (e) in *M. fujisawaense* after feeding of [4,5-$^{13}C_2$]glucose; (f) in *Z. mobilis* after feeding of [U-$^{13}C_6$]glucose; (g) in *E. coli* and *A. acidoterrestris* after feeding of [6-^{13}C]glucose; (h) in *E. coli* and *A. acidoterrestris* after feeding of [1-^{13}C]glucose; (i) in *M. fulvus* after feeding of [1-^{13}C]acetate; and (j) in *E. coli* and *M. fujisawaense* after feeding of [3-^{13}C]pyruvate.

The labeling patterns thus observed in isoprenic units (Figure 3, patterns (a) and (b)) did not fit at all with those expected from the acetate–mevalonate pathway (Figure 1). Exogenous acetate from the culture medium was not directly incorporated into the isoprenoids. The labeling patterns were clear and no scrambling had occurred. Results were reproducible with several bacteria for hopanoid biosynthesis in *Methylobacterium organophilum*, *Rhodopseudomonas palustris*, and *Rhodopseudomonas acidophila*,[23] hopanoid and ubiquinone biosynthesis in *Methylobacterium fujisawaense*, and finally ubiquinone biosynthesis in *E. coli* which does not synthesize hopanoids.[30] When the author started these experiments, there was no reason to reject the classical acetate–mevalonate route. Although the presence of another as yet unknown pathway could not be excluded, the results, like those of all other authors, were interpreted in the framework of the unanimously accepted acetate–mevalonate pathway. Other variants proposed for the formation of mevalonate derived either from leucine or from acetolactate metabolism were also in contradiction with our observations.[31,32] However, the presence of two different and noninterconvertible acetyl CoA pools had to be assumed.[23] This hypothesis, which was rather unsatisfactory in the case of a prokaryote, could not be confirmed by the labeling experiments performed with glucose. These first experiments using ^{13}C-labeled acetate, however, clearly shed light on an unexpected problem. The correct interpretation of the observed labeling patterns was not obvious with these data alone. The true explanation became evident only after incubation experiments with ^{13}C-labeled glucose, once the origin of all carbon atoms in the derived isoprenic units had been determined.

2.03.3 ELUCIDATION OF THE NONMEVALONATE PATHWAY: THE ORIGIN OF THE CARBON ATOMS OF ISOPRENIC UNITS IN BACTERIA

2.03.3.1 Incorporation of ^{13}C-Labeled Glucose into the Isoprenoids of *Z. mobilis* and *M. fujisawaense* via the Entner–Doudoroff Pathway: Origin of the Carbon Atoms of Isoprenic Units

Glucose is metabolized in all *Methylobacterium* species and in *Z. mobilis* via the Entner–Doudoroff pathway. Whereas *M. fujisawaense* is able to utilize a variety of carbon sources, *Z. mobilis* has a very narrow substrate range and utilizes only D-glucose, D-fructose, or sucrose as a carbon source. *Z. mobilis* is an efficient ethanologenic bacterium. It has no other catabolic pathways for hexoses, nor does it convert pyruvate (**18**) into phosphoenol pyruvate, and is therefore unable to synthesize glyceraldehyde 3-phosphate (**19**) from pyruvate (Figure 4). Finally, it possesses an incomplete tricarboxylic acid cycle.[33,34] Feeding *Z. mobilis* with either [1-^{13}C]-, [2-^{13}C]-, [3-^{13}C]-, [5-^{13}C]-, or [6-^{13}C]glucose allowed the determination of the origin of all carbon atoms of the isoprenic units of the hopanoids (**11**) and (**12**) (Figure 3, pattern (c)).[30,35] Only the carbon atoms corresponding to C-4 of IPP were never labeled and could therefore only arise from carbon C-4 of glucose. From the first labeling experiments with *Z. mobilis*, acetyl CoA directly formed from glucose catabolism (Figure 4) could be excluded as precursor; formation of the actual isoprenoid precursor had either to require two distinct pools of unknown C_2 subunits or, more likely, involved a different pathway from the classical one starting from acetyl CoA.[35] These results obtained with *Z. mobilis* were confirmed for the hopanoids and the prenyl chain of ubiquinone in *M. fujisawaense* by incubation of [6-^{13}C]glucose (Figure 3, pattern (d)) and doubly labeled [4,5-$^{13}C_2$]glucose (Figure 3, pattern E). Again, the labeling patterns observed after incorporation of these glucose isotopomers were not consistent with those expected from the acetate–mevalonate pathway. Indeed, in *Z. mobilis*, acetate could only be derived equally from pyruvate decarboxylation either directly from C-2/C-3 of glucose or from C-5/C-6 of glucose via glyceraldehyde phosphate (Figure 4). Such a dual origin for acetate was confirmed by examination of the labeling pattern of *cis*-vaccenic acid, one of the major fatty acids of this bacterium that is synthesized from acetyl CoA. The ^{13}C NMR spectra of its methyl ester showed that odd numbered carbon atoms were labeled from C-2 or C-5 of glucose and even numbered carbon atoms from C-3 or C-6.[36]

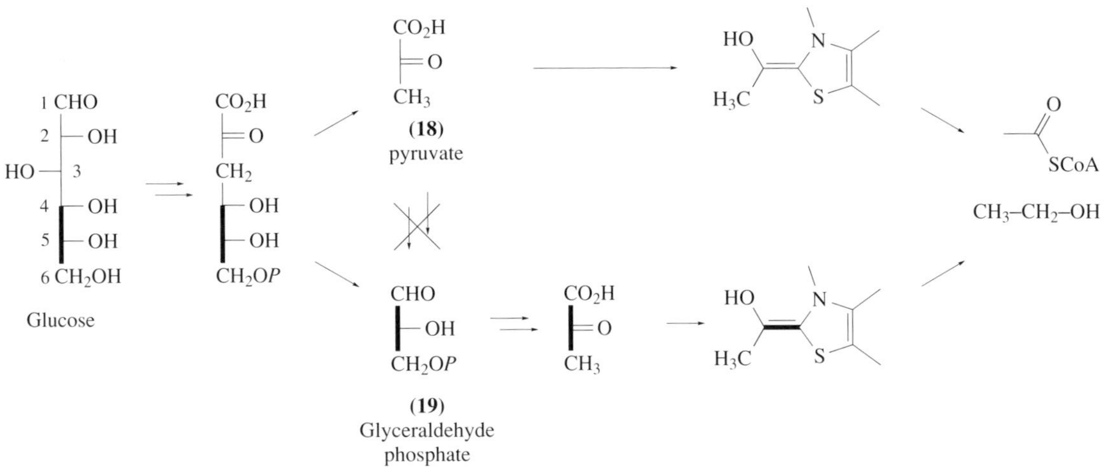

Figure 4 Glucose catabolism in *Z. mobilis* via the Entner–Doudoroff pathway.

From the examination of the labeling patterns of isoprenic units, it was now possible to deduce possible structures for the precursors of isoprenic units. Carbon-3 of IPP arose from C-2 and/or C-5 of glucose, and C-5 of IPP from C-3 and C-6 of glucose (Figure 3, pattern (c)). This pattern was reminiscent of the labeling pattern described above for acetyl CoA in fatty acids and suggested that C-3 and C-5 of IPP were derived from a single two-carbon subunit obtained from pyruvate decarboxylation. The three remaining carbon atoms of IPP (C-1, C-2, and C-4) were each derived from a single carbon atom of glucose (C-6, C-5, and C-4, respectively). The origin of these carbon atoms was the most striking question. Were they introduced via two different precursor molecules (much like in the acetate–mevalonate pathway), or via one single C_3 subunit with insertion of the above-mentioned C_2 subunit between the two IPP carbon atoms derived from C-4 and C-5 of glucose, possibly by a rearrangement?

2.03.3.2 A Rearrangement is Required for Formation of the Isoprenic Skeleton

These two hypotheses concerning the origin of the three carbon atoms derived from C-4, C-5, and C-6 of glucose (i.e., the involvement of two precursor molecules or of one single precursor) could be addressed by incubating bacteria with multiply labeled glucose samples and measurement of the resulting small ^2J and ^3J ^{13}C–^{13}C coupling constants which would represent the signature of the intramolecular rearrangement required in the case of a single precursor. Such couplings are often too small to be easily detected. For the sake of simplicity in the NMR spectra, doubly labeled [4,5-$^{13}C_2$]glucose (diluted in a 1:9 ratio with unlabeled glucose) was therefore preferred for the first experiment in place of uniformly labeled [U-$^{13}C_6$]glucose. The experiment was performed with *M. fujisawaense* which possesses the novel pathway, as shown from the incubation of ^{13}C-labeled acetate and [6-^{13}C]glucose, and produces three series of isoprenoids (hopanoids, 2β-methylhopanoids, and the acyclic prenyl chain of ubiquinone) with different three-dimensional structures, and in concentrations suitable for ^{13}C NMR analysis, thus enhancing the chances of detecting small ^2J ^{13}C–^{13}C coupling constants. After incubation, the signals of nearly all carbon atoms corresponding to C-2 and C-4 of IPP showed enrichments from C-4 and C-5 of glucose and each appeared as doublets with typical ^{13}C–^{13}C ^2J coupling constants between 0.5 and 3 Hz (Figure 3, pattern (e)).[30] This experiment unambiguously established the following. (i) C-4 and C-5 from glucose are introduced simultaneously into isoprenic units from a single precursor molecule. (ii) An intramolecular rearrangement is therefore required for the insertion of the C_2 subunit derived from pyruvate decarboxylation between the C-2 and C-4 carbon atoms of IPP, respectively, derived from C-5 and C-4 of glucose. (iii) Such a transposition in this biogenetic route definitely ruled out the classical pathway involving acetyl CoA and hydroxymethylglutaryl CoA as precursors.

Such conclusions could be confirmed by incorporation of uniformly labeled glucose into ubiquinone (**15**) of *E. coli*[37] and the hopanoids of *Z. mobilis* (**11**, **12**).[38] In experiments with *E. coli*, long-range ^{13}C–^{13}C couplings were observed between carbon atoms derived from C-1 and C-4 of IPP ($^3J = 5$ Hz) and C-2 and C-4 ($^2J = 3$ Hz) adjacent to ^1J couplings (about 40 Hz) between those derived from C-1 and C-2 or C-3 and C-5. In isoprenoids produced by *Z. mobilis*, ^1J couplings were observed for all signals of carbon atoms corresponding to C-3 and C-5 of IPP, indicating that they arose from a single precursor (Figure 3, pattern (f)). This observation was in accordance with a C_2 subunit derived from pyruvate. ^1J Couplings were also always observed between the carbon atoms corresponding to C-1 and C-2 of IPP. Four ^3J couplings between carbon atoms derived from C-4 and C-1 of IPP were found for four isoprenic units of the hopane skeleton, as well as a single ^2J coupling constant between two carbon atoms derived from C-4 and C-2. For other isoprenic units, the presence of weak ^2J and ^3J couplings of the corresponding carbon atoms derived from C-4 of IPP was indicated by broadening of the signals. These data confirmed and extended the results obtained after incorporation of [4,5-$^{13}C_2$]glucose into the isoprenoids of *M. fujisawaense*. They showed indeed that the three carbon atoms C-1, C-2, and C-4 of IPP, respectively, derived in *Z. mobilis* from C-6, C-5, and C-4 of glucose, corresponded to a single C_3 precursor molecule. As *Z. mobilis* does not convert pyruvate into glyceraldehyde phosphate, these three carbon atoms should therefore directly arise from a triose phosphate derivative via glyceraldehyde phosphate (**19**) (Figure 4).

The detection of a rearrangement in the formation of the carbon skeleton of isoprenic units in these bacteria allowed us to propose a consistent interpretation for an odd observation previously reported by Cane and co-workers on the biosynthesis of sesquiterpene antibiotics of the pentalenolactone series in *Streptomyces* UC 5319 (see Chapter 2.06.3.5.1).[39] Indeed, after feeding of [U-$^{13}C_6$]glucose, no long-range couplings had been detected, but in one isoprenic unit the ^1J coupling patterns from the mevalonate pathway were not observed, i.e., doublets for the carbon atoms derived from C-1 and C-2 of IPP and a singlet for that corresponding to C-4. Instead, the signals of carbon atoms corresponding to the C-1 and C-4 of IPP appeared as two doublets, and that from C-2 was a doublet of a doublet. This coupling pattern was the signature of the nonmevalonate route: the first rearrangement involved in the formation of the IPP skeleton separated carbon atoms corresponding to C-2 and C-4 of IPP derived from the same C_3 precursor, whereas the second rearrangement required for the formation of the pentalenolactone carbon skeleton restored the initial carbon atom connectivity of the C_3 precursor. Further evidence for the presence of the mevalonate-independent route in this *Streptomyces* and in other Actinomycetes will be presented in further sections of this chapter.

Together, these experiments fully confirmed the formation of isoprene units from a C_2 and a C_3 subunit. On the one hand, C-1, C-2, and C-4 of IPP were derived from a single C_3 precursor, as shown by the ^1J couplings observed between carbons derived from C-1 and C-2 of IPP and the ^2J

or ^3J couplings found between carbon atoms derived from C-4 and C-2 or C-1 of IPP, respectively. On the other hand, C-3 and C-5 from IPP were derived from a C_2 subunit generated by pyruvate (**18**) decarboxylation (Figure 4).

2.03.3.3 Incorporation of ^{13}C-Labeled Glucose into the Isoprenoids of *E. coli* and *Alicyclobacillus acidoterrestris* via the Embden–Meyerhof–Parnas Pathway

Labeling patterns obtained after incorporation of [1-^{13}C]- or [6-^{13}C]glucose into the ubiquinone (**15**) (Figure 2) of *E. coli* and the hopanoids of *Alicyclobacillus acidoterrestris* differed from those observed in the isoprenoids from *Z. mobilis* and *M. fujisawaense*.[30] They could not be interpreted in the framework of the acetate–mevalonate pathway, but were fully consistent with the interpretation proposed above. Indeed, *E. coli* and *A. acidoterrestris* metabolize glucose mainly via the Embden–Meyerhof–Parnas pathway (glycolysis) (Figure 5) and to some extent via the pentose phosphate pathways. Fructose 1,6-biphosphate (**20**) synthesized from glucose is cleaved by aldolase into dihydroxacetone phosphate (**21**) and glyceraldehyde 3-phosphate (**19**) which are interconverted by triose phosphate isomerase. This pathway allows the incorporation of C-1 as well as C-6 of a hexose into the same position of a triose phosphate derivative (Figure 5). The labeled positions found in the isoprenoids of these two bacteria (Figure 3, patterns (g) and (h)) corresponded qualitatively exactly to those expected from a C_2 subunit derived from pyruvate decarboxylation and from a C_3 moiety arising from a triose phosphate derivative as described above for *Z. mobilis* and *M. fujisawaense*. The less efficient incorporation of C-1 of glucose is most probably due to partial loss of this carbon atom as carbon dioxide when glucose is metabolized via the oxidative pentose phosphate pathway in which C-1 is lost by oxidative decarboxylation of 6-phosphogluconate.

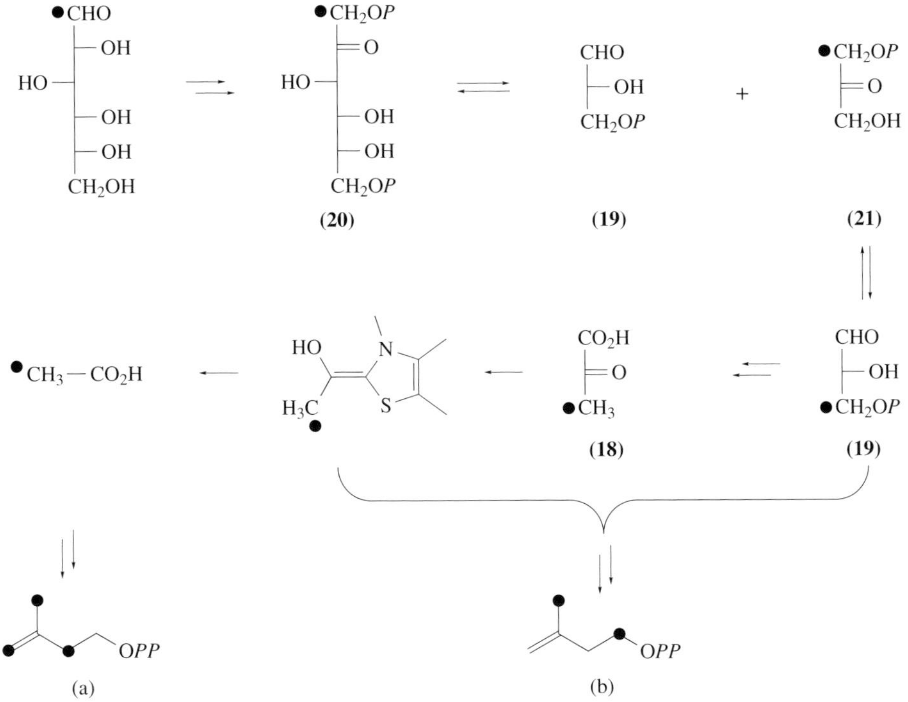

Figure 5 Glycolysis showing labeling patterns of isoprenic units resulting from the mevalonate pathway (a) or the glyceraldehyde phosphate–pyruvate pathway (b) after feeding of [1-^{13}C]glucose.

2.03.3.4 Incorporation of ^{13}C-Labeled Acetate via the Glyoxylate and Tricarboxylic Acid Cycles

When acetate is the only carbon and energy source, it is metabolized in bacteria by the enzymes of the glyoxylate and the tricarboxylic acid cycles (Figure 6). Keeping in mind the interpretation proposed above for the origin of carbon atoms in isoprene units from bacteria showing the unexpected labeling patterns, it is now easy to find a logical interpretation for all labeling patterns observed in the hopanoids and the side chain of ubiquinone after feeding of ^{13}C-labeled acetate.[23,30] Indeed, only the origin of pyruvate and triose phosphate derivatives have to be clarified. The key enzymes are: citrate synthase, yielding citrate (**23**) from oxaloacetate (**22**) and acetyl CoA; isocitrate lyase, cleaving isocitrate (**24**) into glyoxylate (**28**) and succinate (**26**); and malate synthase, condensing acetyl CoA with glyoxylate and yielding malate (**27**). Thus two C_2 units and one C_4 unit yield, via the action of these enzymes, two C_4 dicarboxylic acids, one of which can be utilized as a building block for further syntheses, the other being utilized for starting a new cycle. If only the glyoxylate cycle is involved, C-1 of pyruvate (and therefore also of glyceraldehyde 3-phosphate) is derived from C-1 of acetate, whereas C-2 and C-3 of pyruvate or glyceraldehyde phosphate are derived from C-2 of acetate (Figure 6). However, some scrambling arises via the tricarboxylic acid cycle. Indeed, malate can be obtained in two different ways: either via malate synthase in the glyoxylate cycle, or from the sequence α-oxoglutarate–succinate–fumarate in the tricarboxylic acid cycle. In the latter case, after decarboxylation of α-oxoglutarate (**25**), a carboxylate group of succinate corresponds to the ketone carbonyl group of α-oxoglutarate and is therefore derived from C-2 of acetate (Figure 6). Several recyclings of the C_4 carboxylic acids result in the following average labeling pattern: C-1 of pyruvate (**18**) and glyceraldehyde 3-phosphate (**19**) are equally derived either from C-1 or from C-2 of acetate, whereas the C-2 and C-3 carbon atoms are always derived from the C-2 methyl group of acetate. Such a labeling pattern was in fact observed in the derived hopanoids after feeding of ^{13}C-labeled acetate to *Methylobacterium organophilum*, *Rhodopseudomonas acidophila*, and *Rhodopseudomonas palustris*, in the hopanoids and the ubiquinone from *M. fujisawaense*, and finally in the ubiquinone from *E. coli* (**15**) (Figure 3, patterns (a) and (b)).[23,30]

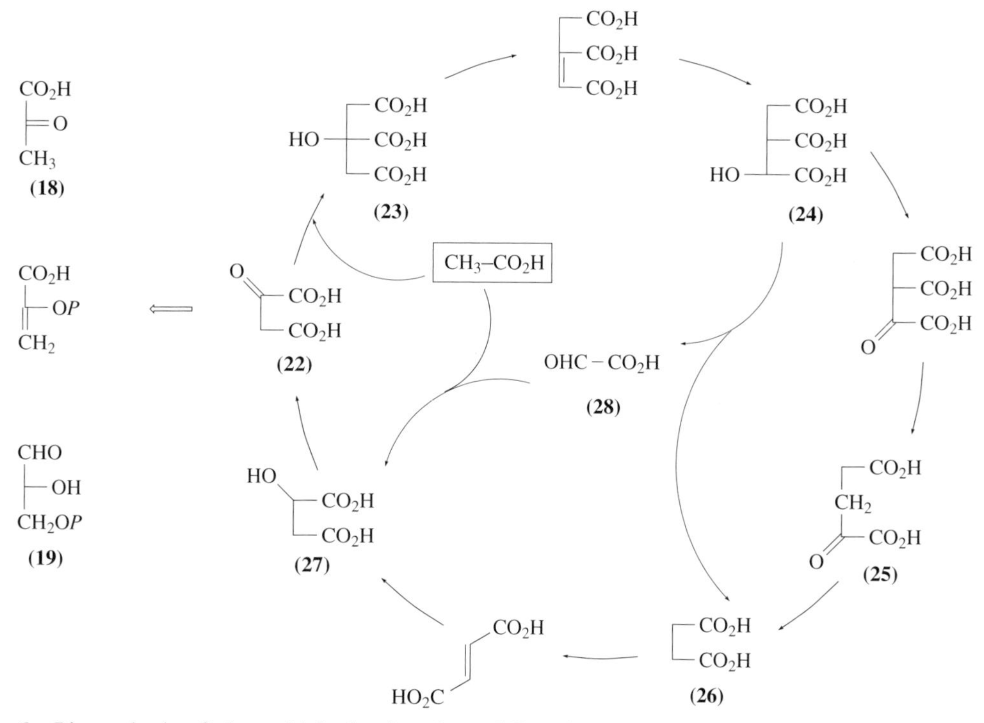

Figure 6 Biosynthesis of glyceraldehyde phosphate (**19**) and pyruvate (**18**) from acetate via the glyoxylate and tricarboxylic acid cycles. For the sake of simplicity, only the carbon skeletons of the intermediates have been represented. All cofactors have been omitted.

2.03.3.5 The Absence of Key Enzymes of the Mevalonate Pathway in Bacteria Possessing the Alternative Pathway

Results of the incubations performed with ^{13}C-labeled precursors definitively ruled out acetyl CoA and hydroxymethylglutaryl CoA as precursors for isoprenoids in the bacteria under investigation. These findings do not exclude mevalonate as precursor, however, since another sequence leading to this metabolite via other intermediates could be imagined. Many studies have reported on the role of mevalonate in the biosynthesis of bacterial isoprenoids. In most cases, incorporation yields were rather modest. This aspect of isoprenoid biosynthesis has been revisited in prokaryotes by Sahm and co-workers in the light of the existence of two distinct pathways.[17] Cell-free systems from six different prokaryotes were investigated for the early steps of isoprenoid biosynthesis. Those of *Halobacterium cutirubrum*, *Lactobacillus plantarum*, *Myxococcus fulvus*, and *Staphylococcus carnosus*, possessing, according to literature data, the acetate–mevalonate pathway, efficiently converted [^{14}C]acetyl CoA or [^{14}C]hydroxymethylglutaryl CoA to [^{14}C]mevalonic acid. Furthermore, [^{14}C]mevalonate, [^{14}C]mevalonate 5-phosphate, or [^{14}C]mevalonate 5-diphosphate were converted into [^{14}C]IPP. The presence of the acetate–mevalonate pathway could later be proved directly *in vivo* in *M. fulvus* by feeding of [1-^{13}C]acetate: the labeling of the prenyl chain of ubiquinone corresponded to isoprene units synthesized from three equivalents of acetyl CoA directly derived from exogenous ^{13}C-labeled acetate via mevalonate (Figure 3, pattern (i)) (Rosa Putra and Rohmer, unpublished results).

In contrast, none of the intermediates of the acetate–mevalonate pathway could be detected when the same experiments were performed with cell-free extracts prepared from *E. coli* and *Z. mobilis*, both of which were shown to possess the alternative pathway. Although negative results may not represent a definitive proof for the absence of a pathway (e.g., because of possibly inappropriate experimental conditions), these data strongly supported two conclusions. On the one hand, mevalonate is not a precursor for isoprenoids in *E. coli* and in *Z. mobilis*, and on the other hand, at least two different pathways for isoprenoid formation occur in bacteria.

2.03.4 THE PRECURSORS OF IPP IN THE MEVALONATE-INDEPENDENT PATHWAY

2.03.4.1 Glyceraldehyde 3-Phosphate and Pyruvate as First Precursors

Based on the similarities observed between isoprenoid biosynthesis and the formation of L-valine from pyruvate, both of which involve a rearrangement (see Section 2.03.4.2), either methylglyoxal or hydroxyacetone could be potential precursors of IPP. Indeed, methylglyoxal is synthesized from dihydroxyacetone phosphate (**21**) (Figure 5) and is present in many bacteria.[40–42] In order to check such a hypothesis, [2-^{13}C]hydroxyacetone (isotopic abundance 99%, 0.12 g L^{-1}) was synthesized and added to a culture of *M. fujisawaense* in the presence of acetate (1 g L^{-1}) as the main carbon source (Seemann and Rohmer, unpublished results). The weak labeling found in the hopanoids and ubiquinone in carbon atoms derived from C-2 and C-3 of IPP (2 and 3%, respectively) most probably did not result from the direct incorporation of hydroxyacetone into these isoprenoids. The observed labeling could instead be explained by successive oxidations of the ketol into methylglyoxal and pyruvate and introduction of pyruvate into triose phosphate metabolism.

The two first precursors in the mevalonate-independent pathway could be identified using *E. coli* mutants each lacking a single enzyme of triose phosphate metabolism (Figure 7): Lin 61 lacking glycerol kinase, Lin 201 lacking glycerol 3-phosphate dehydrogenase, DF 502 defective in triose phosphate isomerase, W3CG missing glyceraldehyde 3-phosphate dehydrogenase, DF 263 lacking 3-phosphoglycerate kinase, and finally DF 261 lacking enolase.[38] When grown on a minimal medium containing only mineral salts, each of these mutants required both pyruvate (**18**) and glycerol (**29**) as carbon sources, being fully unable to interconvert them. Two sets of labeling experiments were performed with each strain: in one case a feeding experiment with [2-^{13}C]glycerol in the presence of unlabeled pyruvate, and in the other, an experiment using pyruvate labeled with ^{13}C either at C-2 or C-3 (depending on the experiment) in the presence of unlabeled glycerol. From the labeling patterns observed in the isoprenoid moiety of ubiquinone 8, the identity of the yet unknown triose phosphate derivative could be determined without ambiguity. ^{13}C-Labeled pyruvate was incorporated in the C_2 subunit of all mutants, whereas ^{13}C-labeled glycerol was incorporated by none of them. This observation was in full agreement with the role of pyruvate as precursor of a C_2 subunit via decarboxylation. Incorporation of ^{13}C label into the C_3 moiety occurred on the one hand from pyruvate only in mutants Lin 61, Lin 201, and DF 502, and on the other hand from glycerol only in mutants DF 261, DF 263, and W3CG. All these results shed light on the unique position of glyceraldehyde 3-phosphate which remained the only precursor possible for the C_3 subunit and excluded as precursor all other compounds derived from triose phosphate metabolism, including methylglyoxal which is formed from dihydroxyacetone phosphate. Direct evidence for the key role of pyruvate was also obtained by the successful incorporation of [3-^{13}C]pyruvate into the hopanoids and ubiquinone of *M. fujisawaense*[30] or of L-[3-^{13}C]lactate into the ubiquinone and menaquinones of *E. coli* (Figure 3, pattern (j)).[37] Indeed, L-lactate is readily converted into pyruvate by the latter bacterium and therefore efficiently introduced into the triose phosphate metabolism.

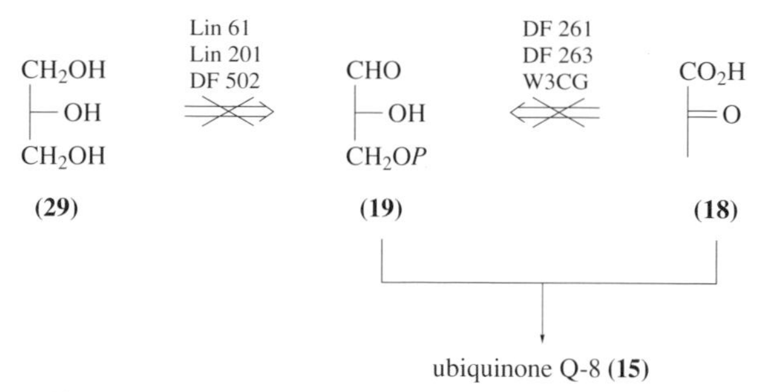

Figure 7 Incorporation of ^{13}C-labeled glycerol or pyruvate into ubiquinone by triose phosphate metabolism mutants of *E. coli*.

2.03.4.2 Hypothetical Biogenetic Scheme for the Mevalonate Independent Pathway

From all data gathered from the above-mentioned labeling experiments, it was now possible to propose a hypothetical biogenetic scheme for the formation of the C_5 branched carbon skeleton of IPP (Figure 8).[38] A C_2 subunit generated by pyruvate (**18**) decarboxylation, most likely hydroxylethylthiamine (**30**), is condensed with the carbonyl group of glyceraldehyde 3-phosphate (**19**) (or possibly glyceraldehyde itself, which however is not a common metabolite). This reaction, resembling the reaction catalyzed by a transketolase, yields a C_5 straight chain carbohydrate, a 1-deoxypentulose phosphate (**31**). The occurrence of this compound formed by such an enzymatic reaction has been previously described many years ago from numerous bacteria and fungi.[43,44] The condensation product between pyruvate and D-glyceraldehyde was identified as D-1-deoxyxylulose which was shown to be directly incorporated into the ubiquinone of *E. coli* (see Section 2.03.4.3). A rearrangement, as pointed out by the labeling experiments using [4,5-$^{13}C_2$]- and [U-$^{13}C_6$]glucose, could yield the branched isoprenoid skeleton from the straight chain carbohydrate. Such a proton-catalyzed acetoin rearrangement is involved in the formation of the skeleton of branched-chain amino acids such as valine, leucine, or isoleucine. Other reactions required in this route would be water eliminations, diol/ketone or aldehyde isomerizations, reduction steps, and at least one phosphorylation, all by an as yet unknown sequence.

As in the classical mevalonate route, the resulting IPP is the precursor of isoprenoids. This could be verified by its successful incorporation into other isoprenoids by cell-free systems from bacteria possessing this pathway. This means that IPP has to be considered as the universal precursor of isoprenoids.[17,45–49] Full elucidation of this new biosynthetic route is hampered by the lack of information about the intermediates, making the identification of intermediates as well as enzymes involved in this pathway tedious, but extremely challenging. Two such intermediates have been identified by comparison of the expected structures for putative intermediates (Figure 8) with those of already-described natural products which had apparently nothing to do with isoprenoid biosynthesis. The role of these two metabolites will be discussed in the following two sections.

2.03.4.3 D-1-Deoxyxylulose 5-Phosphate, a Key Intermediate for Several Biosynthetic Routes

From all the above-mentioned ^{13}C-labeling experiments performed with bacteria, both pyruvate and glyceraldehyde phosphate were shown as the only precursors of the C_5 skeleton of isoprenoid units. Their condensation by a transketolase-like reaction has already been described many years ago by Yokota and Sasajima. A rather unspecific thiamine-dependent enzyme is apparently able to condense a C_2 acetyl unit obtained from pyruvate, acetoin, or methylacetoin on a variety of C_3, C_4, or C_5 aldoses of the D or the L series or on their phosphates.[43,44] This enzyme has been found to be widespread in bacteria as well as in yeasts and lower fungi. From pyruvate and D-glyceraldehyde or D-glyceraldehyde 3-phosphate the enzyme catalyzes the formation of D-1-deoxyxylulose (**31a**) or its 5-phosphate (**31b**) (Figure 8). The free pentulose has already been shown to be involved in other biosynthetic pathways leading to the formation of the carbon skeleton of thiamine in bacteria and

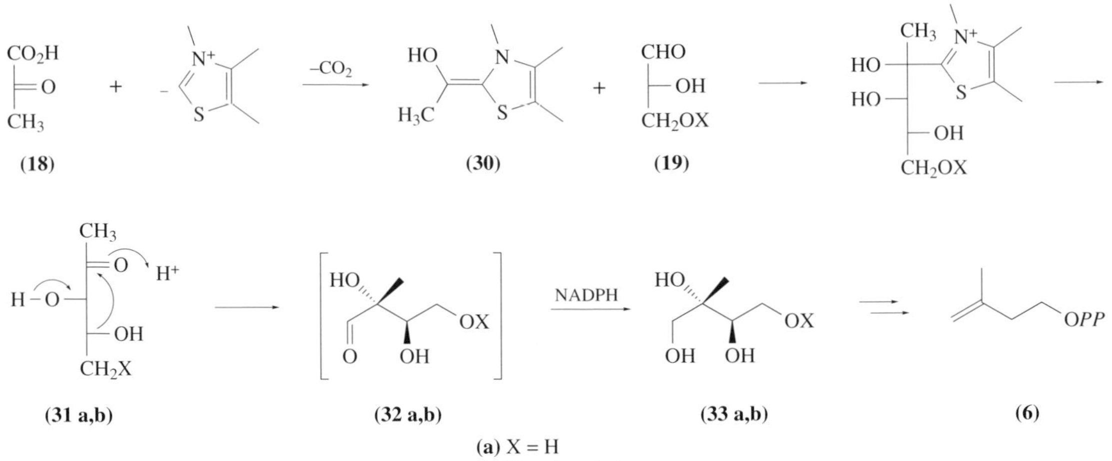

(18) (30) (19)

(31 a,b) (32 a,b) (33 a,b) (6)

(**a**) X = H
(**b**) X = phosphate

Figure 8 Hypothetical biogenetic pathway for isoprenoid biosynthesis from glyceraldehyde phosphate and pyruvate.

higher plants and of pyridoxal phosphate in *E. coli*.[50–52] The participation of (**31a**) in the nonmevalonate route for isoprenoid biosynthesis has been thoroughly checked by Broers and Arigoni.[37] The addition of 1-deoxyxylulose (**31a**) to the culture medium of *E. coli* increased the intracellular concentration of ubiquinone (**15**), menaquinone (**16**), and demethylmenaquinone by a factor of 3 to 4. This allowed successful incorporation of synthetic [1-$^{2}H_{1}$]- or [5,5-$^{2}H_{2}$]-D-1-deoxyxylulose with high yields into the prenyl chains of ubiquinone (**16**) and the menaquinones (**17**) (Figure 1), with labeling on carbon atoms corresponding respectively to C-5 and C-1 of IPP as shown by ^{2}H NMR. These experiments represented the first direct proof for the role of 1-deoxyxylulose as precursor for the C_5 skeleton of isoprene units, most probably after phosphorylation at C-5.

2.03.4.4 2-*C*-Methyl-D-erythritol 4-phosphate, a Putative Intermediate

According to the proposed biogenetic scheme for the nonmevalonate pathway for isoprenoid biosynthesis, rearrangement of the 1-deoxyxylulose skeleton could give the aldehyde (**32a,b**), which could be the precursor of already known natural products. Indeed if reduction immediately follows the rearrangement, analogous to the reaction catalyzed by acetolactate reductoisomerase in valine biosynthesis, the resulting product would be 2-*C*-methyl-D-erythritol (**33a**) or its 4-phosphate (**33b**). This tetrol is found under normal growth conditions as 2,4-cyclodiphosphate in the anaerobic bacterium *Desulfovibrio desulfuricans*.[53] It is also accumulated in high concentrations in the cytoplasm (up to 50 mM) by *Corynebacterium ammoniagenes*, *Staphylococcus aureus*, and *Mycobacterium smegmatis* under oxidative stress conditions induced by treatment during the stationary phase with benzylviologen or the herbicide diquat.[54–56] The free tetrol or erythronolactone resulting from the oxidation of aldehyde (**32a**) (Figure 8) have been also found in higher plants, quite often under oxidative stress conditions (senescent leaves, water stress).[57–63] These observations encouraged us to investigate in detail the biosynthesis of 2-*C*-methyl-D-erythritol and its possible involvement in isoprenoid biosynthesis. For this purpose, labeling experiments were performed with *C. ammoniagenes* treated after its exponential growth phase with benzylviologen.[64] After incubation of [1-^{13}C]- or [6-^{13}C]glucose, the labeling patterns found for 2-*C*-methyl-D-erythritol and the isoprenoid units of the dihydromenaquinones (**17**) (Figure 1) from this bacterium were identical. Both sets of labeling patterns corresponded to those observed in the ubiquinone from *E. coli* or the hopanoids from *A. acidoterrestris* after feeding the same labeled carbon sources (Figure 3, patterns G and H).[30] As Corynebacteria utilize glucose mainly via the Embden–Meyerhof–Parnas pathway and the pentose phosphate pathways as do the former two bacteria,[65,66] these label results strongly indicated that glyceraldehyde 3-phosphate and pyruvate were the precursors of isoprenoid metabolites in these organisms and not acetyl CoA derived from the glycolysis via pyruvate. After incubation of [U-$^{13}C_{6}$]glucose, formation of the C_5 branched skeleton of 2-*C*-methyl-D-erythritol and of the isoprenoid units of dihydromenaquinones via the condensation of C_2 and C_3 subunits and a rearrangement was clearly evident from the ^{13}C–^{13}C couplings, as already reported for the hopanoids of *Z. mobilis*, as well as by the long-range ^1H–^{13}C couplings. All these data showed that the C_5 carbon skeletons of methylerythritol and IPP resulted from the same biosynthetic pathway.[64] The only missing link was the precursor–product relationship between the two compounds. [4,4-$^2H_2$]- and [1,1,4,4-$^2H_4$]-2*C*-methyl-D-erythritol were therefore synthesized and incubated with *E. coli*. Ubiquinone and menaquinone obtained from these cultures were examined by means of mass spectrometry and ^2H NMR spectroscopy.[67] Incorporation rates were lower than those reported by Broers and Arigoni for 1-deoxyxylulose,[37] corresponding to about a 1 to 15 ratio of labeled isoprenoid units. Incorporation of two or four deuteriums could be detected, however, by mass spectrometry after incubation of [4,4-$^2H_2$]- and [1,1,4,4-$^2H_4$]methylerythritol in separate experiments. These results were in accordance with the simultaneous incorporation of all five carbon atoms of intact methylerythritol into IPP without loss of any of the four deuterium atoms present at C-1 and C-4 in the precursor. Only the D enantiomer was incorporated. No labeling appeared in isoprenoids after feeding deuterium-labeled L-methylerythritol. The ^2H NMR spectra obtained after incubation with D-[4,4-$^2H_2$]methylerythritol showed signals in the spectra of ubiquinone and menaquinone corresponding to chemical shifts of deuterium atoms located on carbon atoms derived from C-1 of IPP. Retention of all deuterium atoms excluded any further oxidations of carbon atoms derived from C-4 and C-1 of methylerythritol (i.e., from C-1 and C-4 of IPP) in the reaction sequence yielding IPP. These incorporations of deuterium-labeled methylerythritol completed the partly ambiguous results obtained from labeling experiments using [6,6-$^3H_2$]glucose or L-[3,3,3-$^2H_2$]lactate. Indeed, deuterium labeling was qualitatively found in all feeding experiments at the expected positions in the pentalenic acid derivatives from *Streptomyces* UC 5319,[39] the isoprenoid side chains of the quinones from *E. coli*,[37] or the hopanoids from *Z. mobilis* (Duvold and Rohmer, unpublished results). No deuterium loss occurred for carbon atoms derived from C-1 of IPP, i.e., from C-4 of methylerythritol: as observed in the ^2H NMR spectra, the two deuterium atoms were retained in the ubiquinone of *E. coli* as well as in the hopanoids of *Z. mobilis*. In the methyl groups corresponding to C-5 of IPP and introduced via pyruvate, however, important unexplained losses (up to 75% in *Z. mobilis*) were observed in isoprenoids from all bacteria examined.

These data strongly supported the isoprenoid nature of 2-*C*-methyl-D-erythritol which should now be considered as a hemiterpene and as an intermediate in the mevalonate-independent pathway. This metabolite itself results from the rearrangement of D-deoxyxylulose, representing probably the first committed step of this metabolic route.

2.03.5 DISTRIBUTION OF THE MEVALONATE-INDEPENDENT PATHWAY

2.03.5.1 Distribution Among Prokaryotes

Little is known about the biosynthetic route utilized by prokaryotes for the formation of their isoprenoids. The classical mevalonate route seems present in the very few Archaea examined so far, as shown by the incorporation of ^{13}C-labeled acetate into the biphytanyl chains of the lipids from the thermoacidophilic *Caldariella acidophila*,[68] of ^{14}C-labeled mevalonate and ^{13}C- or ^{14}C-labeled acetate in the phytanyl lipid chains from *Halobacterium cutirubrum*[45,69] and of deuterium-labeled glycerol or glucose into those of *Halobacterium halobium*[70] as well as from the detection of several enzymes of the mevalonate pathway in the former halophile or the conversion of hydroxymethylglutaryl–CoA into mevalonate by enzyme preparations from the latter organism.[71] A small amount of data has indicated that the acetate–mevalonate is operative in some Eubacteria. Although the radioactivity in labeled products had usually not been localized, the incorporation of ^{14}C-, and in one case of ^{13}C-labeled mevalonate, into isoprenoids of different series (C_{30} and C_{40} carotenoids, acyclic polyprenol derivatives, isopentenyl adenosine, sterols) supported the occurrence of the mevalonate pathway in *Staphylococcus* sp.,[47] *Lactobacillus plantarum*,[72–76] *Streptococcus mutans*,[77] *Streptococcus faecium*,[78] *Myxococcus fulvus*,[79] *Nannocystis exedens*,[80] *Chloropseudomonas ethylica*,[81] and in an incompletely identified bacterium reported as *Flavobacterium* species.[82] Unquestionable evidence for this pathway came from characterization of the enzymatic activities of the mevalonate pathway in *L. plantarum*, *Staphylococcus carnosus*, and *M. fulvus*.[17] For the latter bacterium, direct evidence was also obtained from the incorporation of [1-^{13}C]acetate into the ubiquinone with a labeling pattern corresponding to IPP resulting as expected from the condensation of three acetyl CoA units (Rosa Putra and Rohmer, unpublished results). In contrast, evidence against the mevalonate route might be the negative results obtained by attempted incorporation into bacterial isoprenoids of the normal precursors acetate and mevalonate. [U-^{14}C]Acetate was not incorporated into squalene and related hydrocarbons by *Streptomyces hygroscopicus* and *S. griseus*,[83] and ^{14}C-labeled acetate of mevalonate did not label ubiquinone in cells of *Agrobacterium tumefaciens*, *Azotobacter vinelandii*, a "*Pseudomonas*" species, and *E. coli* (which was later shown to possess the alternative route), although the fatty acid-containing lipids were significantly labeled from acetate.[84] Mevalonate or IPP were only weakly incorporated into menaquinones and carotenoids by membrane preparations from *Micrococcus luteus*,[85] and mevalonate was only weakly incorporated into isoprenoids (hopanoids and polyprenol derivatives) by *Alicyclobacillus* (formerly *Bacillus*) *acidocaldarius* cells.[86] Furthermore, ^{14}C-labeled mevalonate was not incorporated into isoprenoids in cell-free systems from *M. fujisawaense* (Knani and Rohmer, unpublished results), and neither ^{14}C-mevalonate, mevalonate phosphate, nor mevalonate diphosphate could be converted by cell-free systems from *Z. mobilis*, *E. coli*, *Corynebacterium glutamicum*, and *Bacillus licheniformis*, whereas ^{14}C-labeled IPP was detected by the same authors when they incubated these precursors in cell-free systems from bacteria possessing the mevalonate route.[17]

Although negative results are always ambiguous, indicating for instance possibly inappropriate conditions for enzymatic tests, the failure to incorporate mevalonate into bacterial isoprenoids, some of which had been shown to possess the alternative route, is in accordance with a rather large distribution of the nonmevalonate pathway among Eubacteria. Finally, the enzymes of the early steps of the mevalonate pathway have almost never been characterized in bacteria. Only bacterial hydroxymethylglutaryl CoA reductase has been thoroughly studied in some *Pseudomonas* species. In the latter case, however, this enzyme is instead involved in the catabolism of mevalonate which can be utilized as carbon source by these bacteria.[87]

When the position of the labeled atoms could be determined, the observed labeling pattern was often not in accordance with that expected from the mevalonate pathway. [1-^{14}C]- and [2-^{14}C]acetate were incorporated into the ubiquinone side chains isolated from *E. coli*, *A. vinelandii*, *Pseudomonas sesami*, and *Rhodopseudomonas capsulata* in a nonclassical way: label from C-1 and C-2 of acetate was incorporated equally into all carbon atoms corresponding to C-1 of IPP, in contradiction with the ratio expected from the known pathway to IPP via acetoacetyl CoA and HMG–CoA.[5] Pandian *et al.* attempted to rationalize these results by proposing that mevalonate is formed from pyruvate, acetaldehyde, and carbon dioxide via the acetolactate pathway, normally utilized for the biosynthesis of L-valine.[31] Their interpretation, at least for isoprenoid biosynthesis in *E. coli*, was not consistent with the glyceraldehyde phosphate–pyruvate route that has been found in this bacterium. In a more recent study on the biosynthesis of the prenyl chain of ubiquinone, again in *E. coli*, Zhou and White excluded this acetolactate route, based on the lack of incorporation of its postulated intermediates.[88] Furthermore, [1,2-$^{13}C_2$]acetate fed in the presence of unlabeled glucose was incorporated into the fatty acids, but not into the isoprenoids, whereas [U-$^{13}C_6$]glucose efficiently labeled both lipid families. These authors showed also that a C_2 unit derived from carbon atoms C-2 and C-3 of pyruvate corresponded to C-3 and C-5 of IPP. Many of their results can now be readily explained in the framework of the glyceraldehyde phosphate–pyruvate pathway. Each of the latter precursors are directly derived from glucose and are therefore efficiently labeled from glucose but not from acetate (especially in the presence of glucose). Acetate is not directly incorporated into the isoprenoids, but instead must enter the tricarboxylic acid and the glyoxylate cycles to yield phosphoenolpyruvate which can be converted to the required precursors (Figure 6). More data for the elucidation of the biosynthetic route for isoprenoids in *E. coli* could not be obtained from this study, most probably because the incorporation of the labeled precursors was hampered by competition with other unlabeled carbon sources contained in the complex culture medium, yielding non-interpretable labeling patterns.

The nonmevalonate pathway is apparently widespread in bacteria (Table 1). It has been detected in many Gram-negative as well as Gram-positive species. Many new taxa can now be added to the list of bacteria that have been more extensively studied for the elucidation of the mevalonate-independent pathway. Incorporation of [1-^{13}C]acetate into ubiquinone allowed the identification of the nonmevalonate biosynthetic route in all other enterobacteria examined in addition to *E. coli*, such as *Citrobacter freundii*, *Salmonella typhimurium*, and *Erwinia carotovora*, in *Pseudomonas aeruginosa* and *Pseudomonas fluorescens*, in *Burkholderia caryophylli* and *Burkholderia gladioli*, *Ralstonia pickettii*, *Acinetobacter calcoaceticus*, in the mycobacteria *Mycobacterium phlei* and in *Mycobacterium smegmatis* (Bravo, Disch, Rosa Putra, and Rohmer, unpublished results). This new pathway is also involved in the formation of β-carotene and the phytyl chain of chlorophylls in the cyanobacterium *Synechocystis* PCC 6714 (Schwender, Lichtenthaler, Disch, and Rohmer, unpublished results). The formation of the hopanoids has not yet been analyzed. The nonmevalonate route is also present in Actinomycetes. The already-mentioned deviations from the mevalonate route for the biosynthesis of the sesquiterpene pentalenolactone and its derivatives in *Streptomyces*

Table 1 Distribution of the glyceraldehyde phosphate–pyruvate pathway for isoprenoid biosynthesis.

Investigated organism	*Isoprenoids*	*Ref.*
Gram-negative Eubacteria		
Rhodopseudomonas acidophila	hopanoids	23
Rhodopseudomonas palustris	hopanoids	23
Synechocystis sp. PCC6714	phytol, carotenoids, hopanoids	a
Zymomonas mobilis	hopanoids	30
Escherichia coli	ubiquinone	30
Citrobacter freundii	ubiquinone	b
Erwinia carotovora	ubiquinone	b
Salmonella typhimurium	ubiquinone	b
Methylobacterium organophilum	hopanoids	23
Methylobacterium fujisawaense	hopanoids, ubiquinone	30
Pseudomonas aeruginosa	ubiquinone	b
Pseudomonas fluorescens	ubiquinone	b
Burkholderia caryophylli	ubiquinone	b
Burkholderia gladioli	ubiquinone	b
Ralstonia pickettii	ubiquinone	b
Acinerobacter calcoareticus	ubiquinone	b
Gram-positive Eubacteria		
Alicyclobacillus acidoterrestris	hopanoids	30
Corynebacterium ammoniagenes	dihydromenaquinones	64
Mycobacterium phlei	dihydromenaquinones	b
Mycobacterium smegmatis	dihydromenaquinones	b
Streptomyces sp. UC 5319	pentalenic acid derivatives (sesquiterpenes)	39
Streptomyces aeriouvifer	terpentecin (monoterpene moiety)	91
Streptomyces argenteolus	cyclized octaprenol derivative	92
Streptomyces exfoliatus	carquinostatin (dimethylallyl moiety)	93
Green algae		
Scenedesmus obliquus	sterols, phytol, carotenoids, plastoquinone ubiquinone	97 and a
Chlorella fusca	sterols, phytol, carotenoids	a
Chlamydomonas reinhardtii	sterols, phytol, carotenoids	a
Red algae		
Cyanidium caldarium	phytol, carotenoids	a
Higher plants		
Daucus carota	phytol, carotenoids	98
Hordeum vulgare	phytol, carotenoids	98
Lemna minor	phytol, carotenoids	98
Ginkgo biloba	diterpenoids (ginkgolides)	99
Salvia miltiorrhiza	diterpenoids (labdane series)	99
Marrubium vulgare	diterpenoids (labdane series)	100
Taxus sinensis	diterpenoids (taxane series)	101
Mentha × piperita	monoterpenoids (menthone)	102
Mentha pulegium	monoterpenoids (pulegone)	102
Thymus vulgaris	monoterpenoids (thymol)	102
Geranium graveolens	monoterpenoids (geraniol)	102
Chelidonium majus	isoprene	103
Populus nigra	isoprene	103
Salix viminalis	isoprene	103

[a] J. Schwender, H. K. Lichtenthaler, A. Disch, and M. Rohmer, unpublished results. [b] A. Disch, S. Rosa Putra, J.-M. Bravo, and M. Rohmer, unpublished results.

UC 5319 were indeed the signature of the glyceraldehyde phosphate–pyruvate pathway.[39,89] Even more striking was the discovery by Seto and co-workers, by feeding experiments with ^{13}C-labeled acetate or glucose, that *Streptomyces aeriouvifer* possessed both the mevalonate and the glyceraldehyde phosphate–pyruvate pathways.[90,91] Indeed dihydromenaquinone, an essential metabolite, was synthesized via the glyceraldehyde phosphate–pyruvate pathway at initial growth stages corresponding to the phase of exponential cell division, whereas the isoprenoid moiety of the composite antibiotic naphterpin was synthesized at a later stage mainly via the acetate–mevalonate route. Similarly, the nonmevalonate route was implied in the formation of the aglycone moiety of longestin, an antibiotic derived from octaprenol and synthesized by *Streptomyces argenteolus*[92] as well as of the dimethylallyl moiety of carquinostatin B from *Streptomyces exfoliatus*.[93] Apparently, all investigated *Streptomyces* species share in common the presence of the mevalonate-independent route (H. Seto, personal communication). The mevalonate route has also been clearly identified in other Actinomycetes by incorporation of ^{13}C-labeled acetate into the isoprenoid moiety of antibiotics such as terpentecins,[94] napyradiomycins,[95] and furaquinocins.[96] Whether the presence of both routes is a general feature in Actinomycetes and might be also found in other prokaryotes should be a matter of further investigation.

2.03.5.2 Distribution amongst Phototrophic Eukaryotes: Algae and Higher Plants

The early steps of isoprenoid biosynthesis in higher plants and their intracellular localization have been debated for several decades. Results from acetate or mevalonate incorporations, using mainly radioisotopes, and inhibition studies favored the formation of IPP from mevalonate in the cytoplasm but suggested the presence of an independent biosynthetic route in the plastids.[104–107] Only a few selected examples will be presented here. Mevalonate was always poorly incorporated, if at all, into chloroplast isoprenoids (e.g., in carotenoids), whereas the same precursor efficiently labeled sterols or sesquiterpenoids in the cytoplasm.[108] Similarly, labeling by mevalonate of monoterpenes or diterpenes, that are also synthesized in plastid-like structures, was also usually extremely low.[109] Furthermore, mevinolin, a strong HMG–CoA inhibitor, completely blocked sterol biosynthesis in the cytoplasm but had practically no effect on the biosynthesis of carotenoids or of chlorophylls containing a phytyl chain.[110,111] Key enzymes of the mevalonate pathway were not found in spinach chloroplasts or in daffodil chromoplasts, and neither hydroxymethylglutaryl CoA, mevalonate, mevalonate phosphate, nor mevalonate diphosphate were incorporated into the corresponding plastidic isoprenoids, whereas IPP was incorporated at high levels.[112,113] All these partly contradictory data were mainly interpreted in terms of compartmentation of isoprenoid biosynthesis between cytoplasm and plastids, the presence of different metabolic channels for the precursors of each isoprenoid series depending on their intracellular localization, and the lack of permeability of the chloroplast membrane towards mevalonate and mevinolin. The presence in both compartments (cytoplasm and chloroplasts) of a unique acetate–mevalonate pathway was postulated by all authors, as expressed by Goodwin: "The failure of labeled acetate and mevalonate to become incorporated into β-carotene in the excised illuminated etiolated seedlings is almost certainly due to their failure to reach the site of carotenoid biosynthesis in the chloroplast rather than to the existence of an entirely novel pathway of biosynthesis."[114,115] This assumption was unanimously accepted and served as background for all research on plant isoprenoid biosynthesis in spite of numerous contradictory results. A thorough critical compilation of the literature data in opposition with the sole occurrence of the mevalonate pathway, and in fact rather in favor of the simultaneous presence of the mevalonate and glyceraldehyde phosphate–pyruvate pathway in higher plants, has been presented by Bach.[106]

Indeed, most odd results concerning isoprenoid biosynthesis in plants could be readily explained if two different biosynthetic routes were present in the cytoplasm and in chloroplasts. The possible occurrence of an as yet unidentified pathway within plastids was not excluded by Lütke-Brinkhaus and Kleinig.[116] After the discovery of the mevalonate-independent glyceraldehyde phosphate–pyruvate route in bacteria, it was therefore tempting to try to investigate directly the biosynthesis of chloroplast isoprenoids using ^{13}C-labeled precursors. This method had apparently never been applied successfully before. Indeed, the main problem to be solved was to identify labeling conditions allowing the incorporation of an exogenous organic carbon source into phototrophic organisms that normally utilize carbon dioxide. Photosynthesis had to be avoided since carbon dioxide is efficiently incorporated into plastid isoprenoids by photosynthesis.[104–107] In the author's first experiments with the unicellular green alga *Scenedesmus obliquus* using ^{13}C-labeled acetate or glucose in the presence of light intensities allowing photosynthesis, all isoprenoids (sterols, carotenoids, phytol) were uniformly labeled, indicating efficient recycling of $^{13}CO_2$ resulting from the catabolism of the carbon source[117] (and Schwender, Lichtenthaler, Seemann, and Rohmer, unpublished results). It was necessary, therefore, to adopt the labeling conditions previously used for the successful bacterial incorporation experiments. Labeling experiments were therefore performed in heterotrophic conditions in the presence of minimal light intensities that were insufficient for photosynthesis but allowed the formation of differentiated chloroplasts and induced the formation of carotenoids and chlorophylls. Only one ^{13}C-labeled carbon source was utilized, as was done for the analogous labeling experiments with bacteria. These conditions allowed facile interpretation of the observed labeling patterns, since reasonable hypotheses could be made concerning the pathways utilized for the metabolism of the labeled carbon source.

For the sake of simplicity, the first experiments were made with the unicellular green alga *S. obliquus*.[97] Labelings were performed either with ^{13}C-labeled acetate or glucose. Surprisingly, the observed labeling patterns in all examined isoprenoids including sterols (**34**) from cytoplasm as well as the phytyl chain (**35**) of chlorophylls, the carotenoids β-carotene (**36**) and lutein (**37**), and plastoquinone (**38**) from chloroplasts (Figure 9) did not correspond to the labeling pattern expected from the acetate–mevalonate route, but corresponded to those resulting exclusively from the glyceraldehyde phosphate–pyruvate route. The patterns were identical to those already observed in bacteria after incorporation of acetate into the glyoxylate and tricarboxylic acid cycles for acetate or glucose main catabolism via glycolysis (Figure 3, patterns A, G, and H). Slight quantitative differences in the observed isotopic abundances were due to a more important contribution of the pentose phosphate pathways. Identical results were obtained with two other unicellular green algae, *Chlorella fusca* and *Chlamydomonas reinhardtii* (Disch, Schwender, Lichtenthaler, and Rohmer, unpublished results).

Cytoplasm

Chloroplasts

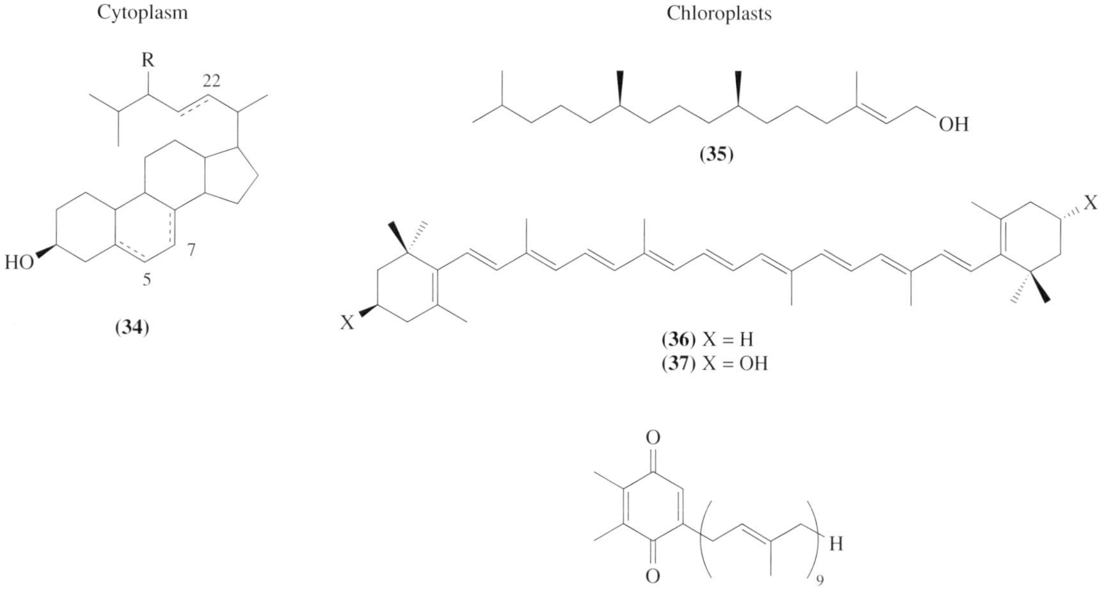

(**34**)

(**35**)

(**36**) X = H
(**37**) X = OH

(**38**)

Figure 9 Isoprenoids from algae and higher plants.

Similar labeling experiments were also performed with higher plant systems: barley seedlings (*Hordeum vulgare*), axenic cultures of a duckweed (*Lemna gibba*), or with green tissue cultures of carrot (*Daucus carotta*).[98] In contrast with the green algae, a clear dichotomy was observed in isoprenoid biosynthesis. Phytosterols (**34**) (Figure 9) showed the expected labeling pattern from the acetate–mevalonate pathway and were only synthesized from acetyl CoA, obtained either directly from acetate when acetate was the carbon source, or from pyruvate decarboxylation via glycolysis when glucose was the carbon source. In contrast, all chloroplast isoprenoids, i.e., phytol (**35**) from the chlorophylls, β-carotene (**36**), lutein (**37**), and plastoquinone (**38**), showed the same labeling as that observed for the isoprenoids in green algae, corresponding exclusively to the glyceraldehyde phosphate–pyruvate pathway. There was no evidence using these labeling conditions for the involvement of the acetate–mevalonate pathway in the formation of chloroplast isoprenoids. The same dichotomy in the formation of sterols and chloroplast isoprenoids was observed in the red alga *Cyanidium caldarium* (Schwender, Lichtenthaler, Disch, and Rohmer, unpublished results). These experiments showed clearly the coexistence in two cell compartments of two biosynthetic routes for isoprenoid biosynthesis in higher plant cells with the acetate–mevalonate route present in the cytoplasm and the glyceraldehyde phosphate–pyruvate route confined to the plastids (Figure 10). As the three plant systems described above were randomly chosen for this study, these results are most probably of general scope, and the conclusions can be safely extended to the biosynthesis of chloroplast isoprenoids in all higher plants.

Data published by other research groups on the biosynthesis of more species-specific terpenoids fully support our observations. Independently of our work on the biosynthesis of bacterial isoprenoids, Schwarz, Cartayrade, and Arigoni identified for the first time this dichotomy for isoprenoid biosynthesis in a higher plant. In embryos of *Ginkgo biloba*[99], sterols were synthesized according to the mevalonate route, whereas the diterpenoid ginkgolides originated from the glyceraldehyde phosphate–pyruvate pathway. Similar conclusions were also found for the biosynthesis of diterpenoids of the labdane series in tissue cultures from the Lamiaceae *Salvia miltiorrhiza*[99] or *Marubium vulgare*[100] and for the formation of a diterpenoid of the taxane series in tissue cultures of *Taxus sinensis*.[101] Since diterpene biosynthesis is supposed to occur in a plastidic compartment, all these observations were fully consistent with the former observations concerning the biosynthesis of the ubiquitous chloroplast isoprenoids.

Monoterpenes are another isoprenoid family synthesized in plastid-like structures. For instance, studies on peppermint secretory cell clusters from glandular trichomes showed that gland cell plastids (leucoplasts) were capable of synthesizing IPP.[118] Furthermore, in 1972 Banthorpe *et al.* pointed out and discussed in detail the paucity of evidence on monoterpene biosynthesis and the generally extremely low incorporation of mevalonate into isoprenoids of this series compared with those observed for sterols.[109] Participation of the glyceraldehyde phosphate–pyruvate has been observed for the formation of monoterpenes. Feeding of [1-^{13}C]- or [U-$^{13}C_6$]glucose to peppermint plantlets afforded low incorporation into menthone resulting from the mevalonate-independent route.[102] Similar results were obtained for the formation of pulegone in *Mentha pulegium*, geraniol in *Pelargonium graveolens*, and thymol in *Thymus vulgaris*.[102] Finally, isoprene emission in higher plants is highly correlated with photosynthesis and chloroplast activity. The biosynthesis of this hydrocarbon from dimethylallyl diphosphate via the glyceraldehyde phosphate–pyruvate pathway was therefore most likely. Indeed, application of deuterium-labeled methyl 1-deoxyxyluloside (which was most likely hydrolyzed by a glycosidase into free deoxyxylulose, a precursor for isoprenoids in the mevalonate-independent pathway) to the leaves of *Chelidonium majus*, *Populus nigra*, or *Salix viminalis* resulted in deuterium labeling of isoprene.[103] The known distribution of the glyceraldehyde phosphate–pyruvate pathway in eukaryotes is summarized in Table 1.

2.03.6 CONCLUSIONS—FURTHER DEVELOPMENTS

The discovery of an alternative mevalonate-independent route for isoprenoid biosynthesis in bacteria removed a major bottleneck in the study of the early steps of isoprenoid biosynthesis. Despite the accumulation of strange, inexplicable, and even contradictory results on the biosynthesis of isoprenoids from plants and bacteria, nobody was ready (including the author's group at the beginning of the work with bacterial triterpenoids[23]) to refute the unanimously accepted role of mevalonate as the universal isoprenoid precursor. Bacterial triterpenoids of the hopane series were the perfect tools for ^{13}C-labeling experiments that allowed the discovery of the novel glyceraldehyde phosphate–pyruvate pathway, owing to their high intracellular concentration, their chemical stability, their well-resolved ^{13}C NMR spectra, with no signals of sp^2 carbon atoms (such as those found in carotenoids and the prenyl chains of quinones and polyprenol derivatives) with long relaxation times lowering the sensitivity of NMR detection, and finally the possibility of producing these metabolites from bacteria grown on culture media of definite composition utilizing a single carbon source. The methodology used for labeling bacterial isoprenoids, as well as the conclusions drawn from these studies, could later easily be extended to studies on isoprenoids from algae or higher plants without major problems, resulting in analogous labeling patterns that could be interpreted in the framework of a common hypothesis. This previously overlooked metabolic route is now a novel tool for microbiologists, plant biochemists, and plant physiologists, as seen from the increasing number of contributions on this topic. One can now safely assume that all of the ubiquitous chloroplast isoprenoids (carotenoids, phytol, plastoquinone), all of the more peculiar isoprenoids synthesized in plastid-related organelles (diterpenoids, monoterpenoids), as well as isoprene itself are mainly, if not exclusively, synthesized via the glyceraldehyde phosphate–pyruvate route. Important plant metabolic regulatory compounds such as gibberellins or abscissic acid are thus most likely derived from this latter pathway. Compartmentation of isoprenoid biosynthesis in plant cells is now also certain with the mevalonate pathway in the cytoplasm for sesquiterpene, ubiquitone, and sterol biosynthesis and the glyceraldehyde phosphate–pyruvate route in the plastids (Figure 10).[98,99] This dichotomy involves, of course, low permeability between the two compartments of all (or most) metabolites, as previously assumed, but is mainly based on different sequences of enzymatic reactions leading to IPP. In fact, this scheme is probably oversimplified. IPP, dimethylallyl diphosphate, geranyl diphosphate, and farnesyl diphosphate are common intermediates of both routes. Exchanges

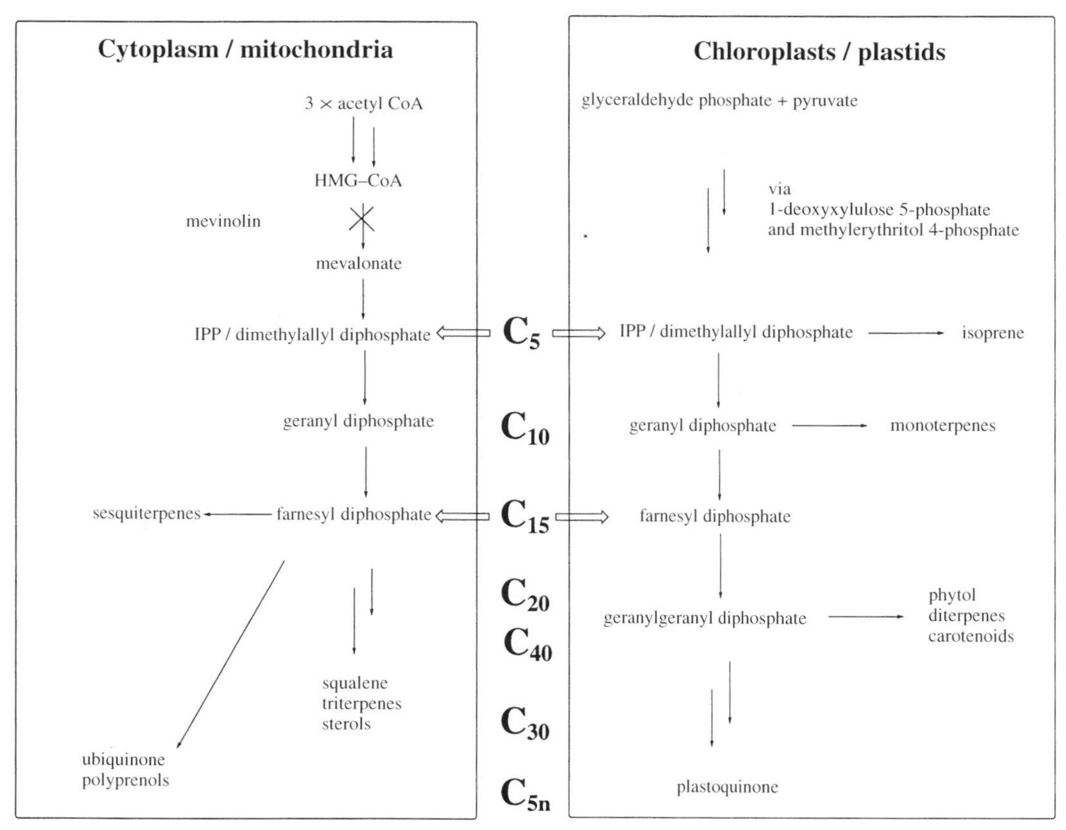

Figure 10 Compartmentation of isoprenoid biosynthesis in higher plants. Arrows at the level of IPP (C_5) and farnesyl diphosphate (C_{15}) indicate possible exchanges between compartments.

have been already detected between both compartments at the level of IPP and farnesyl diphosphate (Figure 10). Feeding of ^{13}C-labeled mevalonate to *G. biloba* embryos resulted in the detection of some transfer of IPP and farnesyl phosphate synthesized in the cytoplasm via the mevalonate pathway into the plastids where they were incorporated into ginkgolides, mainly synthesized from IPP and geranylgeranyl diphosphate arising from the glyceraldehyde phosphate–pyruvate pathway.[99] A similar heterogeneity and a resultant nonequivalent labeling of isoprenoid units were also observed after incorporation of ^{13}C-labeled mevalonate into heteroscyphic acid A, a diterpenoid, and phytol of the liverworts *Heteroscyphus planus* and *Lophocolea heterophylla*.[119–121] The farnesyl diphosphate-derived moiety could be labeled from mevalonate, whereas the fourth isoprenic unit of the diterpene skeleton was not labeled. Although there is presently no evidence for the presence of the glyceraldehyde phosphate–pyruvate route in liverworts, such a heterogeneous labeling pattern most probably reflects the signature of the presence of two pathways involved in isoprenoid biosynthesis. A last feature has to be pointed out. The mevalonate-independent pathway could only be easily detected in plant cells grown heterotrophically, i.e., in rather unnatural physiological conditions for a phototrophic organism. One should verify (and this is not easy) if the clear-cut dichotomy is still present under normal growth conditions when carbon dioxide is the main carbon source and photosynthesis occurs.

Isoprenoid biosynthesis in *Streptomyces* highlighted another striking aspect of isoprenoid biosynthesis, i.e., temporal compartmentation of the metabolic routes.[91] The glyceraldehyde phosphate–pyruvate route is apparently preferentially utilized for the formation of essential metabolites such as menaquinone during the exponential growth phase, whereas the mevalonate route is only induced during the stationary phase (like many other prokaryotic metabolic pathways) for the formation of the isoprenoid moiety of antibiotics. Such changes that depend on the physiological state are to be expected for other organisms, including higher plants.

The recognition of two distinct biosynthetic routes for the formation of IPP requires the thorough reevaluation of many assumptions concerning isoprenoid biosynthesis made *a priori* on the premise that mevalonate would be the only precursor to isoprenoids. For many isoprenoids, for example, the isoprenoids of lower eukaryotes (unicellular algae or protozoa) or even the prenyl chains of ubiquinone or the dolichols in eukaryotes, no clear data on the origin of their isoprenoid units were available. Labeling experiments performed with tobacco tissue cultures using ^{13}C-labeled glucose isotopomers or pyruvate have indicated that ubiquinone from the mitochondria and sterols from the cytoplasm were synthesized via the mevalonate pathway (Disch, Hemmerlin, Bach, and Rohmer, unpublished results) (Figure 10).

There is no simple test to date for the detection of the glyceraldehyde phosphate–pyruvate pathway. Enzymatic tests are not yet available. The only available method is the incorporation of a ^{13}C-labeled precursor (e.g., glucose or acetate, in the latter case in the absence of glucose or triose phosphate derivatives) and determination of the labeling patterns of derived isoprenoids by analysis of their ^{13}C NMR spectra. This method is hampered by its low sensitivity, is rather time-consuming, and requires determination of the proper labeling conditions. This laborious method is, so far, however, the only one that has yielded clear-cut results. Alternative methods based on enzymology and genetics are therefore urgently needed in order to determine more easily the distribution of both pathways and their physiological significance in living organisms.

The occurrence of two different biosynthetic routes leading to the same metabolites, sometimes in the same organism (e.g., the actinomycete *Streptomyces aeriouvifer*, the red alga *Cyanidium caldarium*, or all higher plants) is not a unique feature. Alternative pathways have been recorded for the formation of other metabolites.[121] For example, δ-aminolevulinic acid, the precursor of the porphyrin moiety of heme, chlorophylls, and vitamin B_{12}, is synthesized in higher plants from glutamate in green tissues or from succinate and glycine in the tubers. Dual pathways are also known for the biosynthesis of phosphatidylcholine and nicotinamide. The reason for such an apparent metabolic profligacy is not evident. In the case of isoprenoid biosynthesis, the presence of two separate biosynthetic routes might reflect the possibility of independent regulation of both routes depending on the physiological state of the organism, as in the *Streptomyces*. In higher plants, this would allow the separation of the synthesis of bulk material for membranes (sterols in the cytoplasm) from that of regulatory hormones such as gibberellins or abscissic acid derived from geranylgeranyl diphosphate, whose formation is therefore most likely linked to the chloroplasts. Both pathways to IPP present similar energetic requirements: three ATP and two NADPH for the mevalonate pathway, most likely two ATP (including the ATP required for the formation of glucose 6-phosphate and accordingly of glyceraldehyde 3-phosphate) and three equivalents of a reducing agent such as, for example, NAD(P)H for the glyceraldehyde phosphate–pyruvate pathway. The advantages of one route compared with the other are not obvious. The glyceraldehyde phosphate–pyruvate pathway seems more versatile and less specialized. The two identified C_5 intermediates are involved in roles other than that of isoprenoid precursor. Thus, deoxyxylulose 5-phosphate is a common precursor for isoprenoids as well as for several cofactors (thiamine, pyridoxal), and methylerythritol and its derivatives (cyclodiphosphate or methylerythronolactone) are apparently involved in defense mechanisms such as the response to stress conditions in microorganisms or plants. One could also speculate on the origin and the relative age of both pathways. Since the mevalonate route is present in nearly all eukaryotes (with the exception of the examined green algae which might have lost it), in the few investigated Archaea and in several Eubacteria, it could be the more primitive. The glyceraldehyde phosphate–pyruvate pathway could have appeared later in some Eubacteria, including cyanobacteria, and been transferred into photosynthesizing eukaryotes via endosymbiotic cyanobacteria or prochlorophytes, ancestors of chloroplasts. All these aspects represent a matter of speculation. More information is, of course, required in order to address such questions related to molecular evolution.

We should also point out the work that still has to be done. The problems related to isoprenoid biosynthesis in plants are probably even more complicated, as expected from the presence of two different metabolic routes. Striking labeling patterns have been detected in the prenyl moieties of chalcomoracin, a chalcone derivative from tissue cultures of *Morus alba*, after feeding of several ^{13}C-labeled precursors, while sitosterol showed a fully regular labeling.[124,125] These results have been interpreted in the framework of the mevalonate pathway, assuming that acetyl CoA originated from distinct metabolic routes. Although the interpretation of the labeling patterns was consistent, contradictions were evident: mevalonate did not label the prenyl moiety of chalcomoracin but was readily incorporated into sitosterol, and mevinolin did not block the formation of chalcomoracin but strongly inhibited the biosynthesis of sitosterol. These observations are beyond the scope of this chapter and cannot be discussed in the framework of the two known biosynthetic pathways for isoprenoid biosynthesis. They may open a new area in the regulation of the biosynthesis of plant isoprenoids. Conclusions based on the study of a few model organisms cannot be safely extended to all living organisms. Questions raised by the early steps of isoprenoid biosynthesis still deserve further work.

NOTE ADDED IN PROOF

The gene of the D-1-deoxyxylulose 5-phosphate synthase has been cloned and overexpressed in *E. coli*.[125,126] This enzyme catalyzes the formation of D-1-deoxyxylulose or its 5-phosphate, respectively, from pyruvate and D-glyceraldehyde or its 3-phosphate. Multiply ^{13}C-labeled D-1-deoxyxylulose isotopomers have been synthesized from ^{13}C-labeled pyruvate and D-glyceraldehyde using this enzyme. They were incorporated into phytol or carotenoids by *Catharanthus roseus* cell cultures[127] or into ubiquinone by *E. coli*.[128] Study of the ^{13}C couplings (including long range couplings) showed the identity of the five carbon atoms of the pentulose with those of IPP. D-1-Deoxyxylulose is thus not degraded prior to incorporation into isoprenoids, and the branched isoprenoid skeleton results from a rearrangement of a straight chain precursor as previously highlighted by the incorporation of multiply ^{13}C-labeled glucose isotopomers.

ACKNOWLEDGMENTS

This work was only made possible owing to all the author's co-workers who are cited in the references and whose contribution to the chemistry and biochemistry of bacterial hopanoids since the 1980s made possible the discovery of the mevalonate-independent pathway for isoprenoid biosynthesis. Financial support was continuously provided by the Centre National de la Recherche Scientifique and from 1993 to 1996 by the European Union in the framework of the project "Biotechnology of Extremophiles."

2.03.7 REFERENCES

1. J. D. Connolly and R. A. Hill, "Dictionary of Terpenoids," Chapman and Hall, New York, 1992.
2. S. L. Spurgeon and J. W. Porter, in "Biosynthesis of Isoprenoid Compounds," eds. J. W. Porter and S. L. Spurgeon, Wiley, New York, 1981, vol. 1, p. 1.
3. N. Qureshi and J. W. Porter, in "Biosynthesis of Isoprenoid Compounds," eds. J. W. Porter and S. L. Spurgeon, Wiley, New York, 1981, vol. 1, p. 47.
4. K. Bloch, *Steroids*, 1992, **57**, 378.
5. T. S. Raman, H. Rudney, and N. K. Buzzelli, *Arch. Biochem. Biophys.*, 1969, **130**, 164.
6. S. Fujisaki, T. Nishino, K. Izui, and H. Katsuki, *Biochem. Int.*, 1984, **8**, 779.
7. S. Fujisaki, T. Nishino, and H. Katsuki, *J. Biochem.*, 1986, **99**, 1137.
8. A. A. Kandutsch, H. Paulus, E. Levin, and K. Bloch, *J. Biol. Chem.*, 1964, **239**, 2507.
9. C. M. Allen, W. Alworth, A. Macrae, and K. Bloch, *J. Biol. Chem.*, 1967, **242**, 1895.
10. H. Sagami, K. Ogura, and S. Seto, *Biochemistry*, 1977, **16**, 4616.
11. T. Baba and C. M. Allen, *Arch. Biochem. Biophys.*, 1980, **200**, 474.
12. H. Fujii, T. Koyama, and K. Ogura, *J. Biol. Chem.*, 1982, **257**, 14610.
13. I. Takahashi and K. Ogura, *J. Biochem.*, 1982, **92**, 1527.
14. C. M. Allen, Jr., M. V. Keenan, and J. Saack, *Arch. Biochem. Biophys.*, 1976, **175**, 236.
15. J. G. Christenson, S. K. Gross, and P. W. Robbins, *J. Biol. Chem.*, 1969, **244**, 5436.
16. Y. Shigeri, T. Nishino, N. Yumoto, and M. Tokushige, *Agric. Biol. Chem.*, 1991, **55**, 589.
17. S. Horbach, H. Sahm, and R. Welle, *FEMS Microbiol. Lett.*, 1993, **111**, 135.
18. G. Ourisson, M. Rohmer, and K. Poralla, *Annu. Rev. Microbiol.*, 1987, **41**, 301.
19. G. Ourisson and P. Albrecht, *Acc. Chem. Res.*, 1992, **25**, 398.
20. G. Ourisson and M. Rohmer, *Acc. Chem. Res.*, 1992, **25**, 403.
21. M. Rohmer, *Pure Appl. Chem.*, 1993, **65**, 1293.
22. M. Rohmer, P. Bouvier-Navé, and G. Ourisson, *J. Gen. Microbiol.*, 1984, **130**, 1137.
23. G. Flesch and M. Rohmer, *Eur. J. Biochem.*, 1988, **175**, 405.
24. P. Bisseret and M. Rohmer, *J. Org. Chem.*, 1989, **54**, 2958.
25. P. Zhou, N. Berova, K. Nakanishi, M. Knani, and M. Rohmer, *J. Am. Chem. Soc.*, 1991, **113**, 4040.
26. B. Peiseler and M. Rohmer, *J. Chem. Res. Synop.*, 1992, **9**, 298.
27. N. Zhao, N. Berova, K. Nakanishi, M. Rohmer, P. Mougenot, and U. W. Jürgens, *Tetrahedron*, 1996, **52**, 2777.
28. R. I. Amann, W. Ludwig, and K.-H. Schleifer, *Microbiol. Rev.*, 1995, **59**, 143.
29. M. Rohmer, P. Bouvier, and G. Ourisson, *Proc. Natl. Acad. Sci. USA*, 1979, **76**, 847.
30. M. Rohmer, M. Knani, P. Simonin, B. Sutter, and H. Sahm, *Biochem. J.*, 1993, **295**, 517.
31. S. Pandiank, S. Saengchan, and T. S. Raman, *Biochem. J.*, 1981, **196**, 675.
32. P. Anastasi, I. Freer, K. H. Overton, D. Picken, D. Rycroft, and S. B. Singh, *J. Chem. Soc., Perkin I*, 1987, 2427.
33. S. Bringer-Meyer and H. Sahm, *FEMS Microbiol. Rev.*, 1988, **54**, 131.
34. G. A. Sprenger, *FEMS Microbiol. Lett.*, 1996, **145**, 301.
35. M. Rohmer, B. Sutter, and H. Sahm, *J. Chem. Soc., Chem. Commun.*, 1989, 1471.
36. B. Sutter, Ph.D. Thesis, Nb 91 MULH 0176, Université de Haute Alsace, Mulhouse, France, 1991, p. 11.
37. S. T. J. Broers, Ph.D. Thesis, Nb 10978, Eidgenössische Technische Hochschule, Zurich, Switzerland, 1994.
38. M. Rohmer, M. Seemann, S. Horbach, S. Bringer-Meyer, and H. Sahm, *J. Am. Chem. Soc.*, 1996, **118**, 2564.
39. D. E. Cane, T. Rossi, A. M. Tillman, and J. P. Pachtlako, *J. Am. Chem. Soc.*, 1981, **103**, 1838.

40. R. A. Cooper, *Annu. Rev. Microbiol.*, 1984, **38**, 49.
41. J. P. Richard, *Biochemistry*, 1991, **30**, 4581.
42. R. I. Lindstad and J. S. McKinley-McKee, *FEBS Lett.*, 1993, **330**, 31.
43. A. Yokota and K. I. Sasajima, *Agric. Biol. Chem.*, 1984, **48**, 149.
44. A. Yokota and K. I. Sasajima, *Agric. Biol. Chem.*, 1986, **50**, 2517.
45. S. C. Kushwaha, M. Kates, and J. W. Porter, *Can. J. Biochem.*, 1976, **54**, 816.
46. H. Fujii, T. Koyama, and K. Ogura, *J. Biol. Chem.*, 1982, **257**, 14610.
47. G. Suzue, K. Orihara, H. Morishima, and S. Tanaka, *Radioisotopes*, 1964, **13**, 300.
48. T. Baba, J. Muth, and C. M. Allen, *J. Biol. Chem.*, 1985, **260**, 10467.
49. S. Fujisaki, T. Nishino, and H. Katsuki, *J. Biochem.*, 1986, **99**, 1137.
50. J.-H. Juillard and R. Douce, *Proc. Natl. Acad. Sci. USA*, 1991, **88**, 2042.
51. I. A. Kennedy, T. Hemscheidt, J. F. Britten, and I. D. Spenser, *Can. J. Chem.*, 1995, **73**, 1329.
52. K. Himmeldirk, I. A. Kennedy, R. E. Hill, B. G. Sayer, and I. D. Spenser, *J. Chem. Soc., Chem. Commun.*, 1996, 1187.
53. D. L. Turner, H. Santos, P. Fareleira, I. Pacheco, J. Le Gall, and A. V. Xavier, *Biochem. J.*, 1992, **285**, 387.
54. D. Ostrovsky, A. Shashkov, and A. Sviridov, *Biochem. J.*, 1993, **295**, 901.
55. D. Ostrovsky, E. Kharatian, I. Malarova, I. Shipanova, L. Sibeldina, A. Sashkov, and G. Tantsirev, *BioFactors*, 1992, **3**, 261.
56. D. Ostrovsky, E. Kharatian, T. Dubrovsky, O. Ogrel, I. Shipanova, and L. Sibeldina, *BioFactors*, 1992, **4**, 63.
57. S. W. Shah, S. Brandänge, D. Behr, J. Damén, S. Hagen, and T. Anthonsen, *Acta Chem. Scand.*, 1976, **B30**, 903.
58. R. W. Schramm, B. Tomaszewska, and G. Peterson, *Phytochemistry*, 1979, **18**, 1393.
59. R. Kringstad, A. Olsvik Singsaas, G. Rusten, G. Baekkemoen, B. Smestad Paulsen, and A. Nordal, *Phytochemistry*, 1980, **19**, 543.
60. C. W. Ford, *Phytochemistry*, 1981, **20**, 2019.
61. P. Dittrich and S. J. Angyal, *Phytochemistry*, 1988, **27**, 935.
62. J. Kitajima and Y. Tanaka, *Chem. Pharm. Bull.*, 1993, **41**, 1667.
63. A. A. Ahmed, M. H. Abd El-Razek, E. Abu Mostafa, H. J. Williams, A. I. Scott, J. H. Reibenspies, and T. J. Mabry, *J. Nat. Prod.*, 1996, **59**, 1171.
64. T. Duvold, J.-M. Bravo, C. Pale-Grosdemange, and M. Rohmer, *Tetrahedron Lett.*, 1997, **38**, 4769.
65. T. Conway, *FEMS Microbiol. Rev.*, 1992, **103**, 1.
66. A. Marx, A. A. de Graaf, W. Wiechert, L. Eggeling, and H. Sahm, *Biotechnol. Bioeng.*, 1996, **49**, 111.
67. T. Duvold, P. Cali, J.-M. Bravo, and M. Rohmer, *Tetrahedron Lett.*, 1997, **38**, 6181.
68. M. de Rosa, A. Gambacorta, and B. Nicolaus, *Phytochemistry*, 1980, **19**, 791.
69. N. Moldoveanu and M. Kates, *Biochim. Biophys. Acta*, 1988, **960**, 164.
70. K. Kakinuma, M. Yamagishi, H. Fijimoto, N. Ikekawa, and T. Oshima, *J. Am. Chem. Soc.*, 1988, **110**, 4861.
71. J. A. Cabrera, J. Bolds, P. E. Shields, C. M. Havel, and J. A. Watson, *J. Biol. Chem.*, 1986, **261**, 3578.
72. K. J. I. Thorne and E. Kodicek, *Biochem. J.*, 1966, **99**, 123.
73. A. Peterkofsky, *Biochemistry*, 1968, **7**, 472.
74. D. P. Gough, A. L. Kirby, J. B. Richards, and F. W. Hemming, *Biochem. J.*, 1970, **118**, 167.
75. I. F. Durr and M. Z. Habbal, *Biochem. J.*, 1972, **127**, 345.
76. K. J. I. Thorne, *J. Bacteriol.*, 1973, **116**, 235.
77. K. J. I. Thorne, *Biochem. J.*, 1973, **135**, 567.
78. R. F. Taylor and B. H. Davies, *Can. J. Biochem.*, 1982, **60**, 675.
79. H. Kleinig, *Eur. J. Biochem.*, 1975, **57**, 301.
80. W. Kohl, A. Gloe, and H. Reichenbach, *J. Gen. Microbiol.*, 1983, **129**, 1629.
81. S. E. Moshier and D. J. Chapman, *Biochem. J.*, 1973, **136**, 395.
82. G. Britton, T. W. Goodwin, W. J. S. Lockley, A. P. Mundy, and N. J. Patel, *J. Chem. Soc., Chem. Commun.*, 1979, 27.
83. U. Gräfe, G. Reihardt, F. Hänel, W. Schade, and J. Gumpert, *J. Basic Microbiol.*, 1985, **8**, 503.
84. T. S. Raman, B. V. S. Sharma, J. Jayaraman, and T. Ramasarna, *Arch. Biochem. Biophys.*, 1965, **110**, 75.
85. J. A. Evans and J. N. Pebble, *Microbios Lett.*, 1982, **21**, 149.
86. M. de Rosa, A. Gambacorta, L. Minale and J. Bu'Lock, *Phytochemistry*, 1973, **12**, 1117.
87. M. J. Beach and V. W. Rodwell, *J. Bacteriol.*, 1989, **171**, 2994.
88. D. Zhou and R. H. White, *Biochem. J.*, 1991, **273**, 627.
89. M. Schwarz, *Chimia*, 1996, **50**, 280.
90. K. Shin-ya, K. Furihata, Y. Hayakawa, and H. Seto, *Tetrahedron Lett.*, 1990, **31**, 6025.
91. H. Seto, H. Watanabe, and K. Furihata, *Tetrahedron Lett.*, 1996, **37**, 7979.
92. H. Seto, H. Watanabe, N. Orihara, and K. Furihata, "Symposium Papers, The 38th Symposium on the Chemistry of Natural Products, Tohoku University, Sendai, Japan, 1996," p. 19.
93. N. Orihara, K. Furihata, and H. Seto, *J. Antibiotics*, 1997, **50**, 979.
94. K. Isshiki, T. Tamamura, T. Sawa, H. Naganawa, T. Takeuchi, and H. Umezawa, *J. Antibiotics*, 1986, **39**, 1634.
95. K. Shiomi, H. Iinuba, H. Naganawa, K. Isshiki, T. Takeuchi, and H. Umezawa, *J. Antibiotics*, 1987, **40**, 1740.
96. S. Funayama, M. Ishibashi, K. Komiyama, and S. Omura, *J. Org. Chem.*, 1990, **55**, 1132.
97. J. Schwender, M. Seemann, H. K. Lichtenthaler, and M. Rohmer, *Biochem. J.*, 1996, **316**, 73.
98. H. K. Lichtenthaler, J. Schwender, A. Disch, and M. Rohmer, *FEBS Lett.*, 1997, **400**, 271.
99. M. K. Schwarz, Ph.D. Thesis, Nb ETH 10951, Eidgenössische Technische Hochschule, Zurich, Switzerland, 1994, and references cited therein.
100. W. Knöss, B. Reuter, and J. Zapp, *Biochem. J.*, 1997, **326**, 449.
101. W. Eisenreich, B. Menhard, P. J. Hylands, M. H. Zenk, and A. Bachert, *Proc. Natl. Acad. Sci. USA*, 1996, **93**, 6431.
102. W. Eisenreich, S. Sagner, M. H. Zenk, and A. Bacher, *Tetrahedron Lett.*, 1997, **38**, 3889.
103. J. G. Zeidler, H. K. Lichtenthaler, H. U. May, and F. W. Lichtenthaler, *Z. Naturforsch., Teil C*, 1997, **52**, 15.
104. H. Kleinig, *Annu. Rev. Plant Physiol.*, 1989, **40**, 39.
105. J. Chappel, *Annu. Rev. Plant Physiol.*, 1995, **46**, 521.
106. T. J. Bach, *Lipids*, 1995, **30**, 191.
107. H. K. Lichtenthaler, M. Rohmer, and J. Schwender, *Plant Physiol.*, 1997, **101**, 643.

108. K. J. Treharne, E. I. Mercer, and T. W. Goodwin, *Biochem. J.*, 1966, **99**, 239.
109. D. V. Banthorpe, B. V. Charlwood, and M. J. O. Francis, *Chem. Rev.*, 1972, **72**, 115.
110. T. J. Bach and H. K. Lichtenthaler, in "Ecology and Metabolism of Plant Lipids," eds. G. Fuller and W. D. Nes, American Chemical Society Symposium Series, 1987, **325**, American Chemical Society, Washington, DC, p. 109.
111. T. J. Bach, A. Motel, and T. Weber, *Rec. Adv. Phytochem.*, 1990, **24**, 1.
112. K. Kreuz and H. Kleinig, *Planta*, 1981, **153**, 578.
113. K. Kreuz and H. Kleinig, *Eur. J. Biochem.*, 1984, **141**, 531.
114. T. W. Goodwin, *Biochem. J.*, 1958, **70**, 612.
115. G. D. Braithwaite and T. W. Goodwin, *Biochem. J.*, 1960, **76**, 1.
116. F. Lütke-Brinkhaus and H. Kleinig, *Planta*, 1987, **171**, 406.
117. M. Seemann, Ph.D. Thesis, Nb 95 MULH 0385, Université de Haute Alsace, Mulhouse, France, 1995.
118. D. McCaskill and R. Croteau, *Planta*, 1995, **197**, 49.
119. K. Nabeta, T. Kawae, T. Kikuchi, T. Saitoh, and H. Okuyama, *J. Chem. Soc., Chem. Commun.*, 1995, 2529.
120. K. Nabeta, T. Ishikawa, T. Kawae, and H. Okuyama, *J. Chem. Soc., Chem. Commun.*, 1995, 681.
121. K. Nabeta, T. Ishikawa, and H. Okuyama, *J. Chem. Soc., Perkin Trans. I*, 1995, 3111.
122. K. Bloch, "Blondes in Venetian Paintings, the Nine-Banded Armadillo and other Essays in Biochemistry," Yale University Press, New Haven, CT, USA, 1994, p. 154.
123. Y. Hano, A. Ayukawa, T. Nomura, and S. Ueda, *J. Am. Chem. Soc.*, 1994, **116**, 4189.
124. Y. Hano, T. Nomura, and S. Ueda, *Naturwissenschaften*, 1995, **82**, 376.
125. G. A. Sprenger, U. Schörken, T. Wiegert, S. Grolle, A. A. de Graaf, S. V. Taylor, T. P. Begley, S. Bringer-Meyer, and H. Sahm, *Proc. Natl. Acad. Sci. USA*, 1997, **94**, 12 857.
126. L.-M. Lois, N. Campos, S. Rosa Putra, K. Danielson, M. Rohmer, and A. Boronat, *Proc. Natl. Acad. Sci. USA*, 1998, in press.
127. D. Arigoni, S. Sagner, C. Latzel, W. Eisenreich, A. Bacher, and M. Zenk, *Proc. Natl. Acad. Sci. USA*, 1997, **94**, 10 600.
128. S. Rosa Putra, L.-M. Lois, N. Campos, A. Boronat, and M. Rohmer, *Tetrahedron Lett.*, 1998, **39**, 23.

2.04
Isopentenyl Diphosphate Isomerase and Prenyltransferases

TANETOSHI KOYAMA and KYOZO OGURA
Tohoku University, Japan

2.04.1 INTRODUCTION

2.04.1.1 Scope

In the biosynthesis of isoprenoid compounds, which include numerous structurally different natural products, all of the carbon backbones are derived from linear prenyl diphosphates. These linear prenyl chains are constructed by the action of a group of enzymes commonly called "prenyltransferases." The chain length of prenyl diphosphates varies so widely that it ranges from geraniol (C_{10}) to natural rubber whose carbon chain length extends to several millions. Sixteen enzymes with different catalytic functions have been characterized. "Prenyltransferase" usually refers to any enzymes that catalyze the transfer of prenyl groups to acceptors that include not only isopentenyl diphosphate (IPP) but also aromatic compounds and proteins, etc. In this chapter, however, the term "prenyltransferase" denotes the prenyltransferase whose acceptor is IPP, i.e., "prenyl diphosphate synthase."

Isopentenyl diphosphate : dimethylallyl diphosphate isomerase (IPP isomerase) catalyzes a crucial activation step in the beginning of isoprenoid biosynthesis by converting IPP to its allylic isomer, dimethylallyl diphosphate (DMAPP).

In this chapter we will discuss the progress made in studies on the molecular mechanisms of IPP isomerase and prenyltransferases.

2.04.1.2 History

Studies on isoprenoid biosynthesis were brought to an enzymological level by the discovery of IPP as the true biologically active isoprene unit, which was made independently by the groups of Lynen[1] and of Bloch.[2] The fundamental isoprenoid chain elongation starts with condensation between IPP and DMAPP by the action of a prenyltransferase. The discovery[3] of the "isomerase" for converting IPP to DMAPP, as well as the "prenyltransferase" that catalyzes the head-to-tail condensation between DMAPP and IPP to form farnesyl diphosphate (FPP) by way of geranyl diphosphate (GPP), has led to the recognition that Ruzicka's hypothetical "active isoprene"[4] is the true biological building block for the tremendous number of isoprenoid compounds in nature.

In the ensuing years the stereochemistry of the enzymatic reactions in the biosynthetic pathway from mevalonic acid to squalene have been elucidated by a beautifully designed set of pioneering experiments with supernatant fractions from mammalian liver or yeast by Popják and Cornforth (see Sections 2.04.2.2 and 2.04.3.1.1(ii)). These studies on the stereochemistry of squalene biosynthesis have been reviewed in detail by Bentley[5] and also by Poulter and Rilling.[6]

The first prenyltransferase that was purified to homogeneity is FPP synthase, which was obtained by Eberhardt and Rilling.[7] Reed and Rilling also succeeded in the crystallization of chicken liver FPP synthase.[8] However, determination of the crystal structure had to wait more than 20 years until Tarshis *et al.*[9] determined the structure of recombinant avian enzyme to 2.6 Å resolution, which was the first and only three-dimensional structure for any prenyltransferase.

From the mid-1970s many kinds of substrate analogues for prenyltransferase and IPP isomerase have been synthesized and their susceptibilities as substrates for the enzymes have been examined extensively at Tohoku University in Japan. Some of the artificial substrate homologues have been effectively applied to stereospecific syntheses of biologically active compounds.[10] Poulter and Rilling, at the University of Utah, in the USA, carried out experiments with several fluorinated substrate analogues to explore the reaction mechanisms of prenyltransferases.[11]

In 1987 the cDNA of rat liver FPP synthase was isolated by Clarke *et al.*[12] This was the first introduction of a molecular biological approach to the field of prenyltransferases and related enzymes. Since the late 1980s the genes of many kinds of FPP- and geranylgeranyl diphosphate (GGPP) synthases have been identified and characterized. Additionally, the hexaprenyl diphosphate synthase gene of *Saccharonmyces cerevisiae*[13] and several bacterial genes encoding prenyltransferases that catalyze the synthesis of polyprenyl diphosphates have been cloned.

In 1989, two years after the first cloning of the FPP synthase gene,[12] the gene for IPP isomerase of *S. cerevisiae* was cloned by Anderson *et al.*[14] Six kinds of IPP isomerase genes have been identified, from human to bacterial origin.

2.04.2 ISOPENTENYL DIPHOSPHATE ISOMERASE

2.04.2.1 Enzymology

IPP isomerase (EC 5.3.3.2) catalyzes the isomerization of the double bond from position 3 of IPP to position 2 of DMAPP (Scheme 1), the first and ubiquitous primer substrate for successive prenyl chain elongation reactions catalyzed by prenyltransferases in isoprenoid biosynthesis.

OPP OPP

(IPP) (DMAPP)

Scheme 1

This enzyme was first discovered in baker's yeast by Agranoff *et al.*,[3] then partially purified and characterized from pig liver,[15–17] pumpkin fruit,[18,19] avian liver,[20] *Escherichia coli*,[21] daffodil chromoplasts,[22] and even rubber latex.[23]

Mainly because of its instability, purification to near homogeneity of IPP isomerase was not successful until Banthorpe *et al.* achieved a 200-fold purification of this enzyme from pig liver.[24] Then, it took a further nine years until Bruenger *et al.* reported a purification of the isomerase from *Claviceps* sp. SD 58,[25] and from *Capsicum annuum* chromoplast[26] by Dogbo and Camara. Although the molecular weight of the pig liver enzyme was estimated to be 83 000, consisting of heteromeric proteins,[24] the purified enzymes from the fungus and the plant showed monomeric structures with molecular weights of 35 000 and 33 500, respectively.

IPP isomerase is strongly inhibited by sulfhydryl-directed reagents and is activated by thiol compounds, suggesting that one or more sulfhydryl side chains participate in catalytic function. The enzyme has an absolute requirement for a divalent cation, Mg^{2+} or Mn^{2+}.

2.04.2.2 Stereochemistry of Reaction

The reversible isomerization catalyzed by IPP isomerase leans towards the direction of DMAPP. At equilibrium, the ratio of 13 : 87 has been obtained between IPP and DMAPP, respectively, for pig liver enzyme.[16]

The precise stereochemistry of the isomerization of IPP has been elucidated as follows. In the isomerization catalyzed by IPP isomerase:

(i) the *pro-R* proton at C-2 of IPP is removed to give DMAPP;[27]

(ii) the (*E*)-methyl group in DMAPP is derived from the 4-methylene carbon of IPP;[28,29] and

(iii) the protonation occurs on the *re* face of the double bond during isomerization of IPP to DMAPP.[30]

Hence, the proton abstraction at C-2 and addition to C-4 during isomerization between IPP and DMAPP occur at opposite faces of the isoprene moiety, as illustrated in Figure 1. It should be noted, however, that the stereochemical relation between the methyl groups of DMAPP and the 4-methylene carbon of IPP has been shown to be partially scrambled by analyzing the isomerase reaction in D_2O.[29,31]

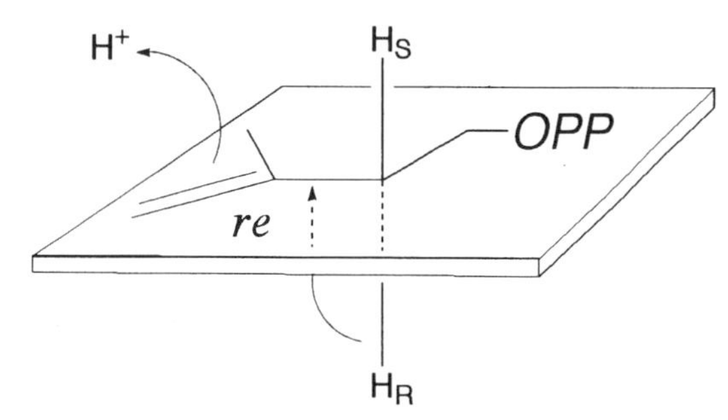

Figure 1 Stereochemistry of the isomerization of IPP by IPP isomerase.

2.04.2.3 Reactions with Artificial Substrate Homologues

Work on the inhibitory effect of several substrate analogues on IPP isomerase from pumpkin fruit[32] has indicated that the allylic diphosphate and even inorganic pyrophosphate (PPi) are inhibitory, whereas the corresponding monophosphates are much less effective. These results led to examination of the substrate specificity of the isomerase with artificial substrate homologues having diphosphate moieties.

Several artificial substrate homologues have been found to react as substrates for IPP isomerase from pig liver. Both 3-ethyl-3-butenyl diphosphate (**1**) and (*E*)-3-methyl-3-pentenyl diphosphate (**2**) act as substrates.[29] The homologue (**2**) is reversibly isomerized to (*E*)-3-methyl-2-pentenyl diphosphate (**3**) in the same stereochemical manner as in the case of the natural substrate, IPP. However, (**1**) is converted almost irreversibly into (**2**). The isomerization of artificial C_6-homologues was analyzed extensively using deuterated water as the reaction medium to elucidate the relationship between the homologues.[33] As a result, two aberrant isomerizations were observed (Scheme 2); one is the isomerization between (**1**) and (**2**), with the equilibrium leaning heavily to the formation of (**2**), and the other is the irreversible isomerization of (**4**) to (**3**), resulting in an apparently peculiar isomerization of (**4**) to (**2**). These observations are explained by presuming at least two binding sites, one for the diphosphate moiety and the other for the methyl group in the catalytic site of the enzyme.

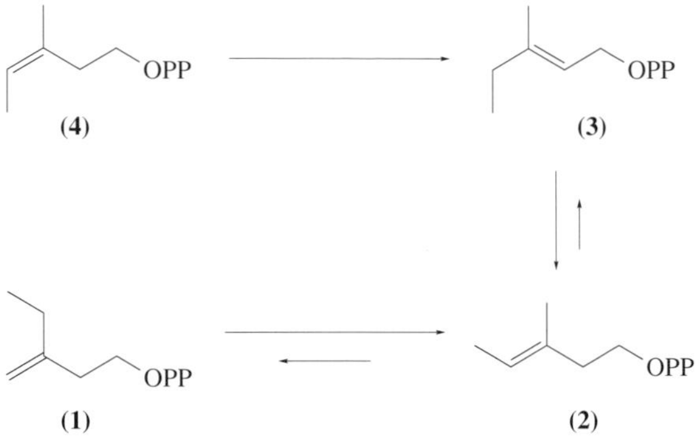

OPP OPP
(**4**) (**3**)

OPP OPP
(**1**) (**2**)

Scheme 2

2.04.2.4 Reaction Mechanism

Shah *et al.*[16] examined the incorporation of tritium from water into IPP and DMAPP during the isomerase reaction, and they proposed a mechanism in which a carbonium ion, formed by protonation, produces a covalent adduct with the enzyme through a thioether linkage. In 1985, two research groups, at Brandeis University,[34] MA and the University of Utah in the USA,[35] independently synthesized a potent inhibitor for IPP isomerase, 2-(dimethylamino)ethyl diphosphate (**5**), which acted as a transition-state/reactive intermediate analogue. By detailed analyses of the inactivation kinetics, substrate protection studies, and labeling experiments using (**5**) (Scheme 3) and some fluorinated substrate analogues, they concluded that isomerization occurs by a protonation–deprotonation mechanism including an intermediate carbonium ion (Scheme 4).[36–39]

Me
H−N+ OPP
Me
(**5**)

F
OPP
(**6**)

Scheme 3

2.04.2.5 Gene Cloning and Biotechnology

Anderson *et al.* purified the IPP isomerase from *S. cerevisiae*.[40] The amino-terminal sequence of the enzyme was used to design oligonucleotide probes for the isolation of clones containing the IPP isomerase gene from a yeast genomic DNA library. The cloned gene encodes a 33 350 Da protein of 288 amino acids, which is the first isolation of the gene for IPP isomerase. Then they constructed a heterologous expression system for the gene and used it to overproduce the isomerase in *E. coli*.[41] The overproduced enzyme constitute 30–35% of the total soluble cell protein and could be purified to homogeneity in two steps. By using radioactive site-directed inhibitor,[37] 3-(fluoromethyl)-3-butenyl diphosphate (**6**) they labeled the yeast enzyme and identified Cys139 as the nucleophile responsible for reaction by the inhibitor.

Site-directed mutageneses exchanging Cys139 and Glu207 with other amino acids resulted in remarkable losses of enzymatic activity.[42] These residues, which were identified as covalently modifiable residues with (**6**), were reasonably assigned as the two nucleophilic groups in the catalytic site of the isomerase, executing the antarafacial stereochemistry of the reaction (Scheme 4).

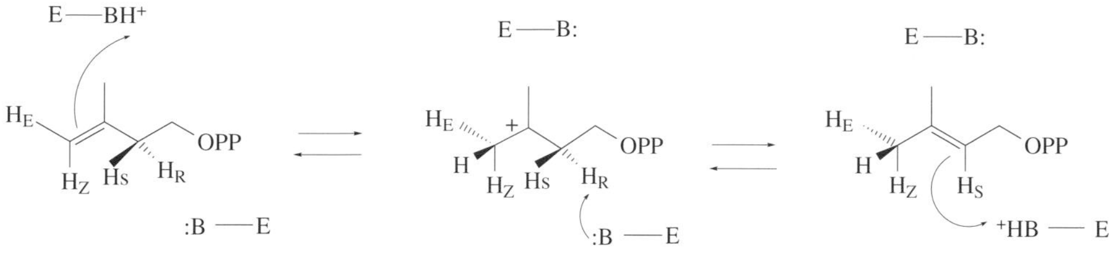

Scheme 4

The *S. cerevisiae* gene for IPP isomerase, *IDI1*, was disrupted with a *LEU2* marker, which revealed that *IDI1* was an essential single-copy gene.[43] By a plasmid shuffle-mediated complementation of the *LEU2* disrupted gene, isolation of the IPP isomerase cDNA clone from *Schizosaccharomyces pombe* was carried out.[44] This clone encoded a 26 864 Da polypeptide of 227 amino acids with a 27% similarity to the *S. cerevisiae* isomerase enzyme containing the essential Cys and Glu catalytic residues.

Xuan *et al.*[45] found a human cDNA sequence in a phorbol-induced library with an open reading frame encoding a protein with a high degree of similarity to both the yeast IPP isomerases. This human cDNA was cloned into an expression plasmid, and the encoded protein was overproduced in *E. coli*, purified to better than 90% homogeneity in two steps by ion exchange- and hydrophobic chromatographies.[46]

Further search for cDNAs showing sequence similarity to the yeast *IDI1* gene yielded a limited sequence of an *Arabidopsis* cDNA clone. By using this clone Blanc and Pichersky[47] identified a cDNA clone from a plant, *Clarkia breweri*. Though the overall sequence identity with the yeast enzyme is slightly less than 50%, regions from Asn104 to Leu146 in the *Clarkia* protein showed 81% identity.

Another thorough database search based on probes from the highly conserved regions in the yeast and human IPP isomerases was carried out by Hahn *et al.*[48] They found a putative IPP isomerase from a worm *Caenorhabditis elegans*, putative bacterial isomerases in the photosynthesis gene cluster of *Rhodobacter capsulatus*, and *E. coli* ORF182, the latter two of which were the first eubacterial IPP isomerase genes to be identified.

Amino acid sequence alignments for the seven IPP isomerases so far identified indicated a high degree of identity (typically 50%) among the four eukaryotic enzymes.[48] However, the identity between the two bacterial enzymes was only 32%. The percentage of conserved amino acids increases substantially in a core region, defined by the residues from Ile74 to Trp255 in the *S. cerevisiae* protein, that encompasses the essential active site Cys139 and Glu207. The length of this core region ranges from 152 amino acids in *R. capsulatus* to 181 in *S. cerevisiae*.

2.04.3 PRENYLTRANSFERASE

The linear prenyl-chain elongation catalyzed by prenyltransferases is interesting in that the reactions are regulated to proceed consecutively and terminate precisely at discrete chain lengths according to the requirements of each organism. The chain length of products varies so widely that it ranges from geraniol to natural rubber. Sixteen kinds of prenyltransferases with different catalytic functions have been characterized and more than five kinds of prenyltransferase genes have been cloned.

Bacteria provide good sources for searching for prenyltransferases, because they do not synthesize secondary metabolites, but produce menaquinones or ubiquinones having polyprenyl side chains whose chain lengths are varied in a species-specific manner.[49] Thus, it was of interest to see whether there would exist respective enzymes corresponding to the variety of the chain length of quinones.

It was also important to explore bacterial enzymes, because *Z,E*-undecaprenyl phosphate is known as a lipid carrier of glycosyl transfer in the biosynthesis of cell wall polysaccharide components in bacteria.

Characterization of many kinds of prenyltransferases, most of which were of bacterial origin, has led us to notice that, in spite of the similarity of the reactions catalyzed by them, the modes of expression of catalytic function are distinctly different according to the chain length and stereochemistry of the reaction products. In fact prenyltransferases can be classified into four groups as depicted in Figure 2.

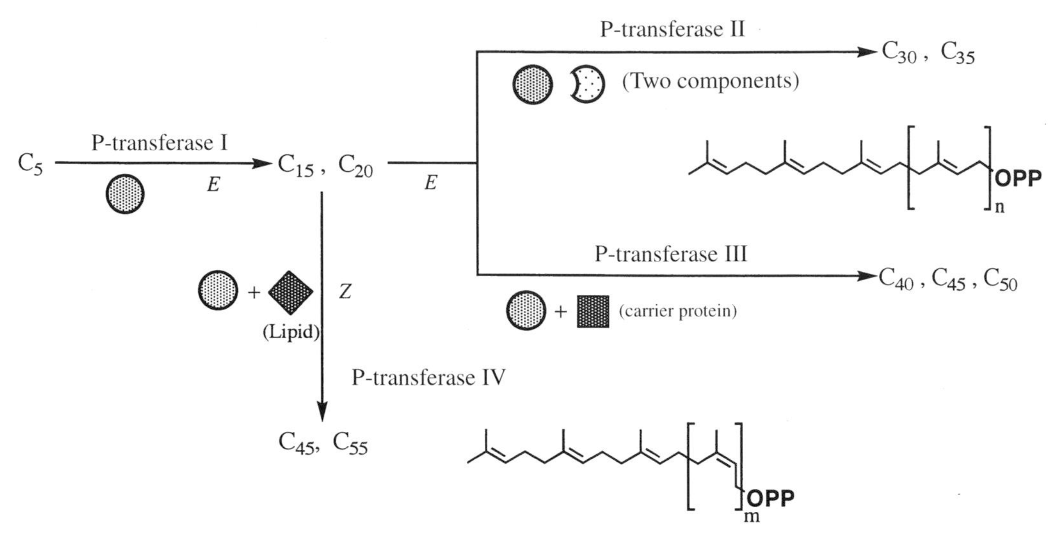

Figure 2 Classification of prenyltransferases.

2.04.3.1 Short-chain Prenyl Diphosphate Synthases (Prenyltransferase I)

Short-chain prenyl diphosphate synthases, including FPP- and GGPP synthases, require no cofactor except divalent metal ions such as Mg^{2+} or Mn^{2+}, which are commonly required by all prenyltransferases.

All living organisms constitutively contain at least one of these short-chain prenyl diphosphate synthases for the production of allylic prenyl diphosphates as the priming substrates for the other groups of prenyltransferases and also as biosynthetic precursors of various isoprenoids including steroids, carotenoids, and prenylated proteins.

2.04.3.1.1 Farnesyl diphosphate synthase

(i) Enzymology

The most widely occurring and the most extensively studied prenyltransferase is FPP synthase (geranyl*trans*transferase (EC 2.5.1.10)). This enzyme occupies a particularly important branch point on the pathway of isoprenoid biosynthesis as shown in Figure 3. The product FPP is the substrate not only for the dimerization to squalene for cholesterol biosynthesis, but also for the steps leading to physiologically important isoprenoid compounds including glycosyl carrier lipids, respiratory quinones, heme a, and prenylated proteins.

The first pure preparation was obtained from *S. cerevisiae* by Eberhardt and Rilling.[7] This group then obtained FPP synthase from chicken liver in a stable crystalline form.[8] Both enzymes were shown to be homodimers with apparent molecular weights of 84–86 kDa.

FPP synthases have been partially purified and characterized from pig liver,[17,50,51] pumpkin fruit,[18] castor bean,[52] *Bacillus subtilis*,[53] silkworm,[54] and *C. annuum*.[55]

Porcine liver FPP synthase exists in two interconvertible forms,[56–58] which are resolved into two active fractions by ion-exchange chromatography. Several lines of evidence have suggested that the state of oxidation of thiol group(s) in the FPP synthase is responsible for the interconversion.[57,59]

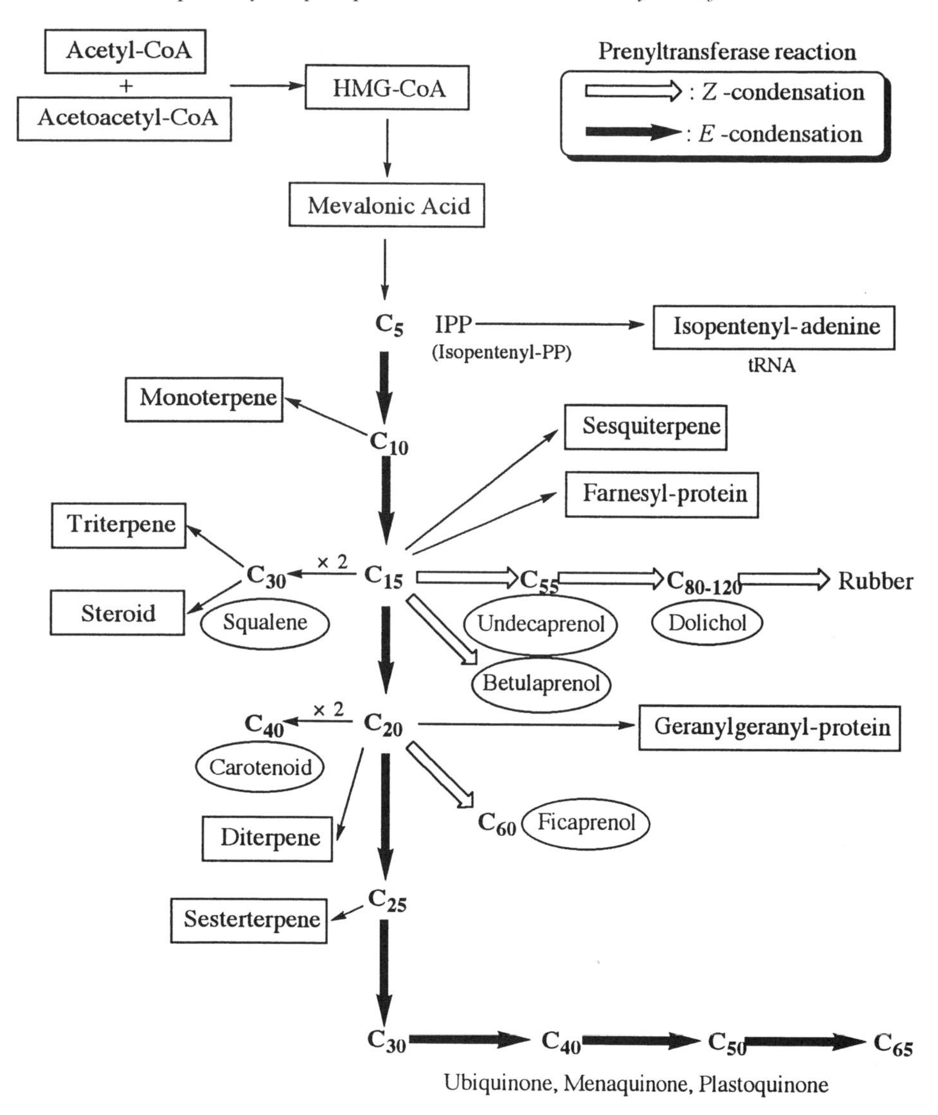

Figure 3 Prenyltransferases in the biosynthesis of isoprenoid compounds.

Similar observations on resolution into multiple forms of FPP synthase have been accumulating on the enzymes from *Ricinus communis*,[52] silkworm,[54] and from a thermophilic bacterium, *B. stearothermophilus*,[60] which will be discussed in the site directed mutagenesis section (2.04.3.1.1(vii)).

(ii) Stereochemistry of reaction

The mode of the reaction catalyzed by prenyltransferases is very unique and interesting from the standpoint of organic reactions constructing carbon–carbon bonds. The 1′-4 condensation reaction between an allylic diphosphate and IPP proceeds, with migration of the double bond at C-1 of IPP, by a unique electrophilic condensation involving carbonium ions as reactive intermediates in such a way that repetition of condensation is possible.

The stereochemistry of the 1′-4 condensation was first studied by Popják and Cornforth[61] during their work on squalene biosynthesis in mammalian liver.

In the 1′-4 condensation catalyzed by FPP synthase:

(i) The condensation occurs with inversion of configuration at the C-1 methylene of the allylic diphosphate, DMAPP, or GPP.[62]

(ii) The 1′-4 bond is constructed by the addition of the allylic moiety to the *si* face of IPP.[63]

(iii) The *pro-R* hydrogen at C-2 of IPP is lost when the new double bond is formed.[62]

Hence, the addition of an allylic prenyl moiety occurs on the same side where the C-2 hydrogen of IPP is eliminated during the 1′-4 condensation. Since the same stereochemical relationships as described above were found in both mammalian liver and yeast, the stereochemistry of 1′-4 condensation is thought to be a highly conserved property of prenyltransferases.

It is interesting that the proton abstraction at C-2 and the addition of a prenyl group to C-4 of IPP during the transferase reaction occur at the *same* face of IPP as depicted in Figure 4, whereas the proton abstraction at C-2 and the addition to C-4 of IPP during the IPP isomerase reaction occur at the *opposite* faces of the IPP molecule as shown in Figure 1.

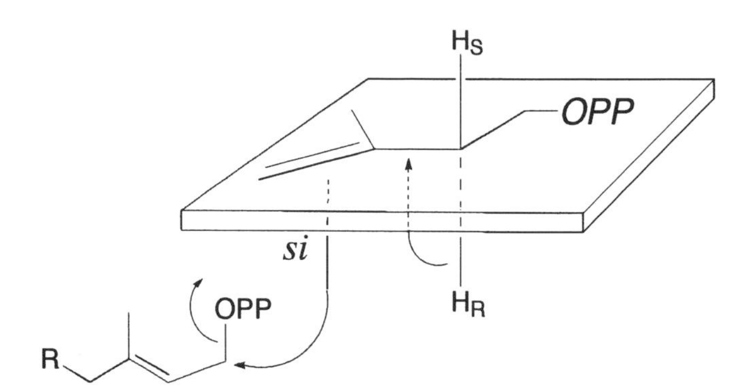

Figure 4 Stereochemistry in the 1′-4 condensation by FPP synthase.

The elegant experimental procedures for the elucidation of the stereochemistry of 1′-4 condensation accomplished by Cornforth and Popják are described in detail by Bentley[5] and by Poulter and Rilling.[6]

(iii) Reactions with artificial substrate homologues

The substrate specificities of pig liver FPP synthase were explored extensively using artificial substrate homologues of allylic diphosphates, DMAPP and GPP, in the authors' laboratory and by Popják's group independently. Popják *et al.* found that 6,7-dihydrogeranyl diphosphate reacted as an artificial substrate for liver FPP synthase.[64] They also examined 16 artificial homologues of DMAPP and found four active DMAPP homologues.[65] Ogura *et al.* explored the substrate specificity of pumpkin fruit enzyme with respect to 3-methyl-2-alkenyl diphosphates.[66] They extended similar studies with various artificial allylic diphosphates including cyclopentylideneethyl- and cyclohexylideneethyl-,[67] 6,7-dihydrogeranyl diphosphate,[68] and (*E*)-3-methyl-2-undecenyl diphosphate.[69] Pig liver FPP synthase showed a broader specificity than the pumpkin enzyme.[70] It is possible to delineate the structural requirements for the allylic diphosphates as substrates for FPP synthase as follows. The hydrogen at C-2 cannot be replaced by a methyl group without loss of activity. However, extension of the methyl groups by a linear hydrocarbon chain can be tolerated to some extent. Nishino *et al.*[69] obtained relative reactivities of 3-methyl-2-alkenyl diphosphates as a function of the linear alkyl chain length extending at either the *E*- or *Z*-position. As shown in Figure 5, there are two optimal chain lengths for the *E*-position, one for DMAPP ($n = 1$) and the other for *n*-pentyl ($n = 5$), which has a similar chain length to the natural substrate, GPP. However, only the chain length for DMAPP ($n = 1$) is optimal for the *Z*-position. Branching of the alkyl group in the 3-methyl-2-alkenyl diphosphate molecule caused a remarkable decrease in the reactivity.[69] Several GPP- or DMAPP- analogues having oxygen atoms in their alkyl moieties were also active as substrates for pig liver FPP synthase to give oxygenated FPP analogues.[71,72] However, the enzyme from a thermophilic bacterium, *B. stearothermophilus*, hardly accepted those analogues having oxygen atoms in their chains.[73]

Although more than 40 artificial allylic homologues of DMAPP or GPP have been shown to act as substrates for FPP synthases, the specificity for the homoallylic substrates is rather stringent with respect to the alkyl homologues of IPP. Ogura and co-workers[74] showed that only 3-ethylbut-3-enyl diphosphate ((**1**), R = Et in (**7**)) was accepted as a substrate for pig liver FPP synthase. This fact led them to examine whether an FPP synthase participates in the biosynthesis of juvenile hormones (JHs) of insects. They showed that the carbon skeletons of all JHs could be constructed by the action of pig liver FPP synthase using (**1**) and (*Z*)-3-methyl-2-pentenyl diphosphate (**8**) (Scheme

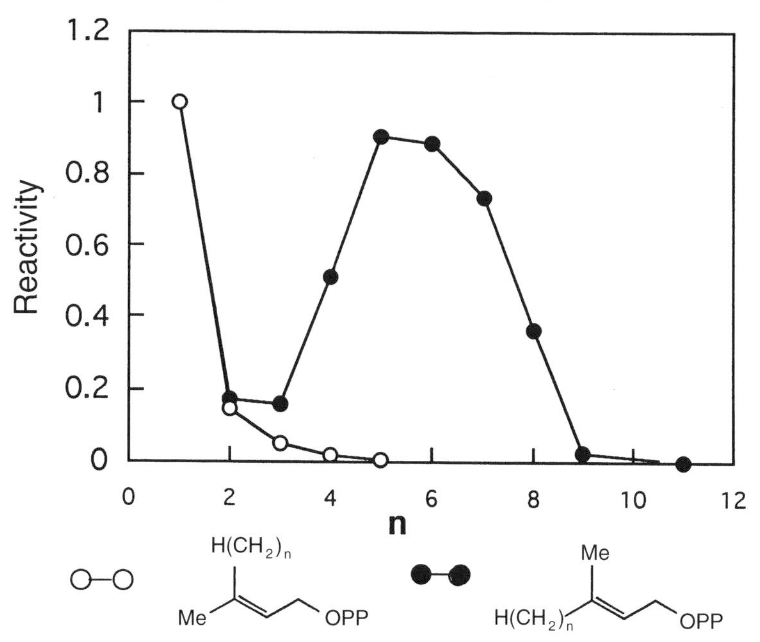

Figure 5 Relative reactivity of 3-methyl-2-alkenyl diphosphates as a function of the linear chain length of the (*Z*)- (○) or (*E*)-alkyl group (●).

5).[75] They also showed that both FPP synthase and IPP isomerase found in silkworm[54] have substrate specificities in favor of the formation of JH structures.[76]

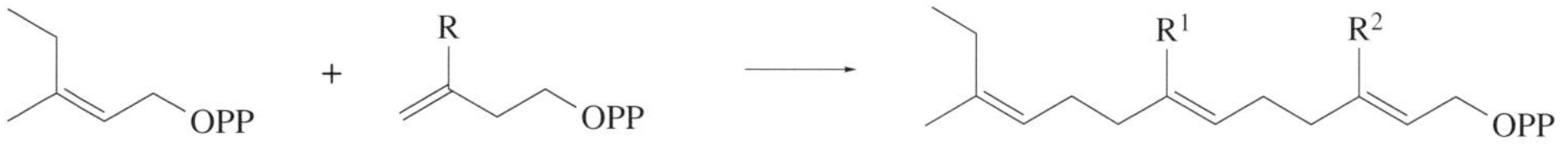

(8) + **(7)** → **(9)**

Scheme 5

Among the IPP homologues in which the number of methylenes between C-1 and C-3 of IPP is changed, only **(10)** was accepted as a substrate for the liver enzyme.[77,78] Surprisingly, the stereochemistry of the newly formed double bond in the product **(11)** or **(12)** was *Z* (Scheme 6). The abnormal 1′-4 condensation with the artificial substrate **(10)** must be a result of conformational distortions in the homologue required to keep the double bond in the proper location for the enzymatic condensation. In the presence of Ni^{2+} instead of Mg^{2+}, however, the reaction between **(10)** and GPP gave an unexpected product, having an exomethylene group, in addition to **(11)**.[79]

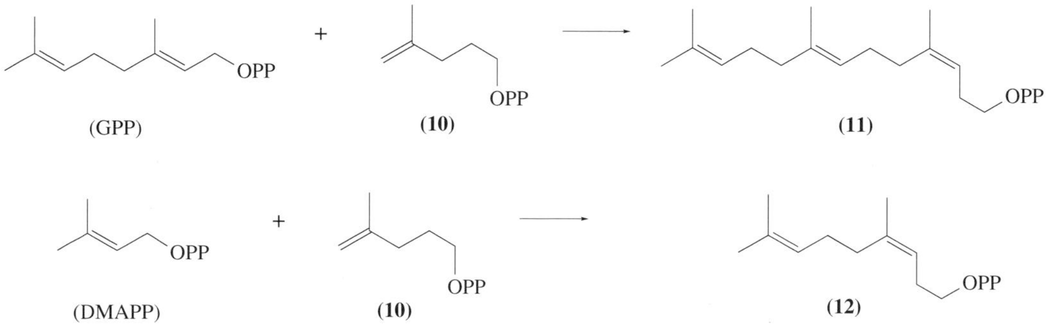

(GPP) + **(10)** → **(11)**

(DMAPP) + **(10)** → **(12)**

Scheme 6

Koyama *et al.* showed that both the (*E*)- and (*Z*)-IPP homologues, each having an extra methyl group at C-4, reacted stereospecifically as substrates for pig liver FPP synthase to give FPP homologues having chiral centers at which the new C—C bonds had been constructed.[80] The (*E*)-(**2**) and (*Z*)-3-methyl-3-pentenyl diphosphates (**4**) were found to condense with DMAPP or GPP in the same stereochemical manner that was demonstrated for the natural substrate IPP by Cornforth *et al.*[63] as depicted in Figure 4. The artificial substrates (**2**) and (**4**) react with DMAPP as well as GPP to give 4,8-dimethylfarnesyl diphosphate ((**14*S***) and (**14*R***)) via 4-methylgeranyl diphosphate ((**13*S***) and (**13*R***)), respectively (Scheme 7).

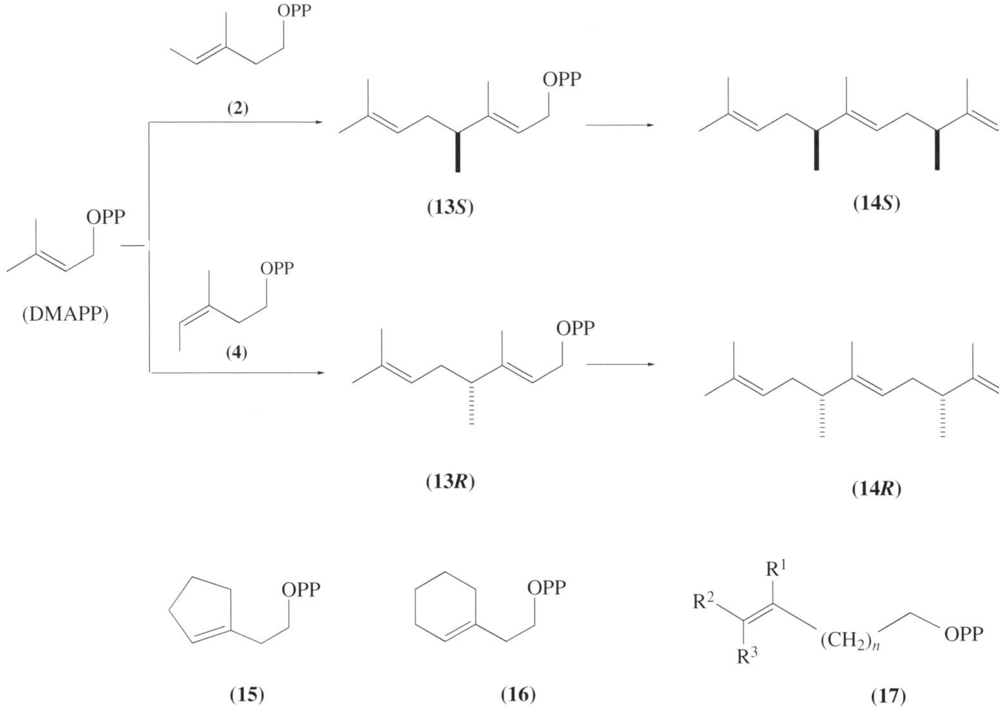

Scheme 7

Furthermore, even the cycloalkene homologues (**15**) and (**16**) are accepted as substrates instead of IPP to give FPP homologues containing cyclopentane and cyclohexane rings with asymmetric carbons, respectively.[81]

To summarize, the structural requirements for the nonallylic diphosphate (**17**) to be accepted instead of IPP by pig liver FPP synthase are: (i) R^1 should be Me or Et; (ii) R^2 and R^3 can be H, Me, or Et; (iii) cyclization between R^1 and R^2 forming a five- or six-membered ring is acceptable; and (iv) *n* should be 1 or 2.[81]

From the standpoint of application of enzymes in organic synthesis, it is of interest that an enzyme can produce artificially the *S* or *R* enantiomer depending on whether it is supplied with the *E* or *Z* isomer as substrate. FPP synthase reactions using (**2**) or (**4**) as substrate were applied effectively in the asymmetric synthesis of some biologically active insect substances.

To determine the absolute configuration of faranal, a trail pheromone of the Pharaoh's ant, the FPP synthase reactions using (**2**) and (**4**) were applied to synthesize the stereoisomers of faranal.[82] Enzymatic condensation of (**2**) or (**4**) with homogeranyl diphosphate (**18**) gave a bishomofarnesyl diphosphate, which had an *S*-(**19*S***) or *R*-(**19*R***) structure, respectively, that was converted to the corresponding alcohol by the action of alkaline phosphatase, followed by chemical oxidation with MnO_2 and reduction with $(Ph_3P)_3RhCl$ and Et_3SiH. The bioassay with the Pharaoh's ant clearly discriminated between the 4*R*- and 4*S*- faranals, showing the stereochemistry of the pheromone to be 3*S*,4*R* (**22*SR***) (Scheme 8). This FPP synthase reaction was further applied to the synthesis of several alkyl chain homologues of faranal to examine the structure–activity relationship of the pheromone.[10] Another application of the FPP synthase reaction was the determination of the absolute configuration of 4-methyl-JH I, one of the juvenoids found in embryos of tobacco hornworm.[83] As shown in Scheme 9, ^3H-labeled (**25*S***) or (**25*R***) was synthesized enzymatically from [1-^3H]bishomogeranyl diphosphate (**23**) and (**2**) or (**4**) by the action of FPP synthase followed by

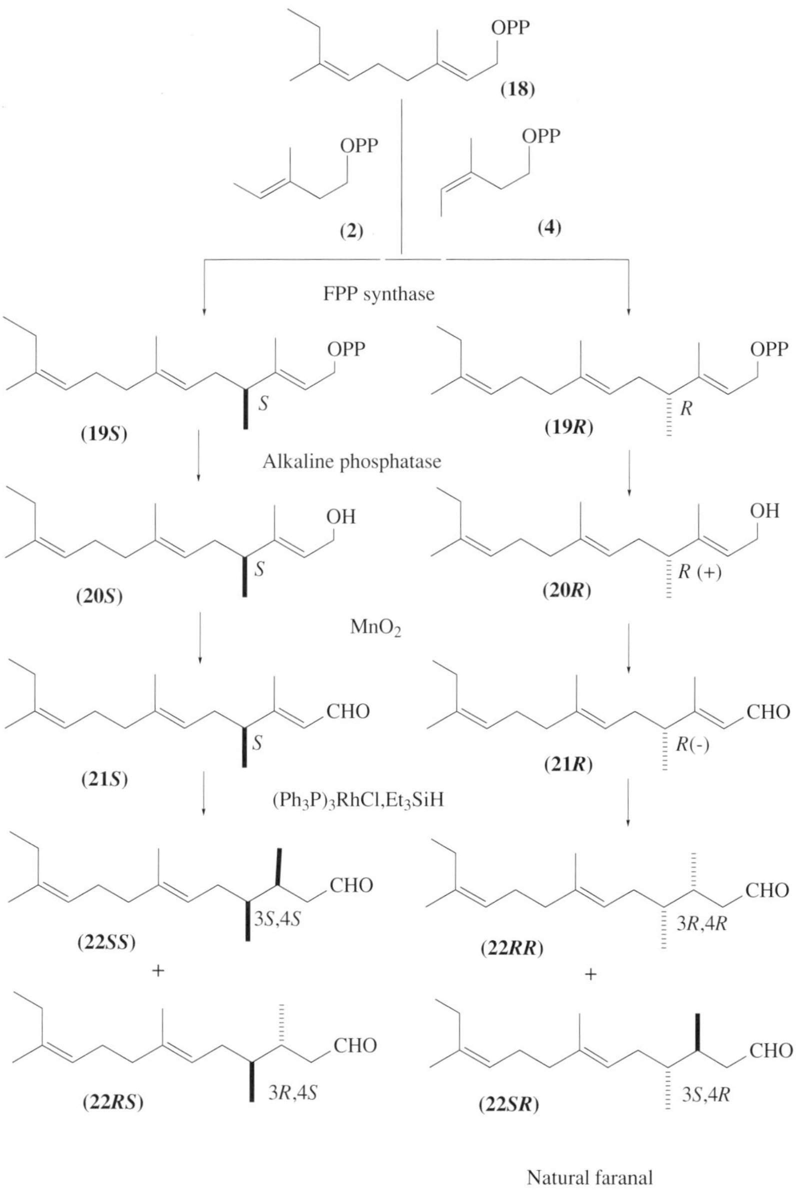

OPP
(18)
OPP
(2)
OPP
(4)
FPP synthase
OPP
S
(19*S*)
OPP
(19*R*)
R
Alkaline phosphatase
OH
S
(20*S*)
OH
(20*R*)
R (+)
MnO_2
CHO
S
(21*S*)
CHO
(21*R*)
R(-)
$(Ph_3P)_3RhCl,Et_3SiH$
CHO
3*S*,4*S*
(22*SS*)
CHO
(22*RR*)
3*R*,4*R*
+
+
CHO
(22*RS*)
3*R*,4*S*
CHO
(22*SR*)
3*S*,4*R*
Natural faranal

Scheme 8

alkaline phosphatase treatment. After biotransformation of (**25*S***) or (**25*R***) in cultured corpora allata from the insect, each tritium-labeled 4-MeJH I was compared with the natural hormone by capillary GLC. As a result the 4*S*-isomer (**26*S***) was identified with the natural 4-MeJH I. It is interesting that 4-MeJH I has the opposite C-4 configuration to that of faranal (**22*SR***), which has the 4*R*-structure.

(iv) Mechanism

From the overall stereochemistry of the prenyltransferase reaction (see Section 2.04.3.1.1(ii)), Cornforth and Popják reached two conclusions on the timing of the individual steps in the FPP synthase reaction:[61,84]

(i) The cleavage of the carbon–oxygen bond at C-1 of the allylic substrate, whose configuration is inverted, and the formation of the new carbon–carbon bond at C-4 is a concerted process.

(ii) The formation of the new carbon–carbon bond is not concerted with elimination of a proton from C-2 of IPP, which occurs at the same side of the addition of an allylic moiety.

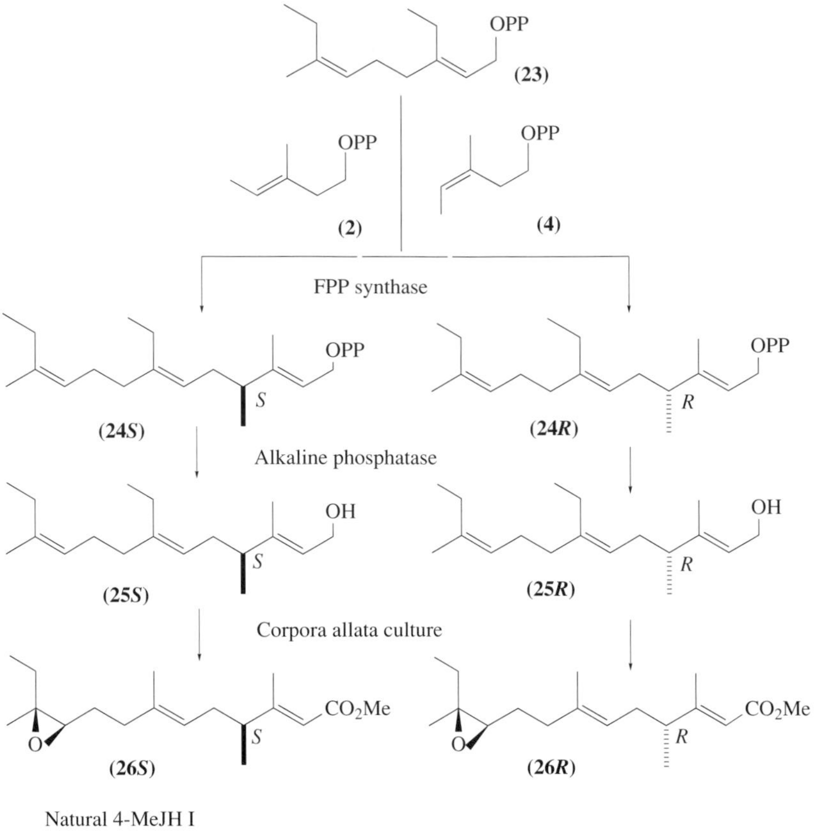

OPP
(23)
OPP
(2)
OPP
(4)
FPP synthase
OPP
S
(24*S*)
OPP
R
(24*R*)
Alkaline phosphatase
OH
S
(25*S*)
OH
R
(25*R*)
Corpora allata culture
CO_2Me
O
S
(26*S*)
CO_2Me
O
R
(26*R*)
Natural 4-MeJH I

Scheme 9

Cornforth *et al.*[85] postulated a two-stage mechanism in which an unknown electron donor (designated X) participates in the 1′-4 condensation by the prenyltransferase. A *trans* addition of the allylic group and X to the double bond of IPP was followed by a *trans* elmination of the *pro-R* proton at C-2 and X, forming the new double bond.

Using highly purified avian FPP synthase,[8] Reed and Rilling developed substrate-binding experiments of the avian prenyltransferase to determine the precise number of substrates bound per mol of the enzyme. As a result, they have indicated that each subunit has a single allylic binding site accommodating DMAPP, GPP, or FPP and one binding site for IPP.[86] Poulter and Rilling also found that hydrolysis of the allylic substrates, DMAPP and GPP was catalyzed by the FPP synthase in the presence of Mg^{2+} or Mn^{2+} and that the rate of this hydrolysis was markedly stimulated by PPi.[87,88]

In the meantime, Poulter and co-workers began to develop mechanistic studies on prenyltansferase using several fluorinated substrate analogues.[89,90] When (*E*)-(**27**) and (*Z*)-3-trifluoromethyl-2-butenyl diphosphates (**28**) were incubated with IPP and pig liver FPP synthase, the rate of condensation between IPP and these analogues was drastically depressed as compared with the normal catalytic rate observed with DMAPP. By assuming that the rate of the reaction varied linearly with the concentration of the enzyme, both of the reactivities of (**27**) and (**28**) were calculated to be 3×10^{-7} of that for DMAPP. Both analogues showed mixed inhibition patterns with respect to DMAPP or GPP. A similar depression was found in the rates of solvolysis for methanesulfonate derivatives of the fluoro analogues. 2-Fluorogeranyl diphosphate (**29**) also acted as a substrate for the liver enzyme, yielding 6-fluorofarnesyl diphosphate upon condensation with IPP.[91] However, the rate of condensation was only 8.4×10^{-4} that of the normal reaction with GPP. A similar rate depression (4.4×10^{-3}) was found for solvolysis of geranyl methanesulfonate and the corresponding 2-fluoro derivative. In addition, (**29**) was shown to be a competitive inhibitor ($K_i = 2.7\ \mu M$) for GPP, and its K_m value ($1.1\ \mu M$) is similar to that of GPP ($0.7\ \mu M$). By accumulating these convincing results, Poulter *et al.* have concluded that the head-to-tail coupling reaction catalyzed by prenyltransferases proceeds by an ionization–condensation–elimination mechanism as depicted in Scheme 10.

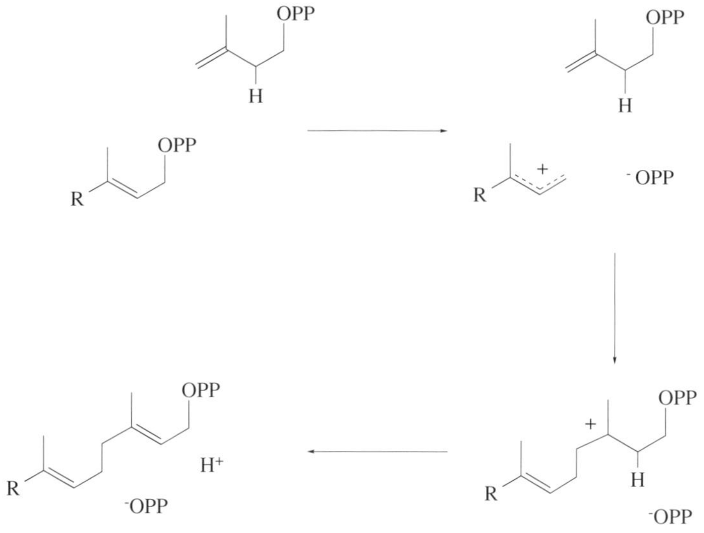

Scheme 10

The participation by an X-group in the 1′-4 condensation is not precluded by the electrophilic mechanism, if the developing positive charge generated at C-3 of IPP during electrophilic alkylation of the double bond is stabilized by covalent attachment of an X-group. Poulter *et al.* probed for the involvement of a covalently bound nucleophile in the reaction by using 2-fluoro- (**30**) and 2,2-difluoroisopentenyl diphosphate (**31**) (Scheme 11).[92] However, all of the experiments with (**30**) or (**31**) failed to obtain any evidence for X-group involvement. Thus it seems unlikely that the 1′-4 condensation involves covalent attachment of a nucleophile at C-3 of IPP.

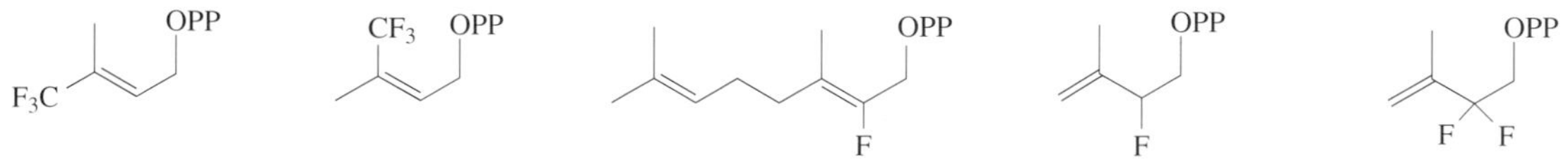

(27) (28) (29) (30) (31)

Scheme 11

The isoprenoid chain-elongation reaction catalyzed by prenyltransferase is unusual, because this reaction is accompanied by liberation of PPi in spite of the irreversibility of the reaction. In most biosynthetic pathways, the high energy involved in the P—O—P bond is ultimately utilized to displace the equilibrium of the reversible reaction in favor of the direction of synthesis. To explore the significance of the diphosphate moiety in the substrates of prenyltransferases, the effect of replacement of the diphosphate (-POP) moiety was investigated.[93] Replacement of the POP moiety of DMAPP and GPP by a methylene-diphosphonate (-PCP) moiety resulted in a decrease of V_{max} (34- and 73-fold, respectively), but the PCP analogues (DMAPCP and GPCP) were both accepted as substrates with K_m values comparable to those of the corresponding substrates (Scheme 12). These results suggest that the high-energy bond of the P—O—P linkage is not essential, but the diphosphate moiety is topologically and kinetically important. This idea suggested that the POP linkage of IPP might be replaced by a PCP linkage without marked loss of reactivity, because PPi is released from allylic substrate but not from IPP. However, the V_{max} value of IPCP in the reaction catalyzed by pig liver FPP synthase was 18-fold lower than that of IPP.[94,95] In addition, markedly different tendencies were observed between IPCP and GPCP with respect to the effects of pH and metal ion concentration. The optimal pH of the reaction of IPCP was around 6.6, whereas that of GPCP was 8.2.

Figure 6 shows the effect of Mg^{2+} concentration on the rate of IPCP reaction at pH 6.6 and 8.2. At pH 8.2 it shows a maximum at about 0.5 mM $MgCl_2$ and declines markedly as the Mg^{2+} concentration is increased. Therefore, the behavior of IPCP is not consistent with the assumption that the IPCP–Mg complex is the true substrate. The optimum pH of the reaction of IPCP was 6.6, which was significantly smaller by 1.1 and 1.4 units than those for IPP and GPCP, respectively. The acidic pH should be unfavorable for substrate–metal complexing. In addition, the reactivity of IPCP at pH 8.2 was strongly inhibited when the concentration of Mg^{2+} was raised, whereas such inhibition was not observed at pH 6.6. These unexpected properties of IPCP in relation to pH and metal ion concentration can be interpreted by assuming that the homoallylic substrate, IPCP, reacts without binding to Mg^{2+}. Thus, a possible mechanism of interaction between the enzyme and the substrates is as follows: the magnesium of the IPP–Mg complex is captured by the IPP-binding site of the enzyme, and then transferred again to the newly formed allylic diphosphate, which is the product of condensation between the allylic diphosphate complexed with Mg^{2+} and the IPP molecule free of Mg^{2+}. During the condensation reaction, the enzyme may dynamically interact with the substrates and the products, including PPi, to maintain an efficient turnover of catalysis. The conformational change caused by the highly charged PPi might also provide a driving force to move the intermediate allylic diphosphate to a position for the next condensation with IPP. The complex formation is favorable for the allylic substrate because it accelerates the release of PPi to form the carbonium ion, but it seems unnecessary for IPP to be complexed with Mg^{2+}, because PPi is not released from IPP during condensation. IPP free of Mg^{2+} must be advantageous, because the negatively charged diphosphate moiety can act as a base to abstract the *pro-R* hydrogen at C-2 of IPP, thereby assisting the nucleophilic attack of the C-4 of IPP on the carbonium ion formed from the allylic substrate.

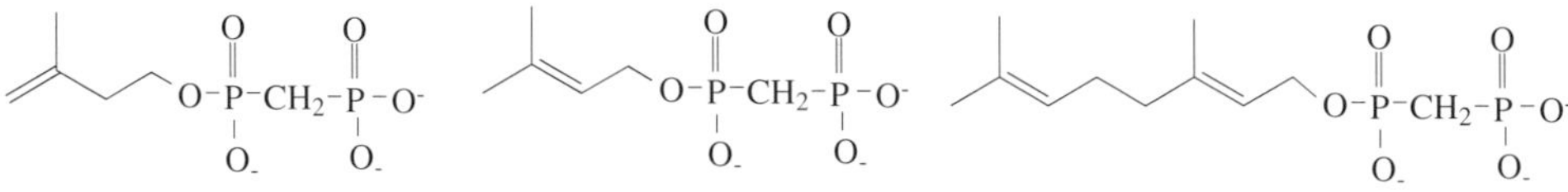

(IPCP) (DMAPCP) (GPCP)

Scheme 12

Stremler and Poulter synthesized GPCP as well as its difluoromethane diphosphonate analogues as alternate substrates and inhibitors inert to hydrolytic activity.[96]

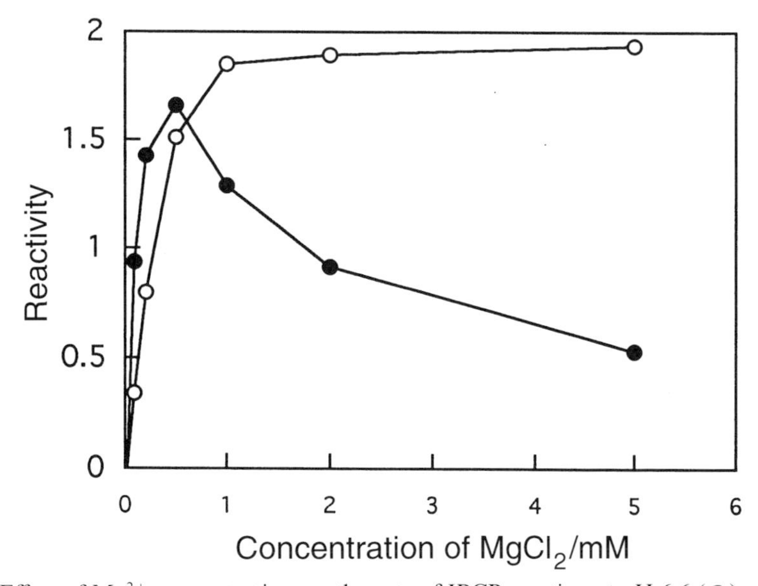

Figure 6 Effect of Mg^{2+} concentration on the rate of IPCP reaction at pH 6.6 (○) and 8.2 (●).

(v) Gene cloning and overproduction

FPP synthase is one of the key enzymes responsible for cholesterol biosynthesis and exhibits changes in the level of its activity in response to the level of cholesterol in mammalian cells. Taking advantage of this fact, Clarke *et al.*[12] succeeded in isolation of a cDNA for rat liver FPP synthase. It is noteworthy that the 30 amino-acid "active-site peptide" of chicken liver FPP synthase, which

Rilling and co-workers[97,98] had identified by photoaffinity labeling with *o*-azidophenylethyl diphosphate, was helpful as the convincing clue for the identification of the gene. Since then, cDNAs or genomic clones encoding FPP synthase have been isolated from various organisms, including human,[99,100] yeast,[101] *E. coli*,[102] *B. stearothermophilus*,[103] white lupin,[104] *Arabidopsis thaliana*,[105] and rubber tree.[106] All these FPP synthases are homodimeric enzymes of subunits ranging in size from 32 to 44 kDa. Koyama *et al.*[103] have succeeded in overproduction of FPP synthase from *B. stearothermophilus* in *E. coli* cells and its purification. Overproducing strains are able to produce thermostable FPP synthase in an amount equivalent to ~20% of the total soluble proteins. This recombinant enzyme is purified easily because of its thermostability and abundance. The enzyme is not inactivated even after treatment at 65 °C for 1 h. Song and Poulter[107] have succeeded in overproduction and purification of a recombinant yeast FPP synthase as a fusion protein having an α-tubulin epitope at the C-terminal, which enabled an efficient purification by immunoaffinity chromatography. Similarly, recombinant FPP synthases of rat liver,[108] human liver,[109] and avian liver[9] were overproduced and purified.

(vi) Crystallization and three-dimensional structure of FPP synthase

Crystallization of FPP synthase was first achieved for avian liver enzyme by Reed and Rilling in 1975.[8] After nearly 20 years, *B. stearothermophilus* FPP synthase was crystallized and a preliminary X-ray diffraction analysis was carried out to about 3 Å resolution.[110] Sacchettini and Poulter and their co-workers[9] succeeded in determining the crystal structure of the avian enzyme to 2.6 Å resolution, which is the first three-dimensional structure for any prenyltransferases. Strikingly, the FPP synthase has a novel folding structure composed of all α-helices joined by connecting loops. According to their analysis, 10 core helices arrange around a large central cavity whose wall carries highly conserved amino acid residues which are assumed to be essential for catalytic function, as described in later sections.

(vii) Comparative primary structures

Comparison of the amino acid sequences not only of FPP synthases from various organisms but also of GGPP synthase from *Neurospora crassa*[111] and HexPP synthase from *S. cerevisiae*[13] revealed the presence of seven conserved regions[103] as shown in Figure 7. The aspartate-rich motifs conserved in region II ($D^1D^2xxD^3xxxxR^1R^2G$) and in region VI ($GxxFQxxD^4D^5xxD^6$) are the most characteristic for prenyltransferases.

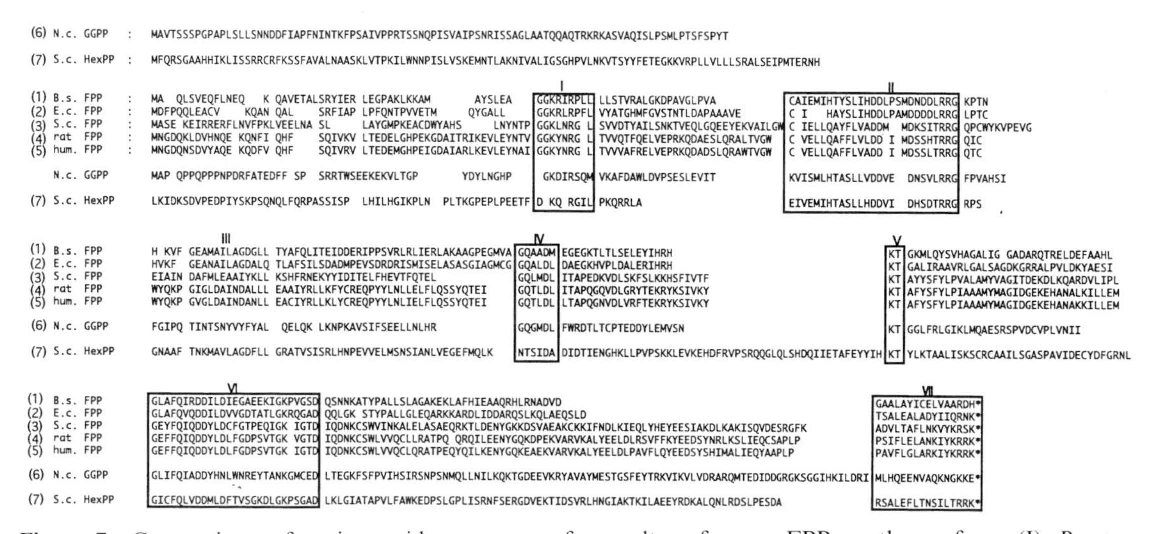

Figure 7 Comparison of amino acid sequences of prenyltransferases. FPP synthases from (I) *B. stearothermophilus*;[103] (II) *E. coli*;[102] (III) *S. cerevisiae*;[101] (IV) rat;[12] (V) human;[99] and (VI) GGPP synthase from *N. crassa*;[111] (VII) HexPP synthase from *S. cerevisiae*.[13] The seven highly conserved regions I to VII are boxed.

(viii) Site-directed mutagenesis studies

To evaluate the roles of the conserved regions in FPP synthases, a number of site-directed mutagenesis experiments have been carried out. The conserved Asp and Arg residues in regions II and VI are crucial for catalytic function. Joly and Edwards[112] showed that the mutagenesis of Asp104, Asp107, Arg112, Arg113, and Asp243 of rat liver FPP synthase, which correspond to D^2, D^3, R^1, R^2, and D^4 in regions II or VI, respectively (Figure 7), resulted in a decreased V_{max} of 1000-fold compared to the wild-type enzyme. However, no significant changes in K_m values for either IPP or GPP were observed in the conservative mutation in which Asp and Arg residues were exchanged with Glu and Lys, respectively. These results suggest that these amino acid residues are involved in catalytic efficiency rather than binding affinity. On the other hand, Asp103 (D^1) and Asp247 (D^6) could be replaced with Glu without marked changes in kinetic properties. Using a recombinant form of yeast enzyme extended by a C-terminal –Glu–Glu–Phe α-tublin epitope, Song and Poulter[107] conducted mutagenesis studies in which the conserved Asp or Arg residue was replaced with Ala residue. They showed that the mutations of the Asp residues and nearby Arg residues in region II (D^1, D^2, R^1, and R^2) and Asp residues in region VI (D^4 and D^5) drastically lowered the catalytic activity of the fused enzyme. However, the third Asp in region VI (D^6) and the Lys254, which is conserved downstream in region VI, were found to be replaced by Ala without marked changes in kinetic properties. Koyama *et al.*[113] also showed that the mutagenesis of Asp224 and Asp225 (D^4 and D^5) of *B. stearothermophilus* FPP synthase resulted in heavy decreases in k_{cat} values of $\sim 10^4$–10^5-fold compared to the wild type. However, replacement of Asp228 (D^6) with Ala resulted in an $\sim$10-fold decrease of k_{cat} value and a 10-fold increase of the K_m value for IPP.

Taken altogether, these results indicate the importance of the six Asp residues in the consensus $D^1D^2xxD^3xxxxR^1R^2G$ (region II) and $GxxFQxxD^4D^5xxD^6$ (region VI) motifs as follows: D^2, D^3, and D^4 are so critical that none of them can be replaced by even Glu without significant decrease in catalytic activity, whereas D^1 can be replaced by Glu but not by Ala. However, D^6 is so tolerant that even Ala can substitute for it.

Replacement of the first Asp residue (D^4, Asp243) in the $GxxFQxxD^4DxxD$ motif in region VI of rat FPP synthase by Glu resulted in a 26-fold increase in the K_m value for IPP.[108] Replacement of the conserved Lys residues (Lys47 and Lys183) in regions I and V of *B. stearothermophilus* FPP synthase with aliphatic amino acids also resulted in marked depressions of the enzyme affinity for IPP, thereby causing a change in the distribution of products.[113] Both K47I and K183A have k_{cat} values $\sim$70-fold lower than those of the wild-type enzyme. Neither of them shows significantly changed K_m values for GPP, but they both show markedly increased K_m values for IPP. These facts indicate that both Lys residues conserved in regions I and V contribute to the binding affinity for IPP as well as the D^6 (or D^4 in case of rat enzyme) in region VI.

Blanchard and Karst[114] isolated a mutant gene *erg20-2* from a yeast strain that had the unusual property of excreting prenyl alcohols such as geraniol. By comparing the nucleotide sequence of the FPP synthase of the mutant, they have shown that the unusual property is attributed to a one-point mutation resulting in a substitution of Glu for Lys197 of *S. cerevisiae* FPP synthase, which is the conserved Lys in region V and which corresponds to Lys183 in the *B. stearothermophilus* enzyme. This is interesting in light of the above mentioned observation by Koyama *et al.*[113] that both K47I and K183A showed markedly increased K_m values for IPP, since a decrease of the affinity for IPP might also cause an accumulation of GPP. Koyama *et al.*[113] examined this possibility by analyzing the relative accumulation of GPP in the FPP synthase reaction by the mutant enzymes, and obtained similar results to those observed by Blanchard and Karst[114] with the yeast mutant extracts. It is noteworthy that K47I and K183A of *B. stearothermophilus* enzyme resemble each other with respect to both kinetic parameters and product distributions. These results raise the possibility that the release of GPP in the reaction catalyzed by the yeast mutant enzyme might also be due to depression of the binding affinity for IPP rather than for GPP.

The significance of the Phe–Gln motif (FQ) in region VI was examined by Koyama *et al.*[115] The highly conserved Phe220 and Gln221 were replaced with Ala and GLu, respectively. These mutageneses resulted in 10^{-5} and 10^{-3} decreases in catalytic activity of the FPP synthase from *B. stearothermophilus*, respectively. Michaelis constants of the Q221E mutant for the allylic substrates (DMAPP and GPP) increased $\sim$25- and two-fold, respectively, compared to the wild type, whereas those for IPP were not altered much. These results suggest that the FQ motif is involved not only in substrate binding but also in catalysis. These facts raise the hypothetical model of the binding of an allylic diphosphate to region VI. However, in view of the dramatic effect on k_{cat} of the replacement of the Phe group by Ala, Phe220 might contribute to the acceleration of catalysis by stabilizing the prenyl cation formed at the beginning of the catalytic reaction through a cation-π interaction[116] as depicted in Figure 8.

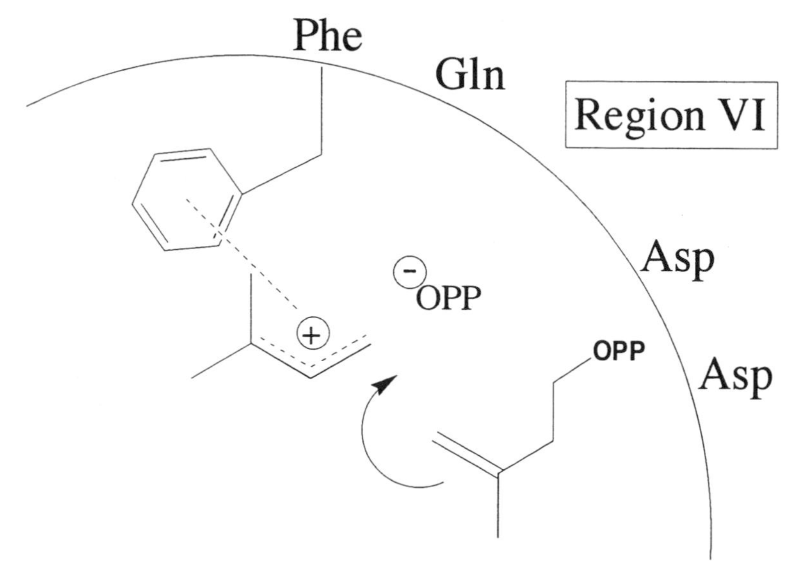

Figure 8 Hypothetical model of the binding of an allylic diphosphate to region VI.

Mammalian FPP synthases share a C-terminal region (region VII) characterized by being crowded with basic amino acids as shown in Figure 7. However, only the Arg residue at the third position from the C-terminus is completely conserved if bacterial FPP synthases are taken into account. To examine the significance of this Arg295 in the *B. stearothermophilus* enzyme, it was mutated to Val.[117] This mutation, however, resulted in no marked decrease of catalytic activity, indicating that the conserved Arg residue in region VII is not essential for catalysis. The R295V mutant showed a three-fold increased K_m value for IPP but showed an unchanged K_m value for GPP as compared with the wild type. Song and Poulter[107] also examined the significance of the Arg350 near the C-terminal of the yeast enzyme with an -EEF epitope. The R350A mutant showed only a slight additional increase in the K_m value for IPP. The addition of the -EEF epitope to the C-terminus of the wild-type enzyme resulted in a 14-fold increase of the K_m value for IPP and a 12-fold decrease of k_{cat}, suggesting that the conserved hydrophilic C-terminus of the enzyme may contribute to substrate binding and catalysis.

It has been suggested that Cys residues play important roles in the catalytic function or the substrate binding of several FPP synthases.[17,118] Moreover, there are two interconvertible forms in FPP synthases from porcine liver, which are attributed to oxidation/reduction of Cys residues of the enzyme.[56–58]

B. stearothermophilus FPP synthase is unique in that it possesses only two Cys residues in contrast to FPP synthases from other sources that have more than four Cys residues. To explore the significance of these Cys residues, Koyama *et al.*[60] examined the effect of replacement of the Cys residues at 73 and 289 of this thermostable enzyme with Ser. All of the mutant enzymes were active as FPP synthase, showing specific activities comparable to that of the wild-type enzyme. These results indicate that neither of the Cys residues is essential for catalytic function. Unexpectedly, the mutant C73S–C289S, in which both Cys residues were replaced with Ser as well as the wild-type enzyme displayed two forms that are separable on MonoQ chromatography. Hence, the two forms of FPP synthase should not be ascribed only to a problem of Cys residues as far as the *B. stearothermophilus* enzyme is concerned.

(ix) Random chemical mutagenesis

Random chemical mutagenesis often provides a powerful method for obtaining a desired mutation product if an effective procedure for screening positive clones is ready. Ohnuma *et al.*[119] have developed an *in vivo* method for detecting GGPP synthase activity, which utilizes carotenoid biosynthesis genes of *Erwinia uredovora*[120] to produce a colored clone expressing GGPP synthase activity.

The FPP synthase gene of *B. stearothermophilus* was subjected to random mutagenesis by $NaNO_2$ treatment to construct libraries of mutated FPP synthase genes. From the libraries the mutants showing GGPP synthase activities were selected by the red-white screening method, and eleven red positive clones were obtained from 24 300 mutants.[121] Each mutant was found to contain a few amino acid substitutions in the FPP synthase, which resulted in acquisition of the catalytic activity synthesizing GGPP as well as FPP. Mutants in which Tyr81 was replaced by His showed the most efficient production of GGPP. From the analysis of the mutations they defined three amino acids that could determine the final chain length of the products. They were Leu34, Tyr81, and Val157. In particular, the mutated enzyme that has a substitution of His for Tyr81, which is situated at the fifth amino acid upstream to the DDxxD motif in region II, produced GGPP most effectively.

To investigate the role of Tyr81 of *B. stearothermophilus* FPP synthase, they constructed 20 mutant enzymes each of which has a different amino acid at position 81.[122] All enzymes except Y81P showed prenyltransferase activities catalyzing condensation of IPP with an allylic diphosphate. When assayed with FPP as the allylic substrate, considerable activities were observed in almost all mutated FPP synthases. These results indicate that the mutated enzymes can catalyze the chain elongation beyond FPP. When DMAPP or GPP was employed as the allylic primer, mutated enzymes, Y81A, Y81G, and Y81S, produced hexaprenyl diphosphate (HexPP) as the longest product. They produced geranylfarnesyl diphosphate (GFPP, C_{25}) as the main product. Y81A and Y81G gave greater amounts of HexPP and GFPP than did Y81S. These mutant enzymes, in which Tyr81 was replaced with Cys, His, Ile, Leu, Gln, Thr, or Val, all produced GFPP as the longest prenyl product. These observations strongly indicate that the amino acid Tyr81 directly contacts the ω-terminal of an elongating allylic product during the catalytic isoprenoid chain elongation. This interaction in the catalytic site must be the critical step determining the chain length of the product of prenyltransferase.

(x) Regulation of product chain length

Tarshis *et al.*[123] have reported strong evidence on the mechanism of regulation of product chain length in the avian FPP synthase reaction by X-ray analyses of some mutated enzymes that acquired the catalytic capability of producing longer chain prenyl diphosphates.

An analysis of the X-ray structure[9] of the avian FPP synthase, coupled with the information about conserved amino acids of many prenyltransferases so far cloned, led them to the idea that the phenyl groups of Phe112 and Phe113 in the avian enzyme were important for determining the ultimate length of the hydrocarbon chains. Then they carried out several site-directed mutations in the avian enzyme with respect to these Phe residues. As a result, enzymes capable of producing GGPP (F112A), GFPP (F113S), and longer chain prenyl diphosphates (F112A/F113S) were obtained. X-ray analyses of the structure of the F112A/F113S mutant in the absence or presence of allylic substrates indicated an alteration of the size of the binding pocket for the growing isoprenoid chain in the active site of the enzyme. The proposed binding pocket in the mutant structure was increased in depth by 5.8 Å compared with that for the wild-type enzyme. Allylic diphosphates were observed in the holo structures, bound through Mg^{2+} to the first two Asp residues in the DDxxD motif (Asp117 ~ Asp121), with the hydrocarbon tails of all the ligands growing down the hydrophobic pocket toward the mutation site.

2.04.3.1.2 Geranylgeranyl diphosphate synthase

(i) Enzymology

It was six years after the discovery of FPP synthase from yeast by Lynen *et al.*[1] that Kandutsoh *et al.*[124] found GGPP synthase in a carotenoid-producing bacterium, *Micrococcus luteus*. This enzyme, farnesyl*trans*transferase (EC 2.5.1.30), catalyzes consecutive $C_5 \rightarrow C_{10} \rightarrow C_{15} \rightarrow C_{20}$ isoprenoid chain elongations to yield all-*E*-GGPP without accumulation of any intermediates. Nandi and Porter obtained and characterized the GGPP synthase from carrot root,[125] which produces carotenoids as well as sterols. Ogura *et al.*[126] succeeded in the separation of FPP synthase and GGPP synthase of pumpkin fruit from each other and demonstrated that these prenyltransferases catalyze the $C_5 \rightarrow C_{10} \rightarrow C_{15}$ and $C_5 \rightarrow C_{10} \rightarrow C_{15} \rightarrow C_{20}$ reactions, respectively. This fact indicates that the biosynthetic pathways to steroids and to carotenoids branch at the stage of C_5 so that the synthesis of GGPP can be independent of the production of FPP by FPP synthase.

Nandi and Porter[125] also reported enzymatic formation of GGPP by incubation of IPP and FPP with a pig liver homogenate. However, it was often assumed implicitly that the formation of GGPP in the liver extracts might be due to the tolerant specificity of FPP synthase, because Reed and Rilling[8] reported that avian liver FPP synthase showed a weak but not negligible GGPP formation. Sagami *et al.*[127] detected a distinct GGPP synthase activity in pig liver and succeeded in separation of the enzyme free of FPP synthase which had been known to occur as a major prenyltransferase in this organ. At that time, the occurrence of this enzyme in mammalian tissue was considered implausible, because mammals lack the ability to synthesize diterpenes or carotenoids. In the late 1990s, however, no one can neglect the significance of this enzyme because of the striking discovery of geranylgeranylated proteins as well as farnesylated proteins.[128]

Although this GGPP synthase was very hard to purify because of its low titres, Sagami *et al.*[129] succeeded in the purification of bovine brain GGPP synthase by use of an affinity gel with a farnesylmethylphosphonyl ligand. This mammalian GGPP synthase is a homooligomer with a molecular mass of the monomer of 37.5 kDa and catalyzes condensations of IPP with GPP and with FPP to synthesize GGPP, but lacks the condensation ability between IPP and GPP unlike the enzymes from plants and bacteria.

Thermostable GGPP synthases were purified from archaea, *Methanobacterium thermoautotrophicum* and *M. thermoformicicum* by Chen and Poulter,[130] and by Tachibana *et al.*,[131] respectively. Both of the enzymes catalyze the sequential addition of IPP to DMAPP, GPP, and FPP by a nonprocessive mechanism which allows substantial accumulation of FPP. As no other short-chain prenyltransferase has been observed in these organisms, these archaeal GGPP synthases seem to have the bifunctional roles to produce FPP and GGPP for further biosyntheses of squalene and membrane lipids, respectively. It is of interest that the enzyme from *M. thermoautotrophicum* is resolved into two interconvertible forms[130] as in the case of several FPP synthases.[56–58,60]

(ii) Gene cloning

In 1990 Armstrong *et al.*[132] determined the nucleotide sequence of three genes from the carotenoid biosynthesis gene cluster of *E. herbicola*, a nonphotosynthetic bacterium, which encode homologues of the CrtB, CrtE, and CrtI proteins of *Rhodobacter capsulatus*, a purple nonsulfur photosynthetic bacterium.[133] These three proteins were thought to be engaged in the early stage in the biosynthesis of carotenoids, because both organisms were expected to possess similar enzymes for synthesis and dehydrogenation of phytoene, the common precursor of all other C_{40} carotenoids.

Meanwhile, Nelson *et al.*[134] isolated the gene encoding GGPP synthase from the filamentous fungus *N. crassa* by complementation of the Albino 3 (*al-3*) mutant which was defective in GGPP synthase. By analyzing the nucleotide sequence of the *al-3* gene, Carattoli *et al.*[111] compared the deduced amino acid sequence with those of FPP synthases, and found that it showed significant homologies in three different regions including the Asp-rich motifs. In addition, the Al-3 protein of 47 kDa is weakly basic and hydrophilic, and does not possess any hydrophobic membrane-spanning regions. These facts confirmed the first isolation of the GGPP synthase gene. They also proposed that the *crtE* gene product of *R. capsulatus* might belong to the prenyltransferase family, because of the impressive homology with all three conserved domains.[111]

The *crtE* gene in *E. herbicola* was assigned for GGPP synthase by Math *et al.*[135] They isolated the *crtE* gene from *E. herbicola* cluster by PCR amplification and cloned the coding region into the *E. coli* expression vector pARC306N. The cell-free extracts from the *E. coli* transformants clearly showed GGPP synthase activity, indicating that *crtE* encodes GGPP synthase.

Similarly the *crtE* gene of *E. uredovora* was assigned as the GGPP synthase gene by Sandmann and Misawa.[136]

Thus, cDNAs or genomic clones encoding GGPP synthase have been isolated from various organisms including *C. annuum*,[137] *M. thermoautotrophicum*,[138] *Sulfolobus acidocaldarius*,[119] *A. thaliana*,[139] white lupin,[140] and *S. cerevisiae*.[141]

(iii) Elucidation of chain length determination mechanisms

One of the most interesting subjects in the catalytic mechanism of prenyltransferases is to understand the mechanism by which individual enzymes determine the chain lengths of their own product during catalysis. The group at the Tohoku University obtained a clue to the solution of the problem.

To obtain mutant prenyltransferases that have changes in the machinery for determination of the product chain length, Ohnuma *et al.*[142] carried out random mutation by $NaNO_2$ treatment on the GGPP synthase gene from *S. acidocaldarius*.[119] The library of mutated GGPP synthase genes on a yeast expression vector were screened for suppression of a *pet* phenotype of yeast C296-LH_3. As *S. cerevisiae* C296-LH_3 is deficient in HexPP synthase, it cannot produce coenzyme Q6, and consequently, the growth of the mutant is limited by the availability of fermentable substrates.[143] If mutated GGPP synthase has HexPP synthase activity, the transformed yeast is expected to be able to grow on a YEPG plate, which contains only glycerol as an energy source instead of glucose. By the selection of 1400 mutants several positive clones were obtained. Three mutant enzymes showed catalytic activity for production of large amounts of GFPP with a concomitant amount of HexPP. Sequence analysis revealed that the mutation of Phe77, which is located at the fifth amino acid upstream from the consensus motif DDxxD in region II of prenyltransferases, is the most effective for elongating the ultimate product. This fact exactly coincides with the results on the FPP synthase mutation[121,122] as described in Section 2.04.3.1.1(ix).

By comparison of amino acid sequences of many kinds of FPP synthases with those of GGPP synthases, Ohnuma *et al.*[144] noticed that there are several homologous regions typical for FPP synthase in the highly conserved region II. Then they introduced mutations into the region II of the GGPP synthase from *S. acidocardarius*. These mutant GGPP synthases had replacements with the corresponding regions of FPP synthases from human,[100] rat,[12] *A. thaliana*,[105] *S. cerevisiae*,[101] *E. coli*,[102] and *B. stearothermophilus*.[103] By analyzing these mutated enzymes, they found that the first aspartate rich motif, $D^1D^2xxD^3$ in region II, is essential for the product specificity of all FPP synthases and that the critical sequence for the specificity in this region of FPP synthase differs between eukaryotic and prokaryotic enzymes, which they designated type I and type II, respectively. Based on these observations they proposed that FPP synthases have evolved from the progenitor corresponding to the archaeal GGPP synthases in two ways.

2.04.3.1.3 *Geranyl diphosphate synthase*

In terms of catalytic reaction, GPP synthase is the simplest prenyltransferase. The existence of this enzyme in higher plants was predicted in relation to the biosynthesis of monoterpenes.[145]

This enzyme, dimethylallyl*trans*transferase (EC 2.5.1.1) exists in many higher plants producing monoterpenes. Partial purification and characterization have been carried out from several plants or plant cell cultures including *Lithospermum erythrorhizon* cell culture,[146] *Salvia officinalis* leaves,[147] and several higher dicotyledon plant leaves.[148] However, because of the unstable feature of this enzyme, a homogeneous preparation has not been obtained. The enzyme from *L. erythrorhizon* cell cultures,[146] which was purified 92-fold, showed a molecular weight of 73 kDa, isoelectric point at pH 4.95, and absolute requirement for divalent cations, Mg^{2+} and Mn^{2+} being the most effective.

2.04.3.1.4 *Geranylfarnesyl diphosphate synthase*

The only report on GFPP synthase was made by Tachibana[149] in 1994. This enzyme was found in the haloalkaliphilic archaeon *Natronobacterium pharaonis*, which has polar lipids, C_{20}, C_{25} diether lipids[150] in addition to the C_{20}, C_{20} diether lipids commonly occurring in Archaea. This enzyme catalyzes a continuous $C_5 \rightarrow C_{10} \rightarrow C_{15} \rightarrow C_{20} \rightarrow C_{25}$ reaction, and it seems to have no two-component system, as described in the following section. Therefore, pentaprenyl diphosphate synthase should be grouped into the class of short-chain prenyl diphosphate synthases in terms of the subunit composition.

2.04.3.2 Medium-chain Prenyl Diphosphate Synthases (Prenyltransferases II)

2.04.3.2.1 *Enzymatic properties and reaction mechanism*

During the search for prenyltransferases in bacteria, which produce menaquinones or ubiquinones having a variety of chain length polyprenyl side chains in a species-specific manner, a peculiar group of prenyltransferases was found. The C_{30} chain elongation is catalyzed by a novel enzyme system, HexPP synthase, *trans*-pentaprenyl*trans*transferase (EC 2.5.1.33). It catalyzes the synthesis of all-

E-HexPP by adding three molecules of IPP to FPP, but it cannot catalyze the synthesis of GPP or FPP from DMAPP and IPP.[151] This enzyme was found in *M. luteus* B-P 26, which produces menaquinone-C_{30}. Similarly heptaprenyl diphosphate (HepPP) synthase, *trans*-hexaprenyl*trans*-transferase (EC 2.5.1.30), has been found in *Bacillus subtilis*, which produces exclusively menaquinone-C_{35}.[152] It catalyzes the synthesis of all-*E*-HepPP by adding four molecules of IPP to FPP, but it is not able to catalyze the synthesis of GPP or FPP from DMAPP and IPP either. FPP is supplied by FPP synthase which occurs in all bacteria.

The purification of the HexPP synthase involved unexpected difficulties, due not only to the unexpected substrate specificities, but also to the fact that the synthesis of HexPP requires two separable proteins, neither of which is catalytically active alone.[151] Therefore, the HexPP synthase had been elusive before this fact was disclosed.

Similarly two different proteins constituting HepPP synthase were separated from *B. subtilis*.[153] The components of HexPP synthase and HepPP synthase are so specific that neither of the two components of one enzyme is exchangeable with that of the other enzyme. Therefore, they appear to be novel heterodimeric enzymes with subunits easily dissociable under physiological conditions.

To obtain insight into the dynamic mechanism on the cooperative interaction between the components A and B of the HexPP synthase, the properties of the heteromeric HexPP synthase on gel filtration were examined.[154] When a mixture of components A and B, which had been incubated in the presence of Mg^{2+} and substrates, was filtered on Superose 12, a small peak of $\sim$50 kDa, which showed HexPP synthase activity, was detected, suggesting that this fraction corresponded to an intermediary state of catalytically active HexPP synthase. Formation of this complex of the two essential components A and B was also observed in the presence of relatively high concentrations of PPi or one of the substrates, IPP and FPP.[155] These results have led us to propose a mechanism of this unusual enzyme system as shown in Figure 9. This might account for the ability of these soluble proteins to show an efficient turnover in the synthesis of amphipathic molecules from water-soluble substrate without association with membrane components. Precise experiments on inhibitory effects of Cys- or Arg-specific reagents suggested that the catalytic site of HexPP synthase is formed by cooperative interaction between the two components, and that Cys and Arg residues on component B play important roles in the catalytic activity.[156]

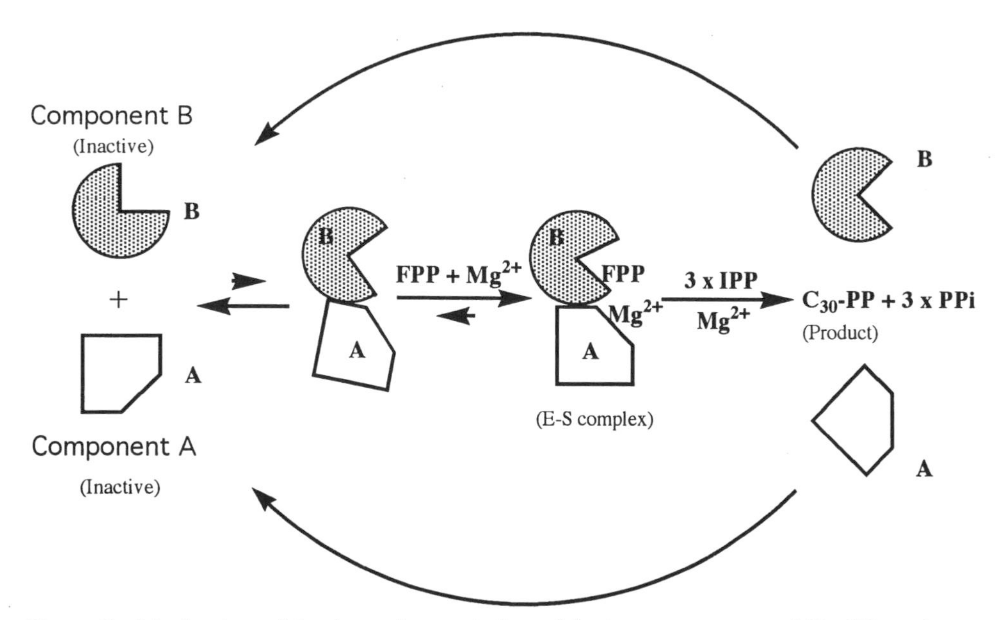

Figure 9 Mechanism of the dynamic association of the two components of HexPP synthase.

A feasible method was developed to determine the stereochemistry of the C—C bond formation with respect to the face of the double bond of IPP in prenyltransferase reactions,[157] and it was applied to the HepPP synthase reaction to examine whether the stereochemical rule in the FPP synthase reaction as described in Section 2.04.3.1.1(ii)[63] holds widely for other prenyltransferases. As a result, the C—C bond formation of the HepPP synthase from *B. subtilis* was found to take place at the *si* face of the double bond with elimination of the *pro-R* hydrogen at C-2 in a *syn* fashion, indicating that the same stereochemistry as that of FPP synthase is involved (Figure 4).

2.04.3.2.2 Gene cloning and expression

Based on the highly conserved amino acid sequences of prenyltransferases as well as the thermostability of the enzyme, Koike-Takeshita *et al.*[158] succeeded in identifying the genes encoding two protein components that constitute a medium-chain prenyl diphosphate synthase of a thermophilic bacterium *B. stearothermophilus*. One of the two proteins constituting an enzyme system for the synthesis of HepPP synthase has 323 amino acid residues and shows a 32% sequence similarity to the FPP synthase of the same bacterium. This protein designated as component II′ has seven highly conserved regions that have been shown typical in prenyltransferases.[103] The other protein (component I′), which is composed of 220 amino acids, has no such similarity within the protein entries in databases. Therefore, it seems likely that the former carries the active sites for substrate binding and catalysis, while the latter plays an auxiliary but essential role in the catalytic function.

Protein database search for amino acid sequences similar to those of the HepPP synthase of *B. stearothermophilus*[158] yielded the GerC proteins of *B. subtilis*, which are encoded in a cluster of three open reading frames, *gerC1*, *gerC2*, and *gerC3*, and have been shown to be involved in vegetative cell growth and spore germination of this bacterium.[159] Two of the *gerC* products, GerC1 and GerC3, showed 37.7% and 65.0% identities to those of the component I′ and component II′, respectively, two of which constitute the HepPP synthase of *B. stearothermophilus*. To examine the hypothesis that the *gerC* gene region encodes the HepPP synthase, the *gerC* gene was amplified by PCR, and the expression of a prenyltransferase activity was examined. When the two proteins were mixed, a distinct HepPP synthase activity was detected, indicating that the HepPP synthase of *B. subtilis* is composed of the two dissociable components, which are encoded by *gerC1* and *gerC3*.[160] Hence, the proteins GerC1 and GerC3 exactly coincide with the two dissociable components I and II of *B. subtilis* HepPP synthase, respectively.

The two genes encoding the components A and B, which are essential for *M. luteus* B-P 26 HexPP synthase have been cloned by Shimizu *et al.*[161] by similar PCR-mediated procedures to those employed for the cloning of *B. stearothermophilus* HepPP synthase.[158]

As shown in Figure 10, the larger components of the HepPP synthase from *B. stearothermophilus* or *B. subtilis*, and the HexPP synthase from *M. luteus* B-P 26, component II′ or -II, and component B, respectively, show high identities between the deduced amino acid sequences. However, the identities between the smaller components of the HepPP synthase and the HexPP synthase is less than 10%. It is very interesting to investigate the molecular mechanism of this kind of novel enzymatic system that expresses the catalytic activity by the dynamic interaction of these heteromeric components.

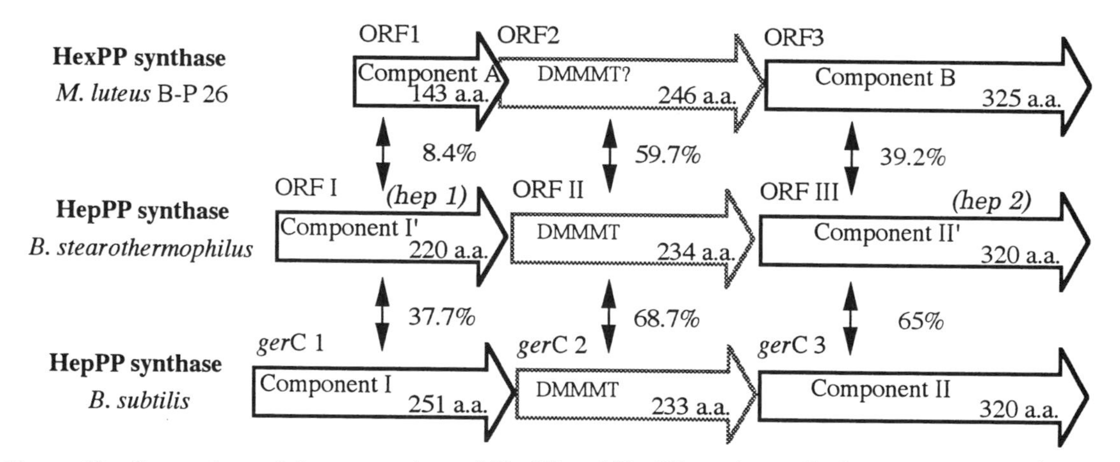

Figure 10 Comparison of the gene regions of HexPP and HepPP synthases. Each arrow corresponds to an open reading frame of each gene region. Amino acid sequence similarities between the genes are described in percentages.

In experiments with a yeast mutant[143] in coenzyme Q biosynthesis, Ashby and Edwards[13] have isolated a gene from a plasmid containing a wild-type genomic DNA fragment that is able to complement the mutant and restore HexPP synthase activity. They have also shown that this gene encodes a 473-amino acid protein having conserved regions characteristic of prenyltransferases. Therefore, this protein seems to correspond to the larger protein component of the two-heteromeric components of bacterial medium-chain prenyl diphosphate synthases as mentioned earlier. However, it is not known whether the yeast 473-amino acid protein acts as HexPP synthase by itself or in association with another gene product similar to the smaller protein. If the latter is the case, the mutant described above must be deficient in one of the two components of the HexPP synthase. It would be interesting to learn whether dissociable heterodimeric systems are common in medium-chain prenyl diphosphate synthases of both prokaryotic and eukaryotic cells.

2.04.3.3 Long-chain *E*-Prenyl Diphosphate Synthases (Prenyltransferases III)

2.04.3.3.1 Enzymology

By analogy with HexPP- and HepPP synthases, it was predicted that the enzymes for the synthesis of polyprenyl diphosphate with chain lengths longer than C_{35} might also be two-component systems. However, this is not the case. *E*-Nonaprenyl (solanesyl) diphosphate (SPP) synthase, which had been characterized by Sagami *et al.*,[162] was isolated from *M. luteus* as an enzyme catalyzing the chain elongation up to C_{45}-PP from GPP (C_{10}), and it was found that this enzyme is a homodimeric protein that is functionally active by itself.[163] In this respect, it resembles the enzymes in the group of prenyltransferase I (4.3.1), but it differs in that it requires a protein factor to maintain an efficient catalytic turnover. This protein was separated as a high-molecular component of a soluble fraction of the same bacterium. It stimulates dramatically the enzymatic synthesis of SPP in a time-dependent manner, but it does not affect the stability of the enzyme.[163] Bovine serum albumin or a detergent such as Tween 80 can substitute for this protein, showing a similar mode of stimulation. These results indicate that the protein facilitates catalytic turnover by removing from the enzyme active site hydrophobic products, which otherwise would inhibit the enzyme reaction (Figure 11). Therefore, this protein should be designated as polyprenyl carrier protein (PCP).

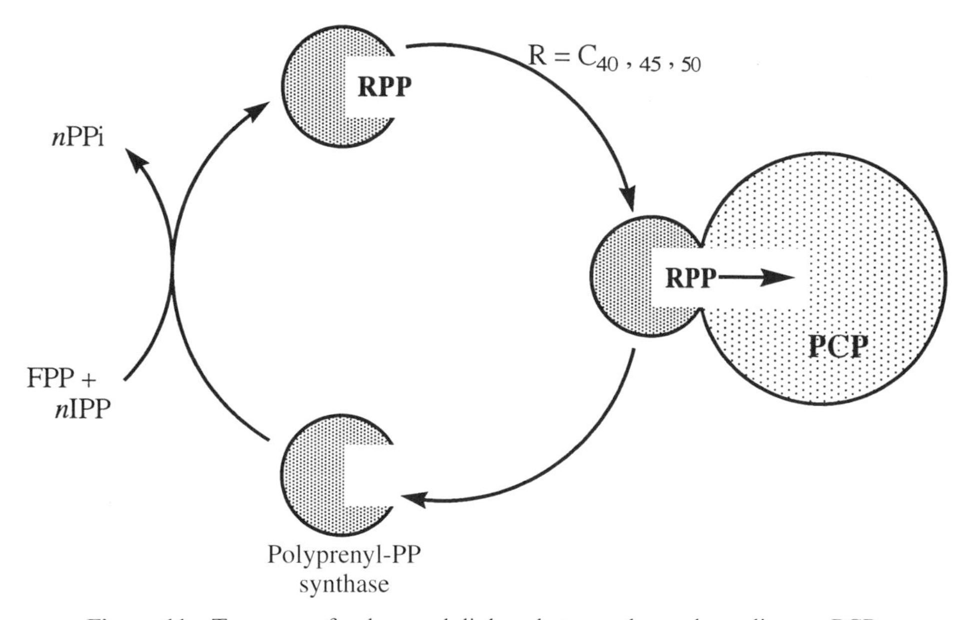

Figure 11 Turnover of polyprenyl diphosphate synthases depending on PCP.

Bacterial all-*E*-octaprenyl PP (C_{40}) synthase and all-*E*-decaprenyl PP (C_{50}) synthase have been found and partially purified from *E. coli*[164] and *Paracoccus denitrificans*,[165] respectively. These polyprenyl diphosphate synthases have similar properties to those of the SPP synthase described above. The PCP from *M. luteus* acted effectively on these synthases as well.[163] However, PCP shows no stimulation on undecaprenyl diphosphate synthase from *E. coli*. Therefore, PCP is effective specifically on all-*E*-polyprenyl diphosphates with chain lengths of C_{40}, C_{45}, and C_{50}.

2.04.3.3.2 Chain-length specificity of SPP synthase

Fujii *et al.*[166] found that the distribution of polyprenyl diphosphates synthesized by SPP synthase from *M. luteus* was dramatically changed depending on the Mg^{2+} concentration in the reaction medium. When the Mg^{2+} concentration was higher than 5 mM, C_{40} and C_{45} products are dominant. However, when the metal ion level is lower than 0.5 mM, a variety of prenyl diphosphates ranging from C_{15} to C_{40} were formed. Ohnuma *et al.*[167] investigated the factors that affect the chain-length determination in the SPP synthase reaction using homogeneously purified *M. luteus* enzyme. As a result, the level of IPP–Mg complex in the reaction mixture is decisive in affecting the chain-length distribution. Thus, the Mg^{2+}-dependent variability of product specificity[166] has been explained in terms of the effect of IPP–Mg concentration.

The effect of other divalent metal ions on the chain-length distribution of polyprenyl diphosphate synthases was also examined.[168] In the presence of Co^{2+} or Mn^{2+}, the octaprenyl-, solanesyl-, and decaprenyl diphosphate synthases gave a variety of polyprenyl products with the longest chains being shifted by one or two isoprene units longer than those of the products formed in the presence of Mg^{2+}.

2.04.3.3.3 Gene cloning and expression

Choi *et al.* found that part of an open-reading frame in the *E. coli* chromosome showed a significant similarity to the *ispA* gene, which encodes FPP synthase of *E. coli.*[102] Jeong *et al.*[170] also determined the whole sequence of this open-reading frame and indicated the high similarity of the gene product to the HexPP synthase of *S. cerevisiae*[13] and GGPP synthases of various organisms. To determine the function of this gene, Asai *et al.*[171] cloned the gene and examined a prenyltransferase activity expressed in transformed *E. coli* cells and they found an increased activity of octaprenyl diphosphate synthase. The deduced amino acid sequence of this enzyme also shows the presence of the seven conserved regions typical of prenyltransferases.

2.04.3.4 *Z*-Polyprenyl Diphosphate Synthases (Prenyltransferases IV)

2.04.3.4.1 Enzymology

Prenyltransferases that catalyze the synthesis of *Z*-prenyl chains occur mostly in the membrane fraction. Their roles are to produce the precursors of polyprenyl lipids required as carbohydrate carrier in the biosynthesis of bacterial cell wall or of glycoproteins in eukaryotic cells.

Undecaprenyl diphosphate (UPP) synthase (*di-trans,poly-cis*-decaprenyl*cis*transferase, EC 2.5.1.31) catalyzes the construction of a *Z*-prenyl chain onto all-*E*-FPP as a primer to yield a C_{55}-prenyl diphosphate with *E,Z*-mixed stereochemistry. It has been described in several bacteria including *M. luteus,*[172] *B. subtilis,*[173] and *E. coli,*[174] but it has been studied most extensively in *Lactobacillus plantarum.*[175–178] This enzyme is so loosely associated with periplasmic membranes that it is easily solubilized, but it requires phospholipid or a detergent such as Triton X-100 for activity. Muth and Allen[178] purified the UPP synthase from *L. plantarum*. The enzyme was shown to be an acidic protein (pI = 5.1) and the estimated molecular weight was 56 kDa. SDS–PAGE (sodium dodecyl sulfate–polyacrylamide gel electrophoresis) analysis indicated that this enzyme was composed of two 30 kDa subunits.

It is interesting that *P. denitrificans*, which is assumed to be an ancestor of mitochondria, has all-*E*-farnesyl-all-*Z*-hexaprenyl (C_{45}) diphosphate synthase[179] instead of UPP synthase, which is common in general bacteria.

Most plants contain large amounts of polyprenols with *E,Z*-mixed stereochemistry, including betulaprenols and ficaprenols. Koyama *et al.*[180] extracted and partially purified a polyprenyl diphosphate synthase from mulberry leaves, *Morus bombysis*. This enzyme catalyzed a consecutive condensation of IPP with GGPP as an allylic primer to produce a series of ficaprenol-type *Z,E*-mixed polyprenyl diphosphates with carbon chain length ranging from C_{40} to C_{60}. FPP and GPP were also accepted as allylic primers, and Triton X-100 stimulated the enzymatic activity.

The *Z*-polyprenyl diphosphate synthase in mammalian cells is dehydrodolichyl diphosphate synthase, which catalyzes much longer chain elongations than do bacterial enzymes. This membrane-bound enzyme has been solubilized and characterized from Ehrlich ascites tumor cell membranes,[181] rat tubular membranes,[182] *S. carlsbergensis,*[183] and rat liver.[184] The chain length composition of the products is coincident with that of the dolichols in each organism, which have saturated α-isoprene units, and which play important roles as the sugar carrier lipids in glycoprotein biosynthesis.

Rubber transferase, rubber *cis*-polyprenyl*cis*transferase (EC 2.5.1.20) has been found and characterized from *Hevea brasiliensis* by Archer *et al.*[185] and from Guayule, *Parthenium argentatum* by Cornish and Backhaus.[186] By analogy with the other prenyltransferases, it has been believed that rubber transferase catalyzes the transfer of *Z*-polyprenyl diphosphates to IPP with elimination of PPi. However, this enzymatic activity occurs only at the surface of the rubber particles, on which the activity is not diminished even after repeated washing, demonstrating the firm association of the enzyme with the particles.

2.04.3.4.2 *Stereochemistry of Z-polyprenyl diphosphate synthase reaction*

The stereochemistry of the hydrogen elimination from C-2 of IPP has been well elucidated with various prenyltransferases since the overall stereochemistry of FPP synthase reaction was first established by Cornforth and co-workers.[62,63] However, pig liver FPP synthase was the only enzyme to have been examined from the viewpoint of the stereochemistry of the C—C bond formation.[63] The stereochemistry of enzymatic C—C bond formation leading to *Z*-prenyl chains was an interesting problem. The success in the separation of medium- and long-chain prenyl diphosphate synthases of *B. subtilis*[53,173] led the authors to determine the absolute stereochemistry involved in the *Z*-prenyl- and *E*-prenyl chain elongations catalyzed by UPP- and HepPP synthases, respectively.[187] The absolute stereochemistry of UPP synthase is illustrated in Figure 12.

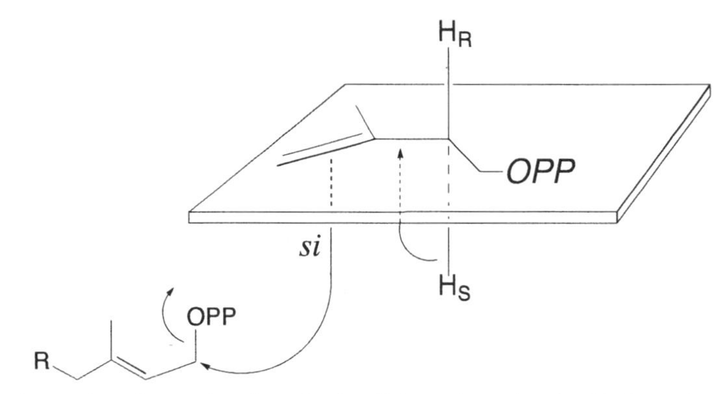

Figure 12 Stereochemistry in the 1′-4 condensation by UPP synthase.

2.04.4 SUMMARY

It is noteworthy that in spite of the similarity of the reactions catalyzed by these prenyltransferases the mode of expression of catalytic functions is surprisingly different according to the prenyl chain length and stereochemistry of reaction products. These prenyltransferases, most of which are of bacterial origin, are summarized and classified into four groups in Figure 2:

(i) *Short-chain prenyl diphosphate synthases* (*prenyltransferases I*). These enzymes require no cofactor except divalent metal ions, Mg^{2+} or Mn^{2+}, which are commonly required by all prenyl diphosphate synthases.

(ii) *Medium-chain prenyl diphosphate synthases* (*prenyltransferases II*). All-*E*-HexPP- and all-*E*-HepPP synthases are unusual because they each consist of two dissociable dissimilar protein components, neither of which alone has catalytic activity.

(iii) *Long-chain all-E*-prenyl diphosphate synthases (prenyltransferases III). Octaprenyl- (C_{40}), nonaprenyl- (C_{45}), and decaprenyl (C_{50}) disphosphate synthases require PCPs that remove the amphipathic polyprenyl products from the catalytic sites of the enzymes to maintain efficient catalytic turnover.

(iv) *Z-Polyprenyl diphosphate synthases* (*prenyltransferases IV*). The enzymes responsible for *Z*-chain elongation including UPP- and *Z*,*E*-nonaprenyl diphosphate synthases require phospholipid for catalytic activity.

2.04.5 REFERENCES

1. F. Lynen, H. Eggerer, U. Henning, and I. Kessel, *Angew. Chem.*, 1958, **70**, 738.
2. S. Chaykin, J. Law, A. H. Phillips, T. T. Tchen, and K. Bloch, *Proc. Natl. Acad. Sci., USA*, 1958, **44**, 998.
3. B. W. Agranoff, H. Eggerer, U. Henning, and F. Lynen, *J. Am. Chem. Soc.*, 1959, **81**, 1254.
4. L. Ruzicka, *Experientia*, 1953, **9**, 357.
5. R. Bentley, in "Molecular Biology, Molecular Asymmetry in Biology," eds. B. Horecker, N. O. Kaplan, J. Marmur, and H. A. Scheraga, Academic Press, New York, 1970, vol. 2, p. 267.
6. C. D. Poulter and H. C. Rilling, in "Biosynthesis of Isoprenoid Compounds," eds. J. W. Porter and S. L. Spurgeon, Wiley, New York, 1981, vol. 1, p. 161.
7. N. I. Eberhardt and H. C. Rilling, *J. Biol. Chem.*, 1975, **250**, 863.
8. B. C. Reed and H. C. Rilling, *Biochemistry*, 1975, **14**, 50.
9. L. C. Tarshis, M. Yan, C. D. Poulter, and J. C. Sacchettini, *Biochemistry*, 1994, **33**, 871.
10. T. Koyama, M. Matsubara, K. Ogura, I. E. M. Brüggemann, and A. Vrielink, *Naturwissenschaften*, 1983, **70**, 469.
11. C. D. Poulter and H. C. Rilling, *Acc. Chem. Res.*, 1978, **11**, 307.
12. C. F. Clarke, R. D. Tanaka, K. Svenson, M. Wamsley, A. M. Fogelman, and P. A. Edwards, *Mol. Cell. Biol.*, 1987, **7**, 3138.
13. M. N. Ashby and P. A. Edwards, *J. Biol. Chem.*, 1990, **265**, 13 157.
14. M. S. Anderson, M. Muehlbacher, I. P. Street, J. Proffitt, and C. D. Poulter, *J. Biol. Chem.*, 1989, **264**, 9169.
15. B. W. Agranoff, H. Eggerer, U. Henning, and F. Lynen, *J. Biol. Chem.*, 1960, **235**, 326.
16. D. H. Shah, W. W. Cleland, and J. W. Porter, *J. Biol. Chem.*, 1965, **240**, 1946.
17. P. W. Holloway and G. Popják, *Biochem. J.*, 1967, **104**, 57.
18. K. Ogura, T. Nishino, and S. Seto, *J. Biochem.*, 1968, **64**, 197.
19. K. Ogura, T. Nishino, T. Koyama, and S. Seto, *Phytochemistry*, 1971, **10**, 779.
20. H. Sagami and K. Ogura, *J. Biochem.*, 1983, **94**, 975.
21. S. Fujisaki, T. Nishino, and H. Katsuki, *J. Biochem.*, 1986, **99**, 1327.
22. M. Lützow and P. Beyer, *Biochim. Biophys. Acta*, 1988, **959**, 118.
23. T. Koyama, D. Wititsuwannakul, K. Asawatreratanakul, R. Wititsuwannakul, N. Ohya, Y. Tanaka, and K. Ogura, *Phytochemistry*, 1996, **43**, 769.
24. D. V. Banthorpe, S. Doonan, and J. A. Gutowski, *Arch. Biochem. Biophys.*, 1977, **184**, 381.
25. E. Bruenger, L. Chayet, and H. C. Rilling, *Arch. Biochem. Biophys.*, 1986, **248**, 620.
26. O. Dogbo and B. Camara, *Biochim. Biophys. Acta*, 1987, **920**, 140.
27. G. Popják and J. W. Cornforth, *Biochem. J.*, 1966, **101**, 553.
28. K. J. Stone, W. R. Roeske, R. B. Clayton, and E. E. van Tamelen, *J. Chem. Soc., Chem. Commun.*, 1969, 530.
29. T. Koyama, K. Ogura, and S. Seto, *J. Biol. Chem.*, 1973, **248**, 8043.
30. K. Clifford, J. W. Cornforth, R. Mallaby, and G. T. Phillips, *J. Chem. Soc., Chem. Commun.*, 1971, 1599.
31. T. Koyama, *Dr. Sci. Thesis, Tohoka University*, 1972.
32. *K. Ogura, T. Koyama, T. Shibuya, T. Nishino, and S. Seto, J. Biochem.*, 1969, **66**, 117.
33. T. Koyama, Y. Katsuki, and K. Ogura, *Bioorg. Chem.*, 1983, **12**, 58.
34. J. E. Reardon and R. H. Abeles, *J. Am. Chem. Soc.*, 1985, **107**, 4078.
35. M. Muehlbacher and C. D. Poulter, *J. Am. Chem. Soc.*, 1985, **107**, 8307.
36. J. E. Reardon and R. H. Abeles, *Biochemistry*, 1986, **25**, 5609.
37. M. Muehlbacher and C. D. Poulter, *Biochemistry*, 1988, **27**, 7315.
38. C. D. Poulter, M. Muehlbacher, and D. R. Davis, *J. Am. Chem. Soc.*, 1989, **111**, 3740.
39. X. J. Lu, D. J. Christensen, and C. D. Poulter, *Biochemistry*, 1992, **31**, 9955.
40. M. S. Anderson, M. Muehlbacher, I. P. Street, J. Proffitt, and C. D. Poulter, *J. Biol. Chem.*, 1989, **264**, 19 169.
41. I. P. Street and C. D. Poulter, *Biochemistry*, 1990, **29**, 7531.
42. I. P. Street, H. R. Coffman, J. A. Baker, and C. D. Poulter, *Biochemistry*, 1994, **33**, 4212.
43. M. P. Mayer, F. M. Hahn, D. J. Stillman, and C. D. Poulter, *Yeast*, 1992, **8**, 743.
44. F. M. Hahn and C. D. Poulter, *J. Biol. Chem.*, 1995, **270**, 11 298.
45. J. W. Xuan, J. Kowalski, A. F. Chambers, and D. T. Denhardt, *Genomics*, 1994, **20**, 129.
46. F. M. Hahn, J. W. Xuan, A. F. Chambers, and C. D. Poulter, *Arch. Biochem. Biophys.*, 1996, **332**, 30.
47. V. M. Blanc and E. Pichersky, *Plant Physiol.*, 1995, **108**, 855.
48. F. M. Hahn, J. A. Baker, and C. D. Poulter, *J. Bacteriol.*, 1996, **178**, 619.
49. L. Jeffries, M. A. Cawthorne, M. Harris, A. T. Diplock, J. Green, and S. A. Price, *Nature*, 1967, **215**, 257.
50. C. R. Benedict, J. Kett, and J. W. Porter, *Arch. Biochem. Biophys.*, 1965, **110**, 611.
51. J. K. Dorsey, J. A. Dorsey, and J. W. Porter, *J. Biol. Chem.*, 1966, **241**, 5353.
52. T. R. Green and C. A. West, *Biochemistry*, 1974, **13**, 4720.
53. I. Takahashi and K. Ogura, *J. Biochem.*, 1981, **89**, 1581.
54. T. Koyama, M. Matsubara, and K. Ogura, *J. Biochem.*, 1985, **98**, 449.
55. P. Hugueney and B. Camara, *FEBS Lett.*, 1990, **273**, 235.
56. L.-S. Yeh and H. C. Rilling, *Arch. Biochem. Biophys.*, 1977, **183**, 718.
57. T. Koyama, Y. Saito, K. Ogura, and S. Seto, *J. Biochem.*, 1977, **82**, 1585.
58. G. F. Barnard, B. Langton, and G. Popják, *Biochem. Biophys. Res. Commun.*, 1978, **85**, 1097.
59. G. F. Barnard and G. Popják, *Biochim. Biophys. Acta*, 1980, **617**, 169.
60. T. Koyama, S. Obata, K. Saito, A. Takeshita-Koike, and K. Ogura, *Biochemistry*, 1994, **33**, 12 644.
61. G. Popják and J. W. Cornforth, *Biochem. J.*, 1966, **101**, 553.
62. J. W. Cornforth, R. H. Cornforth, C. Donninger, and G. Popják, *Proc. R. Soc. London, B.*, 1966, **163**, 492.
63. J. W. Cornforth, R. H. Cornforth, G. Popják, and L. S. Yengoyan, *J. Biol. Chem.*, 1966, **241**, 3970.
64. G. Popják, P. W. Holloway, and J. M. Baron, *Biochem. J.*, 1969, **111**, 325.
65. G. Popják, J. L. Rabinowitz, and J. M. Baron, *Biochem. J.*, 1969, **113**, 861.
66. K. Ogura, T. Nishino, T. Koyama, and S. Seto, *J. Am. Chem. Soc.*, 1970, **92**, 6036.
67. T. Nishino, K. Ogura, and S. Seto, *Biochim. Biophys. Acta*, 1971, **235**, 322.

68. T. Nishino, K. Ogura, and S. Seto, *J. Am. Chem. Soc.*, 1971, **93**, 794.
69. T. Nishino, K. Ogura, and S. Seto, *J. Am. Chem. Soc.*, 1972, **94**, 6849.
70. T. Nishino, K. Ogura, and S. Seto, *Biochim. Biophys. Acta*, 1973, **302**, 33.
71. Y. Maki, H. Satoh, M. Kurihara, T. Endo, G. Watanabe, and K. Ogura, *Chem. Lett.*, 1994, 1841.
72. Y. Maki, M. Kurihara, T. Endo, M. Abiko, K. Saito, G. Watanabe, and K. Ogura, *Chem. Lett.*, 1995, 389.
73. Y. Maki, A. Masukawa, H. Ono, T. Endo, T. Koyama, and K. Ogura, *Bioorg. Med. Chem. Lett.*, 1995, **5**, 1605.
74. K. Ogura, T. Koyama, and S. Seto, *J. Chem. Soc., Chem. Commun.*, 1972, 881.
75. T. Koyama, K. Ogura, and S. Seto, *Chem. Lett.*, 1973, 401.
76. T. Koyama, M. Matsubara, and K. Ogura, *J. Biochem. Tokyo*, 1985, **98**, 457.
77. K. Ogura, A. Saito, and S. Seto, *J. Am. Chem. Soc.*, 1974, **96**, 4037.
78. A. Saito, K. Ogura, and S. Seto, *Chem. Lett.*, 1975, 1013.
79. Y. Maki, G. Watanabe, A. Saito, T. Koyama, and K. Ogura, *Chem. Lett.*, 1986, 1885.
80. T. Koyama, K. Ogura, and S. Seto, *J. Am. Chem. Soc.*, 1977, **99**, 1999.
81. T. Koyama, A. Saito, K. Ogura, and S. Seto, *J. Am. Chem. Soc.*, 1980, **102**, 3614.
82. M. Kobayashi, T. Koyama, K. Ogura, S. Seto, F. J. Ritter, and I. E. M. Brüggemann-Rotgans, *J. Am. Chem. Soc.*, 1980, **102**, 6602.
83. T. Koyama, K. Ogura, F. C. Baker, G. C. Jamieson, and D. A. Schooley, *J. Am. Chem. Soc.*, 1987, **109**, 2853.
84. J. W. Cornforth, *Agnew. Chem., Int. Ed. Engl.*, 1968, **7**, 903.
85. J. W. Cornforth, R. H. Cornforth, C. Donninger, G. Popják, G. Ryback, and G. J. Schroepfer, *Proc. R. Soc. Ser. B*, 1966, **163**, 437.
86. B. C. Reed and H. C. Rilling, *Biochemistry*, 1976, **15**, 3739.
87. C. D. Poulter and H. C. Rilling, *Biochemistry*, 1976, **15**, 1079.
88. D. N. Brems and H. C. Rilling, *J. Am. Chem. Soc.*, 1977, **99**, 8351.
89. C. D. Poulter, D. M. Satterwhite, and H. C. Rilling, *J. Am. Chem. Soc.*, 1976, **98**, 3376.
90. C. D. Poulter and D. M. Satterwhite, *Biochemistry*, 1977, **16**, 5470.
91. C. D. Poulter, J. C. Argyle, and E. A. Mash, *J. Biol. Chem.*, 1978, **253**, 7227.
92. C. D. Poulter, E. A. Mash, J. C. Argyle, O. J. Muscio, and H. C. Rilling, *J. Am. Chem. Soc.*, 1979, **101**, 6761.
93. T. Gotoh, T. Koyama, and K. Ogura, *Chem. Lett.*, 1987, 1627.
94. T. Gotoh, T. Koyama, and K. Ogura, *Biochem. Biophys. Res. Commun.*, 1988, **156**, 396.
95. T. Gotoh, T. Koyama, and K. Ogura, *J. Biochem.*, 1992, **112**, 20.
96. K. E. Stremler and C. D. Poulter, *J. Am. Chem. Soc.*, 1987, **109**, 5542.
97. D. N. Brems and H. C. Rilling, *Biochemistry*, 1979, **18**, 860.
98. D. N. Brems, E. Bruenger, and H. C. Rilling, *Biochemistry*, 1981, **20**, 3711.
99. B. T. Sheares, S. S. White, D. T. Molowa, K. Chan, V. D.-H. Ding, P. A. Kroon, R. G. Bostedor, and J. D. Karkas, *Biochemistry*, 1989, **28**, 8129.
100. D. J. Wilkin, S. Y. Kutsunai, and P. A. Edwards, *J. Biol. Chem.*, 1990, **265**, 4607.
101. M. S. Anderson, J. G. Yarger, C. L. Burck, and C. D. Poulter, *J. Biol. Chem.*, 1989, **264**, 19 176.
102. S. Fujisaki, H. Hara, Y. Nishimura, K. Horiuchi, and T. Nishino, *J. Biochem.*, 1990, **108**, 995.
103. T. Koyama, S. Obata, M. Osabe, A. Takeshita, K. Yokoyama, M. Uchida, T. Nishino, and K. Ogura, *J. Biochem.*, 1993, **113**, 355.
104. S. Attucci, S. M. Aitken, P. J. Gulick, and R. K. Ibrahim, *Arch. Biochem. Biophys.*, 1995, **321**, 493.
105. N. Cunillera, M. Arro, D. Delourme, F. Karst, A. Boronat, and A. Ferrer, *J. Biol. Chem.*, 1996, **271**, 7774.
106. K. Adilwilaga and A. Kush, *Plant Mol. Biol.*, 1996, **30**, 935.
107. L. Song and C. D. Poulter, *Proc. Natl. Acad. Sci., USA*, 1994, **91**, 3044.
108. P. F. Marrero, C. D. Poulter, and P. A. Edwards, *J. Biol. Chem.*, 1992, **267**, 21 873.
109. V. D.-H. Ding, B. T. Sheares, J. D. Bergstrom, M. M. Ponpipom, L. B. Perez, and C. D. Poulter, *Biochem. J.*, 1991, **275**, 61.
110. H. Nakane, T. Koyama, S. Obata, M. Osabe, A. Takeshita, T. Nishino, K. Ogura, and K. Miki, *J. Mol. Biol.*, 1993, **233**, 787.
111. A. Carattoli, N. Romano, P. Ballario, G. Morelli, and G. Macino, *J. Biol. Chem.*, 1991, **266**, 5854.
112. A. Joly and P. A. Edwards, *J. Biol. Chem.*, 1993, **268**, 26 983.
113. T. Koyama, M. Tajima, H. Sano, T. Doi, A. Koike-Takeshita, S. Obata, T. Nishino, and K. Ogura, *Biochemistry*, 1996, **35**, 9533.
114. L. Blanchard and F. Karst, *Gene*, 1993, **125**, 185.
115. T. Koyama, M. Tajima, T. Nishino, and K. Ogura, *Biochem. Biophys. Res. Commun.*, 1995, **212**, 681.
116. D. A. Dougherty, *Science*, 1996, **271**, 163.
117. T. Koyama, K. Saito, K. Ogura, S. Obata, and A. Takeshita, *Can. J. Chem.*, 1994, **72**, 75.
118. G. I. Barnard and G. Popják, *Biochim. Biophys. Acta*, 1981, **661**, 87.
119. S.-I. Ohnuma, M. Suzuki, and T. Nishino, *J. Biol. Chem.*, 1994, **269**, 14 792.
120. N. Misawa, M. Nakagawa, K. Kobayashi, S. Yamano, Y. Izawa, K. Nakamura, and K. Harashima, *J. Bacteriol.*, 1990, **172**, 6704.
121. S.-I. Ohnuma, T. Nakazawa, H. Hemmi, A.-M. Hallberg, T. Koyama, K. Ogura, and T. Nishino, *J. Biol. Chem.*, 1996, **271**, 10 087.
122. S.-I. Ohnuma, K. Narita, T. Nakazawa, C. Ishida, Y. Takeuchi, C. Ohto, and T. Nishino, *J. Biol. Chem.*, 1996, **271**, 30 748.
123. L. C. Tarshis, P. J. Proteau, B. A. Kellogg, J. C. Sacchettini, and C. D. Poulter, *Proc. Natl. Acad. Sci., USA*, 1996, **93**, 15 018.
124. A. A. Kandutsch, H. Paulus, E. Levin, and K. Block, *J. Biol. Chem.*, 1964, **239**, 2507.
125. D. L. Nandi and J. W. Porter, *Arch. Biochem. Biophys.*, 1964, **105**, 7.
126. K. Ogura, T. Shinka, and S. Seto, *J. Biochem.*, 1972, **72**, 1101.
127. H. Sagami, K. Ishii, and K. Ogura, *Biochem. Int.*, 1981, **3**, 669.
128. S. Clarke, *Annu. Rev. Biochem.*, 1992, **61**, 355.
129. H. Sagami, Y. Morita, and K. Ogura, *J. Biol. Chem.*, 1994, **269**, 20 561.

130. A. Chen and C. D. Poulter, *J. Biol. Chem.*, 1993, **268**, 11 002.
131. A. Tachibana, T. Tanaka, M. Taniguchi, and S. Oi, *Biosci. Biotech. Biochem.*, 1993, **57**, 1129.
132. G. A. Armstrong, M. Alberti, and J. E. Hearst, *Proc. Natl. Acad. Sci. USA*, 1990, **87**, 9975.
133. G. A. Armstrong, A. Schmidt, G. Sandmann, and J. E. Hearst, *J. Biol. Chem.*, 1990, **265**, 8329.
134. M. A. Nelson, G. Morelli, A. Carattoli, N. Romano, and G. Macino, *Mol. Cell. Biol.*, 1989, **9**, 1271.
135. S. K. Math, J. E. Hearst, and C. D. Poulter, *Proc. Natl. Acad. Sci. USA*, 1992, **89**, 6761.
136. G. Sandmann and N. Misawa, *FEMS Microbiol. Lett.*, 1992, **69**, 253.
137. M. Kuntz, S. Römer, C. Suire, P. Hugueney, J. H. Weil, R. Schantz, and B. Camara, *Plant J.*, 1992, **2**, 25.
138. A. Chen and C. D. Poulter, *Arch. Biochem. Biophys.*, 1994, **314**, 399.
139. P. A. Scolnik and G. E. Bartley, *Plant Physiol.*, 1994, **104**, 1469.
140. S. M. Aitken, S. Attucci, R. K. Ibrahim, and P. J. Gulick, *Plant Physiol.*, 1995, **108**, 837.
141. Y. Jiang, P. Proteau, C. D. Poulter, and S. Ferro-Novick , *J. Biol. Chem.*, 1995, **270**, 21 793.
142. S.-I. Ohnuma, K. Hirooka, H. Hemmi, C. Ishida, C. Ohto, and T. Nishino, *J. Biol. Chem.*, 1996, **271**, 18 831.
143. A. Tzagoloff, A. Akai, and R. B. Needlemann, *J. Bacteriol.*, 1975, **122**, 826.
144. S.-I. Ohnuma, K. Hirooka, C. Ohto, and T. Nishino, *J. Biol. Chem.*, 1997, **272**, 5192.
145. D. V. Banthorpe, G. A. Bucknall, H. J. Doonan, S. Doonan, and M. G. Rowan, *Phytochemistry*, 1976, **15**, 91.
146. L. Heide and U. Berger, *Arch. Biochem. Biophys.*, 1989, **273**, 331.
147. R. Croteau and P. T. Purkett, *Arch. Biochem. Biophys.*, 1989, **271**, 524.
148. T. Endo and T. Suga, *Phytochemistry*, 1992, **31**, 2273.
149. A. Tachibana, *FEBS Lett.*, 1994, **341**, 291.
150. M. De Rosa, A. Gambacorta, B. Nicolaus, H. N. M. Ross, W. D. Grant, and J. D. BúLock, *J. Gen. Microbiol.*, 1982, **128**, 343.
151. H. Fujii, T. Koyama, and K. Ogura, *J. Biol. Chem.*, 1982, **257**, 14 610.
152. I. Takahashi, K. Ogura, and S. Seto, *J. Biol. Chem.*, 1980, **255**, 4539.
153. H. Fujii, T. Koyama, and K. Ogura, *FEBS Lett.*, 1983, **161**, 257.
154. I. Yoshida, T. Koyama, and K. Ogura, *Biochemistry*, 1987, **26**, 6840.
155. I. Yoshida, T. Koyama, and K. Ogura, *Biochem. Biophys. Res. Commun.*, 1989, **160**, 448.
156. I. Yoshida, T. Koyama, and K. Ogura, *Biochim. Biophys. Acta*, 1989, **995**, 138.
157. M. Ito, M. Kobayashi, T. Koyama, and K. Ogura, *Biochemistry*, 1987, **26**, 4745.
158. A. Koike-Takeshita, T. Koyama, S. Obata, and K. Ogura, *J. Biol. Chem.*, 1995, **270**, 18 396.
159. A. Moir and D. A. Smith, *Annu. Rev. Microbiol.*, 1990, **44**, 531.
160. Y.-W. Zhang, T. Koyama, and K. Ogura, *J. Bacteriol.*, 1997, **179**, 1417.
161. N. Shimizu, T. Koyama, and K. Ogura, *J. Bacteriol.*, in press.
162. H. Sagami, K. Ogura, and S. Seto, *Biochemistry*, 1977, **16**, 4616.
163. S.-I. Ohnuma, T. Koyama, and K. Ogura, *J. Biol. Chem.*, 1991, **266**, 23 706.
164. S. Fujisaki, T. Nishino, and H. Katsuki, *J. Biochem.*, 1986, **99**, 1327.
165. K. Ishii, H. Sagami, and K. Oguri, *Biochim. Biophys. Acta*, 1985, **835**, 291.
166. H. Fujii, H. Sagami, T. Koyama, K. Ogura, S. Seto, T. Baba, and C. M. Allen, *Biochem. Biophys. Res. Commun.*, 1980, **96**, 1648.
167. S.-I. Ohnuma, T. Koyama, and K. Ogura, *J. Biochem.*, 1992, **112**, 743.
168. S.-I. Ohnuma, T. Koyama, and K. Ogura, *Biochem. Biophys. Res. Commun.*, 1993, **192**, 407.
169. Y.-L. Choi, T. Nishida, M. Kawamukai, R. Utsumi, H. Sakai, and T. Komano, *J. Bacteriol.*, 1989, **171**, 5222.
170. J.-H. Jeong, M. Kitakawa, S. Isono, and K. Isono, *DNA Sequence*, 1993, **4**, 59.
171. K.-I. Asai, S. Fujisaki, Y. Nishimura, T. Nishino, K. Okada, T. Nakagawa, M. Kawamukai, and H. Matsuda, *Biochem. Biophys. Res. Commun.*, 1994, **202**, 340.
172. T. Kurokawa, K. Ogura, and S. Seto, *Biochem. Biophys. Res. Commun.*, 1971, **45**, 251.
173. I. Takahashi and K. Ogura, *J. Biochem.*, 1982, **92**, 1527.
174. S. Fujisaki, T. Nishino, and H. Katsuki, *J. Biochem.*, 1986, **99**, 1327.
175. M. V. Keenan and C. M. Allen, Jr., *Arch. Biochem. Biophys.*, 1974, **161**, 375.
176. C. M. Allen, Jr., M. V. Keenan, and J. Sack, *Arch. Biochem. Biophys.*, 1976, **175**, 236.
177. T. Baba and C. M. Allen, Jr., *Biochemistry*, 1978, **17**, 5598.
178. J. D. Muth and C. M. Allen, *Arch. Biochem. Biophys.*, 1984, **230**, 49.
179. K. Ishii, H. Sagami, and K. Ogura, *Biochem. J.*, 1986, **233**, 773.
180. T. Koyama, T. Kokubun, and K. Ogura, *Phytochemistry*, 1988, **27**, 2005.
181. W. L. Adair, Jr., N. Cafmeyer, and R. K. Keller, *J. Biol. Chem.*, 1984, **259**, 4441.
182. Z. Chen, C. Morris, and C. M. Allen, *Arch. Biochem. Biophys.*, 1988, **266**, 98.
183. Y. E. Bukhtiyarov, Y. A. Shabalin, and I. S. Kulaev, *J. Biochem.*, 1993, **133**, 721.
184. S. Matsuoka, H. Sagami, A. Kurisaki, and K. Ogura, *J. Biol. Chem.*, 1991, **266**, 3464.
185. B. L. Archer, B. G. Audley, E. G. Cockbain, and G. P. McSweeney, *Biochem. J.*, 1963, **89**, 565.
186. K. Cornish and R. A. Backhaus, *Phytochemistry*, 1990, **29**, 3809.
187. M. Kobayashi, M. Ito, T. Koyama, and K. Ogura, *J. Am. Chem. Soc.*, 1985, **107**, 4588.

2.05
Monoterpene Biosynthesis

MITCHELL L. WISE and RODNEY CROTEAU
Washington State University, Pullman, WA, USA

2.05.1 INTRODUCTION

2.05.1.1 Historical Perspective

The study of terpenoids has occupied the attention of organic chemists since the early part of the nineteenth century. Such eminent chemists as Kekulé, Perkin, and Boyle dedicated much of their careers to the determination of terpenoid structures and their synthesis. Prior to the advent of chromatographic methods and modern analytical tools, such as mass spectrometry and NMR spectroscopy, the higher terpenes proved far too complex for structural determination; hence monoterpenes were, to a large extent, the focus of much of the early efforts in terpenoid chemistry. An early advance in terpenoid chemistry came from investigations by Wallach on the pyrolysis products of turpentine. His work resulted in the recognition of isoprene as the universal building block of all terpenoids and led him to formulate the "isoprene rule" (i.e., that the terpenoids known at the time could be constructed by head-to-tail joining of isoprene units[1]).

Nearly half a century later, Ruzicka proposed a unifying concept for the formation of nearly all terpenoids in his landmark description of the isoprene rule as it relates to the biogenesis of terpenic compounds.[2] This concept, based on chemically sound rationale, provided a framework upon which to analyze the mechanisms involved in the biosynthesis of cyclic terpenoids and it has, for the most part, withstood the test of time and experiment. At the time Ruzicka proposed his "biogenetic isoprene rule," the actual biosynthetic precursor of the terpenoids (isopentenyl diphosphate) was not known, therefore he illustrated the cyclization reactions using either an acyclic alcohol or alkene (i.e., geraniol (**26**) or myrcene (**22**) in the case of monoterpenes) as substrate. Moreover, the precise nature of the reactions was not established; Ruzicka proposed both cationic as well as radical mechanisms for the ring closures. As will be discussed in Section 2.05.5.3, it was many years before stepwise cationic processes were conclusively demonstrated as the mechanism responsible for most monoterpene cyclization reactions. Nevertheless, Ruzicka's biogenetic isoprene rule presaged the key role that the enzyme must play in bringing about the required "constellation" (= conformation) of the precursor to allow cyclization to occur. Such conformational considerations limit the number of carbon skeletons possible and have accurately predicted the structures actually found most abundantly in nature. In addition, the biogenetic isoprene rule recognizes the inherent facility of these electrophilic reactions to allow hydride shifts and Wagner–Meerwein type rearrangements, thereby providing for the formation of additional skeletal types, including those that do not strictly follow Wallach's isoprene rule in the head-to-tail joining of isoprene units. Ruzicka named geraniol as the parent compound in his monoterpene biogenetic scheme. However, his original paper illustrated the *cis* isomer, nerol (**1**), as the immediate precursor to cyclization (Scheme 1); geraniol, the *trans* isomer, is prevented from direct cyclization due to topological constraints.

Advances in terpenoid biochemistry, particularly as applied to mammalian systems, demonstrated the central role of allylic (prenyl) diphosphates in terpenoid metabolism and it soon became clear that geranyl diphosphate (**16**, Scheme 2) was the first C_{10} intermediate to arise in this universal pathway.[3] Efforts to rationalize the intermediacy of the *cis* isomer, neryl diphosphate (**108**, Scheme 18), in the conversion of geranyl diphosphate to the cyclic monoterpenes resulted in several proposals: (i) direct isomerization of geranyl diphosphate to neryl diphosphate by a specific isomerase;[4] (ii) redox conversion of geraniol to geranial followed by isomerization to neral and, ultimately, reduction to nerol;[5,6] and (iii) direct biosynthesis of neryl diphosphate via the condensation of isopentenyl diphosphate and dimethylallyl diphosphate thereby eliminating geranyl diphosphate as an intermediate.[7,8] Little definitive experimental evidence has been obtained for any of these alternatives. Because of the topological barrier to the direct cyclization of geranyl diphosphate and the ease of cyclization of neryl and linalyl derivatives in chemical model systems (see, for example, Ref. 9), recognition of geranyl diphosphate (**16**) as the true precursor to the cyclic monoterpenes took years to develop. Early biosynthetic studies utilizing crude, cell-free extracts to analyze the substrate specificities of monoterpene cyclases only added to the confusion. Many early reports indicated that

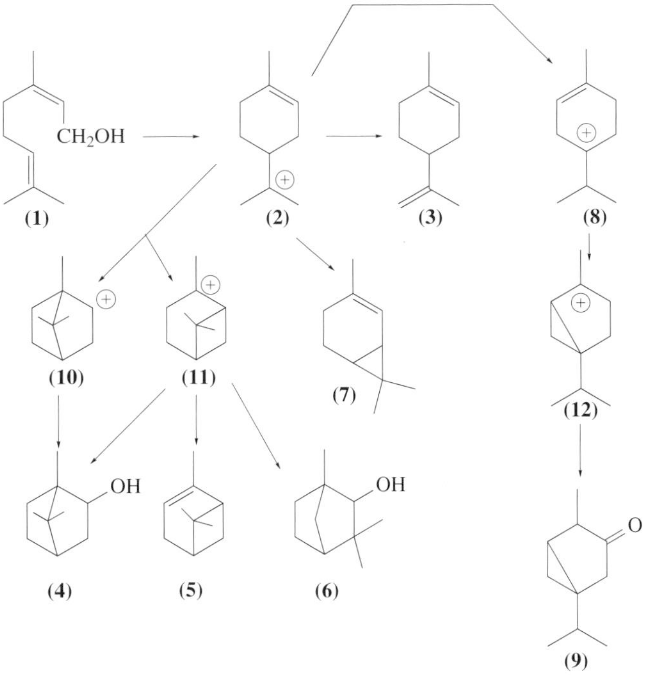

Scheme 1

these enzymes showed a clear preference for neryl diphosphate (**108**) and, in some cases, for linalyl diphosphate (**18**).[10,11] Unfortunately, in most of these studies, the ultimate fate of the precursors was not rigorously determined and the activity of contaminating phospho- and diphosphohydrolases, found ubiquitously in plant cell extracts, was not fully appreciated. In a study specifically designed to evaluate the activity of phosphohydrolases in sage (*Salvia officinalis*) leaves, it was demonstrated that a crude, cell-free homogenate was far more efficient at hydrolyzing geranyl diphosphate than neryl diphosphate.[12] From these experiments, it became obvious that earlier efforts to evaluate the relative incorporation of geranyl, neryl, and linalyl diphosphates into monoterpenes using crude cell-free extracts were, for the most part, seriously flawed. It was only after techniques evolved to separate the monoterpene synthases from competing phosphohydrolases that it became clear that geranyl diphosphate is, in fact, the more efficient substrate (compared to neryl diphosphate) for most (if not all) of the cyclases described to date.[13,14] Linalyl diphosphate is generally the kinetically preferred substrate for these reactions; it is now known that this compound is an essential intermediate in the cyclization of geranyl diphosphate (see below).

The recognition of geranyl diphosphate as the universal precursor of cyclic monoterpenes required a rational mechanism to explain how the steric barrier to cyclization posed by the *trans* double bond is overcome. The postulated involvement of linalyl diphosphate as an intermediate in the biosynthesis of the cyclic monoterpenes from geranyl diphosphate evolved from contemporary studies by Poulter on the mechanism of prenyltransferases. Thus, farnesyl diphosphate synthase had been conclusively demonstrated to act through a delocalized carbocation–diphosphate ion pair in the condensation of dimethylallyl diphosphate (**13**) (and geranyl diphosphate) with isopentenyl diphosphate (**14**, Scheme 2).[3] The cyclization mechanism postulated by Croteau[13] invoked ionization to and collapse of such a geranyl cation–diphosphate ion pair (**17**) to form linalyl diphosphate (**18**), thereby permitting rotation about the newly formed C-2—C-3 single bond from *transoid* to *cisoid* form. This ionization–isomerization step allows juxtaposition of the C-6═C-7 double bond and the delocalized carbocation formed upon subsequent ionization of the enzyme-bound linalyl diphosphate intermediate, thus permitting cyclization of the six-membered ring (Scheme 2). This intramolecular reaction sequence for the coupled isomerization and cyclization of geranyl diphosphate was, in many respects, analogous to the intermolecular reaction catalyzed by prenyltransferase, and was entirely consistent with mechanistic considerations for cyclization of the corresponding C_{15} and C_{20} prenyl diphosphates in the sesquiterpene and diterpene series being simultaneously developed by Cane.[15] Recognition that the monoterpene synthases were capable of isomerizing geranyl diphosphate to form enzyme-bound linalyl diphosphate (or its ion-paired equivalent) within the catalytic

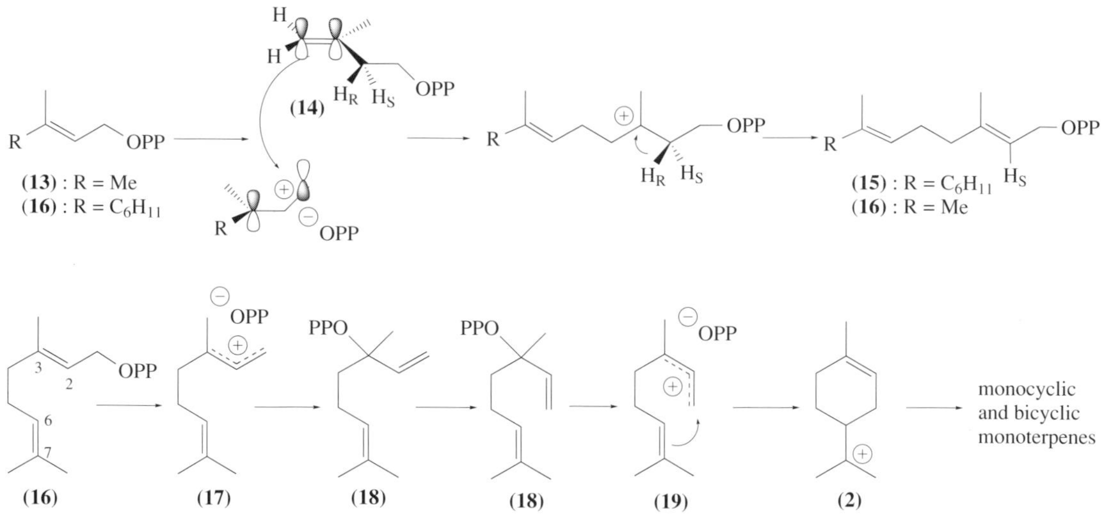

Scheme 2

pocket provided the necessary insight to model a general mechanism by which the common precursor, geranyl diphosphate (**16**), could be cyclized to most monoterpenes in accordance with Ruzicka's biogenetic isoprene rule. This isomerization–cyclization sequence (Scheme 2) rationalizes the deployment of geranyl diphosphate as the universal substrate, accounts for the lack of free neryl or linalyl diphosphates or other free intermediates in the reaction, unifies the mechanistic possibilities for the formation of cyclohexanoid monoterpenes and, as demonstrated throughout this chapter, provides a consistent explanation, in explicit stereochemical terms, for virtually all of the chemical and biochemical data presented to date relating to the biosynthesis of cyclic monoterpenes.

2.05.1.2 Occurrence and Distribution in Nature

In a biosynthetic sense, the monoterpenes represent the second branch of the terpenoid pathway, departing from the central pathway after the formation of isoprene and other hemiterpenes (C_5 family). In being constructed from only 10 carbon atoms, the monoterpenes are somewhat restricted in the diversity of skeletal structures possible relative to the sesquiterpenes (C_{15}) and diterpenes (C_{20}) which allow correspondingly greater reaction flexibility with increasing chain length and number of double bonds. Nevertheless, all of the skeletal types that are mechanistically feasible to form in the monoterpene series are known among the 500 or so naturally occurring compounds. This structural diversity arises not only from the manifold cyclization pathways (with attendant internal electrophilic additions to the remaining double bond, hydride shifts, and Wagner–Meerwein rearrangements, accompanied by terminations via deprotonation or nucleophile capture) but also from a vast array of secondary, largely redox, transformations of these parent compounds (see Section 2.05.6).

Although found in a wide range of organisms, including bacteria, fungi, algae, insects, and even higher animals such as alligators and beavers (where they undoubtedly accumulate from dietary sources or by catabolic processes[16]), monoterpenes are most widely produced by terrestrial plants where they are best known as components of the flower scents, essential oils, and turpentines. The monoterpenes are synthesized and accumulated in significant amounts ($>0.01\%$ fresh tissue weight) by some 100 families of nonvascular and vascular plants ranging from liverworts (Bryophyta), to gymnosperms, to highly evolved members of the angiosperms (e.g., Orchidaceae (monocot), Asteraceae (dicot), and Lamiaceae (dicot)), and they seem to be produced in at least trace amounts by essentially all plants, including crop species. There are no very useful phylogenetic correlations, and, given the range in species of monoterpene producers, it seems possible that the ability to biosynthesize these natural products arose independently several times in the course of plant evolution.

2.05.1.3 Natural Functions and Industrial Applications

Because the monoterpenes do not appear to play a role in the basic metabolic processes of plant growth and development, they have long been considered "secondary metabolites." There is, however, a growing recognition that these natural products play critical roles in the chemical ecology of the producing organisms as defenses against pathogens and herbivores, attractants for pollinators and seed dispersers, and as allelopathic agents (e.g., competitive phytotoxins).[17–19] Owing to the relatively high volatility of monoterpenes, their inherent structural complexity and their hydrophobic character, these metabolites are ideally suited for such communication between species. Determining the ecological roles of monoterpenes is complicated by the fact that these compounds frequently act in concert with each other or other metabolites.[20,21] Moreover, secondary products may be exploited as chemical defenses either directly as toxins or feeding deterrents, or through tritrophic interactions, i.e., by attraction of either predators or parasitoids of the offending herbivores.[22–24] Certain insect species have evolved the ability to detoxify monoterpenes[25] and, in some cases, monoterpenes, which probably developed as allomones (compounds beneficial to the emitting species by deterring pests or inhibiting competing species), have evolved into kairomones (compounds emitted by one species which are beneficial to a receiving species, for example as attractants, but harmful to the emitting species).

The roles of terpenoids in the interactions between bark beetles (Coleoptera: Scolytidea) and conifer species are perhaps the most thoroughly studied and provide an excellent illustration of the complexity of the chemical ecology of terpenoids. Conifer resins are mixtures of monoterpenes, sesquiterpenes, and diterpenoid acids. These resins are secreted as a response to physical wounding as well as fungal infection.[26,27] The efficacy of the oleoresin in determining host selection and repelling bark beetle infestation is dependent on several factors, including the composition and viscosity of the resin, crystallization time, and flow rate.[18] The chemical composition of the resin, particularly the monoterpene constituents, is critical in determining its toxicity, with certain monoterpenes or combinations of monoterpenes being more or less toxic to the invading pests.[18,21] Interestingly, some species of bark beetles, such as the mountain pine beetle (*Dendroctonus ponderosae*), utilize a host monoterpene (pinene) to synthesize the sex attractant pheromone, *trans*-verbenol[28] (**20**). Likewise, species of *Ips* beetles synthesize the aggregation pheromone (+)-ipsdienol (**21**) from the acyclic monoterpene myrcene[29] (**22**). The outcome of these interactions, involving monoterpene-based host selection, insect toxicity, pheromone signaling, and tritrophic-level effects, is highly complex and depends upon the magnitude of the insect infestation, the health of the conifer host, and the vigor of the response, along with numerous other factors related to the precise composition of the host oleoresin.

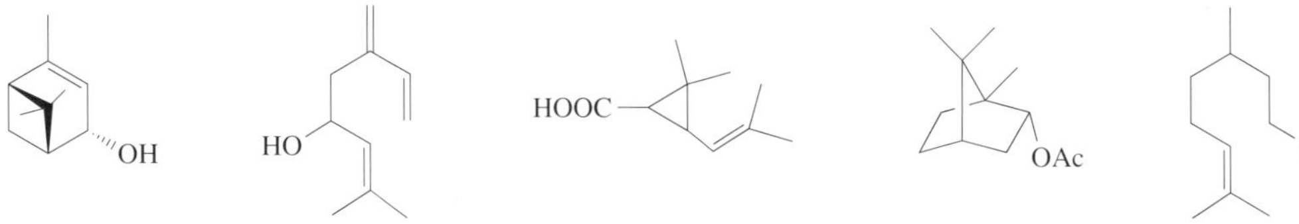

OH HO HOOC OAc OH

(20) **(21)** **(23)** **(24)** **(25)**

Other examples of monoterpene allomones include the pyrethroids (e.g., **23**), bornyl acetate (**24**), citronellol (**25**), and pulegone (**128**, Scheme 33, later). Long recognized for their insecticidal properties, the pyrethroids of chrysanthemum species act as insect neurotoxins, yet have negligible toxicity to birds and mammals, and they now serve as the basis for several commercial insecticides.[30] Bornyl acetate in Douglas fir (*Pseudotsuga menziesii*) needle resin reduces the survival rates of the spruce budworm (*Christoneura occidentalis*), a significant pest of Western US forests.[31] Citronellol, produced by many plant species, is an oviposition deterrent for the leafhopper (*Amrasca devastans*),[32] and pulegone, from pennyroyal mint (*Mentha pulegium*), is an insect feeding deterrent.[33]

In addition to the natural, and in many cases agronomically significant, functions of the monoterpenes, these compounds have been highly valued as medicinals, as flavorants and perfumery materials, and as industrial feedstocks that have been used by humans since antiquity.[34] Camphor (**82**), for example, has been used since prehistoric times for its reputed medicinal properties. Trade between Europe and the Middle East as early as the tenth century featured camphor as one of the principal commodities.[34] During the mid-twentieth century, camphor found extensive use as a plasticizer in the manufacture of celluloid and explosives.[34] During the era of wooden sailing ships, conifer resin and turpentine formed the basis of the "naval stores" industry, and these materials provided the principal feedstocks to the chemical industry until the commercial utilization of petroleum in the 1860s.

From a current industrial perspective, the most common sources of the monoterpenes are the herbaceous plant essential oils and conifer turpentines. Essential oils are those volatile components extracted or distilled from odorous plant materials; monoterpenes generally comprise a large proportion of the essential oils. At present, the most important source of monoterpenes, in terms of production mass, are those found in turpentine. Currently, the global production of turpentine is in the range of 3×10^5 metric tons annually, largely in the form of crude sulfate turpentine[35] as a by-product of the pulp industry. The pinene regioisomers, α-pinene (**5**) and β-pinene (**107**, Scheme 17), are the two principal monoterpenes of turpentine, and they serve as large volume, low cost aroma chemicals. These two compounds also provide industrial starting materials for the synthesis of a variety of other fragrance substances.[35–37]

Other commercial sources of monoterpenes include the essential oils expressed from peels of orange (*Citrus sinensis*), lemon (*C. limon*), and lime (*C. aurantium*), and isolated by steam distillation of peppermint (*Mentha piperita*), spearmint (*M. spicata*), and a broad range of other aromatic plants.[36] Although these more specialized essential oils are produced at a fraction of the scale of the turpentine industry, they usually command substantially higher prices in the marketplace because of their value in flavoring and perfumery. The production of these essential oils, as well as the raw herbs and spices from which they are derived, represent viable agricultural enterprises in many regions of the world.

Although, historically, the main utility of monoterpenes has been as flavorants and odorants, there is a constant search for the application of these renewable materials in other arenas. For example, the use of tall oil turpentine as a diesel fuel additive is under exploration.[38] In a pharmaceutical application, limonene and perillyl alcohol have shown promise as both chemopreventive and chemotherapeutic agents for certain solid tumors.[39,40] The application of monoterpenoid pheromones as insect control agents is also under investigation.[28] In this regard, the significance of monoterpenes, as natural pest deterrents, in protecting valuable timber resources should not be overlooked.[41]

2.05.1.4 Skeletal Types and Chemical Classes

The monoterpenes are generally regarded as occurring in two broad skeletal classifications, regular and irregular. The regular monoterpenes are derived from geranyl diphosphate (**16**), thereby exhibiting a 1′–4 coupling of dimethylallyl diphosphate (**13**) and isopentenyl diphosphate (**14**) and thus the typical "head-to-tail" fusion of isoprene units (Scheme 2). Less common, but nevertheless abundantly represented in nature, are the irregular monoterpenes. These compounds do not fit the typical head-to-tail assembly pattern of isoprene units, and arise from rearrangement of "regular" types or from non-head-to-tail condensations of two molecules of dimethylallyl diphosphate; also generally included within this group are the degraded monoterpenes that contain less than 10 carbons. Within each of these major classes are a number of defined skeletal types. For the sake of completeness, the iridane monoterpenes, as well as the rather unusual halogenated monoterpenes from marine sources, are briefly described here.

2.05.1.4.1 Acyclic monoterpenes

Among the regular monoterpenes, the acyclic forms constitute a relatively small family. These compounds include the alcohols geraniol (**26**), nerol (**1**), and linalool (**27**), and the trienes myrcene (**22**) and the ocimenes (**28** and **29**), all of which are very likely biosynthesized more or less directly from geranyl diphosphate. In the case of geraniol, ample evidence exists for phosphohydrolases which efficiently convert geranyl diphosphate to the alcohol.[12] Nerol may derive via citral (**30**) as the result of redox metabolism,[6,16] and, as mentioned earlier (Section 2.05.1.1), isomerases isolated from several plant species have been found capable of isomerizing geraniol and geranyl phosphate to nerol and neryl phosphate, respectively.[4] A linalool synthase which converts geranyl diphosphate to (*S*)-linalool has recently been isolated from *Clarkia breweri*[42,43] and the corresponding gene has been cloned.[44] This enzyme shows a marked preference for geranyl diphosphate over (*S*)-linalyl diphosphate as precursor and is essentially unreactive with (*R*)-linalyl diphosphate, suggesting direct formation of linalool from the geranyl substrate.[43] The formation of the acyclic alkenes from geranyl diphosphate, often as minor side-products of the cyclization of this precursor[14] has been amply demonstrated. Derivatives formed by simple oxidation (*trans*-citral, **30**) or reduction (citronellol, **25**) of the parent acyclic monoterpenes are well known and widely distributed.

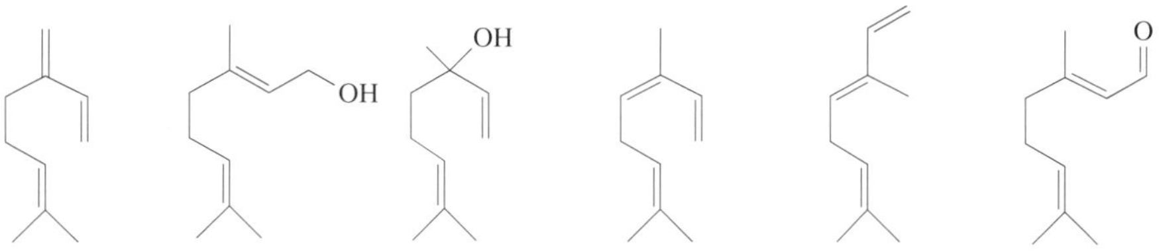

OH OH O

(22) **(26)** **(27)** **(28)** **(29)** **(30)**

2.05.1.4.2 p-*Menthane monoterpenes*

p-Menthane monoterpenes are characterized by the 1-methyl-4-isopropyl cyclohexane skeleton and they represent the largest group of naturally occurring monoterpenes. The *p*-menthadienes, limonene (**3**), terpinolene (**31**), α-terpinene (**32**), γ-terpinene (**33**), α-phellandrene (**35**), and β-phellandrene (**36**), are illustrated in Scheme 3. These alkenes all result from deprotonation of a *p*-menthenyl cation formed either directly upon cyclization or by hydride shift(s) in the cyclic cation. Also included in this structural class are a number of oxygenated compounds, such as 1,8-cineole (**116**, Scheme 31), α-terpineol (**37**), carvone (**133**, Scheme 33), and pulegone (**128**, Scheme 33). Early investigators questioned whether these metabolites arose independently from an acyclic precursor or whether they were formed by subsequent modification of a monocyclic parent compound.[5,16] In the case of 1,8-cineole and α-terpineol, as discussed in Section 2.05.5.4, conclusive evidence has been presented that these monoterpenes result from the direct enzymatic cyclization of geranyl diphosphate.[45] The aromatic *p*-menthane monoterpenes, *p*-cymene (**38**) and the hydroxylated derivatives thymol (**39**) and carvacrol (**40**), always co-occur with α-terpinene, γ-terpinene, and terpinen-4-ol (**41**) and are, in fact, derived by the desaturation of γ-terpinene followed by hydroxylation.[16,46] Secondary transformations in monoterpene metabolism are described in detail in Section 2.05.6.

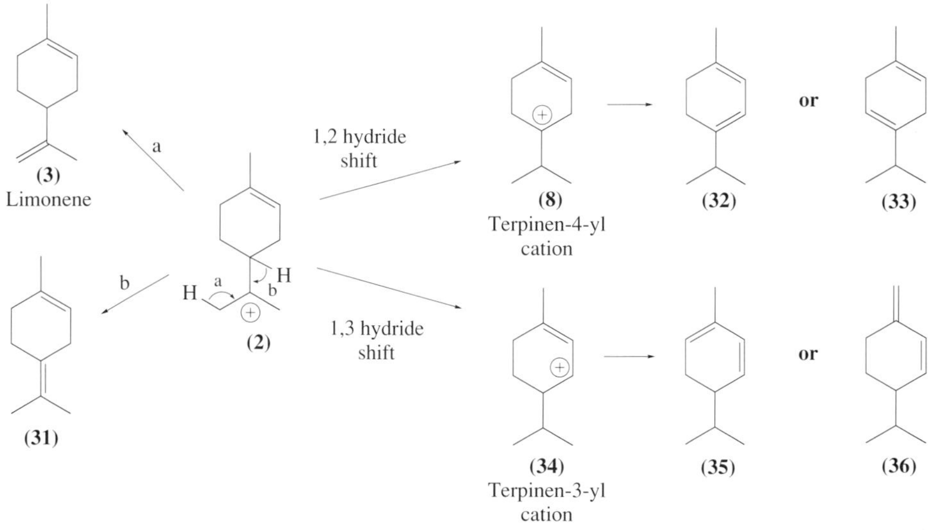

(3) Limonene a b **(31)** H a b H **(2)** 1,2 hydride shift 1,3 hydride shift **(8)** Terpinen-4-yl cation **(32)** or **(33)** **(34)** Terpinen-3-yl cation **(35)** or **(36)**

Scheme 3

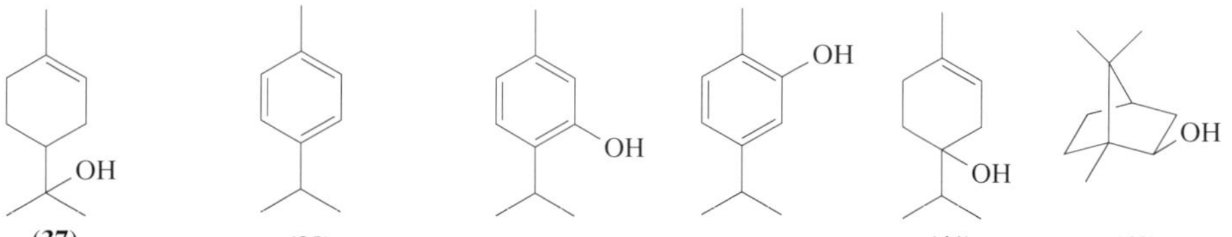

OH OH OH OH OH

(37) **(38)** **(39)** **(40)** **(41)** **(42)**

2.05.1.4.3 *Pinane monoterpenes*

The pinanes are bicyclic monoterpenes resulting from intramolecular Markovnikov addition to the cyclohexenyl double bond of the α-terpinyl cation (**2**), yielding the 3.1.1 bicyclic system with the cationic center at the tertiary position. Subsequent deprotonation of this pinyl cation, either from the adjacent methylene or methyl group, produces α-pinene (**5**) and β-pinene (**107**, Scheme 17), respectively. These two compounds, constituting the bulk of most turpentines, are perhaps the most abundant monoterpenes in nature. Both enantiomers of α-pinene occur naturally (with (+)-(1*R*,5*R*)-α-pinene usually predominating), whereas β-pinene almost always occurs as the nearly optically pure (−)-(1*S*,5*S*)-enantiomer. These observations led to early recognition that at least two enzymatic routes to the pinanes must exist;[5] experimental evidence, obtained using partially purified enzymes[47,48] and, most recently, cloned, heterologously expressed proteins, has borne out these predictions.[49,50] Oxygenated pinane monoterpenes, such as verbenol (**20**), are formed by secondary transformation of α- or β-pinene.[51]

2.05.1.4.4 *Bornane, camphane, and fenchane monoterpenes*

The bornanes, camphanes, and fenchanes are characterized by a 2.2.1 bicyclic carbon skeleton. Prominent members of these families include borneol (**4**), isoborneol (**42**), camphor (**82**), fenchone (**43**), and camphene (**45**). The 2.2.1 bicyclic systems are formed by secondary cyclization of the α-terpinyl cation (**2**) by either of two routes, anti-Markovnikov addition via the C-1–C-2 alkene to form the bornyl cation (**10**) (which by Wagner–Meerwein shift yields the isocamphyl cation (**47**) and by deprotonation yields camphene (**45**)), or Markovnikov addition to generate the pinyl cation (**11**) followed by a Wagner–Meerwein rearrangement. This second alternative could result in formation of either the bornyl or the fenchyl skeleton depending upon how the Wagner–Meerwein shift occurs (Scheme 4). The biosynthesis of fenchone in fennel (*Foeniculum vulgare*) has been shown to occur via rearrangement of the pinyl cation intermediate with water capture[52] to form *endo*-fenchol (**6**), which is subsequently oxidized to fenchone (**43**) by a separate NAD-dependent dehydrogenase.[53] Borneol, however, is produced by the hydrolysis of bornyl diphosphate (**44**) which is derived by cyclization of the α-terpinyl cation to the bornyl cation with internal return of the diphosphate moiety.[54] Enzymatic oxidation of borneol provides camphor.[12]

2.05.1.4.5 *Thujane monoterpenes*

Thujane family monoterpenes bear the a 1-isopropyl-4-methyl-bicyclo[3.1.0]hexane skeleton. Although these compounds possess interesting and characteristic odor properties, thujanes are of relatively little commercial significance. Because the occurrence of cyclopropane rings in natural products is somewhat unusual,[55] the thujanes have found extensive use as models for physical organic studies.[56]

The biosynthesis of most thujane monoterpenes involves a hydride shift in the α-terpinyl cation (**2**) to form the terpinen-4-yl cation[57] (**8**) (Scheme 5) followed by internal addition via the endocyclic double bond[58–60] to create the cyclopropane ring of the sabinyl cation (**12**). α-Thujene (**48**) and sabinene (**49**) are produced via deprotonation at the adjacent methylene or methyl groups, respectively, whereas stereospecific capture of the sabinyl cation by water yields either *cis*- or *trans*-sabinene hydrate[61] (**113**, Scheme 27). Sabinol (**139**, Scheme 35) arises from sabinene by the action of an NADPH/O_2-dependent cytochrome P450 hydroxylase,[62] and oxidation of the alcohol by a specific dehydrogenase produces sabinone[63] (**140**). Subsequent steps in the biosynthesis of C-3-oxygenated thujane monoterpenes (i.e., stereoselective reduction of the sabinone alkenic group to (+)-3-thujone or (−)-3-thujone (**9**), and further reduction of the carbonyl function to the thujyl alcohol isomers) have not been experimentally examined; however, ample evidence exists for similar allylic oxidation–conjugate reduction pathways among other monoterpene families.[14,61] In thyme (*Thymus vulgaris*), α-thujene (**48**) is formed by direct deprotonation at C-6 of the terpinen-4-yl cation[64] (**8**). This unusual route to cyclopropane synthesis likely represents a variant on the normal C-5 deprotonation catalyzed by the cyclase enzyme in the formation of the principal product, γ-terpinene (**33**).

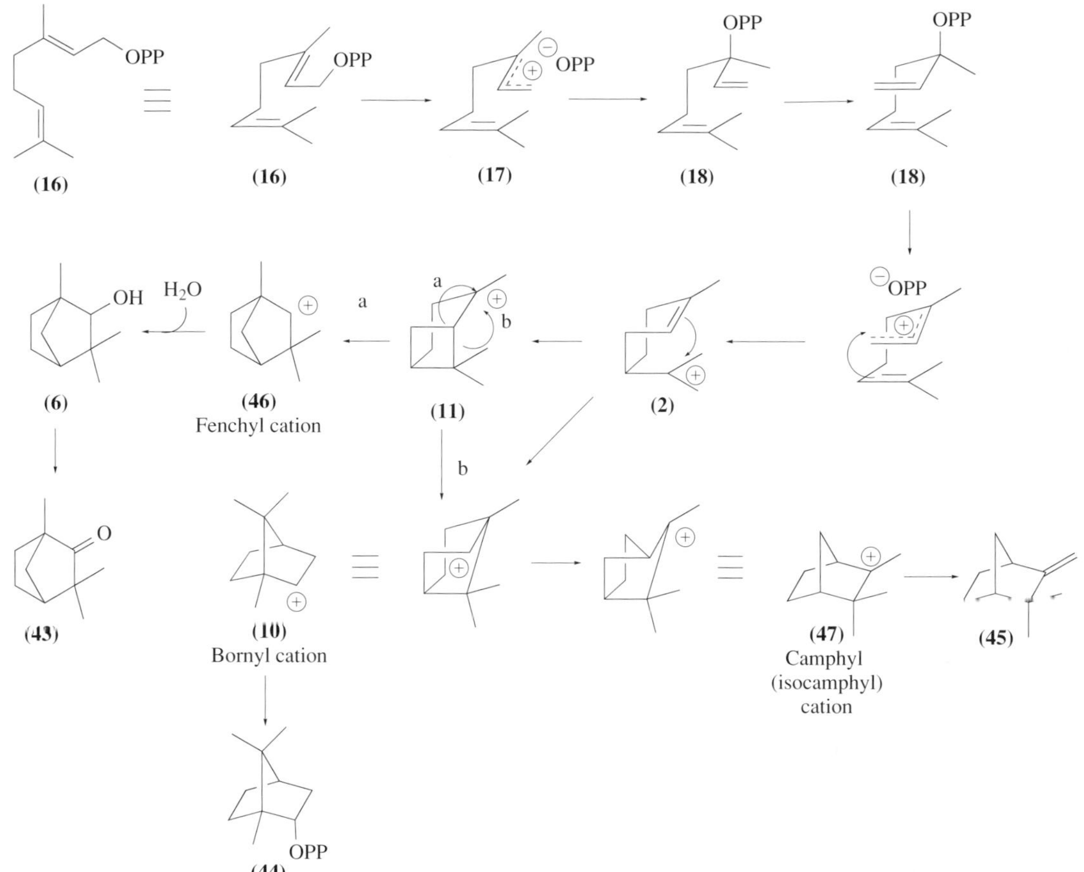

Scheme 4

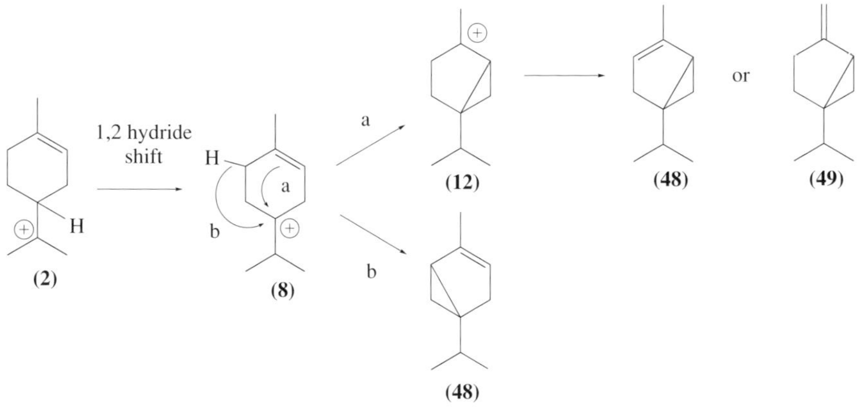

Scheme 5

2.05.1.4.6 Carane monoterpenes

The carane family of monoterpenes represents another example of a cyclopropane functionality, this time in a bicyclo [4.1.0] skeleton. Ruzicka originally hypothesized the involvement of a 1,3 elimination in the α-terpinyl cation to form the carane skeleton (i.e., 3-carene, **7**). The co-occurrence of 3-carene and terpinolene (**31**) (presumably representing two alternatives for proton loss from C-5 or C-4, respectively, of the α-terpinyl cation) was taken as supporting evidence for this mechanism.[5] This proposal came into dispute after detailed degradation studies demonstrated an anomalous labeling pattern in 3-carene (**7**) formed in *Pinus palustris* and *P. sylvestris* from [2-^{14}C]mevalonic acid (**50**).[65] However, subsequent *in vitro* experiments, using [1-^{3}H]geranyl diphosphate with enzyme systems from *Pinus*, have demonstrated the validity of Ruzicka's original biogenetic proposal[66] (see Section 2.05.5.4).

2.05.1.4.7 *Iridane monoterpenes*

The iridane monoterpenes possess a 1,2-dimethyl-3-isopropylcyclopentane carbon skeleton. Originally isolated from ant species of the genus *Iridomyrmex* (from which the name derives),[67,68] several hundred iridanes and related compounds have now been isolated from plants.[69] As precursors to the nontryptophan portions of several medicinally important indole alkaloids, these metabolites have received much attention.[70,71] Unlike the more traditional monoterpenes which tend to be sequestered in anatomically specialized secretory structures (see Section 2.05.2.1), the iridoids frequently occur as relatively water soluble β-D-glucosides that are located throughout the plant tissue.[16]

Numerous tracer studies have demonstrated the precursor role of mevalonate, via geranyl diphosphate and geraniol, in the biosynthesis of loganin (**51**).[16] Early experiments also indicated scrambling of label from [2-^{14}C]mevalonate (**50**) between the C-3 and C-11 positions of loganin[72] (Scheme 6), suggesting equilibration of the *gem*-dimethyl group of geraniol (**26**) at some point during biosynthesis. A general biosynthetic pathway to the iridoids has evolved (Scheme 7) which posits hydroxylation of geraniol to 10-hydroxygeraniol (**52**), catalyzed by a cytochrome P450-type oxygenase,[73,74] as the first committed step. Hydroxygeraniol is converted to the dialdehyde (**53**) by an NADP-dependent oxidoreductase which apparently oxidizes both hydroxyl groups.[75,76] Enzymatic cyclization to iridodial (**54**) is followed by further oxidation to iridotrial (**55**). Glycosylation of the dihydropyran, formed reversibly from the trial, prevents further ring opening, hence isotope scrambling must occur prior to this stage of biosynthesis[69] (probably through enolization of trialdehyde **55**). Although there is still some controversy concerning the step at which cyclopentane ring closure occurs (i.e., whether cyclization occurs at the dial or the trial stage of oxidation), the evidence currently favors iridodial (**54**) as the first cyclized intermediate.[69]

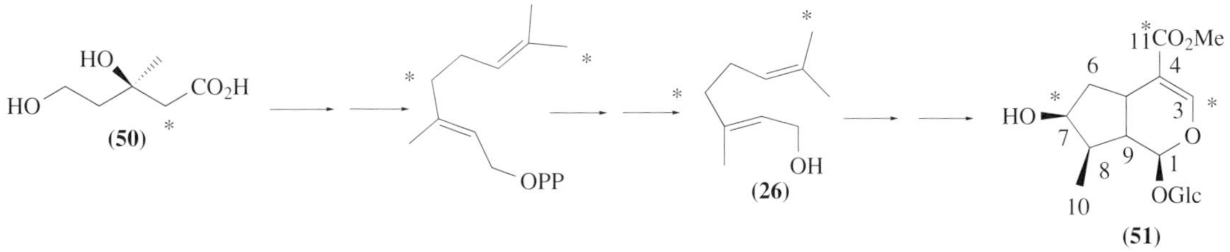

Scheme 6

Madyastha,[77] using partially purified enzyme preparations from *Catharanthus roseus*, has shown the methylation of loganic acid (**56**) to be catalyzed by an *S*-adenosyl-L-methionine-dependent methyltransferase. Hydroxylation at C-7 to form loganin precedes ring cleavage between C-7 and C-8 to yield secologanin (**57**); the enzyme(s) involved in this chemically intriguing cleavage reaction has not yet been described. Incorporation of secologanin into the indole alkaloids is treated elsewhere (Volume 4).

2.05.1.4.8 *Halogenated monoterpenes from marine algae*

Acyclic and cyclic halogenated monoterpenes are produced by macrophytic marine algae. First discovered in the sea hare (*Aplysia californica*),[78] halogenated monoterpenes were subsequently found in species of the red alga *Plocamium*,[79] and it is now generally believed that the metabolites

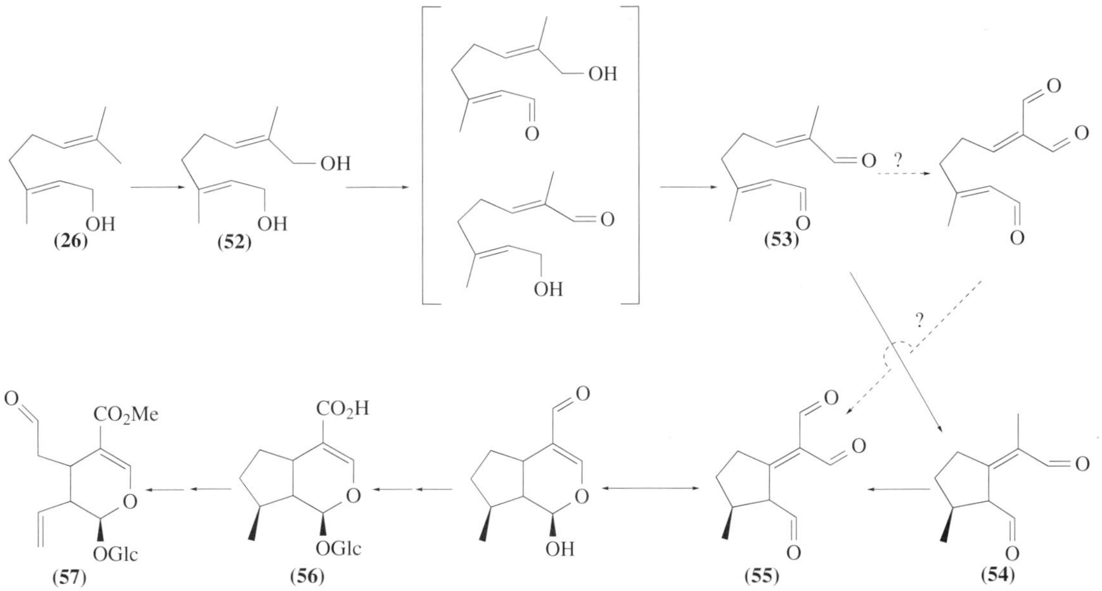

(26) (52) (53) (54) (55) (56) (57)

Scheme 7

found in *A. californica* are of dietary origin.[80] Several dozen halogenated monoterpenes have now been described, and they are found almost exclusively in two genera of Rhodophyta (red algae), namely *Plocamium* and *Chondrococcus*. Reported to be toxic to, and growth inhibitory toward, insects,[81] the physiological role of these metabolites in the producing organisms is unknown.

In addition to the presence of halogen atoms, these monoterpenes possess unusual ring structures. The predominant skeleton types are 1,1-dimethyl-3-ethenylcyclohexane (e.g., **58**), 1,3-dimethyl-1-ethenylcyclohexane (e.g., **59**), and 2,4-dimethyl-1-ethenylcyclohexane (e.g., **60**). Only the 1,1-dimethyl-3-ethenylcyclohexane skeleton has a known counterpart in terrestrial organisms.[82,83] Owing to the co-occurrence of acyclic and cyclic halogenated monoterpenes, the addition of a halonium ion to an alkene to initiate cyclization has been suggested,[82,84] and several chemically plausible mechanisms have been put forward for these processes[85,86] (Scheme 8), although no experimental evidence to substantiate any of these proposals has so far been presented. These biogenetic schemes represent a dramatic departure from the well-established pathways in terrestrial plants in not involving the typical electrophilic reactions of geranyl diphosphate. Efforts to demonstrate the incorporation of geranyl diphosphate in cell-free extracts of the relevant organisms have, so far, been unsuccessful.[85,86] There is, however, ample evidence for the presence of haloperoxidases in other species of macrophytic marine algae that are capable of producing halonium ions,[87] and preliminary evidence suggests that similar enzymes exist in species of *Plocamium*.[88]

2.05.1.4.9 Irregular monoterpenes

Irregular monoterpenes are represented by two general types, those resulting from rearrangement of regular monoterpenes, such as fenchone (**43**), camphene (**45**), eucarvone (**61**), and nezukone (**62**), and thus that obey Ruzicka's biogenetic isoprene rule, and those following decidedly different pathways involving non-head-to-tail condensation of isoprene units. As noted previously, biogenetic schemes for the formation of isocamphane and fenchane monoterpenes have long recognized the likelihood of origin by Wagner–Meerwein rearrangements of either a pinyl cation or, in the case of camphene, a bornyl cation[2,5] (Scheme 4). The use of specifically labeled precursors with cell-free extracts of fennel (*Foeniculum vulgare*) has confirmed the intermediacy of the (+)-(1*R*,5*R*)-pinyl cation (**11**) in the biosynthesis of (1*S*,4*R*)-fenchone.[52] That camphene can arise directly via Wagner–Meerwein rearrangement of the bornyl intermediate (as opposed to indirectly via the pinyl cation) has been deduced from isotopically sensitive branching experiments[89,90] (see Section 2.05.5.5). The

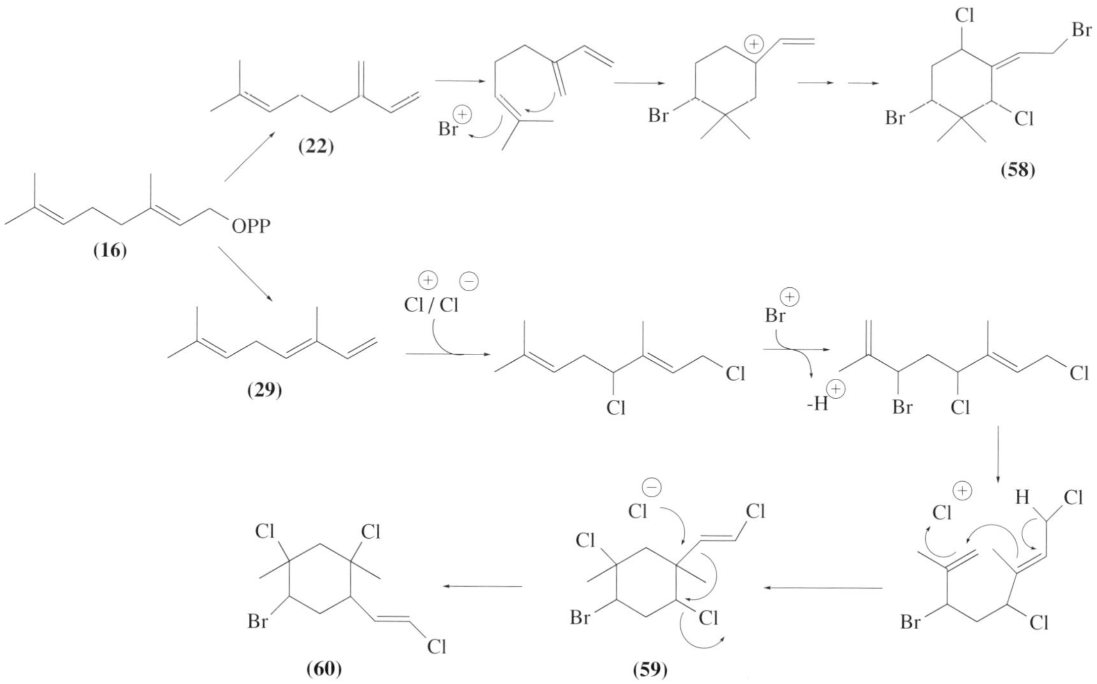

Scheme 8

substituted cycloheptane monoterpenes eucarvone, nezukone, and γ-thujaplicin (**63**) almost certainly arise by ring expansion of cyclohexanoid derivatives, although these reactions have been little explored.

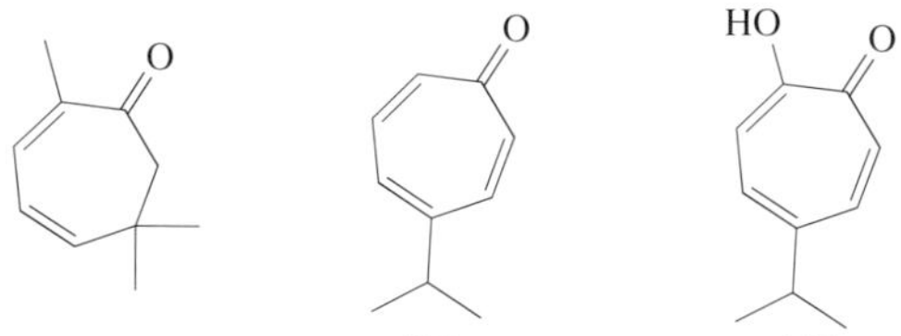

(**61**) (**62**) (**63**)

The other major group of irregular monoterpenes results from non-head-to-tail fusion of dimethylallyl diphosphate units. Compounds of this class include lavandulol (**66**), artemisia ketone (**67**), santolinatriene (**68**), and chrysanthemic acid (**23**). Numerous biogenetic schemes have been proposed for the formation of these irregular monoterpenes.[10] However, based on an earlier proposal,[91] and after detailed analysis of the plant sources as well as consideration of the stereochemistry and mechanistic possibilities, Epstein and Poulter[92] presented a unified model for the formation of these irregular monoterpenes in which the key step is the 1′–2–3 condensation of two dimethylallyl diphosphate molecules to a cyclopropylcarbinyl intermediate.[92] A similar mechanism is well established in the biosynthesis of squalene (**64**) and phytoene (**65**) from farnesyl diphosphate (**15**) and geranylgeranyl diphosphate, respectively.[93] Cleavage or rearrangement of the cyclopropyl ring of the corresponding chrysanthemyl intermediate, as depicted in Scheme 9, yields the various irregular monoterpene skeletons. Model studies have demonstrated the feasibility of the conversion of chrysanthemyl analogues to the artemisane, santolinane, and rothrockane (**69**) systems.[94–96] Formation of the lavandulane system is more likely to result of direct 1′–2 linkage of two molecules of dimethylallyl diphosphate (pathway 1′–2, Scheme 9). This assumption is based on mechanistic considerations as well as the fact that lavandulane-type compounds occur in the plant families Lamiaceae and Umbelliferae, whereas the chrysanthemane, artemisane, and santolinane types are restricted to members of the Asteraceae family.[16]

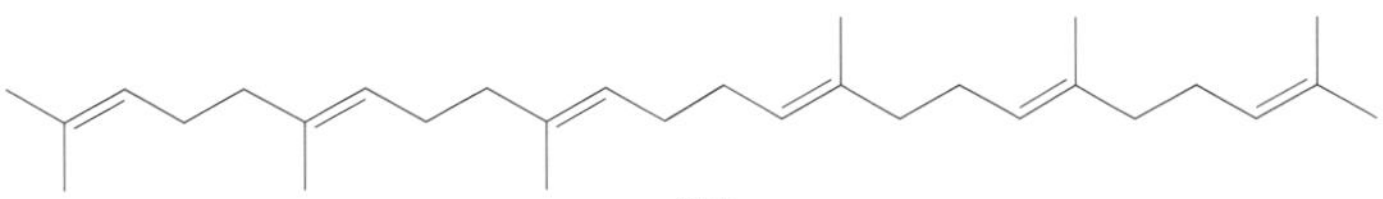

(64)

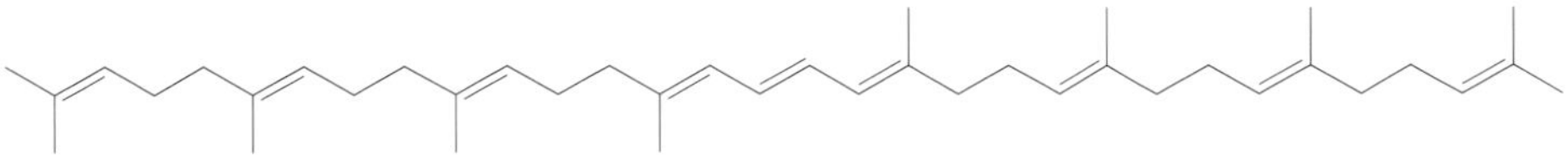

(65)

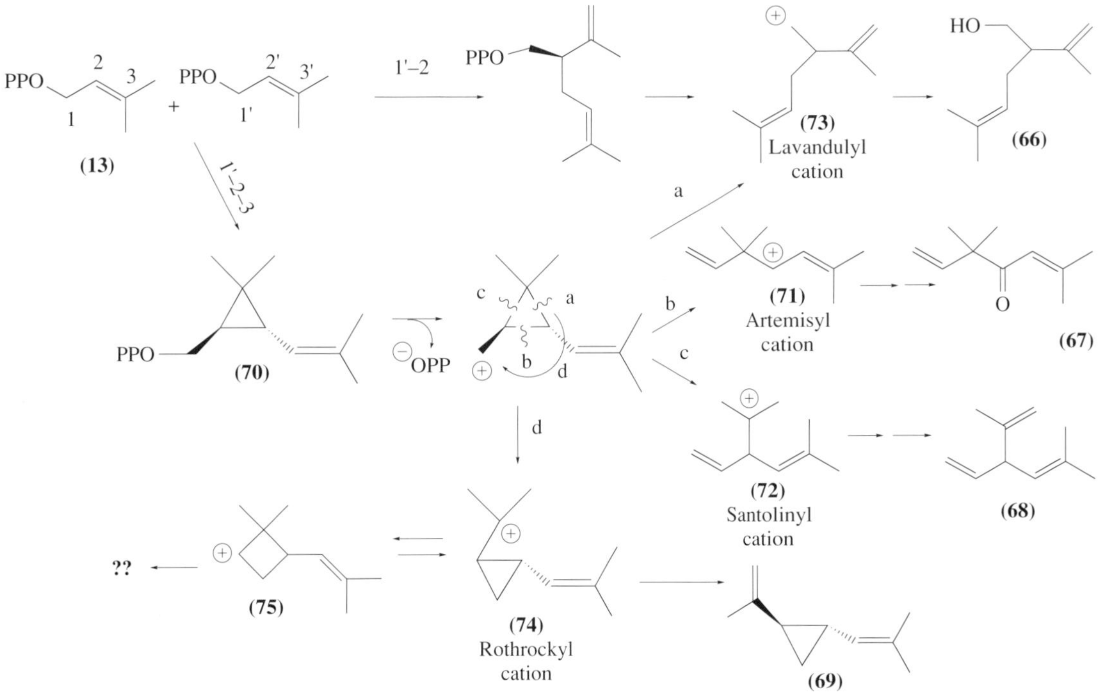

Scheme 9

Incorporation of labeled chrysanthemyl alcohols or chrysanthemates into other irregular monoterpenes by *in vivo* feeding has proven to be relatively unsuccessful;[94,97] however, cell-free extracts from *Santolina chamaecyparissus* and *Artemisia annua* efficiently incorporate both *cis*- and *trans*-chrysanthemyl alcohol and chrysanthemyl diphosphate **(70)** into artemisia alcohol, artemisia ketone, and even trace amounts of lavandulol.[98] More recently, cell-free extracts from callus cultures of *Chrysanthemum cinerariaefolium* were shown to incorporate [1-^{14}C]isopentenyl diphosphate into chrysanthemyl alcohol.[99] There is, as yet, no evidence for a naturally occurring monoterpene analogue of squalene, or of any natural product derived directly from the cyclobutyl cation[100] **(75)**. Nevertheless, the unifying concept of Epstein and Poulter provides an intellectually satisfying paradigm for the biosynthesis of these irregular monoterpenes, with the cyclopropyl carbinyl cation functioning as a key intermediate (similar to the α-terpinyl cation in the biosynthesis of regular monoterpenes[14]); more refined studies with purified enzymes are required to confirm the operation of these proposed pathways.

2.05.1.5 Scope and Emphasis

The remainder of this chapter focuses on the pathways, enzymology, and mechanisms of monoterpene biosynthesis. Because the origins of cyclohexanoid monoterpenes in terrestrial plants have been the most extensively studied, these examples will be emphasized. In as much as the compartmentalization of monoterpene biosynthesis is an important organizational and regulatory feature of metabolism, this biological aspect will be discussed briefly. The formation of the essential precursor, geranyl diphosphate, will also be reviewed; a more detailed discussion of prenyltransferases is presented in Chapter 2.04. Finally, as there are numerous parallels in the biochemistry of monoterpenes, sesquiterpenes, and diterpenes, selected reference to these higher terpenes will be made; comprehensive treatment of sesquiterpene and diterpene biosynthesis is provided in Chapters 2.06 and 2.08. The terms monoterpene and monoterpenoid are used interchangeably throughout this chapter, although in the older literature the former term was generally restricted to the C_{10} hydrocarbons. Nomenclature and numbering systems are based on Erman.[101]

2.05.2 BIOLOGICAL CONSIDERATIONS

2.05.2.1 Organization of Metabolism

The production and storage (or emission) of large quantities of monoterpenes in plants are almost always associated with the presence of anatomically specialized structures (for example, in members of the Lamiaceae, Asteraceae, Pinaceae, Rutaceae, and Umbelliferae[102]). Substantial evidence indicates that monoterpene biosynthesis is restricted to the "secretory" cells of these structures and that various segments of the monoterpene biosynthetic pathways are further confined to specific subcellular locations within the producing cells. Such compartmentalization has important consequences for the regulation of monoterpene metabolism, and the phenomenon is in part responsible for some of the problems traditionally encountered with *in vivo* studies on the formation of these metabolites, such as very low rates of incorporation of exogenous, labeled precursors and anomalous labeling patterns.[10,103]

2.05.2.1.1 Tissue and cellular specialization

Relatively large amounts of monoterpenes are generally synthesized and accumulated in three types of specialized structures, glandular trichomes, secretory cavities, or secretory ducts. These structures are generally nonphotosynthetic and must rely on adjacent cells for carbon substrates, and all are characterized by the presence of an extracytoplasmic storage space. Glandular trichomes are modified epidermal hairs present on the aerial surfaces of many plants (e.g., mint, sage, tansy, geranium). The organization and morphology of these secretory structures, while differing in detail from species to species, tend to have certain common characteristics. The epidermal oil glands of peppermint will be described here as an example; for more comprehensive discussion see Fahn.[104,105]

Peppermint (*Mentha piperita*) leaves bear two types of trichomes, capitate and peltate. The capitate structures consist of a single basal cell surmounted by a single stalk cell. Above the stalk cell is a single secretory cell. Peltate trichomes, likewise, consist of a single basal cell above which lies a single stalk cell. However, in contrast to the capitate structures, the peltate trichomes possess eight radially distributed secretory cells atop the stalk cell, above which lies an extended cuticle (Figure 1(a)). The side walls of the stalk cell are heavily cutinized. This cutinization serves to support the secretory cells and to prevent the apoplastic flow of the secreted oil back into the leaf tissue.[104,105] Secreted monoterpenes (and other lipids) are stored in the subcuticular space outside of the secretory cells, where these materials may accumulate in excess of 1% of dry leaf weight.

Both biochemical and histochemical evidence indicates that monoterpene biosynthesis occurs exclusively in the glandular trichomes of peppermint.[106] Thus, proteins from peltate glandular trichomes were selectively extracted using a mechanized surface abrasion technique,[107] and the activity of enzymes involved in carvone (**133**, Scheme 33) biosynthesis were compared between these extracts and those from the remaining leaf tissue after gland removal. Greater than 99% of the limonene synthase activity and 96% of the limonene hydroxylase activity of the leaf were found in the trichome extract. Stem and root extracts were also devoid of limonene synthase activity, consistent with the lack of oil glands on the surfaces of these tissues. Intact trichome secretory cells with open plasmodesmata (cell boundary openings through which fine threads of cytoplasm, connecting the protoplast of adjacent cells, pass to facilitate intercellular transfer) have been isolated from peppermint leaves and were shown to efficiently incorporate [^{14}C]sucrose as well as [^{3}H]geranyl diphosphate into monoterpenes.[108]

Secretory cavities are typical of the Myrtaceae, Rutaceae, Myoporaceae, Leguminosae, and Hypericaceae families.[104] These structures are formed internally in the leaves or in the exocarp (fruit

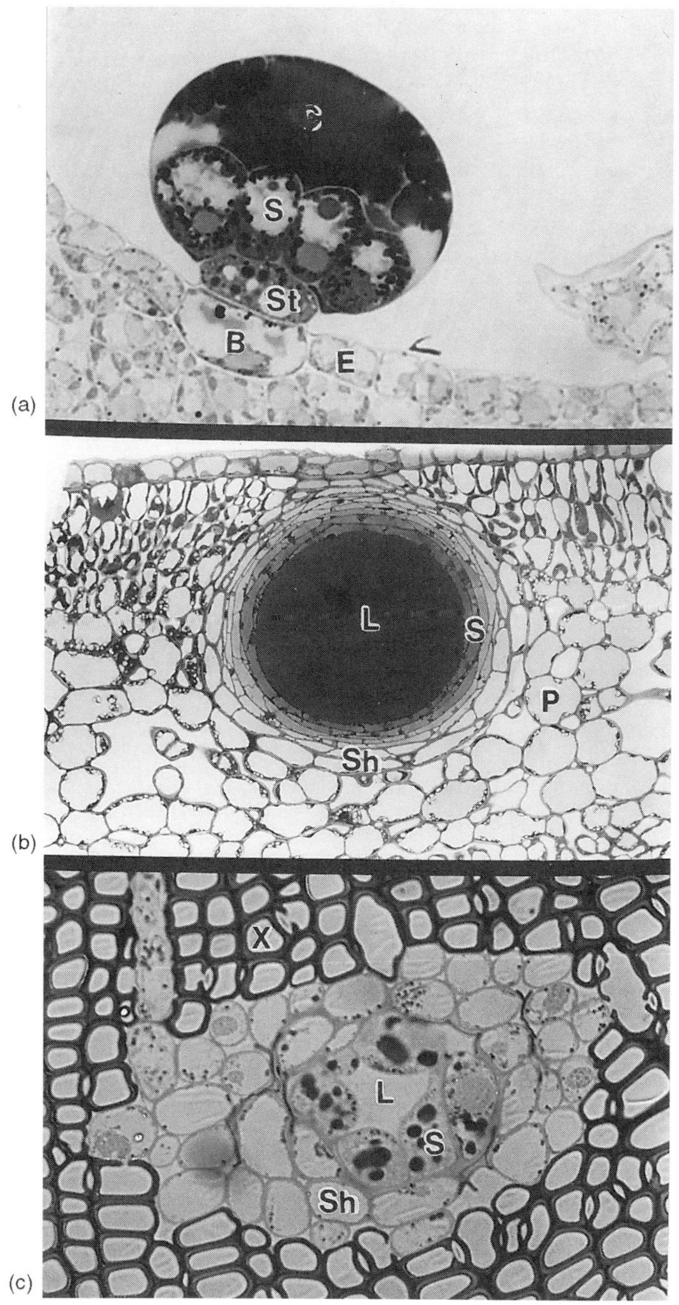

Figure 1 Photomicrographs of secretory structures. (a) Longitudinal section of glandular trichome from spearmint (*M. spicata*) leaf. (b) Transverse section of secretory cavity from lemon (*C. limon*) leaf. (c) Transverse section of resin duct from Jeffrey pine (*P. jeffreyi*) (C, subcuticular space; B, basal cell; E, epidermal cells; L, lumen; S, secretory cells; St, stalk cell; Sh, sheath cells; P, parenchyma cells; X, xylem cells) (Courtesy of Drs. Jonathan Gershenzon, Glenn Turner, and Thomas Savage).

peel) and consist of a lumen surrounded by several layers of specialized epidermal cells (Figure 1(b)). The cells immediately lining the cavity are termed the secretory epithelium and are believed to be responsible for monoterpene biosynthesis.[105] The outermost cells of these secretory structures are usually thick-walled and likely serve as a protective sheathing. Due to the technical difficulties in isolating intact cells from these interior structures,[109] there is no direct evidence that monoterpene biosynthesis occurs in these locations. However, parallels in ultrastructure with other secretory tissues, and the correlations between morphological characteristics and monoterpene accumulation, strongly suggest that they are the sites of monoterpene biosynthesis.[110,111]

Secretory ducts can be likened to elongated secretory cavities (Figure 1(c)). These structures appear in the Pinaceae, Anacardiaceae, Compositae, Hypericaceae, Leguminosae, Umbelliferae, and other families.[104] The conifers are particularly noteworthy for the production of defensive oleoresin (a mixture of roughly equal amounts of monoterpenes and diterpene resin acids) in these structures. Certain species, especially pines, depend almost exclusively on constitutively produced oleoresin as a defense against bark beetles;[21] the site of oleoresin biosynthesis is considered to be the secretory epithelial cells of the resin ducts.[112,113] In contrast, the true firs (*Abies* sp.) depend for defense upon inducible oleoresin produced locally at the infection or wound site by what appear to be nonspecialized (parenchyma) cells.[26,112,114,115] Again, direct experimental evidence for the cellular localization of monoterpene biosynthesis in these organs is lacking.

Extensive investigations of the natural products chemistry of numerous nonglandular plants, including many crop and ornamental species, have indicated that most produce at least trace levels of monoterpenes,[102] some of which are components of pathogen or herbivore defense systems.[116–118] Very little is known about the organization and regulation of monoterpene metabolism in these instances, but the contrast with the "essential oil" plants clearly indicates that the presence of secretory structures in the latter represents a highly specialized adaptation for sequestering high level production of these cytotoxic metabolites.

2.05.2.1.2 Subcellular localization

A fundamental, and seemly widespread, feature of the organization of plant terpenoid metabolism exists at the subcellular level. The sesquiterpenes (C_{15}), triterpenes (C_{30}), and polyterpenes appear to be produced in the cytosol/endoplasmic reticulum compartment, whereas isoprene, the monoterpenes (C_{10}), diterpenes (C_{20}), tetraterpenes (C_{40}), and certain prenylated quinones are primarily of plastidial origin. (Plastids are intracellular organelles surrounded by a double unit membrane. These organelles include the chloroplasts as well as other nongreen plastids which are photosynthetically incompetent. Among this latter group are the leucoplasts and chromoplasts which are generally held to be sites of monoterpene biosynthesis.) The evidence now suggests that the biosynthetic pathways for the formation of the fundamental precursor, isopentenyl diphosphate, differ significantly as well in these two compartments, with the classical acetate/mevalonate pathway (Chapter 2.02) operating in the cytosol/endoplasmic reticulum, and the newly discovered nonmevalonate pathway (the glyceraldehyde phosphate/pyruvate pathway; see Chapter 2.03) operating in the plastid.[119] (There is some circumstantial evidence for a functional mevalonate pathway in plastids that cannot be entirely dismissed.[120,121])

One of the longstanding enigmas to arise from *in vivo* studies of monoterpene biosynthesis is the exceedingly low rates of incorporation of exogenous precursors such as mevalonate and acetate into these metabolites, and the often highly anomalous labeling patterns derived from these basic precursors.[10,103,122] An analysis of the relative incorporation of label from a variety of precursors into sesquiterpenes and monoterpenes of peppermint showed that, with [^{14}C]mevalonate and [^{14}C]acetate, the monoterpenes were dramatically less efficiently labeled than were the less abundant sesquiterpenes.[103] In contrast, [^{14}C]glucose and $^{14}CO_2$ were incorporated into monoterpenes and sesquiterpenes in ratios reflecting their natural abundance. These findings led Loomis and Croteau[103] to suggest that biosynthesis of these two classes of terpenoids occurs at different sites, with monoterpene biosynthesis likely taking place in the plastids of the secretory cells. A range of studies have subsequently supported this proposal. Thus, isolated plastids from *Citrofortunella mitis*,[123] *Narcissus pseudonarcissus*,[124] and from *C. sinensis*[125] have been shown to possess the capability for monoterpene biosynthesis. cDNA clones for limonene synthase from spearmint (*M. spicata*)[126] and perilla (*Perilla frutescens*),[127] and for a number of other monoterpene synthases from common sage (*Salvia officinalis*)[50] and grand fir (*Abies grandis*),[49] encode preprotein forms of these enzymes bearing an amino terminal plastidial targeting sequence. The transit peptide of *in vitro* translated limonene synthase has been verified by plastid uptake and processing experiments. Finally, immunogold labeling has demonstrated the limonene synthase of spearmint to be located specifically in the leucoplasts of the oil gland secretory cells. Related evidence[128–130] suggests that geranyl diphosphate synthase, supplying the immediate precursor of monoterpene biosynthesis, also resides in plastids. The compartmentalization of monoterpene production in plastids, and their probable origin from isopentenyl diphosphate derived from the nonmevalonate pathway, provides a likely explanation for the lack of acetate and mevalonate incorporation, and suggests that the often anomalous labeling patterns observed were the result of degradation of these precursors and subsequent scrambling of the trace amounts of labeled material incorporated.

An exception to the plastidial site of monoterpene biosynthesis may be flower tissue. The linalool synthase of *Clarkia* flowers does not appear to be translated as a preprotein bearing an obvious plastidial transit peptide.[131] The glandular epidermal cells of rose petals possess few plastids but, nevertheless, produce substantial amounts of geraniol, nerol, and related acyclic monoterpenols, mostly as β-D-glucosides,[103] and they incorporate [2-^{14}C]mevalonate into these compounds with exceptionally high efficiency ($>10\%$).[132] This result may suggest that monoterpene biosynthesis in rose petal epidermis also takes place in the cytosol via the acetate/mevalonate pathway or, as an alternative possibility, is localized to plastids but utilizes isopentenyl diphosphate of cytosolic origin. This issue of the source of isopentenyl diphosphate for monoterpene biosynthesis is complicated by observations indicating that, depending on developmental stage or other circumstances, plastids may supply isopentenyl diphosphate for use in cytosolic biosynthesis, and vice versa.[120,121] To determine the origin of the C_5 precursor requires detailed *in vitro* biosynthetic studies or *in vivo* experiments using specifically labeled precursors followed by precise location of label in the derived monoterpene. The interrelationships between the two pathways for isopentenyl diphosphate biosynthesis offer a fertile area for future research.

2.05.2.2 Regulation of Metabolism

Although some degree of control is presumed to exist at the level of isopentenyl diphosphate flux, and perhaps at the level of isopentenyl diphosphate isomerase (see Chapter 2.04), very little is known about the regulation of the early steps of terpenoid metabolism in plants, particularly as regards the partitioning of carbon from primary into secondary metabolism, and the distribution of isopentenyl diphosphate into the various terpenoid types at the subcellular sites of synthesis.[110] The first dedicated steps of monoterpene biosynthesis are considered to be the production of the immediate precursor geranyl diphosphate and the conversion of this intermediate, by the monoterpene synthases, to the parent monoterpene products thereby irreversibly diverting the geranyl substrate from possible elongation by prenyltransferases and thus preventing its utilization in other pathways. Most studies on the developmental or inducible regulation of monoterpene metabolism have focused on these two enzyme types.

Limonene synthase catalyzes the first committed step of monoterpene biosynthesis in *Mentha* and, therefore, is an obvious target of regulatory studies. When limonene synthase activity in oil gland extracts and total monoterpene biosynthetic rate (as determined by $^{14}CO_2$ incorporation *in vivo*) are measured as a function of peppermint leaf development, an essentially coincidental pattern is observed.[133] Both synthase activity and total biosynthetic rate rise rapidly from leaf emergence, peak at about the first week and then decrease to essentially zero by the second week of leaf expansion. Preliminary studies on geranyl diphosphate synthase evidence a very similar pattern of activity. Western immunoblotting, using *anti*-limonene synthase polyclonal antibodies[134] to quantify synthase protein, indicated that enzyme activity parallels enzyme protein. This observation suggests that the limonene synthase activity level directly reflects the amount of synthase present at any given time and that the activity of the enzyme *per se* is unlikely to be regulated by allosterism, covalent modification, or other means. Finally, mRNA blot hybridization studies (northern analysis) demonstrated the time course of appearance and disappearance of limonene synthase transcripts to slightly precede the appearance and subsequent disappearance of the enzyme itself, as might be expected if the newly transcribed message was immediately translated to the encoded protein and was followed by turnover of both. In a similar fashion, camphor production in garden sage (*S. officinalis*), as determined by $^{14}CO_2$ incorporation, closely paralleled the activity of bornyl diphosphate synthase, the monoterpene cyclase catalyzing the first committed step in the biosynthesis of camphor from geranyl diphosphate.[110] The summation of these observations suggests that developmental control of monoterpene biosynthesis (as evidenced by regulation of limonene synthase and bornyl diphosphate synthase in two species of the Lamiaceae) resides at the level of gene expression, and indicates that the oil glands are active in essential oil production only during the first two weeks of leaf development, a period much shorter than previously appreciated. Studies by Banthorpe *et al.* on the seasonal fluctuation of geranyl diphosphate synthase activity in rose geranium (*Pelargonium graveolens*),[135] and by Pauly *et al.*[129] on the biosynthesis of geranyl diphosphate and cyclic monoterpenes in isolated leucoplasts from *Citrofortunella mitis* and *C. unshiu* fruit, also support a role for this prenyltransferase in the control of monoterpene biosynthesis. It is not unreasonable to suggest that the levels of geranyl diphosphate synthase and the monoterpene synthases are coordinately regulated during the production of monoterpenes in oil glands and related secretory structures.

Grand fir (*Abies grandis*) has provided the principal model for studies on the regulation of inducible monoterpene biosynthesis. Mechanical wounding of stem tissue (as a simulation of bark beetle attack) results in a dramatic increase in oleoresin biosynthesis (monoterpenes, sesquiterpenes, and diterpene resin acids).[21,115,136] The wound induction of monoterpene biosynthesis represents both the increase in levels of constitutive monoterpene synthases and the appearance of new monoterpene synthases, together promoting a change in oleoresin turpentine composition[137] from the constitutive mixture to a mixture more toxic to bark beetles and their pathogenic fungal symbionts. Detailed examination of the time-course of wound induction indicated parallel elevation in the levels of both geranyl diphosphate synthase activity and the monoterpene synthase activities, and immunoblot analysis, using a broadly cross-reacting *anti*-pinene synthase antibody,[136] demonstrated the coincidental increase of synthase proteins with synthase activities. Northern blot analysis over the time-course of wound induction, using three respective DNA hybridization probes for grand fir monoterpene, sesquiterpene, and diterpene synthases, confirmed that control of the induced wound response resided at the level of gene expression and suggested that, once translated, the enzymes for oleoresin biosynthesis were stable and functional for many weeks after the corresponding messages had disappeared. Interestingly, the results of enzyme activity assay, immunoblotting, and RNA blot hybridization analysis indicate that monoterpene biosynthesis is up-regulated several days before the coordinated induction of sesquiterpene and diterpene resin acid biosynthesis. These results were interpreted as a reflection of the dual role of monoterpenes in the host response to insect invasion, first as toxins produced immediately in response to attack and later, via sustained production, as a solvent in mobilizing the resin acids for the purpose of wound sealing.[136]

2.05.2.3 Catabolism

The metabolic cost of producing monoterpenes is quite high relative to other secondary metabolites.[138] One potential strategy to recoup the expense of manufacture of these compounds is through catabolism when they are no longer needed to fulfill their ecological function. Unfortunately, most tracer studies on monoterpene metabolism have simply examined the incorporation and subsequent loss of radioactivity from the terpenoid pool; few efforts have been expended to determine the ultimate fate of the labeled products. The mechanisms by which plants might "recycle" terpenoid compounds are, therefore, poorly understood. However, one example is provided by studies on the catabolism of menthone (**76**) in *M. piperita*. In these plants, 50–75% of the monoterpenes that were present in the mature leaves have been lost or degraded by the time the plant completes flowering.[133] Administration of (−)-[^{3}H]menthone (a principal monoterpene of peppermint) results in this substrate being reduced, in approximately equal proportions, to the diastereomeric alcohols (+)-neomenthol (**77**) or (−)-menthol (**78**) (Scheme 10).[139] Some of the menthol is subsequently converted to menthyl acetate (**79**) which remains in the leaf tissue. The neomenthol, however, is largely converted to neomenthyl-β-D-glucoside (**80**) which is transported from the leaf to the rhizome where the glycoside is rapidly hydrolyzed. The aglycone is then oxidized back to menthone which undergoes an unusual lactonization reaction to form 3,4-menthone lactone (**81**).[140] This lactone is apparently subject to β-oxidative metabolism from which the original tritium label could be found in significant yield in acyl lipids, sterols, and carbohydrates.[141]

An analogous mechanism for camphor catabolism appears to be operative in mature sage leaves (*S. officinalis*), where the ketone is lactonized to 1,2-campholide (**83**), glycosylated (**84**), and transported to the roots where it is oxidatively degraded, at least in part, to acetyl-CoA (Scheme 11).[142,143] Camphor catabolism has also been demonstrated in cell suspension cultures of sage. Although the enzymes responsible for the biosynthesis of camphor could be detected in undifferentiated cell cultures, camphor did not accumulate.[144] With the recognition of a catabolic pathway for this compound in the intact plants, it was suggested that the lack of accumulation in cell cultures might be the result of a highly efficient mechanism for degradation; this was found to be the case when exogenous camphor was introduced to sage cell cultures.[145] Interestingly, 1,2-campholide did not appear to be a major intermediate in the catabolism of camphor in these cells, rather camphor was first hydroxylated at the C-6 position followed by oxidation to the β-diketone (6-oxocamphor, **85**) (Scheme 12). The diketone was then subject to a novel hydrolytic ring-opening reaction to form α-campholonic acid (**86**), followed by hydroxylation of the tertiary position of the cyclopentyl ring to

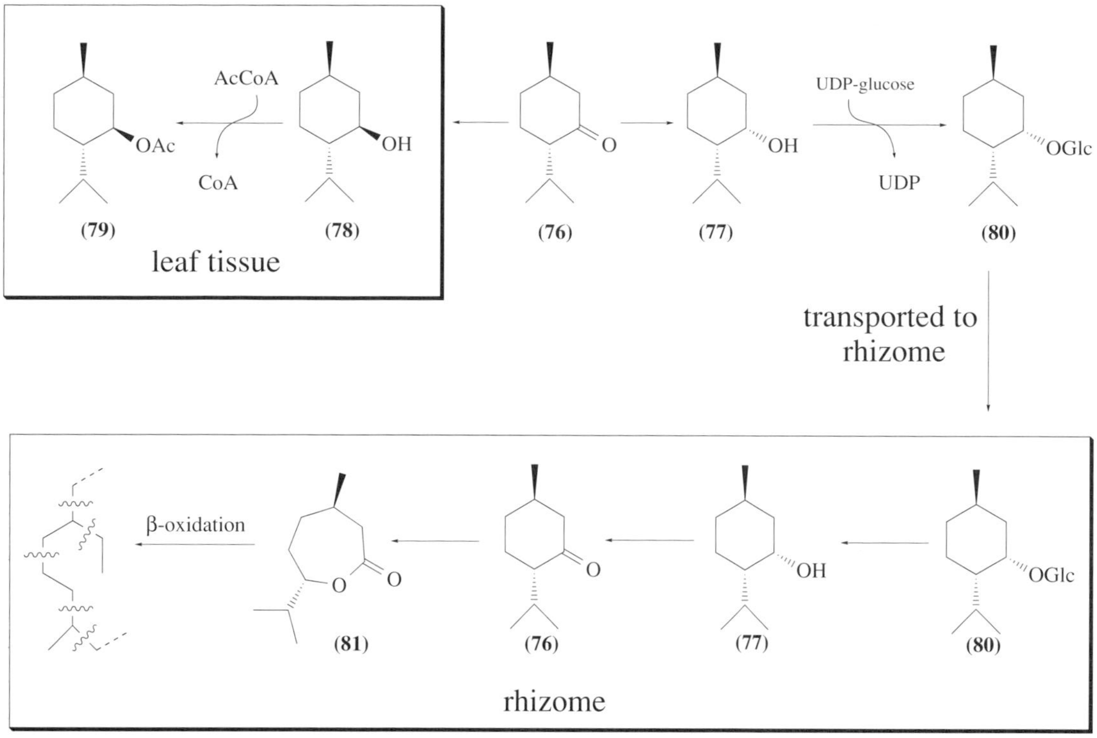

Scheme 10

produce 2-hydroxycampholonic acid (**87**). The latter is then lactonized and undergoes ring opening to yield the branched diacid, isoketocamphoric acid (**88**).[145] Further steps in this degradative sequence have not yet been defined.

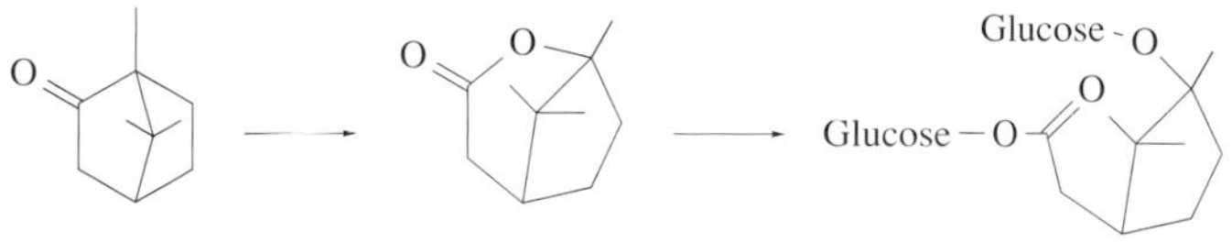

Scheme 11

Both of the systems described above provide evidence for catabolic salvage pathways that are potentially capable of recovering some of the metabolic costs of monoterpene production. However, caution is warranted in the interpretation of many experiments demonstrating rapid turnover of leaf monoterpenes, especially in studies conducted with cuttings or detached stems in which the observed turnover may be an artifact. Thus, intact, rooted plants, in contrast to stem cuttings, do not exhibit rapid turnover of the monoterpene pool, and changes in the monoterpene content of these plants occur only over extended periods of time.[133,146] In the case of the menthone and camphor catabolism studies cited above, it should be noted that the labeled substrates were administered by stem feeding or application to the leaf surface. Since the site of monoterpene synthesis and storage is restricted to specialized secretory tissues designed for the sequestering of these (potentially toxic) metabolites, it is relevant to question whether the catabolism of menthone and camphor observed represents significant processes of the oil glands or detoxification pathways in other cells to which these metabolites gained access.

2.05.2.4 Genetics of Monoterpene Biosynthesis

There are relatively few classical genetic analyses of monoterpene biosynthesis, and these have been performed on a very limited number of plant species. Perhaps the most detailed studies are

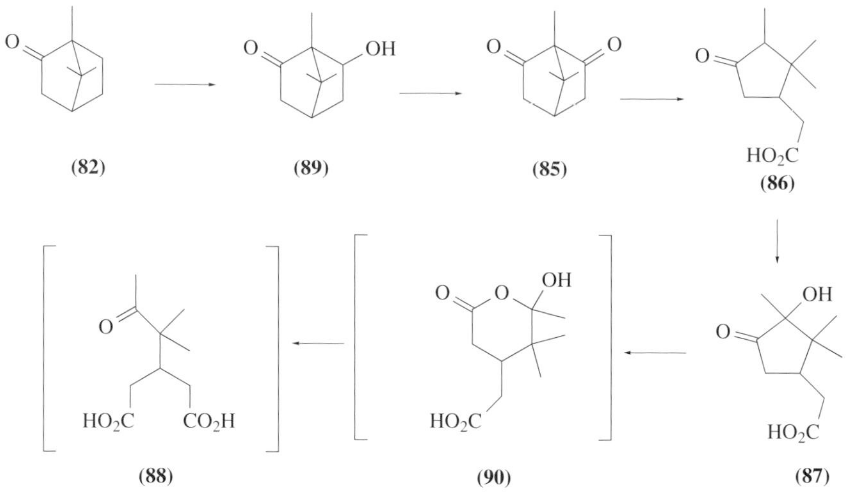

Scheme 12

those conducted by Murray and his co-workers with species of mint (*Mentha*), using traditional hybridization and F_1 backcrossing experiments. Hefendehl and Murray,[147] by correlation of oil composition in these crosses, were able to identify many of the gene loci and allelic dominance patterns associated with the biosynthesis of carvone (**133**), menthol (**78**), and several other monoterpenes of this genus.

With accumulated data on the enzymes involved in monoterpene metabolism in mint, a detailed correlation of specific enzyme activities with proposed gene loci identified by Murray has been proposed.[148] For example, the principal monoterpenes of spearmint and peppermint are distinguished from each other by the position of oxygenation on the *p*-menthane ring of these metabolites. The monoterpenes of spearmint are oxygenated almost exclusively at the C-6 position, whereas with peppermint monoterpenes, the C-3 position of the *p*-menthanes is the specific site of oxygenation. Murray identified the *Lm* and *C* loci as being responsible for the C-3 and the C-6 oxygenations, respectively.[147] The cytochrome P450 oxygenases catalyzing the specific hydroxylations at C-3 and C-6 of the alkene precursor (−)-(4*S*)-limonene (**3**) have been isolated from peppermint and spearmint, respectively,[149,150] and the corresponding cDNAs have been recently cloned and sequenced.[151] These results clearly illustrate the pivotal role of these two specific cytochrome P450 hydroxylases in determining the biosynthetic fate of the common precursor, limonene, in mint species, and suggest a basis for the genetic interpretation. However, whether the *Lm* and *C* loci defined by Murray actually represent such structural genes, or regulatory genes, is not clear.

Other species to which genetic analysis has been applied include *Satureja douglasii*,[152] *Thymus vulgaris*,[153] and several conifer species.[154] Unfortunately, much of the genetic data related to monoterpene biosynthesis in conifers may be seriously flawed by the failure to consider quantitative aspects of oleoresin chemistry.[155] In many cases, however, simple and seemingly valid dominant/recessive allelic relationships have been defined. As will be seen in Section 2.05.5.6, considerable progress in the molecular genetics of monoterpene biosynthesis is being made, and it should soon be possible to better define correlations by determining the specific roles of extant loci as encoding structural or regulatory genes. The combination of classical and molecular genetics can be expected to provide the basis for the transgenic production of new plant species and hybrids selectively altered for improved defense against pathogens and herbivores, improved competition against weeds, and improved flavor and aroma profile, or enriched for the production of specific monoterpenes of perfumery, pharmaceutical, or other commercial value.

2.05.3 EXPERIMENTAL METHODS

The study of monoterpene biosynthesis presents several experimental difficulties that are typically associated with the metabolism of natural products. The enzymes responsible for monoterpene biosynthesis are present in relatively low quantities and, as described in Section 2.05.2.1, are often localized in specialized structures and are under developmental and environmental control. In mint species, for example, monoterpene biosynthesis is restricted to the epidermal oil glands in which biosynthetic capability is expressed only during the first few weeks of leaf expansion. Attention to tissue selection and developmental timing is important in optimizing the enzyme source.

The cyclization enzymes involved in monoterpene biosynthesis do not produce chromophoric products, the only cofactor required is Mg^{2+} or Mn^{2+} (in the case of gymnosperm cyclases, K^+ is also required) and, at the level of activity normally assayed, they do not produce a measurable change in pH. Hence, continuous kinetic monitoring is usually not possible unless inorganic diphosphate release from geranyl diphosphate is measured. The latter methods are cumbersome and expensive.[156,157] Therefore, it is generally necessary to employ either radioisotopically labeled substrates or, if sufficient enzyme is available, to monitor product formation by gas chromatography or other chromatographic means. Many of the secondary transformations of the parent cyclic monoterpenes involve redox reactions employing pyridine nucleotides. However, the levels of activity usually available, except where recombinant enzymes can be employed, are too low to permit routine spectrophotometric monitoring. With the exception of diphosphate esters (e.g., geranyl, linalyl, bornyl diphosphate), the monoterpenes are quite hydrophobic and in most cases relatively volatile. While these characteristics facilitate product isolation and analysis by gas chromatographic methods, caution is necessarily required in trapping these metabolites, or at least in accounting for their loss by volatilization during handling through the use of internal standards. In this section we will discuss some of the specialized experimental methods which have been developed to study the biosynthesis of monoterpenes.

2.05.3.1 *In Vivo* Feeding Experiments

Much of the early work on monoterpene biosynthesis was performed through feeding of radioisotopically labeled substrates to intact plants or plant parts. Administration of labeled precursors to plants can be accomplished by several methods: wicking the precursor from an aqueous solution into the lower part of the plant stem; immersing the root system into an aqueous solution to allow assimilation; directly injecting the precursor into the plant stem or seed capsule; placing excised plant parts into an aqueous solution of the precursor; applying the precursor or solution of precursor onto the leaf surface; or growing the plant in an atmosphere enriched in $^{14}CO_2$.[158] It was found early on that, in most cases, $^{14}CO_2$ or uniformly labeled sugars are incorporated into monoterpenes at reasonably high levels. Unfortunately, incorporation of these precursors is not particularly helpful in elucidating biosynthetic pathways, except where time-course studies can be useful in determining the sequence of formation of biogenetically related metabolites. Incorporation of presumed, more specific precursors, such as [^{14}C]acetate and [^{14}C]mevalonate, which were considered to provide more detailed information,[5] frequently resulted in exceedingly low rates of incorporation and anomalous labeling patterns.[10] It is now known that, under most circumstances, the monoterpenes are not formed by the classical acetate/mevalonate pathway, but rather by the glyceraldehyde-3-phosphate/pyruvate pathway[119] (see Chapter 2.03). Thus, the inefficient and anomalous incorporations of physiologically irrelevant precursors can now be rationalized. Notably, studies on the incorporation of specifically labeled glucose into terpenoid compounds have been exceedingly useful in distinguishing the operation of the traditional acetate/mevalonate pathway and the glyceraldehyde-3-phosphate/pyruvate pathway in the formation of the various classes of metabolites.[119,159,160] Substantial progress in defining monoterpene biosynthesis has only come about with the application of cell-free enzyme systems.

2.05.3.2 Cell-free Enzyme Preparations

2.05.3.2.1 Enzyme isolation from intact plant tissue

Isolation of the native enzymes from plant tissues is complicated by the presence of low molecular weight compounds such as oils, resins, and phenolic materials which (particularly in the presence of phenol oxidases) can inhibit or denature the enzymes of interest.[161] Addition of insoluble polymeric adsorbents such as polyvinyl(poly)pyrrolidone and polystyrene resin (Amberlite XAD-4) to the extraction medium is efficacious in removing the bulk of these inhibitory substances.[161,162] Likewise, the use of thiol-reducing reagents, such as dithiothreitol, is generally essential for preserving enzyme activity. Detailed methodologies for enzyme extraction from intact plant tissues are provided by Alonso, Croteau, and Cane.[163,164]

2.05.3.2.2 Enzyme isolation from glandular trichomes

While whole tissue extracts can provide useful cell-free systems for the study of monoterpene biosynthesis, they present the drawback of gross protein contamination from nonrelevant cell types. This is a particular problem with contaminating phosphohydrolases which compete with the terpenoid synthases by efficiently hydrolyzing the geranyl diphosphate substrate and thereby result in serious errors in the interpretation of kinetic data. Selective enrichment of the enzymes of monoterpene biosynthesis can be achieved with some plants by isolation of the relevant secretory tissues in which monoterpene biosynthesis occurs prior to cell lysis and protein extraction. Thus, the use of only flower petals, the stripped bark of conifer stem, or the peel of *Citrus* fruit, provides tissue sources enriched in the relevant glandular epidermis, resin ducts, and internal glands, respectively, thereby removing much of the bulk proteins and often eliminating most of the potential phosphatase contamination.[165,166] Such an enrichment technique, as applied to the epidermal oil glands of many herbaceous plant species, entails the mechanical abrasion of chilled, turgid leaves with glass beads in a low speed blender. By sieving the abraded leaves through progressively finer nylon mesh screens, the dislodged, intact secretory cell clusters can be isolated from leaf fragments and washed free of other contaminating tissue. Once isolated, the gland cells can be disrupted by sonication or mechanical grinding to extract the protein components, or they can be used intact for *in situ* feeding studies. The isolated, intact secretory cells are nonspecifically permeable to low molecular weight (< 1800 Da), water soluble metabolites (as a consequence of open plasmodesmata) while maintaining a high degree of metabolic competence, including terpenoid biosynthesis.[108] Integrity of organelle membranes is unaffected by the isolation procedure, providing a system to explore numerous aspects of intracellular metabolite transport, effects of various cofactors on terpenoid production and compartmentation of biosynthetic pathways. Disruption of the secretory cells (and the organelles contained therein) by sonication results in complete extraction of the monoterpene biosynthetic machinery.[165] As importantly, prior isolation of the secretory tissues provides a highly enriched source of the RNA messages encoding the enzymes of monoterpene biosynthesis and so is extremely useful in the construction of cDNA libraries.

2.05.3.2.3 Plastid isolation

Plastids are the site of monoterpene biosynthesis, and in some instances it may be possible to isolate these organelles from the relevant tissues prior to biosynthetic studies. Gleizes *et al.* have used the exocarp of calamondin fruits (*C. mitis*)[123] as well as satsuma (*C. unshiu*)[129] as a tissue source for the isolation of leucoplasts involved in monoterpene biosynthesis in *Citrus*. After macerating the exocarp in a buffered 0.3 M sucrose solution containing insoluble polyvinylpyrrolidone, the preparation was filtered and subjected to repeated centrifugation to isolate leukoplasts, the purity of which was determined by light and electron microscopy.[123] The isolated, intact organelles as well as ruptured leucoplasts were capable of converting exogenous [1-^{14}C]isopentenyl diphosphate (**14**) to labeled geranyl diphosphate (**16**) and limonene (**3**), as well as to traces of α- and β-pinene (**5**, **107**) and (somewhat surprisingly) farnesyl diphosphate (**15**). Because of the rigorous cell disruption techniques required with woody tissues (resin ducts) and heavily cutinized structures (oil gland secretory cells), it has not yet been possible to isolate intact plastids from these sources.

2.05.3.3 Enzyme Assay

Although specific reaction conditions need to be determined empirically, some generally useful procedures for the enzymes of monoterpene biosynthesis, especially the synthases, have been developed. For monoterpene synthases, the optimum pH is usually in the range of 6.0–8.0 and buffers of the Good series are commonly employed (phosphate buffers can be inhibitory[163]). A divalent cation (Mg^{2+} or Mn^{2+}) is required for all known synthases and, in the case of several of the conifer-derived synthases, a monovalent cation such as K^+, Rb^+, Cs^+, or NH_4^+ must also be included in the reaction medium.[167] Inclusion of polyols, such as sorbitol or glycerol at 5–20%, and thiol reducing reagents, such as dithiothreitol, in the assay buffer can markedly improve enzyme stability.[164] The monoterpene synthase reaction is initiated by the addition of labeled geranyl diphosphate at 10–50 μM ([1-^{3}H]geranyl diphosphate is the most easily prepared substrate[168]) and the solution is overlaid with approximately 1 mL of pentane to trap the volatile, hydrophobic reaction products (this overlay is not in the least inhibitory). The reaction vessel is sealed with a Teflon cap (anaerobic conditions do not influence the reaction) and the mixture incubated for several minutes to several hours in the 25–32 °C temperature range. The mixture is then chilled on ice and the pentane overlay is used to extract any reaction products from the aqueous phase. Passing the pentane phase through a small column (pasteur pipette) of silica gel, surmounted with a small amount of magnesium sulfate to remove traces of water, effectively separates monoterpene alkenes from any coextracted oxygenated monoterpenes. Subsequent extraction of the reaction mixture with diethyl ether and elution through the original silica gel column will selectively afford the oxygenated monoterpenes, including any geraniol resulting from hydrolysis of the substrate by contaminating phosphohydrolases. Scintillation counting of an aliquot of the organic phases provides an assessment of reaction rate. It should be noted, however, that several different monoterpene synthases may be present in the preparation and that many monoterpene cyclases are multifunctional, in either case resulting in the formation of several reaction products. It is, therefore, necessary to separate the reaction products by radio-GLC (or by GLC–MS if quantities permit) to determine the product profile and to quantitate individual products (see below).

To assay for the formation of phosphorylated products, such as bornyl diphosphate, any residual organic solvent is removed from the sample following extraction of hydrophobic materials (under a stream of nitrogen or under reduced pressure) and 2–3 units of wheat germ phosphatase and apyrase in acetate buffer, pH 5.0, are added to the solution to adjust the pH downward. The vial is again overlaid with pentane, resealed, and incubated for an additional 2–4 h at 31 °C with gentle agitation. The liberated monoterpene alcohols are extracted as described above. Geraniol derived by hydrolysis of residual substrate will also be produced by this treatment; hence, simple scintillation counting of the extract does not afford useful information. It is necessary to chromatographically separate any hydrolyzed reaction products from the geraniol in order to determine conversion rates.

Geranyl diphosphate synthase activity can be monitored by the methods developed by Popják to assay farnesyl diphosphate synthase.[169] The assay involves incubation with dimethylallyl diphosphate and radioisotopically labeled isopentenyl diphosphate in the presence of the divalent cation cofactor. The resulting geranyl diphosphate is measured by the "acid lability" assay, whereby the allylic diphosphate ester is solvolyzed in acid (primarily to linalool). A pentane overlay is generally employed in this assay. Owing to the acid lability of allylic diphosphates and the relative stability of the homoallylic isopentenyl diphosphate (labeled) cosubstrate, the bulk of the radiolabel recovered in this assay will derive from geranyl diphosphate or other prenyl diphosphate esters. Carefully designed controls must be included to account for any labeled dimethylallyl diphosphate formed from isomerase activity or labeled isopentenol resulting from contaminating phosphatases. These complications, and the potential presence of other prenyl diphosphates in the mixture, necessitate the chromatographic separation of reaction products to define the distribution. An alternative assay employs commercially available phosphatases and apyrase to convert the diphosphate esters present to the corresponding alcohols which can be extracted as above. Adjusting the pH of the reaction mixture (following assay) to the range appropriate for either acid or alkaline phosphatase is sufficient to terminate prenyltransferase activity. A disadvantage of this approach is the phosphatase-catalyzed release of labeled isopentenol from residual substrate, thereby requiring a chromatographic step to separate products for rate determination. Both the acid lability assay and the enzyme hydrolysis assay require calibration with known standards; additional details and precautions in using these methods have been described.[163,164,170]

Although secondary transformations of the monoterpenes involve a number of different types of reactions, the same general considerations and precautions apply. Since the substrates for these reactions are generally hydrophobic, a pentane overlay of the assay mixture for trapping products is not used. Unless advantage can be taken of spectrophotometric monitoring of change of a redox cofactor (NAD, NADPH), there is little alternative but to separate product(s) chromatographically from substrate(s) following assay for determination of rates by radiochemical or mass conversion.

2.05.3.4 Radio-GLC and HPLC

Although several chromatographic techniques are available for the separation and quantification of monoterpenes,[164] perhaps the single most powerful tool available is radio-GLC.[171] Interfacing a gas proportional counting chamber with a gas chromatograph allows direct detection of eluting radiolabeled compounds which can be identified by comparison of retention time to an authentic standard. This system is substantially improved by including a thermal conductivity detector in-line with a combustion–reduction chamber or catalytic processor for conversion of labeled compounds to $^{14}CO_2$ or $^{3}H_2$ for proportional counting. This approach allows the introduction of unlabeled standards into the analyte, which are detected by the nondestructive thermal conductivity detector, and eliminates the need for stream splitting as with flame ionization detection. Calibration of the time delay between the thermal conductivity detector and the proportional counting tube allows precise correlation between the retention times of the known standards and the detected radioisotope. Chromatographic conditions can be devised which allow baseline resolution of most monoterpenes with a detection limit as low as 1000 DPM per component.

Although radio-GLC represents a sensitive and reliable technique for the identification of monoterpenes, there are many situations in which identification of a specific enantiomer is desired. Unless sufficient material can be produced for analysis by conventional gas chromatography employing "chiral" capillary columns, it is necessary to convert the monoterpene to a diastereomer which can then be chromatographically resolved by radio-GLC.[172] It should be recognized that chromatographic coincidence is insufficient evidence to conclusively identify a chemical structure and that, in the absence of additional data, it may be necessary to cocrystallize, to constant specific activity, the suspected product or a derivative.[164] Methods for the preparation of crystalline derivative of most monoterpenes are available.[14,164,173] The increased sensitivity of modern GLC–MS systems should allow these techniques to find wider application in the identification of products of cell-free enzyme systems; NMR-based methods are generally only applicable to products of recombinant enzymes of monoterpene metabolism.

Phosphorylated monoterpenes such as geranyl diphosphate and bornyl diphosphate must either be hydrolyzed to their respective alcohols for GLC-based analyses, or they may be analyzed directly by HPLC. Again, due to the relatively small quantities of metabolites typically available, radiochemical detection methods are necessary. The use of flow through scintillation counting chambers interfaced with an HPLC instrument can provide sensitivity equal to or surpassing that of radio-GLC. Reversed-phase ion-paired chromatography (IPC) methods are particularly useful for phosphate esters.[174–176] As with radio-GLC, a nondestructive (UV or refractive index) detector[174] in-line with the scintillation counter allows for the use of authentic internal standards for determination of retention time coincidence.

2.05.3.5 Molecular Biology

Since the early 1990s, it has been possible to isolate cDNAs encoding a number of monoterpene synthases and to express these sequences in *E. coli* as functional enzymes, albeit in the form of preproteins containing a transit peptide (see Section 2.05.2.1). The first monoterpene synthase to be cloned, using the classical reverse genetic strategy, was the (−)-(4*S*)-limonene synthase from spearmint (*M. spicata*).[126] This sequence was subsequently utilized to acquire the corresponding (−)-(4*S*)-limonene synthase from *Perilla frutescens*, a closely related species.[127] More recently, a similarity based, polymerase chain reaction (PCR) strategy has been designed to isolate monoterpene synthases from other plants.[115] Thus, degenerate primers, based on conserved regions of amino acid sequence from spearmint limonene synthase,[126] a monoterpene synthase from sage (*S. officinalis*),[177] a sesquiterpene synthase from tobacco (*Nicotiana tabacum*),[178] and a diterpene synthase from castor bean (*Ricinus communis*),[179] are used to generate specific hybridization probes from an appropriate cDNA library that is subsequently screened. This approach has yielded several monoterpene synthase clones from sage (*S. officinalis*) and grand fir (*Abies grandis*), and is particularly fruitful when combined with the use of enriched cDNA libraries, such as those constructed from mRNA isolated from oil glands (sage) or induced tissue (wound-induced grand fir stems).[49,50,136]

2.05.4 GERANYL DIPHOSPHATE BIOSYNTHESIS

In contrast to the extensive research characterizing farnesyl and geranylgeranyl diphosphate synthase, relatively little is known about geranyl diphosphate synthase, the prenyl transferase that catalyzes the single 1′–4 condensation of dimethylallyl diphosphate (**13**) and isopentenyl diphosphate (**14**) to the C_{10} product (Scheme 2) that serves as the universal precursor of monoterpenes. Although it has been suggested that a specific geranyl diphosphate synthase does not exist, and that the C_{10} intermediate is generated as a by-product of other prenyltransferases,[180] several geranyl diphosphate synthases have been demonstrated that are distinct from farnesyl diphosphate synthase and geranylgeranyl diphosphate synthase. These enzymes appear to be similar in properties to other prenyltransferases and are assumed to operate by a similar reaction mechanism (see Chapter 2.04).

The first reported demonstration of a geranyl diphosphate synthase was from the microorganism *Micrococcus lysodeikticus*[181] which produces solanesyl diphosphate from geranyl diphosphate and hence requires a source of this C_{10} precursor. The isolated geranyl diphosphate synthase was shown to produce geranyl diphosphate exclusively when supplied with dimethylallyl diphosphate and isopentenyl diphosphate as cosubstrates. However, when farnesyl diphosphate was used as the allylic cosubstrate, the partially purified enzyme efficiently synthesized geranylgeranyl diphosphate. Despite the application of a variety of chromatographic separation techniques, these two enzymatic activities could not be resolved.[181] Nevertheless, this enzyme is demonstrably different from a separate geranylgeranyl diphosphate synthase from the same organism and it shows no farnesyl diphosphate synthase activity.[182]

The first specific evidence for a geranyl diphosphate synthase from a plant source was obtained using a cell-free system from rose geranium (*Pelargonium graveolens*).[135] The essential oil of this species consists largely of geraniol (**26**), nerol (**1**), and citronellol (**25**), hence it was selected as a potentially rich source of geranyl diphosphate synthase. Cell-free extracts incorporated high levels of dimethylallyl diphosphate and isopentenyl diphosphate into geranyl diphosphate; however, no efforts were made to purify the responsible enzyme and no evidence was sought for the presence of other prenyltransferases.

Lithospermum erythrorhizon utilizes geranyl diphosphate in the construction of the side chain of the naphthoquinone pigment shikonin (**93**) (Scheme 13) and so was considered a potential source of geranyl diphosphate synthase. Extracts of cell cultures of *L. erythrorhizon* yielded a prenyltransferase with high specificity for geranyl diphosphate as product.[174] The partially purified enzyme was operationally soluble, had a molecular weight of 7.3×10^4 as determined by gel filtration, an isoelectric point of 4.95, and an absolute requirement for a divalent cation (Mg^{2+} or Mn^{2+}, each being effective) as is typical of prenyltransferases.[183] Incubation with isopentenyl diphosphate and dimethylallyl diphosphate yielded geranyl diphosphate as the exclusive product. Only when extremely high levels of geranyl diphosphate were used as the allylic cosubstrate, and in the absence of dimethylallyl diphosphate, were trace amounts of higher isoprenologs formed.

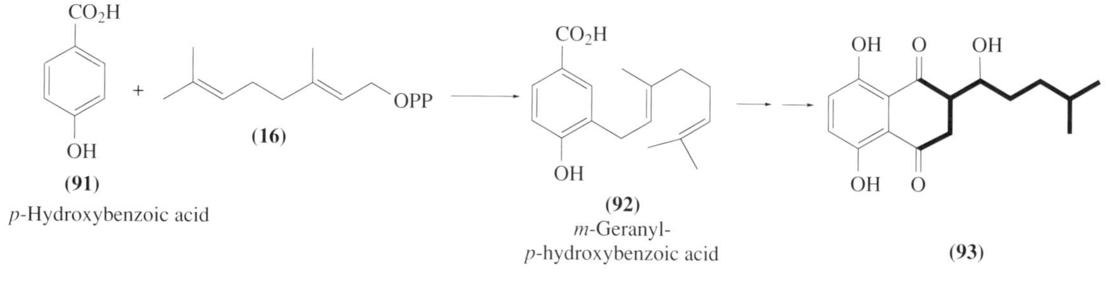

CO_2H OH **(91)** *p*-Hydroxybenzoic acid + OPP **(16)** → CO_2H OH **(92)** *m*-Geranyl-*p*-hydroxybenzoic acid → OH O OH OH O **(93)**

Scheme 13

Another geranyl diphosphate synthase has been reported from garden sage (*S. officinalis*).[14,170] Extraction of the leaf glandular trichomes and of whole leaves, followed by partial purification, allowed the demonstration of three distinct prenyltransferases, geranyl diphosphate synthase, farnesyl diphosphate synthase and geranylgeranyl diphosphate synthase. The preparations from glandular trichomes evidenced a dramatic enrichment of geranyl diphosphate synthase activity relative to the other prenyltransferases when compared to whole leaf extracts,[170] and further purification of the glandular extracts yielded an enzyme fraction largely free of isopentenyl diphosphate isomerase, phosphatases, and competing prenyltransferase activities. Such preparations, when supplied with either dimethylallyl diphosphate, geranyl diphosphate, or farnesyl diphosphate as the allylic cosubstrate, resulted in the corresponding production of geranyl diphosphate, farnesyl diphosphate, or geranylgeranyl diphosphate in the ratio of 100:2:18, respectively, indicating a significant enrichment of the geranyl diphosphate synthase. This enzyme showed a native molecular weight of approximately 10^5 by gel filtration, which is somewhat larger than most prenyltransferases (5.5×10^4–8.5×10^4),[184] gave a pH optimum of 7.0 (similar to most other prenyltransferases), and had an absolute requirement for a divalent cation, with Mg^{2+} preferred.[170] Steady state kinetic analysis of the partially purified geranyl diphosphate synthase demonstrated K_m values of 7.3 μM and 5.6 μM for isopentenyl diphosphate and dimethylallyl diphosphate, respectively; these are in the normal range (1–10 μM) for most prenyltransferases. All of these properties, with the exception of the somewhat higher molecular weight, are typical of prenyltransferases (Chapter 2.04).

More recently, a geranyl diphosphate synthase from cell cultures of Muscat grape (*Vitis vinifera*) has been purified to electrophoretic homogeneity[185] and the enzyme localized to the plastids of this tissue.[128,130] Similarly, a geranyl diphosphate synthase from daffodil (*Narcissus pseudonarcissus*) flowers has been localized to the chromoplasts.[124] Summation of the evidence clearly indicates that geranyl diphosphate synthase, as a product-specific, but otherwise typical, prenyltransferase, does exist, that the enzyme is associated with monoterpene biosynthesis (by the enriched presence in oil glands), and is, in most instances, localized to plastids, consistent with this subcellular site of monoterpene biosynthesis. A gene encoding geranyl diphosphate synthase has been isolated (unpublished, 1997).

2.05.5 CYCLIZATION REACTIONS

The majority of naturally occurring monoterpenes, and nearly all those of commercial significance, are cyclic compounds. Because of this, as well as the intriguing biochemical routes to their formation from geranyl diphosphate, much of the work on monoterpene biosynthesis has focused on cyclization reactions. The enzymes catalyzing these reactions, which establish the basic structures of the monoterpenes, are collectively referred to as monoterpene cyclases, although the more encompassing term, monoterpene synthase, is also used. This section presents an in-depth discussion of the reactions responsible for the formation of the monocyclic and bicyclic carbon skeletons which lead to the myriad monoterpenes found in nature, with a focus on specific examples which best illustrate the underlying principles. A great deal has been learned about the stereochemical and mechanistic features of these reactions. However, relatively little is known about the role of the enzyme in this process. Thus, for example, it is not known how the geranyl substrate is channeled into the appropriate conformation for the precise control of cyclization regio- and stereochemistry, or by what means the enzyme stabilizes the highly reactive carbocationic intermediates while directing them through the reaction coordinate to termination without premature solvent capture or active site alkylation.

2.05.5.1 Model Chemical Reactions

To understand the biochemistry of this class of enzymes, it is helpful to review the chemical principles upon which they depend. In this regard, there are numerous chemical model studies of the monoterpene cyclization reaction.[9] Geranyl derivatives give rise to monocyclic monoterpene products under a variety of solvolytic reaction conditions,[186–188] albeit as minor constituents relative to acyclic products. Because of the topological impediment to direct cyclization presented by the 2-*trans* configuration of the geranyl system, and the formidable energy barrier (>50 kJ mol^{-1}) to rotation of the corresponding allylic cation,[189,190] (**17**) the reaction must involve preliminary conversion to a tertiary (linalyl) intermediate (**18**) or its ion-paired equivalent.[14,188] Formation of a linalyl intermediate allows free rotation about the newly formed C-2–C-3 single bond, thereby allowing C-6–C-1 cyclization (Scheme 2). Reaction conditions which promote the formation and stabilization of the tertiary allylic system favor the production of cyclic products.[191]

In contrast to the geranyl system, solvolysis of neryl and linalyl derivatives affords relatively high yields of cyclic products (e.g., α-terpineol (**37**), terpinolene (**31**)).[186,192] Stereochemical selectivity in the cyclization of linalool (**27**) has been recognized since 1898, when Stephan demonstrated that solvolysis of the (−)-(*R*)-enantiomer afforded optically active α-terpineol.[15] Although the C-6–C-7 double bond of the acyclic precursor is obviously necessary for cyclization, the extent to which the terminal double bond actually participates in the reaction is subject to considerable debate. The enantiomeric excess observed in the formation of α-terpineol from several linalyl esters has been rationalized as the result of anchimeric assistance from the distal π electrons.[188,192,193] However, several other studies present contradictory arguments for the importance of π participation[194,195] and the interpretation of such results is complicated by manifold opposing factors (see ref. 9).

The most definitive model studies to address the stereochemistry and mechanism of monoterpenoid cyclization processes are those of Arigoni and associates,[196] and of Poulter and King.[197,198] Arigoni and his colleagues recognized that cyclization of linalyl *p*-nitrobenzoate (**94**) to α-terpineol must occur either through a *syn, exo* or an *anti, endo* conformation (illustrated for the *R* enantiomer in Scheme 14). Since the alternative *syn, endo* or *anti, exo* reactions do not result in the observed absolute configuration of the product, these investigators reasoned that the two possible stereochemical alternatives could be distinguished by determining the fate of the hydrogens at the C-1 position (H_a and H_b, Scheme 14) of [1,2-$^2H_2$]linalyl *p*-nitrobenzoate in the cyclization.[196] The resulting α-terpineol, derived from the specifically deuterated precursor, was converted to [5,6-$^2H_2$]hydroxycineole and, through NMR analysis, it was determined that 80–90% of the linalyl *p*-nitrobenzoate was cyclized to α-terpineol via the *anti, endo* conformation.[15]

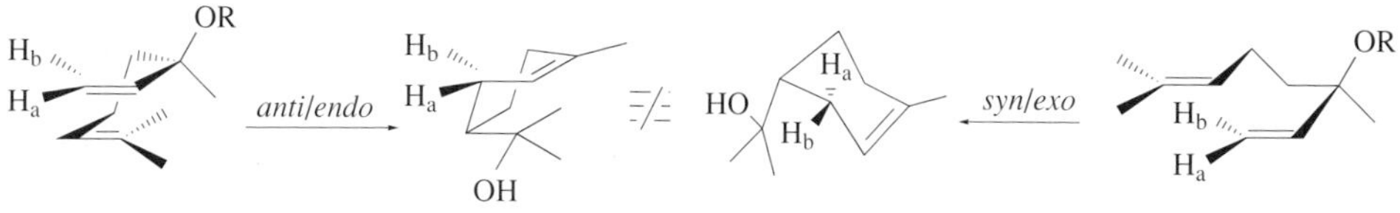

(**94**) : R = *p*-nitrobenzoate (*R*)-(**37**) (*R*)-(**37**) (**94**)

Scheme 14

In an equally insightful series of experiments, Poulter and King used stereospecifically labeled *N*-methyl-(*S*)-4-([1′-^{2}H]neryloxy)pyridinium methyl sulfate (**95**) to demonstrate, by NMR methods, that the solvolytic cyclization of this neryl derivative to α-terpineol (**37**) is stereospecific with respect to C-1 and results in inversion of configuration[197] (Equation (1)). In a subsequent set of experiments, in which rate differentials and proportions of cyclic products resulting from the solvolysis of fluorinated neryl and geranyl methanesulfonates were compared, Poulter and King demonstrated that ionization precedes cyclization.[198] Taken together, these studies provide compelling evidence that the cyclization of acyclic, allylic terpenoid precursors is both stepwise *and* stereospecific.

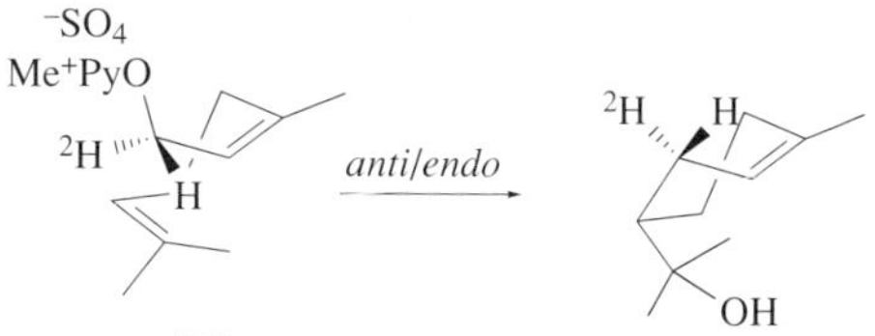

(**95**) (**37**) (1)

Solvolytic formation of bicyclic monoterpenes from acyclic or monocyclic precursors has not been observed as there is no means of enforcing the appropriate conformation for bicyclization under these conditions.[9,14] However, the facile rearrangements of pinyl, bornyl, and camphanyl derivatives are well documented,[9,199] and it was largely investigations on the interconversions of these monoterpenes that led to the concept of the nonclassical bridged carbocation (reviewed by Sargent[200]), a concept that has been vigorously contested.[201] Within the confines of the monoterpene synthase active site, preferential stabilization of particular cationic species during the cyclization process certainly must play a critical role in product distribution. The precise nature and means of stabilizing such enzyme-bound carbocationic intermediates remain speculative.

The stabilizing effect of aromatic groups on carbocations through cation–π interactions has been posited as a factor in enzyme-catalyzed reactions,[202] and the process has been biomimetically modeled using polyaromatic cyclophane "hosts" (**96**) that are able to bind, sequester, and enhance the reaction rates of cationic species.[203,204] The concept that enzymatic stabilization of cationic species is afforded by the electron-rich aromatic rings of proximate phenylalanine, tyrosine, and tryptophan residues has gained general support from X-ray crystallographic studies[205] and sequence analyses[206] of relevant protein catalysts. While direct evidence for stabilization of transient carbocationic intermediates has yet to be provided in the case of monoterpene synthases, it is certainly plausible that protein aromatic residues provide such a function. Sequence data recently obtained from a number of monoterpene synthases do, in fact, reveal substantial conservation of aromatic residues (see Section 2.05.5.6).

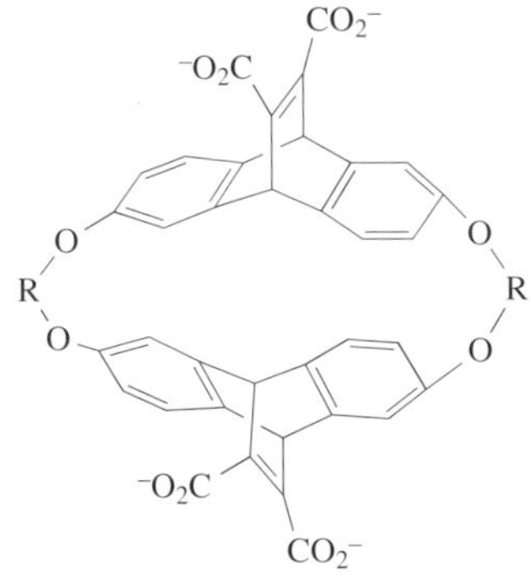

(96) 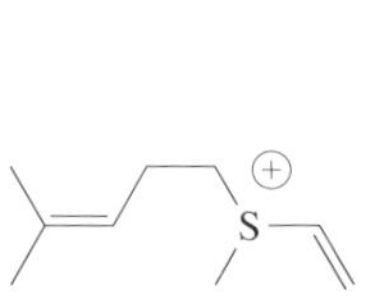 **(97)** 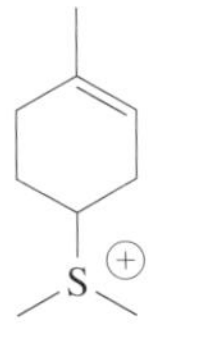 **(98)** 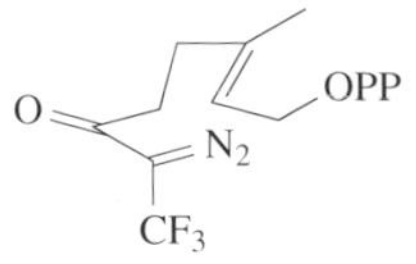 **(99)**

The generation of catalytic antibodies capable of initiating and controlling cationic cyclization reactions has also been reported.[207] Significantly, these biocatalysts show rate accelerations for the cyclization of (**100**) to cyclohexene (**102**) within an order of magnitude of reaction rates observed for monoterpene synthases (note that these enzymes are relatively slow catalysts, see Section 2.05.5.2). Moreover, use of (**101**) as substrate resulted in significant production of a bicyclic [3.1.0] product (**103**) (Scheme 15). It will be of interest to evaluate the role of aromatic residues in these catalysts.

(**100**) : $R_1 = H, R_2 = Si(CH_3)_2C_6H_5$
(**101**) : $R_1 = CH_3, R_2 = Si(CH_3)_2C_6H_5$

(**102**) 90% from **100**

(**103**) 63% from **101**

Scheme 15

The monoterpene synthases exhibit a universal requirement for a divalent metal ion cofactor, usually magnesium or manganese.[14] The metal ion is believed to neutralize the negative charge of the diphosphate moiety, thus promoting ionization. Investigations on the chemistry of allylic diphosphates have revealed that they are quite stable in alkaline solution; however, lowering the pH increases lability to solvolysis, indicating that protonation markedly improves the leaving group potential of the diphosphate and facilitates ionization.[15,208] An investigation of the susceptibility of various alkyl phosphate and diphosphate esters to acid hydrolysis revealed that nonallylic esters undergo P—O bond fission, whereas allylic esters suffer C—O bond cleavage by S_N1 type processes.[209] Solvolysis of geranyl diphosphate at neutral pH is dramatically enhanced in the presence of Mg^{2+} or Mn^{2+} and, based on consideration of dissociation constants[210] and the dependence of rate enhancement on concentration, a dimetal ion–geranyl diphosphate complex appears to be the reactive species.[211,212] Interestingly, ^{31}P- and ^{13}C-NMR spectra of geranyl diphosphate magnesium chelates show a slight (0.14 ppm) downfield shift of the C-1 carbon relative to the unchelated geranyl diphosphate,[213] suggesting a weakening of the C—O bond in the metal ion complex.

While model studies provide a sound chemical rationale for the reactions catalyzed by monoterpene synthases, they cannot provide unequivocal proof that the same processes are operative in the enzymatic reaction. Nevertheless, all of the following evidence is entirely consistent with enzymatic cyclization proceeding by stepwise electrophilic processes involving a series of carbocation–diphosphate anion pairs. The role of the enzyme, then, is to appropriately fold the substrate so as to facilitate metal ion-assisted ionization, and then to stabilize and direct the ensuing cationic intermediates through the electrophilic cascade to termination, and to do so in a hydrophobic environment that excludes water.

2.05.5.2 Enzymology

The enzymes responsible for transformation of geranyl diphosphate to the various monocyclic and bicyclic monoterpene products have been isolated from a variety of plant species of the gymnosperms,[167,214] angiosperms,[14] and bryophytes.[215] Owing to the relatively low tissue levels of these enzymes, as well as to the difficulties inherent to protein isolation from plant sources, the purification of the monoterpene synthases in quantities sufficient for detailed characterization has proven difficult. However, the development of purification techniques, including the preliminary isolation of enriched tissues (such as gland cells), has facilitated sufficient purification of many synthases to allow extensive biochemical characterization (see Section 2.05.3). Purification to electrophoretic homogeneity has been accomplished in the case of limonene synthase from spearmint (*M. spicata*),[216] γ-terpinene synthase from thyme (*Thymus vulgaris*),[217] and (−)-pinene synthase from grand fir (*A. grandis*).[218]

Many plants produce more than one synthase, each of which may generate different products. Separation of the individual enzymes is necessary for accurate biochemical characterization, but this process is complicated by the fact that the monoterpene synthases are very similar in physical properties as reflected by a high level of sequence similarity (see Figure 2). The separation problem is further exacerbated by the fact that many monoterpene synthases produce multiple products from the geranyl substrate. While the total number of unique monoterpene synthases is unknown, it has been estimated at more than 50.[14] Given that mechanistically related multiple-product synthases differ qualitatively and quantitatively in product distribution, and that synthases from different plants that produce similar product profiles are now known to differ significantly in amino acid sequence, this number is undoubtedly a substantial underestimate. The most thoroughly characterized monoterpene synthases have been isolated from *Salvia*, *Mentha*, *Tanacetum*, *Foeniculum*, *Pinus*, *Abies*, and *Citrus* species (these generate (−)-limonene, α-terpinene, γ-terpinene, (+)-sabinene, 1,8-cineole, (+)- and (−)-camphene, (+)- and (−)-α-pinene, (−)-β-pinene, and (+)- and (−)-bornyl diphosphate) and it is largely from this group that illustrative examples will be drawn.

The monoterpene synthases are, for the most part, operationally soluble enzymes with native molecular weights in the range of 50 000–100 000.[164] Based on a limited number of examples, they are either monomeric[216] or homodimeric.[177,217] They tend to be relatively hydrophobic and typically possess low pI values in the 5–6 range.[14,164,167,219] They all require a divalent metal ion for activity, exhibit a pH optimum generally within one unit of neutrality, and turnover rates from 0.01 to 1.0 s^{-1} are typical.[14] Efforts to *N*-terminally sequence native synthases have thus far proven unsuccessful, presumably due to blocked amino termini.[126,167,220]

Immunochemical characterizations of the monoterpene synthases have revealed a high degree of antigenic specificity.[221,222] Polyclonal antibodies raised against denatured (−)-pinene synthase from grand fir (*A. grandis*), for example, exhibited cross-reactivity to other denatured monoterpene synthases from grand fir but not toward any other synthases from other conifer or angiosperm species.[221] Likewise, polyclonal antibodies raised to denatured limonene synthase from spearmint were cross-reactive to other synthases from closely related mint species but not to any other synthases.[222] Neither of these antibody preparations showed demonstrable affinity for the respective native enzyme antigens,[221,222] suggesting that the most highly antigenic regions of the protein are folded into the interior and are exposed only upon denaturation. The inability of the antibody preparations to cross-react beyond the closest phylogenetic relationships is somewhat surprising, in light of the high level of amino acid sequence similarity between these catalysts from various sources (see Section 2.05.5.6).

2.05.5.3 Mechanism

Monoterpene synthases, for reasons indicated previously, require only a divalent metal ion cofactor for activity, usually Mg^{2+} or Mn^{2+}, with K_m values for the metal ion in the range of 0.5–5.0 mM.[14] Most monoterpene synthases can use either geranyl, neryl, or linalyl diphosphate as substrate. Based on comparative V/K_m assessment, the enantiomer of linalyl diphosphate having the appropriate configuration (see below) is the preferred substrate, followed by geranyl diphosphate which is generally a more efficient substrate than neryl diphosphate.[14] Only geranyl diphosphate is thought to arise naturally and so is considered to be the native substrate for these enzymes. Transformation of all these acyclic precursors proceeds without formation of detectable, free intermediates.[13,52,223] Indeed, it was the recognition that geranyl diphosphate was efficiently cyclized without formation of free intermediates that led to the realization that monoterpene synthases catalyze both the isomerization of the geranyl precursor as well as the cyclization step.[13] While it has not been unequivocally established, mixed substrate analyses, using combinations of geranyl, neryl, and linalyl diphosphates, have strongly indicated that all transformations, and thus both isomerization and cyclization steps, occur at the same active site.[223–225] The absence of free intermediates and the mutually competitive inhibition demonstrated by geranyl, linalyl, and neryl diphosphates, further indicate that the isomerization and cyclization steps are tightly coupled and essentially unidirectional.[14] Typical K_m values for geranyl diphosphate are in the low micromolar range.[14]

The monophosphates of geraniol, nerol, and linalool are not substrates for the monoterpene synthases; in fact, these compounds show moderate inhibition toward the normal cyclization.[223,226,227] Inorganic diphosphate is a good inhibitor of monoterpene synthases with a K_i value in the range of 100 μM.[14,228] In a series of studies on the binding characteristics of various substrate analogues, Croteau and co-workers determined that the diphosphate moiety of the acyclic precursor is the principal binding determinant.[228,229] Moreover, when sulfonium analogues of the linalyl (**97**) and α-terpinyl (**98**) cationic intermediates of the cyclization reaction were tested as inhibitors of (+)-bornyl diphosphate synthases and (+)-α-pinene synthase from sage (*S. officinalis*), it was found that the K_i values for these analogues decreased substantially (to submicromolar levels) in the presence of inorganic diphosphate ion, and vice versa.[230] This synergistic inhibition suggests that the paired species are more tightly bound than either partner alone, thereby implicating ion-pairing of intermediates in the course of the normal cyclization reaction. Comparison of the inhibitory properties of geranyl diphosphate analogues modified by reduction or epoxidation of the 2,3- or 6,7-double bonds with that of inorganic pyrophosphate, using the same enzyme systems, confirmed the diphosphate functionality as the primary determinant in substrate recognition and showed that the C-2–C-3 alkene is recognized largely on the basis of geometry whereas the C-6–C-7 alkene is distinguished through electronic interactions.[229] A topological model for the folded substrate was posited in which the planar C-1–C-4 region of the substrate is sandwiched between the diphosphate group and the planar C-5–C-8 region. The isopropylidene group may thus assist in promoting the requisite positioning of the C-1–C-3 allylic functionality to optimize orbital alignment for ionization, diphosphate migration, and cyclization, while simultaneously shielding the reactive site from water.[229]

The substrate analogue, (*E*)-4-[2-diazo-trifluoropropionyloxy]-3-methyl-2-buten-1-ol (DATFP–DMAPP) (**99**) was shown to produce an uncompetitive inhibition pattern with (+)-pinene synthase and (+)-bornyl diphosphate synthase from sage.[231] Pure uncompetitive inhibition, which indicates binding of the inhibitor to the enzyme–substrate complex, is quite rare,[232] particularly in the case of single substrate (geranyl diphosphate–Mg^{2+} complex) enzymes. These results were interpreted as indicating that the inhibitor occupies a bidentate binding site, presumably via the diphosphate functionality while displacing that portion of the normal substrate, or that substrate binding results in a conformational change in the enzyme which unveils an inhibitor binding site.

The electrophilic nature of both the isomerization and the cyclization steps of the reaction sequence has been demonstrated using 2-fluorogeranyl diphosphate and 2-fluorolinalyl diphosphate as substrates.[233] In what could be described as an intramolecular version of the approach used earlier by Poulter to decipher the mechanism of the prenyltransferases,[234] the electron-withdrawing fluorine substituent of these analogues is expected to retard formation of the corresponding carbocationic intermediate. The rates of both the initial ionization–isomerization reaction (with 2-fluorogeranyl diphosphate) and the secondary ionization-cyclization (with 2-fluorolinalyl diphosphate) were suppressed by over two orders of magnitude. Competitive inhibition studies showed that the fluorinated analogues closely resembled their native counterparts in binding characteristics.[233] With the complementary evidence from earlier work that implicated the intermediacy of ion paired intermediates[230] as well as the chemical model studies with geranyl and fluorogeranyl derivatives,[234] these results indicated that the monoterpene synthases catalyze electrophilic isomerization–cyclization reactions in a manner analogous to the prenyltransferase-catalyzed elongation reaction.

Chemical modification of monoterpene synthases has been employed to demonstrate the involvement of specific amino acids in binding and catalysis. Limonene synthase from peppermint (*M. piperita*), for example, is highly sensitive to the thiol-directed reagent *p*-hydroxymercuribenzoate, as well as to the histidine-specific reagents diethylpyrocarbonate (DEPC) and rose bengal.[235] The inactivation by both *p*-hydroxymercuribenzoate and DEPC is significantly reduced in the presence of the substrate or of substrate analogues and, in the case of DEPC inactivation, activity can be restored by treatment with hydroxylamine.[236] Examination of several other monoterpene synthases from herbaceous angiosperms for substrate-protectable inactivation by histidine-directed reagents shows this universal feature.[168,236] These experiments suggest that a histidine residue(s) is located at or near the active site of the monoterpene synthases and could serve as the general base, or may be involved in substrate binding and/or the initial ionization step. More recently, a mechanism-based inhibitor was used to specifically label the active site of monoterpene synthases from sage (*S. officinalis*)[177] and peppermint (*M. piperita*).[237] The inhibitor, 6-cyclopropylidene-(3*E*)-methyl-hex-2-en-1-yl diphosphate (**104**), upon isomerization and cyclization reveals a cyclopropyl cationic intermediate (**105**), analogous to the α-terpinyl cation (**2**) (Scheme 16). This highly reactive species undergoes electrocyclic rearrangement to the more stable allylic cation (**106**) which then alkylates the protein in the active site region.[237] Deployment of this mechanism-based alkylator with limonene synthase demonstrated the requirement for catalytic activation of the inhibitor and the covalent nature of the labeled adduct formed with the protein.[237] When utilized with (+)-pinene synthase and (+)-bornyl diphosphate synthase from sage, the resulting labeled protein, following CNBr digestion, yielded a labeled peptide with the following sequence: L Q L Y E A S F L L X K G E D T X E L A X E (X indicating an unidentified residue).[177] This sequence has since been shown to bear a high level of homology to segments of several other monoterpene synthases (Figure 2). It is noteworthy that this inhibitor shows varying degrees of reactivity with different synthases, being most effective (based on I_{50} values) against limonene synthase from spearmint (*M. spicata*) and γ-terpinene synthase from thyme (*Thymus vulgaris*). Some monoterpene synthases, such as (−)-β-pinene synthase (*Citrus limon*) and (+)-sabinene synthase (*T. vulgare*), are moderately sensitive to inactivation, while others, such as (+)-pinene synthase (*S. officinalis*) and (+)-bornyl diphosphate synthase are discernibly less sensitive.[237] A correlation between decreased susceptibility to inactivation and the ability to convert the analogue to monoterpene products was observed, suggesting that the degree of sensitivity to alkylation reflects the ability of the enzyme to process the substrate analogue through alternative reaction pathways. In the case of limonene synthase and γ-terpinene synthase, the terpinyl cation intermediate in the normal reaction is likely deprotonated by a base proximal to the allylic carbocationic center of the processed analogue; hence, they are highly susceptible to alkylation. Synthases which normally yield bicyclic products, and thus deprotonate from a different position, are likely able to cyclize, or otherwise process, the monocyclic cation generated from the mechanism-based inhibitor and, hence, are more refractory to alkylation.

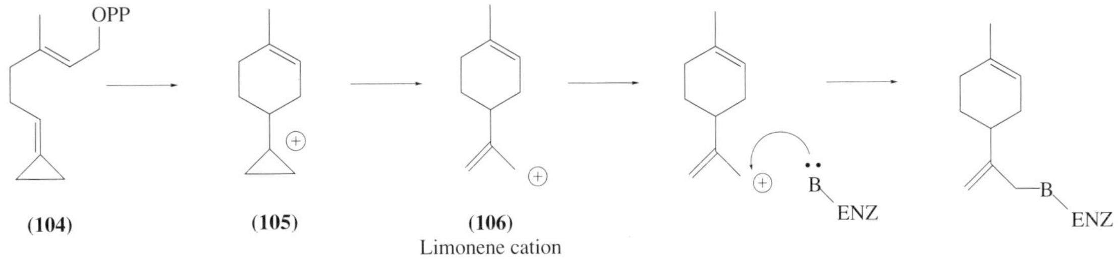

OPP

(**104**) (**105**) (**106**) Limonene cation

B ENZ

B ENZ

Scheme 16

Historically, monoterpene synthases from herbaceous angiosperms have provided the majority of experimental models for the cyclization reaction; however, in the 1990s, enzymes of this type from coniferous species received increased attention. These investigations have revealed that the synthases isolated from gymnosperms differ from those isolated from angiosperms in several respects. For example, the gymnosperm synthases tend to have an exclusive requirement for Mn^{2+} (or Fe^{2+}) as the divalent metal ion cofactor, and Mg^{2+} is an ineffective substitute. Most of these enzymes also require a monovalent cation, such as K^+ (see Table 1).[167] More intriguing, is the difference in susceptibility to amino acid modifying reagents between the two types of synthases. Most of the synthases of coniferous origin are susceptible to arginine-modifying reagents (in a substrate-protectable manner). Thiol- and histidine-modifying reagents also inactivate; however, substrate protection is not observed as with the synthases from angiosperms, thereby indicating differential access.[214] Conversely, angiosperm synthases are inactivated by arginine-directed reagents but they are not protected against inactivation by the substrate–metal ion complex. Thus, while the involvement of the same types of amino acid residues is indicated for both enzyme classes, the ability of the substrate to protect against inactivation differs significantly, suggesting at least differential placement, if not different roles for these residues, in binding and catalysis.

2.05.5.4 Stereochemistry

A historical perspective on the development of the general paradigm for monoterpene cyclization was presented in Section 2.05.1.1 and the analogies to the prenyltransferase reaction described. In

Table 1 Comparison of biochemical characteristics of monoterpene synthases of angiosperm and gymnosperm origin.

Property	*Angiosperms*	*Ref.*	*Gymnosperms*	*Ref.*
pH optimum	6.0–7.2	14,219	7.5–8.0	167,218
pI	4.3–5.25	168,216	~4.75	167
K_m GPP (μM)	1.0–15	14,163,219	3.7–7.8	167,218
Divalent cation requirement	Mg^{2+}, Mn^{2+}	14	Mn^{2+}, Fe^{2+} (Mg^{2+} ineffective)	167, 218
K_m M^{+2} (mM)	0.5–5.0	14,164,219	~0.03	218
Monovalent cation requirement	none		$K^+ > NH_4^+ \gg Na^+, Li^+$	167
K_m M^{+1} (mM)	N/A		5.0–12.0	167
Native size (kDa)	40–100	14,163,219	62–69	167,218
Quaternary structure	monomeric or homodimeric	14,163,219	monomeric	167,218
Active site residues	Cys, His	168,214,236	Arg	214

this section, experimental evidence that defines the stereochemical aspects of the model will be presented based on the mechanistic features of the reaction sequence developed in the prior sections, i.e., that the tightly coupled isomerization–cyclization cascade proceeds stepwise through a series of ion-paired, carbocationic intermediates.

In describing the overall stereochemistry of the isomerization–cyclization, it is important to first note that [1-$^3H_2$, G-^{14}C]geranyl diphosphate is enzymatically converted to a range of different monocyclic and bicyclic monoterpenes without loss of tritium.[238] This observation eliminated the possibility that geranyl diphosphate was isomerized to neryl diphosphate through an aldehyde intermediate, confirmed the central role of geranyl diphosphate as the universal precursor of the monoterpenes, and permitted direct view of the critical stereochemical alterations at C-1 of the substrate without complication by intervening redox processes. In brief overview (see Scheme 17), the geranyl diphosphate–divalent metal complex is initially ionized with *syn*-isomerization to an enzyme-bound linalyl diphosphate intermediate (**18**) (either as the (3*R*)- or (3*S*)-antipode). The stereochemistry of the linalyl intermediate is dictated by the helical folding of the geranyl substrate (**16**) achieved upon binding (the left-handed, screw-sense conformer yielding the (3*R*)-antipode and the right-handed, screw-sense conformer the (3*S*)-linalyl system). Although no direct evidence has been presented that the linalyl cation–diphosphate anion pair formed on initial ionization necessarily collapses to (enzyme-bound) linalyl diphosphate, the high rotational free energy barrier in the allylic cation[189] suggests collapse to the covalent intermediate or sufficiently tight ion-pairing to localize the double bond at C-1–C-2. After rotation about the newly formed C-2–C-3 single bond, a second ionization, this time of the linalyl diphosphate intermediate now folded in the *anti–endo* conformation, promotes cyclization by electrophilic attack by C-1 on the C-6–C-7 alkenic group. The resulting intermediate, the highly reactive (4*R*)- or (4*S*)-α-terpinyl carbocation (**2**) (the specific enantiomer depending on the linalyl antipode) that is common to all cyclization reactions, then undergoes any of a variety of subsequent transformations before terminal quenching of the carbocation (Scheme 17). These further transformations range in complexity from simple deprotonation of one of the adjacent methyl groups of the α-terpinyl cation, yielding limonene (**3**), to additional cyclization reactions involving the endocyclic double bond followed by deprotonation or capture by the cation of an exterior nucleophile (e.g., H_2O or the diphosphate ion). Hydride shifts, either 1,2 from the C-4 methine or 1,3 from the C-3 methylene (carbon numbering based on the terpinyl skeleton), move the charge into the ring and precede further cyclization reactions or deprotonation from one of the adjacent positions, or nucleophile capture.[14] Thus, the overall reaction can be viewed as a stepwise sequence involving ionization, migration of the diphosphate, bond rotation, ionization, cyclization(s), and termination. The role of the enzyme, in addition to catalyzing the required chemistry, is to initially fold the acyclic precursor into a competent conformation and then to direct the ensuing steps by stabilizing and guiding the intermediate cationic species. Although one of the hallmarks of enzyme-catalyzed reactions is the exquisite specificity in substrate recognition and product formation, the monoterpene synthases are quite remarkable in the fact that a single substrate is often transformed into multiple products.

The stereochemical model presented in Scheme 17 dictates that cyclization involving electrophilic attack on the C-6 position by the C-1 carbon would occur from the same face of the π-system from which the diphosphate departed in the prior isomerization step. To establish this, Croteau and co-workers examined the fate of stereospecifically labeled (1*R*)-[1-$^3H_1$; 2-^{14}C] and (1*S*)-[1-$^3H_1$;

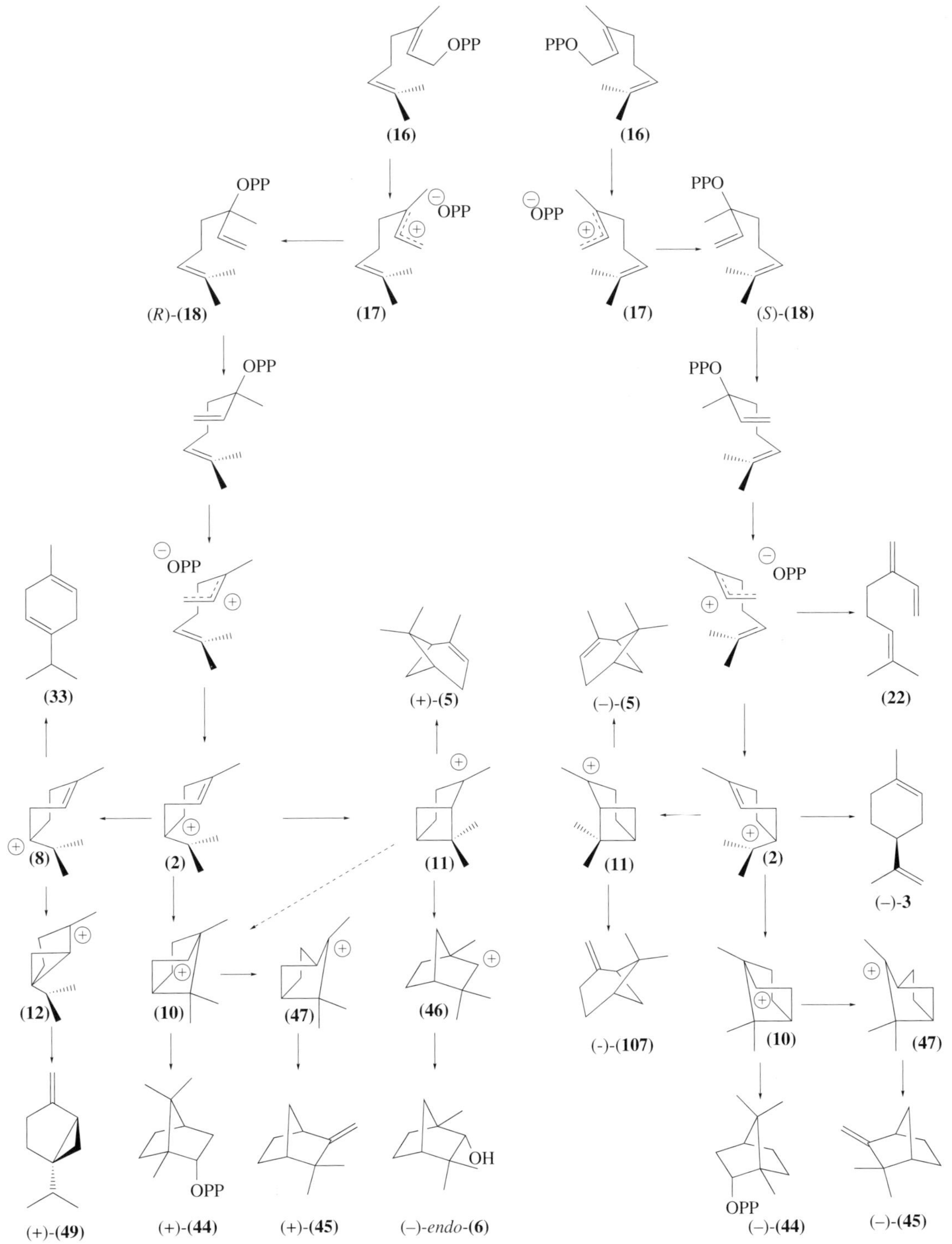

Scheme 17

2-^{14}C]geranyl (**16**) and neryl diphosphate (**108**) (illustrated for the (1*R*)-isomer of geranyl diphosphate and the (1*S*)-isomer of neryl diphosphate in Scheme 18) in the cyclizations catalyzed by (+)-(1*R*,4*R*)-bornyl diphosphate synthase from sage (*S. officinalis*) and (−)-(1*S*,4*S*)-bornyl diphosphate synthase from tansy (*Tanacetum vulgare*).[239] The stereospecifically labeled substrates were separately reacted with each of the enzymes, and the resulting bornyl diphosphate was hydrolyzed and then oxidized to the corresponding (+)- or (−)-camphor (**82**) which was subjected to stereoselective base-catalyzed exchange of the protons at C-3 (derived from C-1 of the acyclic precursor). Because the *exo*-α-hydrogen at C-3 of camphor is approximately 20-fold more labile than the *endo*-α-hydrogen,[240,241] comparison of tritium loss from the derived ketone allows determination of the stereochemistry at C-3 and, thereby, deduction of configuration relative to C-1 of the acyclic progenitor. Thus, it was found that the (+)-bornyl diphosphate (as (+)-camphor) generated from (1*R*)-[1-$^3H_1$]geranyl diphosphate retained significantly more tritium than the (+)-bornyl diphosphate generated from (1*S*)-[1-$^3H_1$]geranyl diphosphate. Conversely, the (−)-bornyl diphosphate retained significantly more tritium from (1*S*)-[1-$^3H_1$]geranyl diphosphate than from (1*R*)-[1-$^3H_1$]geranyl diphosphate. When (1*R*)- and (1*S*)-[1$^3H_1$; 2-^{14}C]neryl diphosphates were used as substrates essentially the opposite results were observed, i.e., (+)-bornyl diphosphate retained tritium from (1*S*)-neryl diphosphate and (−)-bornyl diphosphate retained tritium from the (1*R*)-enantiomer.[239] These results are precisely in accord with the *syn*-isomerization, transoid to cisoid rotation, and *anti*, *endo*-cyclization predicted for the conversion of geranyl diphosphate which results in retention of configuration at C-1. In the case of neryl diphosphate, rotation is unnecessary and so inversion at C-1 is observed, as predicted. Additional experiments, based on this same design but using (+)- and (−)-pinene synthases from sage[242] and (−)-*endo*-fenchol synthase from fennel (*Foeniculum vulgare*),[243] have all proved to be entirely consistent with the stereochemical exemplar discussed above.

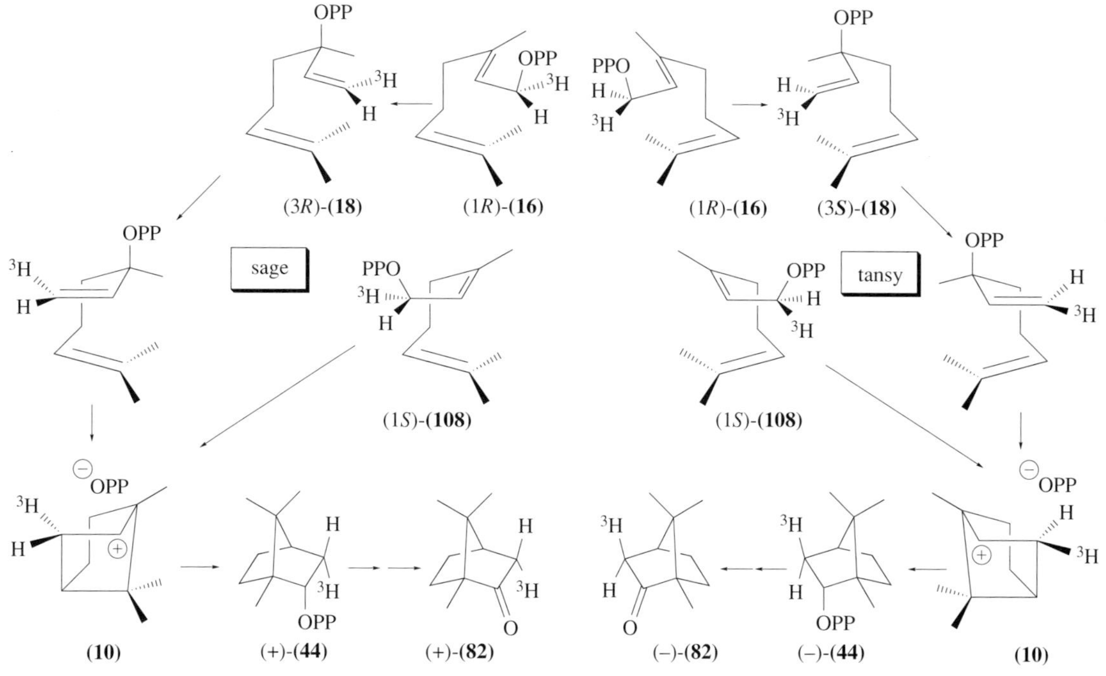

Scheme 18

The coupled isomerization-cyclization model predicts that folding of the substrate into either the right-handed or left-handed helical conformation will, upon ionization and isomerization, result in the appropriate stereoisomer of the bound linalyl diphosphate intermediate for subsequent *anti*, *endo*-cyclization to form products of the correct stereochemistry (see Scheme 17). This prediction can be tested, either by determining if one enantiomer in a racemic mixture of linalyl diphosphate (**18**) is selectively utilized by a particular synthase, or by evaluating the efficiency of utilization of each enantiomer separately. Advantage was again taken of the antipodal cyclizations catalyzed by (+)-bornyl diphosphate synthase from sage (*S. officinalis*) and (−)-bornyl diphosphate synthase from tansy (*F. vulgare*) in examining the transformations of linalyl diphosphate as an alternative substrate. Using both experimental approaches, Croteau and co-workers[244] demonstrated that the enzyme from tansy produced exclusively (−)-(1*S*,4*S*)-bornyl diphosphate from (3*S*)-[1*Z*-^3H]linalyl diphosphate; (3*R*)-[1*Z*-^3H]linalyl diphosphate was not a substrate for this synthase. Additionally, when this enzyme was reacted with a mixture of [(3*R*)-8,9-^{14}C] and [(3*R*,*S*)-1*E*-^3H]linalyl diphosphate (^3H:^{14}C = 5.22) the resulting (−)-bornyl diphosphate gave a ^3H:^{14}C ratio greater than 31, indicating selective use of the (3*S*)-enantiomer.

Interestingly, the (+)-bornyl diphosphate synthase was able to utilize both antipodes of linalyl diphosphate; however, the (3*R*)-enantiomer proved to be a significantly better substrate, with a relative velocity more than four-fold higher than that of the (3*S*)-enantiomer. Cyclization of (3*R*)-linalyl diphosphate resulted in the exclusive formation of (+)-(1*R*,4*R*)-bornyl diphosphate, whereas the (3*S*)-linalyl diphosphate produced, exclusively, (−)-(1*S*,4*S*)-bornyl diphosphate.[244] It should be emphasized that both enzymes show absolute stereochemical fidelity in the cyclization of the achiral precursors geranyl and neryl diphosphate.

In a manner similar to the antipodal biosynthesis of (+)- and (−)-bornyl diphosphate by the related enzymes from sage and tansy, the antipodal (+)- and (−)-pinenes are synthesized by two distinct enzymes in sage.[48] These enzymes, termed cyclase I ((+)-pinene synthase) and cyclase II ((−)-pinene synthase), respectively, are readily separated by size exclusion chromatography, and so can be examined independently. These two synthases were used to evaluate the fate of stereospecifically labeled (1*R*)- and (1*S*)-[1-$^{3}H_1$; 2-^{14}C]geranyl diphosphates in the conversion to the antipodal pinenes.[242] By stereoselective rearrangement of the resulting (+)- and (−)-pinene products to (+)- or (−)-borneol (**4**) and then oxidation to the corresponding camphor (**82**), advantage was again taken of the stereoselective base-catalyzed exchange of the α-*exo*-hydrogen to decipher the stereochemical alteration at C-1 of the acyclic precursor[242] (Scheme 19). Likewise, the stereochemical fates of the C-1 hydrogens of (1*R*)- and (1*S*)-[1-$^{3}H_1$; 2^{14}C]neryl diphosphates were also determined. With the (+)- and (−)-pinene synthases, the results showed retention of configuration at C-1 of geranyl diphosphate and inversion at C-1 of neryl diphosphate in the course of the isomerization–cyclization reaction. Examination of the optically pure (3*S*)- and (3*R*)-enantiomers of linalyl diphosphate as alternative substrates with these two synthases resulted in the predicted stereospecific transformation of (3*R*)-linalyl diphosphate to (+)-α-pinene and of (3*S*)-linalyl diphosphate to (−)-β-pinene. Hence, the reaction was again demonstrated to proceed through *syn* isomerization of the primary allylic diphosphate to the tertiary allylic diphosphate, followed by rotation about the C-2–C-3 single bond and *anti*, *endo*-cyclization.[245]

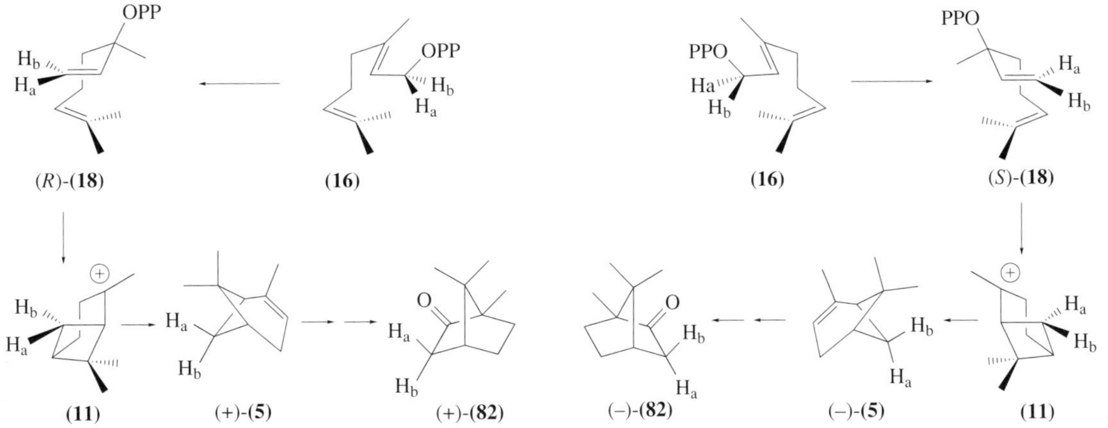

Scheme 19

The (+)-pinene synthase and the (−)-β-pinene synthase from sage both produce multiple products from geranyl diphosphate, a characteristic of several other monoterpene synthases.[14] In the case of (+)-pinene synthase, geranyl diphosphate yields approximately 49% (+)-α-pinene (**5**), 30% (+)-camphene (**45**), and 10% (+)-limonene (**3**), along with minor amounts of myrcene (**22**) and terpinolene (**31**).[246] (−)-Pinene synthase produces approximately 28% (−)-α-pinene, 35% (−)-β-pinene (**107**), and 24% (−)-camphene, as well as minor amounts of (−)-limonene, terpinolene, and myrcene, from geranyl diphosphate.[246] When the alternate substrates linalyl diphosphate or neryl diphosphate are utilized, the product distribution profiles of both enzymes are modified.[246] Particularly significant is the demonstrably higher relative proportion of limonene produced by (+)-pinene synthase from neryl diphosphate; a similar phenomenon is also observed with (−)-pinene synthase using neryl diphosphate as substrate. Even more intriguing is the fact that the product generated from the neryl substrate by (+)-pinene synthase is nearly 90% (−)-(*S*)-limonene, the opposite enantiomer of that expected (from geranyl diphosphate). Even when (3*R*)-linalyl diphosphate is used as substrate with the (+)-pinene synthase, a lesser, but significant, fraction (13%) of the limonene produced is of the (*S*)-configuration. (−)-Pinene synthase generated approximately

65% of the expected (−)-limonene and 45% (+)-limonene from neryl diphosphate, whereas from (3*S*)-linalyl diphosphate, nearly 40% of the limonene produced was of the aberrant (+)-(4*R*)-configuration.

Limonene synthesized from geranyl diphosphate by these two enzymes is essentially optically pure (i.e., (+)-limonene by (+)-pinene synthase, and (−)-limonene by (−)-pinene synthase). Thus, the adulterated stereochemical profiles exhibited in the reactions with neryl diphosphate and linalyl diphosphate clearly belie a fundamental difference in the way these substrates are positioned at the catalytic site. As the diphosphate–divalent metal complex is known to be the primary determinant of substrate binding,[228,229] it is unlikely that this aspect of substrate interaction is significantly altered. Rather, altered orientation of the alkenic side chain, in all likelihood, underlies aberrant product formation. Unlike geranyl diphosphate, which must undergo the relatively slow ionization–isomerization step prior to cyclization, both neryl and linalyl precursors can cyclize directly with possible π-participation from the distal C-6–C-7 group. These stereoelectronic features may be reflected in temporal differences between geranyl diphosphate and the neryl and linalyl diphosphate substrates in achieving optimal binding orientation prior to the initial ionization. Since it is clear that both synthases can cyclize both enantiomers of linalyl diphosphate, the possibility that the neryl substrate is contorted into the abnormal helical conformation during catalysis cannot be dismissed. However, the facts that geranyl diphosphate is converted from a single helical conformer by each of these enzymes, and that the hydrophobic/hydrophilic profiles of the two conformers of neryl diphosphate are considerably different (unlike linalyl diphosphate), argue against this possibility. More likely, it is the "premature" cyclization from extended *anti–exo* conformers of the neryl and linalyl substrates that account for the aberrant stereochemistry in limonene formation (Scheme 20). By contrast, geranyl diphosphate "preassociates" with the isopropylidene group aligned in *endo*-fashion on binding and prior to the initial ionization, thereby resulting in absolute stereochemical fidelity in the cyclization.[246]

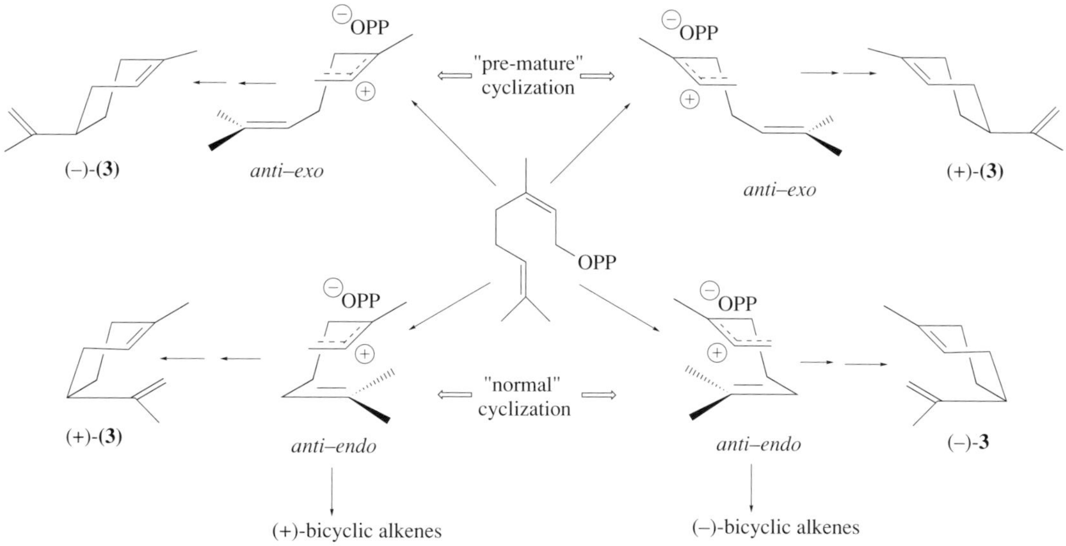

Scheme 20

Reaction of linalyl diphosphate with the (+)- and (−)-pinene synthases from sage also results in the production of substantially higher levels of acyclic alkenes (predominately myrcene (**22**), and *cis*- and *trans*-ocimene (**28**, **29**)) than observed with geranyl diphosphate as substrate.[246] Examination of the potential for alkene formation from linalyl diphosphate (Scheme 17) indicates that only acyclic products can be generated by ionization of this substrate from transoid conformations. There is currently no simple method to probe the precise origin of these acyclic alkenes; however, because the cisoid substrate, neryl diphosphate, does not afford appreciable levels of acyclic alkenes, the derivation of acyclics from transoid conformers of the linalyl substrate seems a plausible rationale. Another significant finding to evolve from study of the enzymatic transformations of neryl diphosphate and linalyl diphosphate with the pinene synthases is the complete lack of oxygenated products from these alternate substrates. Thus, in spite of the markedly different binding interactions that occur with the three substrates, all of the transformations occur in a sufficiently shielded environment to prevent water trapping of the cationic intermediates.

Analysis of the relative efficiencies with which the different substrates are cyclized has indicated that, in nearly all cases, the enantiomer of linalyl diphosphate predicted as an intermediate from geranyl diphosphate exhibits a somewhat lower K_m value than geranyl diphosphate, with a substantially higher relative velocity.[244,245] These data strongly suggest that the initial ionization–isomerization is the slow step in the coupled reaction sequence.[14] Analysis of the β-secondary kinetic isotope effects observed using [4-$^2H_2$; 1-^{3}H] and [10-$^2H_3$; 1-^{3}H]geranyl diphosphate as substrate with the (−)-pinene synthase from grand fir (*A. grandis*) and (−)-phellandrene synthase from lodgepole pine (*Pinus contorta*) provides supporting evidence for this hypothesis[247] (see Section 2.05.5.5).

The bornyl diphosphate synthases are unique among the monoterpene synthases in that they retain the substrate diphosphate in the cyclized product and thus provide an opportunity to evaluate the role of the diphosphate group in monoterpene cyclization reactions.[54] In initial experiments, the (+)- and (−)-bornyl diphosphate synthases were separately incubated with [1-^{3}H; α-^{32}P] and [1-^{3}H; β-^{32}P]geranyl diphosphate. Selective hydrolysis of the resulting bornyl diphosphate products to the corresponding monophosphates showed that the two ends of the substrate diphosphate retained their identities in the cyclization (Scheme 21).[248] With this evidence that the two phosphates were not scrambled in the transformation, deciphering the fate of the C–O–P bridge oxygen became essential. To this end, [8,9-^{14}C; 1-^{18}O]geranyl diphosphate was synthesized and used as substrate with both enzymes, from which the benzoate esters of the derived borneols were analyzed by mass spectrometry and found to have retained the ^{18}O label[248] (Scheme 22). Retention of the phosphate ester bridge oxygen in the transformation from geranyl diphosphate, through the linalyl intermediate, to bornyl diphosphate can be envisioned to formally occur in one of two ways, either by sequential [1,3], [1,2] sigmatrophic shifts or via sequential [3,3], [2,3] sigmatrophic shifts. Using both α- and β-^{32}P-labeled, and 3-^{18}O-labeled, linalyl diphosphates as substrates with both (+)- and (−)-bornyl diphosphate synthases, Croteau and his colleagues again demonstrated that the α- and β-phosphates retained their identity in the cyclizations, and that the bridging ester oxygen was again preserved.[54] In summary, these experiments establish that the diphosphate anion is tightly coupled with its carbocationic partners throughout the course of the reaction and that transfer of the diphosphate involves a formal 1,3- followed by a 1,2-migration.

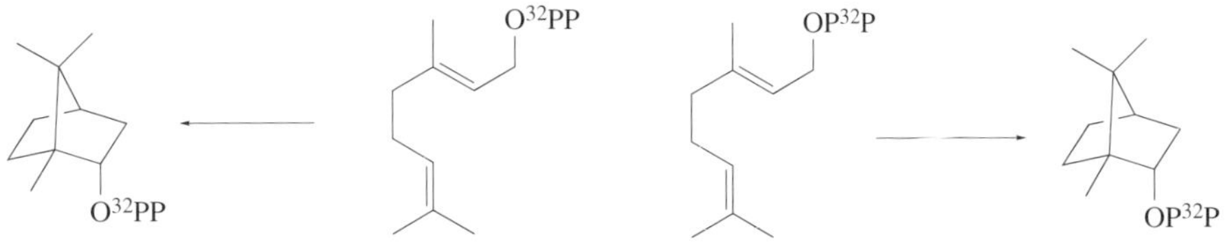

(+)-(**44**) [α-^{32}P](**16**) [β-^{32}P](**16**) (+)-(**44**)

Scheme 21

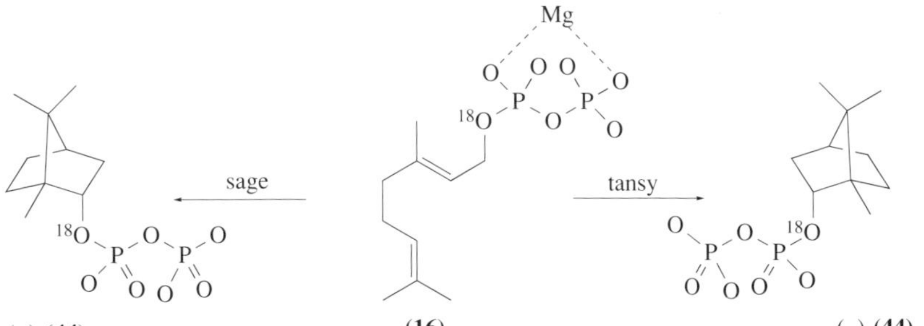

(+)-(**44**) sage (**16**) tansy (−)-(**44**)

Scheme 22

The stereochemical disposition of the *gem*-dimethyl group during formation of the pinenes by the synthases from sage has been investigated using (6*E*)-[8-^{3}H]geranyl diphosphate as substrate,[249] by conversion of the resulting alkene products to diol (**109**) for determination of residual radioactivity (Scheme 23). Both the (+)- and the (−)-pinene synthase displayed complete retention of tritium in the derived diol, indicating that the *E*-methyl of the substrate becomes the *exo*-methyl of the bicyclic products. Thus, cyclization is stereospecific with respect to orientation of the substrate *gem*-dimethyl group, and a least-motion rotation (30° vs. 150°) occurs with the isopropyl group of the α-terpinyl cation upon conversion to the bicyclic pinyl cation.[249] These results are, again, completely consistent with the *anti–endo* cyclization scheme.

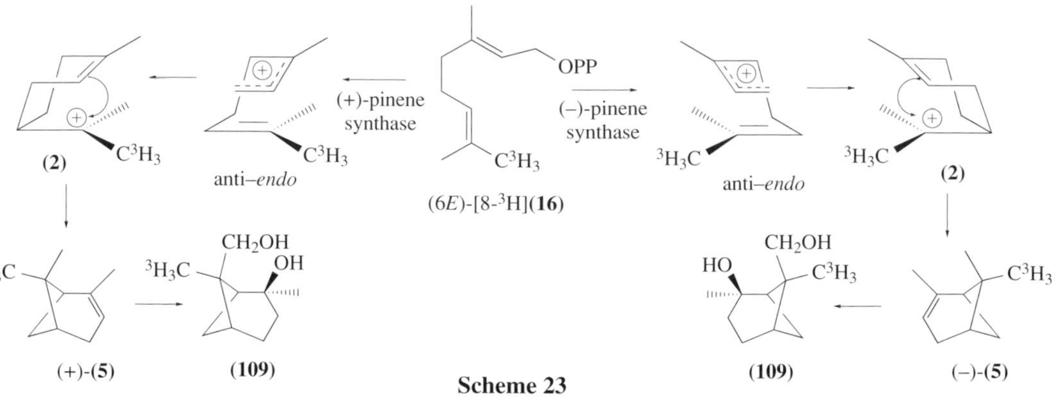

Scheme 23

As a final refinement of the mechanism for the formation of pinene monoterpenes, the stereochemistry of proton elimination from C-3 (of the pinyl intermediate) was examined using stereospecifically deuterated (4*R*)-[4-$^{2}H_{1}$] and (4*S*)-[4-$^{2}H_{1}$]geranyl diphosphates as substrate, with the synthases from sage.[90,250] Mass spectrometric analysis of the (+)-α-pinene produced revealed the loss of 94.3% deuterium from (4*R*)-[4-$^{2}H_{1}$]geranyl diphosphate, with retention of more than 99% deuterium from (4*S*)-[4-$^{2}H_{1}$]geranyl diphosphate. With the (−)-pinene synthase product, the results were somewhat less satisfying in that, from (4*S*)-[4-$^{2}H_{1}$]geranyl diphosphate, only 78.1% of the deuterium was depleted (21.7% retained); however, in the case of the (4*R*)-[4-$^{2}H_{1}$]geranyl diphosphate, the product retained >99% deuterium. These results suggest that it is the axial hydrogens in both (+)- and (−)-enantiomers which are eliminated. Thus, in the course of the cyclization to (+)-α-pinene, the C-4-*proR*-hydrogen of the original substrate occupies the axial position of the pinyl ring (the conformation in which the *proS*-hydrogen is axial is greatly disfavored by 1,4-diaxial interaction with the dimethyl substituted bridge). Axial orientation of the sp^3 orbital of the C—H bond with the periplanar, unoccupied orbital of the cationic center at C-2 would be expected to facilitate elimination.[250] Assuming maintenance of tight ion-pair coupling with the substrate diphosphate anion, this moiety would be expected to reside on the face of the pinyl system opposite the *gem*-dimethyl bridge and, therefore, in close proximity to the eliminated proton (Scheme 24). Thus, the diphosphate anion could, in theory, function as the base in the terminating deprotonation (although an enzyme base could equally serve this role); a related hypothesis has been put forward for the prenyltransferase reaction.[3,251,252]

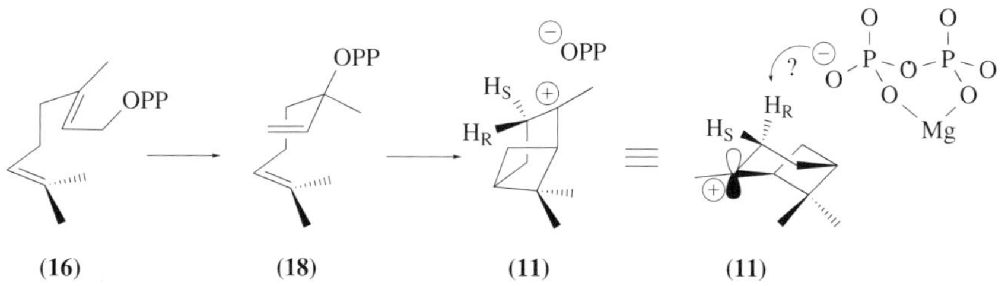

Scheme 24

The regiochemistry of proton elimination from the *gem*-dimethyl groups in the biosynthesis of limonene has been investigated using the stereospecifically deuterated substrates [8,8,8-$^{2}H_{3}$] and [9,9,9-$^{2}H_{3}$]geranyl diphosphates and the (−)-(4*S*)- and (+)-(4*R*)-limonene synthases from spearmint (*M. spicata*) and unshiu orange (*Citrus ushiu*).[253] Using both high resolution mass spectrometry and ^{2}H-NMR analysis of the resulting limonenes, it was found that both enzymes showed pronounced (>94%) regiospecificity for elimination of a hydrogen from the *Z*-methyl group of the substrate. Similarly, Coates, Croteau and associates found that the (+)- and (−)-limonene synthases from *Citrus sinensis* and *Perilla frutescens*, respectively, were also highly regiospecific (>97%) for proton elimination from the *Z*-terminal methyl of geranyl diphosphate.[254] However, analysis of the corresponding limonene antipodes produced by the (+)- and (−)-pinene synthases from sage showed that elimination occurred from both the *Z*- (55–65%) and the *E*- (35–45%) methyl groups. These investigators speculated that the observed preference for elimination from the *Z*-methyl in these reactions, as well as in others involving sesquiterpene and triterpene biosynthesis, results from a minimization of charge separation and an optimization in solvation of the diphosphate leaving group. If a least-motion mechanism is assumed in the conversion of the bound linalyl diphosphate intermediate to the α-terpinyl cation–diphosphate ion pair, then the *Z*-methyl group is positioned significantly closer to the leaving group in what is likely to constitute a more polar region of the reactive site. Thus, charge separation is reduced with better stabilization in the transition state for proton transfer.[254]

The mechanism involved in the biosynthesis of car-3-ene (**7**) is of some historic interest (see Section 2.05.1.4.6). Ruzicka originally proposed a relatively straightforward 1,3 elimination of a C-5 proton in the α-terpinyl cation to account for formation of the cyclopropyl ring (Scheme 25, pathway b). Later, based on *in vivo* feeding experiments, Banthorpe and Ekundayo[65] postulated that the 1,3 elimination occurred by loss of a C-3 proton from the α-terpinyl cation with isomerization of the double bond (Scheme 25, pathway a). However, more recent experiments have demonstrated the validity of Ruzicka's original biogenetic scheme.[66] Thus, stem disks from Douglas fir (*Pseudotsuga menziesii*) were incubated with [1-^{3}H]geraniol and the resulting (+)-3-carene was oxidized to car-3-en-5-one (**110**), resulting in complete loss of tritium (Scheme 26). According to the Banthorpe and Ekundayo proposal,[65] the car-3-ene derived from [1-^{3}H]-geranyl diphosphate would be labeled at C-1 and C-2 of car-3-ene (Scheme 25, path a), neither of these positions will suffer loss of tritium upon oxidation to **110**. Partially purified enzyme preparations from lodgepole pine (*Pinus contorta*) were also utilized to convert [1-^{3}H]geranyl diphosphate to car-3-ene, which was oxidized to 4-isocaranol (**111**) and thence to isocaranone (**112**) without loss of tritium. However, base-catalyzed exchange of **112** to remove the α-hydrogens (Scheme 26) resulted in total loss of tritium, as expected, if the carene product was derived by pathway b, Scheme 25. Finally, incubation of the (+)-3-carene synthase from lodgepole pine with (5*S*)-[5-$^3H_1$; 4-^{14}C]geranyl diphosphate resulted in complete retention of the tritium label in the product, indicating that the carene skeleton originates by an *anti*-1,3 elimination of the 5-*proR* hydrogen of the α-terpinyl intermediate as originally hypothesized by Ruzicka.[66]

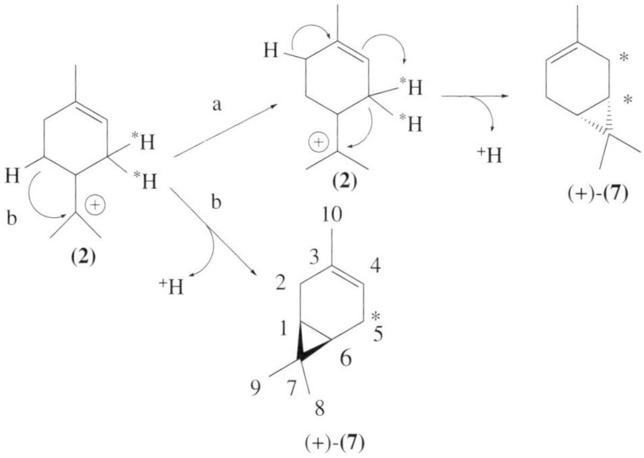

Scheme 25

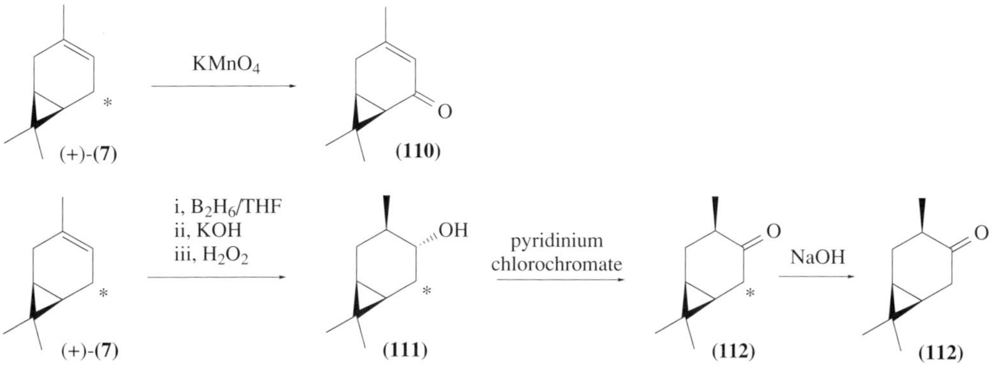

Scheme 26

The biosynthesis of the thujane monoterpenes, which bear the characteristic 1-isopropyl-4-methylbicyclo[3.1.0]hexane skeleton, has been investigated in some detail using partially purified enzyme preparations from sage and marjoram (*Marjorana hortensis*).[61] The (+)-sabinene synthase from sage produces primarily (+)-sabinene (**49**) with trace amounts of (−)-α-thujene (**48**). The sabinene hydrate synthase from marjoram produces approximately 90% (+)-*cis*-sabinene hydrate (**113**) and 10% (+)-*trans*-sabinene hydrate.[60,61] Employment of these enzymes with stereospecifically labeled geranyl diphosphate and linalyl diphosphate substrates, as described earlier with the bornyl diphosphate synthases and pinene synthases, once again provided firm evidence for the *syn*-isomerization-*anti, endo*-cyclization mechanism as a general paradigm for the coupled reaction.[57,61,255] Prior to formation of the cyclopropyl ring, a 1,2-hydride shift from the cyclohexene C-4 to the exocyclic C-8 of the α-terpinyl cation (**2**) is expected to produce the terpinen-4-yl cation (**8**) as an essential intermediate. To test this prediction, [6-^{3}H; 1-^{14}C]geranyl diphosphate was converted to sabinene hydrate (**113**) by the enzyme from marjoram with the result that both *cis* and *trans* isomers bore ^{3}H:^{14}C ratios essentially identical to the starting material, clearly indicating retention of tritium.[57] To confirm that the tritium had been transferred to the C-7 position (carbon number based on the thujane skeleton), the sabinene hydrate products were oxidized to the corresponding diols (**114**, Scheme 27), which resulted in complete loss of ^{3}H for both diastereomers, thereby establishing the predicted 1,2 hydride shift mechanism.

[6-^{3}H](**6**) → (**2**) → (**8**) → (**12**) → *trans*-(**113**) —KMnO$_4$→ *trans*-(**114**); (**12**) → *cis*-(**113**) —KMnO$_4$→ *cis*-(**114**)

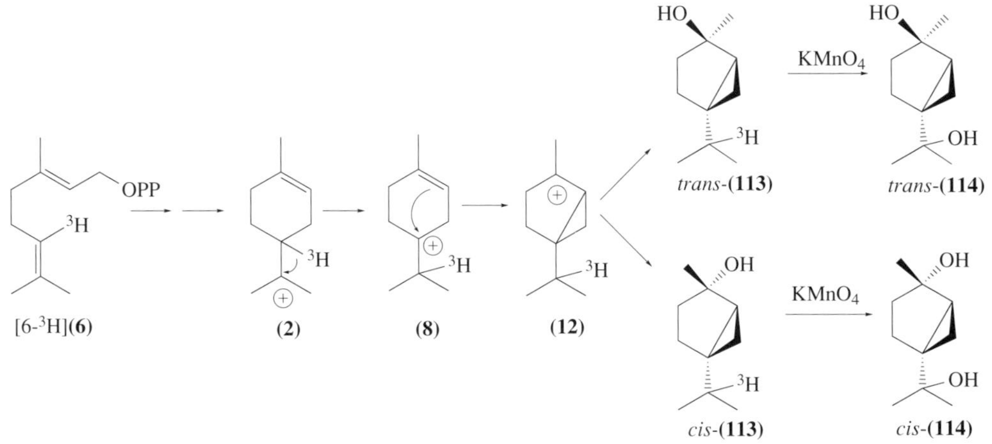

Scheme 27

Similar hydrogen rearrangements are involved in the biosynthesis of several *p*-menthadiene monoterpenes. The nature of these rearrangements, i.e., whether they occur through a 1,3 hydride shift or coupled 1,2 shifts, was explored using α-terpinene synthase from wormseed (*Chenopodium ambrosioides*), γ-terpinene synthase from thyme (*T. vulgaris*), and (−)-β-phellandrene synthase from lodgepole pine (*P. contorta*).[256] The general strategy employed regio- and stereospecifically tritiated geranyl diphosphates as substrates followed by selective oxidation of the product dienes, first to *p*-cymene (**38**) then to methyl-*p*-2-hydroxyisopropyl benzoate (**115**).

Following conversion of [1*RS*-^{3}H]geranyl diphosphate to γ-terpinene (**33**), the specific activity of the diene was reduced by half on oxidation to *p*-cymene, but no further loss of tritium was observed upon oxidation to *p*-2-hydroxyisopropyl benzoate (Scheme 28, A). These results demonstrated the selectivity of the oxidation sequence and the absence of label scrambling. To establish the nature of the hydride shift directly, the synthase was incubated with [6-^{3}H]geranyl diphosphate, and the resulting γ-terpinene was again oxidized to *p*-cymene without change in specific activity. However, conversion of the *p*-cymene to (**115**) resulted in complete loss of tritium label, thereby demonstrating that a 1,2-hydride shift from C-4 to the C-8 position of the α-terpinyl system had occurred (to provide the terpinen-4-yl cation) and eliminating the alternative 1,3-hydride shift mechanism (Scheme 28, B). Conversion of (5*S*)-[5-^{3}H;4-^{14}C]geranyl diphosphate resulted in the complete absence of tritium in the diene product, thereby establishing that the 5-*proS*-hydrogen is lost in the terminating deprotonation step to yield γ-terpinene (Scheme 28, C). A similar strategy was applied to deduce the origin of α-terpinene and, again, a 1,2-hydride shift (in the conversion of the α-terpinyl to the terpinen-4-yl cation) was demonstrated for this synthase (Scheme 29). However, in the case of β-phellandrene synthase, cyclization of [1*RS*-^{3}H]geranyl diphosphate resulted in no change in specific activity of the diene (**36**) upon conversion to *p*-cymene, but 50% loss of ^{3}H upon oxidation to **115**. This result eliminated the possibility of two coupled 1,2-hydride shifts, since this mechanism would have afforded a 50% tritium loss upon conversion of *β*-phellandrene to *p*-cymene. Loss of half the tritium label upon oxidation of *p*-cymene to **115** clearly indicates that a 1,3-hydride shift is operative in the electrophilic cyclization to *β*-phellandrene (Scheme 30).

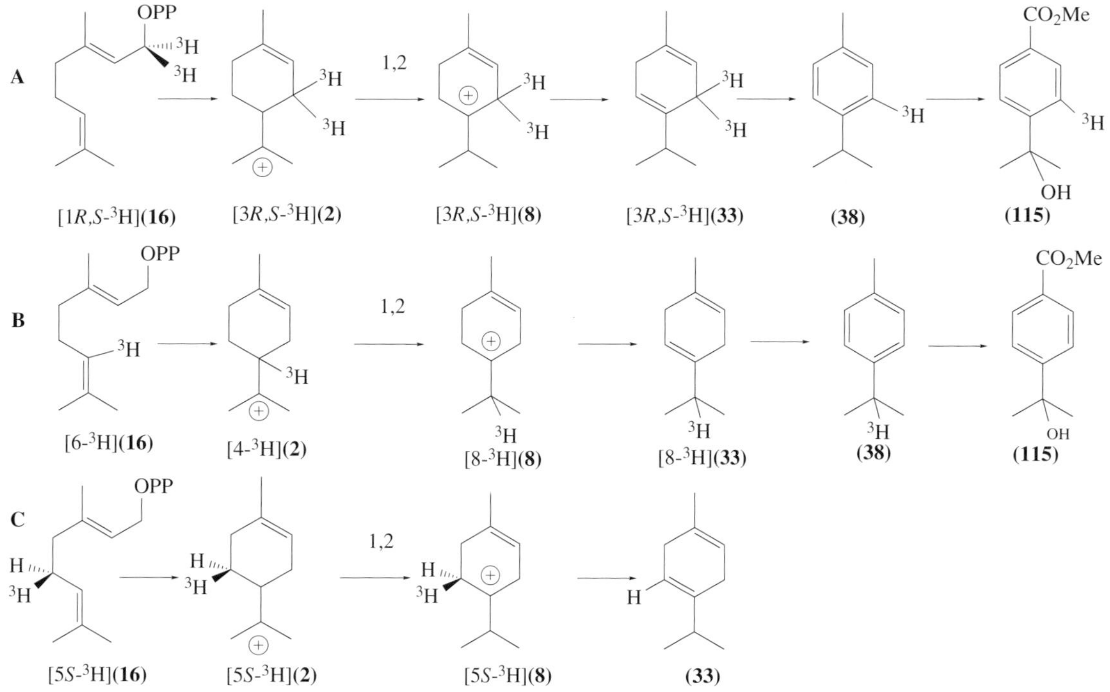

Scheme 28

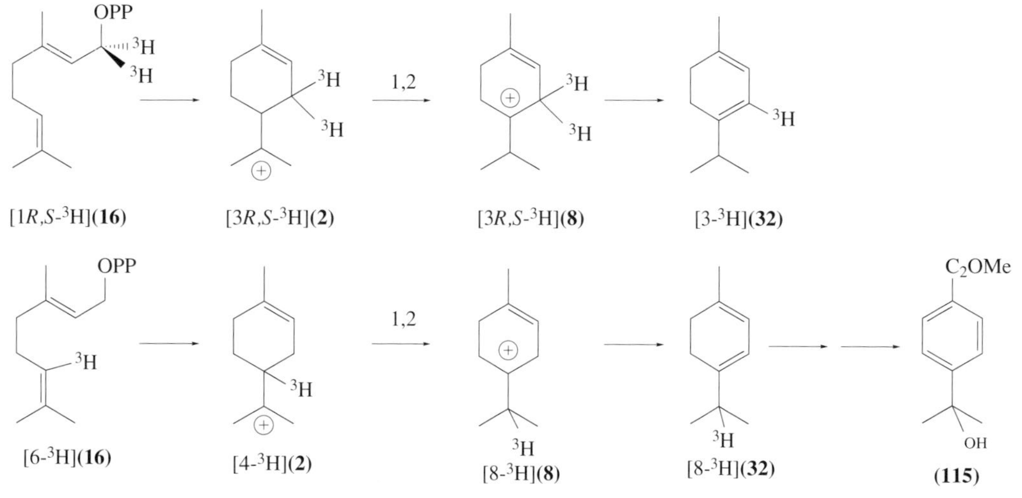

Scheme 29

Studies using $H_2{}^{18}O$ as well as ^{18}O labeled substrates, followed by mass spectrometric analysis of the products, have confirmed H_2O as the source of oxygen in both the cyclic ether 1,8-cineole (**116**)[168] and the alcohol (−)-*endo*-fenchol (**6**).[224] In the case of 1,8-cineole, protonation at C-2 was also shown to involve a solvent-derived hydrogen as determined by experiments run in ^2H_2O.[167] The participation of water in the biosynthesis of these two monoterpenes raises the question of how these enzymes control access of this reactant to the catalytic site. Clearly all intermediate carbocations, other than the terminal species, must be protected from premature capture by the solvent. Folding of the substrate in the *endo*-conformer, with the isopropylidene group positioned toward

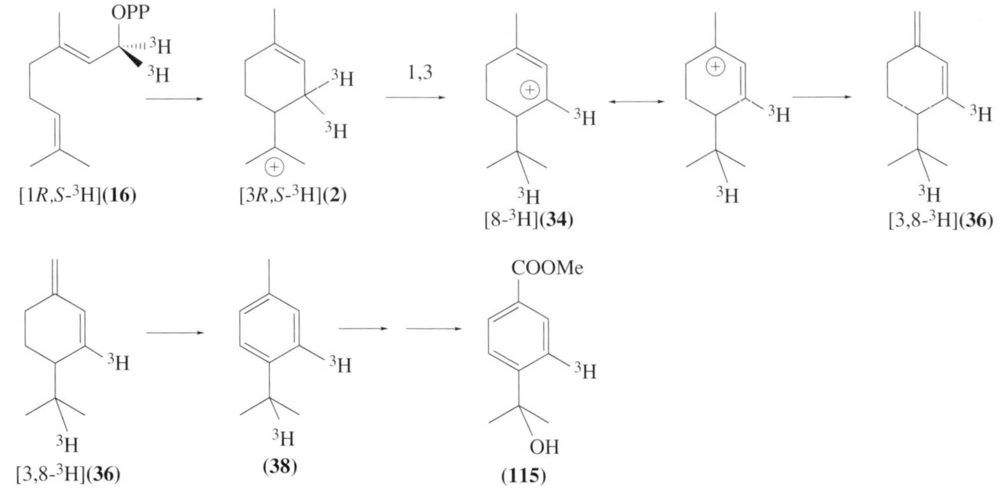

[1*R*,*S*-^{3}H](**16**) [3*R*,*S*-^{3}H](**2**) [8-^{3}H](**34**) [3,8-^{3}H](**36**)

[3,8-^{3}H](**36**) (**38**) (**115**)

Scheme 30

the solvent-exposed region of the reaction cavity, has been proposed as one possible means by which this could be accomplished.[229] However, even if the isopropylidene group functions in shielding the initially formed cation, without further protection, solvent capture of the α-terpinyl cation formed on cyclization would seemingly result, leading to the production of α-terpineol (**37**). In the case of 1,8-cineole, such solvent capture to generate α-terpineol is necessary, but would then require a 180° rotation of the hydroxyisopropyl function prior to the heterocyclic ring closure step (Scheme 31). Furthermore, neither [3-^{3}H]α-terpineol nor the corresponding phosphate or diphosphate esters were converted, by this synthase, to 1,8-cineole. These and other elements of the biosynthesis of 1,8-cineole remain to be defined.

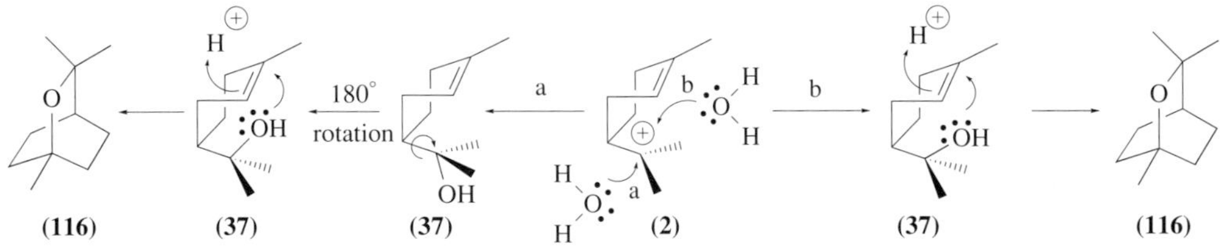

(**116**) (**37**) (**37**) (**2**) (**37**) (**116**)

Scheme 31

In an effort to dissect the cryptic ionization–isomerization step from the tightly coupled ionization–cyclization sequence, 6,7-dihydrogeranyl diphosphate was employed as substrate with the (+)-pinene synthase and (+)-bornyl diphosphate synthase from sage.[257] This substrate analogue, in lacking the C-6–C-7 alkene function, is incapable of cyclization; hence, the reaction is aborted beyond the isomerization step. 6,7-Dihydrogeranyl diphosphate proved to be a reasonably good competitive inhibitor of the natural substrate, and it underwent enzymatic transformation to both acyclic monoterpene alkenes and alcohols, including dihydromyrcene and dihydrolinalool. However, no free dihydrolinalyl diphosphate was detected as a product, indicating that this postulated intermediate (analogous to linalyl diphosphate), if formed, was ionized with subsequent deprotonation or water capture to yield the observed acyclic products. The relatively high proportion of alkenes generated by these enzymes, relative to alcohols, suggested that the reaction took place in a relatively solvent-free environment, as would be expected for a shielded, hydrophobic active site. Moreover, the product distribution generated by the monoterpene synthases closely resembled that derived by solvolysis of 6,7-dihydrolinalyl diphosphate, suggesting that isomerization to the linalyl analogue of the normal reaction had, in fact, occurred prior to product formation.

A somewhat more elegant approach was devised to isolate the isomerization step by using [1-^{3}H]-2,3-cyclopropyl(methano)geranyl diphosphate (**117**) as substrate.[258] This analogue, upon ionization and isomerization, undergoes ring opening to yield the homoallylic analogue of linalyl diphosphate (**119**) which is unreactive to ionization and, thus, to further cyclization (Scheme 32). When utilized as substrate with the sage synthases, the resulting products consisted of 31% mixed trienes (**120**),

(**121**), and (**122**), 58% mixed alcohols (**123**) and (**124**) and 10% of the direct isomerization product, (**119**). Thus, for the first time, direct evidence for the isomerization component of the (normally) tightly coupled isomerization–cyclization sequence was provided.

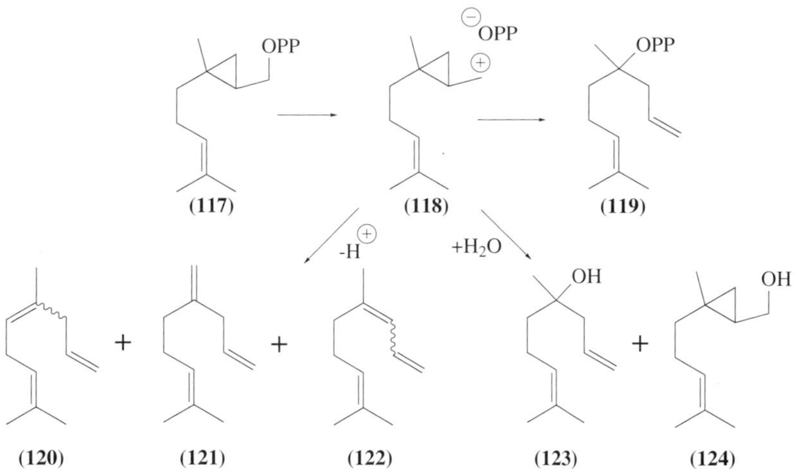

Scheme 32

2.05.5.5 Multiple Product Cyclases

The ability of several of the monoterpene synthases to simultaneously produce multiple products from a single substrate is an unusual and intriguing biochemical phenomenon. The question of adequate isolation of a pure synthase with this capability is a legitimate one; hence, several experimental approaches have been employed to provide unequivocal evidence that certain monoterpene synthases do, in fact, transform geranyl diphosphate into as many as six different products.

Initial efforts to resolve the monoterpene synthases from sage resulted, after four purification steps, in two separate enzyme activities, one producing (+)-α-pinene, (+)-camphene, and (+)-limonene (cyclase I or (+)-pinene synthase), the other producing (−)-α-pinene, (−)-β-pinene, and (−)-camphene (cyclase II or (−)-pinene synthase). The product profiles of each synthase activity remained constant throughout purification.[48] To further substantiate that the multiple products resulted from a single synthase, rather than from copurification of multiple enzymes, the purified enzyme preparations were subjected to differential thermal inactivation, as well as differential inhibition using *p*-hydroxymercuribenzoate and *N*-ethylmaleimide. While the two enzyme preparations showed differential sensitivity to the various treatments relative to each other, no change in product profiles was observed. Although not definitive, these studies provided strong suggestive evidence that both (+)-pinene synthase and (−)-pinene synthase were single enzymes that produced multiple products. This same general approach to differential inactivation was also applied, with similar results, to sabinene hydrate synthase from sweet marjoram (*Majorana hortensis*) which produces both *cis*- and *trans*-sabinene hydrate.[60]

A kinetic strategy, employing the phenomenon of isotopically sensitive branching, has also provided experimental evidence supporting the multifunctional nature of several monoterpene synthases.[89,90,247] Isotopically sensitive branching describes a rate enhancement in the formation of one (or more) products of a multiple product enzyme caused by a suppression of the rate constant for the formation of a second product owing to isotopic substitution in the substrate.[259] This rate enhancement, defined as an induced kinetic isotope effect,[260] indicates that the two (or more) products arise from a common intermediate. Assuming, for example, that a single enzyme catalyzes the biosynthesis of both α- and β-pinene from a common pinyl cation, a simplified kinetic mechanism is illustrated in Equation (2). In this equation, ES I is the enzyme–substrate complex, k_2 is the overall rate constant for the irreversible ionization–isomerization–ionization–cyclization sequence leading to the branch-point intermediates ES II (the pinyl cation), and k_3 and k_4 are the rate constants for the respective deprotonation steps. Thus, in the case of pinene biosynthesis, if product P_2 is taken as β-pinene and [10-$^2H_3$]geranyl diphosphate is used as substrate, the rate constant for the final deprotonation step leading to β-pinene, k_4, will be suppressed due to a primary isotope effect.

Theoretically, this suppression of k_4 will result in an increase in ES II concentration now available for conversion to α-pinene (P_1) and, hence, will lead to an overall increase of α-pinene formation relative to the unlabeled substrate. Alternatively, if α- and β-pinene biosynthesis are mediated by two separate enzymes, then the rate reduction exhibited in β-pinene production will not affect the rate of α-pinene biosynthesis. When a partially purified preparation of (−)-pinene synthase from sage was incubated with [10-^{2}H; 1-^{3}H]geranyl diphosphate, a clear partitioning of intermediate from β-pinene to α-pinene was observed relative to that observed with the nondeuterated substrate (1:3 vs. 1:1.2, respectively). These experiments also revealed an overall rate decrease of 25% with the deuterated substrate, likely due to a secondary kinetic isotope effect on the initial ionization–isomerization step.[89,247] In spite of the overall rate reduction, not only did the relative proportion of (−)-α-pinene increase, but its absolute rate of production also increased relative to the nondeuterated control reaction. These results unambiguously demonstrate that (−)-α-pinene and (−)-β-pinene originate from alternative deprotonations of a common intermediate, and must therefore be products of the same enzyme.

$$\mathrm{E+S} \underset{k_{-1}}{\overset{k_1}{\rightleftharpoons}} \mathrm{ES\ I} \xrightarrow{k_2} \mathrm{ES\ II} \begin{cases} \xrightarrow{k_3} \mathrm{E+P_1} \\ \xrightarrow{k_4} \mathrm{E+P_2} \end{cases} \tag{2}$$

Implicit to the application of isotopically sensitive branching experiments is the requirement that the production of the branch point intermediate be rate-limiting. Recognition that the rate constant of the isotopically *insensitive* reaction, k_3 in the example cited above, does not change is important; it is only by virtue of the increase in concentration of the intermediate (ES II) that the velocity of P_1 formation is increased. Examination of the rate equation for P_1 formation (Equation (3)), derived from the kinetic mechanism described above,[247] reveals that an increase of [ES II], due to a reduction in k_4, depends on the relative size of k_2 compared to (k_3+k_4), and the substrate concentration [S]. In the case of monoterpene synthases, the preponderance of evidence indicates that the initial ionization step is rate-limiting,[14] hence $k_2 \ll (k_3+k_4)$. Therefore, isotopically sensitive branching experiments can be performed under substrate saturating conditions

$$V_1 = k_3[\mathrm{ES\ II}] = \frac{k_3}{k_3+k_4} \cdot \frac{k_2[\mathrm{Eo}]}{1+\dfrac{k_2}{k_3+k_4}+\dfrac{k_{-1}+k_2}{k_1[\mathrm{S}]}} = \rho \cdot V_{\mathrm{total}} \tag{3}$$

Determination of the intrinsic kinetic isotope effect for enzyme-catalyzed reactions is frequently hampered by attenuation from multiple, partially rate limiting transformations along the reaction coordinate.[261] In addition to identification of multiple product enzymes, isotopically sensitive branching experiments can provide a powerful tool for quantifying the lower limit of the intrinsic isotope effect on terminal deprotonation steps.[259] The kinetics of the terminating deprotonations can be separated from the kinetics of earlier steps of the reaction cycle by relating the velocity of formation of one of the alkene products to that of the other as in Equation (4).

$$\frac{V_2}{V_1} = \frac{k_4[\mathrm{ES\ II}]}{k_3[\mathrm{ES\ II}]} = \frac{k_4}{k_3} \tag{4}$$

Thus, a lower limit for the intrinsic kinetic isotope effect ($k_{4\mathrm{H}}/k_{4\mathrm{D}}$) for the production of P_2 can be calculated from the relative velocities of reaction with the deuterated substrate and the control, and since $k_{3\mathrm{D}}$ is the same as $k_{3\mathrm{H}}$ then,

$$\frac{k_{4\mathrm{H}}/k_{4\mathrm{D}}}{k_{3\mathrm{H}}/k_{3\mathrm{D}}} = \frac{k_{4\mathrm{H}}}{k_{4\mathrm{D}}} \cdot \frac{k_{3\mathrm{D}}}{k_{3\mathrm{H}}} = \frac{k_{4\mathrm{H}}}{k_{4\mathrm{D}}} \tag{5}$$

Using this approach, analysis of the data from the above experiments provided an apparent $k_{4\mathrm{H}}/k_{4\mathrm{D}}$ of 2.4 for the deprotonation leading to β-pinene,[89] in close agreement with the intrinsic kinetic isotope effect (2.1) calculated for the formation of β-pinene by *Pinus* species, as determined by natural abundance ^{2}H-NMR.[262]

Analysis of the (−)-pinene synthase from grand fir (*A. grandis*), which produces (−)-α-pinene (**5**) and (−)-β-pinene (**107**), and of the phellandrene synthase from lodgepole pine (*P. contorta*), which produces (−)-α-phellandrene (**35**) and (−)-β-phellandrene (**36**), using both [10-$^2H_3$; 1-^3H] and [4-$^2H_2$; 1-^3H]geranyl diphosphate, has also provided unequivocal evidence for isotopically sensitive branching in the formation of pinenes and phellandrenes by these respective enzymes. Kinetic isotope effects of 3.0 and 2.6 were calculated for the deprotonations leading to (−)-α-pinene and (−)-β-pinene, respectively, and of 4.3 and 3.7 for the deprotonations leading to (−)-α-phellandrene and (−)-β-phellandrene, respectively.[247] Assuming that these values approximate the intrinsic isotope effects, they are considerably lower than the theoretical maximum of 7.0[263] (barring tunneling effects[264]), thereby suggesting a degree of asymmetry in the corresponding transition states. From the proposed reaction mechanism, it seems entirely possible that the transition state for deprotonation more closely resembles the high energy pinyl/phellandryl cation rather than a symmetrical species in which the proton is shared between the cation and the enzyme base. The higher values for the kinetic isotope effects observed in the biosynthesis of the phellandrenes might well reflect resonance stabilization of the phellandryl cation intermediate afforded by the adjacent double bond, thus promoting a more symmetric transition state for proton transfer.[247]

In the evaluation of isotopically sensitive branching by (−)-pinene synthase from sage, the relative rate of camphene production was unperturbed when using the C-10-deuterated substrate. This observation suggests that formation of (−)-camphene (**45**) does not proceed via rearrangement of the pinyl cation but rather through the bornyl cation intermediate that apparently arises from direct (anti-Markovnikov) cyclization of the terpinyl cation[89] (see Scheme 17).

Molecular cloning and heterologous expression of the monoterpene synthases has provided the ultimate confirmation that these enzymes produce multiple products from the same substrate. The limonene synthase from spearmint (*Mentha spicata*), the first monoterpene synthase to be cloned and heterologously expressed, produces a product profile essentially identical to the native enzyme, i.e., 94% limonene with trace amounts (1.8–2.0%) of α-pinene, β-pinene, and myrcene. More recently, a cDNA clone encoding a (−)-pinene synthase from grand fir (*A. grandis*) has been isolated and expressed in *E. coli*, and shown to synthesize α- and β-pinene in a 42:58 ratio,[49] a product distribution essentially identical to that of the electrophoretically pure native enzyme.[218] Perhaps the most intriguing example of a multiple product monoterpene synthase, at least in terms of sheer product diversity, is the (+)-bornyl diphosphate synthase cloned from sage (*S. officinalis*). The recombinant enzyme produces, in addition to (+)-bornyl diphosphate (75%), (+)-camphene (9.9%), (±)-limonene (7.9%), (+)-α-pinene (3.4%), terpinolene (2.1%), and myrcene (1.5%).[50] The availability of cloned and heterologously expressed monoterpene synthases, in addition to unequivocally confirming the ability to synthesize multiple products, provides an abundant source of the enzymes for more detailed analysis of the reaction mechanisms and of structure–function relationships in this family of catalysts.

2.05.5.6 Molecular Biology

The cloning strategies utilized in the isolation of cDNAs encoding monoterpene synthases have employed reverse genetics as well as homology-based polymerase chain reaction (PCR) techniques (see Section 2.05.3.5). Two DNA sequences for limonene synthase,[126,127] one for linalool synthase,[131] and one for chrysanthemyl diphosphate synthase[265] have appeared in the literature, and several more have been obtained in the authors' laboratory.[49,50] An alignment of the deduced amino acid sequences of several of these monoterpene synthases is presented in Figure 2.

The monoterpene synthase cDNAs all appear to express a preprotein bearing a plastidial targeting sequence, consistent with the plastidial location of monoterpene biosynthesis, and as evidenced by protein size considerations and by the relatively poor homology among the first 52–64 residues of the aligned sequences. These amino-terminal targeting sequences also demonstrate several characteristic features of transit peptides, e.g., they are rich in serine, threonine, and small hydrophobic residues, and contain few acidic residues.[266] The fact that the native monoterpene synthases appear to be *N*-terminally blocked has hampered efforts to identify the cleavage site between the transit peptide and the mature protein; this problem has yet to be resolved by *in vitro* processing methods or mass spectrometric sequencing techniques. Attempts to truncate the spearmint limonene synthase based upon predictive methods for chloroplast transit peptide cleavage motifs[267] have all resulted in nonfunctional enzymes.[268] Examination of the aligned sequences, however, reveals that extensive homology begins near a tandem pair of arginine residues that lie approximately 52–64 amino acids downstream from the starting methionine. Truncation of the spearmint limonene synthase at these paired arginines yields a fully functional enzyme, and this "pseudo-mature" synthase displays

Figure 2 Deduced amino acid sequence alignment of myrcene synthase, limonene synthase, and (−)-pinene synthase from grand fir (*A. grandis*); limonene synthase from perilla (*P. frutescens*) and from spearmint (*M. spicata*); and 1,8-cineole synthase, sabinene synthase, and bornyl diphosphate synthase from sage (*S. officinalis*). Sequence alignment performed using the GCG Pileup program. Abbreviations on the left of the figure are the same as in Table 2.

vastly improved expression and solubility characteristics that facilitate purification.[268] Truncation downstream of these tandem arginines severely impairs catalytic function, thereby defining the minimal "mature" enzyme.

Overall sequence homology (deduced amino acid identity) between the monoterpene synthases is quite high (Figure 2), particularly when sequences of angiosperm origin and gymnosperm origin are separately compared (Table 2). Among the synthases of angiosperm origin, the identity ranges from 22 to 73%, and the similarity from 47 to 86%. The lowest identity and similarity scores reflect the comparison to linalool synthase from *Clarkia breweri* which is apparently an outlier. Synthases from the gymnosperm species show 70–77% identity and 83–86% similarity.[49] Comparison of the deduced amino acid sequences of the monoterpene synthases with those of other plant-derived sesquiterpene and diterpene synthases reveals several absolutely conserved residues, including Arg^{161},

Arg[337], and Cys[516] (amino acid placement referred to bornyl diphosphate synthase). In addition to the tandem Arg[55]–Arg[56] pair, several other highly conserved sequence motifs are apparent. These include an Arg[299]–Trp–Trp sequence, an Arg[372]–Trp–Glu/Gln element, a Tyr[384]-Met–Gln/Lys sequence, and a Cys[516]–Tyr–Met–X–Glu/Asp element. The structural or functional significance of these motifs has not been established.

Table 2 Pairwise comparison of deduced amino acid sequence identity of cloned monoterpene synthases. Scored by the GCG Gap program (using the Needleman and Wunsch algorithm).

	Gymnosperms		*Angiosperms*					
	A.g. pine	*A.g. limo*	*C.b. lino*	*P.f. limo*	*M.s. limo*	*S.o. born*	*S.o. cine*	*S.o. sabi*
A.g. myrc	71.2	66.1	24.3	30.2	30.1	31.0	31.2	31.2
A.g. pine		63.5	26.0	30.9	32.3	33.6	31.4	31.6
A.g. limo			24.4	28.4	29.6	31.9	29.6	31.4
C.b. lino				22.3	25.3	22.3	25.6	25.0
P.f. limo					65.3	50.8	54.5	52.5
M.s. limo						49.8	53.3	52.1
S.o. born							50.3	69.7
S.o. cine								53.3

A.g. myrc = *Abies grandis* myrcene synthase; A.g. pine = *A. grandis* ( )-pinene synthase; A.g. limo = *A. grandis* limonene synthase; C.b. lino = *Clarkia breweri* linalool synthase; P.f. limo = *Perillia frutescens* limonene synthase; M.s. limo = *Mentha spicata* limonene synthase; S.o. born = *Salvia officinalis* bornyl diphosphate synthase; S.o. cine = *S. officinalis* 1,8-cineole synthase; S.o. sabi = *S. officinalis* sabinene synthase.

The (I, L, or V)DDXXD motif (350–355 in bornyl diphosphate synthase) found in virtually all terpenoid synthases and prenyltransferases is, as expected, conserved among the monoterpene synthases. This aspartate-rich element was originally recognized as a sequence motif that was absolutely conserved in isoprenyl diphosphate synthases[269] and its functional role has received considerable attention. Based on similar functions of conserved aspartate residues in various ATP binding proteins, it was speculated that the conserved aspartate residues might be involved in binding the Mg^{2+}–diphosphate complex of the diphosphate ester substrates of the prenyltransferases. It is noteworthy that microbial sesquiterpene synthases also display this same aspartate-rich motif, in spite of the lack of overall amino acid sequence similarity between these enzymes and their eukaryotic counterparts.[270] Numerous studies based upon site-directed mutagenesis of prenyltransferases and terpene synthases,[270–274] and upon X-ray structure analysis of avian farnesyl diphosphate synthase[275,276] and of two sesquiterpene synthases from tobacco and *Streptomyces*,[277,278] have confirmed the role of the aspartate-rich motif in Mg^{2+}-diphosphate complex binding, and have suggested possible functions for other amino acid residues in the transformation of prenyl diphosphate substrates (see Chapters 2.04 and 2.06).

The overall similarity in helical core structure between the avian farnesyl diphosphate synthase and the microbial pentalenene synthase is surprising, given the low level of sequence identity between the two, but suggests that general secondary and tertiary structural themes may be conserved among these enzyme types.[277,278] The availability of recombinant monoterpene synthases will soon allow more extensive examination of this possibility, especially via comparison of several mechanistically diverse catalysts.

2.05.6 SECONDARY TRANSFORMATIONS

Secondary modifications of the basic parent carbon skeletons produced by the terpenoid synthases are responsible for generating the many different monoterpenes found in plants. These secondary transformations most often involve oxidation, reduction, isomerization, and conjugation reactions, and it is these modifications of the parent compounds that usually impart the biological activities of significance in ecological interactions[17,19] or the functional properties of commercial value.[35] The reaction types of monoterpene secondary transformations are not unique to terpenoid metabolism and so the responsible enzyme classes will be but briefly defined. Instead, this section will focus on a number of representative pathways to illustrate the general metabolic strategies involved and to demonstrate the role of secondary transformations as a major source of structural diversity among the monoterpenes.

An illustrative example of this type is the pathway for the conversion of (−)-limonene (**3**) to (−)-menthol (**78**) in peppermint (Scheme 33). This pathway has been defined by a series of *in vivo* and *in vitro* experiments[149,279–284] and is initiated by allylic hydroxylation of (−)-limonene, by a microsomal cytochrome P450 limonene-3-hydroxylase, to produce (−)-*trans*-isopiperitenol (**125**).[150] A soluble, NAD-dependent dehydrogenase then oxidizes the alcohol to the ketone ((−)-isopiperitenone **126**), thereby activating the adjacent double bond for reduction by a soluble, NADPH-dependent, stereospecific reductase to afford (+)-*cis*-isopulegone (**127**). An isomerase next moves the remaining double bond into conjugation with the carbonyl group, in a reaction analogous to that catalyzed by keto-steroid isomerase,[285] yielding (+)-pulegone (**128**). Two regiospecific, NADPH-dependent and stereoselective reductases convert (+)-pulegone to (+)-isomenthone (**129**) and (−)-menthone (**76**) ((−)-menthone predominates). Finally, two stereoselective, NADPH-dependent reductases convert (−)-menthone and (+)-isomenthone to (−)-menthol (**78**) and (+)-neoisomenthol (**131**), respectively, and (−)-menthone and (+)-isomenthone to (+)-neomenthol (**77**) and (+)-isomenthol (**130**), respectively. (−)-Menthol greatly predominates among the menthol isomers (to about 40% of the essential oil) and it is primarily responsible for the characteristic flavor and cooling sensation of peppermint. This pathway for allylic oxygenation and conjugate reduction can be seen to multiply the simple parent alkene into 10 distinct derivatives. In fact, when the naturally occurring acetate esters of the various menthol isomers are considered,[286] as well as the minor metabolites generated by isomerization of isopiperitenone prior to reduction,[149] the number of C-3-oxygenated metabolites of limonene produced by peppermint approaches 20.

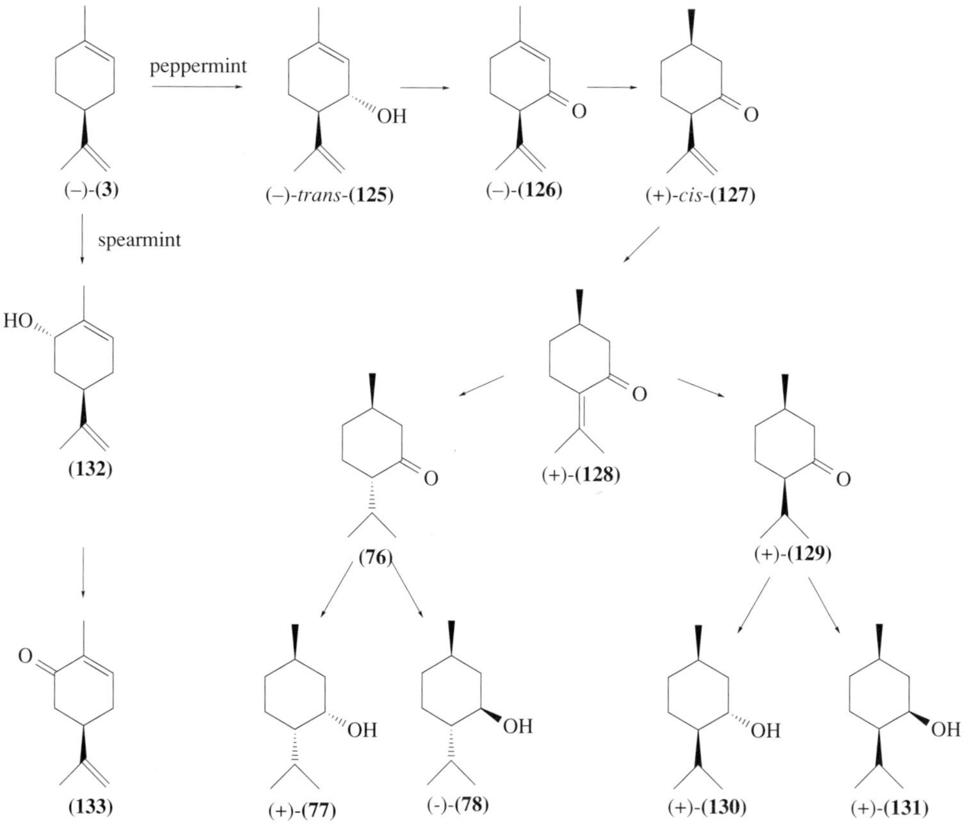

Scheme 33

A very similar pathway operates in spearmint,[149] in this case initiated by hydroxylation specifically at C-6 of (−)-limonene to yield (−)-*trans*-carveol (**132**)[150] which is oxidized to (−)-carvone (**133**). Subsequent metabolism of (−)-carvone occurs to a very limited extent in spearmint such that the carveols, dihydrocarvones, and dihydrocarveols constitute only trace components of the essential oil. Related allylic oxidation–conjugate reduction pathways operate in the metabolism of both pinane (Scheme 34) and thujane (Scheme 35) monoterpenes, although in these cases the enzymology is less well documented.

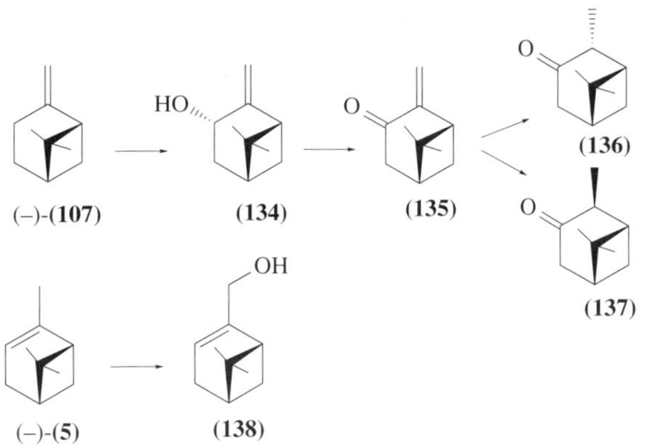

(−)-(**107**) (**134**) (**135**) (**136**) (**137**) (−)-(**5**) (**138**)

Scheme 34

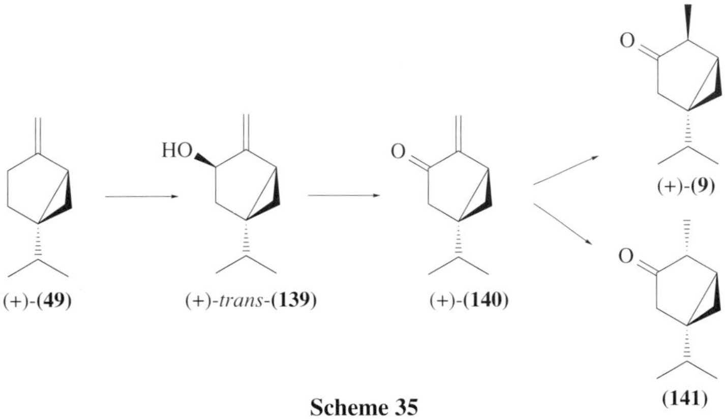

(+)-(**49**) (+)-*trans*-(**139**) (+)-(**140**) (+)-(**9**) (**141**)

Scheme 35

2.05.6.1 Cytochrome P450s

Many of the hydroxylations (or epoxidations) involved in the secondary transformation of monoterpene alkenes are performed by cytochrome P450 mixed function oxidases.[287,288] Of these enzymes, the (−)-limonene hydroxylases of peppermint and spearmint have been studied in the most detail.[150] These enzymes exhibit all of the properties typical of a cytochrome P450, including inhibition by substituted azoles and inactivation by carbon monoxide with reversal by blue light. Both enzymes are highly selective for limonene as substrate and both are absolutely specific in regiochemistry and stereochemistry of oxygen insertion; they are mutually exclusive in the reactions catalyzed (limonene-3-hydroxylation in peppermint and limonene-6-hydroxylation in spearmint). The cDNAs encoding the 3- and the 6-hydroxylases have been isolated from oil gland cDNA libraries from the respective species.[151] The deduced amino acid sequences are very similar and both proteins appear to bear a typical membrane-anchoring sequence rather than an *N*-terminal plastidial targeting sequence, consistent with their location in the endoplasmic reticulum. That limonene originates in plastids, undergoes hydroxylation at the endoplasmic reticulum, and subsequent transformations almost certainly in the cytosol gives an indication of the complex intracellular trafficking that must accompany these secondary metabolic sequences.

In much the same way that the hydroxylation of limonene in mint initiates the production of a broad range of *p*-menthane monoterpenes (Scheme 33), the allylic hydroxylation of the parent alkene (+)-sabinene (**49**) to (+)-*cis*-sabinol (**139**) provides the critical first step in the generation of the C-3-oxygenated thujane monoterpenes thujone (**9**) and isothujone (**141**)[61] (Scheme 35). In some species, reduction of the ketones also occurs, yielding the various thujyl alcohol isomers and often the corresponding esters.[61] The microsomal (+)-sabinene-3-hydroxylase has been isolated from sage (*S. officinalis*), wormwood (*Artemisia absinthium*), and tansy (*T. vulgare*) and shown to catalyze the NADPH- and O_2-dependent hydroxylation to (+)-*cis*-sabinol as the sole product.[62,289] The solubilized enzyme exhibits a typical CO-difference spectrum and type I substrate binding spectrum, and meets most of the established criteria for a cytochrome P450 mixed-function oxygenase.[62] Of the dozen monoterpene alkenes screened as potential substrates, including *β*-pinene (**107**) and *β*-

phellandrene (**36**) that also bear an exocyclic methylene, only (+)-sabinene was hydroxylated. A high level of substrate selectivity is characteristic of this enzyme class from plant sources.

The pattern of allylic oxidation and conjugate reduction in the metabolism of thujane and menthane monoterpenes has also been extended to include oxygenated metabolites of the pinane series. Hyssop (*Hyssopus officinalis*) produces an essential oil containing the saturated bicyclic monoterpene ketones pinocamphone (**136**) and isopinocamphone (**137**), with lesser amounts of myrtenol (**138**) derivatives (Scheme 34). A microsomal preparation from the epidermal oil glands of this species converts the parent bicyclic alkene (−)-β-pinene to the allylic alcohol (+)-*trans*-pinocarveol (**134**) which would then give rise to (−)-pinocamphone and (−)-isopinocamphone by subsequent oxidation (to (+)-pinocarvone **135**) and two stereochemical alternatives for reduction of the conjugated double bond.[51] The pinene hydroxylase from hyssop shares properties with the other monoterpene alkene P450 hydroxylases from leaf oil glands; the same preparation catalyzes the hydroxylation of (−)-α-pinene (**5**) to (−)-myrtenol at a slower rate, but it has not been established whether the same cytochrome species is involved.[51] *In vivo* studies with thyme (*T. vulgaris*) and horsemint (*Monarda punctata*) indicate that the aromatic monoterpenes thymol (**39**) and carvacrol (**40**) are formed by hydroxylation of *p*-cymene (**38**) which is derived by desaturation of γ-terpinene (**33**) (Scheme 36).[64,290] Although these aromatic hydroxylations have not been demonstrated in cell-free systems, such reactions are often catalyzed by cytochrome P450. A novel mode of monoterpene alkene oxygenation is the peroxidase-catalyzed conversion of α-terpinene (**32**) to ascaridole (**142**) (Equation (6))[291] demonstrated in preparations from wormseed (*Chenopodium ambrosiodes*).

OH OH +

(**33**) (**38**) (**39**) (**40**)

Scheme 36

O O (6)

(**32**) (**142**)

2.05.6.2 Redox Reactions

A number of dehydrogenases and reductases involved in the metabolism of acyclic, monocyclic, and bicyclic monoterpenes, such as those responsible for the redox transformations illustrated in Schemes 33–35 have been described,[149,279,281,283,284,292,293] and these are probably representative of metabolic enzymes of this class. These enzymes are operationally soluble, require a pyridine nucleotide, and have ample biochemical precedent in both primary and secondary plant metabolism, as well as in steroidal transformations in animals.

A question often raised with regard to monoterpene interconversions concerns the issue of whether these transformations are carried out by highly specific, or relatively nonspecific, enzymes. A sufficient number of monoterpenol dehydrogenases, and double bond isomerases and reductases, have now been examined to suggest that these enzymes exhibit a significant degree of substrate specificity. Thus, the monoterpenol dehydrogenases of sage, wormwood, and tansy have been examined in detail,[294,295] since the essential oils of these species contain both (+)-camphor (**82**), derived by oxidation of (+)-borneol (**4**), and (+)-3-thujone (**9**) or (−)-3-isothujone (**141**), derived via oxidation of (+)-*cis*-sabinol (**139**) to (+)-sabinone (**140**) en route to the saturated ketones (Scheme 35). The (+)-borneol dehydrogenase and the (+)-*cis*-sabinol dehydrogenase are electrophoretically separable, and each can utilize but a very limited range of alternate, structurally related monoterpenols, such as isoborneol (**42**) or a few thujyl alcohol stereoisomers, respectively, at lower conversion rates and of no apparent metabolic significance.[63] By contrast to these enzymes of redox transformations, the enzymes responsible for the conjugation of monoterpenols, such as acyl transferases and glucosyl transferases, are rather nonselective with regard to their monoterpenoid cosubstrates.[139,286]

A closing example is provided by the enzymes of peppermint which act principally upon C-3-oxygenated *p*-menthane monoterpenes (Scheme 33). The corresponding members of the C-6-oxygenated *p*-menthane series are very poor substrates for these redox enzymes with the exception of the (−)-*trans*-isopiperitenol dehydrogenase which can utilize (−)-*trans*-carveol (**132**) as an alternate, but less efficient substrate.[149] It is thus curious that spearmint and peppermint oil glands contain essentially the same complement of redox enzymes (only the regiospecific limonene hydroxylases differ) but that the machinery in spearmint, beyond the dehydrogenation of (−)-*trans*-carveol, is silent because the specificity of these enzymes is such that (−)-carvone (**133**) is not a substrate.[149] Consequently, carvone, the characteristic component of spearmint, accumulates as the major essential oil component (about 70%). It is notable that spearmint oil also contains 10–20% limonene, depending on leaf age, implying some inefficiency in either transfer of the alkene to the endoplasmic reticulum or in the conversion to carveol at this site. Carveol, on the other hand, does not appreciably accumulate in spearmint oil, indicating that this intermediate is efficiently transformed to carvone prior to oil secretion to the subcuticular storage cavity. This situation contrasts markedly to that of peppermint in which the more extensive pathway operates in menthol biosynthesis. In this instance, neither limonene nor isopiperitenol (**125**) accumulate in the oil ($<1\%$), although other downstream intermediates, such as pulegone (**128**) and menthone (**76**), are present in readily measurable amounts. The exact composition of an essential oil can thereby be seen to depend upon the efficiencies of coupling of the various biosynthetic steps. Given the selectivities of these enzymes of monoterpene secondary metabolism and, thus, their ability to precisely control transformation kinetics, many have become important targets of genetic engineering to alter monoterpene composition and yield toward a particular agronomic or commercial end.[296]

2.05.7 SUMMARY AND PROSPECTS

Considerable insight to the chemical and biological processes involved in the biosynthesis of monoterpenes has developed since Ruzicka originally proposed the biogenetic isoprene rule. The pivotal recognition of geranyl diphosphate as the universal substrate for monoterpene biosynthesis allowed the exploration of the mechanism and the stereochemistry of cyclization reactions leading to the manifold monoterpene skeletal structures and provided the conceptual platform from which to evaluate the transformations of other prenyl diphosphates such as neryl and linalyl diphosphate. In concert with the pioneering studies of Poulter and Rilling on the electrophilic nature of prenyltransferase reactions and with the investigations of sesquiterpene synthases by Cane, a uniform mechanistic model for monoterpene cyclization has emerged. The application of molecular biological techniques to the study of monoterpene biosynthesis has begun and, in the near future, will almost certainly provide answers to many of the yet unresolved questions concerning organization and regulation of metabolism, and structure and mechanism of the responsible catalysts. Molecular techniques will also allow the bioengineering of isoprenoid biosynthesis; some of the potential applications of these techniques are presented here.

2.05.7.1 Enzymology

The availability of cloned, heterologously expressed synthases will allow deciphering the mechanisms of the monoterpene synthases by direct spectroscopic means using levels of enzyme previously unattainable. For example, the availability of highly expressed synthases should, for the first time, allow direct NMR observation of linalyl diphosphate as a reaction intermediate. X-ray crystallographic analysis is in progress with several of these enzymes and will ultimately provide detailed information about overall structure and key residues involved in substrate binding and catalysis. Sequence analysis of the available synthases has already revealed highly conserved residues, sequences and motifs and, as importantly, regions of disparity which undoubtedly impart the different substrate and intermediate conformations that direct alternate reaction pathways affording the various skeletal structures. Site-directed mutagenesis and domain swapping are being applied to these cloned enzymes to identify the function of specific residues and regions involved in catalysis. Indeed, the ability of many of the monoterpene synthases to produce multiple products provides a convenient reporting mechanism whereby alterations in product outcome will communicate those changes in protein structure critical to substrate channeling; these chimeric enzymes may also generate novel products. Also, as more sequence data become available the ability to isolate new, unique monoterpene synthases will become more facile.

2.05.7.2 Bioengineering of Monoterpene Production

With success in the cDNA cloning of monoterpene synthases and hydroxylases, and the development of general cloning strategies, it seems likely that many gene sequences for enzymes of monoterpene metabolism will soon be available. With such genes in hand, the molecular genetic manipulation of monoterpenoid composition and yield in essential oil plants can be contemplated. For example, it might be possible to suppress the undesirable production of menthofuran in peppermint oil through antisense technology. The transformation of microbes for the fermentative production of selected monoterpenoid compounds can be readily realized based on existing technologies, once a geranyl diphosphate synthase gene has been isolated to permit driving these processes from primary metabolism.

It is also conceivable that the monoterpene pathway can be engineered into "non-essential oil" plants, such as fruits and vegetables, and even ornamental species, to impart desirable aroma properties. Given the biological activities and ecological roles of the monoterpenes, their use in the genetic engineering of chemical defenses in field crops against insect pests and pathogens also seems inevitable. The transgenic exploitation of monoterpenes in insect deterrence is of particular significance to the forest products industry where traditional pesticide applications are not feasible and where long-term resource protection is required.

The availability of the structural genes encoding key biosynthetic pathway steps, however, is not in itself sufficient to ensure adequate production of the target metabolites at the appropriate time and location in a transgenic organism. It is clear that successful genetic engineering of monoterpene production will ultimately require a more detailed understanding of the organization and regulation of metabolism, especially as findings related to precursor supply and control of flux,[297] metabolite trafficking, and the secretion/emission processes can be applied to the biosynthesis of monoterpenes in transgenic species. The recent discovery of the nonmevalonate pathway for isopentenyl diphosphate biosynthesis serves as an important reminder that there is still much to be learned about this most basic metabolic pathway.

ACKNOWLEDGMENTS

Work in the authors' laboratory was supported by grants from the NIH, NSF, USDA, and DOE. We also thank Drs. Roy Okuda, William Epstein, David Cane, and Joe Chappell for sharing unpublished data contributing to this chapter. R.C. wishes to acknowledge the contributions of a talented group of colleagues whose names are cited in the references to work from this laboratory, and the productive collaborations with David Cane, Robert Coates, A. C. Oehlschlager, and R. C. Ronald.

2.05.8 REFERENCES

1. O. Wallach, "Terpene und Campher," 2nd edn., Veit, Leipzig, 1914.
2. L. Ruzicka, *Experientia*, 1953, **9**, 357.
3. C. D. Poulter and H. C. Rilling, *Acc. Chem. Res.*, 1978, **11**, 307.
4. W. E. Shine and W. D. Loomis, *Phytochemistry*, 1974, **13**, 2095.
5. B. V. Charlwood and D. V. Banthorpe, *Prog. Phytochem.*, 1978, **5**, 65.
6. D. V. Banthorpe, B. M. Modawi, I. Poots, and M. G. Rowan, *Phytochemistry*, 1978, **17**, 1115.
7. E. Jedlicki, G. Jacob, F. Faini, O. Cori, and C. A. Bunton, *Arch. Biochem. Biophys.*, 1972, **152**, 590.
8. J. A. Attaway, A. P. Pieringer, and L. J. Barabas, *Phytochemistry*, 1966, **5**, 141.
9. R. M. Coates, *Prog. Chem. Org. Nat. Prod.*, 1976, **33**, 73.

10. D. V. Banthorpe, B. V. Charlwood, and M. J. O. Francis, *Chem. Rev.*, 1972, **72**, 115.
11. O. Cori, *Arch. Biochem. Biophys.*, 1969, **135**, 416.
12. R. Croteau and F. Karp, *Arch. Biochem. Biophys.*, 1979, **198**, 523.
13. R. Croteau and F. Karp, *Arch. Biochem. Biophys.*, 1979, **198**, 512.
14. R. Croteau, *Chem. Rev.*, 1987, **87**, 929.
15. D. E. Cane, *Tetrahedron*, 1980, **36**, 1109.
16. R. Croteau, in "Biosynthesis of Isoprenoid Compounds," eds. J. W. Porter and S. L. Spurgeon, Wiley, New York, 1981, vol. 1, p. 225.
17. J. B. Harbone, in "Ecological Chemistry and Biochemistry of Plant Terpenoids," eds. J. B. Harborne and F. A. Tomas-Barberan, Clarendon, Oxford, 1991, vol. 31, p. 399.
18. J. Gershenzon and R. Croteau, in "Herbivores: Their Interaction with Secondary Metabolites," eds. G. A. Rosenthal and M. Berenbaum, Academic Press, New York, 1991, p. 165.
19. J. H. Langenheim, *J. Chem. Ecol.*, 1994, **20**, 1223.
20. J. A. Byers, G. Birgersson, J. Lofqvist, and G. Bergstrom, *Naturwissenschaften*, 1988, **75**, 153.
21. M. Gijzen, E. Lewinsohn, T. J. Savage, and R. Croteau, in "Bioactive Volatile Compounds from Plants," eds. R. Teranishi, R. G. Buttery, and H. Sugisawa, American Chemical Society, Washington, DC, 1993, p. 8.
22. J. B. Harborne, *Nat. Prod. Rep.*, 1993, **10**, 327.
23. M. Dicke, *Chemoecology*, 1995, **6**, 159.
24. M. Baisier, J.-C. Gregoire, K. Delinte, and O. Bonnard, in "Mechanisms of Woody Plant Defenses Against Insects," eds. W. J. Mattson, J. Levieux, and C. Bernard-Dagan, Springer Verlag, New York, 1988, p. 359.
25. L. B. Brattsten, C. F. Wilkinson, and T. Eisner, *Science*, 1977, **196**, 1349.
26. C. Funk, E. Lewinsohn, B. Stofer Vogel, C. L. Steele, and R. Croteau, *Plant Physiol.*, 1994, **106**, 999.
27. R. H. Miller, A. A. Berryman, and C. A. Ryan, *Phytochemistry*, 1986, **25**, 611.
28. J. A. Pickett, in "Ecological Chemistry and Biochemistry of Plant Terpenoids," eds. J. B. Harborne and F. A. Tomas-Barberan, Clarendon, Oxford, 1991, vol. 31, p. 297.
29. S. J. Seybold, D. R. Quilici, J. A. Tillman, D. Vanderwel, D. L. Wood, and G. J. Blomquist, *Proc. Natl. Acad. Sci. USA*, 1995, **92**, 8393.
30. J. E. Casida, in "Pyrethrum, the Natural Insecticide," ed. J. E. Casida, Academic Press, New York, 1973, p. 101.
31. R. G. Cates, C. B. Henderson, and R. A. Redak, *Oecologia*, 1987, **73**, 312.
32. K. N. Saxena and A. Basit, *J. Chem. Ecol.*, 1982, **8**, 329.
33. L. H. Zalkow, M. M. Gordon, and N. Lanir, *J. Econ. Entomol.*, 1979, **72**, 812.
34. R. Croteau, *Perf. Flav.*, 1980, **5**, 35.
35. F. A. Dawson, *Nav. Stores Rev.*, 1994, **104**, 6.
36. R. E. Erickson, *Lloydia*, 1976, **39**, 8.
37. G. Ohloff, "Scent and Fragrances," Springer Verlag, New York, 1994.
38. A. Wong, *Nav. Stores Rev.*, 1991, **101**, 14.
39. M. N. Gould, *J. Cell. Biochem.*, 1995, **22** (Suppl.), 139.
40. M. A. Morse and A. L. Toburen, *Cancer Lett.*, 1996, **104**, 211.
41. T. Savage and R. Croteau, *Nav. Stores Rev.*, 1991, **101**, 6.
42. E. Pichersky, R. A. Raguso, E. Lewinsohn, and R. Croteau, *Plant Physiol.*, 1994, **106**, 1533.
43. E. Pickersky, E. Lewinsohn, and R. Croteau, *Arch. Biochem. Biophys.*, 1995, **316**, 803.
44. V. M. Blanc and E. Pichersky, *Plant Physiol.*, 1995, **108**, 855.
45. R. Croteau and F. Karp, *Arch. Biochem. Biophys.*, 1976, **176**, 734.
46. A. S. Dro and F. W. Hefendehl, *Planta Med.*, 1973, **24**, 353.
47. H. Gambliel and R. Croteau, *J. Biol. Chem.*, 1982, **257**, 2335.
48. H. Gambliel and R. Croteau, *J. Biol. Chem.*, 1984, **259**, 740.
49. J. Bohlmann, C. L. Steele, and R. Croteau, *J. Biol. Chem.*, 1997, **272**, 21 784.
50. M. L. Wise, T. J. Savage, E. Katahira, and R. Croteau, *J. Biol. Chem.*, 1998, submitted for publication.
51. F. Karp and R. Croteau, in "Secondary Metabolite Biosynthesis and Metabolism," eds. R. J. Petroski and S. P. McCormick, Plenum, New York, 1992, p. 253.
52. R. Croteau, M. Felton, and R. C. Ronald, *Arch. Biochem. Biophys.*, 1980, **200**, 524.
53. R. Croteau, M. Felton, and R. C. Ronald, *Arch. Biochem. Biophys.*, 1980, **200**, 534.
54. R. B. Croteau, J. J. Shaskus, B. Renstrøm, N. M. Felton, D. E. Cane, A. Saito, and C. Chang, *Biochemistry*, 1985, **24**, 7077.
55. R. Croteau and F. Karp, in "Perfumes: Art, Science and Technology," eds. P. M. Müller and D. Lamparsky, Elsevier, Amsterdam, 1991, p. 101.
56. V. Hach, *J. Org. Chem.*, 1977, **42**, 1616.
57. T. W. Hallahan and R. Croteau, *Arch. Biochem. Biophys.*, 1989, **269**, 313.
58. D. V. Banthorpe, J. Mann, and K. W. Turnbull, *J. Chem. Soc. (C)*, 1969, 2689.
59. D. V. Banthorpe and K. W. Turnbull, *J. Chem. Soc., Chem. Comm.*, 1966, 177.
60. T. W. Hallahan and R. Croteau, *Arch. Biochem. Biophys.*, 1988, **264**, 618.
61. R. Croteau, in "Recent Developments in Flavor and Fragrance Chemistry," eds. R. Hopp and K. Mori, VCH, Weinheim, 1992, p. 263.
62. F. Karp, J. L. Harris, and R. Croteau, *Arch. Biochem. Biophys.*, 1987, **256**, 179.
63. S. S. Dehal and R. Croteau, *Arch. Biochem. Biophys.*, 1987, **258**, 287.
64. A. J. Poulose and R. Croteau, *Arch. Biochem. Biophys.*, 1978, **191**, 400.
65. D. V. Banthorpe and O. Ekundayo, *Phytochemistry*, 1976, **15**, 109.
66. T. J. Savage and R. Croteau, *Arch. Biochem. Biophys.*, 1993, **305**, 581.
67. G. W. K. Cavill, D. L. Ford, and H. D. Locksley, *Aust. J. Chem.*, 1956, **9**, 288.
68. J. Meinwald, M. S. Chadha, J. J. Hurst, and T. Eisner, *Tetrahedron Lett.*, 1962, 29.
69. S. R. Jensen, in "Ecological Chemistry and Biochemistry of Plant Terpenoids," eds. J. B. Harborne and F. A. Tomas-Barberan, Clarendon, Oxford, 1991, vol. 31, p. 133.
70. V. DeLuca, *Method Plant Biochem.*, 1993, **9**, 345.

71. T. M. Kutchan, *Plant Cell*, 1995, **7**, 1059.
72. S. Escher, P. Loew, and D. Arigoni, *J. Chem. Soc., Chem. Comm.*, 1970, 823.
73. T. D. Meehan and C. J. Coscia, *Biochem. Biophys. Res. Commun.*, 1973, **53**, 1043.
74. D. L. Hallahan, S. M. C. Lau, P. A. Harder, D. W. M. Smiley, G. W. Dawson, J. A. Pickett, R. E. Christoffersen, and D. P. O'Keefe, *Biochim. Biophys. Acta*, 1994, **1201**, 94.
75. D. L. Hallahan, J. M. West, R. M. Wallsgrove, D. W. M. Smiley, G. W. Dawson, J. A. Pickett, and J. G. C. Hamilton, *Arch. Biochem. Biophys.*, 1995, **318**, 105.
76. S. Uesato, H. Ikeda, T. Fujita, H. Inouye, and M. H. Zenk, *Tetrahedron Lett.*, 1987, **28**, 4431.
77. K. M. Madyastha, R. Guarnaccia, C. Baxter, and C. J. Cozcia, *J. Biol. Chem.*, 1973, **248**, 2497.
78. D. J. Faulkner and M. O. Stallard, *Tetrahedron Lett.*, 1973, **14**, 1171.
79. P. Crews and E. Kho, *J. Org. Chem.*, 1975, **40**, 2568.
80. R. E. Moore, *Acc. Chem. Res.*, 1977, **10**, 40.
81. A. San-Martin, R. Negrete, and J. Rovirosa, *Phytochemistry*, 1991, **30**, 2165.
82. S. Naylor, F. J. Hanke, L. V. Manes, and P. Crews, *Prog. Chem. Org. Nat. Prod.*, 1983, **44**, 189.
83. J. H. Tumlinson, R. C. Gueldner, D. D. Hardee, A. C. Thompson, P. A. Hedin, and J. P. Minyard, *J. Org. Chem.*, 1971, **36**, 2616.
84. S. De Rosa, in "Ecological Chemistry and Biochemistry of Plant Terpenoids," eds. J. B. Harborne and F. A. Tomas-Barberan, Clarendon, Oxford, 1991, p. 28.
85. W. Fenical, *Science*, 1982, **215**, 923.
86. K. D. Barrow and C. A. Temple, *Phytochemistry*, 1985, **24**, 1697.
87. A. Butler and J. V. Walker, *Chem. Rev.*, 1993, **93**, 1937.
88. R. K. Okuda, 1997, Personal communication.
89. R. B. Croteau, C. J. Wheeler, D. E. Cane, R. Ebert, and H. J. Ha, *Biochemistry*, 1987, **26**, 5383.
90. K. C. Wagschal, H. J. Pyun, R. M. Coates, and R. Croteau, *Arch. Biochem. Biophys.*, 1994, **308**, 477.
91. R. B. Bates and S. K. Paknikar, *Tetrahedron Lett.*, 1965, no. 20, 1453.
92. W. W. Epstein and C. D. Poulter, *Phytochemistry*, 1973, **12**, 737.
93. C. D. Poulter, *Acc. Chem. Res.*, 1990, **23**, 70.
94. L. Crombie, P. A. Firth, R. P. Houghton, D. A. Whiting, and D. K. Woods, *J. Chem. Soc., Perkin Trans. 1*, 1972, 642.
95. C. D. Poulter, S. G. Moesinger, and W. W. Epstein, *Tetrahedron Lett.*, 1972, no. 1, 67.
96. W. W. Epstein and L. A. Gaudioso, *J. Org. Chem.*, 1982, **47**, 175.
97. K. G. Allen, D. V. Banthorpe, B. V. Charlwood, and C. M. Voller, *Phytochemistry*, 1977, **16**, 79.
98. D. V. Banthorpe, S. Doonan, and J. A. Gutowski, *Phytochemistry*, 1977, **16**, 85.
99. W. S. Zito, V. Srivastava, and E. Abebayo-Olojo, *Planta Med.*, 1991, **57**, 425.
100. W. W. Epstein, 1997, Personal communication.
101. W. F. Erman, "Chemistry of the Monoterpenes. An Encyclopedic Handbook," Marcel Dekker, New York, 1985.
102. R. Croteau, in "Isopentenoids in Plants," eds. D. W. Nes, G. Fuller, and L. S. Tsai, Marcel Dekker, New York, 1984, p. 31.
103. W. D. Loomis and R. Croteau, *Rec. Adv. Phytochem.*, 1973, **6**, 147.
104. A. Fahn, "Secretory Tissues in Plants," Academic Press, London, 1979.
105. A. Fahn, *New Phytol.*, 1988, **108**, 229.
106. J. Gershenzon, M. Maffei, and R. Croteau, *Plant Physiol.*, 1989, **89**, 1351.
107. J. Gershenzon, M. A. Duffy, F. Karp, and R. Croteau, *Anal. Biochem.*, 1987, **163**, 159.
108. D. McCaskill, J. Gershenzon, and R. Croteau, *Planta*, 1992, **187**, 445.
109. W. A. Russin, T. F. Uchytil, and R. D. Durbin, *Plant Science*, 1992, **85**, 115.
110. J. Gershenzon and R. Croteau, *Rec. Adv. Phytochem.*, 1990, **24**, 99.
111. C. Cheniclet and J.-P. Carde, *Isr. J. Bot.*, 1985, **34**, 219.
112. C. Cheniclet, *J. Exp. Bot.*, 1987, **38**, 1557.
113. C. Cheniclet, C. Bernard-Dagan, and G. Pauly, in "Mechanisms of Woody Plant Defenses Against Insects," eds. W. J. Mattson, J. Levieux, and C. Bernard-Dagan, Springer Verlag, New York, 1988, p. 117.
114. F. Lieutier and A. A. Berryman, *Can. J. For. Res.*, 1988, **18**, 1243.
115. C. L. Steele, E. Lewinsohn, and R. Croteau, *Proc. Natl. Acad. Sci. USA*, 1995, **92**, 4164.
116. T. C. J. Turlings, J. H. Loughrin, P. J. McCall, U. S. R. Rose, W. J. Lewis, and J. H. Tumlinson, *Proc. Natl. Acad. Sci. USA*, 1995, **92**, 4169.
117. T. C. J. Turlings, J. H. Tumlinson, R. R. Heath, A. T. Proveaux, and R. E. Doolittle, *J. Chem. Ecol.*, 1991, **17**, 2235.
118. T. C. J. Turlings, J. H. Tumlinson, and W. J. Lewis, *Science*, 1990, **250**, 1251.
119. H. K. Lichtenthaler, J. Schwender, A. Disch, and M. Rohmer, *FEBS Lett.*, 1997, **400**, 271.
120. J. C. Gray, *Adv. Bot. Res.*, 1987, **14**, 25.
121. D. McCaskill and R. Croteau, *Planta*, 1995, **197**, 49.
122. K. Nebeta, T. Ishikawa, and H. Okuyama, *J. Chem. Soc., Perkin Trans. 1*, 1995, 3111.
123. M. Gleizes, G. Pauly, J.-P. Carde, A. Marpeau, and C. Bernard-Dagan, *Planta*, 1983, **159**, 373.
124. U. Mettal, W. Boland, P. Beyer, and H. Kleinig, *Eur. J. Biochem.*, 1988, **170**, 613.
125. L. M. Perez, G. Pauly, J.-P. Carde, L. Belingheri, and M. Gleizes, *Plant Physiol. Biochem.*, 1990, **28**, 221.
126. S. M. Colby, W. R. Alonso, E. J. Katahira, D. J. McGarvey, and R. Croteau, *J. Biol. Chem.*, 1993, **268**, 23016.
127. A. Yuba, K. Yazaki, M. Tabata, G. Honda, and R. Croteau, *Arch. Biochem. Biophys.*, 1996, **332**, 280.
128. G. Feron, M. Clastre, and C. Ambid, *FEBS Lett.*, 1990, **271**, 236.
129. G. Pauly, L. Belingheri, A. Marpeau, and M. Gleizes, *Plant Cell Rep.*, 1986, **5**, 19.
130. E. Soler, G. Feron, M. Clastre, R. Dargent, M. Gleizes, and C. Ambid, *Planta*, 1992, **187**, 171.
131. N. Dudareva, L. Cseke, V. M. Blanc, and E. Pichersky, *Plant Cell*, 1996, **8**, 1137.
132. M. J. O. Francis and M. O'Connell, *Phytochemistry*, 1969, **8**, 1705.
133. J. Gershenzon, *J. Chem. Ecol.*, 1994, **20**, 1281.
134. J. Gershenzon and R. Croteau, in "Lipid Metabolism in Plants," ed. T. S. Moore, Jr., CRC Press, Boca Raton, FL, 1993, p. 340.
135. D. V. Banthorpe, D. R. S. Long, and C. R. Pink, *Phytochemistry*, 1983, **22**, 2459.

136. C. L. Steele, S. Katoh, J. Bohlmann, and R. Croteau, *Plant Physiol.*, 1998, in press.
137. M. Gijzen, E. Lewinsohn, and R. Croteau, *Arch. Biochem. Biophys.*, 1991, **289**, 267.
138. J. Gershenzon, in "Insect–Plant Interactions," ed. E. A. Bernays, CRC Press, Boca Raton, FL, 1994, vol. V, p. 105.
139. R. Croteau and C. Martinkus, *Plant Physiol.*, 1979, **64**, 169.
140. R. Croteau, V. K. Sood, B. Renstrøm, and R. Bhushan, *Plant Physiol.*, 1984, **76**, 647.
141. R. Croteau and V. K. Sood, *Plant Physiol.*, 1985, **77**, 801.
142. R. Croteau, H. El-Bialy, and S. S. Dehal, *Plant Physiol.*, 1987, **84**, 643.
143. R. Croteau, H. El-Bialy, and S. El-Hindawi, *Arch. Biochem. Biophys.*, 1984, **228**, 667.
144. K. L. Falk, J. Gershenzon, and R. Croteau, *Plant Physiol.*, 1990, **93**, 1559.
145. C. Funk, A. E. Koepp, and R. Croteau, *Arch. Biochem. Biophys.*, 1992, **294**, 306.
146. C. A. Mihaliak, J. Gershenzon, and R. Croteau, *Oecologia*, 1991, **87**, 373.
147. F. W. Hefendehl and M. J. Murray, *Lloydia*, 1976, **39**, 39.
148. R. Croteau and J. Gershenzon, *Rec. Adv. Phytochem.*, 1994, **28**, 193.
149. R. Croteau, F. Karp, K. C. Wagschal, D. M. Satterwhite, D. C. Hyatt, and C. B. Skotland, *Plant Physiol.*, 1991, **96**, 744.
150. F. Karp, C. A. Mihaliak, J. L. Harris, and R. Croteau, *Arch. Biochem. Biophys.*, 1990, **276**, 219.
151. S. Lupien, F. Karp, K. Ponnamperuma, M. Wildung, and R. Croteau, *Drug Metabol. Drug. Interact.*, 1995, **12**, 245.
152. D. E. Lincoln and J. H. Langenheim, *Biochem. Syst. Ecol.*, 1981, **9**, 153.
153. P. Vernet, P. H. Gouyon, and G. Valdeyron, *Genetica*, 1986, **69**, 227.
154. E. E. White, *Phytochemistry*, 1983, **22**, 1399.
155. J. S. Birks and P. J. Kanowski, *Silvae Genetica*, 1993, **42**, 340.
156. J. Justesen and N. O. Kjeldgaard, *Anal. Biochem.*, 1992, **207**, 90.
157. G. M. Silver and R. Fall, *J. Biol. Chem.*, 1995, **270**, 13 010.
158. R. B. Herbert, "The Biosynthesis of Secondary Metabolites," 2nd edn., Chapman & Hall, London, 1989.
159. M. Rohmer, M. Knani, P. Simonin, B. Sutter, and H. Sahm, *Biochem. J.*, 1993, **295**, 517.
160. M. K. Schwarz, Ph.D. Diss. Nr. 10951, ETH, Zurich, 1994.
161. W. D. Loomis and J. Battaile, *Phytochemistry*, 1966, **5**, 423.
162. W. D. Loomis, J. D. Lile, R. P. Sandstrom, and A. J. Burbott, *Phytochemistry*, 1979, **18**, 1049.
163. W. R. Alonso and R. Croteau, *Methods Plant Biochem.*, 1993, **9**, 239.
164. R. Croteau and D. E. Cane, *Methods Enzymol.*, 1985, **110**, 383.
165. J. Gershenzon, D. G. McCaskill, J. Rajaonarivony, C. Mihaliak, F. Karp, and R. Croteau, *Rec. Adv. Phytochem.*, 1991, **25**, 347.
166. J. Gershenzon, D. McCaskill, J. I. M. Rajaonarivony, C. Mihaliak, F. Karp, and R. Croteau, *Anal. Biochem.*, 1992, **200**, 130.
167. T. J. Savage, M. W. Hatch, and R. Croteau, *J. Biol. Chem.*, 1994, **269**, 4012.
168. R. Croteau, W. R. Alonso, A. E. Koepp, and M. A. Johnson, *Arch. Biochem. Biophys.*, 1994, **309**, 184.
169. P. W. Holloway and G. Popják, *Biochem. J.*, 1967, **104**, 57.
170. R. Croteau and P. T. Purkett, *Arch. Biochem. Biophys.*, 1989, **271**, 524.
171. D. M. Satterwhite and R. Croteau, *J. Chromatog.*, 1988, **452**, 61.
172. D. M. Satterwhite, C. J. Wheeler, and R. Croteau, *J. Biol. Chem.*, 1985, **260**, 13 901.
173. F. S. Sterrett, in "The Essential Oils," ed. E. Guenther, Krieger, New York, 1975, vol. II (reprinted), p. 769.
174. L. Heide, *FEBS Lett.*, 1988, **237**, 159.
175. D. Zhang and C. D. Poulter, *Anal. Biochem.*, 1993, **213**, 356.
176. E. Tomlinson, T. M. Jefferies, and C. M. Riley, *J. Chromatog.*, 1978, **159**, 315.
177. P. McGeady and R. Croteau, *Arch. Biochem. Biophys.*, 1995, **317**, 149.
178. P. J. Facchini and J. Chappell, *Proc. Natl. Acad. Sci. USA*, 1992, **89**, 11 088.
179. C. J. D. Mau and C. A. West, *Proc. Natl. Acad. Sci. USA*, 1994, **91**, 8497.
180. J. Chappell, *Annu. Rev. Plant Physiol. Plant Mol. Biol.*, 1995, **56**, 521.
181. H. Sagami, K. Ogura, S. Seto, and T. Kurokawa, *Biochem. Biophys. Res. Commun.*, 1978, **85**, 572.
182. H. Sagami and K. Ogura, *J. Biochem.*, 1981, **89**, 1573.
183. L. Heide and U. Berger, *Arch. Biochem. Biophys.*, 1989, **273**, 331.
184. C. D. Poulter and H. C. Rilling, in "Biosynthesis of Isoprenoid Compounds," eds. J. W. Porter and S. L. Spurgeon, Wiley, New York, 1981, vol. 1, p. 558.
185. M. Clastre, B. Bantignies, G. Feron, E. Soler, and C. Ambid, *Plant Physiol.*, 1993, **102**, 205.
186. P. Valenzuela and O. Cori, *Tetrahed. Lett.*, 1967, no. 32, 3089.
187. R. C. Haley, J. A. Miller, and H. C. S. Wood, *J. Chem. Soc. (C)*, 1969, 264.
188. C. A. Bunton, O. Cori, D. Hachey, and J.-P. Leresche, *J. Org. Chem.*, 1979, **44**, 3238.
189. N. C. Deno, R. C. Haddon, and E. N. Nowak, *J. Am. Chem. Soc.*, 1970, **92**, 6691.
190. N. L. Allinger and J. H. Siefert, *J. Am. Chem. Soc.*, 1975, **97**, 752.
191. O. Cori, L. Chayet, L. M. Perez, C. A. Bunton, and D. Hachey, *J. Org. Chem.*, 1986, **51**, 1310.
192. W. Rittersdorf and F. Cramer, *Tetrahedron*, 1968, **24**, 43.
193. S. Winstein, G. Valkanas, and C. F. Wilcox, Jr., *J. Am. Chem. Soc.*, 1972, **94**, 2286.
194. C. A. Bunton, J. P. Leresche, and D. Hachey, *Tetrahedron Lett.*, 1972, **24**, 2431.
195. E. P. Brody and C. D. Gutsche, *Tetrahedron*, 1977, **33**, 723.
196. S. Godtfredsen, J. P. Obrecht, and D. Arigoni, *Chimia*, 1977, **31**, 62.
197. C. D. Poulter and C.-H. R. King, *J. Am. Chem. Soc.*, 1982, **104**, 1420.
198. C. D. Poulter and C.-H. R. King, *J. Am. Chem. Soc.*, 1982, **104**, 1422.
199. D. V. Banthorpe and D. Whittaker, *Q. Rev. Chem. Soc.*, 1966, **20**, 373.
200. G. D. Sargent, *Q. Rev. Chem. Soc.*, 1966, **20**, 301.
201. H. C. Brown, *Acc. Chem. Res.*, 1973, **6**, 377.
202. D. A. Dougherty, *Science*, 1996, **271**, 163.
203. D. A. Dougherty and D. A. Stauffer, *Science*, 1990, **250**, 1558.
204. A. McCurdy, L. Jimenez, D. A. Stauffer, and D. A. Dougherty, *J. Am. Chem. Soc.*, 1992, **114**, 10 314.

205. J. L. Sussman, M. Harel, F. Frolow, C. Oefner, A. Goldman, L. Toker, and I. Silman, *Science*, 1991, **253**, 872.
206. Z. Shi, C. J. Buntel, and J. H. Griffin, *Proc. Natl. Acad. Sci. USA*, 1994, **91**, 7370.
207. T. Li, K. D. Janda, and R. A. Lerner, *Nature*, 1996, **379**, 326.
208. D. S. Goodman and G. Popják, *J. Lipid Res.*, 1960, **1**, 286.
209. B. K. Tidd, *J. Chem. Soc. (B)*, 1971, 1168.
210. H. L. King and H. C. Rilling, *Biochemistry*, 1977, **16**, 3815.
211. D. N. Brems and H. C. Rilling, *J. Am. Chem. Soc.*, 1977, **99**, 8351.
212. L. Chayet, M. C. Rojas, O. Cori, C. A. Bunton, and D. C. McKenzie, *Bioorgan. Chem.*, 1984, **12**, 329.
213. D. I. Ito, S. Izumi, T. Hirata, and T. Suga, *J. Chem. Soc., Perkin Trans. 1*, 1992, 37.
214. T. J. Savage, H. Ichii, S. D. Hume, D. B. Little, and R. Croteau, *Arch. Biochem. Biophys.*, 1995, **320**, 257.
215. K. P. Adam, J. Crock, and R. Croteau, *Arch. Biochem. Biophys.*, 1996, **332**, 352.
216. W. R. Alonso, J. I. M. Rajaonarivony, J. Gershenzon, and R. Croteau, *J. Biol. Chem.*, 1992, **267**, 7582.
217. W. R. Alonso and R. Croteau, *Arch. Biochem. Biophys.*, 1991, **286**, 511.
218. E. Lewinsohn, M. Gijzen, and R. Croteau, *Arch. Biochem. Biophys.*, 1992, **293**, 167.
219. W. R. Alonso and R. Croteau, in "Secondary-Metabolite Biosynthesis and Metabolism," eds. R. J. Petroski and S. P. McCormick, Plenum, New York, 1992, p. 239.
220. R. Croteau, 1997, unpublished.
221. M. Gijzen, E. Lewinsohn, and R. Croteau, *Arch. Biochem. Biophys.*, 1992, **294**, 670.
222. W. R. Alonso, J. E. Crock, and R. Croteau, *Arch. Biochem. Biophys.*, 1993, **301**, 58.
223. G. Portilla, M. C. Rojas, L. Chayet, and O. Cori, *Arch. Biochem. Biophys.*, 1982, **218**, 614.
224. R. Croteau, J. H. Miyazaki, and C. J. Wheeler, *Arch. Biochem. Biophys.*, 1989, **269**, 507.
225. R. Croteau, J. Gershenzon, C. J. Wheeler, and D. M. Satterwhite, *Arch. Biochem. Biophys.*, 1990, **277**, 374.
226. L. Chayet, C. Rojas, E. Cardemil, A. M. Jabalquinto, R. Vicuña, and O. Cori, *Arch. Biochem. Biophys.*, 1977, **180**, 318.
227. R. Croteau and F. Karp, *Arch. Biochem. Biophys.*, 1977, **179**, 257.
228. C. J. Wheeler and R. Croteau, *Arch. Biochem. Biophys.*, 1988, **260**, 250.
229. C. J. Wheeler and R. Croteau, *J. Biol. Chem.*, 1987, **262**, 8213.
230. R. Croteau, C. J. Wheeler, R. Aksela, and A. C. Oehlschlager, *J. Biol. Chem.*, 1986, **261**, 7257.
231. C. J. Wheeler, C. A. Mihaliak, and R. Croteau, *Arch. Biochem. Biophys.*, 1990, **279**, 203.
232. A. Cornish-Bowden, *FEBS Lett.*, 1986, **203**, 3.
233. R. Croteau, *Arch. Biochem. Biophys.*, 1986, **251**, 777.
234. C. D. Poulter, J. C. Argyle, and E. A. Mash, *J. Biol. Chem.*, 1978, **253**, 7227.
235. J. I. Rajaonarivony, J. Gershenzon, and R. Croteau, *Arch. Biochem. Biophys.*, 1992, **296**, 49.
236. J. I. Rajaonarivony, J. Gershenzon, J. Miyazaki, and R. Croteau, *Arch. Biochem. Biophys.*, 1992, **299**, 77.
237. R. Croteau, W. R. Alonso, A. E. Koepp, J. H. Shim, and D. E. Cane, *Arch. Biochem. Biophys.*, 1993, **307**, 397.
238. R. Croteau and M. Felton, *Arch. Biochem. Biophys.*, 1981, **207**, 460.
239. R. Croteau, N. M. Felton, and C. J. Wheeler, *J. Biol. Chem.*, 1985, **260**, 5956.
240. A. F. Thomas, F. A. Schneider, and J. Meinwald, *J. Am. Chem. Soc.*, 1967, **89**, 68.
241. G. A. Abad, S. P. Jindal, and T. T. Tidwell, *J. Am. Chem. Soc.*, 1973, **95**, 6326.
242. R. Croteau, D. M. Satterwhite, C. J. Wheeler, and N. M. Felton, *J. Biol. Chem.*, 1989, **264**, 2075.
243. R. Croteau, D. M. Satterwhite, C. J. Wheeler, and N. M. Felton, *J. Biol. Chem.*, 1988, **263**, 15 449.
244. R. Croteau, D. M. Satterwhite, D. E. Cane, and C. C. Chang, *J. Biol. Chem.*, 1986, **261**, 13 438.
245. R. Croteau, D. M. Satterwhite, D. E. Cane, and C. C. Chang, *J. Biol. Chem.*, 1988, **263**, 10 063.
246. R. Croteau and D. M. Satterwhite, *J. Biol. Chem.*, 1989, **264**, 15 309.
247. K. Wagschal, T. J. Savage, and R. Croteau, *Tetrahedron*, 1991, **47**, 5933.
248. D. E. Cane, A. Saito, R. Croteau, J. Shaskus, and M. Felton, *J. Am. Chem. Soc.*, 1982, **104**, 5831.
249. R. M. Coates, J. F. Denissen, R. B. Croteau, and C. J. Wheeler, *J. Am. Chem. Soc.*, 1987, **109**, 4399.
250. H. J. Pyun, K. C. Wagschal, D. I. Jung, R. M. Coates, and R. Croteau, *Arch. Biochem. Biophys.*, 1994, **308**, 488.
251. C. D. Poulter and H. C. Rilling, *Biochemistry*, 1976, **15**, 1079.
252. L. Jacob, M. Julia, B. Pfeiffer, and C. Rolando, *Tetrahedron Lett.*, 1983, **24**, 4327.
253. T. Suga, Y. Hiraga, M. Aihara, and S. Izumi, *J. Chem. Soc., Chem. Comm.*, 1992, 1556.
254. H.-J. Pyun, R. M. Coates, K. C. Wagschal, P. McGeady, and R. Croteau, *J. Org. Chem.*, 1993, **58**, 3998.
255. R. Croteau, in "Flavor Precursors: Thermal and Enzymatic Conversions," eds. R. Teranishi, G. R. Takeoka, and M. Guntert, American Chemical Society, Washington, DC, 1992, p. 8.
256. R. E. LaFever and R. Croteau, *Arch. Biochem. Biophys.*, 1993, **301**, 361.
257. C. J. Wheeler and R. Croteau, *Arch. Biochem. Biophys.*, 1986, **246**, 733.
258. C.J. Wheeler and R. B. Croteau, *Proc. Natl. Acad. Sci. USA*, 1987, **84**, 4856.
259. J. P. Jones, K. R. Korzekwa, A. E. Rettie, and W. F. Trager, *J. Am. Chem. Soc.*, 1986, **108**, 7074.
260. A. G. Samuelson and B. K. Carpenter, *J. Chem. Soc., Chem. Comm.*, 1981, 354.
261. D. B. Northrop, *Biochemistry*, 1977, **14**, 2644.
262. R. A. Pascal, M. W. Baum, C. K. Wagner, L. R. Rodgers, and D.-S. Huang, *J. Am. Chem. Soc.*, 1986, **108**, 6477.
263. F. H. Westheimer, *Chem. Rev.*, 1961, **61**, 265.
264. M. H. Glickman and J. P. Klinman, *Biochemistry*, 1995, **34**, 14 077.
265. S. R. Ellenberger, G. D. Peiser, G. D. Bell, C. E. Hussey, Jr., D. M. Shattuck-Eidens, and B. D. Swedlund, *Chem. Abstr.*, 1995, **123**, 221 798.
266. K. Keegstra, L. J. Olsen, and S. M. Theg, *Annu. Rev. Plant Physiol. Plant Mol. Biol.*, 1989, **40**, 471.
267. Y. Gavel and G. von Heijne, *FEBS Lett.*, 1990, **261**, 455.
268. D. J. McGarvey, D. C. Williams, and R. Croteau, 1997, unpublished.
269. M. N. Ashby and P. A. Edwards, *J. Biol. Chem.*, 1990, **265**, 13 157.
270. D. E. Cane, Q. Xue, and B. C. Fitzsimons, *Biochemistry*, 1996, **35**, 12 369.
271. L. Song and C. D. Poulter, *Proc. Natl. Acad. Sci. USA*, 1994, **91**, 3044.
272. A. Joly and P. A. Edwards, *J. Biol. Chem.*, 1993, **268**, 26 983.
273. T. Koyama, M. Tajima, H. Sano, T. Doi, A. Koike-Takeshita, S. Obata, T. Nishino, and K. Ogura, *Biochemistry*, 1996, **35**, 9533.

274. P. F. Marrero, C. D. Poulter, and P. A. Edwards, *J. Biol. Chem.*, 1992, **267**, 21 873.
275. L. C. Tarshis, M. Yan, C. D. Poulter, and J. C. Sacchettini, *Biochemistry*, 1994, **33**, 10 871.
276. L. C. Tarshis, P. J. Proteau, B. A. Kellogg, J. C. Sacchettini, and C. D. Poulter, *Proc. Natl. Acad. Sci. USA*, 1996, **93**, 15 018.
277. C. A. Lesburg, G. Zhai, D. E. Cane, and D. W. Christianson, *Science*, 1997, **277**, 1820.
278. C. M. Starks, K. Back, J. Chappell, and J. P. Noel, *Science*, 1997, **277**, 1815.
279. C. Martinkus and R. Croteau, *Plant Physiol.*, 1981, **68**, 99.
280. R. Croteau and J. N. Winters, *Plant Physiol.*, 1982, **69**, 975.
281. R. Kjonaas, C. Martinkus-Taylor, and R. Croteau, *Plant Physiol.*, 1982, **69**, 1013.
282. R. B. Kjonaas and R. Croteau, *Arch. Biochem. Biophys.*, 1983, **220**, 79.
283. R. Kjonaas, K. V. Venkatachalam, and R. Croteau, *Arch. Biochem. Biophys.*, 1985, **238**, 49.
284. R. Croteau and K. V. Venkatachalam, *Arch. Biochem. Biophys.*, 1986, **249**, 306.
285. D. C. Hawkinson, T. C. Eames, and R. M. Pollack, *Biochemistry*, 1991, **30**, 10 849.
286. R. Croteau and C. L. Hooper, *Plant Physiol.*, 1978, **61**, 737.
287. C. Mihaliak, F. Karp, and R. Croteau, *Method Plant Biochem.*, 1993, **9**, 261.
288. F. Karp and R. Croteau, in "Bioflavour '87," ed. P. Schreier, de Gruyter, Berlin, 1988, p. 173.
289. F. Karp and R. Croteau, *Arch. Biochem. Biophys.*, 1982, **216**, 616.
290. A. J. Poulose and R. Croteau, *Arch. Biochem. Biophys.*, 1978, **187**, 307.
291. M. A. Johnson and R. Croteau, *Arch. Biochem. Biophys.*, 1984, **235**, 254.
292. V. H. Potty and J. H. Bruemmer, *Phytochemistry*, 1970, **9**, 1001.
293. A. J. Burbott and W. D. Loomis, *Plant Physiol.*, 1980, **65** (Suppl.), 96.
294. R. Croteau, C. L. Hooper, and M. Felton, *Arch. Biochem. Biophys.*, 1978, **188**, 182.
295. R. Croteau and N. M. Felton, *Phytochemistry*, 1980, **19**, 1343.
296. D. McCaskill and R. Croteau, *Adv. Biochem. Eng. Biotech.*, 1997, **55**, 107.
297. H. Kacser and L. Acerenza, *Eur. J. Biochem.*, 1993, **216**, 361.

2.06 Sesquiterpene Biosynthesis: Cyclization Mechanisms

DAVID E. CANE

Brown University, Providence, RI, USA

2.06.1 INTRODUCTION

Sesquiterpenes are among the most widely occurring, most extensively studied, and best understood families of natural products, from the point of view of chemistry, biochemistry, and biological origin. More than 300 distinct sesquiterpene carbon skeletons have been identified to date, and thousands of naturally occurring oxidized or otherwise modified derivatives have been isolated from marine, terrestrial plant, and microbial sources.[1] These metabolites display a broad range of physiological properties, including antibiotic, antitumor, antiviral, cytotoxic, immunosuppressive, phytotoxic, antifungal, insect antifeedant, and hormonal activities. Numerous sesquiterpene hydrocarbons, alcohols, and derived metabolites found in plant essential oils are also highly valued for their desirable odor and flavor characteristics. Remarkably all of these substances are derived from a single acyclic precursor, farnesyl diphosphate (FPP) (**1**).

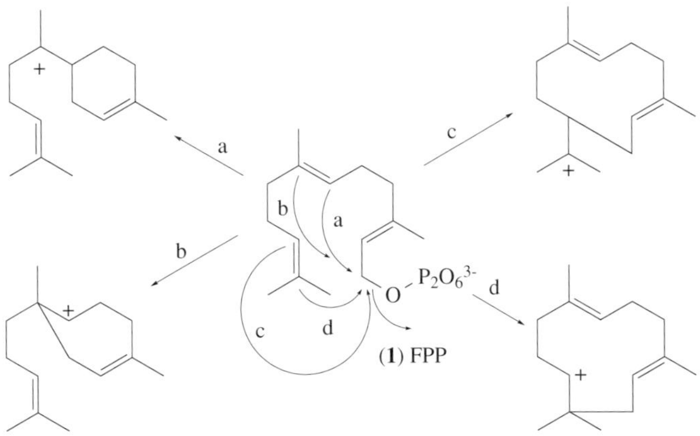

a b c d $P_2O_6^{3-}$ O (**1**) FPP

Scheme 1

Sesquiterpene biosynthesis was the subject of a detailed review published in 1981 covering all important classical precursor incorporation experiments that had been carried out up to that time.[2] Since then, the major development has been the isolation of numerous sesquiterpene synthases, their subsequent purification and characterization, and the eventual cloning, sequencing, and expression of their corresponding structural genes. A 1985 review described methods for the isolation and assay of sesquiterpene and monoterpene synthases[3] and a review in 1990 detailed what was then known about the mechanisms of the enzymatic formation of sesquiterpenes.[4] In 1995, the genetics of isoprenoid antibiotic biosynthesis have also been reviewed.[5] In the mid-1980s, a detailed stereochemical theory of sesquiterpene cyclizations has been articulated.[6] Terpene biosynthesis has also been reviewed on a regular basis.[7] This review focuses mainly on the literature since 1980.

2.06.2 CYCLIZATION MECHANISMS

The role of farnesol as the universal precursor of sesquiterpenes was first recognized by Ruzicka and Stoll in the 1920s.[8] In 1953, the Biogenetic Isoprene Rule was reformulated in explicitly mechanistic terms,[9–11] according to which an activated derivative of farnesol, subsequently recognized as farnesyl diphosphate (**1**), undergoes ionization followed by electrophilic attack of the resultant allylic cation on either the central double bond to form six- or seven-membered ring intermediates (paths a or b) or on the distal double bond to form 10- or 11-membered ring intermediates (paths c or d) (Scheme 1).[12] The derived cationic intermediates can themselves undergo further cyclizations and rearrangements with the reaction being terminated by quenching of the positive charge, either by removal of a proton or capture of an external nucleophile, usually water. Recognizing that there is a geometric barrier to direct cyclization of *trans,trans*-FPP to six-membered rings, as well as to 10- and 11-membered rings containing a *cis* double bond, it was subsequently proposed that FPP would in such cases undergo an initial isomerization to the corresponding tertiary allylic ester, nerolidyl diphosphate (NPP) (**2**), which has the conformational flexibility and appropriate reactivity to allow formation of the derived cyclic products (Scheme 2). Formation of cyclic products therefore involves two distinct but closely related pathways: ionization–cyclization to 10- and 11-membered ring products with all-*trans* double bonds and ionization–isomerization–ionization–cyclization to products with a *cis* double bond. Although the two pathways are superficially distinct, they in fact share a common mechanism—ionization of an allylic diphosphate ester to the corresponding allylic cation—pyrophosphate ion pair followed either by recapture of the pyrophosphate at the allylic position or anti attack on one of the two neighboring double bonds.

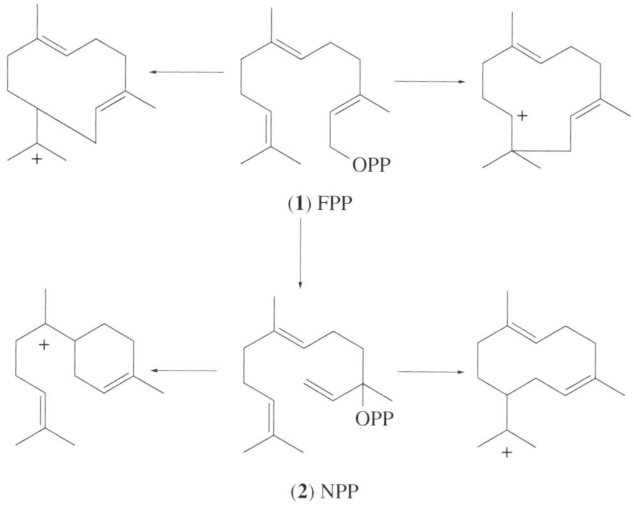

OPP

(**1**) FPP

OPP

(**2**) NPP

Scheme 2

The first explicit stereochemical theory of the formation of sesquiterpenes was elaborated by Arigoni in a landmark IUPAC lecture, based on an exhaustive study of the biosynthesis of a family of cadinane, sativane, and longifolane sesquiterpenes.[13] These ideas were expanded in a later review[6] and have been amply borne out by a wealth of experimental data derived from studies carried out both with intact cells and at the cell-free level. According to this formulation, a major determinant of the structure and stereochemistry of the eventually formed cyclic sesquiterpene is the precise folding of FPP at the active site of a sesquiterpene synthase. Indeed, an extremely fruitful approach to the study of sesquiterpene biosynthesis has been to deduce the folding of the FPP precursor from the relative and absolute configuration of the derived product and then to test these inferences by detailed experiments with stereospecifically labeled substrates. Numerous examples of such experiments are reviewed in Section 2.06.3.

The isolation, characterization, and cloning of many sesquiterpene synthases has made it possible to address explicitly the role of the cyclase in mediating the cyclization of FPP to one or more specific products. Among the most critical questions which must be answered are:

(i) Given that folding of the substrate is a major determinant of the structure and stereochemistry of the ultimately formed product, how does the protein impose a particular conformation on the largely lipophilic C_{15} triene ester substrate, FPP? Given the substantial changes of hybridization, bonding, and stereochemistry which characterize the cyclization of FPP to any one of 300 possible sesquiterpenes, the cyclase active site must be able simultaneously to accommodate and to control an array of reactive intermediates of diverse structure and stereochemistry.

(ii) How is ionization of FPP triggered and how is charge managed—protected from external water, stabilized and directed—during the course of a cyclization? How does the protein control the reactivity of enormously electrophilic cationic intermediates while avoiding self-annihilation which would result from alkylation of nucleophilic side chains or the peptide backbone itself?

(iii) What determines the timing of the eventual quenching of the positive charge, either by deprotonation or controlled admission of water to the active site?

(iv) What is the nature of the active site and the amino acid residues which mediate substrate folding, ionization, intermediate stabilization, and product formation?

2.06.3 SESQUITERPENE BIOSYNTHETIC PATHWAYS

2.06.3.1 Trichothecanes

2.06.3.1.1 Trichodiene synthase

Trichothecanes constitute a large number of mycotoxins and antibiotics produced by a variety of fungi, including *Fusarium*, *Myrothecium*, and *Trichothecium* species, as well as certain plants.[14] Historically, two members of this class, trichothecin (**3**), a product of the apple mold fungus *Trichothecium roseum*, and trichodermin (**4**), a metabolite of *T. sporulosum*, were among the very first sesquiterpenes to be subjected to serious biosynthetic investigation, with the earliest reported experiments dating back to 1959.[15] Interest in this family of metabolites has been further intensified by the serious threats to human and animal health posed by contamination of grain by trichothecane-producing fungi, as well as by fundamental and practical questions regarding the role of the these phytotoxins in fungal virulence.

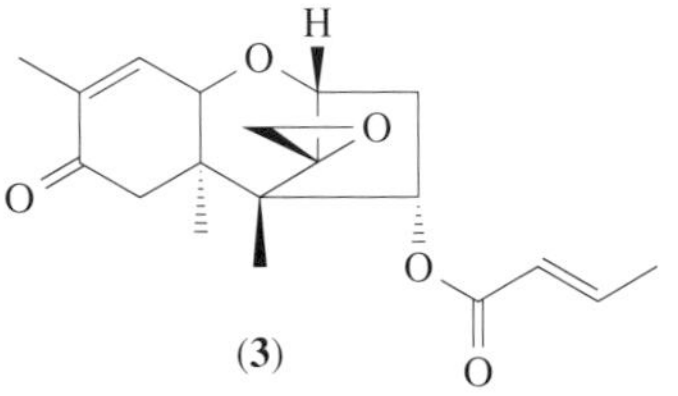

(**3**) 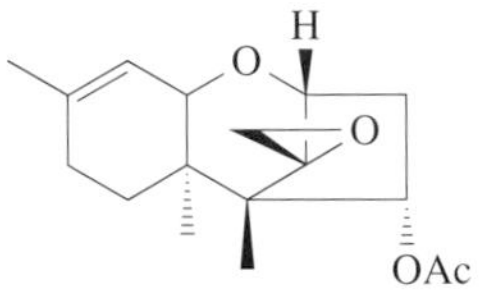 (**4**)

Early experiments involving feeding of labeled acetate and mevalonate precursors to intact cells established the isoprenoid origin of the trichothecanes and demonstrated the operation of a set of mechanistically interesting methyl migrations and hydride transfers.[2] Experiments with ^{13}C-labeled substrates and analysis by ^{13}C NMR largely served to confirm these early conclusions. In two instances, labeled FPP was successfully incorporated into trichothecanes using microbial cultures.[16,17]

In the meantime, systematic investigation of the products of trichothecane-producing cultures resulted in the identification of several metabolites which were recognized as potential intermediates of trichothecane biosynthesis. By far the most important of these latter metabolites was the sesquiterpene trichodiene (**5**), which was shown to be the parent hydrocarbon of the trichothecane family of metabolites.[18] In 1973, Hanson and co-workers reported the cell-free formation of trichodiene itself from farnesyl pyrophosphate catalyzed by a crude extract of *T. roseum*.[19,20] Although several of the experimental claims from this first report were subsequently shown to be erroneous, the isolation of trichodiene synthase opened the way to the modern area of investigation of sesquiterpene biosynthesis. The enzyme has subsequently been isolated from a variety of fungal sources, including *Fusarium sporotrichioides*[21] and *F. sambucinum* (*Gibberella pulicaris*),[22] the source of the mycotoxins T-2 toxin and nivalenol, respectively. Hohn reported the purification to homogeneity of trichodiene synthase from *F. sporotrichioides*, and demonstrated that the native enzyme is a homodimer of M_r 45 000, based on mobility on SDS–PAGE.[21] In common with all other terpenoid cyclases, trichodiene synthase requires no cofactors other than a divalent metal cation, in this case Mg^{2+}. The corresponding structural gene has been cloned from a variety of sources, with the *F. sporotrichioides* gene shown to encode a protein of M_D 43 999 (see Chapter 1.07).[23,24] Cane *et al.* have expressed recombinant trichodiene synthase in *Escherichia coli* as 25–30% of soluble protein.[25]

Detailed insight into the mechanism of trichodiene formation came initially from studies of the enzymatic cyclization of FPP by trichodiene synthase. Incubation of [1-^{3}H,12,13-^{14}C]FPP ((**1**), $H_A = H_B = T$) with a crude extract of trichodiene synthase from *T. roseum* gave labeled trichodiene which was shown by chemical degradation to carry both tritium labels at the expected site, C-11 (Scheme 3).[26] The retention of both of the original C-1 hydrogen atoms of FPP definitively ruled out a previously proposed redox mechanism for the isomerization of the 2,3-double bond of FPP. It was subsequently shown that this cyclization takes place with net retention of configuration at C-1 of FPP.[27] Thus cyclization of (1*S*)- and (1*R*)-[1-^{3}H,12,13-^{14}C]FPP ((**1**), $H_A = T$, $H_B = A$; (**1**), $H_A = H$, $H_B = T$) in separate incubations gave trichodiene, which was shown by stereospecific chemical degradation to be labeled at H-11α and H-11β, respectively. These results, which are in contrast to the well-known inversion of configuration which characterizes the isoprenoid chain elongation reactions catalyzed by the mechanistically related FPP synthase (see Chapter 2.04), are the necessary consequence of the sequential isomerization of FPP to its tertiary allylic isomer, NPP (**2**), which after rotation about the 2,3-single bond undergoes cyclization to the intermediate bisabolyl cation (**6**).

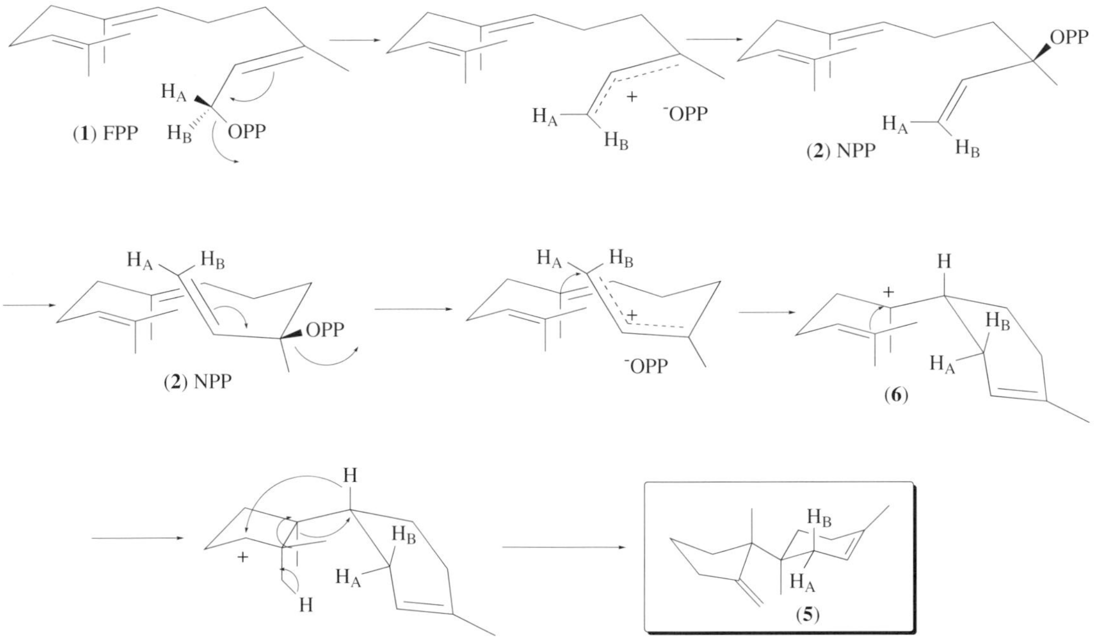

Scheme 3

Further support for the proposed role of NPP in trichodiene formation came from the observation that (3*R*)-NPP can serve as a substrate for trichodiene synthase.[28,29] Incubation of (1*Z*)-[1-^{3}H,12,13-^{14}C]NPP with trichodiene synthase gave labeled trichodiene which was shown by chemical degradation to be labeled exclusively at H-11β, thereby establishing that the cyclization of NPP had occurred by a net *anti* allylic displacement, and leading to the inference that isomerization of FPP to (3*R*)-NPP must take place by a net *syn* migration of the pyrophosphate moiety, consistent with earlier studies of the FPP–NPP isomerization (Scheme 4). Competitive incubation of [1-^{3}H]FPP and [12,13-^{14}C]NPP established that both substrates compete for a common active site in trichodiene synthase. Analysis of labeled FPP and NPP recovered from the latter experiment also established that the isomerization–cyclization sequence takes place without release of enzyme-bound NPP.

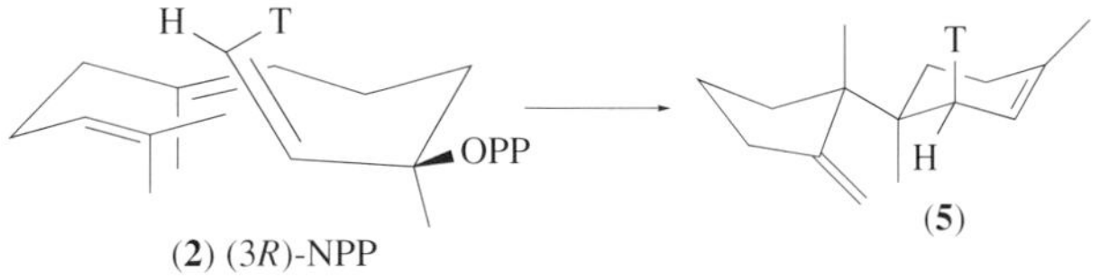

Scheme 4

Comparison of the steady-state kinetic parameters of the natural substrate *trans,trans*-FPP with (3*R*)-NPP as well as *cis,trans*-FPP using both native and recombinant trichodiene synthase was consistent with the proposed intermediacy of NPP, while leading to some unexpected observations.[30] Although the K_m values of *trans,trans*-FPP and (3*R*)-NPP were found to be identical (90 nM), the observed k_{cat} as well as the k_{cat}/K_m for the *trans,trans*-FPP were 1.67 times greater than the corresponding values for the proposed intermediate, (3*R*)-NPP. This apparent kinetic anomaly can be reconciled by the realization that the conformation of free trichodiene synthase which binds exogenous NPP is likely to be different from the conformational state of the protein which binds *enzymatically generated* (3*R*)-NPP. Cases of such closed enzymatic transition states, while rare, have ample precedent,[31,32] and in the case of trichodiene synthase may reflect the necessity of shielding the various cationic intermediates from premature quenching by water from the medium. Transient kinetic studies of the cyclization of FPP to trichodiene, discussed in more detail in Section 2.06.4.2.1, have established that the rate-limiting step of the steady-state reaction is release of the product trichodiene, a process that is 40 times slower than the slowest chemical step, consumption of FPP itself.[33] The product off-rate therefore actually masks the true rate of consumption of exogenously added NPP, which can be calculated to be no more than 1/8 that of FPP and up to 400 times slower than the estimated rate of consumption of enzymatically generated NPP.

Further insight into the mechanism of the trichodiene synthase reaction has come from examination of a group of substrate and intermediate analogues which can variously act as inhibitors[30] or undergo abortive cyclization reactions.[34] Inorganic pyrophosphate itself is a competitive inhibitor with K_I 0.5 μM. The most effective competitive inhibitor tested to date, 10-fluoro-FPP (**7**) exhibited a K_I of 16 nM. The FPP isomer (**8**), in which the distal double bond has been shifted, had a K_I 130 nM, insignificantly higher than the K_m for FPP, while the homologue (**9**), with an extra methylene group, had a K_I of 500 nM. The phosphonophosphate analogue (**10**), in which the ester oxygen of the allylic pyrophosphate moiety has been replaced by a methylene, exhibited mixed noncompetitive inhibition, with K_I for binding to the free enzyme and enzyme–substrate complex found to be 3 and 9 μM respectively. In spite of the somewhat lower binding affinity, this unreactive substrate analogue is an excellent candidate for crystallographic study of cyclase–inhibitor complexes. Interestingly, the FPP ether analogue (**11**), which is one of the most potent inhibitors known for squalene synthase (see Chapter 2.09),[35] an enzyme that also initiates its reaction by ionization of FPP, turns out to be only a modest inhibitor of trichodiene synthase, with a K_I/K_m of 70.

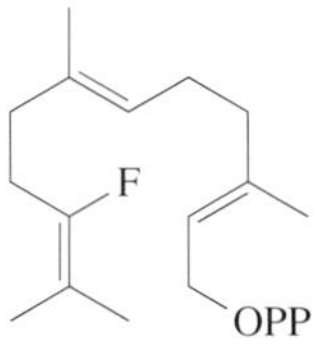

(**7**) 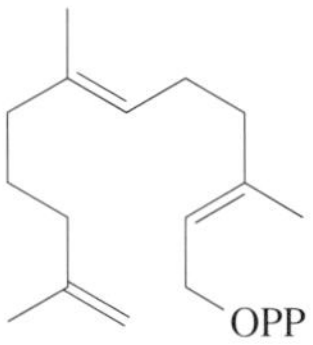 (**8**) 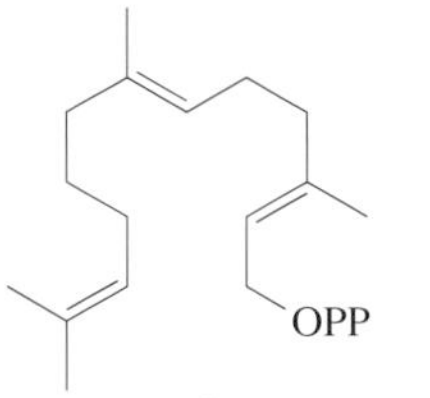 (**9**) 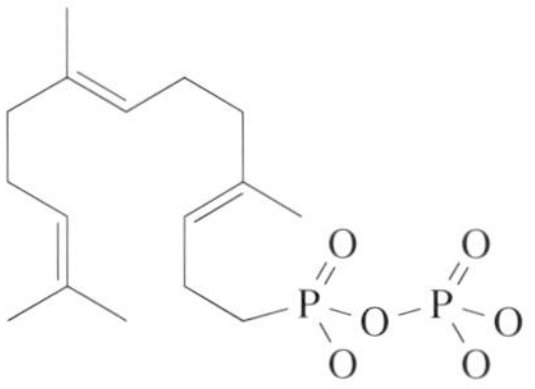 (**10**)

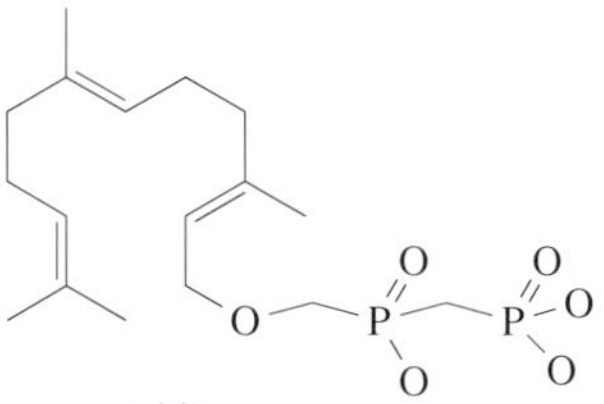

(**11**) 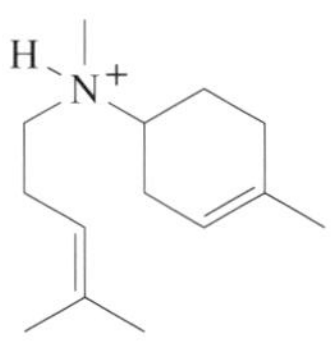 (**12**) 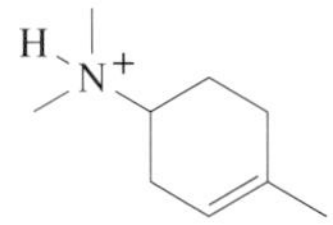 (**13**)

One of the great challenges in studying terpenoid cyclizations is that none of the intermediates are ever released as free intermediates from the cyclase active site. A variety of indirect methods have been used to provide evidence for the existence of chemically reasonable but illusive intermediates. For example, ammonium ion analogues of the proposed bisabolyl cation intermediate act as inhibitors of trichodiene synthase.[36] Although both (*R*)- and (*S*)-(**12**) and (*R*)- and (*S*)-(**13**), as well as trimethylamine, were only weak inhibitors when incubated alone, in the presence of inorganic pyrophosphate, each of these amines showed strongly enhanced, cooperative competitive inhibition. The most significant synergistic effects were observed for the sesquiterpene analogues (**12**), indicative of the importance of both electrostatic and hydrophobic interactions. Remarkably, trichodiene synthase, which normally generates a single enantiomer of its natural product, trichodiene, showed negligible discrimination between the individual enantiomers of each ammonium ion analogue. These observations support the notion that the trichodiene synthase active site, which must accommodate a wide range of structures in converting FPP to trichodiene, is *permissive* rather than *restrictive*, thereby allowing the binding of a family of intermediates ultimately derivable from the initial tightly controlled chiral folding of the substrate FPP. In this sense the active site of trichodiene synthase and related cyclases may be thought of as more like a mitten than a glove, having a high intrinsic specificity and handedness, while simultaneously accommodating a variety of related shapes.

In addition to acting as simple inhibitors, certain substrate analogues can also serve as anomalous substrates for terpenoid cyclases, giving rise to abortive reaction products whose structures can be diagnostic of those of the natural, normally enzyme-bound, reaction intermediates. Thus 6,7-dihydroFPP (**14**) is capable of undergoing the normal allylic isomerization mediated by trichodiene synthase, but the absence of the central double bond prevents further cyclization. In an attempt to study the otherwise cryptic isomerization of FPP to (3*R*)-NPP, both (7*S*)-6,7-dihydroFPP ((7*S*)-(**14**)) and (7*R*)-6,7-dihydroFPP ((7*R*-(**14**)), each of which was a modest competitive inhibitor of trichodiene synthase (K_I/K_m 10–15), were separately incubated with trichodiene synthase, resulting in the formation of a mixture of products consisting of 80–85% of the isomeric trienes (**15**)–(**17**), and 15–20% of the corresponding diastereomer of the allylic alcohols dihydrofarnesol (**18**) and (3*S*)-dihydronerolidol (**19**) (Scheme 5).[34,37] Examination of the water-soluble products from incubation of (7*S*)-(**14**) also established the presence of 25% of the isomeric *cis*-6,7-dihydroFPP (**20**). The net rate of product formation was *ca.* 10% the normal V_{max} of trichodiene formation from FPP. These results can be explained by initial enzyme-catalyzed isomerization of dihydroFPP to the corresponding tertiary allylic isomer, (3*R*)-6,7-dihydroNPP (**21**). Since (**21**) is incapable of further cyclization, ionization of (**21**) and quenching of the resultant allylic cation–pyrophosphate ion pair by deprotonation, recapture of the pyrophosphate ion to give the isomeric primary diphosphate esters, or backside attack by adventitious water will give the observed family of abortive cyclization products.

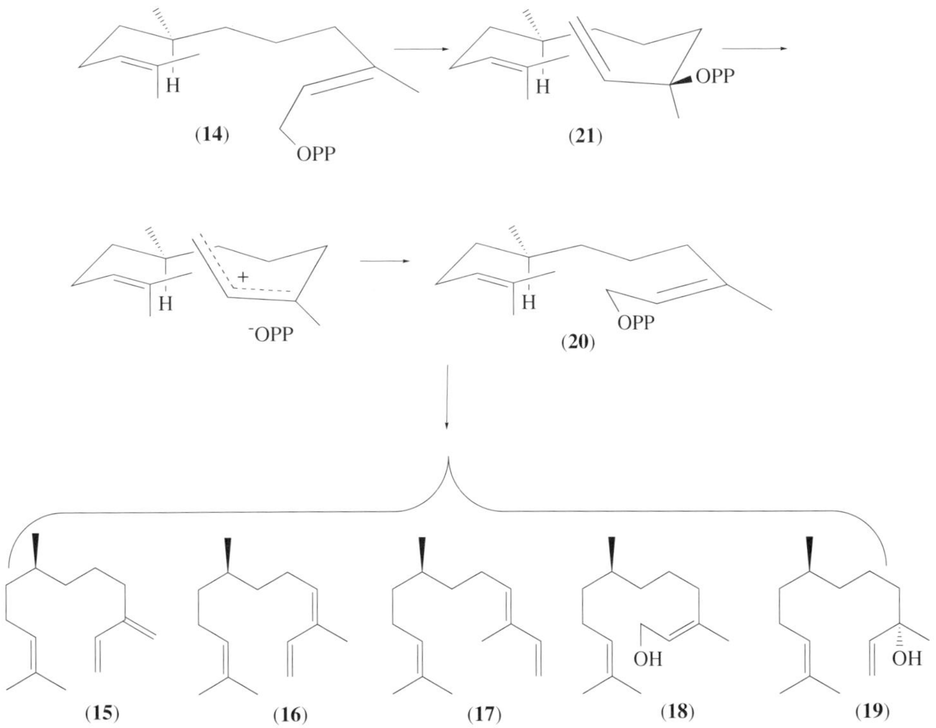

(14) OPP H → (21) OPP H →

⁻OPP H → (20) OPP H

(15) (16) (17) (18) OH (19) OH

Scheme 5

The genes for several trichodiene synthases of fungal origin have been cloned and sequenced (see Chapter 2.07).[23,24,38] Although these synthases show a high degree of amino acid sequence conservation, trichodiene synthase shows no discernible overall sequence similarity to any other known protein. Classical experiments with amino acid-specific reagents had suggested the presence of two cysteine residues in or near the active site.[39] Consistent with this observation, the corresponding C146A mutant of *F. sporotrichioides* cloned in *E. coli* proved to be essentially inactive. The precise mechanistic role of this cysteine residue, however, remains completely obscure. Although initial results had suggested that the C190A mutant had only partial activity, repetition of these experiments has revealed that this mutant has kinetic parameters essentially identical to the wild-type protein.[40]

Trichodiene synthase from *F. sporotrichioides* has a short basic amino acid-rich region, corresponding to amino acids 302–306 (DRRYR), while the closely related sequence DHRYR is found in both the *G. pulicaris* and *F. poae* enzymes. An analogous base-rich region in FPP synthase, EERYK, had previously been identified, by use of a photoaffinity analogue of isopentenyl diphosphate, as a potential pyrophosphate binding region.[41] Recognizing the apparent similarity of these two motifs, site-directed mutagenesis was used to systematically modify the DRRYR region in trichodiene synthase.[39] Although the D302N, D302E, R303I, R303E, and R306K mutants were found to be essentially fully active, the R304K, Y305F, and Y305T mutants showed interesting changes in activity and were selected for further study. The importance of Arg304 for catalysis was established by the finding that the R304K mutant showed a 200-fold reduction in k_{cat} and 25-fold increase in K_m compared to wild-type trichodiene synthase. Although replacement of Tyr305 with the aromatic residue Phe resulted in a modest seven- to eightfold increase in K_m, there was little effect on k_{cat}. By contrast, the Y305T mutant had a k_{cat} reduced by 120-fold accompanied by an 80-fold increase in K_m. Curiously, crystallographic data on FPP synthase have subsequently shown that the putative active site Arg is actually positioned on the exterior of the protein on a strand distinct from, but possibly buttressing, the active site itself.[42,43] Indeed replacement of this arginine with lysine in rat FPP synthase has negligible effect on the activity of the enzyme.[44] The structural basis for the apparent role of the DRRYR domain in the trichodiene synthase reaction remains to be established.

Although wild-type trichodiene synthase produces trichodiene (**5**) as the exclusive product, the three mutants were found to produce not only trichodiene, but varying proportions of several sesquiterpene hydrocarbons, including (−)-(*Z*)-α-bisabolene (**22**), β-bisabolene (**23**), and cuprenene (**24**) (Scheme 6).[45] The formation of mixtures of products by site-directed mutants is startling and has provided rich insights into the trichodiene synthase cyclization mechanism. Binding of FPP (**1**) to these mutants may involve small but significant changes in the precise folding of the substrate and its orientation relative to active site residues. Ionization of the aberrantly bound FPP can then result in generation of abortive cyclization products by premature deprotonation of the normal cationic cyclization intermediates. Analysis of the products themselves opens a window on the nature of these normally enzyme-bound, cryptic intermediates and raises the possibility that the active base, which would normally carry out the final deprotonation step leading to trichodiene, may be responsible for some or all of the anomalous deprotonation events in the mutants. This possibility has been investigated by the use of isotope effects to perturb the relative proportions of trichodiene and cuprenene. Thus incubation of (1*S*)-[1-^{2}H]FPP with the Y305F mutant shifted the ratio of (**5**) to (**24**) from 3:1 to >7:1, due to the isotope effect on removal of the H-11α deuterium from the common bicyclic cation intermediate (**25**). The intrinsic isotope effect for this deprotonation was calculated to be >2.3.

Cane *et al.* have extended these results by analysis of a second series of mutants modified in the highly conserved aspartate-rich region that is found in all terpenoid cyclases characterized to date.[46] In FPP synthase, both site-directed mutagenesis[44,47–49] and direct crystallographic evidence[42,43] have established that the corresponding aspartate-rich domains in this protein are involved in binding of the catalytically essential divalent Mg^{2+} ions. In fact, the D100E and D101E mutants of *F. sporotrichioides* trichodiene synthase each exhibited modest increases in K_m and decreases in k_{cat}. By contrast the D104E mutant was virtually identical to wild-type enzyme in both k_{cat} and K_m. Transient kinetic analysis has revealed that the effect on the rate of FPP consumption by the D101E mutant is considerably greater than is first apparent from steady state kinetic measurements; this retardation is partially masked by the slow intrinsic rate of product release (see Section 2.06.4.2.1).[33] Thus FPP ionization was reduced by a factor of 100 compared to wild-type enzyme, while the net effect on the overall observed k_{cat} was only a factor of 3. Once again, all three mutants produced mixtures of sesquiterpene hydrocarbons, with trichodiene (**5**) being accompanied by varying amounts of β-farnesene (**26**), (*Z*)-α-bisabolene (**22**), β-bisabolene (**23**), cuprenene (**24**), and a new sesquiterpene with a previously unknown skeleton which was assigned structure (**27**) and named isochamigrene (Scheme 6).[50] All of the aberrant cyclization products generated by the D100E and other mutants, except for (**27**), can be derived by direct deprotonation of the normal cationic cyclization intermediates. Formation of isochamigrene, an "unnatural natural product", on the other hand, requires a diversion from the normal cyclization pathway by ring expansion of the bicyclic intermediate (**25**), followed by loss of a proton from C-12.

During the ionization of FPP by typical terpenoid cyclases, chelation with Mg^{2+} can aid both substrate binding, by proper orientation of the pyrophosphate moiety, and catalysis, by promoting ionization of the allylic diphosphate ester through neutralization of two of the three negative charges. Substitution of Mn^{2+} for Mg^{2+} decreased the k_{cat}/K_m for both the wild-type and D104E mutants, but had little net effect on the k_{cat}/K_m for both the D100E and D101E mutants, due to compensating, 4-fold reductions in both k_{cat} and K_m for each of the latter two mutant enzymes. More interestingly, Mn^{2+} enhanced the relative proportions of aberrant sesquiterpenes generated by each mutant, while having no effect on the purity of the trichodiene produced by the wild-type trichodiene synthase. These results shed light on the role of the aspartate-rich region and its interaction with divalent metals.

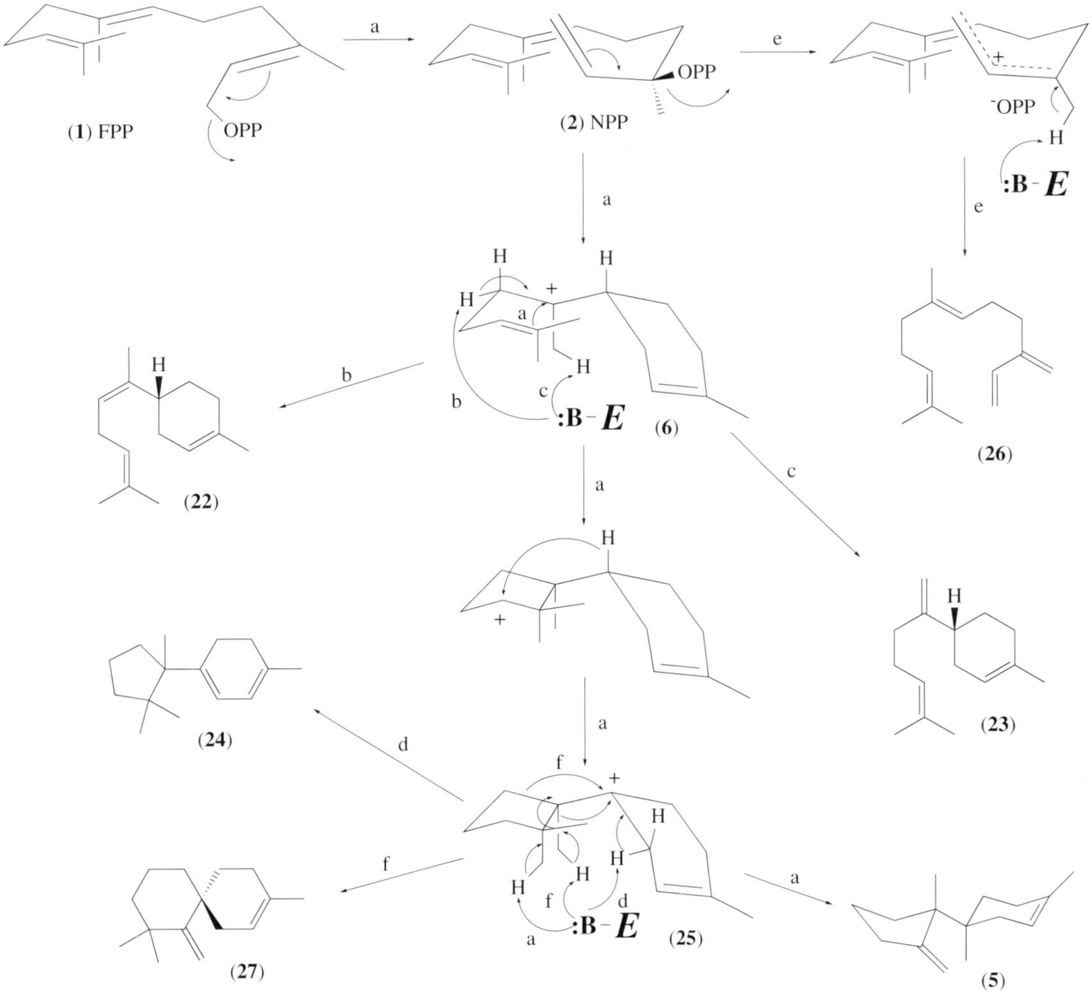

Scheme 6

In related work, Nabeta *et al.* have studied the formation of cuparene (**28**) by callus cultures of *Perilla frutescens* (Shiso).[51] Incorporation of [4,4-$^2H_2$]mevalonate and [2,2-$^2H_2$]mevalonate followed by detailed GC–MS analysis of the fragmentation patterns of the derived cuparene led to a deduced labeling pattern essentially in agreement with the accepted mechanism of formation of cuprenene, and oxidized cuparene derivatives, including the operation of the 1,4-hydride shift (Scheme 7).[2] The claim, however, that this 1,4-shift also competes with a set of sequential 1,3-hydride shifts is completely inconsistent with the results of earlier experiments on the formation of biogenetically related trichothecanes in which the positions and distribution of isotopic labels were determined beyond question by rigorous chemical degradation.[2] It may be that results based exclusively on the use of MS to determine sites of deuterium isotopic labeling must be viewed with caution due to the possibility of rapid hydride rearrangements during the fragmentation of molecular ions and daughter ions.

2.06.3.1.2 *Metabolism of trichodiene*

Trichodiene is the apparent precursor of the several dozen known trichothecene metabolites. The conversion of trichodiene to oxygenated trichothecenes has been studied by a variety of techniques, including kinetic pulse labeling, isotopic labeling, precursor incorporation, and isolation of presumptive intermediates. These methods have revealed the basic outlines of the oxidative metabolism of trichodiene.[52] However, very few studies have been carried out at the enzyme or genetic level.

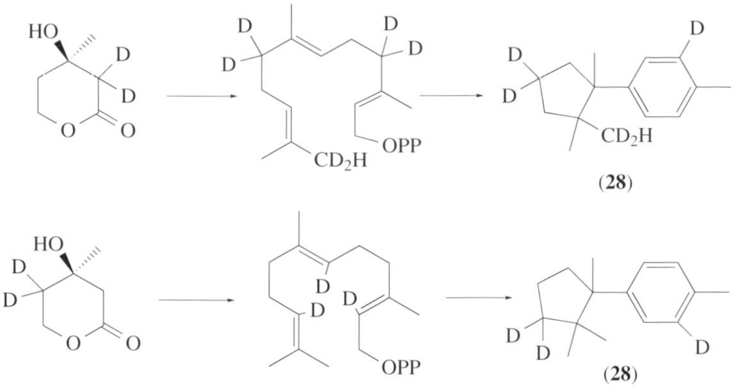

Scheme 7

Labeled trichodiene was first incorporated into trichothecolone in *T. roseum*[18] and has been subsequently shown to be a precursor of various trichothecenes in *Fusarium* species. The oxygen atoms of the epoxide and hydroxyl groups of various trichothecenes are derived from molecular oxygen, presumably due to the action of several dedicated P450 mono-oxygenases.[53] Indeed, incubations of three different species of *Fusarium* with oxygenase inhibitors, such as xanthotoxin or ancymidol, blocked trichothecene production and resulted in accumulation of trichodiene itself.[54–56] The first identified post-trichodiene intermediates are 12,13-epoxy-9,10-trichoene-2-ol (**29**) and isotrichodiol (**30**) (Scheme 8).[55,56] Each of these metabolites was detected by kinetic pulse labeling and shown to be converted to more highly-oxidized trichothecenes such as 12,13-epoxytrichothec-9-ene (EPT) (**31**), 3-acetyldeoxynivalenol (3-ADN) (**32**), and calonectrin (CAL) (**33**), in *F. culmorum*.[57–59] Interestingly, the isomeric 9β-trichodiol (**34**), originally thought to be an intermediate in trichothecene formation, is now recognized as a shunt metabolite.[60] Hohn *et al.* have isolated the *Tri4* gene of *F. sporotrichioides*, part of a larger cluster of three or more trichothecene biosynthetic genes, by complementation of a *Tri4*$^-$ mutant which accumulates trichodiene.[61] *Tri4*$^-$ mutants, generated by targeted gene disruption, were shown to be able to convert isotrichotriol to T-2 toxin (**35**), indicating the involvement of the Tri4 gene product in one or more early oxidation steps.[62] Sequence comparisons indicated that Tri4 is a cytochrome P450 which has been placed in a new gene family, *CYP58*. The *Tri4* coding region contains an open reading frame of 1560 bp encoding 520 amino acids with a deduced M_D of 59 056 Da. Although the protein itself has not been further characterized, it may possibly correspond to the enzyme activity observed by Bycroft and co-workers, who reported that a crude extract of *F. culmorum* could convert 9,10-epoxytrichodiene, a semisynthetic trichodiene derivative, to 9,10,12,13-diepoxytrichodiene in a reaction requiring NADPH and molecular oxygen.[63] On the other hand, the latter extract could not epoxidize trichodiene itself, nor did it catalyze oxidation at C-3, casting doubt on the physiological relevance of the crude *F. culmorum* extract. This issue will be best addressed by expression of Tri4 and systematic *in vitro* analysis of its substrate specificity and product spectrum.

The trichothecene pathways branch at isotrichodiol (**30**), with isotrichodiol either undergoing direct cyclization to EPT (**31**) or C-3 oxidation to isotrichotriol (**36**) prior to cyclization and acetylation to give isotrichermol (**37**) and isotrichodermin (**38**), respectively. Precursor incorporation and kinetic pulse labeling methods have established that EPT (**31**) is converted to trichothecenes such as sambucinol (**39**) in *F. sambucinum*, whereas isotrichodermin (**38**) is the precursor of 3-ADN (**32**) and CAL (**33**) and various oxygenated derivatives of *F. culmorum*.[57–59] Hydroxylation of the latter metabolites at C-15 appears to be the first step in further oxygenation of isotrichodermin. No interconversion occurs between EPT and isotrichodermin. None of the enzymes or genes mediating steps downstream of isotrichodiol have as yet been identified.

2.06.3.2 Bisabolene

(*Z*)-γ-Bisabolene synthase activity has been reported in crude extracts of *Andrographis paniculata* tissue cultures.[64] Early results suggesting a possible redox mechanism for the isomerization–cyclization of FPP have been shown to be in error by the finding that [1-^{3}H,12,13-^{14}C]FPP is cyclized to γ-bisabolene (**40**) without loss of tritium label, as judged by the absence of change in the observed

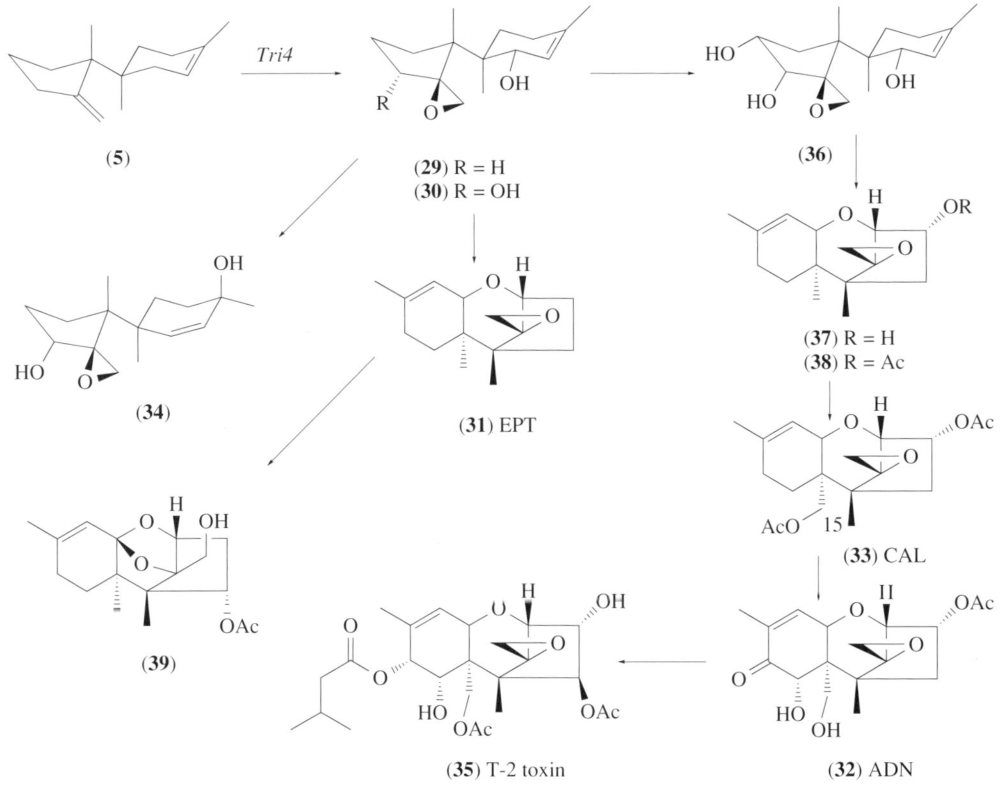

Tri4

(5)

(29) R = H
(30) R = OH

(36)

(34)

(31) EPT

(37) R = H
(38) R = Ac

(33) CAL

(39)

(35) T-2 toxin

(32) ADN

Scheme 8

$^3H/^{14}C$ ratio (Scheme 9).[65] The latter results are therefore consistent with the expected intermediacy of NPP, although no direct information is available on the stereochemical course of the cyclization. The formation of *β*-bisabolene by *P. frutescens* has also been studied by Nabeta *et al.* using deuterated mevalonates, and confirms the retention of all six deuteriums from [5,5-$^2H_2$]mevalonate.[51]

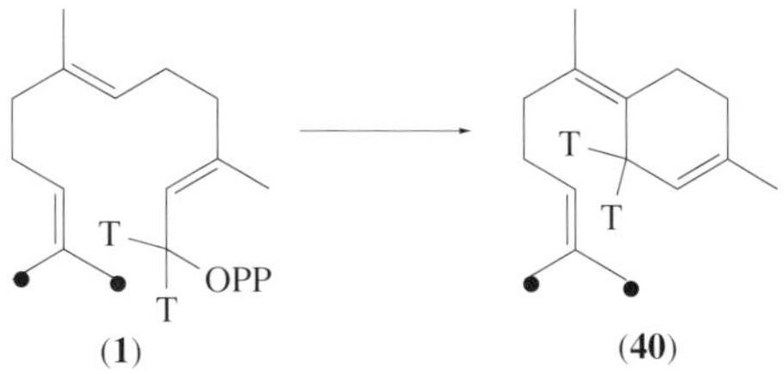

(1)

(40)

Scheme 9

2.06.3.3 Ovalicin and Bergamotene

2.06.3.3.1 Bergamotene synthase

The immunosuppressive antibiotic ovalicin (**41**) and the related fungal metabolite fumagillin (**42**) share an unusual sesqui-*o*-menthane skeleton. The parent hydrocarbon for this family of antibiotics is *β*-*trans*-bergamotene (**43**). The sesquiterpene has itself been isolated from the mycelia of the ovalicin-producing organism *Pseudeurotium ovalis*[66] and the fumagillin producer *Aspergillus fumigatus*,[67] and is also present in the essential oils of a variety of higher plants. Crude extracts of *P. ovalis* were shown to contain bergamotene synthase, based on their ability to convert [12,13-^{14}C]FPP to bergamotene, in which the sites of labeling were unambiguously determined by oxidative degradation of the side chain.[68] Consistent with these results was the finding that cyclization of [1-$^2H_2$]FPP ((**1**), $H_A = H_B = D$) gave bergamotene carrying two deuterium atoms at C-6, as determined by ^{2}H NMR (Scheme 10).

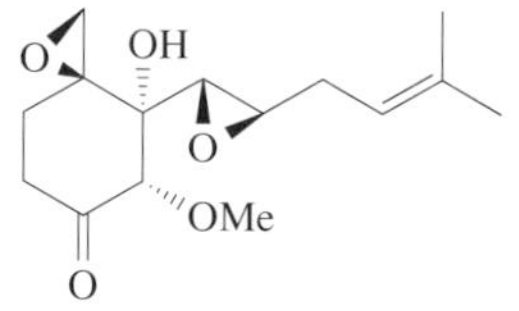

(**41**) 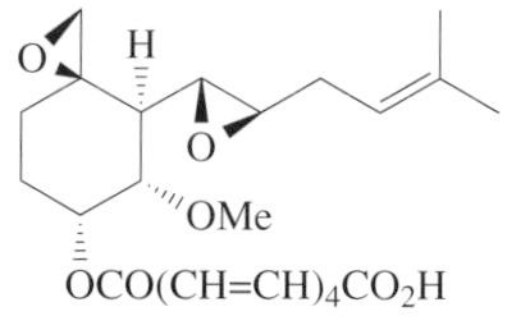 (**42**)

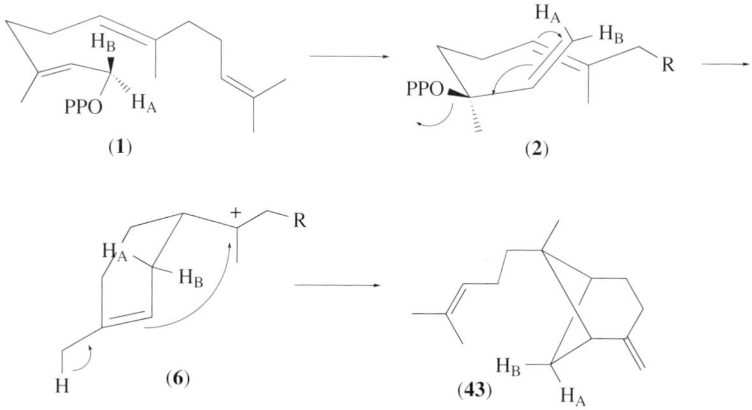

(**1**) (**2**) (**6**) (**43**)

Scheme 10

The absolute configuration of the enzymatically generated (−)-*β*-*trans*-bergamotene was established by a novel combination of enzymatic and NMR spectroscopic methods.[69] Separate incubations of (4*S*)- and (4*R*)-[4-^{2}H]FPP with crude bergamotene synthase gave labeled samples of bergamotene, which were each analyzed by ^{2}H NMR. The sample of (**43**) derived from (4*S*)-[4-^{2}H]FPP ((**1**), $H_C = D$, $H_D = H$) displayed a ^{2}H NMR signal corresponding to H-3_{endo} (H-3_{si}), while the complementary sample of (**43**) obtained from (4*R*)-[4-^{2}H]FPP (**1**), ($H_C = H$, $H_D = D$) showed a ^{2}H NMR signal corresponding to H-3_{exo} (H-3*re*) (Scheme 11). Since the cyclization itself does not disturb the configuration of the chirally deuterated carbon, the relative stereochemistry of deuteration in the product could be used to deduce the absolute configuration of the bicyclic sesquiterpene.

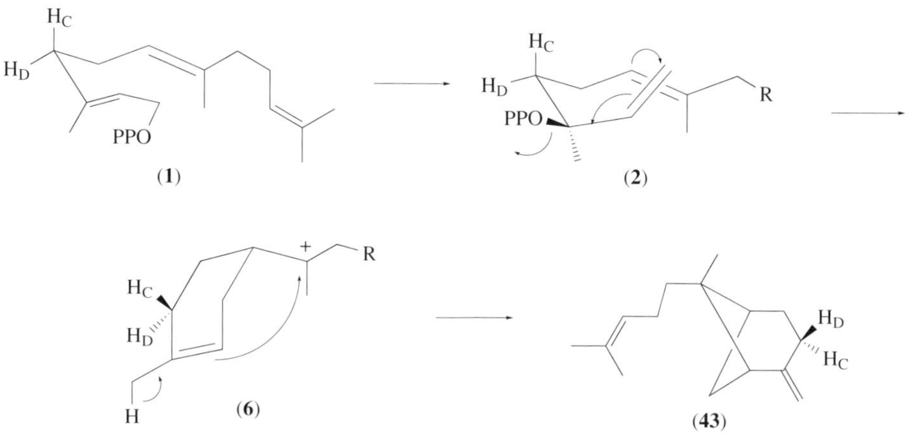

(**1**) (**2**) (**6**) (**43**)

Scheme 11

Knowing the absolute configuration of bergamotene, it was then possible to determine the stereochemistry of the enzymatic cyclization of FPP. Separate incubations of (1*S*)- and (1*R*)-[1-^{2}H]FPP ((**1**), $H_A = D$, $H_B = H$; (**1**) $H_A = H$, $H_B = D$) with bergamotene synthase gave labeled bergamotene carrying deuterium at H-6_{exo} and H-6_{endo}, respectively (Scheme 10).[68] These results established that cyclization occurs with net retention of configuration at C-1 of FPP, consistent with an isomerization–cyclization mechanism involving the intermediacy of NPP (**2**) and the bisabolyl cation (**6**).

Bergamotene synthase has been found to be a membrane-bound protein,[70] an unexpected observation since all other known sesquiterpene synthases are operationally soluble although markedly hydrophobic. Bergamotene synthase activity could be solubilized from the 250 000 g pellet of a *P. ovalis* extract by treatment with a detergent. The resolubilized protein could be further purified some 30–45-fold over three steps, but with substantial losses in enzyme activity. The partially purified protein cyclized (1*Z*)-[1-^{3}H]NPP to bergamotene, as measured by detection of radioactive product and confirmed by GC–MS detection of bergamotene. The configuration of the active enantiomer of NPP is under investigation.

2.06.3.3.2 *Ovalicin biosynthesis*

Feeding of [12,13,-^{13}C]-β-*trans*-bergamotene to cultures of *P. ovalis* gave ovalicin (**41**) labeled at C-12 and C-13, as established by ^{13}C NMR (Scheme 12).[71] Little is known about the cleavage of the bicyclo[3.3.1]heptane ring, nor is there any information available about the details of the pathway linking bergamotene to ovalicin, other than what can be inferred from whole-cell labeling studies of ovalicin derived from [5-^{3}H]-, (5-^{2}H]- and [2-^{2}H]mevalonates.[2,72]

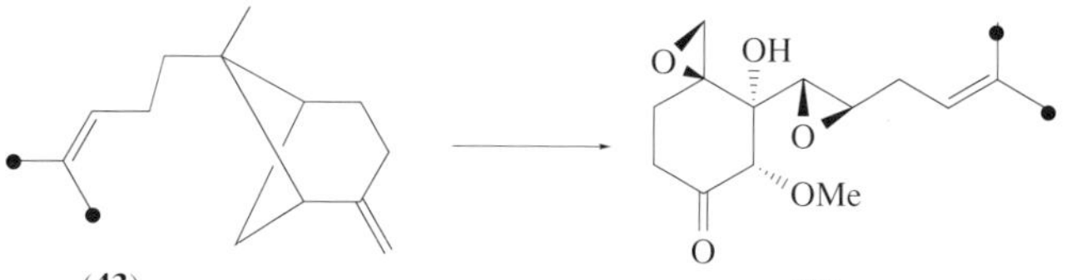

(**43**) (**41**)

Scheme 12

2.06.3.4 Humulene and Caryophyllene

Both humulene (**44**) and β-caryophyllene (**45**) are widely occurring sesquiterpenes found in numerous higher plants. The formation of each product can be accounted for by a common humulyl intermediate, itself generated by ionization of FPP and electrophilic attack at C-11 of the distal double bond. Direct deprotonation from C-9 will yield humulene (**44**), while further cyclization by attack on the 2,3-double bond and removal of a proton from the attached methyl group can lead to β-caryophyllene (**45**) (Scheme 13). In support of this mechanism was the observation that cyclization of [9-^{3}H,12,13-^{14}C]FPP gave humulene (**44**) which had lost half the original tritium label, as well as caryophyllene (**45**), in which the ^{3}H/^{14}C ratio was unchanged, as predicted.[73] Furthermore cyclization of [1-^{3}H]FPP by the 105 000 g supernatant of a crude extract of sage (*Salvia officinalis*) leaves gave a 2:1 mixture of labeled humulene (**44**) and caryophyllene (**45**), in which the sites of labeling were established by chemical degradation (Scheme 14).[74]

Using extracts of sage leaf oil glands, Croteau and Gundy have achieved a resolution of the humulene and β-caryophyllene synthase activities and partially purified each of the two cyclase activities.[73] Both proteins had an apparent M_r of 58 000, based on gel filtration, and exhibited the usual requirement for Mg^{2+}. Inactivation by *p*-hydroxymercuribenzoate and by diethyl pyrocarbonate indicated a possible role for cysteine and histidine residues, respectively.

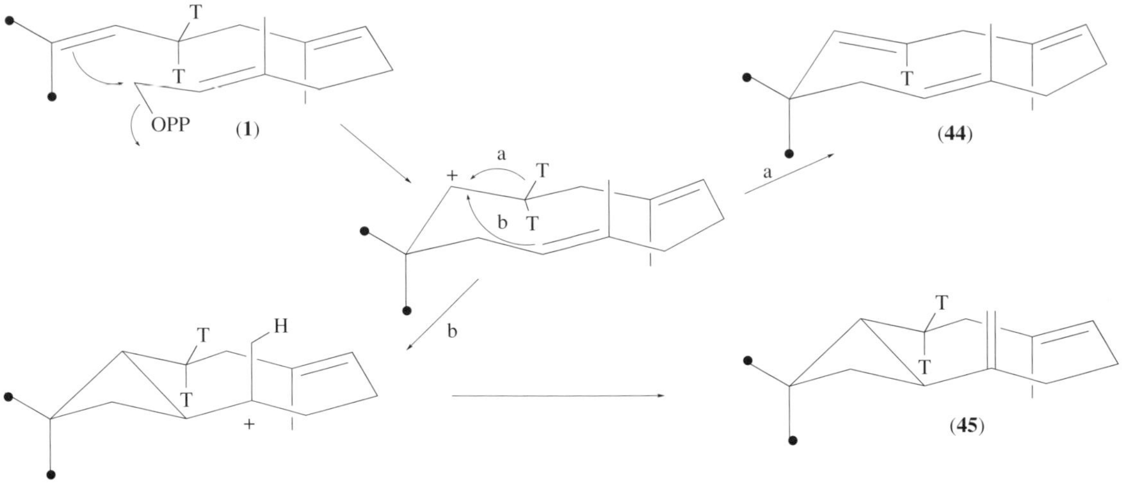

Scheme 13

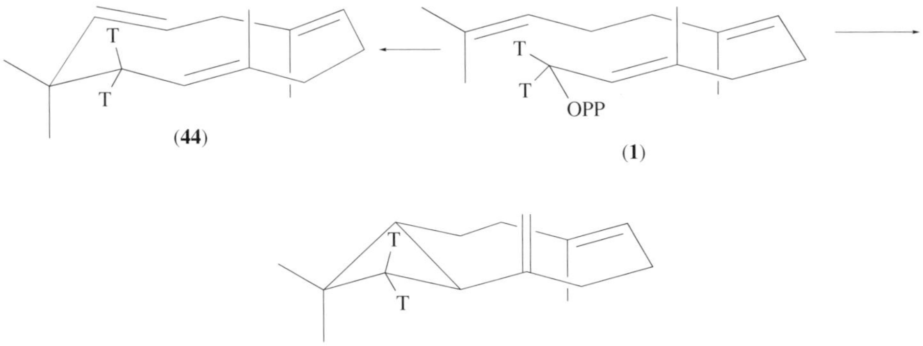

Scheme 14

2.06.3.5 Pentalenolactones

2.06.3.5.1 Pentalenene synthase

The pentalenolactones are a family of antibiotics which have been isolated from a variety of *Streptomyces* species. Early work on the biosynthesis of these metabolites involved feedings of [UL-$^{13}C_6$]glucose to *Streptomyces* UC5319.[75] The labeled glucose was designed as an *in vivo* precursor of [1,2-$^{13}C_2$]acetyl-CoA. Indeed, analysis of the pattern of enhancements and couplings in the derived pentalenolactone (**46**) and pentalenic acid (**47**) appeared to be consistent with the predicted mevalonoid origin of this group of metabolites, in spite of the fact that all attempts to directly incorporate either labeled acetate or mevalonate had been met with failure (Scheme 15). Although the derivation of the pentalenenes from FPP is now beyond question, it has recently become apparent that the producing *Streptomyces* does not utilize mevalonate but instead employs the nonclassical pyruvate–glyceraldehyde route to isopentenyl diphosphate,[76] a pathway which fortuitously results in a labeling pattern in the derived (**47**) identical to that predicted by the better-known acetate–mevalonate pathway, while the pattern of enhancements and couplings in labeled pentalenolactone (**46**) is as predicted by the acetate–mevalonate route, except for a previously unexplained additional coupling between C-1 and the attached C-14 methyl group that has migrated from C-2. As it has now turned out, the apparently anomalous coupling is fully accounted for by the non-classical origin of isopentenyl diphosphate. In support of these conclusions, Arigoni and co-workers have demonstrated the direct incorporation of deoxyxylulose (**48**) into (**46**) and (**47**) (Scheme 16),[76] and Seto *et al.* have independently reported the results of labeling experiments which confirm operation of the nonclassical pathway to FPP in *Streptomyces*.[77]

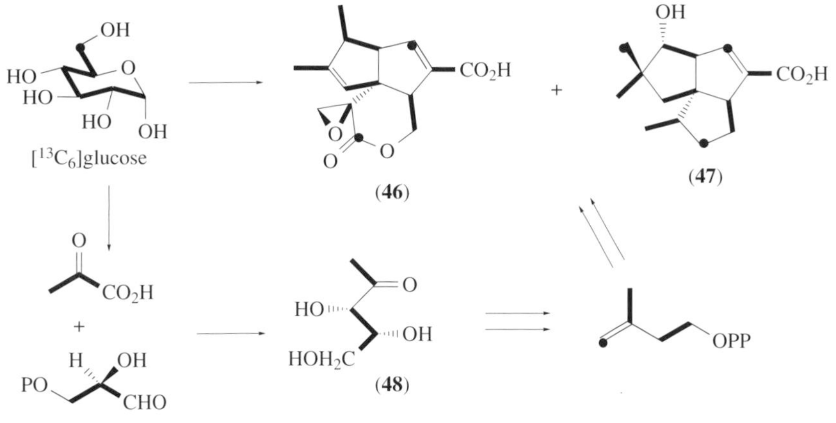

Scheme 15

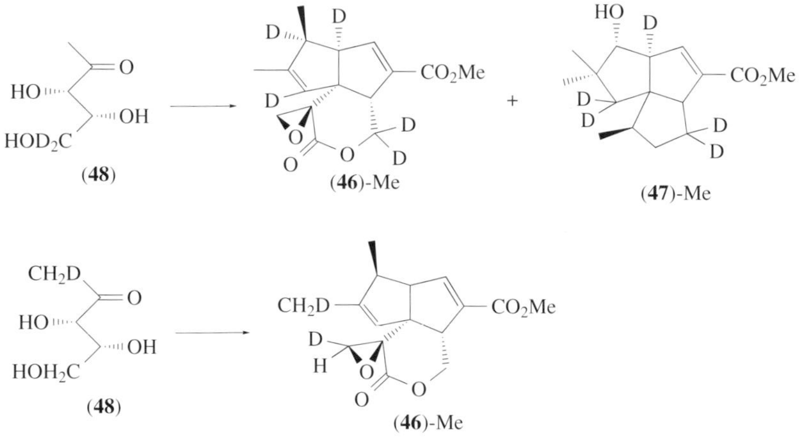

Scheme 16

The parent hydrocarbon of the pentalenolactone family of metabolites, pentalenene (**49**), was first isolated in 1980.[78] Shortly thereafter, it was reported that a crude cell-free extract of *Streptomyces* UC5319 would convert FPP to pentalenene.[79] Cane *et al.* subsequently purified pentalenene synthase to homogeneity, cloned the corresponding structural gene, and overexpressed recombinant pentalenene synthase in *E. coli*.[80] In the mid to late 1990s, the structure of this protein has been solved by X-ray crystallography (see Section 2.06.4.3.1).[81,82] Pentalenene synthase is a monomer of M_D 38 002. The enzyme requires Mg^{2+}, with the K_m for FPP 0.3 μM and a k_{cat} of 0.3 s^{-1}.

The formation of pentalenene can be explained by the mechanism illustrated in Scheme 17 in which the substrate FPP is folded as shown. Following ionization and cyclization of FPP, the proposed intermediate humulene (**44**) is reprotonated at C-10 to initiate cyclization to the protoilludyl cation (**50**). Following hydride shift and further cyclization, a final deprotonation will generate pentalenene (**49**). This mechanism is supported by extensive labeling experiments, which have revealed a wealth of details concerning the stereochemical course of the cyclization.

The cyclization of FPP was shown to take place with inversion of configuration at C-1, as expected for a direct cyclization without intervening isomerization of the 2,3-double bond (Scheme 17).[83,84] Thus separate incubations of (1*S*) and (1*R*)-[1-^{2}H]FPP ((**1**), $H_A = D$, $H_B = H$; (**1**), $H_A = H$, $H_B = D$) gave pentalenenes labeled with deuterium in place of H-3*si* and H-3*re*, respectively, as determined by ^{2}H NMR. When [9-^{3}H,12,13-^{14}C]FPP ((**1**), $H_C = H_D = T$) was incubated with pentalenene synthase, the resulting labeled pentalenene carried one equivalent of tritium at the bridgehead, C-8, and a second which had migrated to the adjacent site C-1, as established by a combination of chemical and microbial degradation (Scheme 18).[85] Since formation of humulene requires removal

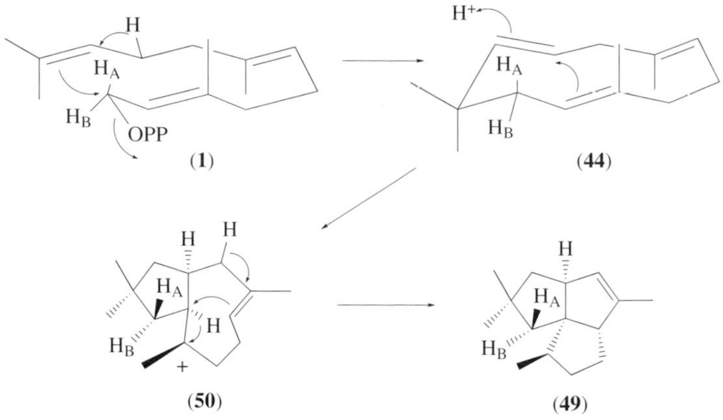

Scheme 17

of one of the original H-9 protons of FPP, the retention of both labels implies that the proton which is removed is redonated to humulene at C-10 without significant exchange with the medium or indeed with any other protons on the base itself.

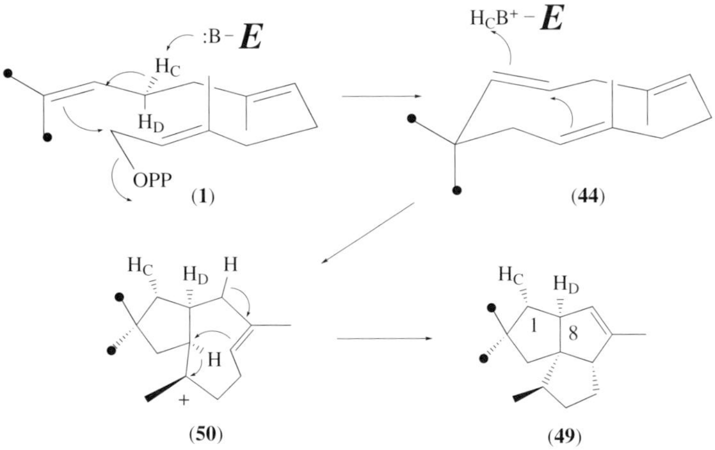

Scheme 18

The stereochemical course of the deprotonation step was established by incubation of (9*S*) and (9*R*)-[9-^{3}H,12,13-^{14}C]FPP ((**1**), H_C = T, H_D = H; (**1**), H_C = H, H_D = T) with pentalenene synthase and unambiguous determination of the sites of labeling.[83,86] In this manner it was established that H-9*re* of FPP becomes H-8 of pentalenene, while H-9*si* is transferred to C-1 of (**49**). Coupled with the fact that cyclization of FPP involves attack on the *si* face of the 10,11-double bond, these results established that the overall electrophilic allylic addition–elimination reaction that generates humulene takes place with net *anti*stereochemistry. The stereochemistry of the subsequent reprotonation of humulene was determined by incubation of [10-^{2}H, 11-^{13}C]FPP with pentalenene synthase and analysis of the resultant labeled pentalenene by ^{2}H NMR and ^{13}C NMR (Scheme 19).[83] In this manner it was demonstrated that the deuterium label resided at the H-1*si* (H-1β) position of pentalenene (**49**), indicating that the proton which is removed form H-9*si* of FPP is redonated to the 10*re* face of the humulene double bond, in accord with the predicted folding of FPP and the derived humulene.

Incubation of [8-^{3}H,12,13-^{14}C]FPP with pentalenene synthase confirmed the expected loss of half the tritium label and the retention of the remainder at C-7 of pentalenene, as determined by chemical degradation.[79] By incubating both (4*S*,8*S*)- and (4*R*,8*R*)-[4,8-^{3}H$_2$,4,8-^{14}C$_2$]FPP ((**1**), H_E = T, H_F = H; (**1**), H_E = H, H_F = T) and analyzing the derived labeled pentalenenes, it was found that it is H-8*si* that is lost in the final deprotonation step (Scheme 20).[87] Furthermore, analysis of the geometry of folding of the FPP substrate and intermediate humulene has suggested that a single base at the active

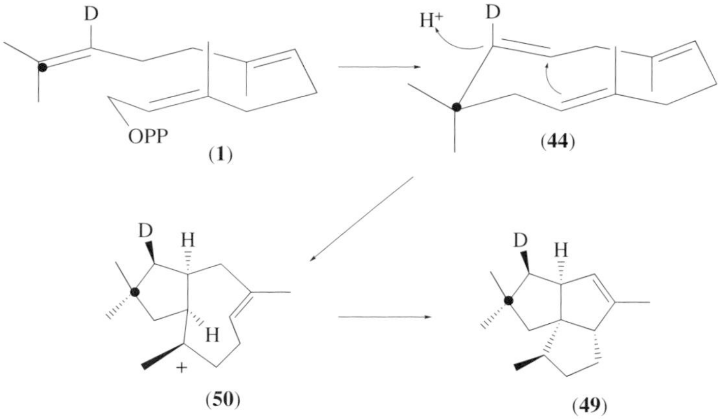

Scheme 19

site of pentalenene synthase may be responsible for the successive deprotonation–reprotonation–deprotonation steps. Crystallographic analysis (see Section 2.06.4.3.1) has revealed that this base may be His309.[82]

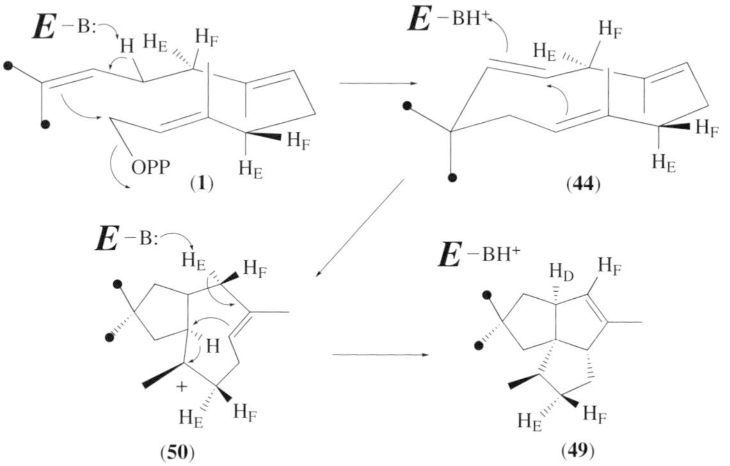

Scheme 20

Pentalenene synthase is inhibited by inorganic pyrophosphate with K_I 3.2 μM, about 10 × the observed K_m for FPP. The FPP analogue (**10**) is an effective competitive inhibitor (K_I 0.34 μM), in contrast to its lower efficacy and mixed competitive inhibition behavior with trichodiene synthase.[88] The enzyme was inactivated by incubation with the sulfhydryl-directed reagent PCMB.[89] The mechanistic significance of this observation is unclear since crystallographic analysis has indicated the absence of Cys residues at the pentalenene synthase active site (see Section 2.06.4.3.1).

2.06.3.5.2 Conversion of pentalenene to pentalenolactone

Various pentalenolactone-producing *Streptomyces* strains produce a number of oxidized metabolites that are presumptive intermediates or shunt metabolites in the conversion of pentalenene to pentalenolactone. Feeding of labeled pentalenene (**49**) to cultures of *Streptomyces* UC5319 gave labeled pentalenolactone (**46**) and pentalenic acid (**47**), confirming the role of pentalenene as the parent hydrocarbon of this family of sesquiterpenes (Scheme 21).[79] Although no direct feeding experiments have been carried out with the more oxidized metabolites, the results of refeeding labeled pentalenene samples, obtained by cyclization of appropriately labeled FPP precursors, have established several key stereochemical features of the metabolism of pentalenene and the stereochemistry of pentalenolactone formation. Formation of pentalenic acid (**47**) was shown to involve hydroxylation with retention of configuration by stereospecific removal of H-1α of pentalenene.[83] By contrast, this same proton is *retained* in pentalenolactone, thereby ruling out the intermediacy of pentalenic acid in the formation of (**46**). Consistent with this data is the parallel finding that H-1β of pentalenene is retained in pentalenic acid but lost in pentalenolactone (see Scheme 16), thereby requiring that the migration of the β-methyl group to C-1 takes place with retention of configuration at C-1, possibly by way of a cyclopropyl intermediate such as pentalenolactone P (**51**).[90] In this model, pentalenolactones A (**52**) and B (**53**) would be shunt metabolites of this same rearrangement.[91] It has also been shown that the migration of the 14-methyl group in pentalenolactone is accompanied by stereospecific loss of the *anti* H-3α proton.[83]

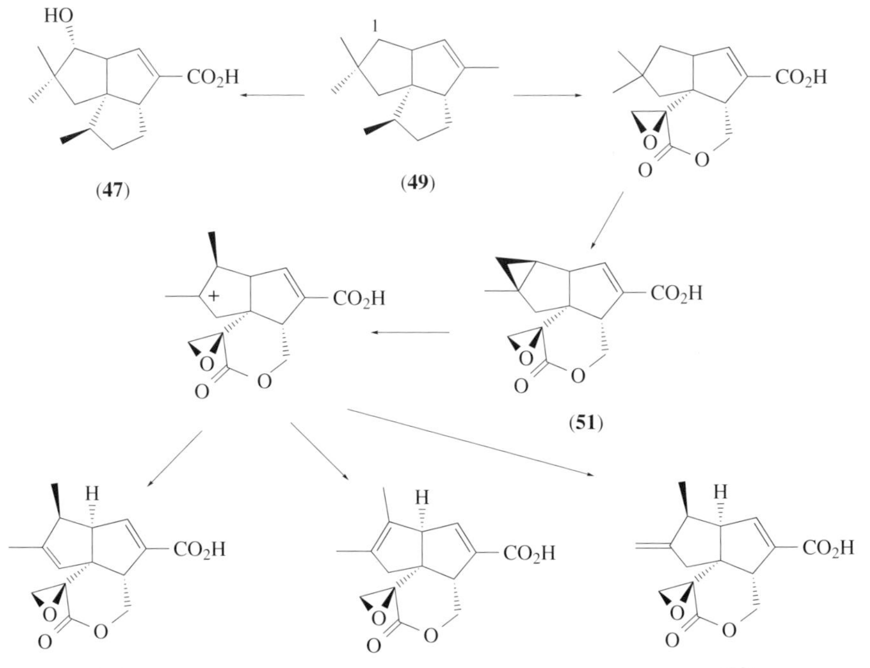

HO CO$_2$H (**47**) 1 (**49**) CO$_2$H O O O

CO$_2$H + O O O CO$_2$H O O O (**51**)

H CO$_2$H O O O (**46**) H CO$_2$H O O O (**52**) H CO$_2$H O O O (**53**)

Scheme 21

2.06.3.6 Sterpurenes

The tricyclic sesquiterpenes (**54**) and (**55**), as well as the parent hydrocarbon sterpurene (**56**), are produced by the fungus *Stereum purpureum*.[92] These metabolites are derivable from FPP by way of humulene (**44**) and the protoilludyl cation (**50**) (Scheme 22). Support for this proposal initially came from incorporation of [1-^{13}C]- and [2-^{13}C]acetate into (**54**) and (**55**), which revealed that a rearrangement had occurred in the formation of the cyclobutyl ring.[93] These findings were confirmed and expanded by the observation that incorporation of [1,2-^{13}C]acetate into (**57**) and (**58**) gave rise to ^{13}C NMR spectra exhibiting five pairs of enhanced and coupled doublets in addition to five enhanced singlets.[94,95] Of particular interest was the finding of three adjacent singlets arising from the cyclobutyl ring. Consistent with the proposed rearrangement pathway was the detection of a long-range $J_{1,3}$ coupling between C-3 and C-5, indicating that these two carbon atoms arose from a common acetate precursor. The fact that C-15 also appeared as a singlet, combined with the known pattern of labeling of the intermediate FPP, led to the inference that cyclization of FPP involves electrophilic attack on the *si* face of the C-10,11 double bond.

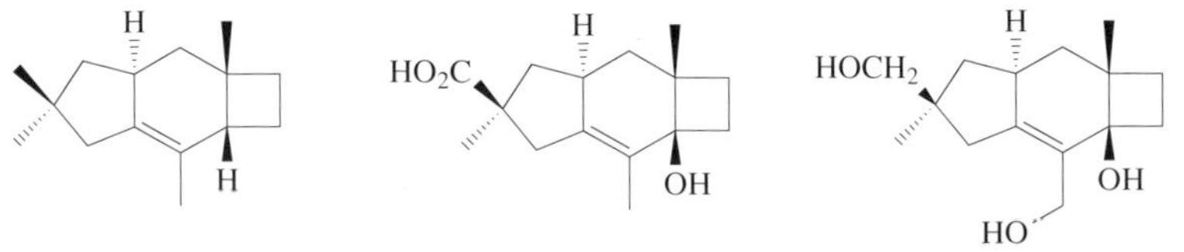

H H (**56**) H HO$_2$C OH (**54**) H HOCH$_2$ OH HO (**55**)

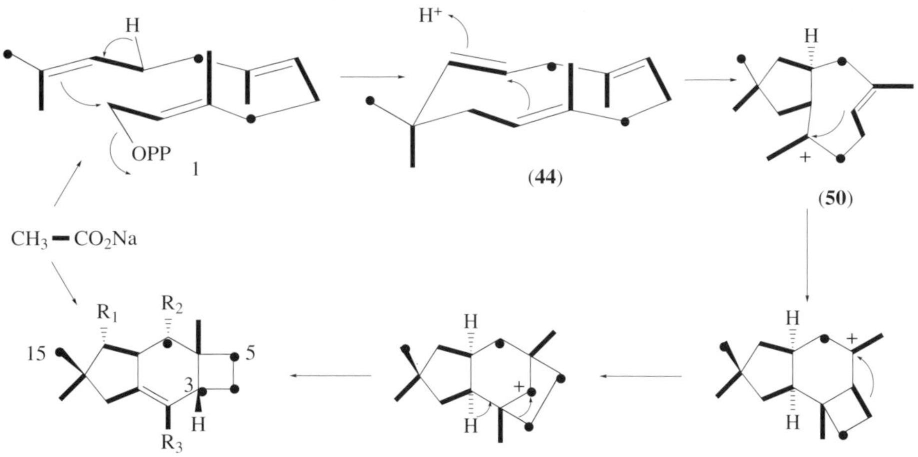

(**44**) (**50**)

(**56**) $R_1=R_2=R_3=H$
(**57**) $R_1=H$, $R_2=OH$, $R_3=CH_2OH$
(**58**) $R_1=OH$, $R_2=H$, $R_3=CH_2OH$

Scheme 22

2.06.3.7 Botrydials

Dihydrobotrydial (**59**) and botrydial (**60**) are phytotoxic metabolites of the plant pathogenic fungus *Botrytis cinerea*.[96] The sesquiterpene origin of the lactone (**61**) was confirmed by incorporation of [1-^{14}C]FPP.[97] The mode of formation of dihydrobotrydial from FPP has been elucidated by incorporation of labeled acetate and mevalonate. Thus incorporation of [1,2-$^{13}C_2$]acetate gave dihydrobotrydial (**59**), which displayed five pairs of enhanced and coupled doublets in its derived ^{13}C NMR spectrum, as well as five enhanced singlets, indicating that one of the original acetate units has undergone cleavage in the course of the cyclization of FPP (Scheme 23).[97] The nature of this cleavage was revealed by the observation that incorporation of [4,5-$^{13}C_2$]mevalonate gave rise to two pairs of coupled doublets (C-4,5 and C-1,10) as well as two enhanced singlets (C-7 and C-9).[98] Further insight came from incorporation of [1-^{13}C]acetate at sufficiently high levels so as to give rise to two sets of induced couplings between C-6,7 and between C-7,8.[97,98] The juxtaposition of these sets of acetate C-1-derived carbons, as well as the earlier labeling results, could be explained by the mechanism illustrated in Scheme 24, in which an initially formed humulyl cation cyclizes to a caryophyllenyl cation. Ring expansion, cyclization, and a 1,3-hydride shift will give the tricyclic cation (**62**). Capture of water will generate the parent tricyclic alcohol (**63**) which can be converted to botrydial and related metabolites by oxidative modification. As an alternative to the 1,3-hydride shift, one might postulate a geometrically more favorable deprotonation–reprotonation sequence via a cyclopropane intermediate, without exchange of the proton with the external medium.

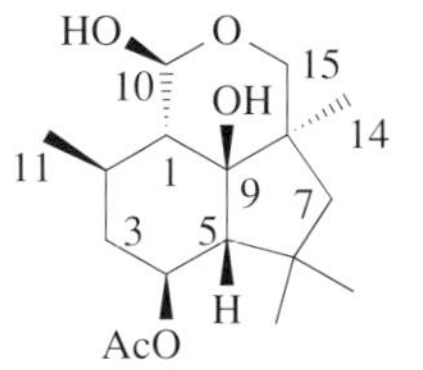

(**59**) 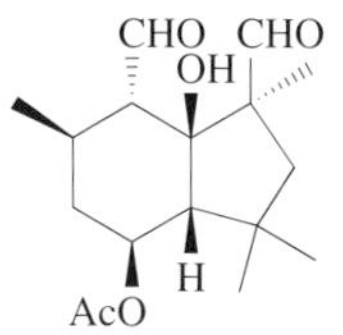 (**60**) 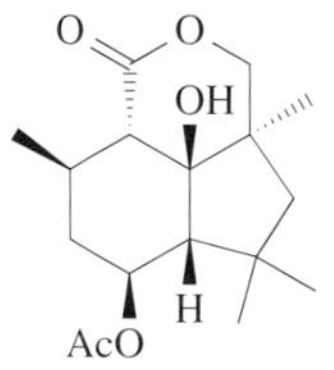 (**61**)

Direct evidence for the 1,3-hydride (or proton) shift came from incorporation of [4-^{2}H,4-^{13}C] mevalonate (Scheme 25).[99] The ^{2}H NMR spectrum of the derived botryaloic acid methyl ester (**64**) showed ^{2}H–^{13}C couplings for the signals corresponding to D-1 and D-5, indicating preservation of an intact ^{2}H–^{13}C bond, whereas the signal for D-2 appeared as a singlet. Since the corresponding C-4 of mevalonic acid has become C-9 of the botryaloic acid skeleton, the originally attached deuterium has undergone a 1,3-shift, thus unambiguously ruling out an alternative set of consecutive 1,2-shifts.

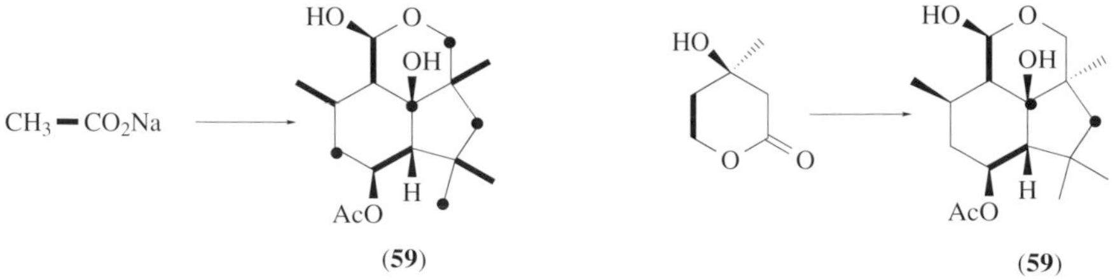

Scheme 23

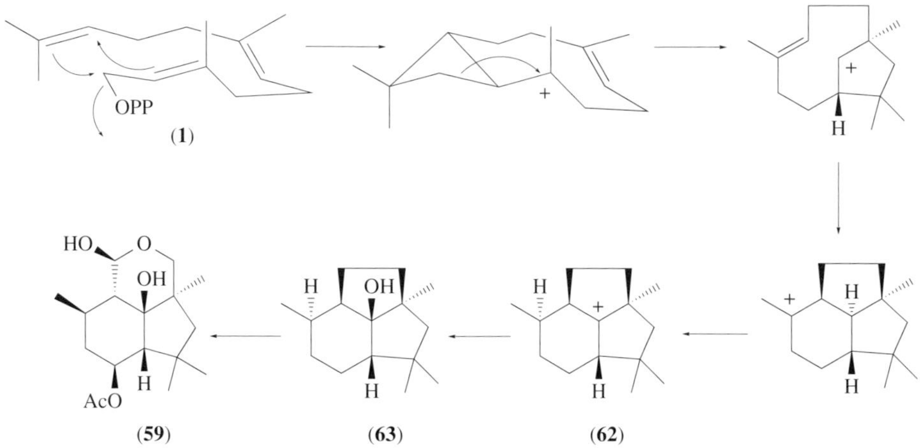

Scheme 24

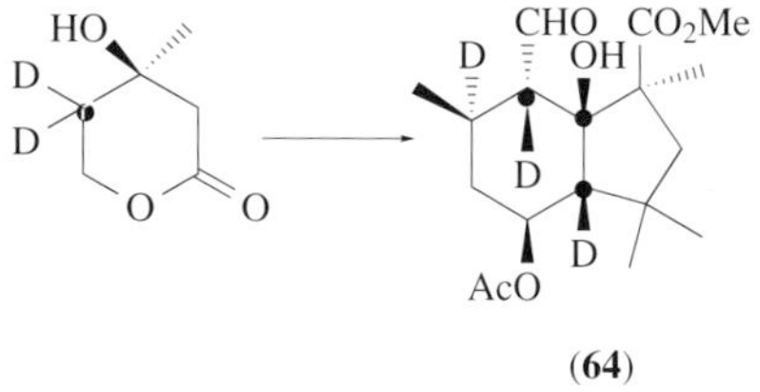

Scheme 25

2.06.3.8 Quadrone and Terrecyclic Acid

Cultures of *Aspergillus terreus* produce two novel rearranged sesquiterpenes, the tetracyclic lactone quadrone (**65**) and the closely related terrecyclic acid (**66**). The isoprenoid origin of these metabolites was independently investigated by three groups through incorporation of various combinations of labeled acetate and mevalonate. Incorporation of [1,2-$^{13}C_2$]acetate gave rise to quadrone and terrecyclic acid, which each displayed a pattern of three enhanced singlets and six pairs of enhanced and couple doublets in the corresponding ^{13}C NMR spectra, consistent with the preservation of the six intact acetate units of the intermediate FPP (Scheme 26).[100–103] Incorporations of [1-^{13}C]- and [2-^{13}C]acetate confirmed these observations and established the orientation of each acetate building block. Further insight came from incorporation of [2-^{13}C]- and [5-^{13}C]mevalonate.[103]

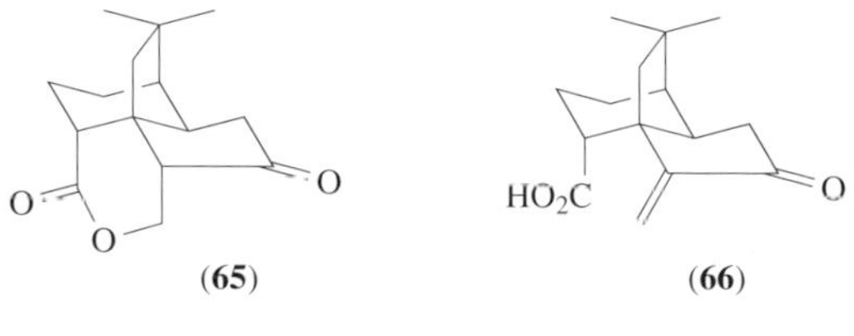

(**65**) (**66**)

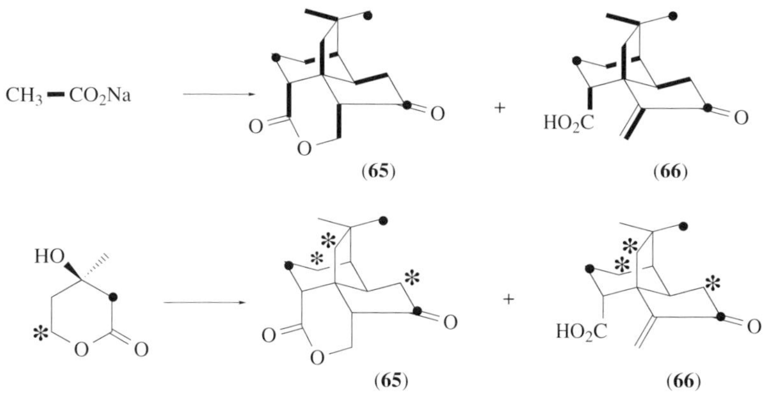

CH_3—CO_2Na → (65) + (66)

(65) (66)

Scheme 26

Two competing explanations were originally offered for these labeling results, illustrated in Schemes 27a and 27b. Analysis of the ^{13}C NMR spectra of terrecyclic acid and quadrone derived from incorporation of [3,4-$^{13}C_2$]mevalonate revealed two pairs of enhanced and coupled doublets corresponding to the C-1,5 and C-11,13 pairs (Scheme 28), consistent with the postulated cleavage of the 6,7-double bond of FPP, but without distinguishing between the two competing mechanisms.[101] Mechanism 27a was firmly excluded, however, by the experiments of Hirota *et al.*, who found that the methyl protons of acetate are retained at C-2 and C-8 of terrecyclic acid (**66**), based on incorporations of [2-$^2H_3$]- and [2-$^2H_3$,2-^{13}C]acetate and analysis of the derived dihydroterrecyclic acid (**67**) by ^{2}H NMR and ^{13}C NMR.[104]

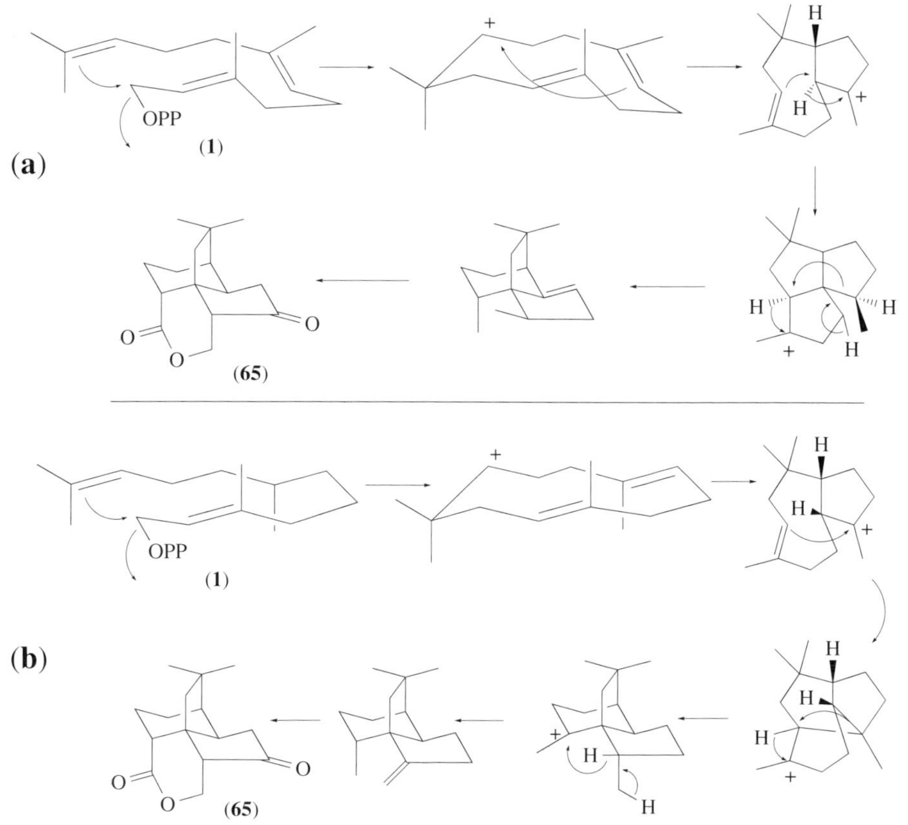

(a) (1) (65)

(b) (1) (65)

Scheme 27

Although Scheme 27b can account for the cumulative labeling data, the 1,3-hydride shift postulated for the penultimate step of the cyclization is problematical on purely stereoelectronic grounds. An alternative mechanism has been suggested by Coates and co-workers based on an

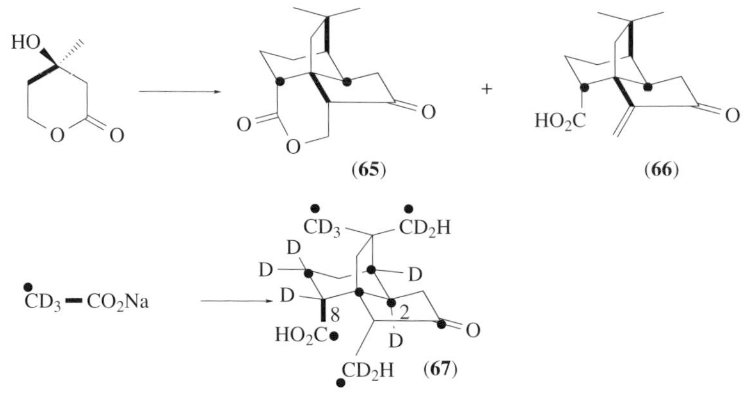

Scheme 28

ingenious set of model reactions.[105] Solvolysis in formic acid of synthetically prepared silphiny-2α-yl mesylate (**68**) gave an 81% yield of the rearranged tricyclic hydrocarbon (**69**), which was named α-terrecyclene (Scheme 29). From these results, Coates and co-workers have extended a previous suggestion of Bohlmann *et al.*[106] for the origin of the silphinyl carbo-cation to account for the formation of terrecyclic acid and quadrone (Scheme 30). Cyclization of FPP to the tricyclic carbocation (**62**), previously implicated in the formation of botrydial (see Scheme 24), can be followed by ring contraction to give the silphinyl cation (**70**) which can then rearrange to α-terrecyclene (**69**) or its exocyclic β-isomer. Although neither (**69**) nor its isomer have as yet been detected in mycelial extracts of *A. terreus*, the Coates hypothesis not only accounts elegantly for all the available labeling data, but also provides a unified biogenetic mechanism for the origin of several families of sesquiterpenes.

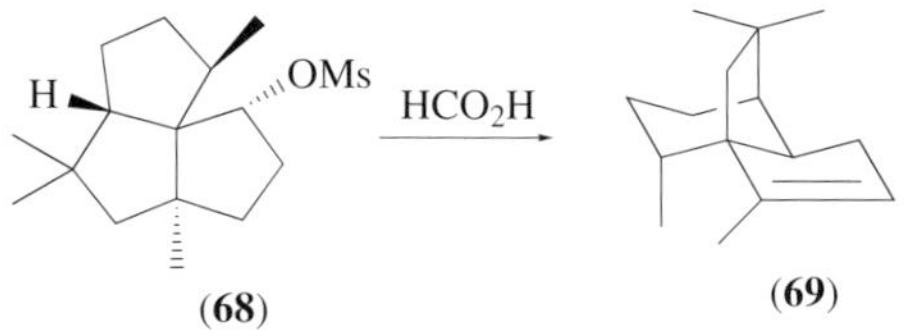

Scheme 29

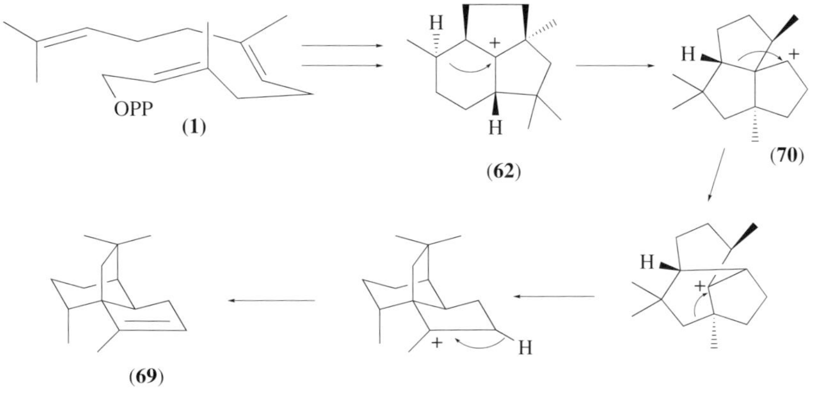

Scheme 30

2.06.3.9 Alliacolide

Alliacolide (**71**) is a novel sesquiterpene produced by the Basidiomycete *Marasmius alliaceus*. Incorporation of [1-^{13}C]- and [1,2-$^{13}C_2$]acetate established the isoprenoid origin, giving rise to the labeling pattern illustrated in Scheme 31.[107,108] Of particular interest was the observation that the C-12 methyl in the side chain of (**71**) was derived from C-2 of mevalonate, as evidenced by the appearance of a singlet in the ^{13}C NMR spectrum of (**71**) derived from [1,2-$^{13}C_2$]acetate. This conclusion was supported by incorporation of [2-$^2H_2$]mevalonate, which gave rise to ^2H NMR signals corresponding to two deuteriums at C-12, as well as two each at C-15 and C-2.[108] Evidence for a hydride shift during the cyclization of FPP came from the finding that dehydroalliacolide retained two deuteriums from [4-$^2H_2$]mevalonate, located at C-1 and C-6 by ^2H NMR. An additional hydride shift or proton transfer was evident from the observation that incorporation of [5-$^2H_2$]mevalonate results in labeling of C-8 (1D), C-3 (2D), and C-6 (1D).[108] No deuterium was found at C-11 of alliacolide derived from the same feeding experiment, thereby ruling out a 1,3-hydride shift analogous to that involved in the biosynthesis of cadinene and related metabolites (see Sections 2.06.3.15 and 2.06.3.16). When [5-$^2H_2$,4-^{13}C]mevalonate was fed to *M. alliaceus*, the ^2H NMR spectrum of the resultant labeled sample dehydroalliacolide displayed a doublet $J = 21$ Hz for D-6, consistent with a 1,2-shift.[109]

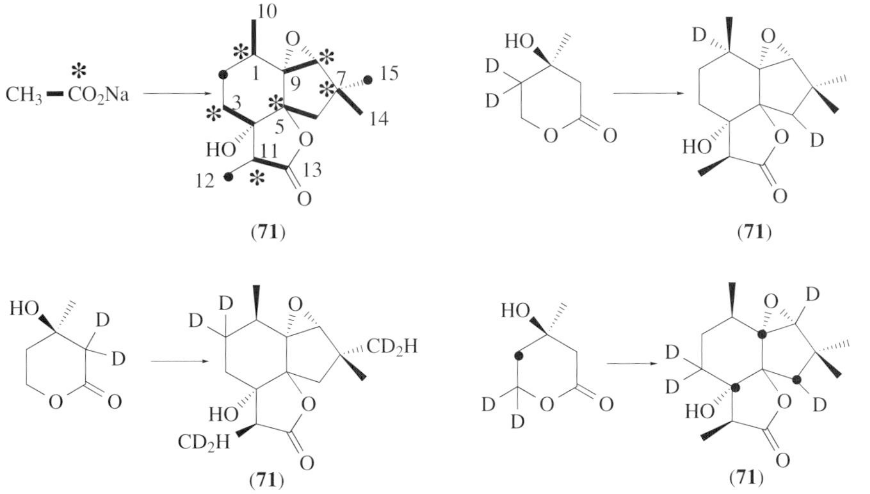

Scheme 31

These results can be explained by the cyclization mechanism illustrated in Scheme 32, in which the bicyclic cation (**50**) is generated from FPP via humulene (**44**), analogous to the pathway leading to pentalenene (see Scheme 17). An additional 1,2-hydride shift and cyclization will generate the cyclobutyl cation (**72**), which can undergo ring opening and quenching by water to give the hypothetical alcohol (**73**). Conversion of (**73**) to alliacolide and related metabolites could involve allylic oxidation at C-13, lactonization, isomerization of the C-11,12 double bond, and hydration of the derived C-4,11 double bond.

2.06.3.10 Aristolochene

Aristolochene (**74**) is a bicyclic hydrocarbon belonging to the eremophilene class of sesquiterpenes. The (+)-enantiomer has been isolated from *A. terreus*[110] and is believed to be present in numerous other fungi, where it is the presumed precursor of a variety of fungal metabolites including PR toxin (**75**) (*Penicillium roqueforti*)[111] and sporogen-AO 1 (**76**) (*Aspergillus oryzae*).[112] (−)-Aristolochene has been identified in several plant species, including *Aristolochia indica* and *Bixa orellana*,[113] as well as in the defensive secretions of *Syntermes* soldier termites.[114] Aristolochene synthase has been purified from *P. roqueforti*[115] and the gene isolated[116] and expressed at high levels in *E. coli*.[117] The cyclase, which is a monomer of M_D 39 200, catalyzes the Mg^{2+}-dependent cyclization of FPP to aristolochene by a mechanism believed to involve the intermediacy of the 10-membered ring hydrocarbon, germacrene A (**77**) (Scheme 33). Aristolochene synthase harbors the cyclase signature sequence LIDDVLE, beginning at amino acid 113. The *A. terreus* enzyme has also been purified to homogeneity and has a M_r of 39 000.[118] Sequencing of the cDNA has indicated that the two proteins share an approximate 70% identity at the amino acid sequence level and that the *A. terreus*

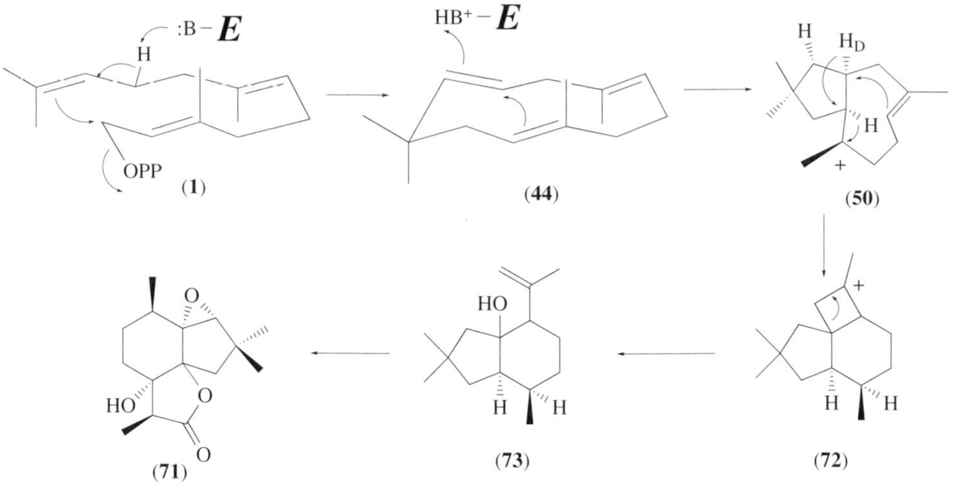

Scheme 32

aristolochene synthase also has a conserved aspartate-rich sequence. Interestingly, however, antibodies to the *P. roqueforti* protein do not cross-react with the *A. terreus* enzyme.

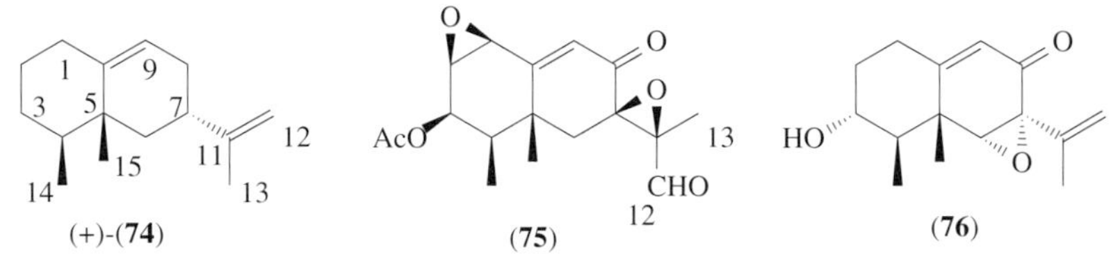

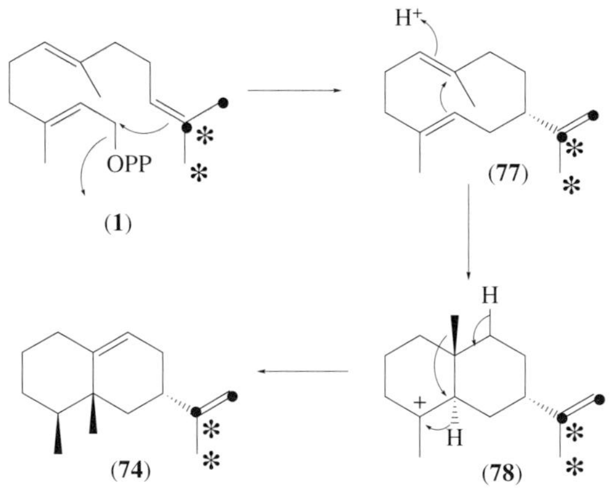

Scheme 33

Interest in the biogenesis of eremophilene sesquiterpenes dates back to the time of Robinson, who first proposed that the formation of the characteristic skeleton could be reconciled with Ruzicka's Isoprene Rule by a simple methyl migration.[119,120] Experimental evidence in favor of this suggestion came some 40 years later when it was shown that incorporation of [1,2-$^{13}C_2$]acetate gave PR toxin (**75**), in which the ^{13}C NMR signals for C-10 and C-15 appeared as singlets, in addition to the usual three singlets corresponding to the mevalonate C-2-derived carbons, C-12, C-9, and C-3 (Scheme 34).[121,122] Also consistent with the proposed mechanism of formation, incorporation of [2,3-$^{13}C_2$]mevalonate led to enhanced and coupled doublets in the ^{13}C NMR spectrum arising from the C-3,4, C-9,10, and C-11,12 pairs.[123] In support of the proposed 1,2-hydride migration from C-5 to C-4, ^{2}H NMR was used to show that [4-$^2H_2$]mevalonate labeled PR toxin at the predicted sites,

D-1 and D-4. PR toxin itself can be derived from aristolochene by a series of oxidation and acetylation reactions.

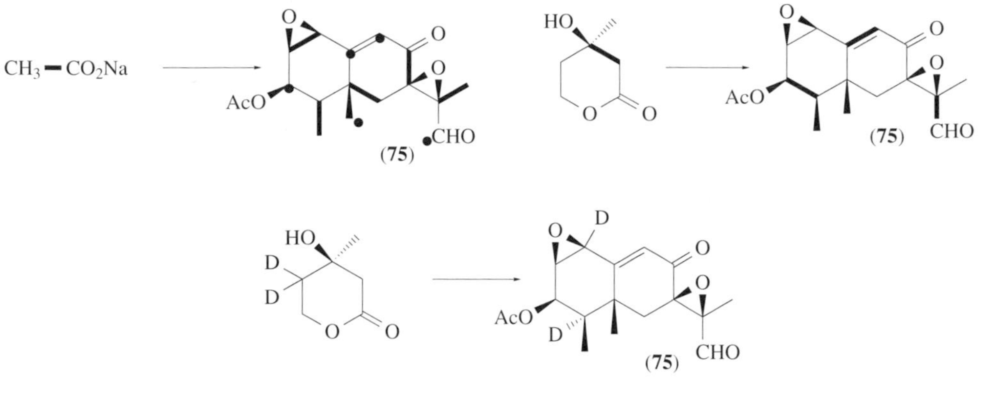

Scheme 34

Incubation of the *A. terreus* aristolochene synthase with a mixture of [11,12-$^{13}C_2$]- and [11,13-$^{13}C_2$]FPP gave rise to labeled aristolochene which displayed the predicted pair of enhanced and coupled doublets in the ^{13}C NMR spectrum (see Scheme 33).[124] By incubation of [12,12,12-$^2H_3$]FPP ((**1**), $H_A = D$, $H_B = H_C = H_D = H$) with aristolochene synthase, it was established that it is the *cis*-(C-12) methyl group of FPP which undergoes deprotonation, based on the ^{2}H NMR observation of deuterium at the C-12 methylene in the product, and confirmed by the conversion of [13,13,13-$^2H_3$]FPP ((**1**), $H_B = D$, $H_A = H_C = H_D = H$) to [13,13,13-$^2H_3$]aristolochene (**74**) (Scheme 35).[125] No transfer of the original H-12 deuterium to the ultimate site of reprotonation, H-1 of aristolochene, could be detected, suggesting that if a common base and its conjugate acid should mediate both steps, exchange with the medium would be more rapid than reprotonation of the intermediate germacrene A (**77**). The exclusive loss of the C-12 proton of FPP has also been observed in formation of several other germacrene A-derived sesquiterpenes, including capsidiol and lubimin (see Sections 2.06.3.11.2 and 2.06.3.12.2). It will also be noted that the C-12 methylene of aristolochene must eventually become the C-13 methyl of PR toxin, based on their common derivation from the *cis*(C-12) methyl of FPP. This conversion may involve two-step allylic oxidation of C-12 to the aldehyde, followed by isomerization of the exomethylene double bond.

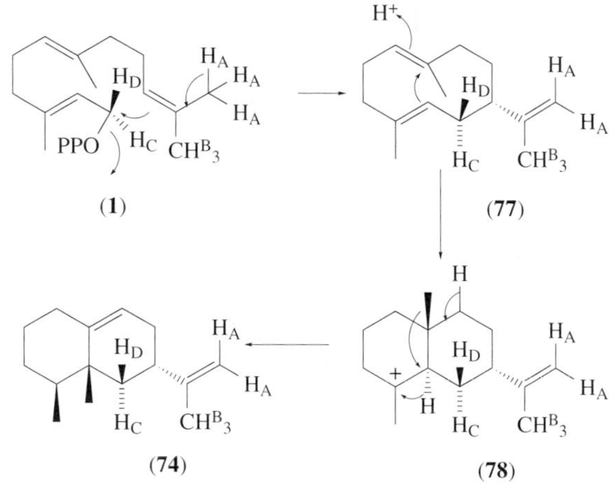

Scheme 35

The cyclization of FPP to aristolochene was shown to take place with inversion of configuration at C-1 of FPP, as expected for a direct displacement without isomerization of the 2,3-double bond.[124] Thus separate incubations of (1*R*)- and (1-*S*)-[1-^{2}H]FPP with the *A. terreus* synthase and analysis of the resulting samples by ^{2}H NMR established that aristolochene derived from (1*R*)-[1-^{2}H]FPP ((**1**), $H_D = D$, $H_A = H_B = H_C = H$) was labeled with deuterium exclusively at H-6_{eq} (H-6*re*), while aristolochene obtained from (1-*S*)-[1-^{2}H]FPP ((**1**), $H_C = D$, $H_A = H_B = H_D = H$) was labeled at H-6_{ax} (H-6*si*) (see Scheme 35). Further incubations involving (4*R*,8*R*)- and (4*S*,8*S*)-[4,8-$^2H_2$]FPP ((**1**), H_E = D, H_F = H; (**1**), H_F = D, H_E = H) and analysis by ^{2}H NMR established that it is the original H-8*si* (H_F) of FPP which is lost in the formation of the 9,10-double bond of aristolochene (**74**) (Scheme 36).[125] Based on the known relative and absolute configurations of (+)-aristolochene, these results imply that the substrate FPP and the derived intermediate germacrene A (**77**) are initially folded in a chair–boat conformation. Furthermore the sequential 1,2-hydride and methyl migrations take place with *anti*stereochemistry from the eudesmane intermediate (**78**), while the final deprotonation involves loss of the proton *syn* to the migrating methyl. This latter geometry is consistent with, but does not require, the involvement of a single base which mediates the series of deprotonation–reprotonation–deprotonation steps, similar to what has been proposed for the pentalenene synthase reaction. Further tests of this idea will have to await the results of ongoing X-ray crystallographic analysis of aristolochene synthase.

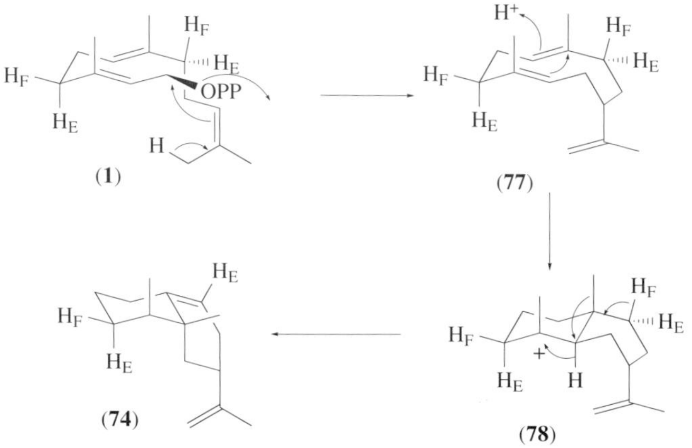

Scheme 36

The proposed intermediate germacrene A (**77**) is never released from the aristolochene active site and has therefore never been directly detected during the formation of aristolochene. Indirect evidence for the intermediacy of the otherwise cryptic germacrene A was obtained by incubation of the FPP analogue, (7*R*)-6,7-dihydroFPP (**14**) with *A. terreus* aristolochene synthase, giving rise to formation of a new product, which was identified as 6,7-dihydrogermacrene (**79**) by GC–MS comparison with a synthetic sample (Scheme 37).[126] Although (7*R*)-6,7-dihydroFPP can undergo cyclization to (**79**), the absence of the 6,7-double bond prevents further cyclization and dihydrogermacrene is ejected from the active site. (**14**) is itself an effective competitive inhibitor of aristolochene synthase (K_I 0.18 μM, K_m FPP 3.3 μM) and is converted to dihydrogermacrene (**79**) at a rate of about 1.5% that of the normal rate of turnover of FPP.

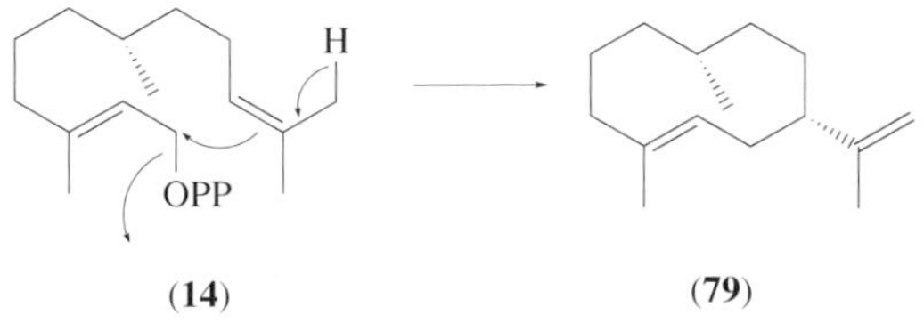

Scheme 37

The development of mechanism-based inhibitors of terpenoid synthases presents a formidable challenge since these enzymes have evolved to process highly reactive electrophilic intermediates without suicide inactivation by their natural substrate. To address this problem, it was envisaged that use of a substrate analogue which, upon cyclization, could undergo rearrangement or delocalization of charge, might bring the reactive electrophilic centers into contact with regions of the protein that do not normally encounter such charged species and which, therefore, might become covalently modified by alkylation. Indeed, the vinyl analogue of FPP, 12-methylideneFPP (**80**), proved to be just such an effective mechanism-based inhibitor, exhibiting pseudo-first-order, time-dependent inactivation of recombinant *P. roqueforti* aristolochene synthase (k_{inact} 0.33 min^{-1} ($t_{1/2}$ 2.10 min), K_I 0.46 μM) (Scheme 38).[127] These inactivation parameters are comparable to the steady-state parameters for the normal substrate FPP (k_{cat} 1.95 min^{-1}, K_m 1.2 μM). Co-incubation with FPP retarded the observed rate of cyclase inactivation. Inactivation showed an absolute requirement for Mg^{2+} and was accompanied by stoichiometric loss of the pyrophosphate moiety, as evidenced by the absence of radioactivity in aristolochene synthase that had been inactivated with [^{32}P]-12-methylideneFPP (**80**). [1-^{3}H]-(**80**) was covalently bound to aristolochene synthase with essentially 1:1 stoichiometry and radioactivity co-migrated with inactivated enzyme on SDS–PAGE. Future studies are directed at establishing the site of covalent modification of aristolochene synthase.

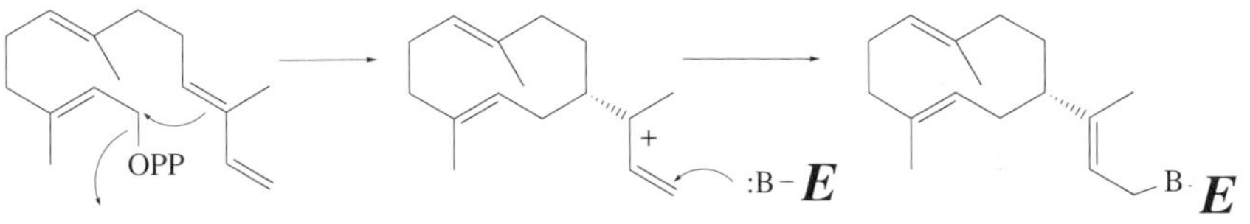

Scheme 38

2.06.3.11 *epi*-Aristolochene and Capsidiol

2.06.3.11.1 epi-*Aristolochene synthase*

5-*epi*-Aristolochene (**81**), a diastereomer of aristolochene, is the precursor of capsidiol (**82**), a sesquiterpene phytoalexin produced by several members of the Solanaceae, including tobacco (*Nicotiana tabacum*),[128] jimson (*Datura stramonium*),[129] and sweet pepper (*Capsica annum*),[130] in response to tissue injury or exposure to fungal elicitors. *epi*-Aristolochene itself was first identified as the C_{15}-hydrocarbon product of an inducible, Mg^{2+}-dependent, FPP cyclase from tobacco tissue cultures.[131–134] Chappell and co-workers purified *epi*-aristolochene synthase, which proved to be a monomer, M_r 63 500, with a typical k_{cat} of 0.05 s^{-1} and K_m 2–5 μM.[135] The same group subsequently showed that the corresponding structural gene was present in *N. tabacum* as a gene family of some 12–15 members of closely-related sequences.[136] Sequencing the cDNA and genomic DNA for two of these alleles, which map within 5 kb of one another, revealed that each contains a 1479-bp open reading frame containing five introns and encoding a protein of M_D 56 828 Da. Not only was *epi*-aristolochene synthase the first plant terpenoid cyclase to be cloned and sequenced, but its regulation has been studied in detail. Although the deduced amino acid sequence showed no similarity to the functionally closely-related fungal aristolochene synthase, *epi*-aristolochene synthase displayed the conserved aspartate-rich motif, IVDDTFD, beginning at amino acid residue 299. The protein has also been expressed in *E. coli* using a variety of T7-lac and tac-based expression vectors.[137] Detailed studies of the presteady-state kinetics,[138] as well as an X-ray crystallographic determination of the three-dimensional structure of the recombinant enzyme,[139] have been carried out (see Sections 2.06.4.2 and 2.06.4.3).

The cyclization of FPP to *epi*-aristolochene is believed to occur by a mechanism closely related to that established for the formation of aristolochene (Scheme 39). Following initial formation of a germacrene A (**77**) intermediate, folded in the manner illustrated, reprotonation and cyclization will generate a eudesmane intermediate (**83**) diastereomeric to that postulated to lead to aristolochene. Sequential *syn* 1,2-hydride and methyl migrations, followed by deprotonation will generate *epi*-aristolochene (**81**). To date, all labeling studies supporting this mechanism have been carried out not on *epi*-aristolochene itself, but on the derived capsidiol.

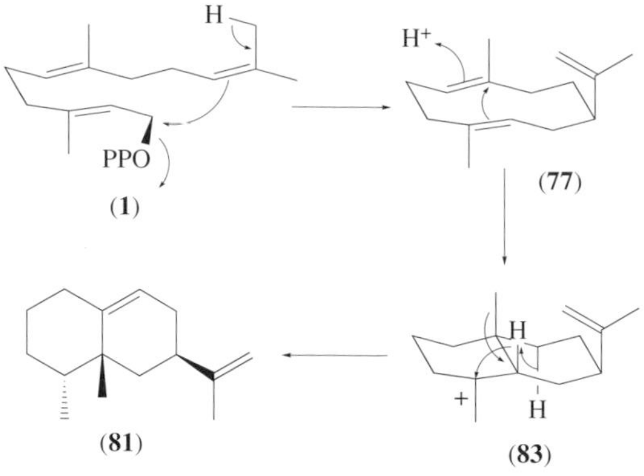

Scheme 39

2.06.3.11.2 Capsidiol

Using cellulase-elicited cultures of *N. tabacum*, Whitehead and Threllfall demonstrated the oxidative conversion of [^{14}C]-*epi*-aristolochene (**81**) to capsidiol (**82**) by way of 1-deoxycapsidiol (**84**) (Scheme 40).[132] Analysis of the ^{13}C NMR spectrum of capsidiol (**82**) obtained by feeding [1,2-$^{13}C_2$]acetate to *C. annum* revealed the presence of five pairs of enhanced and coupled doublets, consistent with the expected isoprenoid origin, while the appearance of enhanced singlets corresponding to C-10 and C-15 confirmed the proposed methyl migration (Scheme 41).[140,141] Analogous to aristolochene, the C-12 methylene gave rise to a doublet coupled to C-11, consistent with loss of a proton from the *cis* (*Z*)-methyl of FPP in the formation of the intermediate germacrene A. Three additional enhanced singlets corresponding to the predicted sites C-3, C-9, and C-13, were also evident. Evidence for the proposed 1,2-hydride migration from C-5 to C-4 came from incorporation of [4,4-$^2H_2$]mevalonate into (**82**) by elicited sweet peppers.[142] The ^{2}H NMR spectrum of the resulting (**82**) established the presence of deuterium at each of the predicted sites, D1, D-7, and the terminus of the migration, C-4.

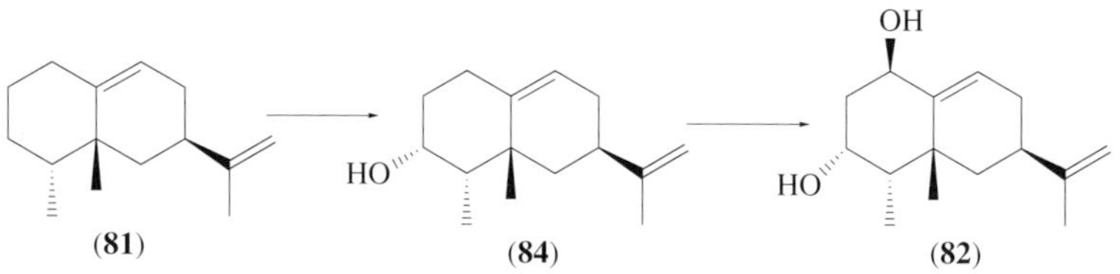

Scheme 40

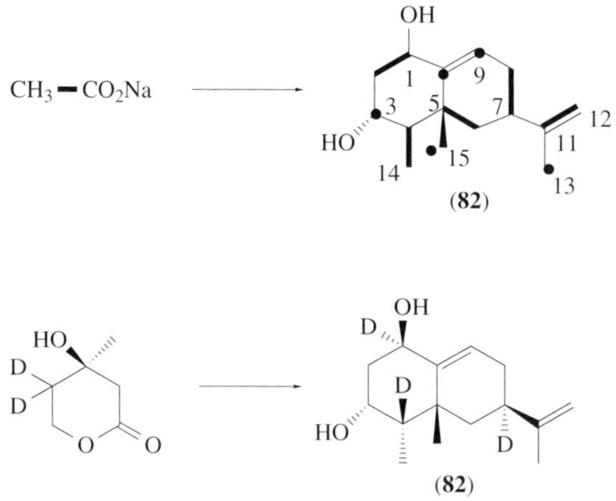

Scheme 41

2.06.3.12 Vetispiradiene, Lubimin, and Rishitin

2.06.3.12.1 Vetispiradiene synthase

Several solanaceous plants produce bicyclic stress metabolites derived from the parent sesquiterpene hydrocarbon, vetispiradiene. Using a PCR probe derived from *N. tabacum epi*-aristolochene synthase, Chappell and Back have isolated genomic and cDNA clones for vetispiradiene synthase from henbane (*Hyoscyamus muticus*) corresponding to an elicitor-inducible Mg^{2+}-dependent FPP-cyclase activity.[143] Southern hybridization indicated a gene family of 6–8 members and bacterial expression gave a protein which could cyclize FPP to a single sesquiterpene product corresponding to vetispiradiene (**85**). The His-tagged recombinant enzyme showed a K_m for FPP of 4.4 μM.[139]

Not only did the *H. muticus* vetispiradiene synthase share the same IVDDTFD motif with *epi*-aristolochene synthase, the two enzymes had an overall 77% identity and 81% similarity at the amino acid level. Moreover, comparison of the corresponding genomic sequences for the two structural genes revealed a common intron–exon organization between the two cyclase genes. This similarity could be extended to two unrelated plant terpenoid synthases, the monoterpene cyclase, limonene synthase, from peppermint, and the diterpene cyclase, casbene synthase, from castor bean. Comparison of the sequences has suggested a set of three conserved His dispersed over the N-terminal half of the protein and a conserved Cys residue near the C-terminus.

The formation of vetispiradiene from FPP (**1**) is believed to involve cyclization to the same germacrene A (**77**) and eudesmane cation (**83**) intermediates as for *epi*-aristolochene (Scheme 42). Instead of the previously described methyl migration, the pathway to vetispiradiene diverges by a 1,2-Wagner–Meerwein shift resulting in a net ring-contraction, followed by deprotonation to generate the characteristic spiro[6.5]bicyclodecane skeleton of (**85**). Based on the close parallels between the cyclization mechanisms leading to (**81**) and (**85**), as well as the marked similarity in amino acid sequence and intron–exon organization between the two cyclases, Chappell has explored the possibility that individual steps in the overall cyclization process might be controlled by explicit domains within each cyclase. To test these ideas, Chappell and Back carried out a clever set of experiments in which a series of chimeric proteins were constructed from systematically varied combinations of different *epi*-aristolochene and vetispiradiene domains.[144] In fact, in several cases the chimeric enzymes produced *both* sesquiterpenes. To explain these results, Chappell and Back postulated that there are so-called product specificity domains in each protein, as well as a third domain, termed a "ratio-determinant domain", which influences the observed proportion of products generated by each chimeric protein. According to this model, each domain would have an explicit function. An alternative explanation is that each native cyclase folds the substrate and derived intermediates in a precisely determined manner, resulting in the formation of a single, characteristic product. Transgenic substitution of domains from one terpene cyclase to another would result in a perturbation of the natural shape, hydrophobicity, and polarity of the corresponding region of the active site, giving rise to a hybrid enzyme which no longer perfectly accommodates a given substrate folding or intermediate conformation, but instead is *permissive* enough to allow two cyclization pathways to compete. Such relaxed specificity is reminiscent of the generation of mixtures of biogenetically related products as a consequence of point mutations in active site residues of trichodiene synthase. A distinction between these active and passive models of domain function will require the construction of additional chimeric mutants as well as detailed analysis of protein structure. Significant progress in structural analysis of the native and mutant proteins has already been made (see Section 2.06.4.3.2).

2.06.3.12.2 Lubimin and rishitin

Although no explicit labeling experiments have been carried out with the vetispiradiene synthase, considerable information on the formation of this class of sesquiterpenes has been obtained by studies of the origins of derived phytoalexins. Feeding of [1,2-$^{13}C_2$]acetate to fruit capsules of *D. stramonium* elicited with spores of *Monilinia fructicola* gave labeled samples of lubimin (**86**) and hydroxylubimin (**87**), which each exhibited the expected set of six pairs of enhanced and coupled droplets and three enhanced singlets in the derived ^{13}C NMR spectra, consistent with the proposed mechanism of formation of the parent sesquiterpene hydrocarbon, vetispiradiene (Scheme 43).[145] Murai and Masamune demonstrated that (−)-[8,8-$^2H_2$]solavetivone (**88**) is a precursor of (+)-(**86**),

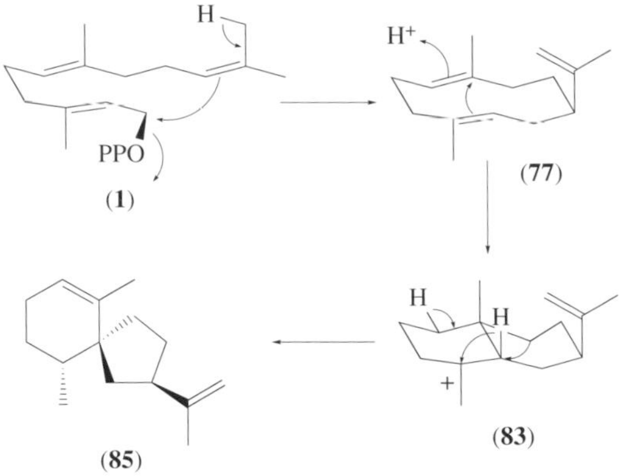

Scheme 42

(+)-(**87**), and the rearranged sesquiterpene (−)-rishitin (**89**) in slices of potato (*Solanum tuberosum*) (Scheme 44).[146] Independently, Stoessl and Stothers confirmed the predicted 1,2-hydride shift by incorporation of [2-$^2H_3$,2-^{13}C]acetate and observation of a β-deuterium isotope shift in the ^{13}C NMR signal corresponding to C-5 of rishitin, arising from deuterium at C-4 which has originated from a common acetate precursor.[147] Additional deuterium labels were evident from the observed α-deuterium isotope shift and [^{13}C-^2H]-coupling at each of the remaining predicted sites in (**89**). Finally, growth in the presence of [$^{18}O_2$] gave rise to labeled oxygen at C-2 of lubimin (**86**) and C-2 and C-3 of rishitin (**89**), consistent with the operation of mixed-function oxygenases in the late stages of the biosynthetic pathway.[148] Any additional isotopic label in the aldehyde oxygen of (**86**) would presumably have been lost by rapid exchange.

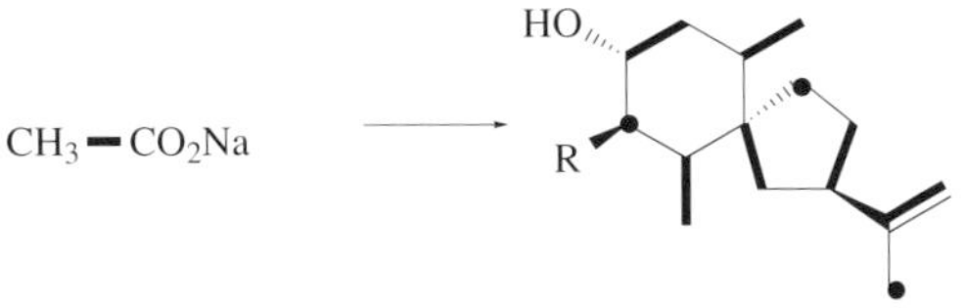

CH_3—CO_2Na

(**86**) R=H
(**87**) R-OH

Scheme 43

2.06.3.13 Patchoulol and Patchoulene

Patchoulol (**90**) is the major component of the essential oil of patchouli (*Pogestemon cablin*), where it is produced and accumulated in specialized glandular trichomes, accompanied by an array of biogenetically related sesquiterpenes, including α-, β-, and γ-patchoulene (**91**)–(**93**), α-bulnesene (**94**), and α-guaiene (**95**).[149] A free-extract of patchouli leaf homogenates was found by Croteau *et al.* to catalyze the cyclization of [1-^3H,12,13-^{14}C]FPP to patchoulol, chemical degradation of which located the tritium label at C-1 (Scheme 45).[150] Isotopic dilution experiments with unlabeled sesquiterpene hydrocarbons failed to influence the formation of patchoulol, and labeled samples of (**91**), (**92**), (**94**), and (**95**) were not converted to (**90**). Further insight into the mechanism of cyclization came from the finding that cyclization of [6-^3H,12,13-^{14}C]FPP gave patchoulol (**90**) in which the tritium label had undergone a net 1,3-migration to C-3, as established rigorously by chemical degradation (Scheme 46). Consistent with this result was an independent report that incorporation of (4*R*)-[4-^3H,2-^{14}C]mevalonate into α- and γ-patchoulene ((**91**) and (**93**)) by feeding to *P. cablin*, resulted in retention of all three tritium atoms, as judged by the lack of change in ^3H:^{14}C isotope ratio, while formation of β-patchoulene (**92**) was accompanied by loss of one-third of the original tritium (Scheme 47).[151] No effort was made to locate the label in the latter samples. The results of cyclization of [6-^3H,12,13-^{14}C]FPP[150] also ruled out an alternative set of sequential 1,2-hydride migrations suggested by Akhila *et al.*[151]

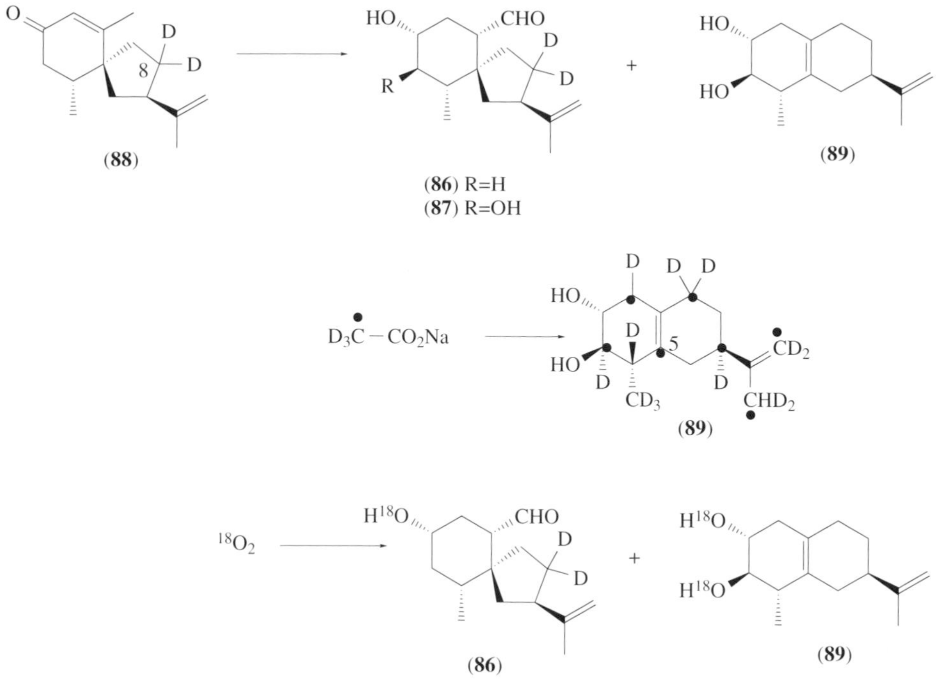

Scheme 44

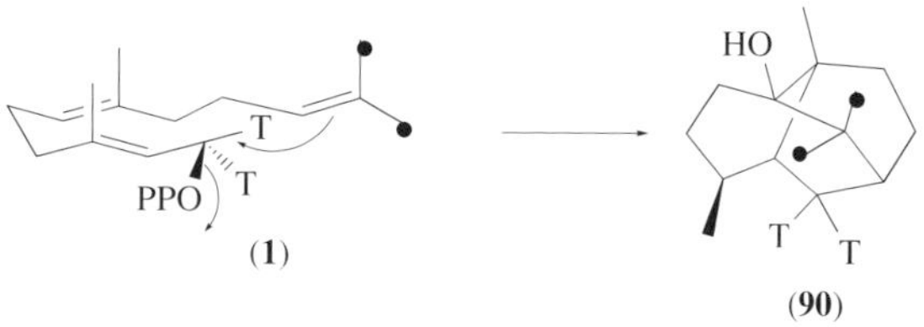

Scheme 45

These labeling results can be rationalized by the cyclization mechanism illustrated in Scheme 46, key features of which are the electrophilic attack of the 2-propyl cation of **(96)** on the original 6,7-double bond of FPP followed by the 1,3-hydride migration and a series of backbone rearrangements. An alternative to the 1,3-hydride migration would be a deprotonation–reprotonation sequence involving an intermediate cyclopropane. Diversion of several of the cyclization intermediates can lead to formation of the various sesquiterpene hydrocarbon co-metabolites. Indeed, purification of patchouli synthase to homogeneity failed to resolve the sesquiterpene alcohol and alkene cyclase activities, suggesting that all these biogenetically related products are produced at a common active site. Although other examples of multiple product formation by sesquiterpene synthases are known, this trait is more commonly observed among plant monoterpene synthases.

The homogeneous patchouli synthase is a homodimer of subunit M_r 40 000 with a K_m for FPP of 6.5 μM and a typical terpene synthase k_{cat} of 0.03 s^{-1}.[152] The cyclization requires Mg^{2+} (K_m 1.7 mM); other divalent cations are ineffective. The cyclase was inhibited by the cysteine-specific reagents *N*-ethylmaleimide and methyl methanethiolsulfonate, while diethylpyrocarbonate, a histidine-directed reagent, was also inhibitory. Partial inactivation did not affect the proportion of hydrocarbon and alcohol products.

2.06.3.14 β-Selinene

β-Selinene **(97)** is the major sesquiterpene hydrocarbon of calamondin orange (*Citrofortunella mitis*) fruits. The bicyclic skeleton can be formed by cyclization of FPP **(1)** to the eudesmane cation

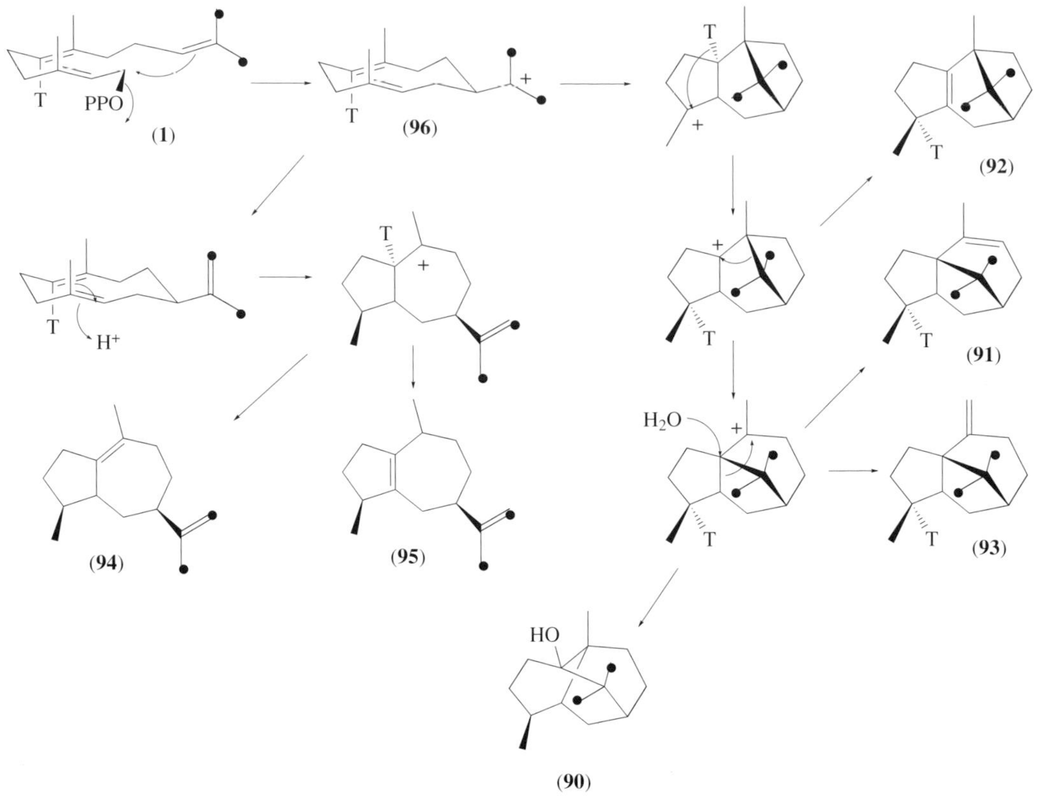

Scheme 46

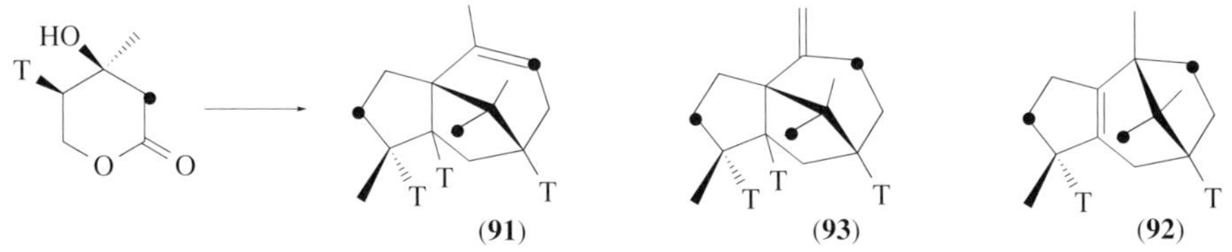

Scheme 47

(**78**) by way of germacrene A (**77**) (Scheme 48). Although no studies of this mechanism have as yet been reported, β-selinene synthase has been purified more than 80-fold from the 100 000 g supernatant of an extract of the flavedo of immature calamondin.[153] The protein, which appeared to be associated with the endoplasmic reticulum, had a native M_r 67 000 and catalyzed the Mg^{2+}-dependent cyclization FPP to β-selinene. The K_m for FPP, 45 μM, was relatively high for this type of enzyme.

2.06.3.15 Cadinenes and Gossypol

2.06.3.15.1 δ-Cadinene synthase

The hydrocarbon δ-cadinene (**98**) is one of the most widely occurring plant sesquiterpenes. Cadinene is found in several species of cotton where it is the parent hydrocarbon of a variety of inducible phytoalexins, including 2,7-dihydroxy cadalene (**99**), lacinilene C (**100**), and gossypol (**101**). For example, inoculation of glandless cotyledons of upland cotton (*Gossypum hirsutum* L.) with the bacterial pathogen *Xanthomonas campestris* pv. *malvacearum* or oligogalacturonide elicitors

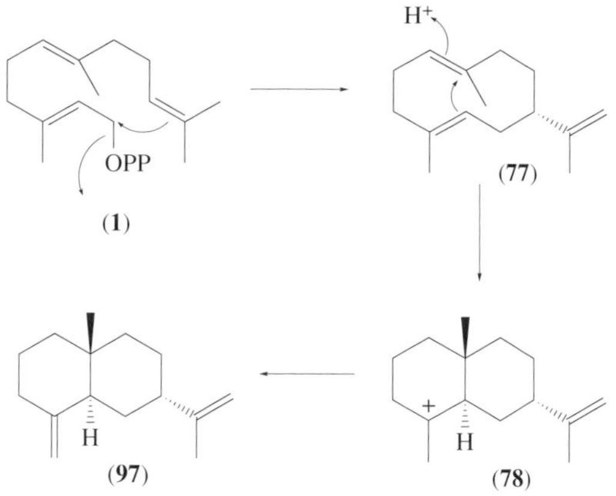

(1) (77) (97) (78)

Scheme 48

led to accumulation of (+)-δ-cadinene as the major inducible sesquiterpene.[154] Essenberg and Davis have isolated the corresponding δ-cadinene synthase from similarly elicited cotyledons[154] and purified the enzyme to apparent homogeneity.[155] The synthase, a M_r 64 000–65 000 protein, catalyzed the Mg^{2+}-dependent cyclization of [1-^{3}H]FPP to (+)-δ-cadinene. The sequences of three tryptic peptides obtained from the purified *G. hirsutum* cadinene synthase proved to be identical to homologous sequences of cadinene synthase cloned from *G. arboreum*.

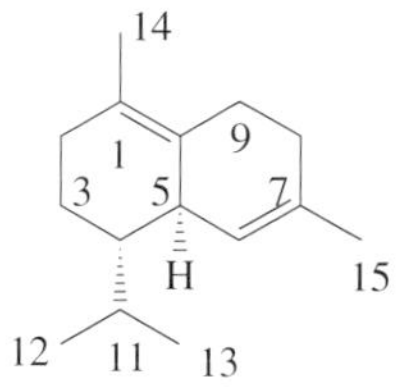

(98) 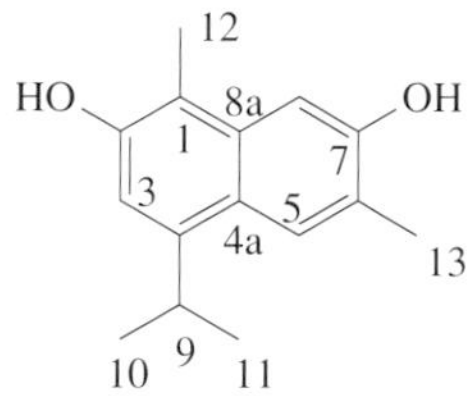 (99) 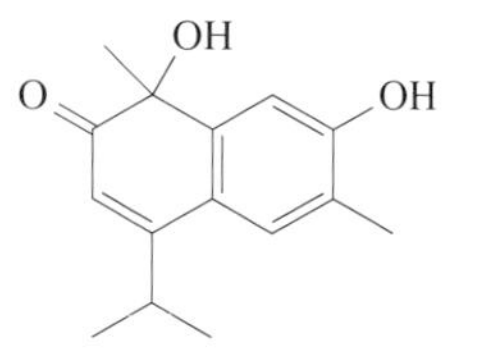 (100) 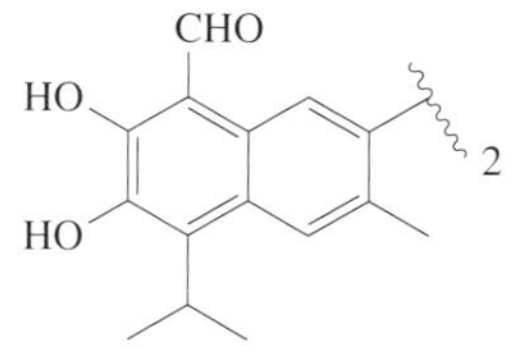 (101)

Heinstein, Davisson and co-workers have isolated three independent cDNA clones for δ-cadinene synthase by screening a cDNA library from *G. arboreum* cell suspension cultures elicited with the phytopathogenic fungus *Verticillium dahliae*, using as primary probes PCR products initially generated using PCR primers based on conserved regions of the *epi*-aristolochene sequence.[156,157] Two of the corresponding synthases, CAD1-C1 and CAD14, were nearly identical to one another (95% at the nucleic acid and 97% at the amino acid sequence level),[156] while the third, CAD1-A, showed 80% amino acid sequence identity to the first two.[157] All three proteins could be expressed in *E. coli* as M_D 64 kDa His-tagged fusion proteins with K_m for FPP in the 10–17 μM range and k_{cat} 0.03 s^{-1}. Excision of the polyhistidine leader sequence by thrombin had negligible effect on the observed K_m with only a minor (1.8-fold) increase in k_{cat}. The consensus aspartate-rich motif, IVDDTYD, was present at amino acids 305–311. These cadinene synthases also showed significant sequence homologies to other plant terpenoid synthases, including *epi*-aristolochene synthase, ranging from 32–48% identity and 53–66% similarity. Among all these sequences there were two conserved histidines and a single conserved cysteine.

The formation of δ-cadinene is thought to take place by isomerization of FPP (**1**) to nerolidyl diphosphate (**2**), followed by cyclization to the *cis,trans*-helminthogermacradienyl cation (**102**) (Scheme 49). A 1,3-hydride shift leads to the corresponding allylic cation (**103**), which upon further cyclization and deprotonation yields δ-cadinene (**98**). In support of this mechanism are the findings of Benedict *et al.* who reported that incubation of [1-$^2H_2$]FPP with an extract of cotton stems (*G. barbadense*) that had been elicited with *V. dahliae* gave δ-cadinene which was shown by GC-MS to retain both deuterium atoms, one of which had been transferred to C-11 of the isopropyl side-chain.[158] The cyclization mechanism, which is consistent with the results of extensive investigations of the biosynthesis of related cadinene-type metabolites carried out by Arigoni and his co-workers,[13] is further supported by labeling studies on derived cadalene sesquiterpenes.

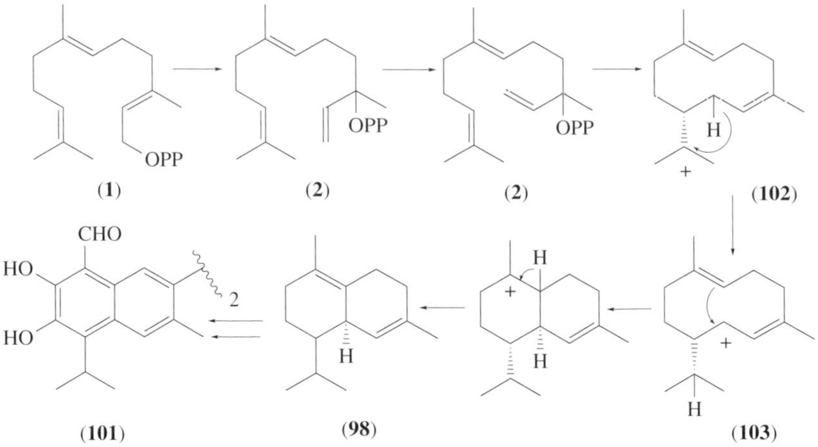

Scheme 49

2.06.3.15.2 2,7-Dihydroxycadalene and gossypol biosynthesis

Feeding of $[^{3}\mathrm{H}]$-δ-cadinene (**98**) to elicited cotyledons of *G. hirsutum* gave labeled 2,7-dihydroxycadalene (**99**), lacinilene C (**100**), and a variety of closely-related sesquiterpenoid phytoalexins (Scheme 50).[154] A report in 1970 had claimed one such cadalene, represented by gossypol (**101**), was formed from *cis,cis*-farnesyl diphosphate,[159] but this was shown to be erroneous by feeding of $[2\text{-}^{14}\mathrm{C}]$- and $[4\text{-}^{14}\mathrm{C}]$mevalonate to *G. herbaceum* and chemical degradation of the derived samples of labeled gossypol (Scheme 51).[160] The deduced distribution of label from each precursor was found to be inconsistent with the alleged *cis,cis*-FPP isomer and supported the cyclization mechanism illustrated in Scheme 49. Consistent with these conclusions was the demonstration of the proposed 1,3-hydride shift, as evidenced by the retention of tritium from $[5\text{-}^{3}\mathrm{H},2\text{-}^{14}\mathrm{C}]$mevalonate in the side chain methine of (**101**). Similar observations were subsequently reported for the incorporation of tritium from $[5\text{-}^{3}\mathrm{H},2\text{-}^{14}\mathrm{C}]$mevalonate into 2,7-dihydroxycadalene (**99**) from elicited *G. hirsutum* and the location of the label in the isopropyl side chain.[161] Independently, $[1,2\text{-}^{13}\mathrm{C}_2]$acetate was incorporated in separate feedings into gossypol[162] (**101**) and 2,7-dihydroxycadalene (**99**),[163] giving rise to a pattern of six enhanced and coupled doublets and three enhanced singlets in the $^{13}\mathrm{C}$ NMR spectra of each metabolite, fully consistent with the proposed FPP folding pattern.

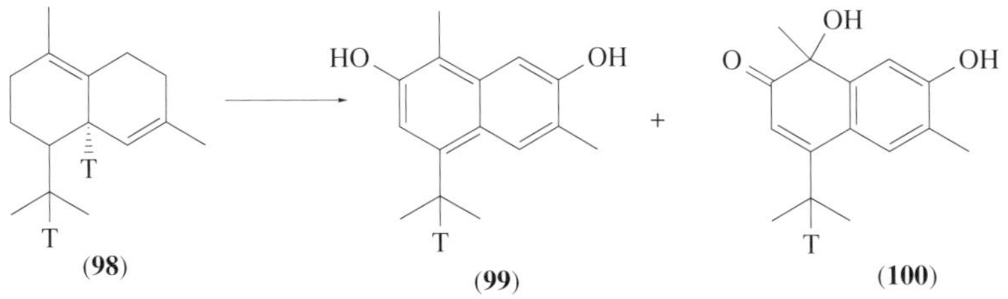

Scheme 50

Nabeta *et al.* have carried out a detailed $^{13}\mathrm{C}$ NMR study of the origins of (1*S*)-7-methoxy-1,2-dihydrocadalenene (**104**) produced by suspension cultures of the liverwort, *Heteroscyphus planus*.[164] Incubation of $[2\text{-}^{13}\mathrm{C}]$mevalonate gave (**104**) labeled at C-2, C-7, and C-10, while incorporation of $[4\text{-}^{13}\mathrm{C}]$mevalonate led to labeling at the predicted sites, C-4, C-5, and C-8a (Scheme 52). Formation of (**104**) must involve not only a 1,3-hydride shift analogous to that involved in the formation of δ-cadinene, but a 1,2-hydride shift as well to generate the secondary methyl group at C-1 (Scheme 53). To test the latter possibility, $[4\text{-}^{2}\mathrm{H}_2]$mevalonate was administered to *H. planus* cultures and the derived (**104**) was analyzed by $^{2}\mathrm{H}$ NMR, thereby confirming the presence of deuterium at the predicted site, C-1 as well as at C-5. Evidence for the 1,3-hydride shift came from incorporation of $[5\text{-}^{2}\mathrm{H}_2,5\text{-}^{13}\mathrm{C}]$mevalonate. In the $^{2}\mathrm{H}$ NMR of the derived (**104**), the signals for D-3 and D-8 appeared as doublets, due to coupling to the directly attached $^{13}\mathrm{C}$, while the signal for D-9, the deuterium

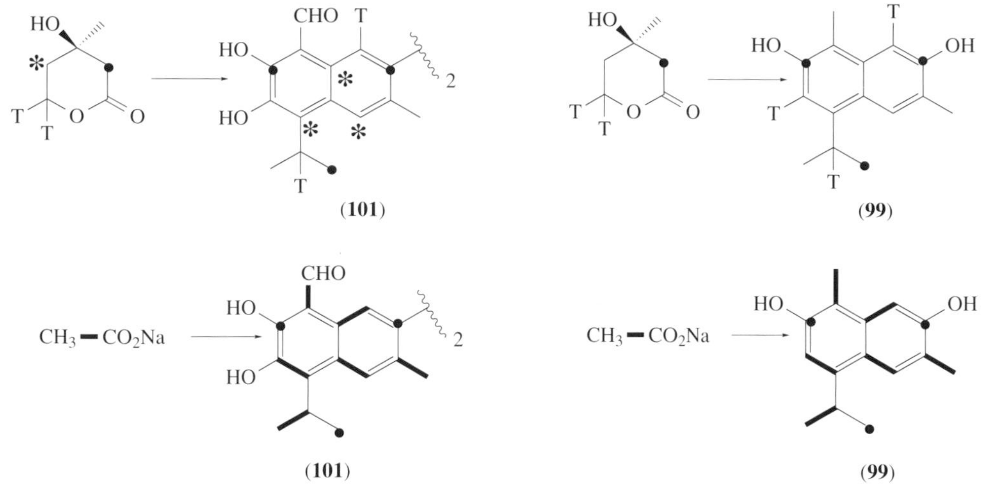

(101)

(99)

(101)

(99)

Scheme 51

that had undergone migration, appeared as a singlet. In the corresponding ^{13}C NMR spectrum, C-3 and C-8 appeared as isotopically shifted triplets, while C-4a, which had lost its attached deuterium, appeared as an enhanced singlet.

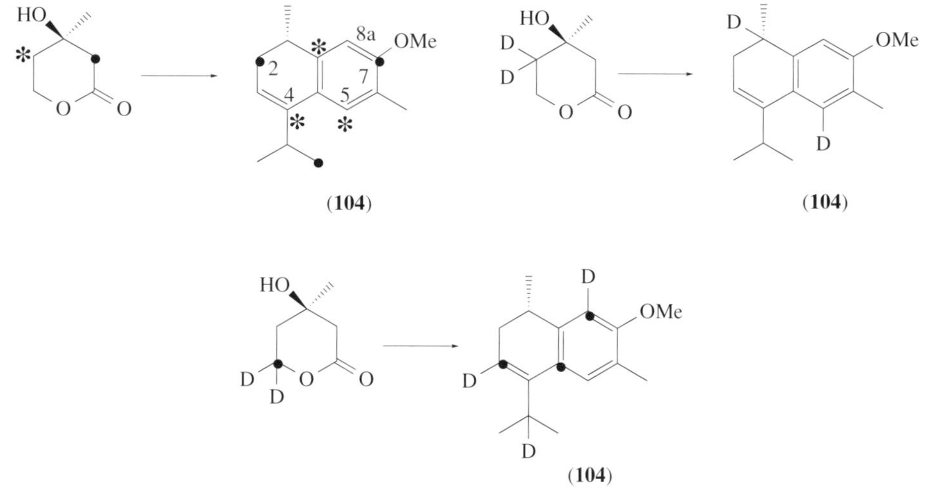

(104)

(104)

(104)

Scheme 52

2.06.3.16 Epicubenol

(+)-Epicubenol (**105**) is a cadinane alcohol which has been isolated from a variety of organisms, including *Streptomyces* sp. LL-B7[165] and the liverwort *H. planus*.[166] The (−)-enantiomer has also been isolated from cubeb oil and other plants. A cell-free extract of *Streptomyces* sp. LL-B7 was shown to catalyze the Mg^{2+}-dependent cyclization of [1-^{3}H]FPP to epicubenol, the identity of the product being confirmed by dilution with synthetic (±)-cubenol and recrystallization of a diol derivative to constant activity.[167] Incubation of [13,13,13-$^2H_3$]FPP (**1**) with crude epicubenol synthase gave epicubenol (**105**) which was shown by ^{2}H NMR to be labeled in one of the diastereotopic methyls of the side chain isopropyl group (Scheme 54).

The cyclization of FPP to epicubenol (**105**) is thought to occur by a mechanism closely analogous to the formation of δ-cadinene (**98**) and (1*S*)-7-methoxy-1,2-dihydrocadalenene (**104**). Thus isomerization of FPP (**1**) to NPP (**2**) followed by ionization and cyclization will give the *cis,trans*-helminthogermacradienyl cation (**102**) (Scheme 55). Following a 1,3-hydride shift, the resultant

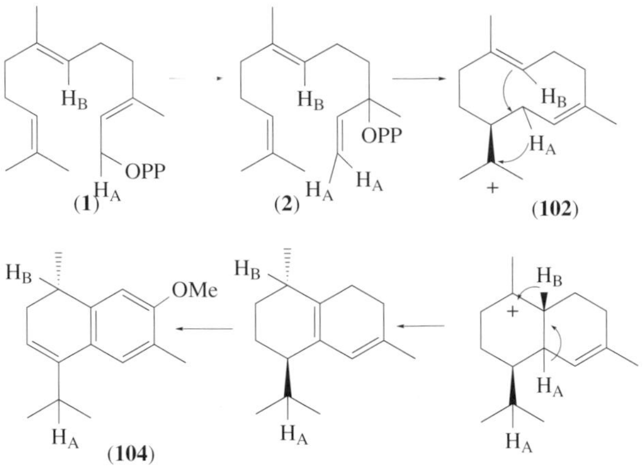

Scheme 53

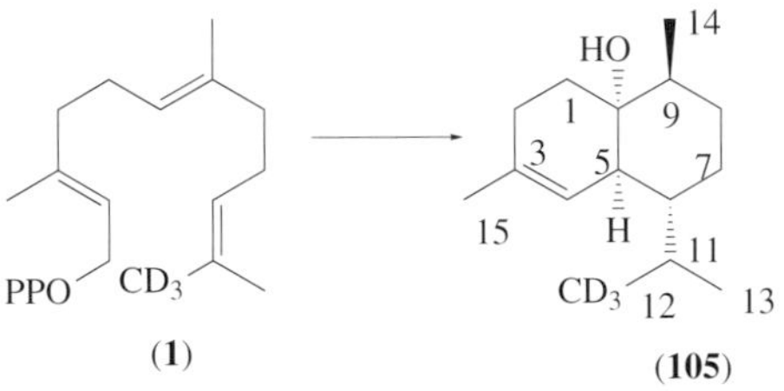

Scheme 54

allylic cation can cyclize to the bicyclic cation common to all three pathways. A 1,2-hydride shift and capture of water will generate epicubenol (**105**).

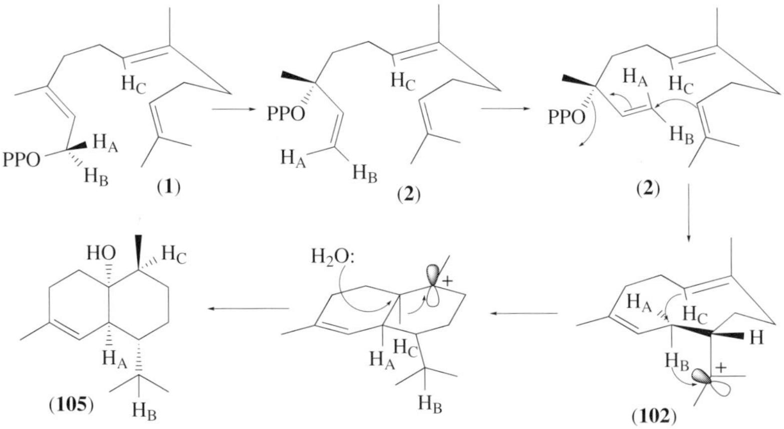

Scheme 55

Evidence for the proposed 1,3-hydride shift came from incubation of [1,1-$^2H_2$]FPP ((**1**), $H_A = H_B = D$, $H_C = H$) with epicubenol synthase from *Streptomyces* sp. LL-By (see Scheme 55).[167] The resultant labeled epicubenol (**105**) displayed two signals in the ^2H NMR corresponding to D-5 (δ 1.62), the original site of deuteration, and D-11 (δ 1.93), the terminus of the proposed 1,3-hydride migration. Incubation of stereospecifically deuterated FPP and ^2H NMR analysis established the stereochemical course of this hydride shift. Thus cyclization of (1*R*)-[1-^2H]FPP ((**1**), $H_A = D$, $H_B = H_C = H$) gave epicubenol (**105**) labeled at C-5 while epicubenol derived from (1*S*)-[1-^2H]FPP ((**1**), $H_B = D$, $H_A = H_C = H$) gave epicubenol deuterated at C-11 on the isopropyl side chain.

The role of nerolidyl diphosphate in the cyclization was first demonstrated by cyclization of (3*R*)-(1*Z*)-[1,^{3}H]NPP (**2**) to epicubenol.[168] Under identical conditions (3*S*)-(1*Z*)-[1-^{3}H]NPP did not undergo cyclization. When (3*R*)-(1*Z*)-[1-^{2}H]NPP ((**2**), H_A = D, H_B = H_C = H) was incubated with partially purified epicubenol synthase, GC–MS analysis established the presence of deuterium exclusively at the C-5 ring junction. Since cyclization of (1*R*)-[1-^{2}H]FPP had given the identical labeling pattern in the product epicubenol, it is evident that the isomerization of FPP to (3*R*)-NPP that initiates the cyclization reaction must take place with exclusive suprafacial stereochemistry, in complete agreement with the stereochemical course of the FPP to NPP isomerization catalyzed by trichodiene synthase. Rotation about the C-2,3 double bond of the NPP intermediate and reionization of the cisoid conformer will generate the allylic cation–pyrophosphate anion pair which is captured by the distal 10,11-double bond with net *anti*stereochemistry, in accord with the stereochemistry of all known S_N' displacements of allylic diphosphates.[169] In the resulting *cis*-germacradienyl cation (**102**), the original H-1*si* hydrogen of FPP (H_B) is able to overlap the vacant *p*-orbital of the C-11 cation, thereby accounting for the observed stereochemistry of the hydride rearrangement. Such results are fully in agreement with the stereochemical model of such cyclizations originally proposed by Arigoni.[13]

Cyclization of the allylic cation gives the cadinanyl cation, which can undergo a 1,2-hydride shift before capture of water from the same face. Direct confirmation of the proposed hydride migration was obtained by cyclization of [6-^{2}H]FPP ((**1**), H_C = D, H_A = H_B = H) with the *Streptomyces* cyclase and location of the deuterium at C-9 by ^{2}H NMR (see Scheme 55).[170]

The formation of epicubenol (**105**), along with that of the closely related hydrocarbon (+)-cubenene (**106**), has also been investigated by Nabeta *et al.* using cell-free extracts of *H. planus*.[171] By GC–MS analysis, *trans,trans*-[1,1-$^2H_2$]FPP was found to act as a precursor of both (**105**) and (**106**), while the corresponding (2*Z*,6*E*)-FPP was not cyclized (Scheme 56). The mass spectra also confirmed the predicted 1,3-hydride shift, based on the presence of deuterium in the isopropyl group of both cubenene and epicubenol. Examination of the derived cubenene (**106**) by ^{2}H NMR located the deuterium atoms at C-5 and C-11. In a similar manner, cyclization of [6-^{2}H]FPP gave rise to cubenene (**106**) labeled at D-9, consistent with the expected 1,2-hydride shift.

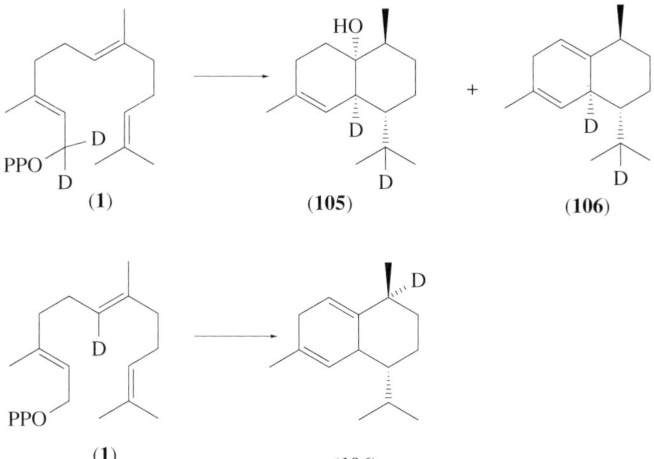

PPO D D HO D D D D
(**1**) (**105**) (**106**)

PPO D D
(**1**) (**106**)

Scheme 56

2.06.3.17 Miscellaneous Sesquiterpenes

Labeling studies with $^{18}O_2$ have established that the three oxygen atoms in ipomeamarone (**107**), a major sesquiterpenoid stress metabolite of sweet potato (*Ipomoea batatas*), are all derived from molecular oxygen.[172] Deuterated mevalonates have been used in conjunction with GC–MS analysis to study the formation of several sesquiterpenes, including α-cedrene (**108**), produced by *Lerix leptolepis* callus.[173]

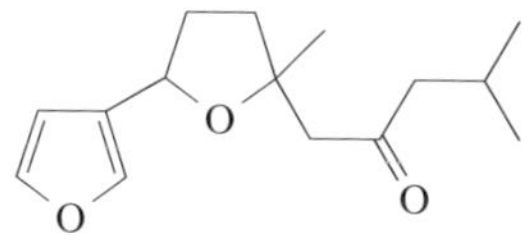

(107) 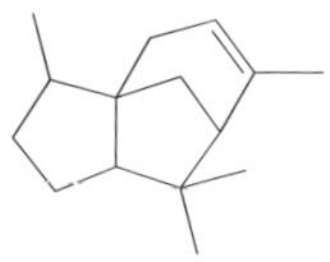 (108)

Marine invertebrates are well known as an exceptionally rich source of sesquiterpenes. However, numerous technical problems, ranging from difficulty in laboratory culturing of marine organisms to apparent severe barriers to uptake of exogenous precursors, have hindered biosynthetic investigations of novel marine terpenes. The results of recent ^{13}C incorporation experiments with the dorid nudibranch, *Acanthodoris nanaimoensis*, are therefore of considerable interest and are likely to stimulate further efforts in this area.[174] Injection of [1,2-$^{13}C_2$]acetate into specimens of *A. nanaimoensis* over a period of 16 days was followed by extraction of the resulting aldehydes nanimoal (**109**), acanthodoral (**110**), and isoacanthodoral (**111**), reduction with $NaBH_4$, and HPLC purification. ^{13}C NMR analysis of (**112**), revealed a pattern of six enhanced and coupled doublets and three enhanced singlets corresponding to the labeling pattern illustrated in Scheme 58. The ^{13}C NMR spectrum of (**113**), however, displayed five singlets, in addition to five sets of enhanced and coupled doublets, indicating additional cleavage of one of the original acetate units (Scheme 58). (**112**) and (**113**), therefore, cannot be derived from a common cyclobutane intermediate. Instead, it has been proposed that the pathways branch immediately after formation of the monocyclofarnesyl diphosphate intermediate (see Scheme 57).

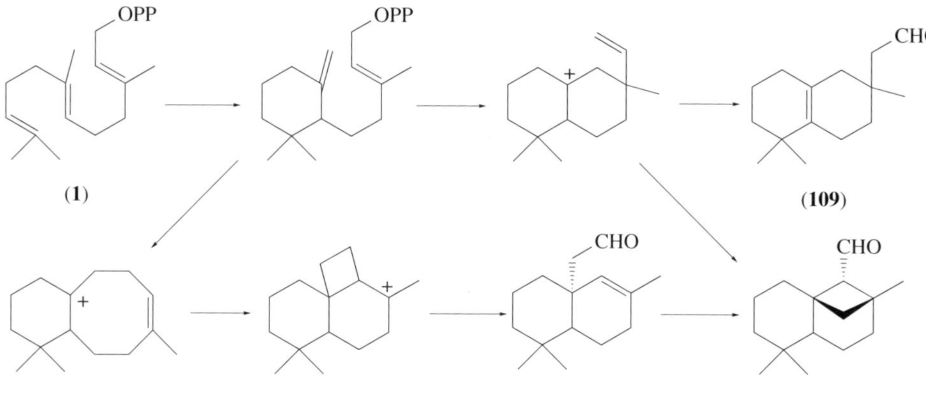

OPP OPP CHO

(1) (109)

CHO CHO

(111) (110)

Scheme 57

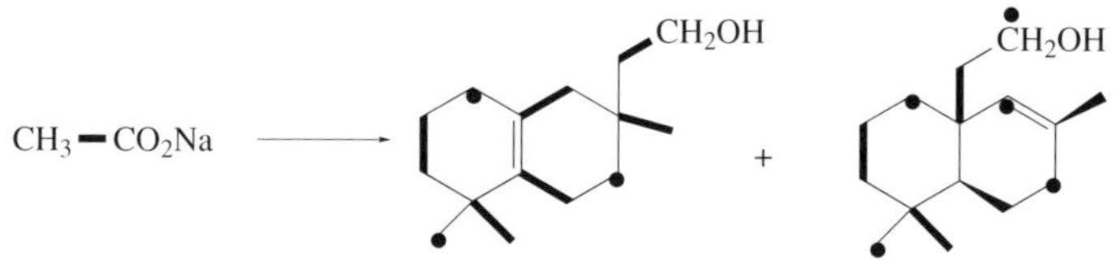

CH_3—CO_2Na → CH_2OH + CH_2OH

(112) (113)

Scheme 58

2.06.4 SESQUITERPENE SYNTHASES

2.06.4.1 Protein Chemistry

Although only a minor fraction of the potential several hundred sesquiterpene synthases have been investigated in any way, several clear patterns have already emerged.[3–5] All the enzymes characterized up until the 1990s are monomers or homodimers of subunit molecular weight *ca.* 40 000–70 000. All require a divalent cation, Mg^{2+} almost always being preferred, and the majority exhibit K_m values for FPP in the range of 0.1–10 μM, with turnover numbers, k_{cat}, of 0.03–0.5 s^{-1}. Recent evidence suggests that the latter rates may be dominated by rate-limiting product release and that the rate of enzyme-catalyzed FPP consumption is actually 1–2 orders of magnitude faster. Although nothing is known about the subcellular distribution of the sesquiterpene synthases, it is reasonable to speculate that they may be loosely membrane-associated, based not only on their pronounced hydrophobicity, but on the fact that they provide a metabolic link between the detergent-like substrate farnesyl diphosphate, produced by a cytosolic enzyme, and hydrocarbon or alcohol products which frequently undergo oxidative processing by presumably membrane-bound cytochrome P450s. Interestingly all such synthases have been obtained in crude extracts in soluble form, save for bergamotene synthase, which is initially found in the microsomal fraction. The synthases are broadly similar in molecular weight, enzymology, divalent metal ion dependence, and mechanism of action to FPP synthase and related prenyl transferases. Although there is no evident sequence identity between the two classes of enzymes, they all share a common aspartate-rich motif, (I,L,V)XDDXXD, believed to play a critical role in binding the pyrophosphate moiety of the substrate by chelation of a bridging Mg^{2+} ion or ions.

Among the plant sesquiterpene synthases there is a striking degree of sequence similarity, indicating a common evolutionary origin, as well as the presence of multiple genes for individual synthases within a given plant species, suggesting the existence of synthase isozymes and possibly complex patterns of differential expression. These sequence similarities are discussed in more detail in Chapter 2.07. Although several conserved histidine and cysteine residues have been observed in the plant sesquiterpene synthases, a result consistent with the inhibitory action of amino-acid specific reagents, the actual catalytic role of these residues remains obscure. By contrast, none of the plant sesquiterpene synthases bears any significant sequence identity to the microbial cyclases, nor do the latter enzymes show homologies either to one another or to any other known protein.

2.06.4.2 Presteady-state Kinetics

2.06.4.2.1 Trichodiene synthase

The availability of several cloned sesquiterpene synthases has allowed a detailed look at the presteady-state kinetics of sesquiterpene formation. In collaboration with Anderson, Cane *et al.* have used rapid chemical quench methods to study the transient kinetics of the trichodiene synthase reaction.[33] Using a rapid-mixing apparatus, [1,2-^{14}C]FPP was reacted with a five-fold excess of trichodiene synthase at 15 °C and the reaction was quenched with KOH at time intervals ranging from 3 ms to 6 s. A typical time course for FPP consumption and trichodiene formation is shown in Figure 1(a). The calculated single turnover rate was found to be 3.5–3.8 s^{-1}, some 40 times faster than the steady-state k_{cat} of 0.09 s^{-1} for trichodiene (**5**) formation measured at the same temperature. Multiturnover experiments, using an excess of FPP, showed biphasic behavior, with a burst phase, $k_b = 4.2\ s^{-1}$, for initial consumption of FPP and accumulation of trichodiene on the enzyme, followed by a slower, linear release of products, $k_{lin} = 0.086\ s^{-1}$, which was in close agreement with the steady-state k_{cat} (Figure 1(b)). These results establish that the overall rate-limiting step in the trichodiene synthase reaction is k_4, the release of the product trichodiene, while ionization of FPP, k_2, is the slowest enzyme-catalyzed chemical step (Scheme 59). Interestingly, product release has also been shown to be rate-limiting for FPP synthase, with actual FPP formation on the enzyme (4.7 s^{-1}) being some 47 times faster than the off-rate (0.1 s^{-1}). The behavior of the trichodiene synthase D101E mutant was also examined. Although the D101 mutant exhibits only an approximate threefold decrease in the observed steady-state k_{cat}, the slow rate of release of trichodiene turns out to mask a more profound, 100-fold decrease in the actual rate of FPP consumption.

In an attempt to directly observe the proposed intermediate, nerolidyl diphosphate (**2**), single-turnover reactions were carried out at 4 °C, 15 °C, and 30 °C, but under none of these conditions could transient accumulation of NPP be observed, nor was there any measurable lag in the formation of trichodiene on the enzyme. Numerical simulation of the kinetics leads to the conclusion that k_3, the rate of breakdown of NPP, must be greater than 200 s^{-1}.

Deuterium isotope effects on the rate of FPP ionization can be detected under steady-state conditions only indirectly by measuring the effect on k_{cat}/K_m, since the rate of consumption of FPP does not affect the overall rate of turnover, k_{cat}. Using rapid-chemical quench methods on [1-^{2}H,1,2-^{14}C]FPP the deuterium isotope effect on k_2, the rate of consumption of FPP, was found to be 1.11 ± 0.06, consistent with the expected secondary deuterium isotope effect, which destabilizes the allylic cation.

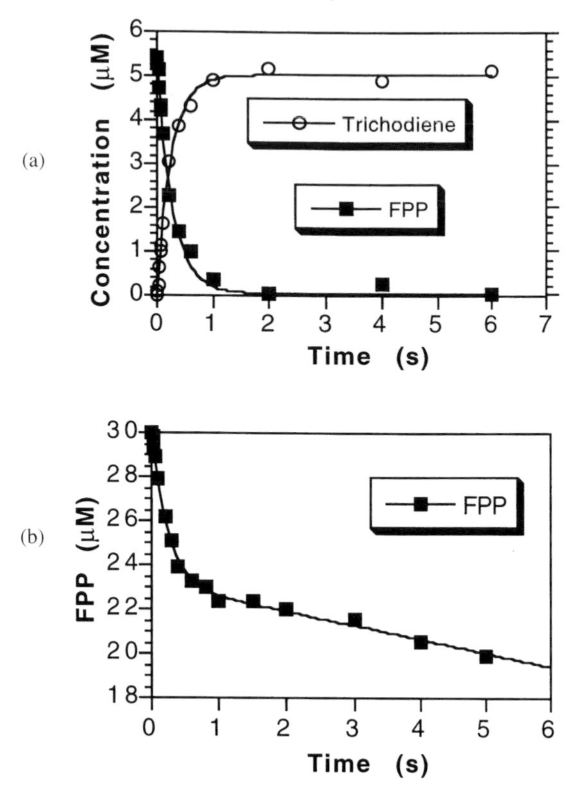

Figure 1 (a) Time course for a single turnover of FPP and production of trichodiene at the active site of trichodiene synthase monitored by rapid chemical quench methods. (b) Time course of presteady-state multiturnover of FPP in 4:1 excess over trichodiene synthase, showing initial burst in consumption of FPP followed by slower linear phase.

$$\mathrm{FPP + E_1} \underset{k_{-1}}{\overset{k_1}{\rightleftharpoons}} \mathrm{FPP{\bullet}E_1} \xrightarrow{k_2} \mathrm{NPP{\bullet}E_2} \xrightarrow{k_3} \mathrm{Trichodiene{\bullet}E} \xrightarrow{k_4} \mathrm{Trichodiene + E_1}$$

Scheme 59

2.06.4.2.2 epi-*Aristolochene synthase and vetispiradiene synthase*

In parallel with the above studies, Chappell, Poulter, Noel and co-workers have examined used rapid quench and isotope trapping experiments to study the presteady-state behavior of recombinant *epi*-aristolochene synthase, vetispiradiene synthase, and a chimeric protein which generates mixtures of *epi*-aristolochene (**81**) and vetispiradiene (**85**).[138] All three cases exhibited burst kinetics, with FPP ionization proceeding 10–70 times faster than ultimate product release.

By adding unlabeled FPP at different time intervals following initial mixing of enzyme and substrate, then allowing the incubation to proceed an additional 30 s before quenching with KOH/EDTA, labeled FPP, which had been released from the enzyme•FPP complex before catalysis, could be trapped by dilution. No difference in the burst rates of product formation were observed between the rapid quench and isotope trapping experiments, leading to the conclusion that there is a rapid preequilibration between free FPP and the enzyme•FPP complex prior to a slower chemical step, the ionization of FPP. Eventual release of the sesquiterpene hydrocarbon product is an order of magnitude or more slower still. By carrying out rapid quench experiments with varying amounts (6.25, 12.5, and 25.0 nM) of FPP and a fixed excess (190 nM) of cyclase, the differences in burst amplitude could be used to calculate the K_D for FPP for both vetispiradiene synthase (69 ± 25 μM) and the chimeric cyclase (18 ± 2 μM). Numerical integration of the differential equations corresponding to the simple kinetic model of Scheme 60 allowed calculation of k_2 (0.3–1.6 s^{-1}), the rate of FPP consumption, and k_3 (0.01–0.03 s^{-1}), the rate of product release, in each case.

$$\text{FPP} + \text{E} \underset{k_{-1}}{\overset{k_1}{\rightleftharpoons}} \text{FPP}\bullet\text{E} \xrightarrow{k_2} \textit{epi}\text{-Aristolochene}\bullet\text{E} \xrightarrow{k_3} \textit{epi}\text{-Aristolochene} + \text{E}$$

Scheme 60

2.06.4.3 Protein Structure

2.06.4.3.1 Pentalenene synthase

In spite of years of mechanistic investigations on terpenoid biosynthesis, including determination of the amino acid sequences of a significant number of enzymes of terpenoid biosynthesis, including prenyl transferases, as well as monoterpene, sesquiterpene, diterpene, and triterpene cyclase, the protein structural basis for the catalysis of these electrophilic cyclizations has remained a mystery. Now two research teams have independently determined the structures of a set of sesquiterpene synthases of both microbial and plant origin, opening the way to a detailed and comprehensive understanding of the mode of action of this synthetically versatile and widely occurring class of enzymes.

In collaboration with Christianson, the structure of pentalenene synthase has been determined to 2.7 Å resolution.[81,82] The recombinant protein crystallized as hexagonal prisms belonging to space group $P6_3$, with hexagonal unit cell dimensions of $a = b = 179.8$ Å, $c = 56.6$ Å. The asymmetric unit contained two 38 kDa monomers. Pentalenene synthase has a globular structure consisting of 11 α helices connected by polypeptide loops of 3–10 residues in length (Figure 2). Remarkably, comparison with the structure of avian farnesyl diphosphate synthase[42,43] revealed a striking similarity in the overall secondary structure and folding patterns of the two enzymes, particularly in the active site region, in spite of the lack of any significant amino acid sequence identity ($<17\%$). Indeed the aspartates 80, 81, and 84 of pentalenene synthase at the C-terminal end of helix C were nearly identically positioned with the corresponding aspartates 117, 118, and 121 of FPP synthase when the two structures are superimposed. The latter three residues have been shown by site-directed mutagenesis and by direct crystallographic analysis to chelate the required Mg^{2+} which bind and activate the allylic diphosphate substrate.

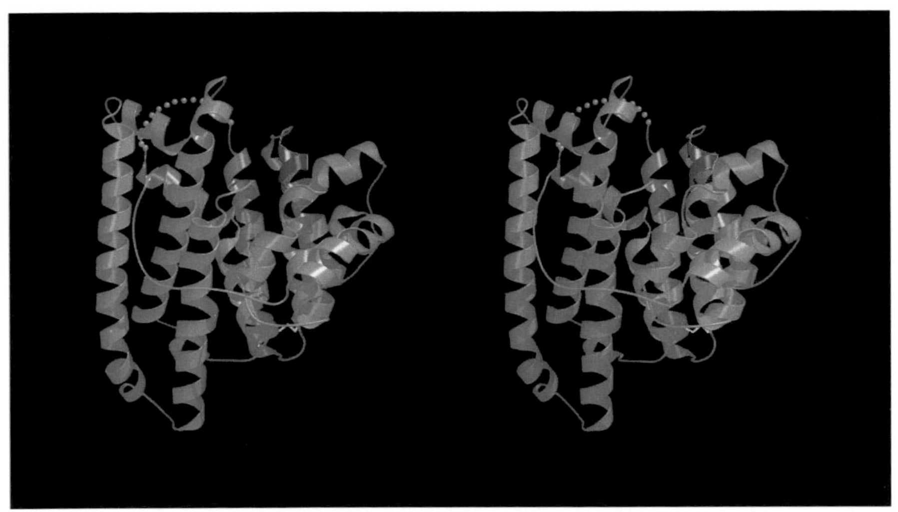

Figure 2 Ribbon plot of pentalenene synthase showing 11 helical domains. The aspartate-rich region with Asp-80, Asp-81, and Asp-84 is colored red.

Flanking the upper regions of the pentalenene synthase active site are several basic residues—Arg-157, Arg-173, Lys-226, and Arg-230, which are well positioned to orient and stabilize the pyrophosphate moiety of FPP (Figure 3). Lining the actual active site cavity are several aromatic amino acids, including Phe-76 and Phe-77 whose π-faces are ideally positioned to stabilize positive charge in the initially generated allylic cation and derived intermediates by strong quadrupole–cation interactions, as predicted by Dougherty.[175] Additional aromatic residues are evident at Trp-308 and Phe-57, located near the C-10 and C-11 carbons of FPP at which positive becomes located at various stages in the cyclization of FPP to pentalenene. A dipole–charge interaction with

Asn-219 may provide additional stabilization of carbo-cation intermediates. This Asn residue may also play an important role in guiding the proper folding of the FPP substrate and the derived humulene.

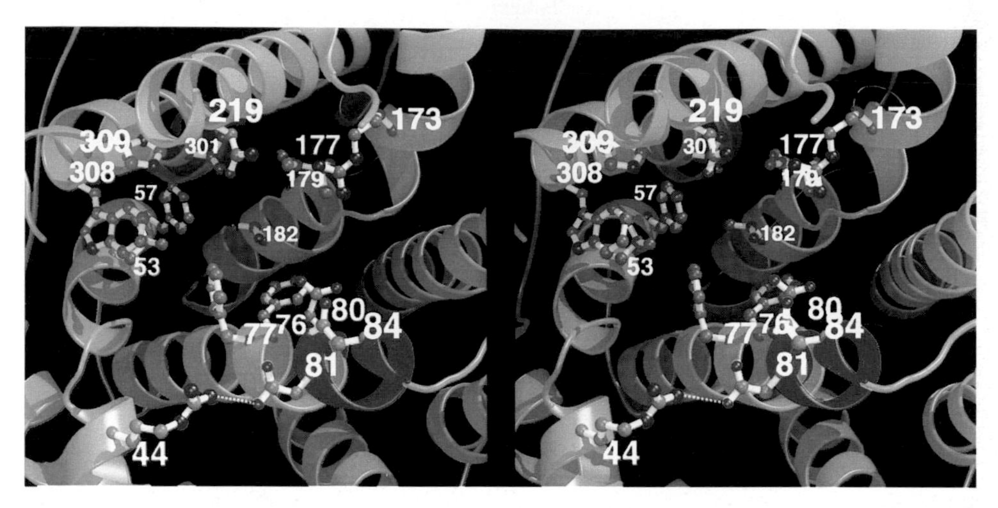

Figure 3 Close up of active site of pentalenene synthase, viewed from top as seen in Figure 2, showing conserved aspratate-rich region and key aromatic residues, as well as the putative active-site base, His-309.

Finally, examination of the crystal structure of pentalenene synthase reveals the presence of a single basic residue, His-309, which is geometrically well placed to serve as the critical base which mediates the proposed deprotonation–reprotonation–deprotonation sequence. The possible role for this histidine is evident from Figure 4, in which FPP and several cyclization intermediates have been modeled into the active site.

2.06.4.3.2 **epi-*Aristolochene synthase***

Chappell, Noel and co-workers have also completed their own study of the crystal structures of *epi*-aristolochene and related plant cyclases.[139] The structures of these enzymes show many of the same features as pentalenene synthase and exhibit a pronounced structural homology to the overall structure of the bacterial protein. The active site of *epi*-aristolochene synthase contains numerous aromatic residues that are proposed to play a role in the stabilization of positively charged intermediates. Interestingly, unlike pentalenene synthase, no histidine residues were evident at the active site. Chappell and Noel have made the intriguing suggestion that the indole moiety of the strategically placed Trp-273 acts as the base that is responsible for the first deprotonation step in the formation of *epi*-aristolochene. In support of this unprecedented but completely plausible suggestion was the finding that replacement of Trp-273 with either aromatic or nonaromatic residues abolishes cyclase activity. Interestingly, a tryptophan residue is also present at position 308 of pentalenene synthase, immediately adjacent to the proposed active site base, His-309.

The determination of the structures of pentalenene synthase and of *epi*-aristolochene synthase has pulled aside the veil which has long hidden the secrets of how terpenoid synthases catalyze the marvelously intricate and precisely controlled cyclization of FPP to complex sesquiterpenes. Armed with the powerful tools of modern molecular genetics and mechanistic enzymology, we now find ourselves able to address fundamental questions of molecular recognition and catalytic control.

ACKNOWLEDGMENTS

Work carried out in the author's laboratory was supported by NIH grants GM30301 and GM22172. I would like to acknowledge the contributions of a talented group of co-workers whose names can be found in the references cited to work from this laboratory, as well as fruitful and enjoyable collaborations with Karen Anderson, Rodney Croteau, Robert Coates, David Christianson, and Thomas Hohn. I would also like to thank Duilio Arigoni, Joe Chappell, Joe Noel, and Dale Poulter for sharing results from their own laboratories prior to publication.

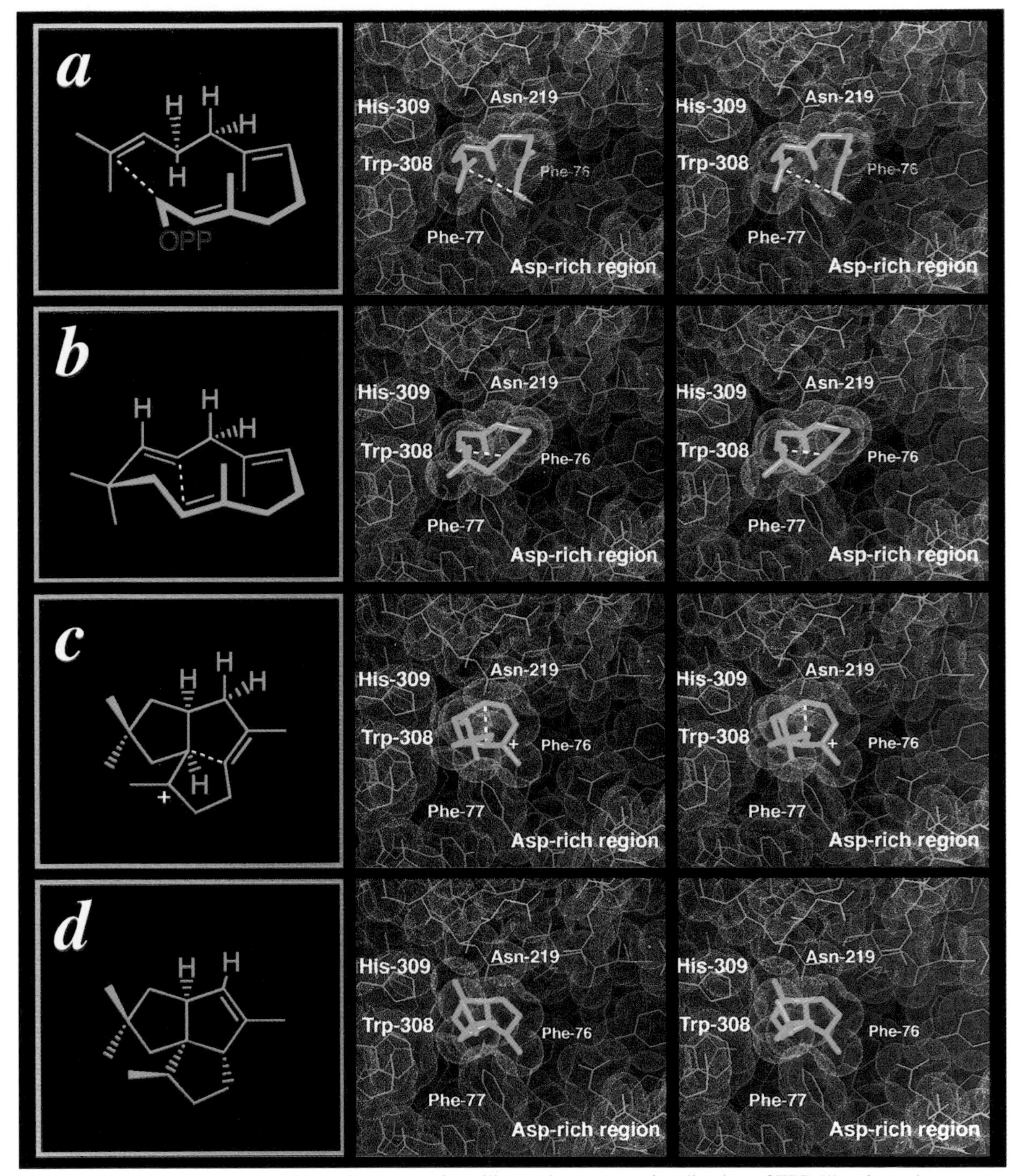

Figure 4 Model of active site of pentalenene synthase illustrating proposed cyclization of FPP (**1**) to humulene (**44**), the protoilludyl cation (**50**), and pentalenene (**49**). Positions of the inorganic pyrophosphate and the bridging MG^{2+} chelated by Asp-80 and Asp-84 are omitted for clarity. Note positions of Phe-76, Phe-77, Trp-308, and Asn-219, as well as active site base, His-309.

2.06.5 REFERENCES

1. J. S. Glasby, "Encyclopedia of Terpenoids," Wiley, Chichester, 1982.
2. D. E. Cane, in "Biosynthesis of Isoprenoid Compounds," eds. J. W. Porter and S. L. Spurgeon, Wiley, New York, 1981, p. 283.
3. R. Croteau and D. E. Cane, in "Methods in Enzymology. Steroids and Isoprenoids," eds. J. H. Law and H. C. Rilling, Academic Press, New York, 1985, p. 383.
4. D. E. Cane, *Chem. Rev.*, 1990, **9**, 1089.

5. D. E. Cane, in "Genetics and Biochemistry of Antibiotic Production," eds. L. C. Vining and C. Stuttard, Butterworth-Heinemann, Boston, MA, 1995, p. 633.
6. D. E. Cane, *Acc. Chem. Res.*, 1985, **18**, 220.
7. P. M. Dewick, *Nat. Prod. Rep.*, 1995, **12**, 507.
8. L. Ruzicka and M. Stoll, *Helv. Chim. Acta*, 1922, **5**, 923.
9. L. Ruzicka, A. Eschenmoser, and H. Heusser, *Experientia*, 1953, **9**, 357.
10. L. Ruzicka, *Proc. Chem. Soc.*, 1959, 341.
11. L. Ruzicka, *Pure Appl. Chem.*, 1963, **6**, 493.
12. W. Parker, J. S. Roberts, and R. Ramage, *Q. Rev., Chem. Soc.*, 1967, **21**, 331.
13. D. Arigoni, *Pure Appl. Chem.*, 1975, **41**, 219.
14. C. Tamm and W. Breitenstein, in "The Biosynthesis of Mycotoxins," ed. P. S. Steyn, Academic Press, New York, 1980, p. 69.
15. J. Fishman, E. R. H. Jones, G. Lowe, and M. C. Whiting, *Proc. Chem. Soc.* London, 1959, 127.
16. B. Achilladelis and J. R. Hanson, *Phytochemistry*, 1968, **7**, 589.
17. D. Arigoni, D. E. Cane, B. Mueller, and C. Tamm, *Helv. Chim. Acta*, 1973, **56**, 2946.
18. S. Nozoe and Y. Machida, *Tetrahedron*, 1972, **28**, 5105.
19. R. Evans, A. M. Holton, and J. R. Hanson, *J. Chem. Soc., Chem. Commun.*, 1973, 465.
20. R. Evans and J. R. Hanson, *J. Chem. Soc., Perkin Trans. 1*, 1976, 326.
21. T. M. Hohn and F. VanMiddlesworth, *Arch. Biochem. Biophys.*, 1986, **251**, 756.
22. T. M. Hohn and M. N. Beremand, *Appl. Environ. Microbiol.*, 1989, **55**, 1500.
23. T. M. Hohn and P. D. Beremand, *Gene*, 1989, **79**, 131.
24. T. M. Hohn and R. D. Plattner, *Arch. Biochem. Biophys.*, 1989, **275**, 92.
25. D. E. Cane, Z. Wu, J. S. Oliver, and T. M. Hohn, *Arch. Biochem. Biophys.*, 1993, **300**, 416.
26. D. E. Cane, S. Swanson, and P. P. N. Murthy, *J. Am. Chem. Soc.*, 1981, **103**, 2136.
27. D. E. Cane, H. Ha, C. Pargellis, F. Waldmeier, S. Swanson, and P. P. N. Murthy, *Bioorg. Chem.*, 1985, **13**, 246.
28. D. E. Cane and H. Ha, *J. Am. Chem. Soc.*, 1986, **108**, 3097.
29. D. E. Cane and H. Ha, *J. Am. Chem. Soc.*, 1988, **110**, 6865.
30. D. E. Cane, G. Yang, Q. Xue, and J. H. Shim, *Biochemistry*, 1995, **34**, 2471.
31. W. W. Cleland, *Biochemistry*, 1990, **29**, 3194.
32. R. Kluger and T. Smyth, *J. Am. Chem. Soc.*, 1981, **103**, 1214.
33. D. E. Cane, H.-T. Chiu, P.-H. Liang, and K. S. Anderson, *Biochemistry*, 1997, **36**, 8332.
34. D. E. Cane, J. L. Pawlak, R. M. Horak, and T. M. Hohn, *Biochemistry*, 1990, **29**, 5476.
35. S. A. Biller, M. J. Sofia, B. Delange, C. Forster, E. M. Gordon, T. Harrity, L. C. Rich, and C. P. Ciosek, *J. Am. Chem. Soc.*, 1991, **113**, 8522.
36. D. E. Cane, G. Yang, R. M. Coates, H. Pyun, and T. M. Hohn, *J. Org. Chem.*, 1992, **57**, 3454.
37. D. E. Cane and G. Yang, *J. Org. Chem.*, 1994, **59**, 5794.
38. R. H. Proctor, T. M. Hohn, and S. P. McCormick, *Mol. Plant–Microbe Interact.*, 1995, **8**, 593.
39. D. E. Cane, J. H. Shim, Q. Xue, B. C. Fitzsimons, and T. M. Hohn, *Biochemistry*, 1995, **34**, 2480.
40. D. E. Cane, J. H. Shim, Q. Xue, B. C. Fitzsimons, and T. M. Hohn, *Biochemistry*, 1997, **36**, 9636.
41. D. N. Brems, E. Bruenger, and H. C. Rilling, *Biochemistry*, 1981, **20**, 3711.
42. L. C. Tarshis, M. J. Yan, C. D. Poulter, and J. C. Sacchettini, *Biochemistry*, 1994, **33**, 10 871.
43. L. C. Tarshis, P. J. Proteau, B. A. Kellogg, J. C. Sacchetini, and C. D. Poulter, *Proc. Natl. Acad. Sci.* USA, 1996, **93**, 15 018.
44. A. Joly and P. A. Edwards, *J. Biol. Chem.*, 1993, **268**, 26 983.
45. D. E. Cane and Q. Xue, *J. Am. Chem. Soc.*, 1996, **118**, 1563.
46. D. E. Cane, Q. Xue, and B. C. Fitzsimons, *Biochemistry*, 1996, **35**, 12 369.
47. P. F. Marrero, C. D. Poulter, and P. A. Edwards, *J. Biol. Chem.*, 1992, **267**, 21 873.
48. T. Koyama, M. Tajima, H. Sano, T. Doi, A. Koike-Takeshita, S. Obata, T. Nishino, and K. Ogura, *Biochemistry*, 1996, **35**, 9533.
49. L. S. Song and C. D. Poulter, *Proc. Natl. Acad. Sci.* USA, 1994, **91**, 3044.
50. D. E. Cane, Q. Xue, J. E. Van Epp, and T. S. Tsantrizos, *J. Am. Chem. Soc.*, 1996, **118**, 8499.
51. K. Nabeta, K. Kawakita, Y. Yada, and H. Okuyama, *Biosci. Biotechnol. Biochem.*, 1993, **57**, 792.
52. A. E. Desjardins, T. M. Hohn, and S. P. McCormick, *Microbiol. Rev.*, 1993, **57**, 595.
53. A. E. Desjardins, R. D. Plattner, and F. VanMiddlesworth, *Appl. Environ. Microbiol.*, 1986, **51**, 493.
54. A. E. Desjardins, R. D. Plattner, and M. N. Beremand, *Appl. Environ. Microbiol.*, 1987, **53**, 1860.
55. A. R. Hesketh, L. Gledhill, B. W. Bycroft, P. M. Dewick, and J. Gilbert, *Phytochemistry*, 1992, **32**, 93.
56. A. R. Hesketh, L. Gledhill, D. C. Marsh, B. W. Bycroft, P. M. Dewick, and J. Gilbert, *J. Chem. Soc., Chem. Commun.*, 1990, 1184.
57. L. O. Zamir, K. A. Devor, N. Morin, and F. Sauriol, *J. Chem. Soc., Chem. Commun.*, 1991, 1033.
58. L. O. Zamir, K. A. Devor, A. Nikolakakis, and F. Sauriol, *J. Biol. Chem.*, 1990, **265**, 6713.
59. L. O. Zamir, K. A. Devor, and F. Sauriol, *J. Biol. Chem.*, 1991, **266**, 14 992.
60. A. R. Hesketh, B. W. Bycroft, P. M. Dewick, and J. Gilbert, *Phytochemistry*, 1992, **32**, 105.
61. T. M. Hohn, S. P. McCormick, and A. E. Desjardins, *Curr. Gene.*, 1993, **24**, 291.
62. T. M. Hohn, A. E. Desjardins, and S. P. McCormick, *Mol. Gen. Genet.*, 1995, **248**, 95.
63. L. Gledhill, A. R. Hesketh, B. W. Bycroft, P. M. Dewick, and J. Gilbert, *FEMS Microbiol. Lett.*, 1991, **81**, 241.
64. K. H. Overton and D. J. Picken, *J. Chem. Soc., Chem. Commun.*, 1976, 105.
65. P. Anastasis, I. Freer, C. Gilmore, H. Mackie, K. Overton, and S. Swanson, *J. Chem. Soc., Chem. Commun.*, 1982, 268.
66. D. E. Cane and G. G. S. King, *Tetrahedron Lett.*, 1976, 4737.
67. S. Nozoe, H. Kobayashi, and N. Morisaki, *Tetrahedron Lett.*, 1976, 4437.
68. D. E. Cane, D. B. McIlwaine, and P. H. M. Harrison, *J. Am. Chem. Soc.*, 1989, **111**, 1152.
69. D. E. Cane, D. B. McIlwaine, and J. S. Oliver, *J. Am. Chem. Soc.*, 1990, **112**, 1285.
70. D. E. Cane and X. Lu, unpublished.

71. D. E. Cane and D. B. McIlwaine, *Tetrahedron Lett.*, 1987, **28**, 6545.
72. D. E. Cane and S. L. Buchwald, *J. Am. Chem. Soc.*, 1977, **99**, 6132.
73. S. S. Dehal and R. Croteau, *Arch. Biochem. Biophys.*, 1988, **261**, 346.
74. R. Croteau and A. Gundy, *Arch. Biochem. Biophys.*, 1984, **233**, 838.
75. D. E. Cane, T. Rossi, A. M. Tillman, and J. P. Pachlatko, *J. Am. Chem. Soc.*, 1981, **103**, 1838.
76. D. Arigoni, M. Schwarz, and S. Eppacher, personal communication, 1996.
77. H. Seto, H. Watanabe, and K. Furihata, *Tetrahedron Lett.*, 1996, **37**, 7979.
78. H. Seto and H. Yonehara, *J. Antibiot.*, 1980, **33**, 92.
79. D. E. Cane and A. M. Tillman, *J. Am. Chem. Soc.*, 1983, **105**, 122.
80. D. E. Cane, J. Sohng, C. R. Lamberson, S. M. Rudnicki, Z. Wu, M. D. Lloyd, J. S. Oliver, and B. R. Hubbard, *Biochemistry*, 1994, **33**, 5846.
81. C. A. Lesburg, M. D. Lloyd, D. E. Cane, and D. W. Christianson, *Protein Sci.*, 1995, **4**, 2436.
82. C. A. Lesburg, G. Zhai, D. E. Cane, and D. W. Christianson, *Science*, 1997, **277**, 1820.
83. D. E. Cane, J. S. Oliver, P. H. M. Harrison, C. Abell, B. R. Hubbard, C. T. Kane, and R. Lattman, *J. Am. Chem. Soc.*, 1990, **112**, 4513.
84. P. H. M. Harrison, J. S. Oliver, and D. E. Cane, *J. Am. Chem. Soc.*, 1988, **110**, 5922.
85. D. E. Cane, C. Abell, and A. M. Tillman, *Bioorg. Chem.*, 1984, **12**, 312.
86. D. E. Cane, C. Abell, R. Lattman, C. T. Kane, B. R. Hubbard, and P. H. M. Harrison, *J. Am. Chem. Soc.*, 1988, **110**, 4081.
87. D. E. Cane and S. W. Weiner, *Can. J. Chem.*, 1994, **72**, 118.
88. D. E. Cane and G. Zhai, unpublished.
89. D. E. Cane and M. D. Lloyd, unpublished.
90. H. Seto, T. Sasaki, H. Yonehara, S. Takahashi, M. Takeuchi, H. Kuwano, and M. Arai, *J. Antibiot.*, 1984, **37**, 1076.
91. D. E. Cane, J. K. Sohng, and P. G. Williard, *J. Org. Chem.*, 1992, **57**, 844.
92. W. A. Ayer and M. H. Saeedi-Ghomi, *Can. J. Chem.*, 1981, **59**, 2536.
93. W. A. Ayer, T. T. Nakashima, and M. H. Saeedi-Ghomi, *Can. J. Chem.*, 1984, **62**, 531.
94. C. Abell and A. P. Leech, *Tetrahedron Lett.*, 1987, **28**, 4887.
95. C. Abell and A. P. Leech, *Tetrahedron Lett.*, 1988, **29**, 4337.
96. H.-W. Fehlhaber, R. Geipal, H.-J. Mercker, R. Tschesche, and Welmar, *Chem. Ber.*, 1974, **107**, 1720.
97. A. P. W. Bradshaw, J. R. Hanson, and R. Nyfeler, *J. Chem. Soc., Perkin Trans. 1*, 1981, 1469.
98. J. R. Hanson and R. Nyfeler, *J. Chem. Soc., Chem. Commun.*, 1976, 72.
99. A. P. W. Bradshaw, J. R. Hanson, R. Nyfeler, and I. H. Sadler, *J. Chem. Soc., Chem. Commun.*, 1981, 649.
100. D. E. Cane, Y. G. Whittle, and T. Liang, *Tetrahedron Lett.*, 1984, **25**, 1119.
101. D. E. Cane, Y. G. Whittle, and T. Liang, *Bioorg. Chem.*, 1986, **14**, 417–428.
102. A. Hirota, M. Nakagawa, H. Sakai, and A. Isogai, *Agric. Biol. Chem.*, 1984, **48**, 835.
103. J. M. Beale, R. L. Chapman, and J. P. N. Rosazza, *J. Antibiot.*, 1984, **37**, 1376.
104. A. Hirota, M. Nakagawa, H. Sakai, A. Isogai, K. Furihata, and H. Seto, *Tetrahedron Lett.*, 1985, **26**, 3845.
105. M. Klobus, L. J. Zhu, and R. M. Coates, *J. Org. Chem.*, 1992, **57**, 4327.
106. F. Bohlmann, C. Zdero, J. Jakupovic, H. Robinson, and R. King, *Phytochemistry*, 1981, **20**, 2239.
107. A. P. W. Bradshaw, J. R. Hanson, and I. H. Sadler, *J. Chem. Soc., Chem. Commun.*, 1981, 631.
108. A. P. W. Bradshaw, J. R. Hanson, and I. H. Sadler, *J. Chem. Soc., Perkin Trans. 1*, 1982, 2787.
109. A. P. W. Bradshaw, J. R. Hanson, and I. H. Sadler, *J. Chem. Soc., Chem. Commun.*, 1982, 292.
110. D. E. Cane, B. J. Rawlings, and C. Yang, *J. Antibiot.*, 1987, **40**, 1331.
111. S. Moreau, J. Biguet, A. Lablache-Combier, F. Baert, M. Foulon, and C. Delfosse, *Tetrahedron*, 1980, **36**, 2989.
112. S. Tanaka, K. Wada, S. Marumo, and H. Hattori, *Tetrahedron Lett.*, 1984, **25**, 5907.
113. T. R. Govindachari, P. A. Mohamed, and P. C. Parthasarathy, *Tetrahedron*, 1970, **26**, 615.
114. R. Baker, H. R. Cole, M. Edwards, D. A. Evans, P. E. Howse, and S. Walmsley, *J. Chem. Ecol.*, 1981, **7**, 135.
115. T. M. Hohn and R. D. Plattner, *Arch. Biochem. Biophys.*, 1989, **272**, 137.
116. R. H. Proctor and T. M. Hohn, *J. Biol. Chem.*, 1993, **268**, 4543.
117. D. E. Cane, Z. Wu, R. H. Proctor, and T. M. Hohn, *Arch. Biochem. Biophys.*, 1993, **304**, 415.
118. D. E. Cane and I. Kang, unpublished.
119. A. R. Penfold and J. L. Simonsen, *J. Chem. Soc.*, 1939, 87.
120. R. Robinson, "The Structural Relations of Natural Products," Clarendon, Oxford, 1955, p. 12.
121. A. A. Chalmers, A. E. De Jesus, C. P. Gorst-Allman, and P. S. Steyn, *J. Chem. Soc., Perkin Trans. 1*, 1981, 2899.
122. S. Moreau, A. Lablache-Combier, and J. Biguet, *Phytochemistry*, 1981, **20**, 2339.
123. C. P. Gorst-Allman and P. S. Steyn, *Tetrahedron Lett.*, 1982, **23**, 5359.
124. D. E. Cane, P. C. Prabhakaran, E. J. Salaski, P. H. M. Harrison, H. Noguchi, and B. J. Rawlings, *J. Am. Chem. Soc.*, 1989, **111**, 8914.
125. D. E. Cane, P. C. Prabhakaran, J. S. Oliver, and D. B. McIlwaine, *J. Am. Chem. Soc.*, 1990, **112**, 3209.
126. D. E. Cane and Y. S. Tsantrizos, *J. Am. Chem. Soc.*, 1996, **118**, 10 037.
127. D. E. Cane and C. Bryant, *J. Am. Chem. Soc.*, 1994, **116**, 12 063.
128. J. A. Bailey, R. S. Burden, and G. G. Vincent, *Phytochemistry*, 1975, **14**, 597.
129. E. W. B. Ward, C. H. Unwin, G. L. Rock, and A. Stoessl, *Can. J. Bot.*, 1976, **54**, 25.
130. A. Stoessl, C. H. Unwin, and E. W. B. Ward, *Phytopathol. Z.*, 1972, **74**, 141.
131. I. M. Whitehead, D. F. Ewing, and D. R. Threlfall, *Phytochemistry*, 1988, **27**, 1365.
132. I. M. Whitehead, D. R. Threlfall, and D. F. Ewing, *Phytochemistry*, 1989, **28**, 775.
133. I. M. Whitehead, P. C. Prabhakaran, D. F. Ewing, D. E. Cane, and D. R. Threlfall, *Phytochemistry*, 1990, **29**, 479.
134. U. Vögeli and J. Chappell, *Plant Physiol.*, 1988, **88**, 1291.
135. U. Vögeli, J. W. Freeman, and J. Chappell, *Plant Physiol.*, 1990, **93**, 182.
136. P. J. Facchini and J. Chappell, *Proc. Natl. Acad. Sci.* USA, 1992, **89**, 11 088.
137. K. Back, S. Yin, and J. Chappell, *Arch. Biochem. Biophys.*, 1994, **315**, 527.
138. J. R. Mathis, K. Back, C. Starks, J. Noel, C. D. Poulter, and J. Chappell, *Biochemistry*, 1997, **36**, 8340.
139. C. M. Starks, K. Back, J. Chappell, and J. P. Noel, *Science*, 1997, **277**, 1815.

140. F. C. Baker, C. J. W. Brooks, and S. A. Hutchinson, *J. Chem. Soc., Chem. Commun.*, 1975, 293.
141. F. C. Baker and C. J. W. Brooks, *Photochemistry*, 1976, **15**, 689.
142. Y. Hoyano, A. Stoessl, and J. B. Stothers, *Can. J. Chem.*, 1980, **58**, 1894.
143. K. Back and J. Chappell, *J. Biol. Chem.*, 1995, **270**, 7375.
144. K. W. Back and J. Chappell, *Proc. Natl. Acad. Sci.* USA, 1996, **93**, 6841.
145. G. I. Birnbaum, C. P. Huber, M. L. Post, J. B. Stothers, J. R. Robinson, A. Stoessl, and E. W. B. Ward, *J. Chem. Soc., Chem. Commun.*, 1976, 330.
146. A. Murai, S. Sato, A. Osada, N. Katsui, and T. Masamune, *J. Chem. Soc., Chem. Commun.*, 1982, 32.
147. A. Stoessl and J. B. Stothers, *J. Chem. Soc., Chem. Commun.*, 1982, 880.
148. P. A. Brindle, T. Coolbear, P. J. Kuhn, and D. R. Threlfall, *Phytochemistry*, 1985, **24**, 1219.
149. P. Weyerstahl, H.-D. Splittgerber, J. Walteich, and T. Wolny, *J. Essential Oil Res.*, 1989, **1**, 1.
150. R. Croteau, S. L. Munck, C. C. Akoh, H. J. Fisk, and D. M. Satterwhite, *Arch. Biochem. Biophys.*, 1987, **256**, 56.
151. A. Akhila, P. K. Sharma, and R. S. Thakur, *Phytochemistry*, 1987, **26**, 2705.
152. S. L. Munck and R. Croteau, *Arch. Biochem. Biophys.*, 1990, **282**, 58.
153. L. Belingheri, A. Cartayrade, G. Pauly, and M. Gleizes, *Plant Sci.*, 1992, **84**, 129.
154. G. D. Davis and M. Essenberg, *Phytochemistry*, 1995, **39**, 553.
155. E. M. Davis, J. Tsuji, G. D. Davis, M. L. Pierce, and M. Essenberg, *Phytochemistry*, 1996, **41**, 1047.
156. X. Chen, Y. Chen, P. Heinstein, and V. J. Davisson, *Arch. Biochem. Biophys.*, 1995, **324**, 255–266.
157. X. Y. Chen, M. S. Wang, Y. Chen, V. J. Davisson, and P. Heinstein, *J. Nat. Prod.*, 1996, **59**, 944.
158. C. R. Benedict, I. Alchanati, P. J. Harvey, J. Liu, R. D. Stipanovic, and A. A. Bell, *Phytochemistry*, 1995, **39**, 326.
159. P. F. Heinstein, D. L. Herman, S. B. Tove, and F. B. Smith, *J. Biol. Chem.*, 1970, **245**, 4658.
160. R. Masciadri, W. Angst, and D. Arigoni, *J. Chem. Soc., Chem. Commun.*, 1985, 1573.
161. G. D. Davis, E. J. Eisenbraun, and M. Essenberg, *Phytochemistry*, 1991, **30**, 197.
162. R. D. Stipanovic, A. Stoessl, J. B. Stothers, D. W. Altman, A. A. Bell, and P. Heinstein, *J. Chem. Soc., Chem. Commun.*, 1986, 100.
163. M. Essenberg, A. Stoessl, and J. B. Stothers, *J. Chem. Soc., Chem. Commun.*, 1985, 556.
164. K. Nabeta, T. Ishikawa, T. Kawae, and H. Okuyama, *J. Chem. Soc., Perkin Trans. 1*, 1994, 3277.
165. N. N. Gerber, *Phytochemistry*, 1971, **10**, 185.
166. K. Nabeta, K. Katayama, S. Nakagawara, and K. Katoh, *Phytochemistry*, 1993, **32**, 117.
167. D. E. Cane, M. Tandon, and P. C. Prabhakaran, *J. Am. Chem. Soc.*, 1993, **115**, 8103.
168. D. E. Cane and M. Tandon, *J. Am. Chem. Soc.*, 1995, **117**, 5602.
169. D. E. Cane, *Tetrahedron*, 1980, **36**, 1109.
170. D. E. Cane and M. Tandon, *Tetrahedron Lett.*, 1994, **35**, 5355.
171. K. Nabeta, K. Kigure, M. Fujita, T. Nagoya, T. Ishikawa, H. Okuyama, and T. Takasawa, *J. Chem. Soc., Perkin Trans. 1*, 1995, 1935.
172. L. T. Burka and A. Thorsen, *Phytochemistry*, 1982, **21**, 869.
173. K. Nabeta, Y. Ara, Y. Aoki, and M. Miyake, *J. Nat. Prod.*, 1990, **53**, 1241.
174. E. I. Graziani and R. J. Andersen, *J. Am. Chem. Soc.*, 1996, **118**, 4701.
175. D. A. Dougherty, *Science*, 1996, **271**, 163.

2.07 Cloning and Expression of Terpene Synthase Genes

THOMAS M. HOHN
National Center for Agricultural Utilization Research, USDA/ARS Peoria, IL, USA

2.07.1 INTRODUCTION

Terpenoid biosynthesis is widespread in nature and occurs in bacteria, fungi, plants, and several phyla of invertebrates. Terpenoid structures have an enormous range of complexity and perform equally diverse biological functions.[1] Biosynthetic pathways for mono- (C_{10}), sesqui- (C_{15}), and diterpenoids (C_{20}) branch off from the general isoprenoid pathway, beginning with the linear isoprenoid pathway intermediates geranyl diphosphate (GPP), farnesyl diphosphate (FPP), and geranylgeranyl diphosphate (GGPP), respectively. Research since the 1970s has demonstrated the existence of a group of mechanistically related enzymes, known as terpene synthases, that catalyze the first step in terpenoid pathways.[2,3] Evidence for terpene synthase involvement in terpenoid biosynthesis is now sufficient to consider these enzymes a defining feature of terpenoid pathways. Because terpene synthases are located at the branch points of terpenoid and isoprenoid pathways, these enzymes are frequently the focus of studies aimed at understanding the regulation and function of terpenoid pathways.

Isolation of terpene synthase genes has played an increasingly important role in studies of terpenoid biosynthesis and function. Terpene synthase genes have now been isolated from actinomycete,[4] fungal,[5,6] and plant[7–15] sources, providing insights into both enzyme structure and evolution. Expression of terpene synthase genes in *Escherichia coli* has provided a means for obtaining large quantities of enzyme for research purposes, and permitted the use of various molecular tools such as site-directed mutagenesis for studies of structure–function relationships.[16,17] This chapter will discuss the progress in the cloning and heterologous expression of terpene synthase genes with an emphasis on fungal enzymes.

2.07.2 CLONING OF TERPENE SYNTHASE GENES

One of the most important discoveries in terpene synthase research is the observation that all of the known plant terpene synthases appear to be closely related.[10,12] Plant terpene synthase similarities include regions of sequence conservation and the positioning of intron sequences.[18] Recognition of plant terpene synthase sequence similarities has had important implications for studies of these enzymes and efforts to isolate plant terpene synthase genes. In contrast, no such relationships were observed among the three microbial terpene synthase sequences that have been described.[4,6] Terpene synthase gene families may exist within fungi and actinomycetes, but evidence of these relationships must await the isolation of additional terpene synthase genes. Because the basic strategies for gene isolation currently differ in plants and microorganisms, discussions of terpene synthase gene cloning are organized according to enzyme source.

2.07.2.1 Fungi

Fungi produce thousands of different sesqui-, sester-, and diterpenoids but remarkably few monoterpenoids. Although fungi have been reported to produce low levels of monoterpenes,[19,20] there is no evidence that fungi possess monoterpene synthases. Interest in fungal terpenoids has been driven largely by the fact that several fungal sesquiterpenoids are mycotoxins. Sesquiterpenoid mycotoxin biosynthesis has been investigated in pathways beginning with either trichodiene (**1**) or aristolochene (**2**). Trichodiene-derived mycotoxins consist primarily of trichothecenes which represent the greatest problem for agriculture among the sesquiterpenoid mycotoxins,[21] while aristolochene-derived mycotoxins include PR toxin (**3**) and appear to be of limited agricultural importance. In this section, research on trichodiene and aristolochene synthases will be discussed along with strategies for fungal terpene synthase gene isolation.

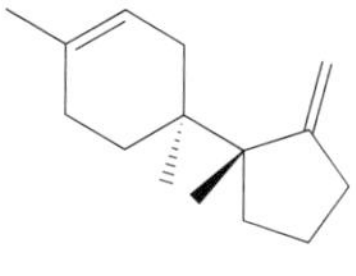

AcO O O O H O

(1) 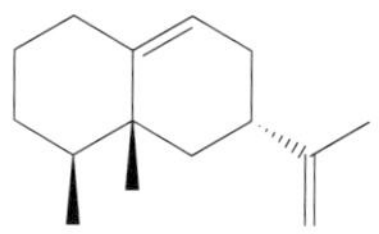 **(2)** 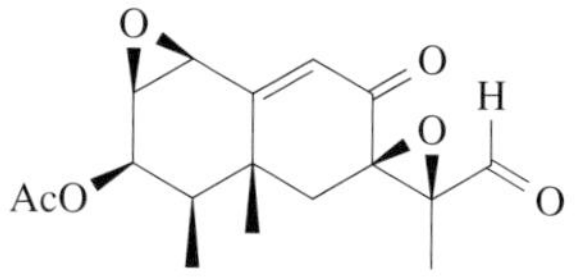 **(3)**

2.07.2.1.1 Trichodiene synthase

Trichothecenes are sesquiterpenoid toxins produced by members of at least 10 genera of fungi[22,23] and are also known to accumulate at high levels in two species of the plant genus *Baccharis*.[24] Between 80 and 100 different trichothecenes have been identified from natural sources, all of which share the core trichothecene structure (**4**). Macrocyclic trichothecenes, such as roridin E (**5**), represent the most structurally complex trichothecenes and are characterized by the presence of a macrocycle ring esterified through hydroxy groups at C-4 and C-15.

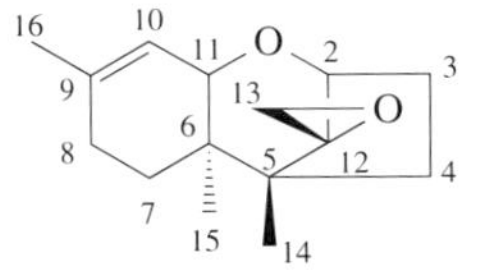

(4) 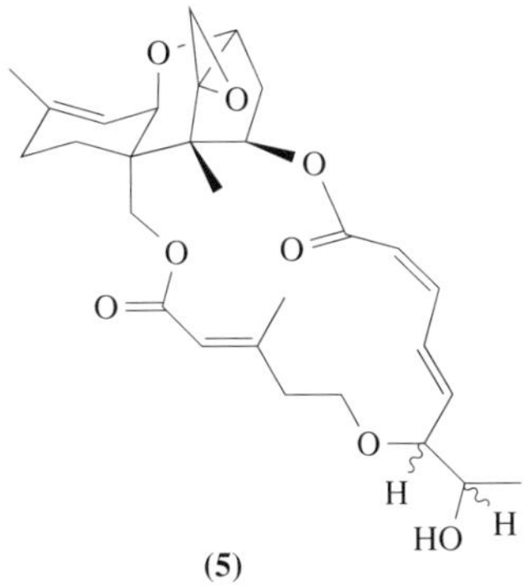 (5)

Trichothecenes such as T-2 toxin (**6**) and deoxynivalenol (**7**) are typical of the trichothecenes produced by *Fusarium* species in grains. Accumulation of trichothecenes in grains is an important agricultural problem because these sesquiterpenoids are highly toxic to vertebrates. Trichothecenes are also toxic to plants, and considerable evidence now points to trichothecene production as a virulence factor in *Fusarium* diseases of wheat,[25,26] and maize (L. Harris, A. E. Desjardins, T. M. Hohn, R. P. Proctor, and R. Plattner, unpublished results). Trichothecenes are not the only trichodiene metabolites produced by fungi. A number of apparent shunt metabolites that include apotrichothecenes[27] (**8**) and sambucinol[28] (**9**) are also commonly produced by *Fusarium* species. The function and agricultural significance of these metabolites are presently unknown.

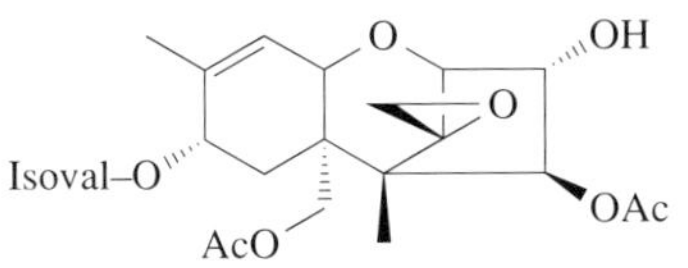

(6) 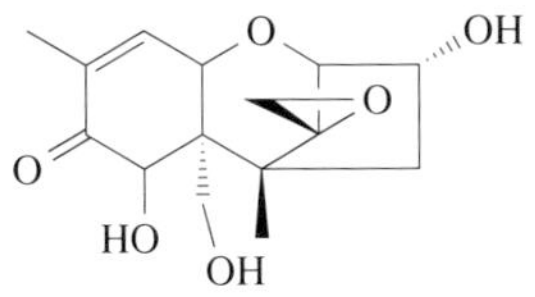 (7)

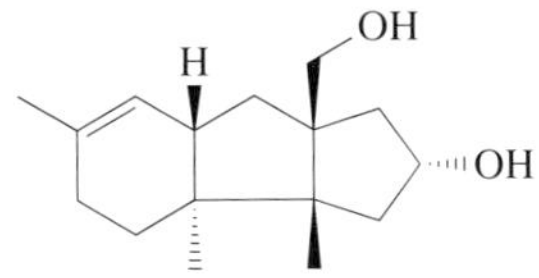

(8) 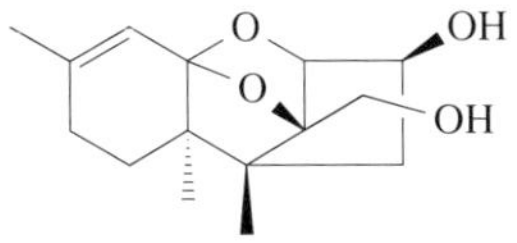 (9)

Trichodiene synthase (TS) catalyzes the formation of the trichothecene precursor, trichodiene[29] (**1**). TS has been characterized from both *Fusarium*[30] and *Trichothecium*[31] species. Purification of TS has only been reported for *F. sporotrichioides*.[30] A hybrid enzyme containing the N terminal 309 amino acids (aa) of the *F. sporotrichioides* TS and the C terminal 74 aa of the *Gibberella pulicaris* enzyme has been purified following expression in *E. coli*.[16] *F. sporotrichioides* cultures grown in complex media with nitrogen as a growth-limiting nutrient[32] yield cell homogenates containing levels of TS approaching 0.5% of the total soluble protein.[30] TS activity is first detected under these growth conditions at about 21 h postinoculation and increases to maximum levels within 2–3 h.[33] The appearance of TS is closely coupled with the initial detection of trichothecenes. TS was purified from cell homogenates from *F. sporotrichioides*[30] and shown to consist of an M_r 45 000 polypeptide by sodium dodecyl sulfate–polyacrylamide gel electrophoresis (SDS–PAGE) (see Chapter 2.06).

The TS gene (*FSTri5*) was subsequently isolated from *F. sporotrichioides* using a TS-specific antibody to screen a λgt11 genomic expression library.[5] TS gene identification was confirmed by comparisons between the *FSTri5* deduced amino acid sequence and TS peptide sequences obtained from purified TS. Further confirmation of TS gene identity was obtained following expression of the *FSTri5* coding sequence in *E. coli*.[34] Additional genes encoding TS have been isolated from several different fungal sources using either the *FSTri5* as a hybridization probe[35,36] or the polymerization chain reaction (PCR).[26]

Comparisons between the *Tri5* nucleotide and amino acid sequences from *Fusarium* species show that they share approximately 90–95% identity (Figure 1). The most notable features are the positional conservation of the single intron that interrupts codon 157 and the sequence divergence that occurs within the C terminal 20–25 aa of each sequence. It appears that the sequence of a particular sesquiterpene synthase can change significantly in distantly related fungi. When the *Tri5* gene from *Myrothecium roridum* (*MRTri5*) is compared to *Fusarium* TS gene sequences, the amount of sequence identity drops to between 70% for nucleic acid sequences and 75% for amino acid sequences.[36] This level of identity was still sufficient to permit identification of *MRTri5* using the *FSTri5* gene as a hybridization probe, but required hybridization conditions employing reduced stringency. Thus, the use of terpene synthases as hybridization probes to identify the same terpene synthase in a distantly related fungus may require altered hybridization conditions to overcome differences in the two sequences.

```
Fs M E N F P T E Y F L N T T V R L L E Y I R Y R D S N Y T R E E R I E N L H Y A Y 40
Gp M E N F P T E Y F L N T S V R L L E Y I R Y R D S N Y T R E E R I E N L H Y A Y 40
Gz M E N F P T E Y F L N T S V R L L E Y I R Y R D S N Y T R E E R I E N L H Y A Y 40
Fp M E N F P T E Y F L N T S V R L L E Y I R Y R D S N Y T R E E R I E N L H Y A Y 40
Mr M D E F P T E Y F L G T A V R L L E N V K Y R D S N Y T R E E R I E N L S Y A N 40

Fs N K A A H H F A Q P R Q Q Q L L K V D P K R L Q A S L Q T I V G M V V Y S W A K 80
Gp N K A A H H F A Q P R Q Q Q L L K V D P K R L Q A S L Q T I V G M V V Y S W A K 80
Gz N K A A H H F A Q P R Q Q Q M L K V D P K R L Q A S L Q T I V G M V V Y S W A K 80
Fp N K A A H H F A Q P R Q Q Q L L K V D P K R L Q A S L Q T I V G M V V Y S W A K 80
Mr N K A A A H F A Q E R Q Q R I L K V N P K R L E A S L R T I V G M V V Y S W V K 80

                                         *
Fs V S K E C M A D L S I H Y T Y T L V L D D S K D D P Y P T M V N Y F D D L Q A G 120
Gp V S K E C M A D L S I H Y T Y T L V L D D S S D D P Y P A M M N Y F N D L Q A G 120
Gz V S K E C M A D L S I H Y T Y T L V L D D S S D D P H P A M L N Y F D D L Q A G 120
Fp V S K E C M A D L S I H Y T Y T L V L D D S S D D P Y A A M M N Y F N D L Q A G 120
Mr V S K E L M A D L S I H Y T Y T L I L D D S E D D P H N N M L T F F D D L Q A G 120

Fs R E Q A H P W W A L V N E H F P N V L R H F G P F C S L N L I R S T L D F F E G 160
Gp R E Q A H P W W A L V N E H F P N V L R H F G P F C S L N L I R S T L D F F E G 160
Gz R E Q A H P W W A L V N E H F P N V L R H F G P F C S L N L I R S T M D F F E G 160
Fp R E Q A H P W W A L V N E H F P N V L R H F G P F C S L N L I R S T L D F F E G 160
Mr R E Q K H P W W M L V N E H F P N V L R H F G P F C S L N L I R S T L D F F E G 160

Fs C W I E Q Y N F G G F P G S H D Y P Q F L R R M N G L G H C V G A S L W P K E Q 200
Gp C W I E Q Y N F G G F P G S H D Y P Q F L R R M N G L G H C V G A S L W P K E Q 200
Gz C W I E Q Y N F G G F P G S D D Y P Q F L R R M N G L G H C V G A S L W P K D L 200
Fp C W I E Q Y N F G G F P G S H D Y P Q F L R R M N G L G H C V G A S L W P K E Q 200
Mr C W I E Q Y N F H G F P G S Y D F P G F L R R M N G L G H C V G G S L W P K E L 200

Fs F N E R S L F L E I T S A I A Q M E N W M V W V N D L M S F Y K E F D D E R D Q 240
Gp F D E R G L F L E I T S A I A Q M E N W M V W V N D L M S F Y K E F D D E R D Q 240
Gz F D E R K H F L E I T S A V A Q M E N W M V W V N D L M S F Y K E F D D E R D Q 240
Fp F D E R S L F L E I T S A I A Q M E N W M V W V N D L M S F Y K E F D D E R D Q 240
Mr F D E Q K H F L E I T S A V A Q M E N W M V W V N D L M S F Y K E F D D P R D Q 240

Fs I S L V K N Y V V S D E I S L H E A L E K L T Q D T L H S S K Q M V A V F S D K 280
Gp I S L V K N Y V V S D E I T L H E A L E K L T Q D T L H S S K Q M V A V F S E K 280
Gz I S L V K N F V T C H E I T L D E A L E K L T Q E T L H S S K Q M V A V F S D K 280
Fp I S L V K N Y V V S D E I S L H E A L E K L T Q D T L H S S K Q M V A V F S D K 280
Mr T S L V K N Y V V C D E I S L T Q A L E K L T V D T L T S S E Q M M E V F S D K 280

                                           **
Fs D P Q V M D T I E C F M H G Y V T W H L C D R R Y R L S E I Y E K V K E E K T E 320
Gp D P Q V M D T I E C F M H G Y V T W H L C D H R Y R L N E I Y E K V K G Q K T E 320
Gz D P Q V M D T I E C F M H G Y V T W H L C D A R Y R L H E I Y E K V K D Q D T E 320
Fp D P Q V M V T I E C F M H G Y V T W H L C D H R Y R L N E I S E K V K E Q K T E 320
Mr D A K L M E T I E C F M H G Y I T W H L C D H R Y R L K E V Y E G T M H I E T E 320

Fs D A Q K F C K F Y E Q A A N V G A V S P S E W A Y P P V A Q L A N V R S K D V K 360
Gp D A E K F C K F Y E Q A A N V G A V S P S E W A Y P P I A Q L A N I R T K D V K 360
Gz D A K K F C K F F E Q A A N V G A V A P S E W A Y P Q V A Q L A N V R A K D - - 359
Fp D A Q K F C K F Y E Q A A N V G A V S P S E W A Y P P V A Q L A N V R S K D V - 360
Mr D A I K F R K F Y G Q A A K V G A I E H E E W A F P T V A E R I E V R L A E E K 360

Fs E - - - - - - - - - V Q K P F L S - - S I E L V E                           374
Gp D L K D V K D L K E I Q K P L L S - - S I E L V E                           383
Gz - - - - - - D V K E A Q K P I L S - - S I E L V E                           375
Fp - - - - - K N V K Q I E K P L L S - - S I E L V E                           377
Mr A A K D G Q A V L T S A E P A V P Q A A Q E V L A                           385
```

Figure 1 Alignment of predicted amino acid sequences for five trichodiene synthase genes. Key to abbreviations for enzyme sources together with GenBank accession numbers: Fs, *F. sporotrichioides* (M27246); Gp, *Gibberella pulicaris* (M64348); Gz, *Gibberella zeae* (U22464); Fp, *Fusarium poae* (U15658); Mr, *Myrothecium roridum* (AF009416). The locations of the DDXXD (*) and RYR (**) motifs are indicated.

Trichothecene production within the genus *Fusarium* appears restricted to a group of approximately 10–15 closely related species. Attempts have been made to identify *Tri5*-related genes in other *Fusarium* species such as *F. solani*, *F. oxysporum*, and *F. moniliforme* that do not produce trichothecenes using both Southern blotting with *FSTri5* as the probe under stringent hybridization conditions and PCR with degenerate primer pairs capable of amplifying TS gene fragments (T. M. Hohn, unpublished results). These efforts have been unsuccessful, suggesting that trichothecene pathway genes are absent from nonproducing species of *Fusarium*. The failure of these experiments also argues against the presence of other closely related fungal terpene synthases. Many *Fusarium* species are known to produce terpenoids other than trichothecenes, including cyclonerodiol[37] and gibberellins.[38]

Additional genes in the trichothecene pathway are closely linked to *FSTri5* in *F. sporotrichioides*. Clustering for fungal natural product pathway genes has only been recognized since the early 1990s.[39] Trichothecenes are the first example of a fungal terpenoid pathway gene cluster. At least eight pathway-related genes are clustered around *FSTri5* within a 25 kilobase region of DNA[40] (Figure 2, T. M. Hohn, unpublished results). In addition to *FSTri5*, five genes encoding enzymes involved in trichothecene biosynthesis have been identified. Two genes, *FSTri4*[41] and *FSTri11*,[42,43] encode cytochrome-P450 monooxygenases involved in the addition of hydroxy groups, while of the remaining three genes, *FSTri3* appears to encode an acetyltransferase[44] and the enzymatic functions of *FSTri7* and *FSTri8* have not yet been determined (T. M. Hohn and S. P. McCormick, unpublished results). Genes encoding a positive-acting transcription factor (*FSTri6*) that is required for pathway gene expression[45] and a probable transport protein (*FSTri12*) (N. J. Alexander, T. M. Hohn, and S. P. McCormick, unpublished results) have also been identified. Several lines of evidence indicate that the expression of trichothecene pathway genes is coordinately regulated.[33,45] Efforts to characterize additional trichothecene pathway genes within the cluster are in progress.

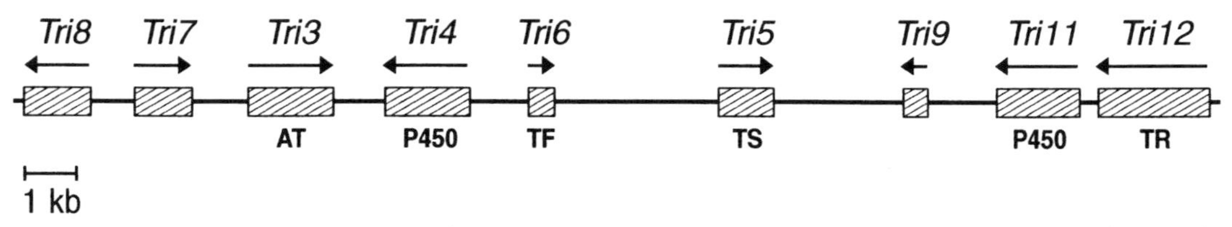

Figure 2 Map of the trichothecene pathway gene cluster in *F. sporotrichioides*. Boxes represent the locations of gene coding regions and arrows indicate the direction of transcription. AT, acetyltransferase; P450, cytochrome-P450 monooxygenase; TF, transcription factor; TS, trichodiene synthase; TR, transport protein.

Macrocyclic trichothecene pathway genes in *M. roridum* were also found to be clustered. A single cosmid clone carrying the *MRTri5* gene also carries homologues of *FSTri4* and *FSTri6*. The *MRTri4* and *MRTri6* genes share between 45% and 70% identity with their *Fusarium* counterparts. The functional identities of *MRTri4* and *FSTri4* have been demonstrated by successful complementation of an *FSTri4* mutant in *F. sporotrichioides* with *MRTri4*.[36] Interestingly, gene orientations and the distances between genes differ significantly in the *F. sporotrichioides* and *M. roridum* clusters. These differences may be due in part to the presence of macrocyclic pathway-specific genes within the *M. roridum* gene cluster but may also reflect the phylogenetic distance between these two fungi.

2.07.2.1.2 Aristolochene synthase

PR toxin (**3**) is a sesquiterpenoid toxin produced by *Penicillium roqueforti*.[46] PR toxin was originally identified as a mycotoxin from its presence in samples of silage that were found to cause animal mycotoxicoses.[47] Because it occurs rarely in agricultural products, PR toxin is currently not recognized as a problem mycotoxin. Investigations of *P. roqueforti* in cheeses have shown that PR toxin does not accumulate.[48,49] PR toxin appears to be the most toxic member of several related sesquiterpenoids produced by *P. roqueforti*.[50] Based on structural considerations, PR toxin is one of numerous fungal toxins that appear to be derived from aristolochene[6,51] (**2**).

An aristolochene synthase (AS) has been isolated from *P. roqueforti* cultures grown under the same conditions previously described for trichothecene production in *Fusarium* species.[52] AS activity is detected 60–80 h postinoculation, and can account for up to 0.5% of total cellular protein; however, little PR toxin accumulates under these growth conditions (T. M. Hohn, unpublished results). AS has been purified from *P. roqueforti* and shown to consist of an M_r 37 000 polypeptide by SDS–PAGE. AS has also been partially purified from *Aspergillus terreus*, and was reported to have a molecular weight of 58 000.[53] However, the *A. terreus* enzyme has now been purified, and its molecular weight is nearly identical to the *P. roqueforti* enzyme (D. E. Cane, personal communication). In *A. terreus*, aristolochene has been found to accumulate under certain growth conditions,[51] although no aristolochene metabolites have been reported.

Isolation of the AS gene (*Ari1*) in *P. roqueforti* was accomplished using PCR with degenerate primers based on sequences obtained from isolated peptide fragments.[6] The *Ari1* gene specifies a 342 aa polypeptide with a predicted molecular weight of 39 200. Identification of the *Ari1* gene was confirmed by expression of active AS in *E. coli*. Sequence analysis of the cloned *A. terreus* enzyme has revealed an approximate 70% identity between the *A. terreus* and *P. roqueforti* proteins at the amino acid level (D. E. Cane, personal communication).

2.07.2.1.3 Sesquiterpene synthase purification

The AS and TS genes are the only terpene synthase genes that have been isolated from fungi. Comparisons between the AS and TS sequences indicate that these enzymes share no significant similarities and are at best distantly related. At least one TS structural feature of apparent mechanistic importance is not present in AS. In TS, there is evidence that the sequence RYR at aa 304–306 is located near the active site based on structure–function studies employing site-directed mutagenesis[16,54] (see Figure 1). Conservative changes in this sequence result in either a reduction in k_{cat} or an increased value for K_m and the production of anomalous sesquiterpene products. No RYR sequence is observed in AS, although an RYR motif is present in the actinomycete sesquiterpene synthase, pentalenene synthase.[4] Evidence for another catalytically important TS sequence, DDXXD, has also been reported.[55,56] DDXXD is located at aa 100–104 and conforms to the consensus for similar sequences in other terpene synthases and prenyltransferases,[55,57] where it is proposed to coordinate the Mg^{2+} ion. Conservative changes in the DDXXD sequence due to Glu substitutions for Asp100 (EDXXD) and Asp101 (DEXXD) resulted in enzymes with reduced k_{cat} and increased K_m values. The Glu substitution for Asp104 (DDXXE) had only modest effects on the activity of the altered TS enzyme. Interestingly, AS does not contain the DDXXD sequence; however, the related sequence DDXXE is present. The functional importance of the DDXXE sequence in AS has not been investigated.[6] Differences in the apparent active site residues for AS and TS may reflect their distinctive catalytic properties or, alternatively, may mean that these enzymes have arrived at different protein structural solutions to some common mechanistic requirements such as FPP binding.

In the absence of conserved structural features among terpene synthases, enzyme purification remains an important part of gene isolation strategies. Purified enzyme preparations can lead to the development of terpene synthase gene probes either by providing structural information for the design of oligonucleotides or by serving as antigens for the generation of specific antibodies. Successful terpene synthase purifications depend on many of the same methods that have proven useful in the isolation of other types of fungal natural product pathway enzymes. Strain selection is particularly important, since the same terpenoid products are often produced by several closely related species and even by members of several different genera. Criteria used for strain selection should include both terpenoid production levels and stability of the production trait in culture. It has been observed that trichothecene production levels appear to be positively correlated with TS activity levels in trichothecene-producing species of *Fusarium*. Comparisons between *F. sporotrichioides* NRRL 3299 and *F. sambucinum* (telemorph, *G. pulicaris*) R-6380[33] have shown that under certain growth conditions, *F. sporotrichioides* accumulates approximately 2.5 times more trichothecenes than *G. pulicaris*, and this difference in trichothecene accumulation is reflected in correspondingly higher levels of TS activity. Increasing the enzyme levels in the cell homogenates used as starting materials for terpene synthase isolations is important because recoveries of purified terpene synthases are frequently low. Changes in starting material enzyme levels of several-fold can determine whether or not a purification protocol is successful. Stability of the terpenoid production trait is also an important consideration, since fungi can lose the ability to produce natural products in culture. With respect to trichothecene biosynthesis, *F. graminearum* is an example of an organism that undergoes spontaneous loss of the terpenoid production trait in culture.[58] Loss of trichothecene production by *F. graminearum* is associated with distinctive changes in colonial morphology that appear as sectors on plates of agar-containing media. In contrast, the production of trichothecenes by cultures of *F. sporotrichioides* NRRL 3299 remains at high levels even when stock cultures are repeatedly subcultured and stored at 4 °C for extended periods of time. Changes in production due to subculturing can be minimized by storing sufficient quantities of inoculum using appropriate long-term storage methods such as freezing or desiccation.[59]

Culture growth conditions also play important roles in terpene synthase purification schemes. Although time-consuming, optimizing growth conditions to maximize terpene synthase expression levels can be critically important. Usually the analysis of terpenoid production levels serves as the best means for evaluating the effectiveness of different growth media and various growth parameters. However, there are exceptions to this generalization; for example, AS was found to peak at high levels in *P. roqueforti* cultures in the absence of significant PR toxin (**3**) accumulation (T. M. Hohn, unpublished results). Optimizing growth conditions involves the systematic manipulation of media composition and culture incubation conditions. Production of fungal natural products is influenced by a number of nutritional factors including differences in medium carbon:nitrogen ratios and the types of carbon and nitrogen sources employed. Growth parameters such as temperature, light, pH, and the degree of culture aeration must also be optimized. Systematic approaches to optimizing growth media for fungal natural product production have been reported.[60] Details such as the size and shape of the flask or the type of flask closure can significantly alter culture aeration and have large effects on terpenoid production levels. Production of trichothecenes by *F. sporotrichioides* can vary by as much as 100-fold depending on the type of flask closures used (T. M. Hohn, unpublished results).

Although the structures of AS and TS differ significantly, it is interesting to note that similar purification strategies proved effective with both enzymes. This is presumably due to the fact that, despite their structural differences, both enzymes are cytosolic proteins with similar isoelectric points and subunit molecular weights. Both AS and TS were eluted with a KCl gradient from anion exchange columns at about the same salt concentration (190–200 mM KCl), and both enzymes underwent significant purification when eluted from an anion exchange column using a counter gradient method that employed increasing KCl and decreasing sodium pyrophosphate concentrations.[30,52] Similarities between the AS and TS purification protocols suggest that they can serve as a starting point for the development of other terpene synthase purification schemes.

Gene isolation efforts based on the purification of terpene synthases require either amino acid sequence data to generate nucleic acid probes or specific antibody preparations. Peptides have been isolated and sequenced from AS[6] and TS[5] using either CNBr treatment or digestion with the protease Lys-C, while specific antibody preparations have been generated from TS.[33]

2.07.2.1.4 Other gene isolation strategies

The discovery that many fungal natural product pathway genes are organized in clusters[39] suggests possible indirect strategies for isolating terpene synthase genes. For example, if all pathway genes are clustered, then the entire pathway can serve as a genetic target for gene isolation efforts. In some instances, it may be possible to exploit this situation by first isolating a closely linked gene as a preliminary step toward isolation of the terpene synthase gene itself. Following the isolation of the closely linked pathway gene, analysis of flanking DNA sequences could then be used to identify the terpene synthase gene. This approach is beneficial when the closely linked pathway genes lend themselves to sequence-based gene isolation methods but may also present advantages if the properties of the protein encoded by a closely linked gene suggest that affinity chromatography methods can be used for its purification. Indirect approaches are particularly attractive when research goals are not limited to terpene synthase gene isolation but include the characterization of other pathway genes.

Many terpenoid pathways employ cytochrome-P450 monooxygenases and *o*-acetyltransferases, both of which might be useful for the development of indirect terpene synthase gene-cloning strategies. Trichothecene pathways can have as many as six different cytochrome P450-catalyzed steps.[61] The two trichothecene pathway cytochrome P450 genes that have been characterized, *Tri4*[41] and *Tri*11,[42] are distantly related to each other, although both appear to be of fungal origin. Because cytochrome P450s comprise a superfamily of enzymes with common structural and mechanistic features,[62,63] it may be possible to design degenerate primers to amplify P450 gene fragments; however, it is likely that fungi contain numerous cytochrome P450 genes, and identifying a pathway-specific gene fragment could be problematic. Purification of cytochrome P450s involved in fungal natural product pathways has proven extremely difficult and seems unlikely to facilitate terpene synthase gene isolation efforts. On the other hand, *o*-acetyltransferases are often cytosolic enzymes and could offer advantages over terpene synthase purification. Preliminary studies indicate that the trichothecene pathway acetyltransferase TRI3[44] is stable in cell homogenates and can be conveniently assayed. TRI3 catalyzes the acetylation of the C-15 hydroxy group and is the best characterized acetyltransferase in the pathway. TRI3 is unrelated to two other suspected pathway acetyltransferases, TRI7 and TRI8 (T. M. Hohn and S. P. McCormick, unpublished results), and other fungal *O*-acetyltransferases. In contrast, *cefG* from *Cephalosporium acremonium* encodes an acetyltransferase involved in cephalosporin biosynthesis[64] that is similar to several other fungal *O*-acetyltransferases. Isolation of additional fungal acetyltransferase genes may reveal sequence information useful for gene isolation efforts, but there are currently insufficient data to pursue this approach.

Terpene synthase and terpenoid pathway genes can also be isolated using established fungal gene isolation methods. Several different methods, including the complementation of pathway gene mutations and the use of various subtractive cDNA methods, have been successfully employed to isolate fungal natural product pathway genes. Complementation of mutants lacking a functional pathway gene has been used to isolate genes from several pathways.[65–67] However, in most instances these efforts were successful largely because complementation resulted in phenotypic changes such as altered mycelial pigmentation that could be easily monitored. Mutant complementation approaches usually involve the transformation of pathway mutants with DNA from cosmid libraries constructed with wild-type genomic DNA. If ordered libraries are used, then the complementing cosmid can be easily identified;[66] otherwise, isolation of the complementing gene is accomplished using plasmid rescue methods.[67] For terpenoid pathways, this approach is limited by difficulties both in obtaining the appropriate pathway mutants[68] and in screening for transformants with restored pathway function. Other approaches that have been used to isolate fungal natural product genes are based on differential gene expression[69] and use various techniques involving cDNA subtractive hybridization.[70] cDNA clones that are identified within a terpenoid pathway subtractive library represent genes specifically expressed under conditions supporting terpenoid biosynthesis. Functional analysis of the corresponding genes can be performed using gene disruption (see below). The success of this approach depends upon finding conditions that maximize changes in pathway gene expression relative to changes in the expression of nonpathway genes. Changes in the growth medium or growth parameters, such as incubation temperature, can often be used to induce terpenoid pathway expression; however, they can also produce high backgrounds of differentially expressed transcripts, thereby making terpenoid pathway cDNA isolation difficult. Ideally, changes in pathway gene expression should occur over a short time period without abrupt changes in growth conditions. Trichothecene pathway gene expression changes 50- to 100-fold over a 3 h time period for *F. sporotrichioides* cultures[33] in early stationary phase growth.

Most fungi are amenable to gene disruption,[71] and this molecular genetic technique has proven valuable for characterizing natural product pathway genes.[40] Gene disruption permits the construction of genetically defined mutants and can provide clues as to gene function through analysis of the resulting terpenoid production phenotype. There are two general types of gene disruption techniques: gene replacement and integrative gene disruption (Figure 3). Of these, gene replacement is the most commonly used disruption method. Gene replacement involves the replacement of an existing functional copy of a gene with the altered nonfunctional copy residing on the transformation vector. Typically, a DNA fragment containing a selectable marker, such as a drug resistance gene, is inserted into the cloned copy of the target gene with or without the simultaneous deletion of coding sequences. Deletion of the target gene coding region ensures that no partial gene products or hybrid proteins will be expressed that might later interfere with phenotypic analysis. Gene replacement is irreversible and requires two recombination events involving the homologous target gene sequences present on both sides of the selectable marker. Integrative gene disruption is performed using a portion of the gene coding region that has been truncated at both ends.[72] Integration of the doubly truncated coding region results in the generation of two nonfunctional copies of the target gene, each with a single truncation. This type of gene disruption requires only a single recombination event which is potentially reversible. Because it does not require additional sequencing outside the gene coding region, it can be usefully coupled with PCR or other approaches which yield only portions of gene coding regions. In addition, integrative disruption vectors can be prepared in a single step and may produce disruptants more efficiently when the amount of available sequence is small. While nearly all fungi can be transformed by molecular means, transformation efficiencies are often very low so that the disruption efficiency becomes an important factor. For gene identification efforts involving a number of different DNA fragments, integrative gene disruption could present advantages over gene replacement in some fungi.

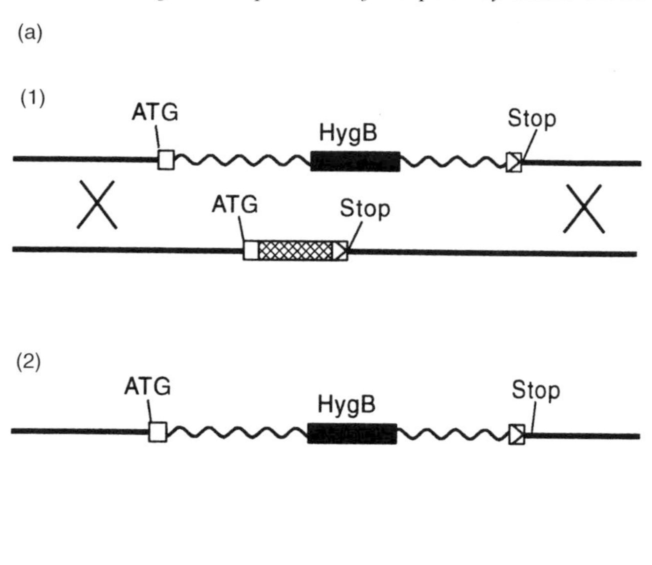

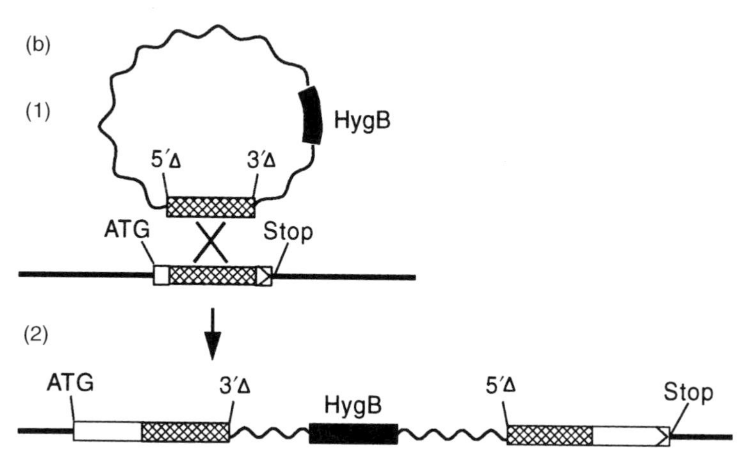

Figure 3 Gene disruption methods: (a) gene replacement/gene disruption; (b) integrative gene disruption. Steps in gene disruption: (1) recombination between disrupter fragment (top) and target gene (bottom); (2) product of gene disruption. HygB refers to the hygromycin phosphotransferase gene which is frequently used as a selectable marker for fungal transformation, 5′ and 3′ indicate the respective truncations of the gene coding region, wavy lines represent the plasmid sequences in the DNAs used for gene disruption, ATG indicates the start of translation, and Stop is the end of translation.

2.07.2.2 Plants

Plants are the most prolific producers of terpenoids and have been reported to produce thousands of different mono-, sesqui-, and diterpenoids.[1] Purification of plant terpene synthases is often complicated by factors such as localized expression within specific tissues and subcellular organelles, low levels of enzyme, and enzyme hydrophobicity. Despite these obstacles, several plant terpene synthases have been purified from plant tissues[73–77] and, in one instance, from plant cell cultures.[78] Initial efforts to isolate terpene synthase genes from plants employed probes based on either amino acid sequence data from purified enzymes[10,11] or specific antibody preparations.[79]

Comparisons between the first three available terpene synthase sequences established that all three enzymes, and by implication many other plant terpene synthases, are related.[1,80] These enzymes represented mono- (limonene synthase),[10] sesqui- (5-*epi*-aristolochene synthase),[11] and diterpene synthases (casbene synthase)[79] from spearmint, tobacco, and castor bean, respectively. Identities in global alignments between the three enzymes ranged from 31% to 42%, although higher levels of identity (77%) have since been observed between enzymes from closely related plant species.[13] Degenerate primers based on sequence identities among terpene synthases were used to isolate a taxadiene synthase from Pacific yew[9] by PCR. Primers for the isolation of the taxadiene synthase gene were designed using two different alignment strategies involving either the three terpene synthase sequences discussed above or a set of four diterpene synthases from either angiosperm[12,14,15] or gymnosperm (abietadiene synthase, GenBank #U50768) sources. A total of 11 regions of homology were identified in the two different alignments, and the resulting 20 primers generated from these sequences were used in all pairwise combinations with DNA from a cDNA library as the template. Only a single primer pair resulted in the amplification of a DNA fragment encoding a terpene synthase-like polypeptide. Primers making up the successful primer pair were based on sequences obtained from each of the two alignment strategies. This suggests that efforts to design degenerate primers for specific terpene synthase gene isolations should use conserved sequence information obtained from several different alignment strategies, and that factors such as phylogenetic relationships and enzyme substrate requirements should be taken into account.

Plant sesquiterpene synthase genes have also been isolated using PCR methods employing perfect primers with genomic DNA templates. Primers consisting of sequences from tobacco 5-*epi*-aristolochene synthase (EAS) genes[11] were used to amplify portions of a vetispiradiene synthase (VS) gene from *Hyoscyamus muticus*[13] and a cadinene synthase (CS) gene from cotton.[7] Selection of the VS primer sequences for CS gene amplification was guided by the identification of conserved regions resulting from alignments between the limonene synthase, EAS, and casbene synthase genes. VS was subsequently found to be 77% identical and 81% similar to EAS, however, CS was considerably less similar to EAS with 48% identity and 66% similarity. Higher levels of similarity between VS and EAS could reflect the fact that tobacco and *Hyoscyamus muticus* are more closely related and that the reaction mechanisms of these enzymes are remarkably similar.[17] Based on available terpene synthase sequence information, mono-, sesqui-, and diterpene synthases constitute a gene superfamily within plants and, in many instances, the similarities between enzymes are sufficient to permit the application of sequence-based gene cloning techniques.

2.07.2.3 Actinomycetes

Actinomycetes appear to make limited numbers of terpenoids, although it is unclear how rigorously this question has been investigated. Sesquiterpenoid production by actinomycetes has been described,[81] and two different sesquiterpene synthases have been characterized.[4,82] Only the pentalenene synthase (PS) from *Streptomyces* UC5319 has been purified to homogeneity.[4] PS has a molecular weight of 41–42 kDa as determined by SDS–PAGE. The gene encoding PS has been isolated from *Streptomyces* UC5319, and its identity confirmed by expression in *E. coli*.[4]

Sequence comparisons between the predicted amino acid sequence of the PS gene and other terpene synthases do not reveal significant similarities with either plant or fungal enzymes. Currently, there are no sequence-based gene isolation strategies for actinomycete terpene synthase genes. Efforts to isolate additional terpene synthase genes from actinomycetes will likely need to rely on enzyme purification to obtain amino acid sequence information or preparations of specific antibody.

2.07.3 HETEROLOGOUS EXPRESSION OF TERPENE SYNTHASE GENES

Difficulties associated with the isolation of terpene synthases were seen as limiting factors for studies of many enzymes. Expression of terpene synthase genes in *E. coli* has greatly reduced this barrier and promises to facilitate all aspects of terpene synthase research.

2.07.3.1 Expression in *E. coli*

2.07.3.1.1 Microbial terpene synthases

All three of the available microbial sesquiterpene synthase genes have been expressed to high levels in *E. coli*.[4,83,84] High-level expression of these enzymes was accomplished using an expression vector containing a bacteriophage T7 promoter and resulted in enzyme levels of 10–15% of soluble protein. In each case, milligram quantities of the recombinant enzymes were obtained in homogeneous form using purification protocols greatly simplified over those employed for purifying the native enzymes. The recombinant enzymes were similar or identical to purified native enzyme preparations with respect to kinetic parameters such as K_m and k_{cat}, and molecular weight as determined by SDS–PAGE. The fact that three distantly related microbial enzymes have been successfully expressed to high levels in *E. coli* suggests that this same expression strategy will work with other microbial terpene synthases. The utility of site-directed mutagenesis as a means for investigating structure–function relationships has been demonstrated with TS.

The fungal sesquiterpene synthases have also been expressed in *E. coli* using alternative expression strategies. Both TS and AS were expressed as fusion proteins consisting of the entire TS and AS coding regions with a 23 kDa polypeptide containing the IgG-binding domain of protein A fused at the N terminus[6] (T. M. Hohn, unpublished results). The resulting fusion proteins made the appropriate sesquiterpene products, but enzyme activities in the cell homogenates were low. TS has also been expressed at low levels using an expression vector with a trp–lac promoter.[34] *E. coli* cultures expressing TS from a chimeric gene regulated by the trp–lac promoter were found to accumulate low levels of trichodiene.

2.07.3.1.2 *Plant terpene synthases*

Several plant terpene synthases have been expressed in *E. coli.*[7,9,10,13,14,85,86] Expression of plant terpene synthase genes has resulted in high-level expression for two sesquiterpene synthases, but high-level expression has yet to be demonstrated for monoterpene and diterpene synthases. In all cases, sufficient enzyme was obtained to confirm gene identity through GC–MS analysis of the resulting enzyme reaction products. A variety of expression strategies has been employed including the use of expression vectors containing the bacteriophage T7 promoter[14,85,86] and the expression of N terminal fusions with the *lacZ* gene.[9,10] Expression of monoterpene and diterpene synthase genes is often complicated by the presence of chloroplast transit sequences at the N terminus. Despite the presence of chloroplast transit sequences, a monoterpene[10] and two diterpene synthases[9,85] have been successfully expressed in *E. coli.* An additional consideration for plant terpene synthase expression in *E. coli* is that plant genes are known to contain the arginine codons AGA and AGG which are used infrequently in *E. coli.*[87] AGA and AGG codons are particularly troublesome when they occur consecutively. In some instances, the use of plasmids containing a copy of the $tRNA_{arg4}$ gene has been shown to greatly improve plant gene expression[88] in *E. coli.* Casbene synthase expression levels increased two- to threefold when performed in an *E. coli* strain carrying a $tRNA_{arg4}$ gene.[85]

Identification of terpene synthase genes isolated from plants by molecular methods can be problematic. Because plants frequently possess numerous terpene synthase genes and many terpene synthases are closely related, it is difficult to know the identity of a particular gene without characterizing the enzyme product. The situation is further complicated by the fact that individual terpene synthase genes are frequently members of gene families within which sequence differences can occur.[7,11] For these reasons, heterologous expression of putative terpene synthase genes is often critical for gene identification. While terpene synthase gene expression in *E. coli* has been shown to generate sufficient product from reaction mixtures for GC–MS analysis, identification of the reaction product is difficult if the appropriate reference terpenes are not available. In the case of unknown terpene synthase genes, structural analysis of enzyme products requires the application of NMR methods. If the unknown enzyme is expressed at low levels in *E. coli*, then difficulties in isolating sufficient quantities of the enzyme product may be encountered. One solution to this problem is the development of expression systems that result in both terpene synthase production and in the accumulation of terpene synthase products. Such an expression system was demonstrated with the kaurene synthase A from *Arabidopsis thaliana.*[14] *E. coli* was cotransformed with plasmids containing either the kaurene synthase A coding region or the *CrtE* gene (GGPP synthetase) from *Erwinia uredova*. Extracts from cultures expressing kaurene synthase A contained both GGPP and the diterpene synthase product copalyl pyrophosphate. Expression of plant terpene synthases in an appropriate fungal host could also provide a means for accumulating enzyme products.

2.07.3.2 Expression in Plants

Heterologous expression of terpene synthases in plants has been reported for a single terpene synthase. The TS gene (*FSTri5*) of *F. sporotrichioides* has been successfully expressed in transgenic tobacco plants.[89] Expression of *Tri5* was regulated by the CaMV 35S promoter and resulted in the production of immunodetectable TS polypeptide and active enzyme in the leaves of transgenic plants. Low levels of the enzyme product trichodiene were also detected in leaves. Further studies of TS expression were carried out with cell suspension cultures derived from transgenic plants.[90] The resulting cell cultures were also shown to produce TS polypeptide and active enzyme. Expression of TS in cell cultures was constitutive, and the levels of activity did not change in response to elicitors of sesquiterpenoid phytoalexins such as cellulase. Production levels of trichodiene were low in uninduced cultures but increased following elicitor treatment. Significantly, a novel trichodiene metabolite, 15-hydroxytrichodiene[91] (**10**), was identified following elicitor treatment. Levels of 15-hydroxytrichodiene were comparable to that of the tobacco sesquiterpenoid phytoalexin capsidiol.

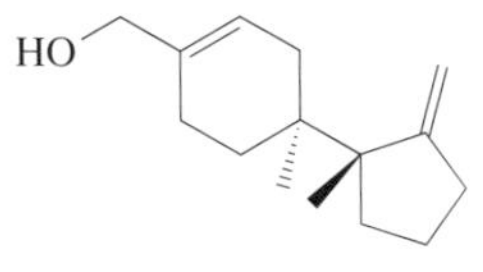

HO

(10)

2.07.4 IMPLICATIONS FOR FUTURE RESEARCH

2.07.4.1 Understanding Terpenoid Function

Considerable information is available concerning the structural complexity of terpenoids, however, the specific biological functions of most terpenoids remain obscure. Terpenoids are comprised of functionally diverse natural products that play important roles in both producer-organism physiology and ecology.[92] Terpenoids are known to function as hormones, pheromones, defensive chemicals, pollinator attractants, antibiotics, and phytotoxins.[1] Efforts to understand terpenoid function as it relates to plant–microbe, plant–insect, and plant–plant interactions is of particular interest to agricultural research. Terpene synthase genes present opportunities for studying terpene function within complex biological systems involving the participation of numerous natural product pathways. Using molecular genetic techniques such as gene disruption or antisense mRNA expression, it is possible to isolate the functions of specific terpenoids.[21] Because terpene synthases are located at the branch points of terpenoid pathways, terpene synthase genes are excellent targets for molecular efforts to disable these pathways.

In fungi, gene disruption has been used to investigate the role of trichothecene production by *Fusarium* species in certain plant diseases. Disruption of the TS gene in different *Fusarium* species resulted in TS-deficient mutants unable to produce trichothecenes. Analysis of these mutants indicated that their virulence was reduced in some disease interactions but not in others.[25,93] Other plant pathogenic fungi produce terpenoid phytotoxins, and a number of these are thought to enhance fungal virulence.[21] Understanding the role of fungal phytotoxins in plant diseases may provide useful information for efforts to develop resistant crop plant varieties.

While molecular approaches have not yet been applied to plant terpenoid pathways as a means for determining terpenoid function, they undoubtedly will be in the near future. Techniques involving either antisense mRNA expression or sense suppression techniques could be used to study the role of terpenoid phytoalexins and other plant terpenoids thought to function as defensive chemicals. These approaches are currently being applied to the functional analysis of several plant natural product pathways.[94,95] Several crop plants including potatoes, tomatoes, and cotton are known to produce sesquiterpenoid phytoalexins in response to a variety of elicitors, but the contribution of these compounds to plant defense against microorganisms and insect pests remains to be proven. Molecular approaches focused on terpene synthase genes could provide a means for demonstrating the role of sesquiterpenoid phytoalexins in specific disease interactions. Similarly, molecular approaches could also be applied to studies of terpenoids thought to function as pollinator attractants, insect antifeedants, and pheromone mimics.

2.07.4.2 Applications

Those terpenoids with recognized commercial applications are principally of plant origin. Terpenoids are major constituents of essential oils and are important as fragrances and flavorings. Only a few terpenoids have made significant contributions as pharmaceuticals (e.g., taxol), which is surprising given the structural complexity of this group of natural products and the fact that they are produced in large numbers by both plants and microbes. Terpene synthase genes could be used as part of strategies to genetically alter the types of terpenoids produced by plants. Expression of foreign plant terpenoid pathways could produce desirable changes in plant fragrance and flavor properties in addition to the possibility of improving plant chemical defenses. This approach requires the heterologous expression of plant terpene synthase genes in plant hosts, something which has not yet been demonstrated. However, the fact that most plant terpene synthases appear to be structurally related suggests that heterologous expression of plant terpene synthase genes is feasible.

Expression of the fungal sesquiterpene synthase TS in tobacco demonstrates that fungal terpene pathways could also be used to alter plant terpenoid production, although the utility of this approach is presently limited by the small number of available fungal terpene synthase genes. The accumulation of 15-hydroxytrichodiene in transformed cell suspension cultures expressing TS indicates that foreign terpene synthase products can be metabolized in unpredictable ways and that this is a potential problem for efforts to genetically alter plant terpenoid production.

Finally, the genetic engineering of plant terpene synthase genes in fungal hosts may also represent possible commercial applications. Fungi are more amenable than plants to genetic manipulation, and a number of fungi are known to produce large quantities of terpenoids in liquid culture. If plant terpenoids can be produced in fungi, the likely applications would include specialty terpenoids related to fragrances and flavors. Plant sesquiterpenoid synthases and other cytosolic plant terpene synthases are the best candidates for fungal expression, although it may also be possible to express the chloroplast-localized terpene synthases.

ACKNOWLEDGMENTS

I thank R. Plattner and R. Proctor for reading the manuscript and providing helpful comments.

2.07.5 REFERENCES

1. D. J. McGarvey and R. Croteau, *Plant Cell*, 1995, **7**, 1015.
2. D. E. Cane, *Chem. Rev.*, 1990, **90**, 1089.
3. R. Croteau, *Chem. Rev.*, 1987, **87**, 929.
4. D. E. Cane, J. K. Sohng, C. R. Lamberson, S. M. Rudnicki, Z. Wu, M. D. Lloyd, J. S. Oliver, and B. R. Hubbard, *Biochemistry*, 1994, **33**, 5846.
5. T. M. Hohn and P. D. Beremand, *Gene*, 1989, **79**, 131.
6. R. H. Proctor and T. M. Hohn, *J. Biol. Chem.*, 1993, **268**, 4543.
7. X. Y. Chen, M. S. Wang, Y. Chen, V. J. Davisson, and P. Heinstein, *J. Nat. Prod.*, 1996, **59**, 944.
8. X. Y. Chen, Y. Chen, P. Heinstein, and V. J. Davisson, *Arch. Biochem. Biophys.*, 1995, **324**, 255.
9. M. R. Wildung and R. Croteau, *J. Biol. Chem.*, 1996, **271**, 9201.
10. S. M. Colby, W. R. Alonso, E. J. Katahira, D. J. McGarvey, and R. Croteau, *J. Biol. Chem.*, 1993, **268**, 23016.
11. P. J. Facchini and J. Chappell, *Proc. Natl. Acad. Sci. USA*, 1992, **89**, 11088.
12. C. J. D. Mau and C. A. West, *Proc. Natl. Acad. Sci. USA*, 1994, **91**, 8497.
13. K. W. Back and J. Chappell, *J. Biol. Chem.*, 1995, **270**, 7375.
14. T. P. Sun and Y. Kamiya, *Plant Cell*, 1994, **6**, 1509.
15. R. J. Bensen, G. S. Johal, V. C. Crane, J. T. Tossberg, P. S. Schnable, R. B. Meeley, and S. P. Briggs, *Plant Cell*, 1995, **7**, 75.
16. D. E. Cane, J. H. Shim, Q. Xue, B. C. Fitzsimons, and T. M. Hohn, *Biochemistry*, 1995, **34**, 2480.
17. K. W. Back and J. Chappell, *Proc. Natl. Acad. Sci. USA*, 1996, **93**, 6841.
18. J. Chappell, *Plant Physiol.*, 1995, **107**, 1.
19. J. L. Laseter, J. D. Weete, and C. H. Walkinshaw, *Photochemistry*, 1973, **12**, 387.
20. H.-P. Hanssen, V. Sinnwell, and W.-R. Abraham, *Z. Naturforsch. Teil C*, 1986, **41**, 825.
21. T. M. Hohn, in "The Mycota", vol. V, part A. "Plant Relationships," eds. G. Carroll and P. Tudzynski, Springer, Berlin, 1997, p. 129.
22. Committee On Protection Against Mycotoxins, "Protection Against Trichothecene Mycotoxin", National Academy Press, Washington, DC, 1983, p. 22.
23. Y. Ueno, *Adv. Nutr. Res.*, 1980, **3**, 301.
24. B. B. Jarvis, N. Mokhtari-Rejali, E. P. Schenkel, C. S. Barros, and N. I. Matzenbacher, *Phytochemistry*, 1991, **30**, 789.
25. A. E. Desjardins, R. H. Proctor, G. Bai, S. P. McCormick, G. Shaner, G. Buechley, and T. M. Hohn, *Mol. Plant–Microbe Interact.*, 1996, **9**, 775.
26. R. H. Proctor, T. M. Hohn, and S. P. McCormick, *Mol. Plant–Microbe Interact.*, 1995, **8**, 593.
27. R. Greenhalgh, D. A. Fielder, L. A. Morrison, J. P. Charland, B. A. Blackwell, J. D. Miller, M. E. Savard, and J. W. ApSimon, in "Mycotoxins and Phycotoxins '88," eds. S. Natori, K. Hashimoto, and Y. Ueno, Elsevier, Amsterdam, 1989, p. 223.
28. P. Mohr, C. Tamm, W. Zurcher, and M. Zehnder, *Helv. Chim. Acta*, 1984, **67**, 406.

29. D. E. Cane, H. Ha, C. Pargellis, F. Waldmeir, S. Swanson, and P. P. N. Murthy, *Bioorg. Chem.*, 1985, **13**, 246.
30. T. M. Hohn and F. Vanmiddlesworth, *Arch. Biochem. Biophys.*, 1986, **251**, 756.
31. D. E. Cane, S. Swanson, and P. P. N. Murthy, *J. Am. Chem. Soc.*, 1981, **103**, 2136.
32. Y. Ueno, M. Sawano, and K. Ishii, *Appl. Microbiol.*, 1975, **30**, 4.
33. T. M. Hohn and M. N. Beremand, *Appl. Environ. Microbiol.*, 1989, **55**, 1500.
34. T. M. Hohn and R. D. Plattner, *Arch. Biochem. Biophys.*, 1989, **275**, 92.
35. T. M. Hohn and A. E. Desjardins, *Mol. Plant–Microbe Interact.*, 1992, **5**, 249.
36. S. C. Trapp, T. M. Hohn, and B. B. Jarvis, *Mol. Gen. Genet.*, in press.
37. D. E. Cane, R. Iyengar, and M. Shiao, *J. Am. Chem. Soc.*, 1981, **103**, 914.
38. E. Cerdaolmedo, R. Fernandezmartin, and J. Avalos, *Anton. Leeuwenhoek*, 1994, **65**, 217.
39. N. P. Keller and T. M. Hohn, *Fungal Genet. Biol.*, 1997, **21**, 17.
40. T. M. Hohn, A. E. Desjardins, S. P. McCormick, and R. H. Proctor, in "Molecular Approaches to Food Safety," eds. M. Eklund, J. L. Richard, and M. Katsutoshi, Alaken, Fort Collins, CO, 1995, p. 239.
41. T. M. Hohn, A. E. Desjardins, and S. P. McCormick, *Mol. Gen. Genet.*, 1995, **248**, 95.
42. N. Alexander, T. M. Hohn, and S. P. McCormick, *Appl. Environ. Microbiol.*, 1998, **64**, 221.
43. S. P. McCormick and T. M. Hohn, *Appl. Environ. Microbiol.*, 1997, **63**, 1685.
44. S. P. McCormick, T. M. Hohn, and A. E. Desjardins, *Appl. Environ. Microbiol.*, 1996, **62**, 353.
45. R. H. Proctor, T. M. Hohn, S. P. McCormick, and A. E. Desjardins, *Appl. Environ. Microbiol.*, 1995, **61**, 1923.
46. R. D. Wei, H. K. Schnoes, E. B. Smalley, S. Lee, Y. Chang, and F. M. Strong, in "Animal, Plant, and Microbial Toxins," Vol. 2. "Chemistry, Pharmacology, and Immunology," eds. A. Ohsaka, K. Hayashi, and Y. Sawai, Plenum, New York, 1976, p. 137.
47. P. E. Still, R. D. Wei, E. B. Smalley, and F. M. Strong, *Fed. Proc.*, 1972, **31**, 733.
48. L. Polonelli, G. Morace, F. delle Monache, and R. A. Samson, *Mycopathologia*, 1978, **66**, 99.
49. P. M. Scott and S. R. Kanhere, *J. Assoc. Off. Anal. Chem.*, 1979, **62**, 141.
50. R. Wei, Y. W. Lee, and Y. Wei, in "Trichothecenes and Other Mycotoxins," ed. J. Lacey, Wiley, Chichester, 1985, p. 337.
51. D. E. Cane, B. J. Rawlings, and C.-C. Yang, *J. Antibiotics*, 1987, **40**, 1331.
52. T. M. Hohn and R. D. Plattner, *Arch. Biochem. Biophys.*, 1989, **272**, 137.
53. D. E. Cane and C. Bryant, *J. Am. Chem. Soc.*, 1994, **116**, 12063.
54. D. E. Cane and Q. Xue, *J. Am. Chem. Soc.*, 1996, **118**, 1563.
55. D. E. Cane, Q. Xue, and B. C. Fitzsimons, *Biochemistry*, 1996, **35**, 12369.
56. D. E. Cane, Q. Xue, and J. E. Van Epp, *J. Am. Chem. Soc.*, 1996, **118**, 8499.
57. M. N. Ashby, D. H. Spear, and P. A. Edwards, in "Molecular Biology of Atherosclerosis," ed. A. D. Attie, Elsevier, Amsterdam, 1990, p. 27.
58. J. D. Bu'Lock and S. Chulze-de-Gomez, *Mycol. Res.*, 1990, **94**, 851.
59. K. O'Donnell and S. W. Peterson, in "Biotechnology of Filamentous Fungi, Technology and Products," eds. D. B. Finklestein and C. Ball, Butterworth-Heineman, Boston, MA, 1992, p. 7.
60. R. L. Monaghan, J. D. Polishook, V. J. Pecore, G. F. Bills, M. Nallinomstead, and S. L. Streicher, *Can. J. Bot.*, 1995, **73**, S925.
61. A. E. Desjardins, T. M. Hohn, and S. P. McCormick, *Microbiol. Rev.*, 1993, **57**, 595.
62. D. R. Nelson and D. W. Nebert, *DNA Cell Biol.*, 1993, **12**, 1.
63. D. R. Nelson and H. W. Strobel, *Biochemistry*, 1989, **28**, 656.
64. L. Mathison, C. Soliday, T. Stepan, T. Aldrich, and J. Rambosek, *Curr. Genet.*, 1993, **23**, 33.
65. N. Kimura and T. Tsuge, *J. Bacteriol.*, 1993, **175**, 4427.
66. G. A. Payne, G. J. Nystrom, D. Bhatnagar, T. E. Cleveland, and C. P. Woloshuk, *Appl. Environ. Microbiol.*, 1993, **59**, 156.
67. P. Chang, C. D. Skory, and J. E. Linz, *Curr. Genet.*, 1992, **21**, 231.
68. M. N. Beremand, *Appl. Environ. Microbiol.*, 1987, **53**, 1855.
69. M. Ehrenshaft and R. G. Upchurch, *Appl. Environ. Microbiol.*, 1991, **57**, 2671.
70. L. Diatchenko, Y. F. Lau, A. P. Campbell, A. Chenchik, F. Moqadam, B. Huang, S. Lukyanov, K. Lukyanov, N. Gurskaya, E. D. Sverdlov, and P. D. Siebert, *Proc. Natl. Acad. Sci. USA*, 1996, **93**, 6025.
71. J. R. S. Fincham, *Microbiol. Rev.*, 1989, **53**, 148.
72. D. Shortle, J. E. Haber, and D. Botstein, *Science*, 1982, **217**, 371.
73. W. R. Alonso, J. I. M. Rajaonarivony, J. Gershenzon, and R. Croteau, *J. Biol. Chem.*, 1992, **267**, 7582.
74. P. Moesta and C. A. West, *Arch. Biochem. Biophys.*, 1985, **238**, 325.
75. S. L. Munck and R. Croteau, *Arch. Biochem. Biophys.*, 1990, **282**, 58.
76. S. S. Dehal and R. Croteau, *Arch. Biochem. Biophys.*, 1988, **261**, 346.
77. E. Pichersky, E. Lewinsohn, and R. Croteau, *Arch. Biochem. Biophys.*, 1995, **316**, 803.
78. U. Vogeli, J. W. Freeman, and J. Chappell, *Plant Physiol.*, 1990, **93**, 182.
79. A. F. Lois and C. A. West, *Arch. Biochem. Biophys.*, 1990, **276**, 270.
80. J. Chappell, *Annu. Rev. Plant Physiol.*, 1995, **46**, 521.
81. N. N. Gerber, *Phytochemistry*, 1971, **10**, 185.
82. D. E. Cane and M. Tandon, *J. Am. Chem. Soc.*, 1995, **117**, 5602.
83. D. E. Cane, Z. Wu, R. H. Proctor, and T. M. Hohn, *Arch. Biochem. Biophys.*, 1993, **304**, 415.
84. D. E. Cane, Z. Wu, J. S. Oliver, and T. M. Hohn, *Arch. Biochem. Biophys.*, 1992, **300**, 416.
85. A. M. Hill, D. E. Cane, C. J. D. Mau, and C. A. West, *Arch. Biochem. Biophys.*, 1996, **336**, 283.
86. K. W. Back, S. H. Yin, and J. Chappell, *Arch. Biochem. Biophys.*, 1994, **315**, 527.
87. R. Mattes, in "Biotechnology," eds. H. J. Rehm and G. Reed, VCH, Weinheim, 1993, vol. 2, p. 233.
88. P. M. Schrenk, S. Baumann, R. Mattes, and H. Steinib, *Biotechniques*, 1995, **19**, 196.
89. T. M. Hohn and J. B. Ohlrogge, *Plant Physiol.*, 1991, **97**, 460.
90. M. Zook, T. Hohn, A. Bonnen, J. Tsuji, and R. Hammerschmidt, *Plant Physiol.*, 1996, **112**, 311.
91. M. Zook, K. Johnson, T. Hohn, and R. Hammerschmidt, *Phytochemistry*, 1996, **43**, 1235.
92. J. B. Harbone, in "Ecological Chemistry and Biochemistry of Plant Terpenoids," eds. J. B. Harborne and F. A. Tomas-Barberian, Clarendon Press, Oxford, 1991, vol. 31, p. 399.

93. A. E. Desjardins, T. M. Hohn, and S. P. McCormick, *Mol. Plant–Microbe Interact.*, 1992, **5**, 214.
94. R. A. Dixon, C. J. Lamb, S. Masoud, V. J. H. Sewalt, and N. L. Paiva, *Gene*, 1996, **179**, 61.
95. N. Courtney-Gutterson, in "Genetic Engineering of Plant Secondary Metabolism," eds. B. E. Ellis, G. W. Kuroki, and H. A. Stafford, Plenum, New York, 1994, vol. 28, p. 93.

2.08 Diterpene Biosynthesis

JAKE MacMILLAN and MICHAEL H. BEALE
University of Bristol, UK

2.08.1 INTRODUCTION

This chapter concentrates on the cyclic diterpenes. Their formation can be rationalized by considering the different types of cyclization of geranylgeranyl diphosphate (GGPP) that have been revealed by studies on the enzymes. To date, four different kinds of cyclases (synthases) have been cloned and their functional proteins have been sequenced. These synthases are:

Casbene synthase[1–3]123 catalyzes the formation of (1*S*,3*R*)-casbene by ionization of the diphosphate of GGPP and attack on the resultant allylic 1-carbonium by the terminal double bond (Equation (1)).

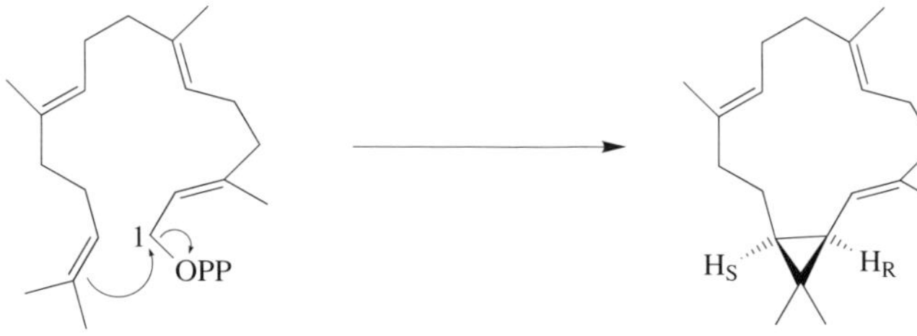

GGPP Casbene (1)

ent-Copalyl diphospate synthase[4,5] catalyzes the formation of *ent*-copalyl diphosphate (*ent*-CPP) by proton-induced cyclization (Equation (2)).

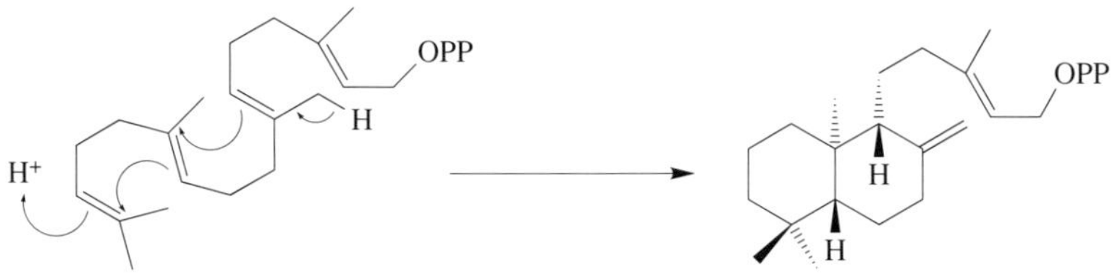

GGPP *ent*-copalyl diphosphate (*ent*-CPP) (2)

Taxadiene synthase[6,7] catalyzes the formation of taxa-4,11-diene by a series of cyclization steps initiated by ionization of the diphosphate and continued by proton-induced cyclization (Equation (3)).

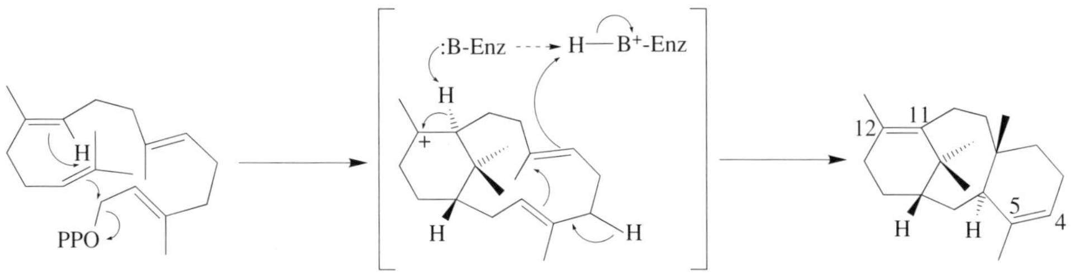

GGPP taxa-4,11-diene (3)

Abietadiene synthase[8,9] catalyzes the formation of abieta-7,13-diene by a series of cyclization steps, initiated by proton-induced cyclization and continued by ionization of the diphosphate (Scheme 1).

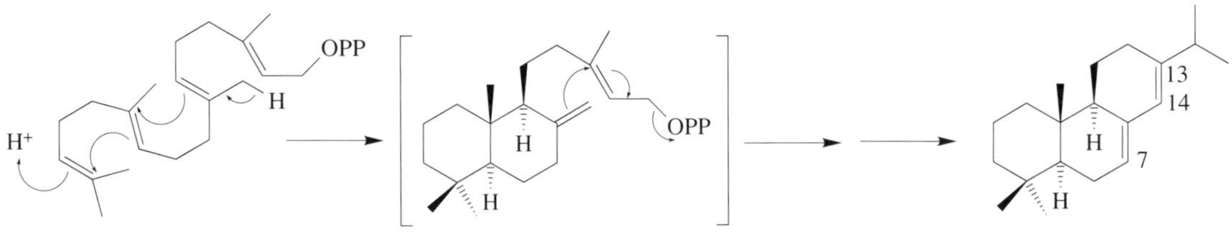

GGPP copalyl diphosphate (CPP) abieta-7,13-diene

Scheme 1

ent-Kaurene synthase[10,11] is not a GGPP cyclase, but catalyzes the formation of *ent*-kaur-16-ene from *ent*-CPP by diphosphate-induced cyclization. Thus, in contrast to abietadiene, *ent*-kaur-16-ene is formed from GGPP in two distinct steps catalyzed by two enzymes: *ent*-CPP synthase and *ent*-kaurene synthase (Scheme 2). However, in 1997 it was shown that GGPP is converted to *ent*-kaur-16-ene via *ent*-CPP in the fungus *Phaeosphaeria* sp. L487 by a single enzyme.[12] The amino acid sequence, deduced from the cloned gene, contains both motifs **a** and **b**.

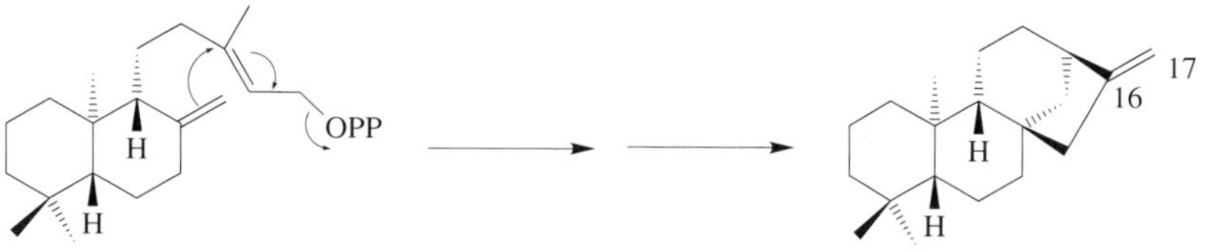

ent-CPP → → *ent*-kaur-16-ene

Scheme 2

Alignment of the amino acid sequences of these synthases reveals two characteristic conserved motifs, **a** and **b** (Table 1). Casbene and *ent*-kaurene synthases, in which cyclization is initiated by ionization of the diphosphate, contain only motif **a**, which is the suggested binding site for a diphosphate–divalent metal complex. *ent*-CPP synthase contains only motif **b**, nearer the N terminus, and may stabilize the incipient positive charge, generated by protonation of the 14,15-double bond of GGPP. Abietadiene synthase catalyzes both types of cyclization and contains both motifs. Taxadiene synthase also contains motif **a** but motif **b** is modified and does not align with motif **b** of the other cyclases; this modified motif **b** may be associated with internal deprotonation and reprotonation, catalyzed by taxadiene synthase (see Section 2.08.3.1).

Table 1 Conserved regions in geranylgeranyl diphosphate cyclases.

Cyclase	*Motif* ***a*** *(I,L,V)****DD****XXD*	*Motif* ***b*** *D(I,V)****DD****TAM*	*Ref.*
Casbene synthase	LI**DD**TI**D**	—	1
ent-Kaurene synthase (pumpkin)	VV**DD**FY**D**	—	5
ent-CPP synthase (arabidopsis)	—	**D**I**DD**TAM	11
ent-CPP synthase (maize)	—	**D**V**DD**TAM	13
Abietadiene synthase	IL**DD**LY**D**	**D**I**DD**TAM	9
Taxadiene synthase	LF**DD**MA**D**	(**D**SY**DD**)	6

The individual cyclases are described in more detail later. On the basis of the general properties of the GGPP cyclases and *ent*-kaurene synthase, the origins of the diverse structures of the cyclic diterpenes are discussed under two general headings: Type A, initiation by ionization of the diphosphate; and Type B, initiation by protonation at the 14,15-double bond.

Type A may be followed by Type B, catalyzed by the same or separate enzymes. Likewise, Type B may be followed by Type A, catalyzed by the same or separate enzymes. Further transformations of the initial cyclization products are catalyzed by oxidases, some of which have been characterized and/or cloned. These oxidases are discussed where appropriate.

2.08.2 TYPE A CYCLIZATION

2.08.2.1 Casbene

Casbene ((**2**), Scheme 3)[14,15] is a phytoalexin, elicited in seedlings of castor bean (*Ricinus communis* L.) by the fungus *Rhizopus stolonifer*,[16] or oligogalacturonides from cell wall fragments of this fungus.[17] Its formation has been rationalized[18] by cyclization of GGPP to a nonclassical carbocation at C-1, C-14, and C-15. However, the intermediate (**1**) is shown in Scheme 3 as a classical carbonium ion to illustrate that formation of the cyclopropane ring proceeds by suprafacial approach of the *re,re*-face of the 14,15-double bond of GGPP to C-1 and stereospecific loss of the pro-*S* hydrogen at C-1 of GGPP.[19] The overall conversion is catalyzed by a single enzyme, casbene synthase. The native enzyme (M_r, ca. 53 000) has been purified (700-fold) from castor bean seeds and shows a preference for Mg^{2+} over Mn^{2+}.[2,3] A near full-length cDNA clone for casbene synthase has been obtained from castor bean seedlings,[1] encoding a 601 amino acid protein with a predicted M_r of 68 690. Although casbene synthase is located in proplastids of germinating seedlings,[20] the deduced amino acid sequence contains no clear proplastid targeting signature. However, it does contain features, common to mono- and sesquiterpene cyclases of Type A (see Chapter 2.07), including the DDXXD motif **a** (Table 1). Expression of the gene is increased during elicitation by pectic fragments.[21]

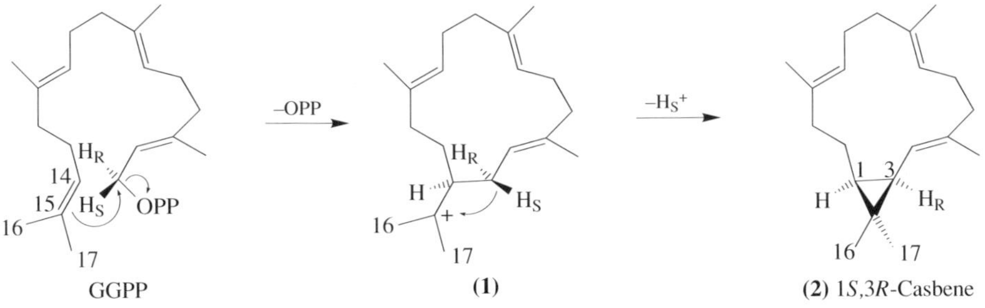

GGPP —OPP→ **(1)** $-H_S^+$→ **(2)** 1*S*,3*R*-Casbene

Scheme 3

2.08.2.2 Cembrenes

The cembrenes (Scheme 4) are 14-membered ring diterpenes that occur in the turpentine gum of pines and gum from the trichomes of tobacco leaves. *neo*-Cembrene (cembrene A, **(4)**) represents the simplest product of the Type A cyclization of GGPP and, in principle, may be derived from GGPP via the carbonium ion ((**3**), Scheme 4). 1*S*-Cembrene (**7**),[22] the parent compound of the cembrene 4-ols (**5**) and cembrene 4,6-diols (**8**), may, likewise, be formed from (**3**) via (**6**). However, there is no experimental evidence for these steps, nor for the postulate[23] that 1*S*-cembrene (**7**) is formed from the (1*R*,3*S*)-enantiomer of (1*S*,3*R*)-casbene ((**2**), Scheme 3). The biosynthesis of the cembrene-diols (**8**) has been examined in tobacco by two groups. Using excised calyces, Crombie *et al.*[23] observed the incorporation of [2-^{14}C]geranylgeraniol and 1*R*,*S*-cembrene into the 4α and 4β-cembrenediols (**8**). In contrast, Guo and Wagner[24] found that a purified enzyme preparation from trichome glands catalyzed the formation of both cembrene 4α and 4β-ols (**5**) from [1-^{3}H]GGPP. They concluded that the steps from GGPP to (**5**) were catalyzed by a single, soluble polypeptide of 58 kDa and that the cembrene diols (**8**) were probably formed from the cembrene 4-ols (**5**) by a cytochrome P450-dependent monooxygenase. However, no definitive enzymology has been reported.

2.08.3 TYPE A–TYPE B CYCLIZATION

2.08.3.1 Taxanes

More than 100 taxanes occur in *Taxus* species. The most detailed biosynthetic studies have concerned taxol (paclitaxel, (**16**), Scheme 5), a potent anticancer agent from the bark of Pacific yew (*Taxus brevifolia* Nutt). These studies, summarized in Scheme 5 and discussed later, show that the first committed step is the cyclization of GGPP to taxa-4,11-diene (**13**). A cDNA clone for taxadiene synthase has been obtained from stems of *T. brevifolia* and functionally expressed in *Escherichia coli*.[6] The expressed protein catalyzes all steps from GGPP to the taxadiene (**13**). The deduced amino acid sequence of the polypeptide contains a presumptive plastidial targeting sequence and has an M_r of 98 303 compared to 79 000 for the mature native enzyme.[7] Sequence comparison with other plant terpene synthases showed similarities, including the motifs **a** and **b** (Table 1).

Details of the cyclization of GGPP to taxa-4,11-diene (**13**) have been established by metabolic studies. Using a cell-free enzyme preparation from the bark of young saplings of *T. brevifolia*, Koepp *et al.*[25] established the cyclization of [1-^{3}H]GGPP to taxa-4,11-diene (**13**) and the further conversion of the derived [^{3}H]-diene by stem discs into taxol (**16**), cephalomannine (**17**), and 10-desacetyl baccatin III (**18**). That the first intermediate was the 4,11-diene (**13**) and not the 4(20),11-

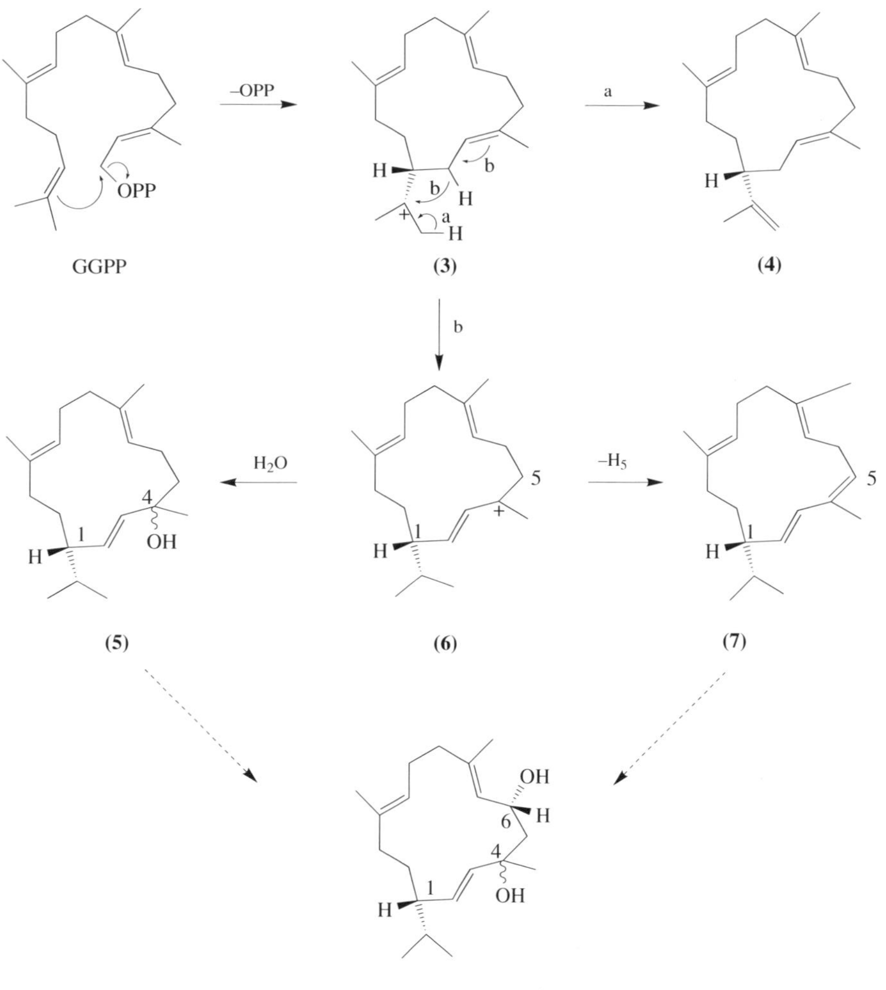

Scheme 4

diene (**14**) was unexpected. However, confirmation of this result, and further details of the formation of taxa-4,11-diene (**13**) from GGPP, were obtained from an elegant series of experiments by Lin *et al.*[26] These authors used a $\sim$600-fold purified enzyme preparation[7] from the bark and adhering cambium of *T. brevifolia* to obtain the following information. First, the conversion of [20-$^2H_3$]GGPP to taxa-4,11-diene (**13**), without loss of deuterium, eliminated the intermediate formation of the 4(20),11-diene (**14**). Second, the conversion of [1-$^2H_2$,20-$^2H_3$]GGPP to taxa-4,11-diene (**13**) without loss of deuterium eliminated the intermediate formation of casbene ((**2**), Scheme 3). Third, incorporation of label from [10-^{2}H]GGPP into taxa-4,11-diene (**13**) showed that intramolecular proton transfer occurred in the cyclization. These results are incorporated in the mechanism shown in Scheme 5. Thus, ionization of GGPP promotes bond formation in the substrate from C-1 to C-14, followed by ring closure via *re*-face attack at C-15 to give cation (**9**). Deprotonation of (**9**) by removal of its 11α-H gives 1*S*-verticillene (**10**); rapid reprotonation at C-7 by the same enzyme base initiates transannular cyclization of (**11**) by *re*-face attack of C-3 at C-8 to give the taxenyl 4-cation (**12**) which is deprotonated to (**13**). Thus, these results establish that taxa-4,11-diene (**13**) is formed from GGPP by a single enzyme, catalyzing Type A then Type B cyclization.

Following the cyclization of GGPP to taxa-4,11-diene (**13**), the latter is hydroxylated to the 5α-ol (**15**) by a microsomal enzyme preparation from stems of *T. brevifolia* or cultured cells of *T. cuspidata*.[27] The enzyme preparation has the characteristic properties of a cytochrome P450-dependent monooxygenase. No isotope effect was observed for the [20-$^2H_3$]-labeled substrate and no mechanistic details of this hydroxylation with allylic rearrangement are known (cf. the formation of gibberellin A_3 from gibberellin A_5, discussed in Section 2.08.6.1.2). The intermediacy of the dienol

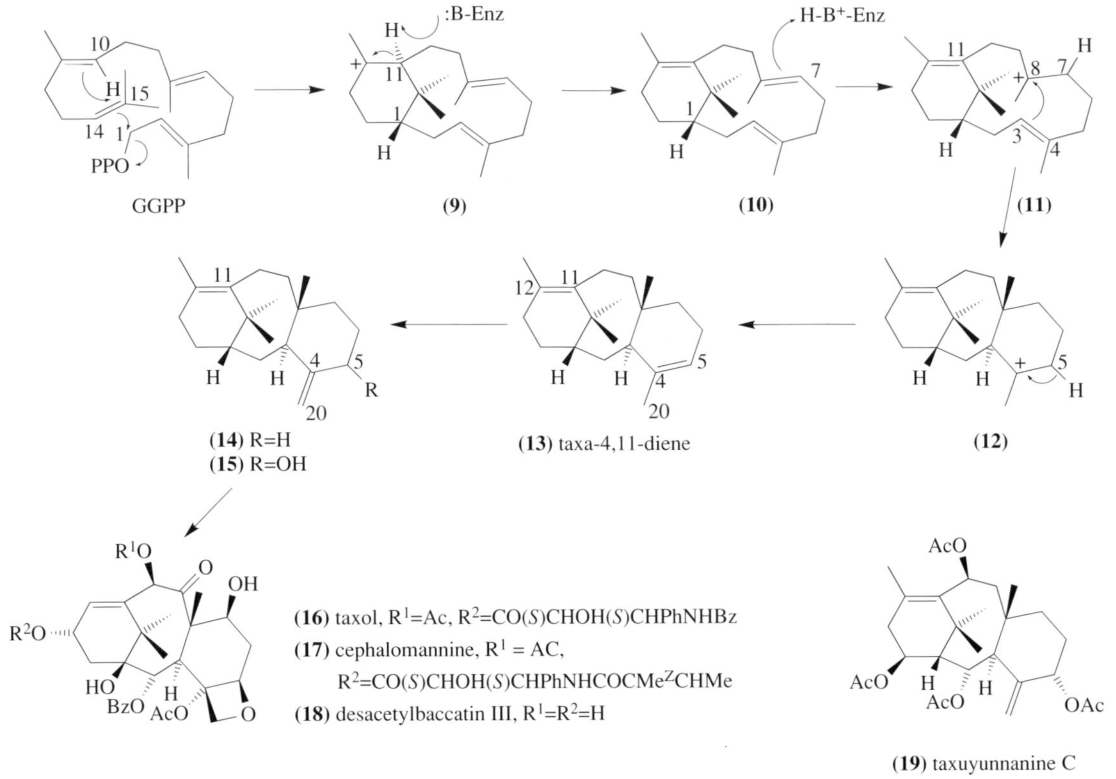

GGPP (9) (10) (11)

(14) R=H
(15) R=OH

(13) taxa-4,11-diene

(12)

(16) taxol, R^1=Ac, R^2=CO(*S*)CHOH(*S*)CHPhNHBz
(17) cephalomannine, R^1 = AC, R^2=CO(*S*)CHOH(*S*)CHPhNHCOCMe^{Z}CHMe
(18) desacetylbaccatin III, R^1=R^2=H

(19) taxuyunnanine C

Scheme 5

(**15**) was demonstrated by its occurrence in bark of *T. brevifolia* and its conversion into taxol (**16**), cephalomannine (**17**), and 10-desacetylbaccatin III (**18**) by stem discs of *T. brevifolia*.[27]

Although there has been much speculation on the mechanism of formation of the oxetane ring, no metabolic studies have been reported. Regarding the N-containing side chain, *β*-phenylalanine, phenylisoserine, and the intact *N*-benzoylphenylisoserine are incorporated into taxol; phenylalanine–ammonia lyase does not seem to be involved since cinnamic acid is not incorporated.[28]

A nonmevalonate pathway to taxuyunnanine C (**19**) in cell cultures of *T. chinensis* has been reported[29] and is discussed in Chapter 2.03.

2.08.3.2 Fusicoccin and Transversial

Fusiccocin ((**24**), Scheme 6) is the major component of a related family of phytotoxic metabolites from cultures of *Fusicoccum amygdali* Del. No enzymological studies have been reported. However, its biosynthesis from GGPP (Scheme 6) by Type A–Type B cyclization can be proposed from the results of labeling experiments.[30,31] Label was incorporated from [3-^{13}C]MVA into (**24**) at the C-3, C-7, C-11, C-15 in the rings and C-24 in the side chain, and only four of the expected five ^{3}H labels were incorporated from [2-^{14}C,4*R*-^{3}H]MVA.[30] In addition, it was shown[31] that the intensities of the ^{13}C-signals of C-7 and C-15 were suppressed, compared to those of C-3, C-11 and C-24, in the NMR spectrum of fusicoccin (**24**), obtained from (4*R*)-[3-^{13}C,4-$^2H_2$]MVA. These results support pathway a (Scheme 6), from GGPP via (**23**) to (**20**), then C-6 to C-2 bond formation, initiated by protonation at the *re*-face at C-3 of (**20**), to give (**21**) which is converted to (**22**) by two consecutive 1,2-hydride shifts, from C-2 to C-6 and from C-6 to C-7, rather than one 1,3-hydride shift from C-2 to C-7.

In contrast, biosynthesis of the related transversial (**27**) from *Cercospora tranversiana* does not involve hydride shifts from C-2, C-10, or C-14 as shown by NMR spectrometry of (**27**), obtained from [2,2,2-$^2H_3$,1-$^{13}C_1$] and [2,2,2-$^2H_3$,2-$^{13}C_1$]acetate.[32] The proposed pathway b, shown in Scheme 6, includes the suggestion that the stereochemical differences between fusicoccin and transversial at

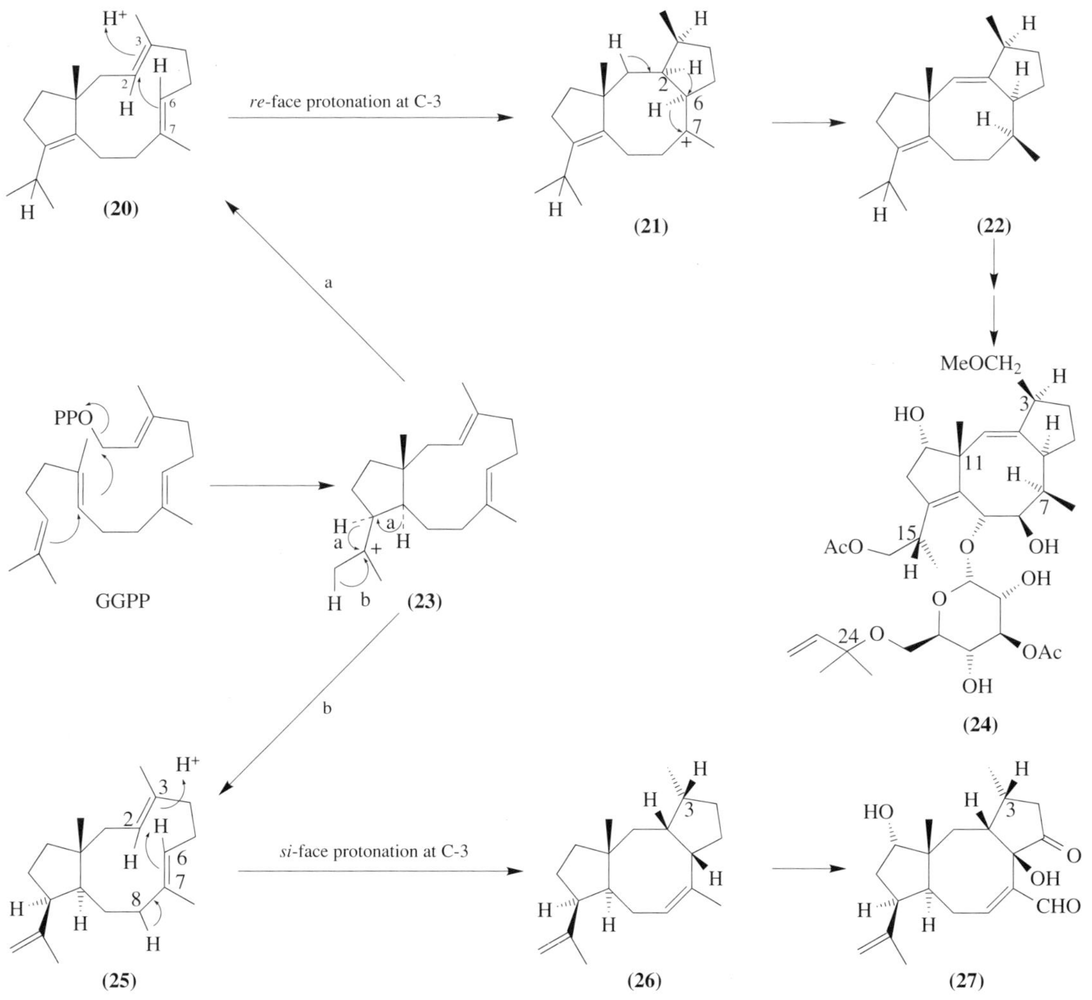

Scheme 6

C-3 may be the result of the protonation of the intermediates (**20**) and (**25**) on the *re*- and *si*-faces of C-3, respectively. The proposed intermediate, transversadiene (**26**), is a metabolite of *C. transversiana*[33] but its conversion to (**27**) has not been reported.

2.08.3.3 Miscellany

The jatrophane, latherane, tigliane, daphnane, and ingenane skeleta are close structural relatives of casbane and cembrenes. No biosynthetic studies have been reported but a putative biosynthetic relationship has been proposed.[34] The verrucosanes are also putative products of Type A cyclization, followed by Type B.[35]

2.08.4 TYPE B CYCLIZATION

As illustrated in Scheme 7, protonation of the 14,15-double bond of *E,E,E*-GGPP, followed by the attack of C-10 on C-15, then C-7 on C-11, gives four possible products, depending on the conformation of the *pro*chiral substrate. The chair–chair conformation (**28**) gives the 8-carbonium ion (**29**) of copalyl diphosphate (CPP, (**30**)) with the "normal" *anti,anti* absolute stereochemistry (Type B-1). The antipodal chair–chair conformation (**31**) of GGPP gives the 8-carbonium ion (**32**) of *ent*-copalyl diphosphate (*ent*-CPP, (**33**)) with the enantiomeric *anti,anti* absolute stereochemistry

(Type B-2). The chair–boat conformation (**34**) gives the 8-carbonium ion (**35**) of the "normal" *syn*-CPP (**36**) (Type B-3). The chair–boat conformation (**37**) gives the 8-carbonium ion (**38**) of *syn-ent*-CPP ((**39**); Type B-4).

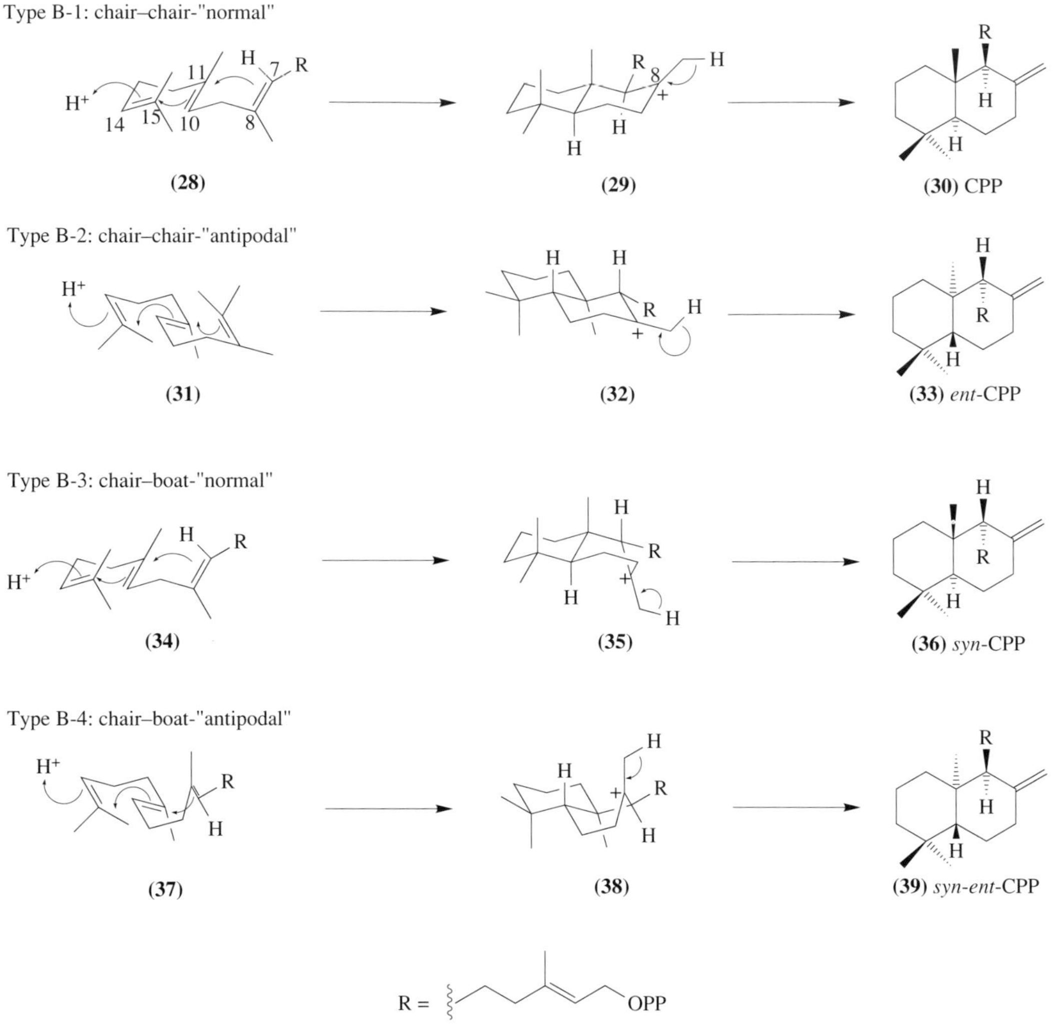

Type B-1: chair–chair-"normal"

(**28**) (**29**) (**30**) CPP

Type B-2: chair–chair-"antipodal"

(**31**) (**32**) (**33**) *ent*-CPP

Type B-3: chair–boat-"normal"

(**34**) (**35**) (**36**) *syn*-CPP

Type B-4: chair–boat-"antipodal"

(**37**) (**38**) (**39**) *syn-ent*-CPP

R = OPP

Scheme 7

Each of these cyclization types can be inferred from studies on the diterpenes that are formed by further elaboration of the immediate products (see later sections). However, the only cyclase that has been characterized is *ent*-CPP (**33**) synthase (Type B-2). The formation of *ent*-[^{14}C]CPP (**33**) from [^{14}C]GGPP was first established[36] using a soluble enzyme preparation from the mycelia of the gibberellin-producing fungus, *Gibberella fujikuroi*. The enzyme was partially purified from this fungus[36] and from the endosperm of *Marah macrocarpus*.[37] However, detailed information on *ent*-CPP synthase comes from the cloning[4] of the *GA1* locus in *Arabidopsis thaliana*, using the GA-responding dwarf mutant *ga1* and a genomic subtraction technique. This gene has been shown[5] to encode an *ent*-CPP synthase with a molecular mass of 86 kDa which is imported into pea chloroplasts and processed to a 76 kDa protein. The amino acid sequence of *ent*-CPP synthase contains the motif b (Table 1) and is similar to those of other plant cyclases (for a review, see ref. **37**). The An1 gene from *Zea mays* (maize) contains motif b and probably also encodes an *ent*-CPP synthase.[13]

Type B cyclization of *Z*-isomers of GGPP has not been studied in detail. However, the *E,Z,E*-GGPP was excluded as a precursor of 9,10-*syn*-diterpenes in cell cultures of *Oryza sativa*.[39]

The immediate products of these Type B cyclizations of *E,E,E*-GGPP are modified by functional group transformations to yield labdanes and their rearrangement products, the clerodanes. These 6,6-bicyclic products are discussed in the following sections.

2.08.4.1 Labdanes and *ent*-Labdanes

Labdanes, derived via (**29**), and *ent*-labdanes, derived via (**32**), are of widespread occurrence but few definitive biosynthetic studies have been reported. Indeed, it has been suggested that they may be derived from GGPP by a two-step process via monocyclic intermediates on the basis that both labdane and retinanes co-occur in different populations of *Bellardia trixago*.[40] Another point of biosynthetic interest is the reported co-occurrence of labdanes and *ent*-labdanes. For example (Scheme 8), the enantiomers (**40**) and (**41**) occur in the resin of *Eperua purporea*[41] and *ent*-sclarene (**42**) and sclarene (**43**) have been isolated from different specimens of *Dacridium intermedium*.[42] The co-occurrence of enantiomeric labdanes is discussed by Carman and Duffield[43] but no enzymological studies on this point have been reported.

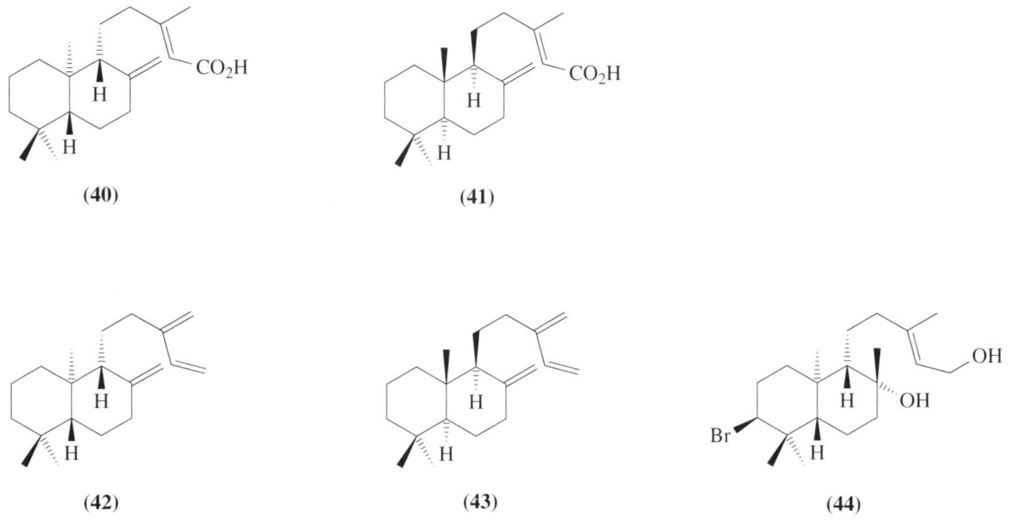

Scheme 8

In addition to the previously discussed formation of *ent*-CPP (**33**) from GGPP, the biosynthesis of "normal" labdanes has been investigated in tobacco (*Nicotania tobaccum* L.).[44,45] Cyclization of GGPP to *cis*-abienol (**45**), as shown in Scheme 9, occurs in extracts from trichome glands of the leaves of the cultivar T.I. 1068.[44] The synthase activity was not present in the epidermal or sub-epidermal tissue from which the glands were taken. It was soluble, Mg^{2+} stimulated, and unaffected by conditions that inhibit cytochrome P450 oxygenases. From these results it was suggested that the enzyme-bound intermediate, CPP 8-carbonium ion (**29**), is hydrated at C-8, followed by elimination of the elements of HOPP as shown in Scheme 9. It may be that capture of the 8-carbonium ion by H_2O diverts a Type A cyclization, resulting in the elimination of 12-H from the 13-cation. In a related study,[45] labdenediol (**46**) and sclareol (**47**) were both formed from GGPP in cell-free extracts of another cultivar, 24A, of tobacco. The enzyme activities for the formation of each product were not separated after protein purification, or by thermal inactivation or in product inhibition studies. It was therefore concluded that labdenediol (**46**) and sclareol (**47**) are the direct products of one synthase, operating on GGPP. Although these products appear to require hydrolysis of the diphosphate, they could also arise from an aborted Type A cyclization whereby the allylic carbonium ion is captured by H_2O; thus a diphosphatase would not be required. It remains to be determined if the two cultivars, T.I. 1068 and 24A, of tobacco contain two distinct GGPP cyclases.

The occurrence of the 3-bromo-*ent*-labdane, aplysin-20 ((**44**), Scheme 8), from the mollusc, *Aplysia kurodai*, suggests[46] that Type B cyclization can be initiated by Br^+ as well as H^+.

2.08.4.2 Clerodanes

The clerodanes[47] comprise a large family, considered to be derived from rearrangement of the 8-carbonium ions (**29**) and (**32**). As shown in Scheme 10, *trans*-clerodanes can, in principle, be the products of concerted 1,2-shifts including C-19 to C-5 (sequences a). In the case of the *cis*-clerodanes, a similar series of shifts cannot be concerted; for a stereoelectronic shift of C-18 to C-5, sequences

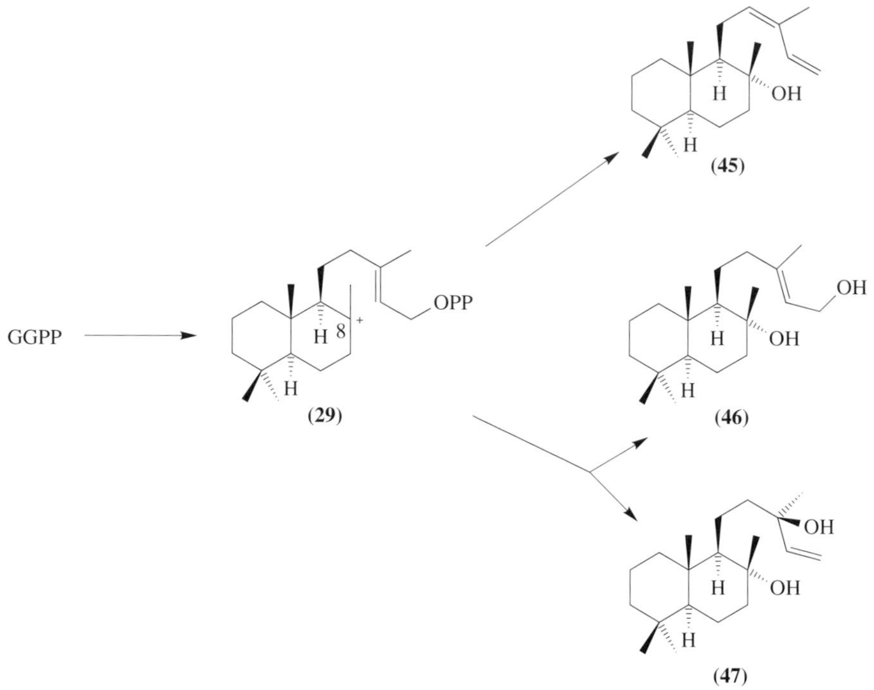

Scheme 9

b would require a conformational change in ring A. A consequence would be the prediction that the 3β-H of the bicyclic intermediates is lost in the formation of the 3,4-dehydro-*trans*-clerodanes and that the 3α-H is lost in the formation of the 3,4-dehydro-*cis*-clerodanes. However, this point has not been examined experimentally. Evidence for the proposed rearrangements come from the isotope ratios in a series of furanoclerodanes, obtained from *Tinospora cordifolia*, after administering [4*R*-^{3}H,2-^{14}C]MVA.[48] Particularly instructive is the compound ((**48**) or its enantiomer, Scheme 10) in which C-18, derived from the 2-^{14}C of MVA, was shown to be retained at C-5 from the degradative evidence that C-19 was unlabeled. Indirect evidence for the shift of C-18 from C-4 to C-5 is provided by the results of the labeling of heteroscyphic acid A ((**49**) or its enantiomer, Scheme 10) in cultured cells of the liverwort *Heteroscyphus planus*.[49] However, the most interesting result from this study is the nonequivalent labeling, indicating that the GGPP precursor is formed from MVA-derived FPP and non-MVA-derived IPP (see Chapter 2.03).

2.08.5 TYPE B–TYPE A CYCLIZATIONS

The initial 6,6-bicyclic products of Type B cyclization of GGPP can undergo Type A cyclization by two different stereochemical routes. Following ionization of the diphosphate, the 8(17)-double bond can attack the *si*- or *re*-face of C-13 and give two series of tri-, tetra-, and pentacyclic systems which are then further modified by oxidation and/or by rearrangement. The outcome from each of the 6,6-bicyclic products CPP (**30**), *ent*-CPP (**33**), *syn*-CPP (**36**), and *syn*-*ent*-CPP (**39**) (see Scheme 7) is discussed separately.

2.08.5.1 Tricycles from CPP

The possible pathways to known 6,6,6-tricyclic diterpenes are outlined in Scheme 11. The *re*-face and *si*-face cyclization of CPP (**30**) on C-13 gives the 8-carbonium ions (**54**) and (**63**), respectively. Deprotonation of (**54**) leads to virescene (**51**) and sandaracopimaradiene (**55**), and deprotonation

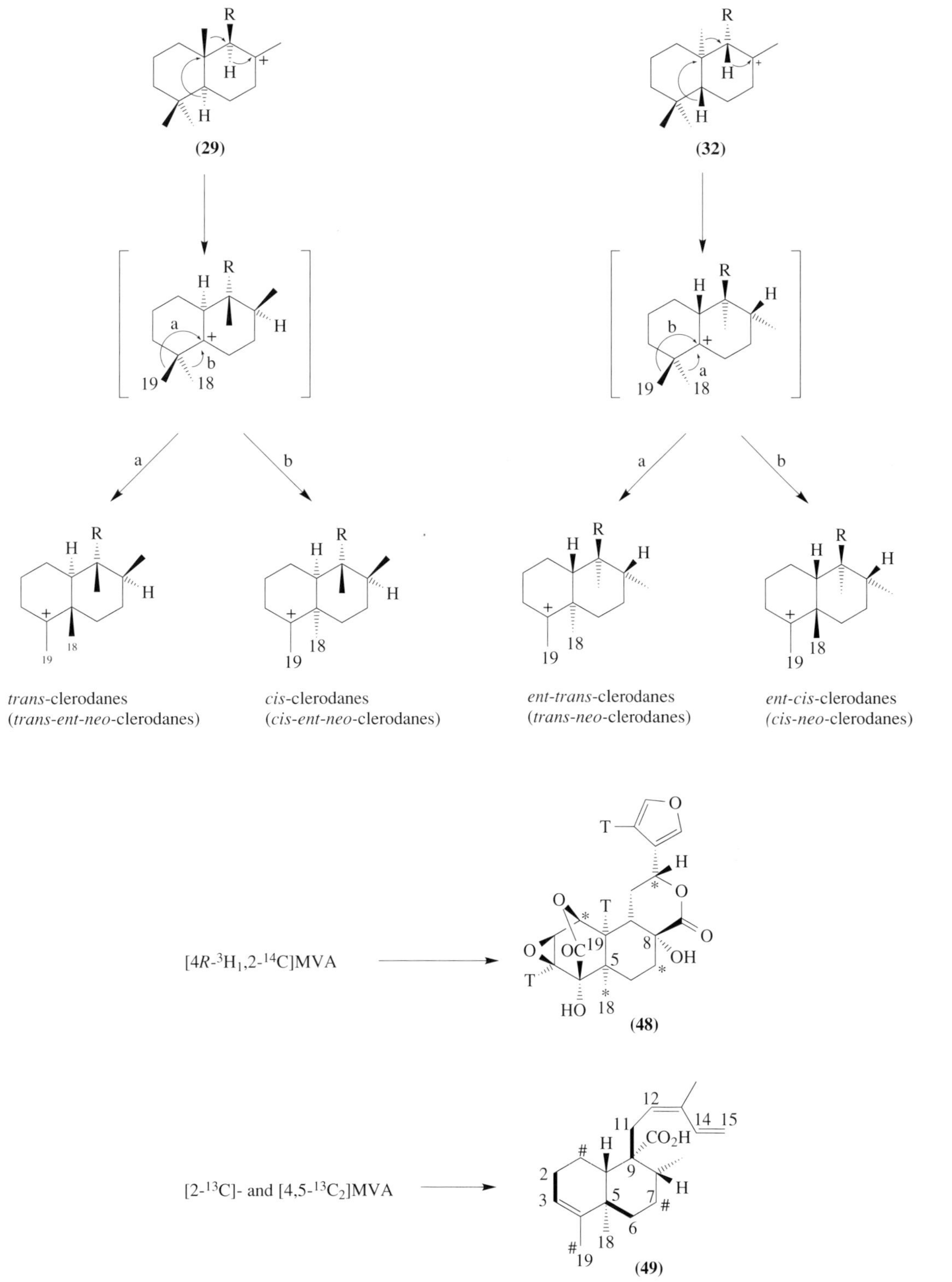

R H + H **(29)**

R H + H **(32)**

H R H a + b 19 18

H R H b + a 19 18

a b a b

H R H + 19 18 H R H + 18 19 H R H + 18 19 H R H + 18 19

trans-clerodanes (*trans-ent-neo*-clerodanes)

cis-clerodanes (*cis-ent-neo*-clerodanes)

ent-trans-clerodanes (*trans-neo*-clerodanes)

ent-cis-clerodanes (*cis-neo*-clerodanes)

$[4R\text{-}^{3}H_1,2\text{-}^{14}C]$MVA →

T O H * O O T O * OC 19 8 OH 5 * O T HO 18 *

(48)

$[2\text{-}^{13}C]$- and $[4,5\text{-}^{13}C_2]$MVA →

12 11 14 15 CO_2H # H 9 2 3 5 7 H # 6 # 19 18

(49)

* denotes carbon from $[2\text{-}^{14}C]$MVA; # denotes carbon from $[2\text{-}^{13}C]$MVA; thick lines show $^{13}C\text{–}^{13}C$ coupling from$[4,5\text{-}^{13}C_2]$MVA

Scheme 10

of (**63**) leads to pimara-8(14),15-diene (**64**) and pimara-8,15-diene (**67**). Metabolic evidence for these steps includes the conversion of [^{14}C]CPP (**30**) to the [^{14}C]pimaradiene (**67**) in cell-free extracts of *Tricothecium roseum*[50] and the NMR spectrum of virescenol B (**50**), derived from [1,2-^{13}C]acetate in *Oospora virescens*, shows ^{13}C–^{13}C coupling, consistent with its derivation via GGPP and (**54**).[51] In the step (**30**) to (**63**) an overall *anti*-S_N2' displacement of the diphosphate anion has been established in the biosynthesis of rosenonolactone (see later) in which the 16*Z*-H and the 16*E*-H (see (**63**)) are respectively derived from the 5-pro*R* and 5-pro*S*-H of MVA.[52] Also, the 1,2-shift (a) of the methyl from C-13 to C-15 in (**55**) or (**64**) to form (**59**) is stereoselective, at least for cryptotanshinone (**68**) (**52**) and ferruginol (**69**).[54] The expected ^{13}C-isotopic labeling (see structures) was observed for each compound, derived from [U-$^{13}C_6$]glucose in cell cultures of *Salvia miltiorrhiza*. This information, together with determination of the absolute stereochemistry at C-15 in (**68**), established that the C-17 methyl group migrates to the *si*-face of C-15.[53] Confirmation of this result was obtained[54] by showing that it is the migrated methyl group that becomes the pro*R*-methyl in the isopropyl group in ferruginol (**69**).

Enzymological studies provide detailed evidence for the biosynthesis of abieta-7,13-diene (**59**) and abietic acid (**62**) by the steps shown by bold arrows in Scheme 11. Abietadiene synthase is constitutive in stems of lodgepole pine (*Pinus contorta*) and both constitutive and wound-inducible in stems of grand fir (*Pinus grandis*).[55,56] Partial purification[9] of the soluble extract from grand fir gave a native protein of 80 kDa that catalyzed the cyclization of [$^3H_1$]GGPP to abietadiene ((**59**), 90%) and sandaracopimaradiene ((**55**), 9%) in the presence of a divalent cation, preferably Mg^{2+}. Using cDNA from wound-induced stem and PCR degenerate primers, based on the amino acid sequences of tryptic digests from the native protein, three cDNA clones were obtained and functionally expressed in *E. coli*,[8] thereby confirming that a single protein catalyzed the conversion of GGPP to abietadiene in a multistep sequence of Type B–Type A cyclizations. The clone that yielded the highest levels of cyclase activity encoded a protein of 99.5 kDa, including a putative plastid targeting sequence, rich in serine and threonine. The deduced size (88 kDa) of the mature protein agrees with the molecular mass assigned to the partially purified, native protein. The deduced amino acid sequence also contained both motifs a and b (Table 1), consistent with the catalytic properties of a Type B–Type A cyclase. A possible mechanism is shown in the bold arrow sequence of Scheme 11 whereby the enzyme-bound intermediate CPP (**30**) is transformed either by 13*re*-face cyclization to (**54**) and sandaracopimaradiene (**55**) or by 13*si*-face cyclization to (**63**) and the pimaradiene (**64**). Rearrangement of (**55**) or (**64**) to (**58**), followed by deprotonation, yields abietadiene (**59**). The isolation of sandaracopimaradiene (**55**) as a minor product from GGPP, using both the purified native protein and the cloned cyclase, indicates that 13*re*-face cyclization of (**30**) may be the preferred route.

The formation of abietic acid (**62**) from abietadiene (**59**) by stepwise oxidation at C-18 has been demonstrated[55] in cell-free extracts from stems of lodgepole pine and grand fir. The first two steps to the 18-ol (**60**) and the 18-al (**61**) are catalyzed by a microsomal fraction with the expected properties of cytochrome P450 oxygenases but showing different sensitivities to known inhibitors of plant P450s. The third step from the 18-al (**61**) to the acid (**62**) is catalyzed by soluble fractions; the activity is not inhibited by CO, does not require O_2, and uses NAD^+ as a cofactor, indicating a soluble aldehyde dehydrogenase.

Simple variations of the position of deprotonation of the intermediate (**58**), followed by stepwise oxidation, would account for the origin of the common resin acids, isopimaric acid (**52**), palustric acid (**53**), *neo*-abietic acid (**57**), and laevopimaric acid (**66**), but no experimental details have been published.

Rearrangement of the tricyclic carbonium ion ((**63**), Scheme 11) accounts for the origins of rosenonolactone ((**71**), Scheme 12). This metabolite of *Trichothecium roseum* has been shown to be formed from CPP (**30**) by migration of the C-9 hydrogen in (**63**), derived from the pro-4*R* hydrogen of MVA, to C-8 and of the C-10 methyl to C-9.[57,58] These two shifts (Scheme 12) would lead to (**70**), analogous to the formation of the clerodanes (Scheme 10). However, the exact timing of lactone formation from (**70**) is not known. Carbon-19, not C-18, is derived from C-2 of MVA.[59] This point is returned to in discussing the gibberellins and the evidence that Type-B cyclization of *E,E,E*-GGPP is all-*trans*.

The biosynthesis of pleuromutillin ((**74**), Scheme 12), a metabolite of *Pleurotus mutilus*, has been studied in detail.[60] The various rearrangements, summarized in Scheme 12, were proposed from the results of labeling in (**74**) from [1-^{14}C]acetic acid, [2-^{14}C]MVA, [1-^{14}C]GGPP, (3*R*,4*R*)-[4-$^3H_1$]MVA, [5-$^3H_2$]MVA, and (3*R*,5*R*)-[5-$^3H_1$]MVA. Since the absolute stereochemistry of (**74**) is known,[61] GGPP appears to be cyclized to the *syn*-CPP 8-carbonium ion (**35**) which is then transformed to (**72**) by a series of nonconcerted 1,2-shifts. Type A cyclization of (**72**) to (**73**), followed by a *trans*-annular hydride shift, leads to (**74**).

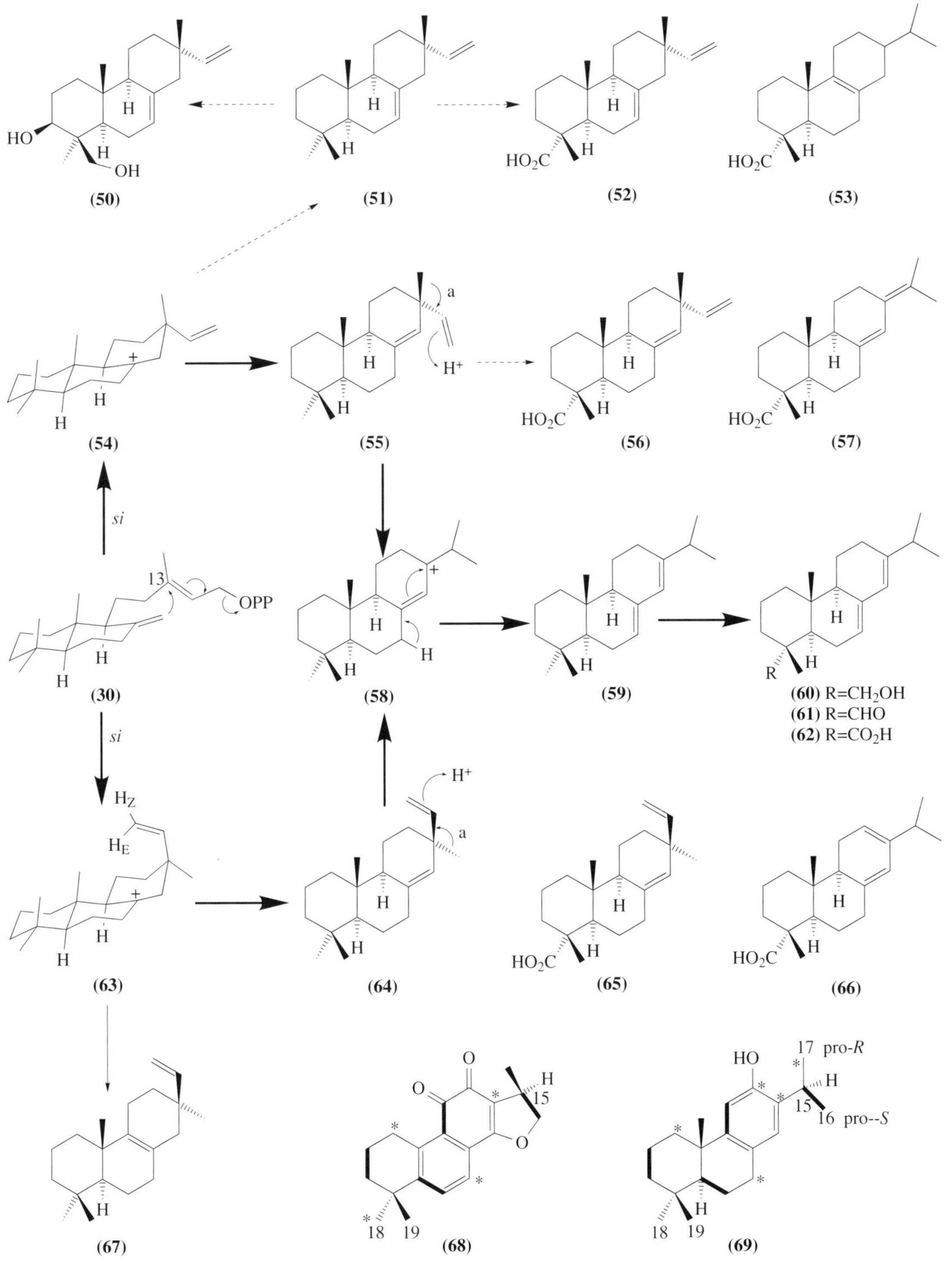

Scheme 11

2.08.5.2 Tricycles from *ent*-CPP and *syn*-CPP

In cell cultures of *Oryza sativa* (rice) that have been exposed to either UV light[62] or chitin,[39] *ent*-CPP (**33**) is cyclized to *ent*-sandaracopimaradiene (**77**), presumably by *si*-face attack of C-17 at C-13 via ((**75**), Scheme 13). In the same system, *syn*-CPP (**36**) is cyclized to 9β-pimara-7,15-diene (**78**),

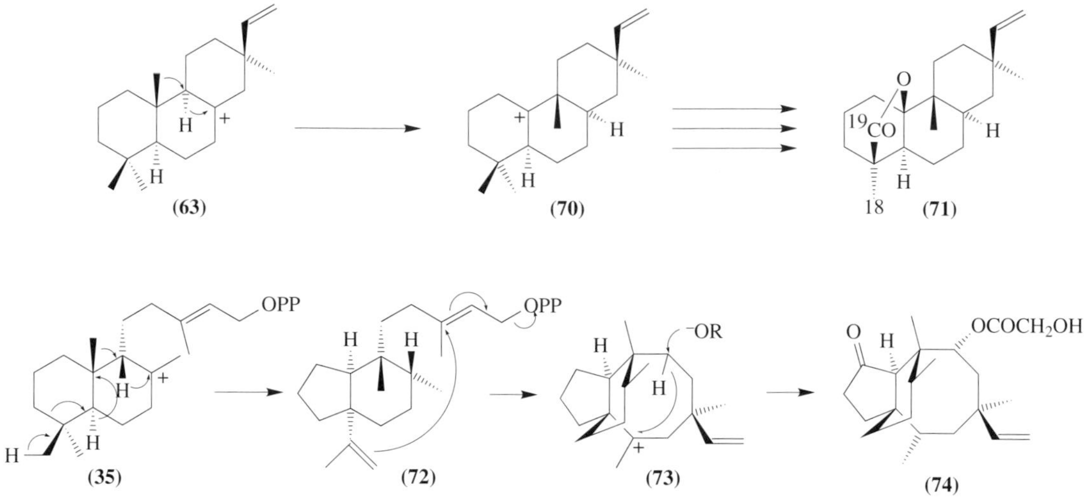

Scheme 12

presumably via (**76**). Neither *ent*-sandaracopimaradiene (**77**) nor 9β-pimara-7,15-diene (**78**) are formed in nonelicited cell cultures. The elicitation of the enzymes for the formation of (**77**) and (**78**) is of interest vis-à-vis the phytoalexins, such as the oryzalexins (e.g., **79**)[63] and momililactones (e.g., **81**)[64] which are formed in rice plants that have been exposed to UV light or fungal infection. However, no biosynthetic studies on the formation of (**79**) from (**77**) or (**81**) from (**78**) have been reported. *ent*-Sandaracopimaradiene (**77**) is also formed from GGPP in soluble enzyme preparations from seedlings of castor bean.[18]

The *ent*-abieta-7,13-diene ((**80**) Scheme 13), reported to occur in *Helichrysum chionosphaerum*,[65] may be formed from *ent*-CPP (**33**) via (**77**) by a sequence similar to the formation of abieta-7,13-diene ((**59**), Scheme 11).

There are no reports of the cyclization of *ent-syn*-CPP (**39**) to tricyclic diterpenes. The absolute stereochemistry, assigned to the *ent*-9β-pimara-7,15-dienes, reported to occur in *Calceolaria* species,[66] has not been established.

2.08.6 TETRACYCLES AND PENTACYCLES

As illustrated in Scheme 7, Type B cyclization of GGPP gives four bicyclic products ((**30**), (**33**), (**36**), and (**39**)). Further cyclization of each of these 6,6-bicycles (**82**) is shown in general terms in Scheme 14. Type A cyclization by attack of C-17 at the *re*- or *si*-face of C-13 gives rise to eight possible stereoisomers of the tricyclic 8-carbonium ion (**83**). These tricycles can undergo cyclization from C-15 to C-8 ((**83**) to (**84**)) or from C-15 to C-9 after migration of 9H to C-8 ((**83**) to (**85**)) to give 16 possible tetracyclic carbonium ions that can undergo further rearrangement to give more than 75 different tetra- and pentacyclic ring systems. Some, but not all, of these ring systems have been shown to occur naturally. Those that are known, and for which biosynthetic information is available, are discussed in the following sections.

2.08.6.1 From *ent*-CPP by *re*-Face Cyclization on C-13

This group is the most abundant and the most studied. Scheme 15 shows possible intermediates from the 8-carbonium ion (**87**) from *ent*-CPP (**33**) to the natural *ent*-kaur-16-ene (**86**), *ent*-beyerene (**88**), *ent*-trachylobane (**89**), *ent*-atisir-16-ene (**90**), and *ent*-atisir-15-ene (**91**). Early evidence for these pathways came from the observation that MVA and GGPP were converted to (**86**), (**88**), (**89**), and

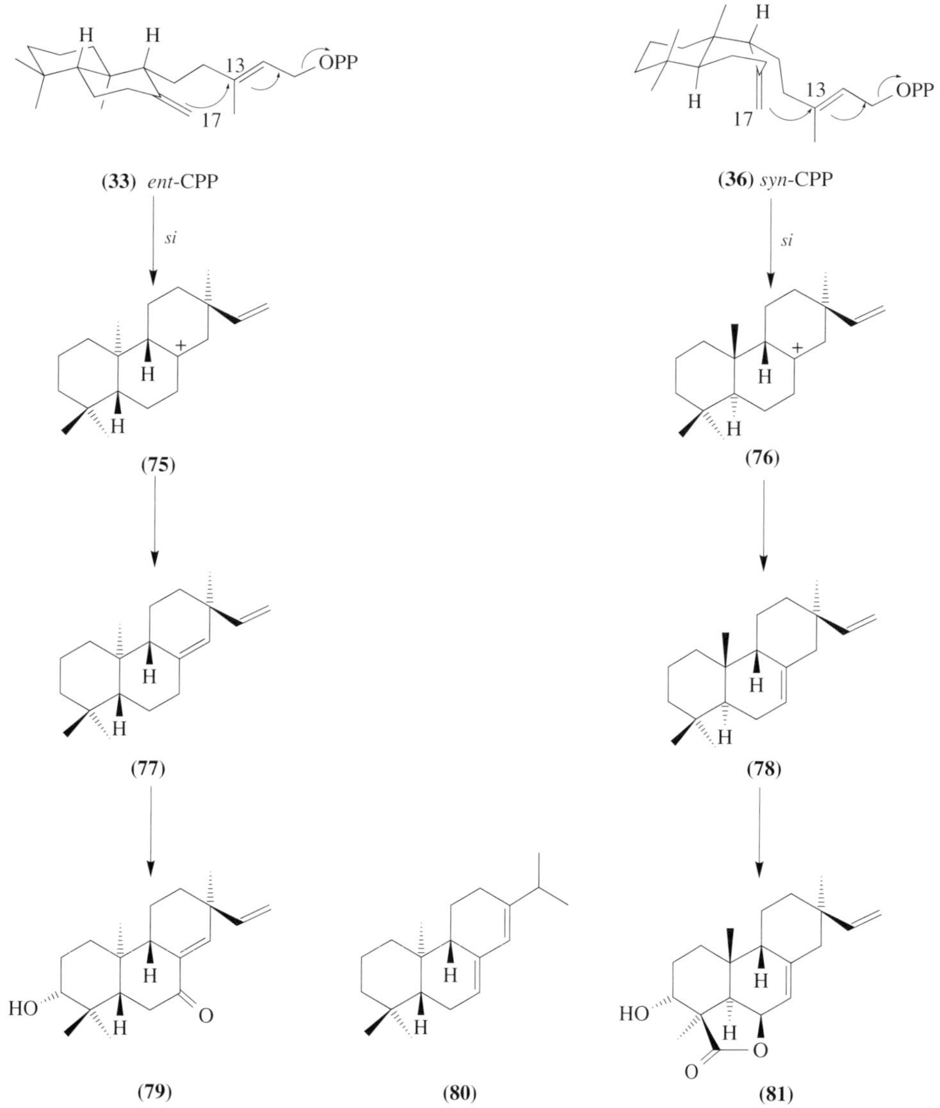

(33) *ent*-CPP **(36)** *syn*-CPP

si *si*

(75) **(76)**

(77) **(78)**

(79) **(80)** **(81)**

Scheme 13

(90) in addition to casbene (**(2)**, Scheme 3) and *ent*-sandaracopimaradiene (**(77)**, Scheme 13) by soluble enzyme preparations from castor been seedlings.[14,18] Partial purification of the crude enzyme preparation indicated that separate GGPP cyclases were present for the formation of each of the diterpenes except for *ent*-kaur-16-ene **(86)** and *ent*-trachylobane **(89)**. However, subsequent studies have been directed to the biosynthesis of *ent*-kaur-16-ene **(86)** and its conversion to the gibberellin family of plant hormones and these studies are now discussed separately.

2.08.6.1.1 **ent-*Kaur-16-ene and* ent-*kaurenoids***

The Type A cyclization of ent-CPP **(33)** to *ent*-kaur-16-ene (**(86)**, Scheme 15) was first demonstrated in a cell-free system from the fungus, *Gibberella fujikuroi*, and in homogenates from seeds of *M. macrocarpus* and seedlings of *R. communis*.[67] The enzyme concerned, *ent*-kaur-16-ene synthase,

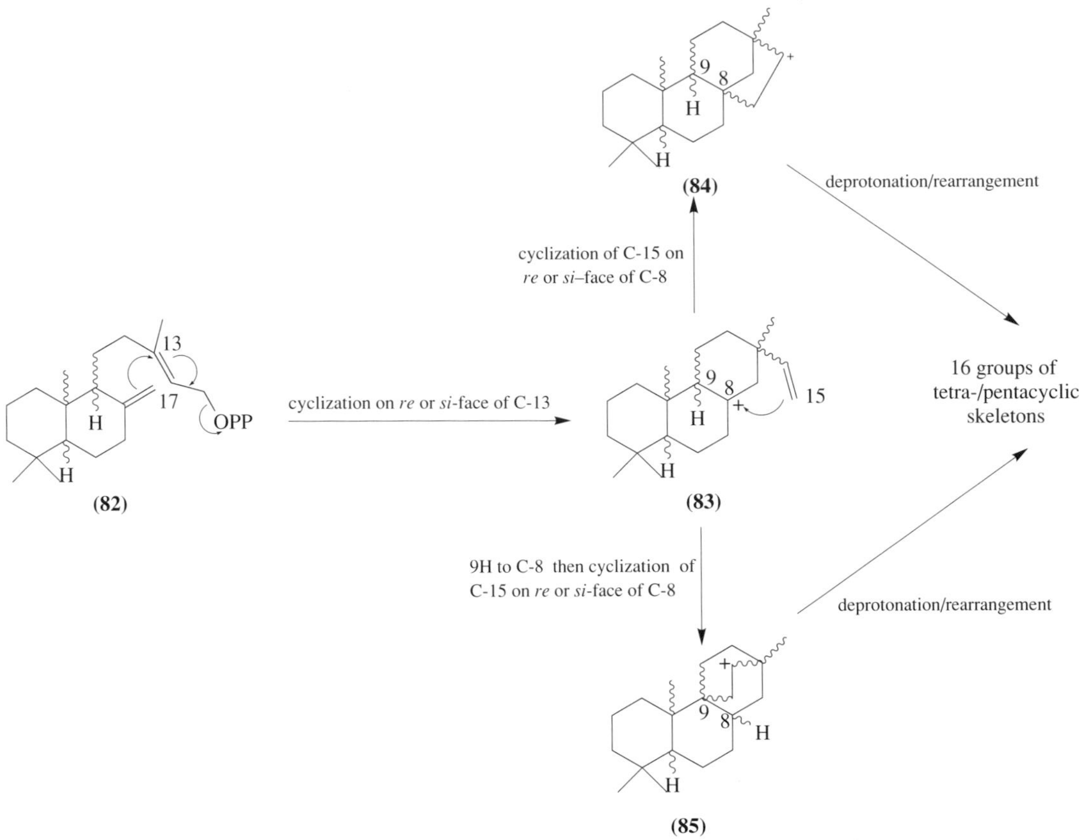

(84)

cyclization of C-15 on *re* or *si*–face of C-8

deprotonation/rearrangement

16 groups of tetra-/pentacyclic skeletons

(82)

cyclization on *re* or *si*-face of C-13

(83)

9H to C-8 then cyclization of C-15 on *re* or *si*-face of C-8

deprotonation/rearrangement

(85)

Scheme 14

has been partially purified from *M. macrocarpus*[37] and, more recently, purified to near homogeneity (M_r, 81 000) from *Cucurbita maxima*.[10] The gene encoding the *C. maxima* enzyme has been cloned and expressed in *E. coli* and the recombinant protein converts *ent*-CPP to (**86**).[11] The gene product has an M_r of 89 000 and the sequence contains a possible plastid-targeting N terminal sequence and motif **a** (Table 1), characteristic of a Type A cyclase. Some details of the stereochemistry of the cyclization in *M. macrocarpus* were provided in earlier investigations with [17-*E*-$^3H_1$]-*ent*-CPP which gave rise to *ent*-kaur-16-ene (**86**) labeled with tritium at the C-15 *endo*-position. Similarly, [1(*S*)-$^3H_1$]GGPP gave *ent*-kaur-16-ene with tritium at the 14β-position.[68] Support for the migration of C-12 to C-16 comes from feeding [3,3′-$^{13}C_2$]MVA to *G. fujikuroi*, where retention of ^{13}C–^{13}C coupling was observed at C-16 and C-17 in *ent*-kaur-16-ene ((**86**), Scheme 15).[69]

Oxidative metabolism of *ent*-kaur-16-ene (**86**) gives rise to the plethora of naturally occurring *ent*-kaurenoid diterpenes and also the gibberellin plant hormones (see Section 2.08.6.1.2). Oxidation at C-19 to the acid is a feature common to many kaurenoids and is also the first transformation on the route to the gibberellins. Because of this, oxidative metabolism at rings A and B of *ent*-kaur-16-ene has been much studied (Scheme 16). The oxidation of (**86**) to the alcohol (**92**), and of the aldehyde (**93**) to *ent*-kaur-16-en-19-oic acid (**94**) is catalyzed by microsomal enzymes with the properties of P450 monooxygenases.[70] The conversion of the 19-alcohol (**92**) to the 19-aldehyde (**93**) is also catalyzed by a microsomal activity and involves the loss of the 19-pro-*R* hydrogen atom.[71]

Studies in a cell-free system from *G. fujikuroi* have shown that the hydroxylation of *ent*-kaur-16-en-19-oic acid (**94**) to *ent*-7α-hydroxykaur-16-en-19-oic acid (**96**) is also catalyzed by a microsomal P450 monooxygenase activity.[72] It is likely, but not proven, that the 6,16-dienoic acid (**97**) can also be a product of this enzyme. The biosynthesis of the kaurenolides (e.g. **101**) has been established to proceed from (**97**),[73,74] probably via the unstable epoxide (**99**) which undergoes spontaneous intramolecular cyclization to (**101**) under aqueous conditions.[74] Further oxidation of *ent*-7α-hydroxykaur-16-en-19-oic acid (**96**) at C-6 gives rise to the *ent*-6α,7α-*diol* (**98**) and the ring-B contraction product, gibberellin A_{12}-aldehyde (**100**). Both (**98**) and (**100**) appear to be products of the same P450-type of activity in the *C. maxima* system,[75] and thus can be formed from the same

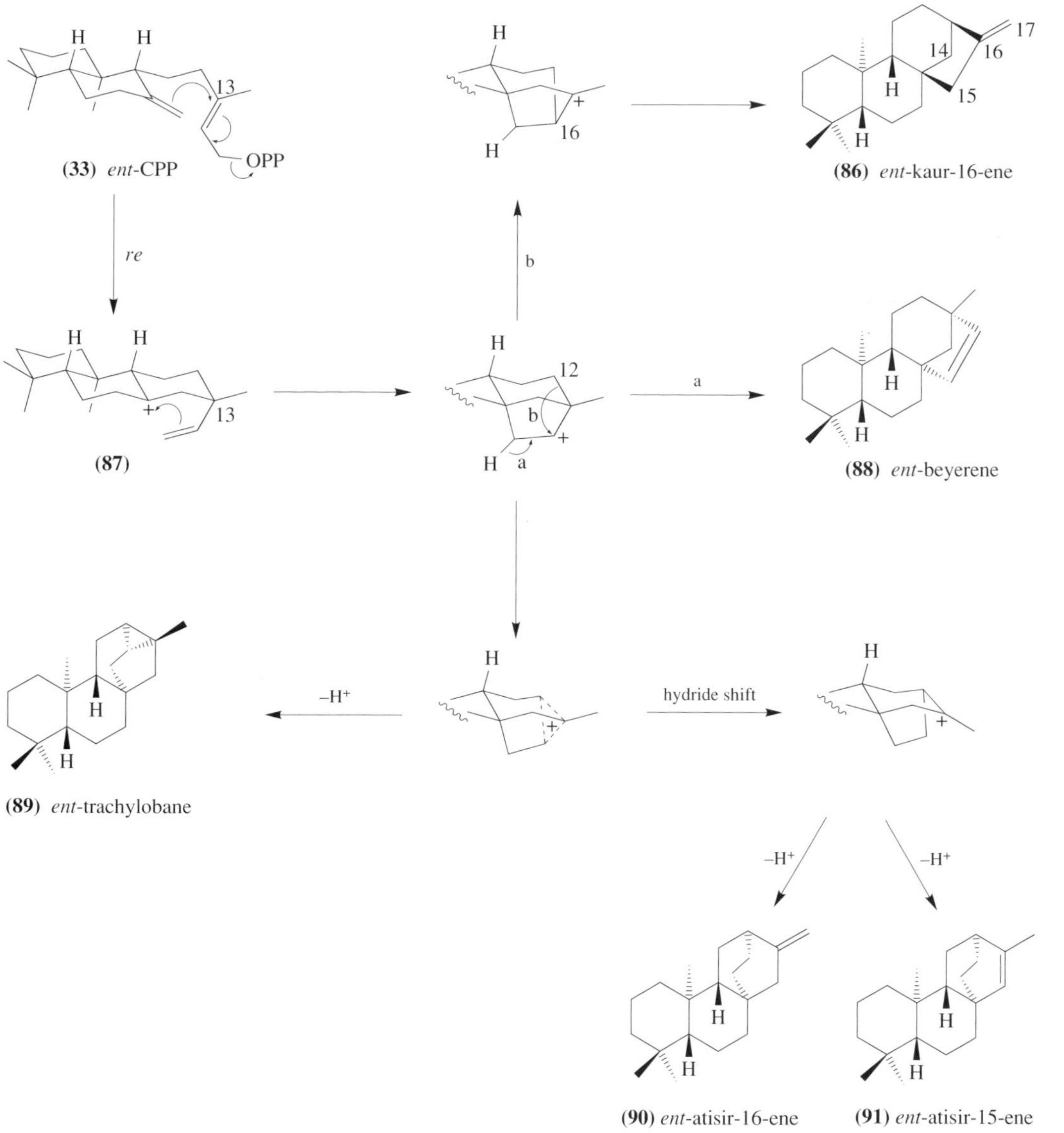
H
H
13
OPP
(33) ent-CPP
re
H
H
+
13
(87)
H
+
16
H
b
H
12
b
+
H a
H
14
16
17
H
15
H
(86) ent-kaur-16-ene
a
H
H
(88) ent-beyerene
H
H
(89) ent-trachylobane
–H+
H
+
hydride shift
H
+
–H+
–H+
H
H
(90) ent-atisir-16-ene
H
H
(91) ent-atisir-15-ene

Scheme 15

C-6 free radical (**95**) from which a 1,2 radical shift results in the extrusion of C-7 (Scheme 16). Studies with *ent*-kaur-16-en-19-oic acid (**94**), stereospecifically labeled with deuterium at C-6 and C-7, have demonstrated that the C-6 and C-7 hydroxylations proceed with retention of configuration, while the dienoic acid (**97**), and hence the kaurenolides (**101**), are formed with stereoselective loss of the *ent*-7α-hydrogen and nonstereoselective loss of the *ent*-6α and 6β hydrogen.[76] This study also revealed that ring contraction to (**100**) proceeds with stereoselective loss of the *ent*-6α-hydrogen, complementing earlier results with [6,6-$^3H_2$,^{14}C]-*ent*-kaur-16-en-19-oic acid (**94**).[77]

2.08.6.1.2 Gibberellins

The gibberellins (GAs) are a family of over 100 tetracyclic diterpenoids biosynthetically derived from gibberellin A_{12}-aldehyde ((**100**), Scheme 16). They are produced by several fungi, the most notable being *G. fujikuroi*, from which the GAs were first isolated. They also occur in higher plants, where they have an important role as hormones involved in many aspects of normal growth and development. As a result of their biological significance there is much literature on the biosynthesis

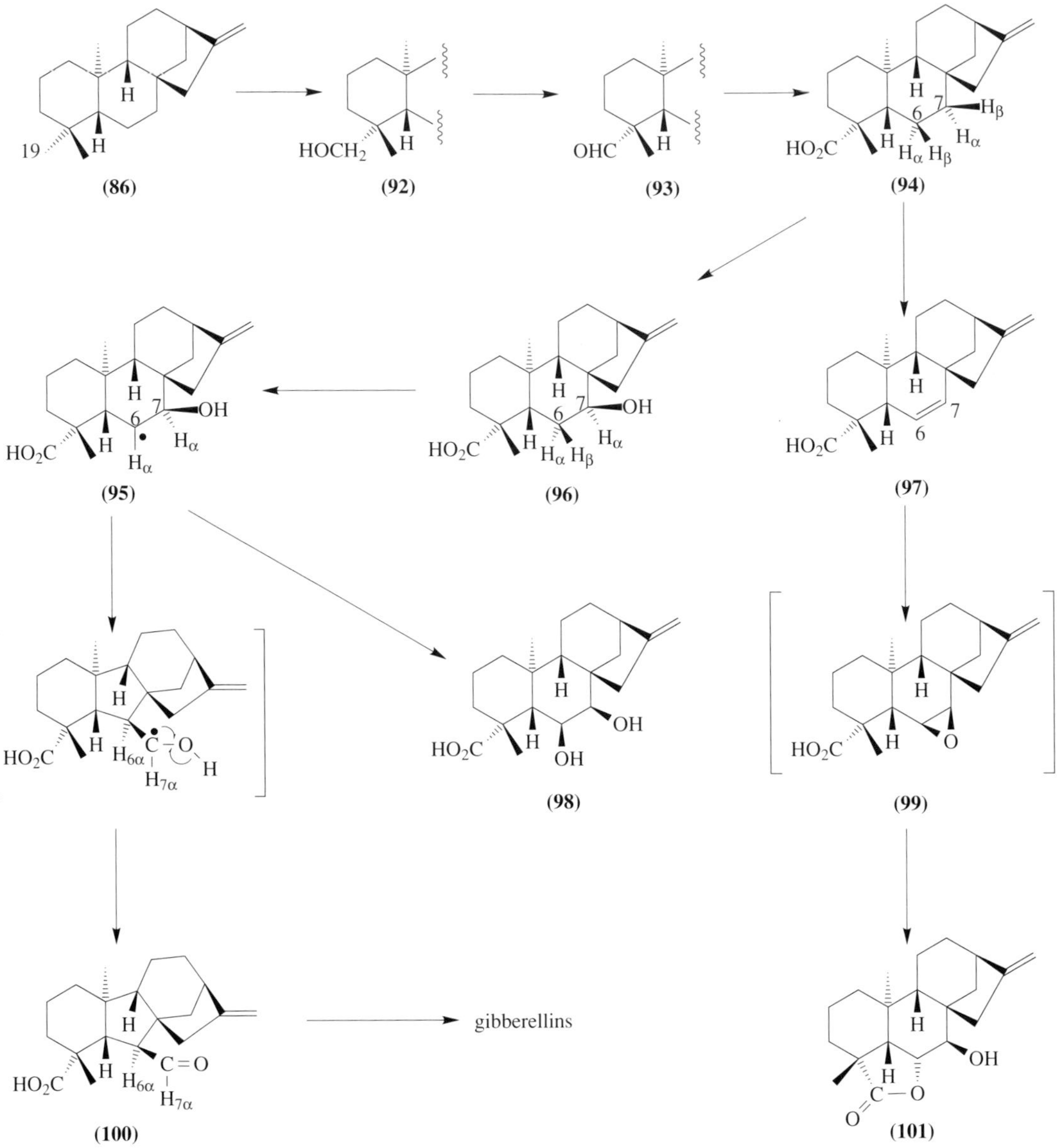

Scheme 16

of GAs, comprehensively reviewed in several recent articles.[78–80]787980 Here discussion will be confined to the basic pathway from (**100**) to biologically active GAs, with the emphasis on information arising out of the recent cloning of some of the enzymes involved. The pathways from GA_{12}-aldehyde (**100**) are shown in Scheme 17. Several parallel pathways exist as a result of the different timing of hydroxylation steps. The pathways in higher plants (shown in Scheme 17 by bold arrows) differs from that in *G. fujikuroi* (plain arrows) and they are discussed first. Oxidation of (**100**) at C-7 to the acid, with or without hydroxylation at C-13, gives rise to two pathways, via GA_{12} (**102**) and GA_{53} (**103**). The sequence of subsequent events involves progressive oxidation and eventual loss of C-20 with the respective formation of the γ-lactones, GA_9 (**113**) or GA_{20} (**114**), the parents of the biologically active GAs. Stepwise oxidation at C-20 proceeds via the alcohols GA_{15} (**105**) or GA_{44} (**106**) and the aldehydes GA_{24} (**108**) or GA_{19} (**109**). The stereochemistry of the oxidation of the 20-alcohols to the 20-als is known from deuterium-labeling studies to involve loss of the 20-pro*R* hydrogen atom.[81] The next step involves loss of C-20 from the aldehydes (**108**) or (**109**) with the formation of the lactones, GA_9 (**113**) and GA_{20} (**114**), respectively. The biologically inactive carboxylic acids, GA_{25} (**111**) or GA_{17} (**112**), are also metabolites of the aldehydes, but are not precursors of the lactones.

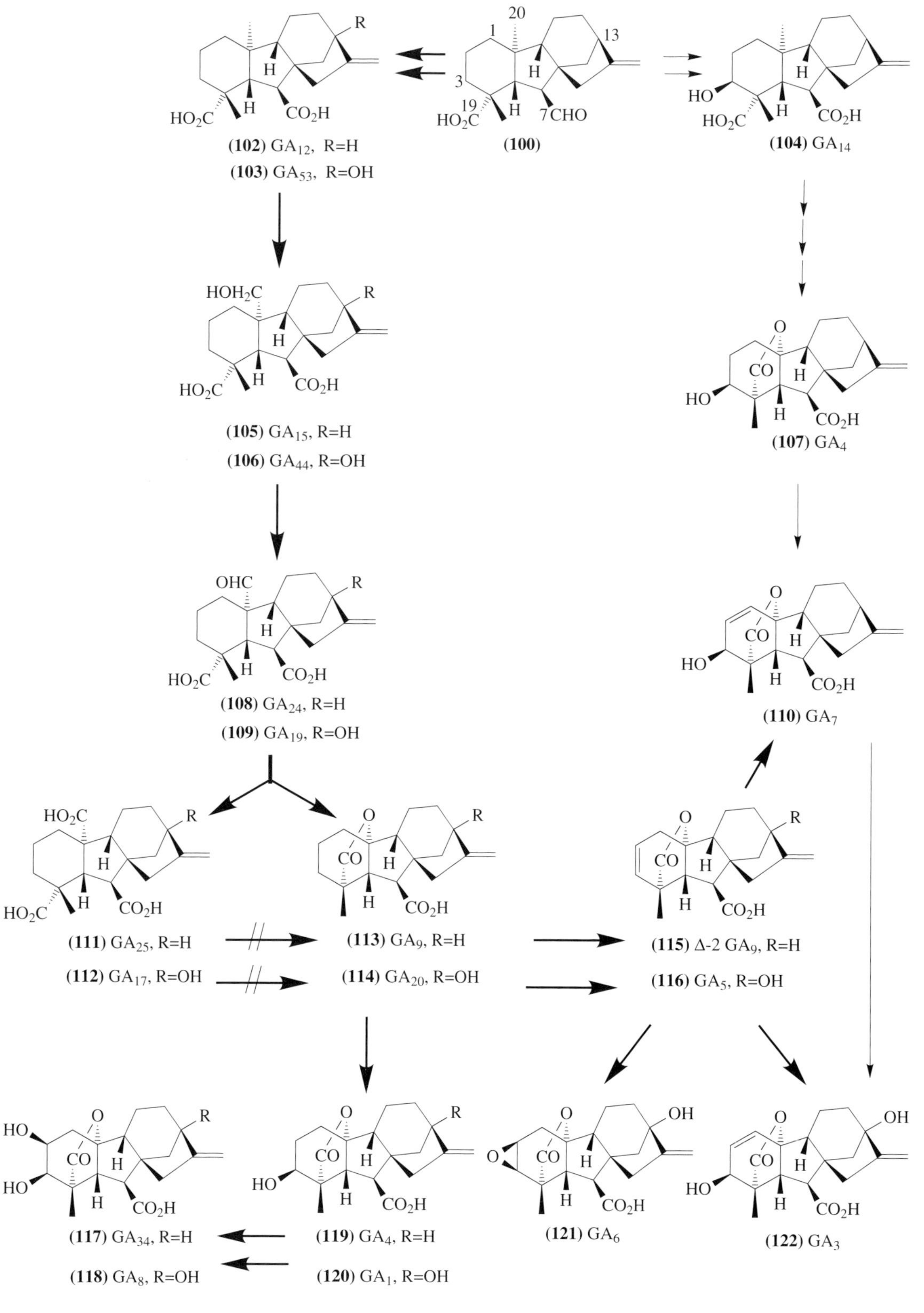

(102) GA_{12}, R=H
(103) GA_{53}, R=OH

(100)

(104) GA_{14}

(105) GA_{15}, R=H
(106) GA_{44}, R=OH

(107) GA_4

(108) GA_{24}, R=H
(109) GA_{19}, R=OH

(110) GA_7

(111) GA_{25}, R=H
(112) GA_{17}, R=OH

(113) GA_9, R=H
(114) GA_{20}, R=OH

(115) Δ-2 GA_9, R=H
(116) GA_5, R=OH

(117) GA_{34}, R=H
(118) GA_8, R=OH

(119) GA_4, R=H
(120) GA_1, R=OH

(121) GA_6

(122) GA_3

Scheme 17

The enzyme (GA-20-oxidase) involved in the oxidation of C-20 in *C. maxima* seeds has been purified[82] and cloned.[83] Related GA-20-oxidases have also been cloned from *Arabidopsis thaliana*,[84] *Spinacea oleracea*,[85] *Pisum sativum*,[86] and *Marah macrocarpus*.[87] These enzymes are members of the 2-oxoglutarate-dependent dioxygenase family. They are soluble enzymes requiring Fe^{2+}, ascorbate, and oxygen as well as 2-oxoglutarate, which acts as cosubstrate with the GA. The deduced sequences of these GA-20-oxidases have 52–81% identity and contain highly conserved regions. Two motifs (e.g., $H_{243}XD$ and $H_{298}R$ in the *C. maxima* enzyme) may be associated with the metal binding site; two histidine residues and an aspartic acid residue have also been identified as metal binding sites in the related dioxygenase, isopenicillin-*N*-synthase, by X-ray crystallographic studies[88] and site-directed mutagenesis.[89] There is also an NYYPXCXXP motif that may identify the 2-oxoglutarate binding site and a LPWKET motif that may be associated with binding of the GA substrate. In many plant species there are multiple enzymes indicating tissue specific regulation of GA biosynthesis. However, it is now evident from heterologous expressing studies in *E. coli* that each of the known GA-20-oxidases catalyze all the steps for the conversion of the C-20 methyl in (**102**) and (**103**) through to the γ-lactones (**113**) and (**114**) and/or to the carboxylic acids (**111**) and (**112**).[83,84]

The mechanism of oxidative loss of C-20 in the GAs is unusual. In other systems, removal of methyl groups usually occurs via oxidation to the carboxylic acid followed by decarboxylation, often involving activation by β-carbonyl functions (cf. loss of 4-methyl groups in cholesterol biosynthesis) or via oxidation to the aldehyde, then double bond formation from elimination of vicinal functions (cf. aromatase in estrogen biosynthesis). In GA biosynthesis, involvement of the 1,3-*trans* diaxially disposed carboxylic acid function at C-19 appears to lead to a novel mechanism, the details of which are still unknown. Studies with specifically-labeled mevalonate and *G. fujikuroi* indicate that adjacent hydrogen atoms at C-1, 5, and 9 are retained.[90,91] In the same system, ^{18}O-labeling has shown that both oxygen atoms of the γ-lactone arise from the 19-carboxylic acid.[92] Results from experiments using cell-free systems of *P. sativum* indicate that carbon-20 is released ultimately as CO_2 in the overall conversion of GA_{12} (**102**) to γ-lactones.[93] If a 20-oxidase were the only enzyme involved, this result would require two cycles of 2-oxyglutarate turnover, and thus an intermediate, between the aldehyde (**108**) and the γ-lactone (**113**). The nature of this intermediate, which could be enzyme-bound, is not yet known. The recent availability of recombinant enzyme should give further insight into this mechanism.

In higher plants the next step in the biosynthesis of bioactive GAs is hydroxylation at C-3β to give GA_4 (**119**) or GA_1 (**120**). This step is also catalyzed by 2-oxoglutarate-dependent dioxygenases. Such enzymes have been cloned from *A. thaliana*[94] and from *P. sativum*[95] and their function has been confirmed by heterologous expression in *E. coli*.[95,96] Both recombinant enzymes convert (**113**) to (**119**) and, less efficiently, (**114**) to (**120**); the fusion protein from *A. thaliana* can also oxidize GA_5 (**116**) to the epoxide, GA_6 (**121**).[96] Previous experiments with partially purified 3β-hydroxylase from *P. vulgaris* embryos and GA_{20} (**114**) as substrate had shown that the 2,3-olefin, GA_5 (**116**) is also a product along with the hydroxy compound (**120**).[97] The multifunctionality of a 3β-hydroxylase in *Zea mays* has also been suggested from metabolic studies with the *dwarf-1* mutant in which the conversion of GA_{20} (**114**) to GA_1 (**120**) and of GA_5 (**116**), and of GA_5 (**116**) to GA_3 (**122**), is blocked.[98] Deactivation of the biologically active GA_4 (**119**) and GA_1 (**120**) in plants is accomplished by further hydroxylation at C-2β, by 2-oxoglutarate-dependent dioxygenases,[99] to give the biologically inactive GA_{34} (**117**) and GA_8 (**118**).

The biosynthesis of the majority of the 100 or so gibberellins found in plants follows the basic linear pathway described above. Structural variation is achieved by hydroxylation at various positions at different stages in the pathway. Common positions for hydroxylation additional to those in Scheme 17 are at C-1, 11, 12, and 15. The result is the occurrence of parallel pathways, often in the same plant, with cross-over points, creating metabolic grids of intermediates formed probably by a small number of enzymes acting on a number of structurally related substrates. One unusual feature of GA biosynthesis in plants is the formation of GA_3 (gibberellic acid, (**122**)) and GA_7 (**110**). These GAs are formed, respectively, by an unusual *ene*-hydroxylation of the 2,3-olefins, GA_5 (**116**) and 2,3-dehydroGA_9 (**115**), initiated by abstraction of the 1β-hydrogen demonstrated using substrates specifically labeled with deuterium.[100] The pathway to GA_3 and GA_7 in plants is different from that in the fungus, *G. fujikuroi*. In the fungal pathway, shown in Scheme 17 by normal arrows, 3β-hydroxylation occurs early, at the GA_{12}-aldehyde stage, then 7-oxidation gives GA_{14} (**104**). Oxidation at C-20 then occurs, presumably by the same sequence that occurs in plants, to give GA_4 (**107**), a compound common to both plant and fungal pathways. In the fungus, metabolism to GA_3 then occurs by 1,2-dehydrogenation to give GA_7 (**110**), a process in which the 1α and 2α-H atoms are eliminated,[100] followed by hydroxylation at C-13. Little is known about the enzymology of fungal GA biosynthesis. Preparation of cell-free systems that take precursors beyond GA_{14} has not been achieved, and at the present time it is not clear if the enzymes concerned are 2-oxoglutarate-dependent dioxygenases as in plants.

Although the majority of GAs are derived from *ent*-kaurenoid precursors by variations in the basic pathways shown in Scheme 17, the recent discoveries of ring C/D variants in higher plants[101,102] and ferns[103–105]103104105 (Scheme 18) indicates that other tetracyclic and pentacyclic hydrocarbons may give rise to gibberellin-like structures. Indeed, the fungal enzymes do show a remarkable degree of nonspecificity and convert a range of "unnatural" *ent*-kaurenonids to "unnatural" fungal GAs; this topic and the fungal biosynthesis of GAs is comprehensively reviewed by Bearder.[106] However, as shown in Scheme 18 by dashed arrows, formation of the ring C/D structures present in 9,11 dehydro-GAs (e.g., (**124**)), 12,15 cyclo-GAs (e.g., (**125**), (**126**)), and antheridic acid (**127**), can be rationalized as rearrangements of the 9-radical (**123**) from GA_9 (**113**). An alternative route to (**125**) could also be from the C-15 radical of (**113**). The only pertinent experimental information comes from the observed metabolism of (**125**) to (**126**) and (**126**) to (**127**) in *Anemia phyllitidis* cultures.[105]

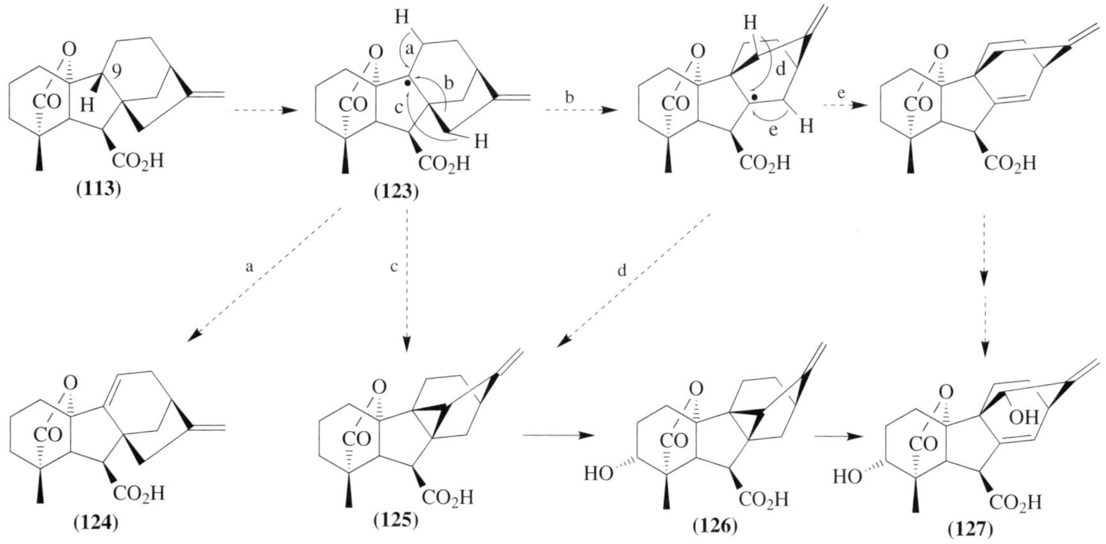

Scheme 18

The all-*trans* cyclization of *E,E,E*-GGPP to *ent*-CPP has been established by: (i) the exclusive labeling[59] of C-18, and not C-19, in GA_3 from [2-^{14}C]MVA (see also rosenolactone (**71**), Scheme 12); (ii) retention[93] of the 4-pro*R*-hydrogen from MVA at C-3 and C-9; and (iii) the demonstration[107] that the 4-pro*R*-hydrogen from MVA at C-3 in *ent*-kaur-16-ene is lost by hydroxylation with retention of configuration. This is one of the few documented cases of the generally accepted all-*trans*-cyclization of GGPP.

2.08.6.2 From *ent*-CPP by *si*-Face Cyclization on C-13

Cyclization of the 13-epimer of the 8-carbonium ion ((**87**), Scheme 15), derived from *ent*-CPP (**33**) by *si*-face attack on C-13, would give ring C/D enantiomers of the structures shown in Scheme 15. None are known.

2.08.6.3 From CPP by *re*-Face Cyclization on C-13

Cyclization of the 8-carbonium ion ((**54**), Scheme 11), derived from CPP (**30**) by *re*-face attack on C-13, gives the carbon skeletons corresponding to those in Scheme 15 but with the "normal" absolute stereochemistry at C-5, -9, and -10. Only phyllocladene ((**128**), Scheme 19) is known and no biosynthetic studies have been reported.

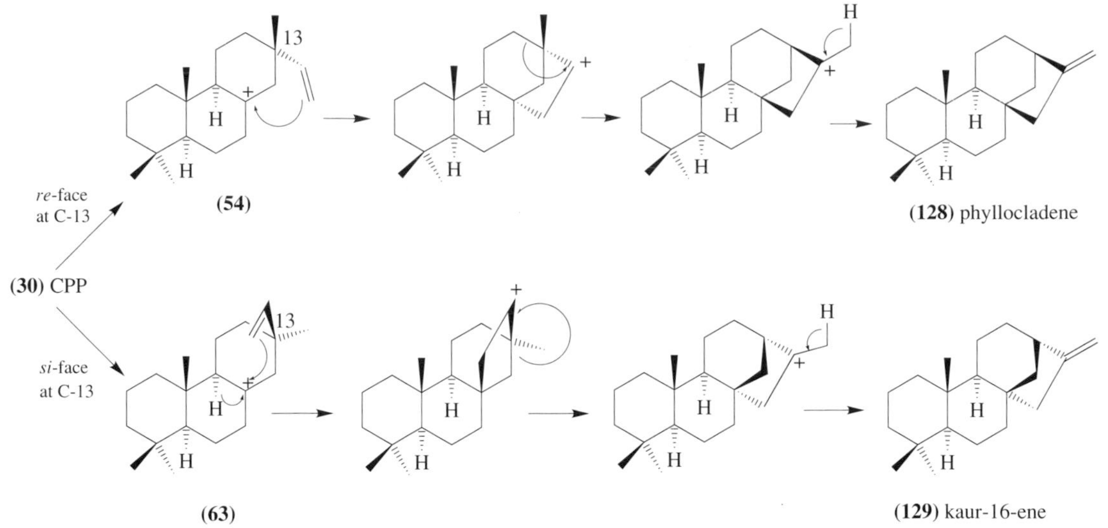

re-face at C-13
(54)
(128) phyllocladene
(30) CPP
si-face at C-13
(63)
(129) kaur-16-ene

Scheme 19

2.08.6.4 From CPP by *si*-Face Cyclization on C-13

Cyclization of the 8-carbonium ion ((**63**), Scheme 19), derived from CPP (**30**) by *si*-face attack on C-13, gives the enantiomers of the structures shown in Scheme 15. Of these enantiomers only kaur-16-ene ((**129**), Scheme 19) is known but no biosynthetic studies have been reported.

2.08.6.5 From *ent*-CPP by *re*-Face Cyclization on C-13, then 9H → C-8

Cyclization of the 8-carbonium ion ((**130**), Scheme 20), derived from *ent*-CPP (**33**) by *re*-face attack on C-13, then 9H → C-8 in (**87**), provides a possible origin for helifulvanic acid (**131**), isolated from *Helichrysum chionosphaerum*.[64] This plant also contains *ent*-atisir-16-ene ((**90**), Scheme 15), *ent*-kaur-16-ene ((**86**), Scheme 15), and *ent*-abieta-7,13-diene (the enantiomer of (**59**), Scheme 11), indicating the presence of at least two different GGPP cyclases in *H. chionosphaerum*. The 11α-hydroxy derivative (**132**) of helifulvanic acid (**131**) and *ent*-kaur-16-en-19-oic acid ((**94**), Scheme 16) also co-occur in *H. fulvum*,[108] indicating that cyclization of the unrearranged ion (**87**) and the rearranged ion (**130**) takes place in the same plant.

Other products that may arise from the ion (**130**) have not been reported.

2.08.6.6 From *ent*-CPP by *si*-Face Cyclization on C-13, then 9H → C-8

The postulated cyclization (Scheme 20) of the 8-carbonium ion (**133**), derived from *ent*-CPP (**33**) by *si*-face attack on C-13, then 9H → C-8 in (**75**), gives the *ent*-8a-stemarenes, e.g., the villanovanes (**134**), which have been isolated from *Villanova titicaenis*.[109] No experimental evidence for this pathway is available.

2.08.6.7 From *syn*-CPP by *re*-Face Cyclization on C-13, then 9H → C-8

Cyclization of the 8-carbonium ion (**136**), derived from *syn*-CPP (**36**) by *re*-face attack on C-13 to (**135**), then 9H → C-8, is shown in Scheme 21. Pathway a from (**137**) via (**138**) is the probable pathway to the known stemodanes, stemodin (**139**) and stemodinone (**140**). Pathway b, from (**137**) via (**141**), gives stemar-16-ene (**142**) which is formed from *syn*-CPP (**36**) in rice suspension cultures that have been treated with chitin.[39] Stemar-16-ene (**142**) is the presumed precursor of the phytoalexin, oryzalexin S (**143**), found in UV-irradiated rice leaves.[110] It is also the probable precursor of stemarin (**144**), occurring in *Stemodia maritima* L.,[111] and of the stemodanes that occur in *S. chiensis*.[112]

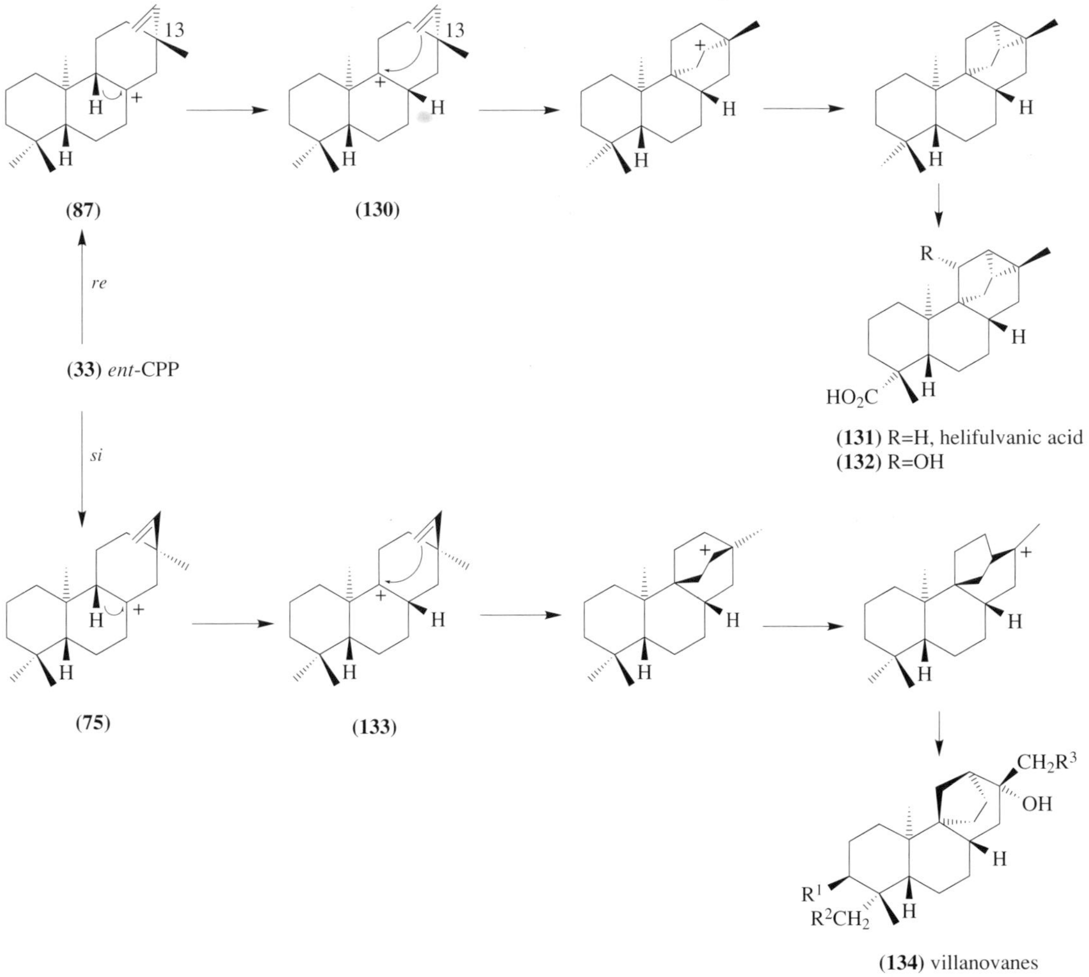

(87)

re

(33) *ent*-CPP

si

(130)

(131) R=H, helifulvanic acid
(132) R=OH

(75)

(133)

(134) villanovanes

Scheme 20

2.08.6.8 From *syn*-CPP by *si*-Face Cyclization on C-13, then 9H → C-8

Cyclization of the 8-carbonium ion (**76**), derived from *syn*-CPP (**36**) by *si*-face attack on C-13, then 9H → C-8, is shown in Scheme 21. The sequence via (**148**), (**149**), and (**150**) leads to scopadulin (**145**) and aphidicolin (**147**). The sequence via (**151**) gives the related scopadulcic acids (e.g., **152**) and the thyrsiflorins (**153**).

2.08.6.8.1 Aphidicolin

Aphidicolin (**147**) is the major metabolite of the fungus *Cephalosporium aphidicola*. Its origin from *syn*-CPP (**36**) has not been directly established but can be inferred from the established 8βH-stereochemistry[113] and the generation of ^{2}H–^{13}C coupling at C-8 in the ^{2}H-NMR spectrum of aphidicolin, biosynthesized from [4-$^{2}H_2$,^{13}C]MVA.[114] These findings implicate the intermediate (**148**), derived from (**76**) by a 9β-H → C-8 shift. The 16-OH is derived from H_2O in the first step from the 16-carbonium ion (**150**).[115] Subsequent hydroxylation of (**146**), in the main pathway to aphidicolin (**147**), appears to proceed in the order, C-18, C-3, C-17, on the basis of the observed metabolism of intermediates.[116] However, the addition of cytochrome P-450 inhibitors indicate that C-3 hydroxylation may be the last step.[117]

2.08.6.8.2 Scopadulcic acids and thyrsiflorins

The biosynthesis of scopadulcic acid B (**152**) from the herb *Scoparia dulcis* L. has not been studied but a speculative pathway has been suggested[118] from the aphidicolin intermediate (**150**) via (**151**)

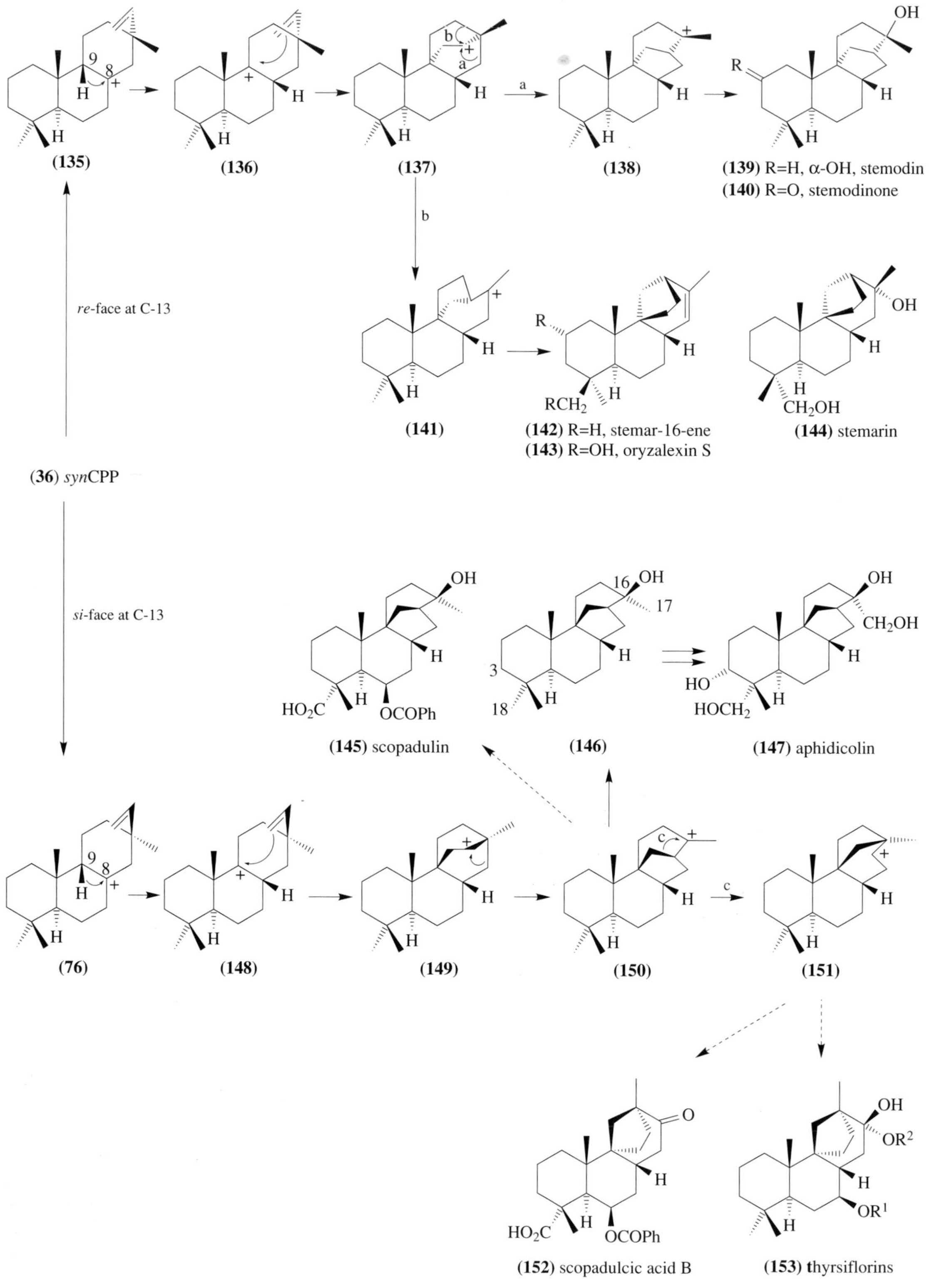

(135) (136) (137) (138)

(139) R=H, α-OH, stemodin
(140) R=O, stemodinone

re-face at C-13

(141)

(142) R=H, stemar-16-ene
(143) R=OH, oryzalexin S

(144) stemarin

(36) *syn*CPP

si-face at C-13

(145) scopadulin

(146)

(147) aphidicolin

(76) (148) (149) (150) (151)

(152) scopadulcic acid B

(153) thyrsiflorins

Scheme 21

as shown in Scheme 21. The co-occurrence of (**145**) and (**152**) in *S. dulcis* lends credence to their common origin.[119] The thrysiflorins (**153**) from the herb *Calceolaria thyrsiflora*[120] have the same presumptive origin.

2.08.6.9 Miscellany

Cyclization of *syn-ent*-CCP (**39**) by *re-* or *si*-face attack on C-13, then 9H → C-8, leads to the enantiomers of the compounds shown in Scheme 21 but none are known. *ent*-Stemarenes have been reported to occur in *Calceolaria* species although the evidence for their absolute stereochemistry is not definitive.[121]

No tetra- or pentacyclic diterpenes are known from the cyclization of CPP by *re-* or *si*-face of C-17 on C-13, then 9H → C-8; from the cyclization of *syn*-CPP by *re-* or *si*-face on C-13; or from the cyclization of *syn-ent*-CPP by *re-* or *si*-face on C-13.

2.08.7 SUMMARY AND FUTURE PROSPECTS

It has been shown that the diverse structures of the cyclic diterpenes arise by different modes of cyclization of GGPP, then by oxidation of the cyclization products. Details of these processes are emerging through the purification and cloning of the enzymes involved.

Four types of GGPP cyclases have been cloned and the function of their gene products has been determined. Thus, four types of cyclization of GGPP have been established, depending on the initiation of cyclization (protonation or diphosphate ionization). Furthermore, each of the four classes of cyclases are multifunctional, catalyzing a series of steps on the enzyme surface. Since these enzymes can now be overexpressed *in vitro*, detailed study of the mechanisms of cyclization can be anticipated. Rapid advances in the characterization of other cyclases can also be anticipated using conserved sequences of the known genes and PCR. Questions of particular interest to be addressed are the existence of enantiomeric products of the cyclization of GGPP and the diastereotopic cyclization of the 6,6-bicyclic products from the proton-initiated cyclization of GGPP.

The oxidative metabolism of the products of cyclization of GGPP has only been studied in detail for a few groups of diterpenes. Details information on the soluble 2-oxoglutarate-dependent dioxygenases is emerging from studies on gibberellin biosynthesis. Characterization of the microsomal P450 oxidases, however, has been less tractable.

The progress that has been made sets the scene for the genetic engineering of diterpene biosynthesis as a means of studying the regulation of pathways and flux of precursors into them, and also to manipulate diterpene synthesis to give desirable properties. Uses may range from the production of rare medicinal terpenes in amenable plants or microorganisms to the manipulation of plant development by sense or antisense expression of gibberellin biosynthesis genes, a goal that has already been achieved in Arabidopsis, or to the manipulation of the plant defense response via terpenoid phytoalexins and antifeedant compounds.

2.08.8 REFERENCES

1. C. J. D. Mau and C. A. West, *Proc. Natl. Acad. Sci. USA*, 1994, **91**, 8497.
2. M. T. Dueber, W. Adolf, and C. A. West, *Plant Physiol.*, 1978, **62**, 598.
3. P. Moesta and C. A. West, *Arch. Biochem. Biophys.*, 1985, **238**, 325.
4. T.-p. Sun, H. M. Goodman, and F. M. Ausubel, *Plant Cell*, 1992, **4**, 119.
5. T.-p. Sun and Y. Kamiya, *Plant Cell*, 1994, **6**, 1509.
6. M. R. Wildung and R. Croteau, *J. Biol. Chem.*, 1996, **271**, 9201.
7. M. Hezari, N. G. Lewis, and R. Croteau, *Arch. Biochem. Biophys.*, 1995, **322**, 437.
8. B. Stofer Vogel, M. R. Wildung, G. Vogel, and R. Croteau, *J. Biol. Chem.*, 1966, **271**, 23262.
9. R. E. LaFever, B. Stofer Vogel, and R. Croteau, *Arch. Biochem. Biophys.*, 1994, **313**, 139.
10. T. Saitò, H. Abe, H. Yamane, A. Sakurai, N. Murofushi, K. Takio, N. Takahashi, and Y. Kamiya, *Plant Physiol.*, 1995, **109**, 1239.
11. S. Yamaguchi, T. Saito, H. Abe, H. Yamane, N. Murofushi, and Y. Kamiya, *Plant J.*, 1996, **10**, 203.
12. H. Kawaide, R. Imai, T. Sassa, and Y. Kamiya, *J. Biol. Chem.*, 1997, **272**, 21 706.
13. R. J. Bensen, G. S. Johal, V. C. Crane, J. T. Tossberg, P. S. Schnable, R. B. Meeley, and S. P. Briggs, *Plant Cell*, 1995, **7**, 75.
14. D. R. Robinson and C. A. West, *Biochemistry*, 1970, **9**, 70.
15. L. Crombie, G. Kneen, G. Pattenden, and D. Whybrow, *J. Chem. Soc., Perkin Trans. 1*, 1980, 1711.

16. D. Sitton and C. A. West, *Phytochemistry*, 1975, **14**, 1921.
17. M. Stekoll and C. A. West, *Plant Physiol.*, 1978, **61**, 38.
18. D. R. Robinson and C. A. West, *Biochemistry*, 1970, **9**, 80.
19. W. J. Guilford and R. M. Coates, *J. Am. Chem. Soc.*, 1982, **104**, 3506.
20. M. W. Dudley, M. T. Dueber, and C. A. West, *Plant Physiol.*, 1986, **81**, 335.
21. A. F. Lois and C. A. West, *Arch. Biochem. Biophys.*, 1990, **276**, 270.
22. W. G. Dauben, W. E. Thiessen, and P. R. Resnick, *J. Org. Chem.*, 1965, **30**, 1693.
23. L. Crombie, D. McNamara, D. F. Firth, S. Smith, and P. C. Bevan, *Phytochemistry*, 1988, **27**, 1685.
24. Z. Guo and G. J. Wagner, *Plant Sci.*, 1995, **110**, 1.
25. A. E. Koepp, M. Hezari, M. Zajicek, B. Stofer Vogel, R. E. LaFever, N. G. Lewis, and R. Croteau, *J. Biol. Chem.*, 1995, **270**, 8686.
26. X. Lin, M. Hezari, A. E. Koepp, H. G. Floss, and R. Croteau, *Biochemistry*, 1996, **35**, 2968.
27. J. Hefner, S. M. Rubinstein, R. E. B. Ketchum, D. M. Gibson, R. M. Williams, and R. Croteau, *Chem. Biol.*, 1996, **3**, 479.
28. P. E. Fleming, U. Mocek, and H. G. Floss, *J. Am. Chem. Soc.*, 1993, **115**, 805.
29. W. Eisenreich, B. Menhard, P. J. Hylands, M. H. Zenk, and A. A. Bacher, *Proc. Natl. Acad. Sci. USA*, 1996, **93**, 6431.
30. A. Banerji, R. B. Jones, G. Mellows, L. Phillips, and K.-Y. Sim, *J. Chem. Soc., Perkin Trans. 1*, 1976, 2221.
31. A. Banerji, R. Hunter, G. Mellows, K.-Y. Sim, and D. H. R. Barton, *J. Chem. Soc., Chem. Commun.*, 1978, 843.
32. A. Stoessl, G. L. Rock, J. B. Stothers, and R. C. Zimmer, *Can. J. Chem.*, 1988, **66**, 1084.
33. A. Stoessl, G. L. Rock, and J. B. Stothers, *Can. J. Chem.*, 1989, **67**, 1302.
34. W. Adolf and E. Hecker, *Isr. J. Chem.*, 1977, **16**, 75.
35. M. Toyota, E. Nahaishi, and Y. Asakawa, *Phytochemistry*, 1996, **43**, 1057.
36. R. R. Fall and C. A. West, *J. Biol. Chem.*, 1971, **246**, 6913.
37. J. D. Duncan and C. A. West, *Plant Physiol.*, 1981, **68**, 1128.
38. D. J. McGarvey and R. Croteau, *Plant Cell*, 1995, **7**, 1015.
39. R. S. Mohan, N. K. N. Yee, R. M. Coates, Y.-Y. Ren, P. Stamenkovic, I. Mendez, and C. A. West, *Arch. Biochem. Biophys.*, 1996, **330**, 33.
40. A. F. Barrero, J. F. Sanchez, and F. G. Cuenca, *Phytochemistry*, 1988, **27**, 3676.
41. V. De Santis and J. D. Medina, *J. Nat. Prod.*, 1981, **44**, 370.
42. N. B. Perry and R. T. Weavers, *Phytochemistry*, 1985, **24**, 2899.
43. R. M. Carman and A. R. Duffield, *Aust. J. Chem.*, 1993, **46**, 1105.
44. Z. Guo, R. F. Severson, and G. J. Wagner, *Arch. Biochem. Biophys.*, 1994, **308**, 103.
45. Z. Guo and G. J. Wagner, *Planta*, 1995, **197**, 627.
46. W. Fenical, in "Marine Natural Products: Chemical and Biological Perspectives," ed. P. J. Scheuer, Academic Press, New York, 1978, vol. 2, p. 173.
47. A. T. Merritt and S. V. Ley, *Nat. Prod. Rep.*, 1992, **9**, 243.
48. A. Akhila, K. Rani, and R. S. Thakur, *Phytochemistry*, 1991, **30**, 2573.
49. K. Nabeta, T. Ishikawa, and H. Okuyama, *J. Chem. Soc., Perkin Trans. 1*, 1995, 3111.
50. B. Dockerill and J. R. Hanson, *J. Chem. Soc., Perkin Trans. 1*, 1997, 324.
51. J. Polonsky, G. Lukacs, N. Cagnoli-Bellavita, and P. Ceccherelli, *Tetrahedron Lett.*, 1975, 481.
52. D. E. Cane and P. P. N. Murthy, *J. Am. Chem. Soc.*, 1977, **99**, 8327.
53. Y. Tomita and Y. Ikeshiro, *J. Chem. Soc., Chem. Commun.*, 1987, 1311.
54. Y. Tomita, M. Annaka, and Y. Ikeshiro, *J. Chem. Soc., Chem. Commun.*, 1989, 108.
55. C. Funk and R. Croteau, *Arch. Biochem. Biophys.*, 1994, **308**, 258.
56. C. Funk, E. Lewinsohn, B. Stofer Vogel, C. L. Steele, and R. Croteau, *Plant Physiol*, 1994, **106**, 999.
57. B. Achilladelis and J. R. Hanson, *J. Chem. Soc. (C)*, 1969, 2010.
58. B. Dickerill and J. R. Hanson, *Phytochemistry*, 1978, **17**, 1119.
59. A. J. Birch, R. W. Richards, H. Smith, A. Harris, and W. B. Whalley, *Tetrahedron*, 1959, **7**, 241.
60. D. Arigoni, *Pure Appl. Chem.*, 1968, **17**, 331.
61. G. Bonavia, *Diss. ETH (Zürich)*, 4189, 1968.
62. K. A. Wickham and C. A. West, *Arch. Biochem. Biophys.*, 1992, **293**, 320.
63. Y. Kono, S. Takeuchi, O. Kodama, H. Sekido, and T. Akatsuka, *Agric. Biol. Chem.*, 1985, **49**, 1675.
64. D. W. Cartwright, P. W. Langcake, R. J. Pryce, D. P. Leworthy, and J. P. Ride, *Phytochemistry*, 1981, **20**, 535.
65. F. Bohlmann, W.-R. Abraham, and W. S. Sheldrick, *Phytochemistry*, 1980, **19**, 869.
66. M. Piovano, V. Gambaro, M. C. Chamy, J. A. Garbarino, M. Nicoletti, J. Guilhem, and C. Pascard, *Phytochemistry*, 1988, **27**, 1145.
67. I. Shechter and C. A. West, *J. Biol. Chem.*, 1969, **244**, 3200.
68. R. M. Coates and P. L. Cavender, *J. Am. Chem. Soc.*, 1980, **102**, 6358.
69. K. Honda, T. Shishibori, and T. Suga, *J. Chem. Res. (S)*, 1980, 218.
70. C. A. West, in "The Biochemistry of Plants," ed. D. D. Davies, Academic Press, New York, 1980, vol. 2, p. 317.
71. P. F. Sherwin and R. M. Coates, *J. Chem. Soc., Chem. Commun.*, 1982, 1013.
72. J. C. Jennings, R. C. Coolbaugh, D. A. Nakata, and C. A. West, *Plant Physiol.*, 1993, **101**, 925.
73. P. Hedden and J. E. Graebe, *Phytochemistry*, 1981, **20**, 1011.
74. M. H. Beale, J. R. Bearder, G. H. Down, M. Hutchison, J. MacMillan, and B. O. Phinney, *Phytochemistry*, 1982, **21**, 1279.
75. J. E. Graebe and P. Hedden, in "Biochemistry and Chemistry of Plant Growth Regulators," eds K. Schreiber, H. R. Schutte, and G. Sembdner, Acad. Sci. German Democratic Republic, Inst. Plant Biochem., Halle, Germany, 1974, p. 1.
76. S. J. Castellaro, S. C. Dolan, P. Hedden, P. Gaskin, and J. MacMillan, *Phytochemistry*, 1990, **29**, 1833.
77. J. E. Graebe, P. Hedden, and J. MacMillan, *J. Chem. Soc., Chem. Commun.*, 1975, 161.
78. J. MacMillan, *Nat. Prod. Rep.*, 1997, **14**, 221.
79. P. Hedden and Y. Kamiya, *Annu. Rev. Plant Physiol. Plant Mol. Biol.*, 1977, **48**, 431.
80. J. E. Graebe, *Annu. Rev. Plant Physiol.*, 1987, **38**, 419.

81. J. L. Ward, G. J. Jackson, M. H. Beale, P. Gaskin, P. Hedden, L. N. Mander, A. L. Phillips, H. Seto, M. Talon, C. L. Willis, T. M. Wilson, and J. A. D. Zeevaart, *J. Chem. Soc., Chem. Commun.*, 1997, 13.
82. T. Lange, *Planta*, 1994, **195**, 108.
83. T. Lange, P. Hedden, and J. E. Graebe, *Proc. Natl. Acad. Sci. USA*, 1994, **91**, 8552.
84. A. L. Phillips, D. A. Ward, S. Ucknes, N. E. J. Appleford, T. Lange, A. K. Huttly, P. Gaskin, J. E. Graebe, and P. Hedden, *Plant Physiol.*, 1995, **108**, 1049.
85. K. Wu, L. Li, D. A. Gage, and J. A. D. Zeevaart, *Plant Physiol.*, 1996, **110**, 547.
86. D. N. Martin, W. M. Proebsting, T. D. Parks, W. G. Dougherty, T. Lange, M. J. Lewis, P. Gaskin, and P. Hedden, *Planta*, 1996, **200**, 159.
87. J. MacMillan, D. A. Ward, A. L. Phillips, M. J. Sanchez-Beltran, P. Gaskin, T. Lange, and P. Hedden, *Plant Physiol.*, 1997, **113**, 1369.
88. P. L. Roach, I. J. Clifton, V. Fulop, K. Harlos, G. J. Barton, J. Hajdu, I. Andersson, C. J. Schofield, and J. E. Baldwin, *Nature (London)*, 1995, **375**, 700.
89. H. Borovok, O. Landman, R. Kreisberg-Zakarin, Y. Aharonowitz, and G. Cohen, *Biochemistry*, 1996, **35**, 1981.
90. R. Evans, J. R. Hanson, and A. F. White, *J. Chem. Soc. (C)*, 1970, 2601.
91. J. R. Hanson and A. F. White, *J. Chem. Soc. (C)*, 1969, 981.
92. J. R. Bearder, J. MacMillan, and B. O. Phinney, *J. Chem. Soc., Chem. Commun.*, 1976, 834.
93. Y. Kamiya, N. Takahashi, and J. E. Graebe, *Planta*, 1986, **169**, 524.
94. H.-H. Chiang, I. Hwang, and H. M. Goodman, *Plant Cell*, 1993, **7**, 195.
95. D. M. Martin, W. M. Proebsting, and P. Hedden, *Proc. Natl. Acad. Sci. USA*, 1997, **94**, 8907.
96. J. Williams, A. L. Phillips, P. Gaskin, and P. Hedden, *Plant Physiol.*, 1998, in press.
97. V. A. Smith, P. Gaskin, and J. MacMillan, *Plant Physiol.*, 1990, **94**, 1390.
98. C. R. Spray, M. Kobayashi, Y. Suzuki, B. O. Phinney, P. Gaskin, and J. MacMillan, *Proc. Natl. Acad. Sci. USA*, 1996, **93**, 10515.
99. V. A. Smith and J. MacMillan, *Planta*, 1986, **167**, 9.
100. K. S. Albone, P. Gaskin, J. MacMillan, B. O. Phinney, and C. L. Willis, *Plant Physiol.*, 1990, **94**, 132.
101. P. Hedden, G. V. Hoad, P. Gaskin, M. J. Lewis, J. R. Green, M. Furber, and L. N. Mander, *Phytochemistry*, 1993, **32**, 231.
102. N. Oyama, T. Yamauchi, H. Yamane, N. Murofushi, M. Agatsuma, M. Pour, and L. N. Mander, *Biosci. Biotech. Biochem.*, 1996, **60**, 305.
103. H. Yamane, Y. Satoh, K. Nohara, M. Nakayama, N. Murofushi, N. Takahashi, K. Takeno, M. Furuya, M. Furber, and L. N. Mander, *Tetrahedron Lett.*, 1988, **29**, 3959.
104. M. Furber, L. N. Mander, J. E. Nester, N. Takahashi, and H. Yamane, *Phytochemistry*, 1989, **28**, 63.
105. T. Yamauchi, N. Oyama, H. Yamane, N. Murofushi, N. Takahashi, H. Schraudolf, M. Furber, L. N. Mander, G. L. Patrick, and B. Twitchen, *Phytochemistry*, 1991, **30**, 3247.
106. J. R. Bearder, in "The Biochemistry and Physiology of Gibberellins," ed. A. Crozier, Praeger, New York, 1983, vol. 1, p. 251.
107. R. M. Dawson, P. R. Jefferies, and J. R. Knox, *Phytochemistry*, 1975, **14**, 2593.
108. F. Bohlmann, C. Zdero, R. Zeisberg, and W. S. Sheldrick, *Phytochemistry*, 1979, **18**, 1359.
109. F. Bohlmann, C. Zdero, R. M. King, and H. Robinson, *Liebigs Ann. Chem.*, 1984, 250.
110. S. Tamogami, M. Mitani, O. Kodama, and T. Akatsuka, *Tetrahedron*, 1993, **49**, 2025.
111. P. S. Manchand and J. F. Blount, *J. Chem. Soc., Chem. Commun.*, 1975, 894.
112. M. C. Chamy, M. Piovano, J. A. Garbarino, and V. Gambaro, *Phytochemistry*, 1991, **30**, 1719.
113. W. Dalziel, B. Hesp, K. M. Stevenson, and J. A. J. Jarvis, *J. Chem. Soc., Perkin Trans. 1*, 1973, 2841.
114. M. J. Ackland, J. R. Hanson, and A. H. Ratcliffe, *J. Chem. Soc., Perkin Trans. 1*, 1984, 2751.
115. M. J. Ackland, J. Gordon, J. R. Hanson, B. L. Yeoh, and A. H. Ratcliffe, *Phytochemistry*, 1988, **27**, 1031.
116. M. J. Ackland, J. F. Gordon, J. R. Hanson, B. L. Yeoh, and A. H. Ratcliffe, *J. Chem. Soc., Perkin Trans. 1*, 1988, 1477.
117. H. Oikawa, A. Ichihara, and S. Sakamura, *Agric. Biol. Chem.*, 1989, **53**, 299.
118. T. Hayashi, K. Okamura, M. Kawasaki, and N. Morita, *Phytochemistry*, 1991, **30**, 3617.
119. T. Hayashi, M. Kawasaki, Y. Miwa, T. Taga, and N. Morita, *Chem. Pharm. Bull.*, 1990, **38**, 945.
120. M. C. Chamy, M. Piovano, J. A. Garbarino, C. Miranda, V. Gambaro, M. L. Rodriguez, C. Ruis-Perez, and I. Brito, *Phytochemistry*, 1991, **30**, 589.
121. M. C. Chamy, M. Piovano, J. A. Garbarino, and V. Gambaro, *Phytochemistry*, 1991, **30**, 3365.

2.09 Squalene Synthase

ISHAIAHU SHECHTER and GUIMIN GUAN
Uniformed Services University of the Health Sciences, Bethesda, MD, USA

and

BRIAN R. BOETTCHER
Novartis Pharmaceuticals Corporation, Summit, NJ, USA

2.09.1 INTRODUCTION

Squalene synthase (farnesyl diphosphate:farnesyl diphosphate farnesyltransferase, EC 2.5.1.21) is a key regulatory enzyme in isoprenoid biosynthesis (Scheme 1) and occupies an important position as the first enzyme in the pathway entirely committed to cholesterol biosynthesis (Scheme 2). As such it represents an attractive target enzyme for the identification of novel agents to reduce total and low-density lipoprotein (LDL)[1] cholesterol levels. The approach of reducing LDL cholesterol levels through upregulation of the LDL receptor has been demonstrated with the development and clinical use of a number of hydroxymethylglutaryl-Co enzyme A (HMG-CoA) reductase inhibitors. The statin class of HMG-CoA reductase inhibitors (fluvastatin, lovastatin, pravastatin, and simvastatin are currently in widespread clinical use) effectively reduce LDL cholesterol levels.[1] Moreover, as observed by coronary angiography, the statins both slow and reverse the progression of atherosclerotic lesions in patients with coronary heart disease as well as in individuals with elevated LDL cholesterol levels.[2] Therefore, it is expected that squalene synthase inhibitors will also have utility as agents to reduce LDL cholesterol levels and thereby reduce cardiovascular events.[3]

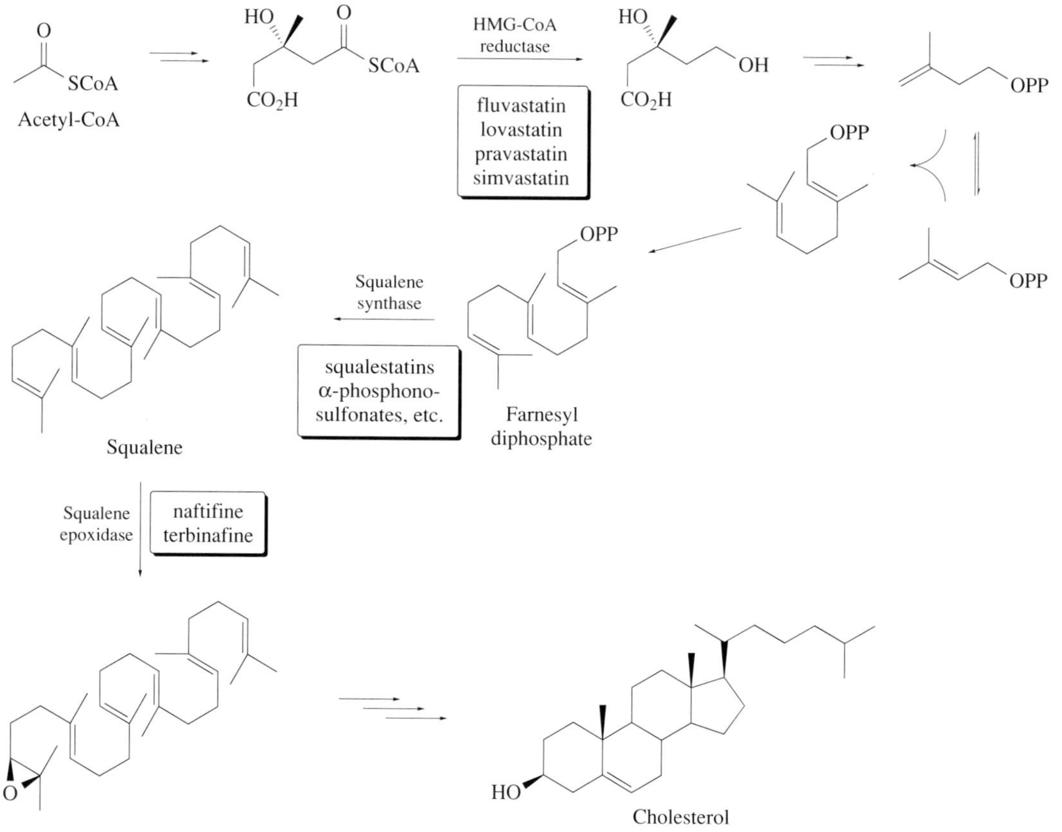

Scheme 1

Interest in the study of squalene synthase has also been stimulated by the unique chemistry of carbon–carbon bond formation catalyzed by the enzyme. The formation of squalene from farnesyl diphosphate (FPP) is catalyzed by squalene synthase in two separate reactions that represent unique and important steps in the formation of lipophilic cholesterol intermediates (Scheme 3). In the first reaction, the condensation of two molecules of FPP occurs to form the cyclopropyl-containing intermediate, presqualene diphosphate (PSPP) (**1**), through a unique 1′–2–3 prenyl transferase reaction. The intermediate is then rearranged and reduced by NADPH to form squalene (**2**)[4–7] Scheme 4 illustrates the conversion of two FPP molecules to squalene in more detail with hypothetical intermediates shown. Although, at present, there is no direct evidence for some of the intermediates shown (e.g. (**4**)), they are shown to provide a more complete description of possible

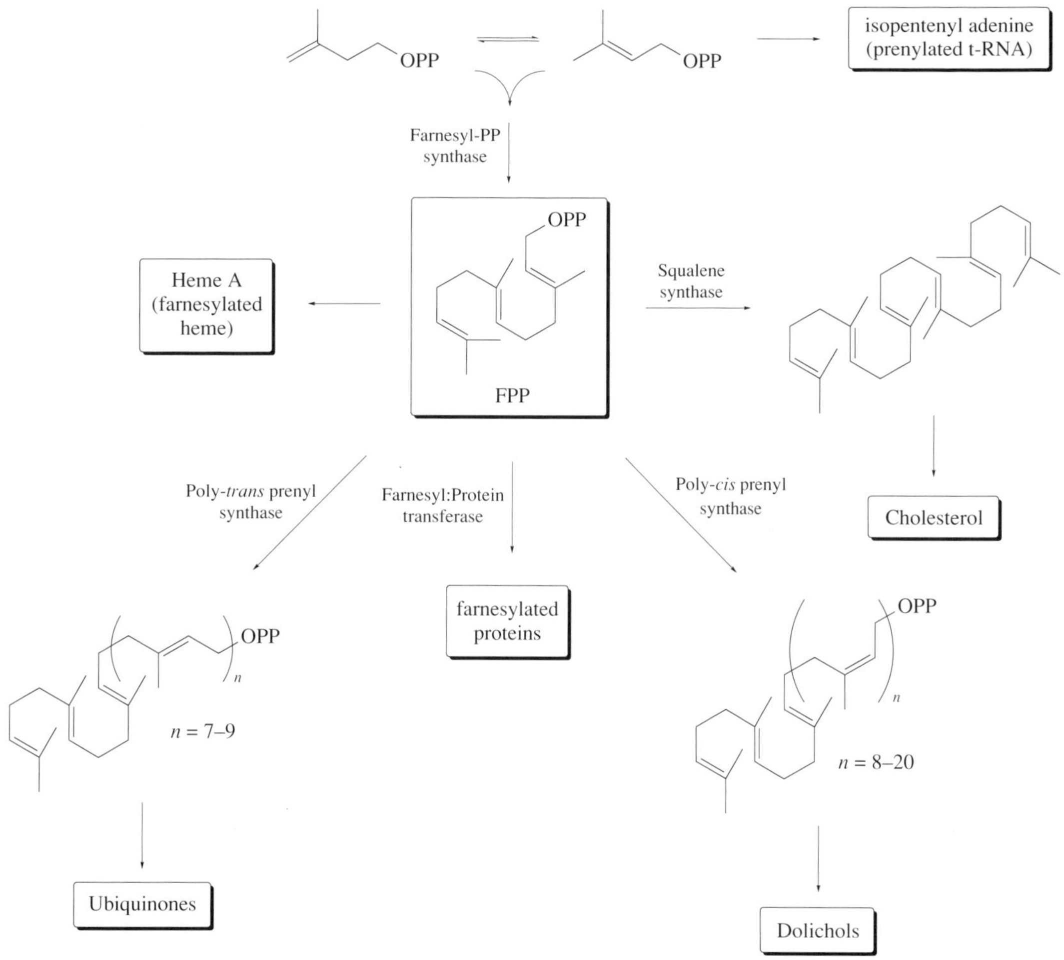

Scheme 2

reaction intermediates. An understanding of the mechanism and inhibitor design principles for this enzyme should provide prototypical structures for the development of other prenyl transferase inhibitors. The recent cloning, overexpression, and purification of the enzyme provide the tools needed to address this unique chemistry through structural, mutagenic, and mechanistic studies. The enhanced understanding of inhibitor design principles for this class of enzymes may in turn lead to the development of novel therapies for cancer[8,9] and dermatology.[10]

The substrate for squalene synthase is FPP, which also serves as a metabolic intermediate in the formation of sterols, dolichols, ubiquinones, and farnesylated proteins[11,12] (Scheme 2). Repetitive *cis* additions of isopentenyl diphosphates yields 2,3-dehydrodolichyl diphosphate and *trans* polymerization gives nonaprenyl diphosphate as an intermediate in ubiquinone synthesis. The identification of the farnesylated yeast a-mating factor, rhodotorucine *A*,[13] and the observation that mevalonate-derived products are incorporated into cellular proteins,[14] led to the identification of thioether-linked farnesyl and geranylgeranyl moieties with proteins.[11,12] These prenylated proteins such as rab and ras play a major role in the control of normal cellular processes, for example, intracellular translocation of proteins during maturation and oncogenesis. The position of squalene synthase at this central branch point of the mevalonate pathway and the observation that its activity can be suppressed in the presence of low-density lipoproteins led to the hypothesis that its regulation plays a major role in the direction of intermediates to the sterol or nonsterol pathways.[15]

Inhibitors of other enzymes in the sterol biosynthetic pathway also have significant value in clinical use (Scheme 1). The inhibitors of squalene epoxidase, naftifine and terbinafine, represent an important new class of topical and oral antifungal agents.[16,17] Thus, it is possible that squalene synthase inhibitors may also find utility as a new class of antifungal agents. The identification of

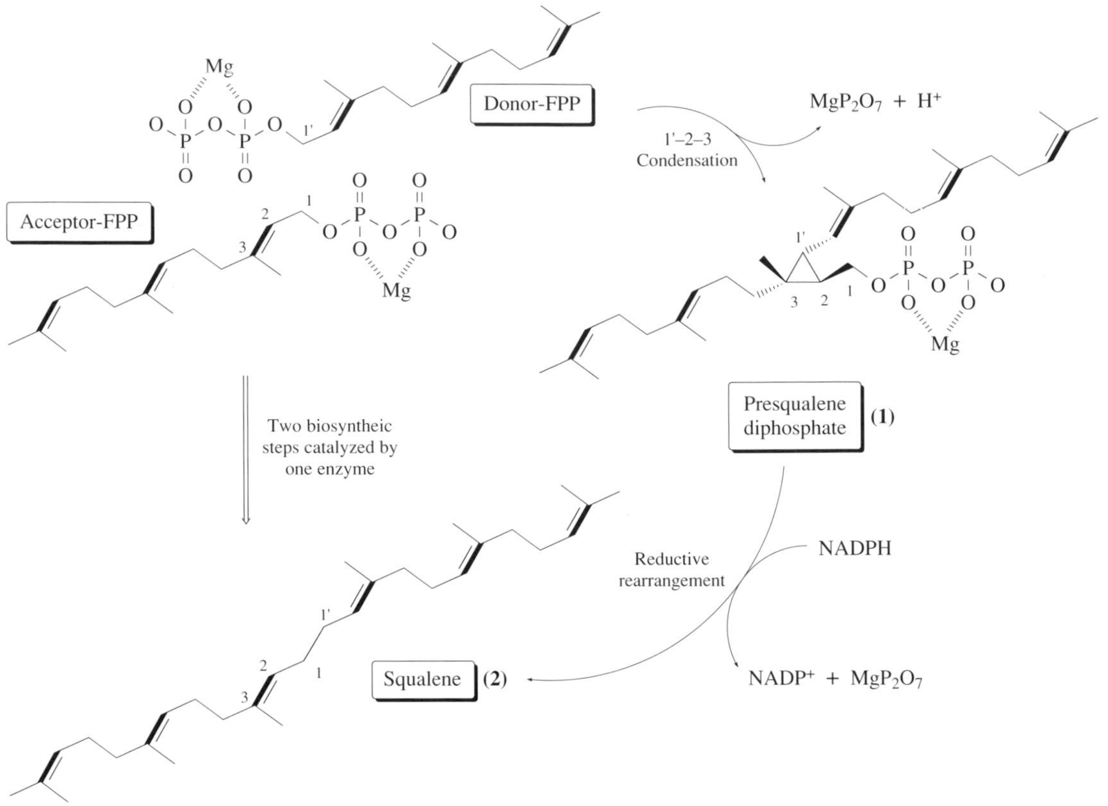

Scheme 3

inhibitors of squalene synthase as agents to lower LDL cholesterol and as antifungal agents have been reviewed recently by a number of groups.[8,18–20] Readers interested in more detailed information regarding squalene synthase inhibitors should consult these excellent and comprehensive review articles. This chapter presents a brief overview of the natural product and synthetic squalene synthase inhibitors which have been prepared.

Finally, a better understanding of the regulation of squalene synthase at the molecular level was possible through advances in cloning of the squalene synthase gene. Thus, the human gene promoter structure and DNA sequence elements involved in the sterol-control of transcription have been described. In addition, chromosomal mapping studies have been carried out and the subcellular localization of the enzyme has been determined. These findings are also reviewed in this chapter.

2.09.2 PURIFICATION

2.09.2.1 Nonrecombinant Squalene Synthases

Squalene synthase has been shown to be an intrinsic microsomal protein that is resistant to solubilization and purification.[21–25] Yeast cells (*S. cerevisiae*) have been used most often for studies and purification of squalene synthase. Early attempts to solubilize the enzyme used the ionic detergent deoxycholate.[21,23] The yeast enzyme was finally purified to homogeneity by using a mixture of nonionic detergents and chromatographic purification.[24,25]

Early attempts to obtain the mammalian enzyme by use of the yeast enzyme sequence data were unsuccessful due to insufficient overall sequence homology between the yeast and mammalian enzymes. However, a soluble truncated form of the enzyme was purified from fluvastatin-treated rat liver microsomes (to induce squalene synthase expression) after limited proteolysis with trypsin. The proteolytic truncation enabled the enzyme to be purified using standard chromatographic

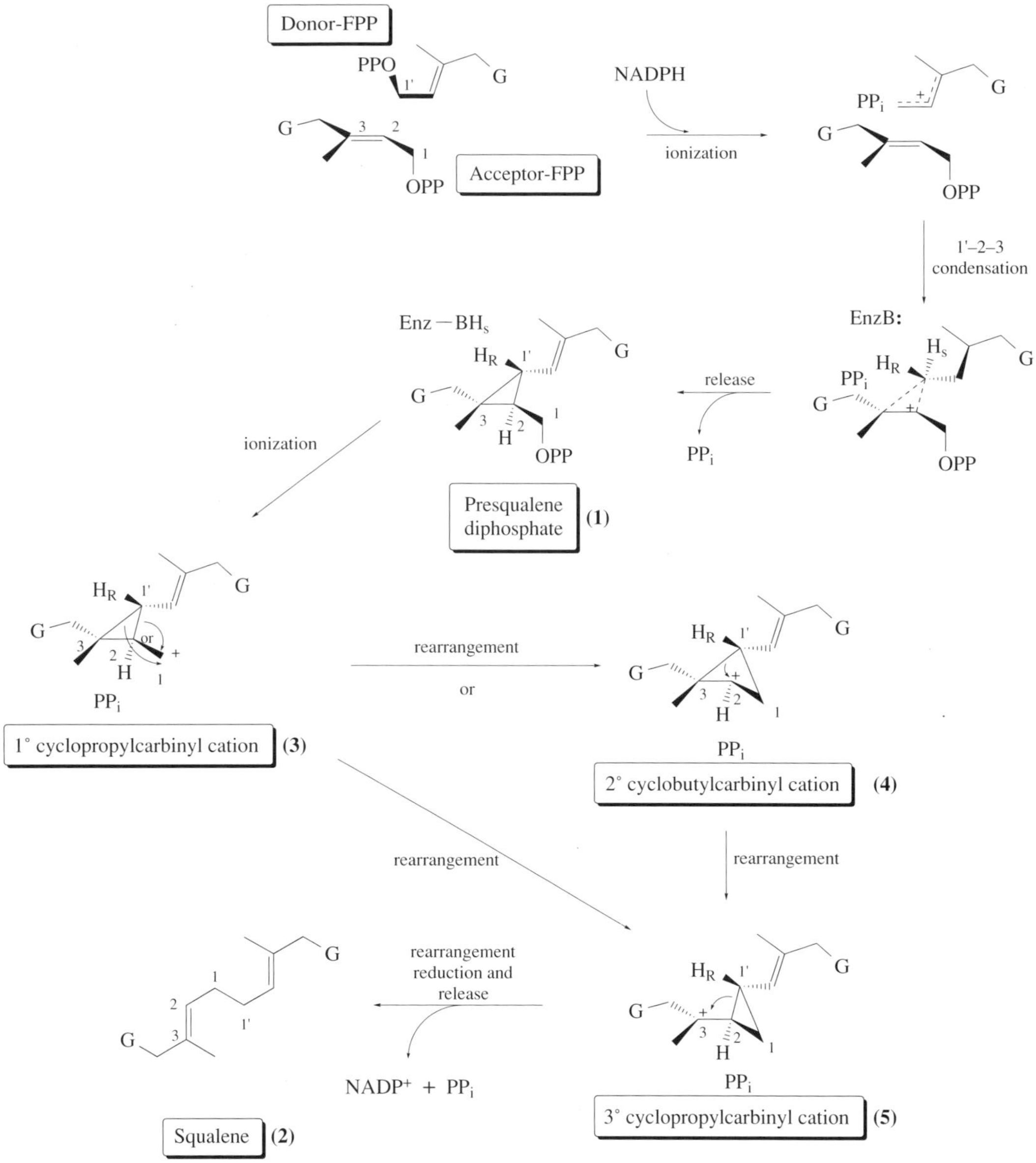

Scheme 4

methods without the use of detergents.[26] The amino acid sequence of the purified protein proved to be essential to the first successful cloning of the gene for the mammalian enzyme[27] and the subsequent isolation of the promoter of the human squalene synthase gene.[28]

2.09.2.2 Recombinant Squalene Synthases

The yeast (*S. cerevisiae*) enzyme was cloned by functional complementation of a squalene synthase-deficient mutant and by screening a genomic library.[29–31] The enzyme from *S. pombe* has also been cloned.[32] A carboxy-terminal truncated form of the yeast enzyme in which the putative membrane anchor was removed has also been overexpressed in *E. coli* and purified to homogeneity using standard purification methods.[33,34] This overexpression provides access to sufficient quantities of the enzyme for structural and mechanistic studies as well as a convenient expression system for site-directed mutagenic studies.

Rat squalene synthase was cloned from cDNA libraries using amino acid sequence data from the purified rat enzyme.[27] Subsequently the human enzyme was cloned by identifying conserved sequences in rat and yeast squalene synthases or in a mechanistically related plant enzyme, phytoene synthase.[32,35–37] The full-length human enzyme was overexpressed and purified from Sf9 insect cells in sufficient quantities for structural and mechanistic studies using a baculoviral expression system.[37] The cDNAs for squalene synthases have also been cloned from mouse,[38] *Arabidopsis thaliana*,[39] and *Nicotiana benthamiana*[40] using sequence data obtained from the earlier studies.

Prokaryotic organisms also produce a variety of steroid-like molecules from squalene. Some of the compounds, like the hopanoids, have functions similar to the steroids in eukaryotic organisms. Until recently the synthesis of squalene from FPP in bacterial species appeared to occur through the action of more than a single enzyme, that is no bacterial squalene synthase had been identified. However, Poralla and co-workers have recently cloned a cDNA for squalene synthase from *Methylococcus capsulatus*.[41] Therefore, this unique biochemical transformation by a single enzyme also occurs in some bacterial species.

2.09.3 COMPARISON BETWEEN YEAST, PLANT, AND MAMMALIAN ENZYMES

2.09.3.1 Sequence Comparisons

A comparison of the squalene synthase encoding genes derived from various sources[27] has revealed conserved motifs which may be involved in the catalytic function of this enzyme. A model based on the predicted amino acid sequence deduced from cDNA sequence data and on secondary structure prediction programs is shown in Figure 1.[27] The truncation site in this model was based on N-terminal amino acid analysis of the cytosolic, active form of the enzyme following proteolytic release with trypsin. The three sections of homology indicated in this model also extended to the plant (*Arabidopsis*) enzyme.[39] Sections A and B exhibit homology to prephytoene diphosphate synthase and phytoene synthase (pTOM5 gene product) in plants.[27] Section B is homologous to geranylgeranyl diphosphate synthase and farnesyl diphosphate synthase. The aspartate-rich motif found in Section B of this model was proposed to be the polyprenyl diphosphate binding site in a number of distinct enzymes of this class.[42] Section C is one of the most highly conserved regions of the enzyme and does not appear to exhibit homology with other enzymes that utilize prenyl diphosphates. This may indicate uniqueness with respect to the type of catalysis (1′–1 carbon–carbon bond formation), the size of the prenyl moiety (C-15), or involvement of the NADPH cofactor. A hydrophobic region at the carboxy terminus was suggested to be important for the membrane binding.[32,43] However, in the case of the rat enzyme, such a binding motif could not be identified.[27] Finally, an additional conserved region of the enzyme near the amino terminus may also play an important role.[40]

It was not possible to clone the human squalene synthase gene by expression cloning strategies in *S. cerevisiae*. This difficulty might have been due to the 3′-end of the coding region.[32] This hypothesis is supported by the finding that expression of a chimeric human/*ERG9* protein containing the carboxy terminus of the yeast protein was sufficient to rescue growth of an *erg9* mutant strain.[32] Additional possible reasons for the inability of the full-length human enzyme to complement the *erg9* mutant include protein instability, incorrect membrane localization, or translational inefficiencies due to the use of mammalian codons.[32] However, the inability to complement the *erg9* disruption mutant with human squalene synthase appeared not to be a problem of protein stability or insufficient protein synthesis since substantial levels of protein were detected by immunoblotting.[37] Moreover, the detection of increased enzymatic activity, following induction with galactose, in detergent-solubilized yeast cell extracts demonstrated that the expressed human enzyme was capable of catalysis when extracted from the cells.[37] Nevertheless, the expression of the full-length human enzyme was insufficient to rescue growth of spores defective in *ERG9* gene function.[37] These results, in total, support the hypothesis that structural differences in the carboxy termini of the yeast and human enzymes affect the localization or folding of the protein within intracellular membranes, thereby preventing the full-length human squalene synthase from reconstituting sufficient sterol synthesis to support the growth of yeast cells.[37]

2.09.3.2 Steady-state Kinetic and Inhibition Studies

Comparisons between the yeast and mammalian enzymes have been carried out using steady-state kinetic analysis with substrates and inhibitors. The steady-state kinetic experiments suggest

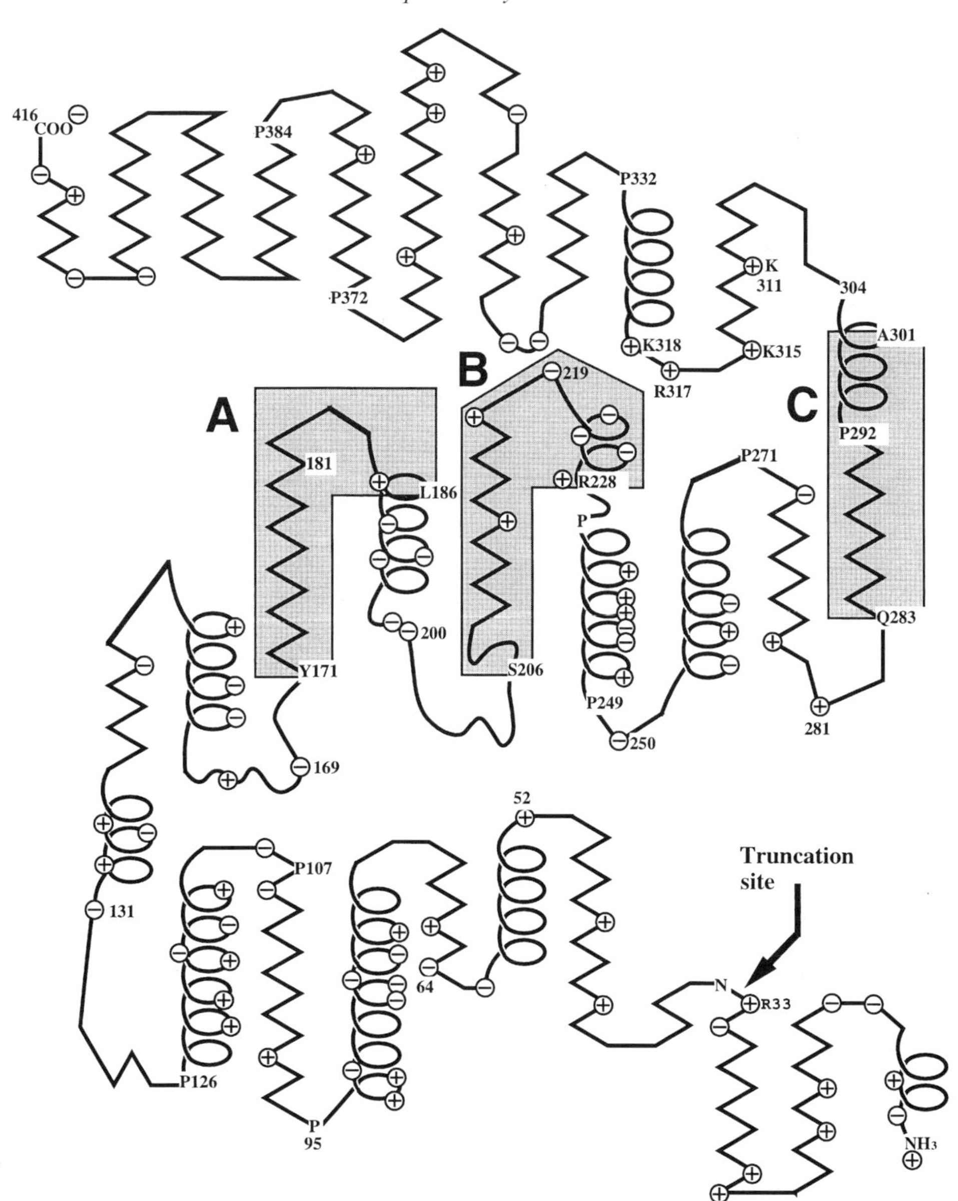

Figure 1 Model for secondary structure of rat squalene synthase. The model was based on the predicted amino acid sequence deduced from cDNA sequence data and on secondary structure prediction programs as described by McKenzie *et al.*[27] The truncation site was based on N-terminal amino acid analysis of the cytosolic, active form of the enzyme following proteolytic release with trypsin. Shaded areas A, B, and C contain sequences with high homology to other mammalian, yeast and plant squalene synthases (reproduced by permission of The American Society for Biochemistry and Molecular Biology from *J. Biol. Chem.*, 1992, **267**, 21 368).

that the mammalian enzymes exhibit significant structural differences from the yeast enzyme in both the NADPH binding site as well as in the FPP binding site(s). These studies are summarized below.

LoGrasso *et al.*[44] compared the purified yeast, rat, and human enzymes with respect to inhibition by two FPP analogues ((**6**) and (**7**)) and zaragozic acid C(**8**). Whereas (**6**) inhibited both the yeast and human enzymes similarly (IC_{50} = 48 μM and 55 μM, respectively), (**7**) inhibited the human enzyme more potently (IC_{50} = 68 nM for the human enzyme and IC_{50} = 1000 nM for the yeast enzyme). These results indicated substantial differences in the binding of these FPP analogues to the two forms of the enzyme, although FPP itself interacted similarly with all forms of the enzyme when measured under similar assay conditions (see below). These results also indicate the importance of using the relevant form of the enzyme for identification of therapeutic agents. Zaragozic acid C (**8**) exhibited similar degrees of inhibition for all three forms of the enzyme (IC_{50} = 0.5–2 nM).

Fulton *et al.*[45] also compared the inhibition of tobacco (*Nicotiana tabacum*) and rat liver enzyme extracts with a series of squalestatin analogues. As observed by LoGrasso *et al.*[44] for inhibition of the yeast and mammalian enzymes by a structurally related inhibitor, zaragozic acid C (**8**), similar degrees of inhibition were observed for all squalestatin analogues with both the rat and tobacco forms of the enzyme. It would be of interest to compare the inhibition of plant squalene synthase by (**6**) and (**7**). Since the sequence of plant (*Arabidopsis*) squalene synthase is more closely related to the yeast enzyme than mammalian squalene synthases,[39] it could be expected that (**7**) would also be a more effective inhibitor of the mammalian enzyme than of the plant enzyme.

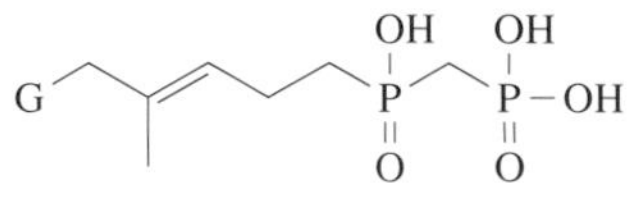

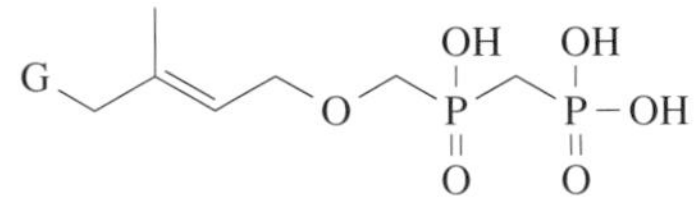

(**6**) G = geranyl ($C_{10}H_{17}$) (**7**)

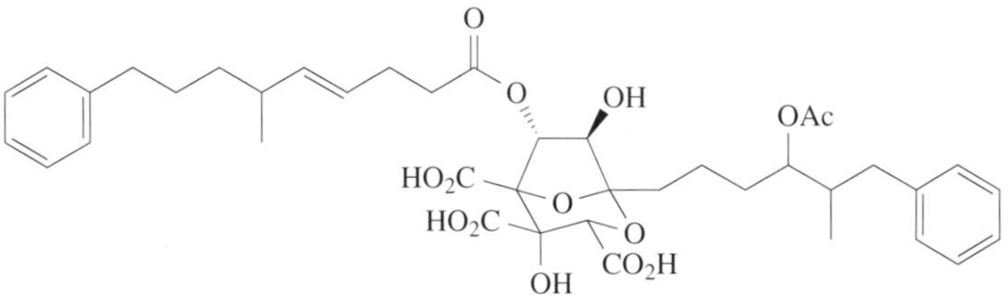

zaragozic acid C (**8**)

The reported steady-state kinetic values for different forms of purified squalene synthases are summarized in Table 1. For the human enzyme, the K_m value for NADPH was 430 μM and the $S_{0.5}$ value for FPP was 2.3 μM.[37] The corresponding values for the trypsinized (carboxy terminal and amino terminal truncated)[27] rat liver enzyme were 40 μM for NADPH and 1.0 μM for FPP.[26] The reported values for the carboxy terminal truncated yeast enzyme were 530 μM, 100 μM, and 180 μM for NADPH and 2.5 μM, 40 μM, and ~1 μM for FPP.[33,34,46] Thus, the enzyme from yeast, rat, and human sources had similar $S_{0.5}$ values for FPP (1–3 μM) when measured under similar assay conditions (see below). Finally, the K_m values for NADPH for the full-length human and carboxy terminal truncated yeast enzymes were 3–12 fold higher than the value for the trypsinized rat liver enzyme.

Table 1 Steady-state kinetic parameters for various forms of purified squalene synthases.

Squalene synthase source	$S_{0.5}$ *for FPP* (μM)	K_m *for NADPH* (μM)	V_{max} (nmol min^{-1} mg^{-1})	*Ref.*
Trypsinized rat	1.0	40	1200	26
C-terminal truncated yeast	2.5	530	1250	33
C-terminal truncated yeast	40	100	4100	34
C-terminal truncated yeast	~1	180	9200	46
Full-length human	2.3	430	4900	37

$S_{0.5}$ is substrate concentration giving 50% of V_{max} activity.

The full-length human enzyme had a V_{max} value approximately fourfold higher than the trypsinized rat enzyme[26] and the carboxy terminal truncated yeast enzyme[33] when measured with similar assay conditions. Zhang *et al.*[34] have also reported kinetic constants for the identical carboxy terminal truncated form of the yeast enzyme reported by LoGrasso *et al.*[33] The V_{max} value reported by Zhang *et al.*[34] was threefold higher than the value reported by LoGrasso *et al.*[33] Moreover, the $S_{0.5}$ value for FPP observed by Zhang *et al.*[34] was 40 μM or more than 10-fold higher than the value for carboxy terminal truncated squalene synthase reported by LoGrasso *et al.*[33] The reason for these differences may be related to variations in the assay conditions, for example, use of different detergents and other additives such as bovine serum albumin (BSA). Zhang *et al.*[34] included BSA in the assay while LoGrasso *et al.*[33] did not use BSA. Zhang *et al.*[34] reported that BSA stimulated the apparent enzymatic activity 5–6-fold; this additional stimulation of the enzymatic activity could account for the difference in the V_{max} values reported by the two groups. Moreover, the substantially higher $S_{0.5}$ value for FPP reported by Zhang *et al.*[34] may also be a result of FPP binding to the BSA in the assay mixture. BSA can bind a wide variety of amphipathic molecules such as fatty acids. In subsequent studies, in which the 0.5 mg^{-1} ml^{-1} BSA was replaced by 0.1 mg^{-1} ml^{-1} gelatin, 10% methanol, and 10% glycerol, the $S_{0.5}$ value for FPP was approximately 1 μM.[46]

While the full-length yeast enzyme exhibited similar V_{max}/K_m values for both NADH and NADPH,[25] the carboxy terminal truncated yeast enzyme showed a moderate preference for NADPH. Thus, the V_{max}/K_m value for NADPH was sevenfold greater than the value for NADH.[33] However, the trypsinized rat enzyme had a marked preference for NADPH. In this case, the V_{max}/K_m value for NADPH was 50-fold greater than the value for NADH.[47] The tobacco enzyme was also reported to exhibit a preference for NADPH over NADH.[48]

2.09.4 STRUCTURE AND MECHANISM

2.09.4.1 Mechanistic Studies

Many of the recent experiments directed at defining the chemical mechanism of squalene synthase have been carried by Poulter and co-workers.[7] Scheme 4 illustrates the conversion of presqualene diphosphate (**1**) to squalene (**2**) and indicates the involvement of various hypothetical carbocationic intermediates. In order to access the possible involvement of the carbocation intermediates in squalene synthase catalysis, Poulter and co-workers[7] prepared ammonium ion compounds (e.g., (**9**)) as reaction intermediate analogues of the primary (**3**) and tertiary (**5**) cyclopropylcarbinyl cations (Scheme 4). These analogues were tested as inhibitors of the enzyme and, somewhat surprisingly, were found not to inhibit the enzyme ($IC_{50} > 170$ μM). However, when they were tested in the presence of inorganic pyrophosphate, these ammonium ion compounds were found to be much better inhibitors ($IC_{50} = 3$ μM). Poulter and co-workers[7] suggested that the inorganic pyrophosphate may combine with the ammonium ion compound in the active site and thereby generate a potent ion-pair inhibitor. Another analogue (**10**) , in which a diphosphate moiety was tethered to (**9**), was prepared and found to be an effective inhibitor of the enzyme ($IC_{50} = 5$ μM), giving additional support to the ion-pair hypothesis. In addition Poulter and co-workers determined that (**10**) inhibited both partial reactions similarly and concluded that squalene synthase has a single active site or two overlapping sites catalyzing both partial reactions. Other groups have obtained similar results with analogues designed to mimic the carbocation species.[8] For example, Prashad *et al.*[49] prepared ammonium ion analogue inhibitors such as (**11**) and also observed increased inhibition in the presence of inorganic pyrophosphate. This result may be rationalized by hypothesizing that (**11**) plus the added inorganic pyrophosphate forms an ion-pair species. Moreover, Prashad[50] prepared a diphosphate tethered analogue (**12**) and found that (**12**) was also a potent inhibitor of the enzyme ($IC_{50} = 20$ nM). Studies such as these provide support for the involvement of carbocationic species in the conversion of PSPP to squalene.

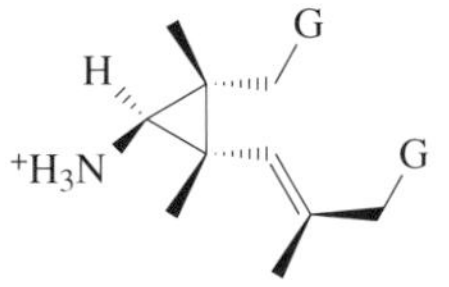

(9) 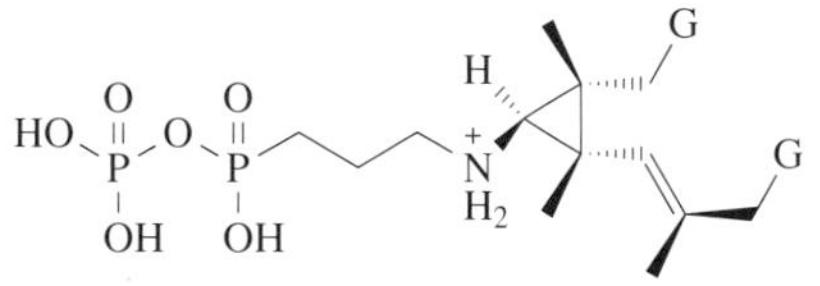 **(10)**

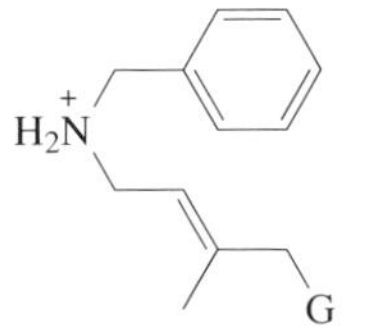

(11) 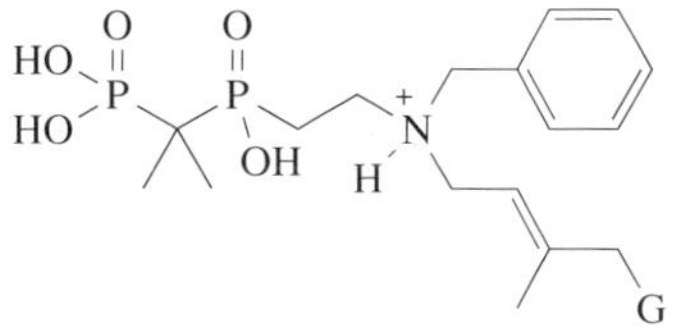 **(12)**

G = geranyl ($C_{10}H_{17}$)

Zhang and Poulter[51] have used the purified recombinant squalene synthase to characterize the reaction products formed by the enzyme in the absence of NADPH. Based on their observation of hydroxybotryococcene formation, they proposed a common mechanism for both squalene synthase (1′–1 isoprenoid) and botryococcene synthase (1′–3 isoprenoid). These authors suggested that the regiochemistry of the two reactions is determined by the location of the hydride group from NADPH relative to carbocation intermediates formed from PSPP. In the case of squalene synthase, the hydride moiety from NADPH is transferred to the 1′-position of the cyclopropyl ring of the tertiary carbocation intermediate ((**5**) in Scheme 4). They also suggested that botryococcene synthase may have evolved from squalene synthase by a mutation that shifted the location of the nicotinamide cofactor in the active site to allow hydride transfer to the 3′-position of the carbocationic intermediate.

Poulter and co-workers[52] have also obtained chemical evidence for the formation of the tertiary carbocationic intermediate ((**5**) in Scheme 4). When recombinant squalene synthase was incubated with FPP and a nonredox active dihydro derivative of NADPH, they observed the formation of three products. One of the products was a previously unobserved alcohol and was identified as a tertiary cyclopropylcarbinol, rillingol (**13**). This finding suggested that the corresponding tertiary cyclopropylcarbinyl cation ((**5**) in Scheme 4) is an intermediate in the conversion of PSPP to squalene.

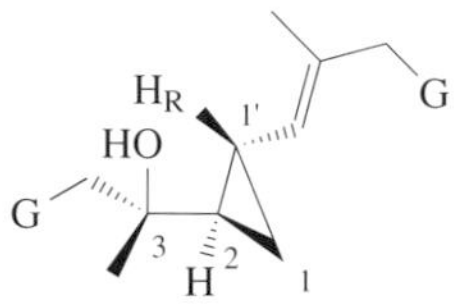

Rillingol (**13**)
G = geranyl ($C_{10}H_{17}$)

2.09.4.2 Effects of NADPH on Substrate and Inhibitor Binding

Early studies on the role of NADPH in the reaction indicated that PSPP synthesis appears to be stimulated by NADPH.[21] Later studies by Popjak and Agnew[4] suggested an additional role for NADPH in the formation of squalene rather than merely acting as a reducing agent. The studies of Mookhtiar *et al.*[46] suggested that at least part of the effect of NADPH on stimulation of PSPP synthesis may be due to slower release of PSPP from the enzyme relative to the release of squalene. However, observations of synergistic effects of NADPH (see below) with squalene synthase inhibitors suggest that NADPH plays an additional role in the reaction beyond simple hydride transfer and allowance for more rapid release of product from the enzyme. It is of interest to note that inclusion of an unreactive dihydro analogue of NADPH in a reaction mixture with purified recombinant squalene synthase and FPP led to the isolation of a novel tertiary cyclopropylcarbinol (**13**) as discussed in Section 2.09.4.1.[52] It is possible that occupancy of the NADPH binding site stabilized formation of (**5**) (Scheme 4) and/or allowed the tertiary cationic intermediate (**5**) to be more accessible to solvent.

The effect of NADPH on inhibitor binding was measured with the trypsinized rat liver enzyme by determining the IC_{50} values at various NADPH concentrations with an FPP concentration of 5 μM ($\sim 5 \times S_{0.5}$ value).[47] The K_m value for NADPH is 0.04 mM for the trypsinized rat enzyme and NADPH concentrations were chosen above and below this value. Therefore, inhibitor effectiveness was determined under relatively low and high degrees of saturation with NADPH. If an inhibitor were to bind partly in the NADPH binding site, one would expect a higher apparent IC_{50} value at higher NADPH concentrations than at lower NADPH concentrations. This situation did not occur with the squalene synthase inhibitors tested. If there was no binding of the inhibitor in the NADPH site, then the IC_{50} values should be similar at the low and high levels of NADPH. This situation occurs with zaragozic acid C (see below and Section 2.09.4.3). The third possibility is that the apparent IC_{50} value is lower in the presence of higher concentrations of NADPH. This is the least likely situation, but as previously described, an analogous situation occurs with the ammonium ion inhibitors of squalene synthase and inorganic pyrophosphate (see Section 2.09.4.1).

The effects of NADPH on the IC_{50} values of (**6**) and (**14**) are illustrated in Figure 2. The effect of NADPH on the IC_{50} value of (**14**) is substantial. The IC_{50} for (**14**) in the presence of 0.025 mM NADPH was 17 μM, and the IC_{50} value decreased to 0.12 μM in the presence of 1.2 mM NADPH (i.e., 140-fold decrease in the apparent IC_{50} value). This is a minimal estimate for the difference since the IC_{50} value in the absence of NADPH was not estimated. When 1.0 mM inorganic pyrophosphate was included, the IC_{50} value decreased further to 30 nM (data not shown). This effect of inorganic pyrophosphate has previously been described for (**14**) with the rat microsomal enzyme.[18] For (**7**), approximately a 10-fold decrease in the IC_{50} value was observed between 0.025 mM and 1.2 mM NADPH (data not shown), while for (**6**), less than a threefold decrease in the IC_{50} values were observed over the range of NADPH concentrations used. Note that the changes in the IC_{50} values occurred near the K_m level of NADPH (0.04 mM), consistent with a role of NADPH in influencing the binding of carbocation intermediate analogues. The NADPH effect on inhibitor binding indicated that strong synergies occur between the substrates of the reaction. In particular, the effects were most pronounced for analogues which mimic carbocation intermediate species such as (**14**), although smaller effects were still observed for the ground state substrate (FPP) mimetic analogues such as (**6**) and (**7**). However, zaragozic acid C (**8**), which appears to mimic the intermediate PSPP (**1**),[8,53] does not exhibit such synergistic binding behavior with NADPH and the human enzyme.[37]

(**14**)
G = geranyl ($C_{10}H_{17}$)

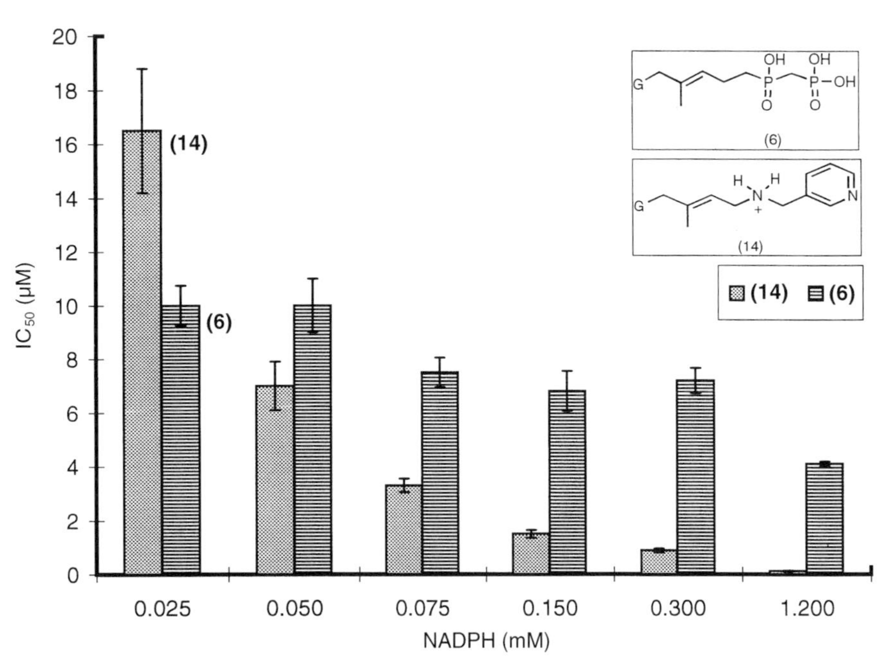

Figure 2 Effect of NADPH on the apparent IC_{50} values for inhibition of trypsinized rat squalene synthase with compounds (**6**) and (**14**). The effect of NADPH on inhibitor binding was measured with the trypsinized rat liver enzyme by determining the apparent IC_{50} values at various NADPH concentrations.[47] The NADPH concentrations were chosen above and below the K_m value for NADPH (0.04 mM). The IC_{50} for (**14**) in the presence of 0.025 mM NADPH was 17 μM, and the IC_{50} value decreased to 0.12 μM in the presence of 1.2 mM NADPH (i.e., 140-fold decrease in the apparent IC_{50} value). When 1.0 mM inorganic pyrophosphate was included, the apparent IC_{50} value for (**14**) decreased further to 30 nM (data not shown).

2.09.4.3 Studies with Zaragozic Acid Analogues

Studies with the zaragozic acids have been reviewed extensively.[8,18–20] However, in this section we discuss studies which have used these potent squalene synthase inhibitors to characterize the active site of the enzyme. Zaragozic acid A was reported to be a competitive inhibitor of squalene synthase with a K_i value of 80 pM[53] and 1.6 nM.[54] In addition, zaragozic acid A also inhibited both partial reactions.[54] Because these inhibitors were so potent, it was possible to have an excess of enzyme over inhibitor in the assay, thereby resulting in apparently higher K_i and IC_{50} values. Thus, it was necessary to dilute the microsomal protein until the inhibition was independent of the protein concentration in order to obtain an accurate estimate of the K_i value.[53] Zaragozic acid C (**8**) was a competitive inhibitor of the purified recombinant human enzyme with respect to FPP and had a K_i value of 250 pM. It should be noted that 30 pM squalene synthase was used in these experiments and therefore the inhibitor was in excess over the enzyme.[37] This K_i value was moderately higher than the value of 45 pM reported for the microsomal rat liver enzyme.[53] In addition, NADPH, at concentrations below and above the K_m value (at $\sim 0.5\times$ and $\sim 10\times$ the K_m value for NADPH), had little effect on the IC_{50} value for zaragozic acid C. This result indicated that the NADPH binding site was separate from the binding site for zaragozic acid C.

Lindsey and Harwood[55] observed that zaragozic acid A appeared to irreversibly inactivate squalene synthase. Although this may be a plausible interpretation of the experiments, it will be necessary to demonstrate covalent bond formation with the enzyme directly before concluding that irreversible inhibition occurred. For example, methionine sulfoximine inhibited glutamine synthetase by formation of a complex of methionine sulfoximine phosphate and ADP in the active site. Although the inhibition was reversible, in practice this was not observed, since the half-life for release from glutamine synthetase was 10^5–10^6 years.[56] Given the availability of highly purified squalene synthases,[26,33,34,37] it should be possible to demonstrate covalent bond formation with zaragozic acid A and squalene synthase, for example, using electrospray mass spectrometry. If this is the case, zaragozic acid A should be a useful tool for identifying active site residues. However, as pointed out by Biller *et al.*,[8] this class of compounds (zaragozic acids/squalestatins) are generally reversible inhibitors of the enzyme.

2.09.4.4 Steady-state Kinetic Analysis

The availability of the purified carboxy terminal truncated yeast enzyme[33,34] has allowed an analysis of the kinetic mechanism[46] as a prelude to further studies of the kinetic and chemical mechanisms. These studies indicated that the kinetic mechanism was sequential with NADPH binding to the enzyme following an irreversible step. Stimulation of the rate of PSPP formation by NADPH appeared to result, at least partially, from alleviation of PSPP inhibition. In this case, the rate of PSPP dissociation was rate limiting in the absence of NADPH. The K_m value for NADPH was approximately 180 μM and the estimated K_m values for the donor and acceptor FPP molecules were less than 100 nM. Previously, it was suggested that PSPP formation and squalene formation may occur at separate sites on the enzyme(s).[4] However, the studies by Poulter and co-workers[7] with reaction intermediate analogue inhibitors and steady-state kinetic analysis[46] using the purified recombinant enzyme suggest that both reactions occur at a single site or two overlapping sites.

Additional steady-state kinetic experiments have been carried out to determine how well the nicotinamide adenine dinucleotide product of the reaction ($NADP^+$) binds to squalene synthase. The oxidized product of the reaction, $NADP^+$, binds poorly to the trypsinized rat enzyme, relative to the binding of NADPH as estimated by the K_m value. The IC_{50} value for $NADP^+$ with the trypsinized rat enzyme measured in the presence of 0.1 mM NADPH is 6.5 mM.[47] Thus, assuming that $NADP^+$ is competitive vs. NADPH, the calculated K_i value for $NADP^+$ is ~ 2 mM or ~ 50-fold higher than the K_m value for NADPH. These results indicated that mimetics of the reduced form of the nicotinamide cofactor were preferred in the design of bisubstrate squalene synthase inhibitors.

2.09.5 INHIBITORS

Inhibitors have been extensively reviewed by various groups.[8,18–20] Identification of squalene synthase inhibitors has been based on rational drug design and random screening of synthetic compound collections and natural products. Both strategies were successful and illustrate the value of using rational design and random screening approaches in drug discovery programs. We briefly summarize results with some of the more potent and selective inhibitors of the enzyme which have been identified, e.g., squalestatins/zaragozic acids, bisphosphonates, and α-phosphonosulfonates. The utility of these compounds as valuable research tools for selective inhibition of sterol biosynthesis and as potential therapeutic agents is discussed.

2.09.5.1 Natural Products Inhibitors

The structure of one of the natural product inhibitors, zaragozic acid C (**8**), is shown. The squalestatins/zaragozic acids are characterized by a citric acid-like 'core' structure (2,8-dioxabicyclo-[3.2.1]octane-3,4,5-tricarboxylic acid). The squalestatins/zaragozic acids appear to be mimetics of PSPP based on modeling studies.[8,53] A wide variety of related structures have been found from many different fungal sources.[19] Other examples of natural product inhibitors of squalene synthase include the viridiofungins (**15**),[18] nonadrides (**16**),[8] and schizostatin (**17**).[57]

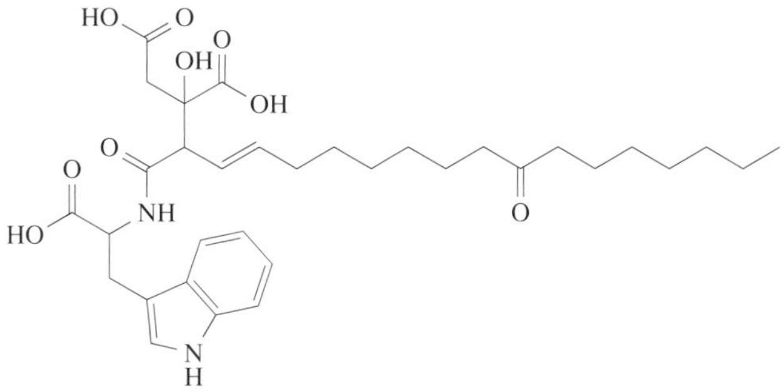

Viridiofungin C (**15**)

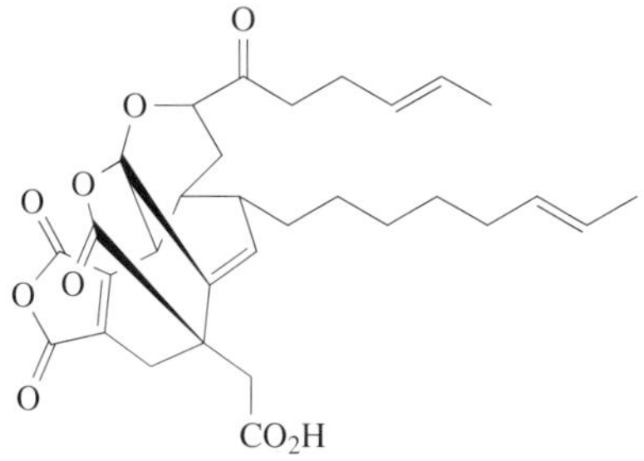

(**16**)

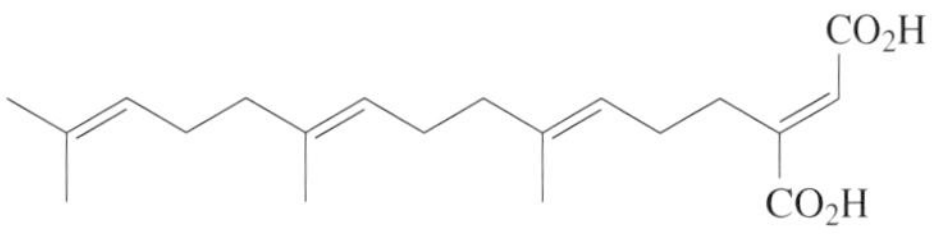

Schizostatin (**17**)

The function of these natural product inhibitors is not known. However, it is possible that these natural products may offer a competitive advantage to the producer organism through inhibition of the squalene synthases of competitive species. A similar situation may occur with the natural product inhibitors of glutamine synthetase, such as phosphinothricin and taboxinine-β-lactam.[58,59] In these cases, the glutamine synthetase inhibitors are produced as inactive tripeptide precursor forms of the inhibitors. Some of the natural product squalene synthase inhibitors may also be produced in inactive precursor forms, thereby preventing inhibition of their own sterol biosynthetic pathway.

The sterol biosynthetic pathway appears to be a particularly sensitive pathway for inhibition by these natural products. The HMG-CoA reductase inhibitors, such as compactin, are additional examples of natural product inhibitors of sterol biosynthesis.[1] Moreover, potent inhibition of squalene synthase has been observed using plant extracts, suggesting that some plants, in addition to some fungi, may produce squalene synthase inhibitors for protective purposes.[47]

The squalestatins/zaragozic acids are effective inhibitors of cholesterol biosynthesis and cholesterol lowering agents *in vivo*. In rats treated with zaragozic acid A, FPP-derived dicarboxylate metabolites appear transiently in the liver and are excreted in the urine. In some cases the build-up of farnesol-derived dicarboxylic acids is toxic. However, the toxicity is species dependent.[8]

2.09.5.2 Synthetic Inhibitors

Much of the early work was based on designing mimetics of the substrate FPP.[60,61] These compounds were used in mechanistic studies to distinguish the donor and acceptor sites of the enzyme.[61] Poulter and co-workers[7] prepared reaction intermediate analogues such as (**9**) and (**10**). These analogues were useful in supporting the hypothesis that carbocationic intermediates were involved in the reaction as well as indicating the formation of ion-pair combinations with inorganic pyrophosphate (see Section 2.09.4.1).

Other groups have prepared additional analogues as mimetics of substrates or intermediates of the reaction. For example, Vedananda *et al.*[62] prepared a series of cyclic amino alcohols (e.g., (**18**)) as replacements for the diphosphate moiety of FPP and found such structures to be moderately effective inhibitors of the enzyme. Shechter *et al.*[63] prepared a series of sulfobetaine zwitterionic inhibitors of squalene synthase (e.g., (**19**)). These neutrally charged compounds are mimetics of putative ion-pair reaction intermediates (i.e., analogous to (**10**) and (**12**)) and were found to be moderately effective enzyme inhibitors. It would be of interest to see if additional hydrophobic substitution on the tertiary amine leads to increased potency for this series of inhibitors. The arylalkylamine analogues of the carbocation intermediates (e.g., (**11**) and (**14**)) were effective squalene synthase inhibitors, but suffered from lack of selectivity since they also inhibited 7-dehydrocholesterol reductase. Thus, *in vivo* treatment with these compounds led to accumulation of the undesirable product, 7-dehydrocholesterol.[64] Likewise, a series of cycloalkylamine squalene synthase inhibitors (e.g., (**20**)) also inhibit 7-dehydrohydrocholesterol reductase and led to the formation of 7-dehydrocholesterol.[65] A number of groups also prepared 3-substituted quinuclidines (e.g., (**21**)) as squalene synthase inhibitors.[8] Some of the 3-substituted quinuclidines appear to be selective for squalene synthase as no build-up of postsqualene metabolites occurred.[66] A series of cyclobutyldicarboxylic acids (e.g., (**22**)) was prepared and some of these compounds were reported to be effective squalene synthase inhibitors.[8] This class of compounds, like the squalestatins/zaragozic acids, may also be mimics of PSPP.

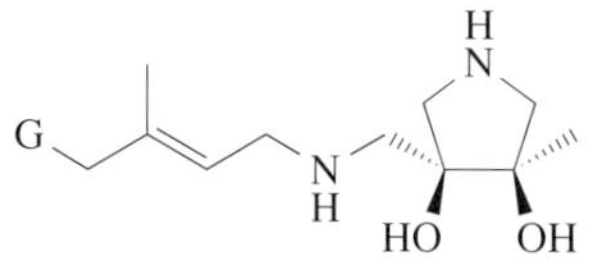

(**18**) 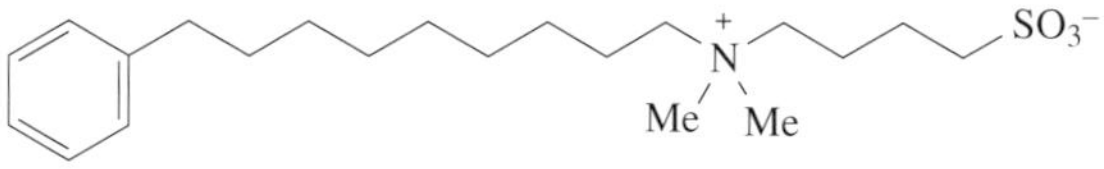 (**19**)

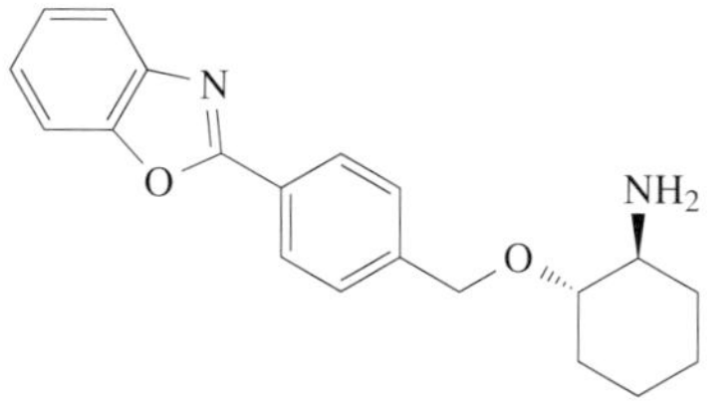

(**20**) 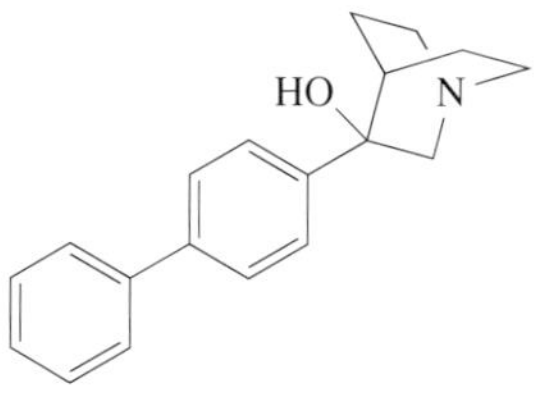 (**21**)

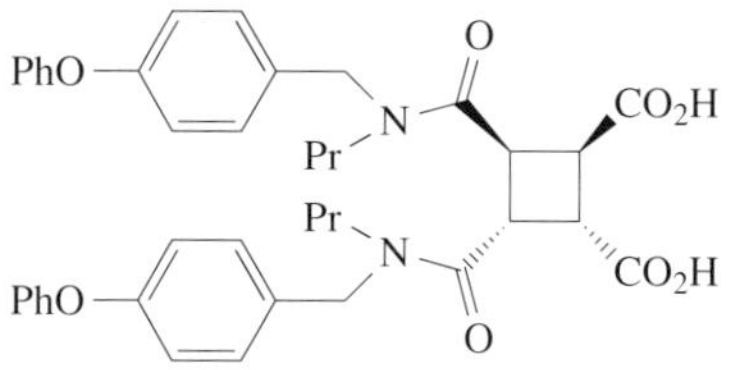

(**22**)

G = geranyl ($C_{10}H_{17}$)

The phosphonate, bisphosphonate, and α-phosphonosulfonate inhibitors appear to be among the most investigated series of synthetic squalene synthase inhibitors. The incorporation of an ether oxygen into an FPP analogue ((**7**)) led to greater than 1000-fold improvement in binding[67] and indicated potential interaction between the ether oxygen of (**7**) and an active site residue of the enzyme. However, some of the most potent (IC_{50} values in the picomolar range) synthetic inhibitors came from the bisphosphonate (e.g., (**23**)) series of structures.[18] The conversion of these potent enzyme inhibitors into potential drug candidates by investigators at Bristol-Myers Squibb is a good example of a multidisciplinary team approach to overcome problems of absorption, distribution, and metabolism.[8] The bisphosphonate group was replaced with an α-phosphonosulfonate moiety to eliminate retention in liver and bone tissues. Next the farnesyl moiety was replaced with a diphenyl ether group to block the farnesyl side chain metabolism. Finally, a prodrug strategy was used with bioreversible bis(pivaloyloxy)methyl esters to overcome absorption issues leading to structures such as (**24**). Compounds such as (**24**) are effective cholesterol lowering agents in rats, hamsters, and primates. Compound (**24**) (BMS-188494) was reported to have entered Phase I clinical trials.[68]

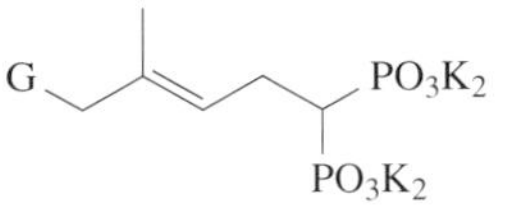

(23) 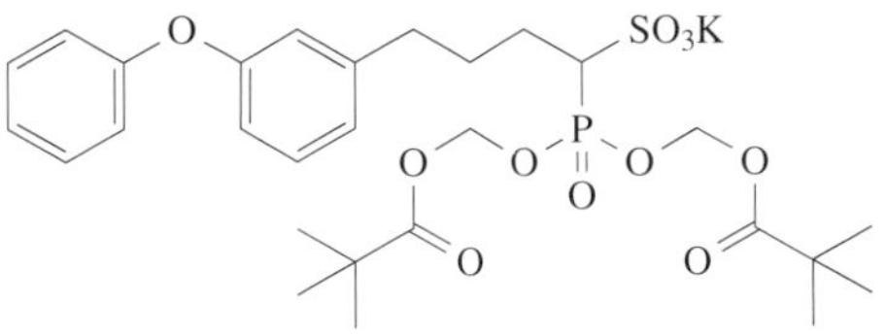 (24)

G = geranyl ($C_{10}H_{17}$)

In general, the carbocation mimetic analogues appear to exhibit synergistic binding behavior with both inorganic pyrophosphate and with the nicotinamide cofactor NADPH (see Sections 2.09.4.1 and 2.09.4.2). Therefore, the *in vivo* potency and specificity of these analogues will also be determined by local concentrations of both inorganic pyrophosphate and NADPH. In contrast, the putative mimetics of PSPP and FPP such as the squalestatins/zaragozic acids, cyclobutyldicarboxylic acids, bisphosphonates, and α-phosphonosulfonates may not be as sensitive to local concentrations of inorganic pyrophosphate and NADPH and therefore may be better therapeutic candidates. *In vivo* pharmacology and toxicology studies with the various classes of squalene synthase inhibitors will be required to determine if this is the case.

2.09.6 REGULATION OF THE MAMMALIAN ENZYME

2.09.6.1 Regulation of the Rat Enzyme

Early observations, related to the regulation of squalene synthase, have shown that the activity of this enzyme is suppressed in cultured fibroblasts grown in the presence of low-density lipoproteins and have led to the hypothesis that this enzyme plays a major role in the regulation of the flux of terpenoids to both sterol and nonsterol end products.[15] Later reports, using animal experimentation, showed that rat hepatic squalene synthase (RSS) activity decreased in animals fed diet supplemented with cholesterol, whereas feeding with a low cholesterol diet or with a diet containing HMG-CoA reductase inhibitors such as fluvastatin caused more than 25-fold increase in hepatic enzyme activity.[26] Similarly, Stamellos *et al.* showed, by Western blot analysis, variations in the amount of hepatic enzyme protein in response to dietary manipulations.[69] In addition, they demonstrated a significant decrease in hepatic enzyme in response to fenofibrate feeding and an increase in response to gemfibrozil.

These studies of the regulation of both the enzyme protein level and its activity, together with the successful cloning of the RSS cDNA, enabled examination of the transcriptional regulation of the RSS gene. Regulation of the rat hepatic enzyme in response to dietary manipulation is transcriptionally controlled. There are two size classes of mRNA for the rat enzyme at 1.9 kb and 3.4 kb. Figure 3 demonstrates the variation in the amount of the mRNA in response to supplementation of the diet with cholesterol or addition of cholestyramine or lovastatin. Both size classes of mRNA show a significant decrease in response to cholesterol feeding and a marked increase in the presence of either cholestyramine or lovastatin in the diet. This variation in mRNA levels is very similar to the variations in the amount of enzyme proteins in response to these diets (Figure 4). Comparison between relative rate of transcription, as measured by nuclear runoff studies, relative mRNA levels, and relative amount of RSS protein showed similar variations in response to the different dietary manipulations[70] (see Table 2). Such synchronous variations strongly suggest that the regulation of RSS is primarily transcriptionally controlled.

2.09.6.2 Regulation of the Human Enzyme

The human hepatic squalene synthase (HSS) was reported to be transcriptionally regulated in cultured hepatic cells similar to that observed for RSS in rat livers.[35] There are three size classes of HSS mRNA at 1.4 kb, 1.6 kb, and 2.1 kb. All three sizes of mRNA are regulated in the human hepatoma cell line, HepG2. Jiang *et al.*[35] have shown (Figure 5) that growing of cells in the presence of 5 μg ml^{-1} lovastatin resulted in an increase in the HSS mRNA level, whereas the presence of 5 μg ml^{-1} 25-hydroxycholesterol almost completely suppressed the level of this mRNA. These

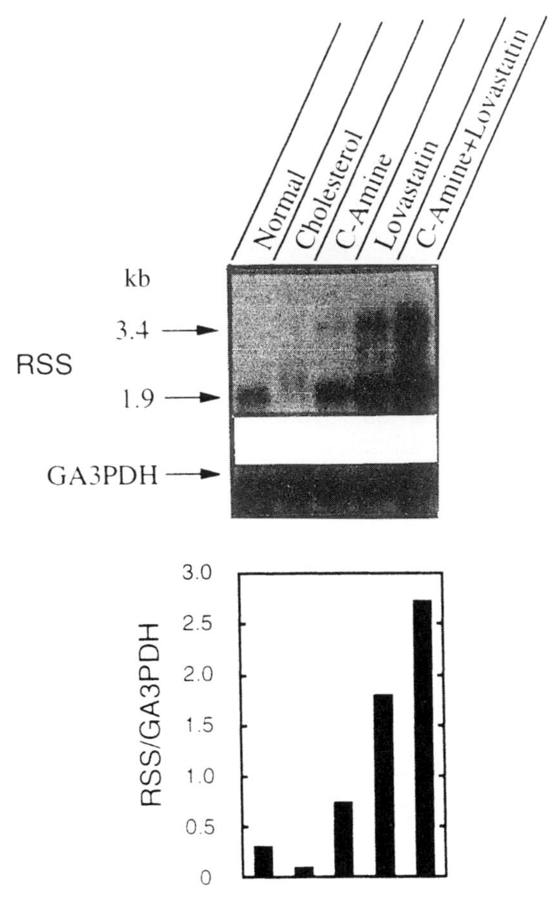

Figure 3 Northern blot analysis of rat Poly(A^+) RNA from livers of animals fed various diets. Poly(A^+) RNA, obtained from livers of animals fed with the labeled diets (C-Amine represents cholestryamine), was twice purified on oligo(dT) cellulose and separated by 0.8% (w/v) agarose formaldehyde gel electrophoresis.[70] After being transferred to a nitrocellulose membrane, the mRNA was sequentially hybridized to a ^{32}P-labeled RSS cDNA probe and a rat GA3PDH probe. RSS mRNA shows at two size transcripts of 1.9 and 3.4 kb. Quantitation of the signals of the probed livers mRNAs was done by PhosphorImager scanning. The relative amount of RSS mRNA in the various lanes is expressed by the ratio of the signals RSS/GA3PDH.

variations in mRNA level in response to lovastatin and 25-hydroxycholesterol can be observed in cells grown in either lipid-depleted serum or in media containing 10% normal serum (see Figure 6). Based on these observations, additional studies were focused on the HSS gene promoter and its regulatory sequences.

Table 2 Comparison of the effect of different diets on rate of transcription, mRNA level, and amount of RSS protein.

Diets[a]	*Relative rates of transcription*	*Relative mRNA level*	*Relative amount of RSS protein*
Normal	1.0 ± 0.07	1.0 ± 0.02	1.0 ± 0.03
2% Cholesterol	0.8 ± 0.05	0.3 ± 0.1	0.3 ± 0.02
5% Cholestyramine	2.3 ± 1.1	1.2 ± 0.2	2.1 ± 0.16
0.1% Lovastatin	2.9 ± 1.2	3.0 ± 0.04	3.2 ± 0.05
5% Cholestryamine + 0.1% lovastatin	4.1 ± 0.5	4.0 ± 1.27	4.3 ± 0.05

[a] Percent additions refers to (w.w). The table summarizes the relative rate of transcription, mRNA level and the amount of RSS immunoreactive protein in liver of rats fed with the indicated die.[70] Relative rates of transcription were assayed in nuclear run-off experiments, relative mRNA levels were measured in slot blot analyses assays using poly(A^+) RNA, and RSS proteins were determined by immunoreactivity. Values for different diets were normalized for normal diet. Each value is a mean of two experiments.

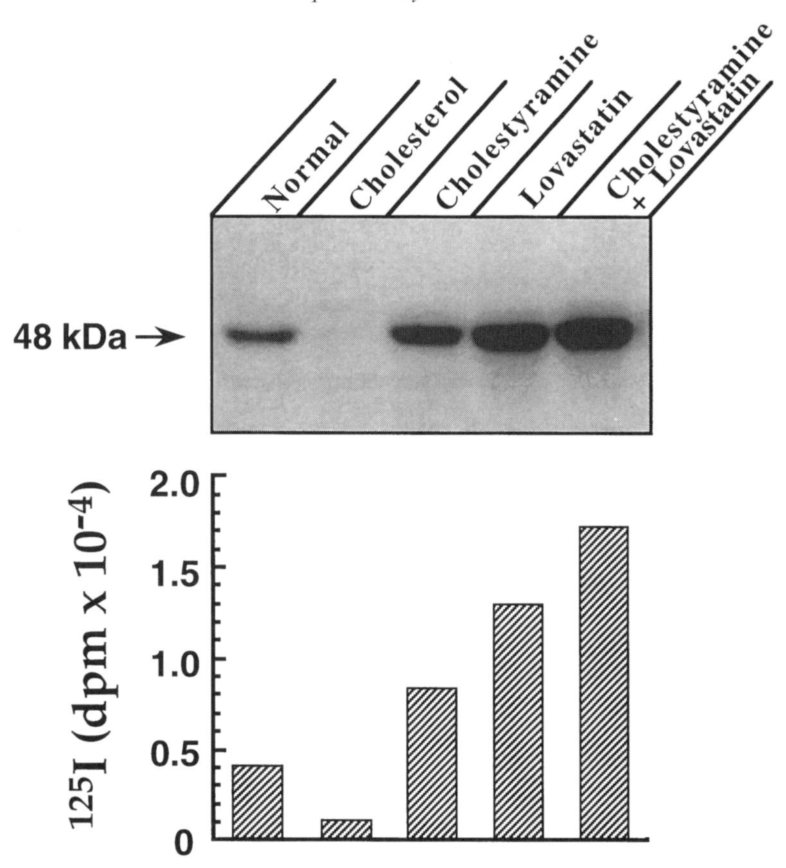

Figure 4 Immunoblot analysis of RSS from livers of rats fed different diets. Rat liver microsomes (50 μg protein per lane), obtained from animals fed different diets, were analyzed by Western blots analysis.[70] Detection of immunoreactive proteins was with ^{125}I-Protein A and autoradiography. Quantitation of radioactivity was done by cutting each of the labeled bands and counting for ^{125}I radioactivity. The figure shows the autoradiogram obtained (top) and the quantitation of radioactivity corresponding to RSS (bottom).

2.09.6.3 HSS Gene Promoter Structure and the Presence of DNA Sequence Elements Involved in Transcriptional Regulation

The 5′ flanking region (1.5 kb) of the gene encoding human squalene synthase was cloned from a fixII human placenta genomic library and the promoter activity of successively 5′ truncated sections of 1 kb of this region were measured with a luciferase chimeric reporter gene transfected into HepG2 cells.[28] A sterol-mediated promoter activity was located primarily in a DNA segment of 200 bp 5′ to the transcription start site. The lowering of the cells' sterols, by the presence of 5 μg ml^{-1} lovastatin in the growth media, resulted in an approximately 50-fold induction of luciferase activity in transiently transfected HepG2 cells for fusion constructs containing sections of 200, 459, and 934 bp of the HSS promoter, compared to cells grown in the presence of 5 μg ml^{-1} 25-hydroxycholesterol. Loss of promoter activity and response to sterols was localized to a 69 bp section located 131 nucleotides 5′ to the transcription start site (Figure 7). Sequence analysis of this region showed that it contained a sterol regulatory element 1 (SRE-1) previously identified in other sterol regulated genes,[71] a sterol regulatory element 3 (SRE-3), and a Y-Box sequence previously reported to function in concert in the sterol-mediated transcriptional regulation of the FPP synthase gene,[72] two potential NF-1 binding sites and a CCAAT box. An additional 8 out of 10 bp SRE-1 sequence elements and two Sp1 sites are present 3′ to this section. It is of interest that the SRE-3 and the Y-Box in the HSS promoter are both in an inverted orientation compared to those found in FPP synthase (hence the terminology Inv-Y-Box and Inv-SRE-3). Their relative position in the promoter is reversed as well. In FPP synthase, the Y-Box is 5′ to the SRE-3 sequence and they are separated by a 21 bp spacer. In HSS, the Inv-SRE-3 is 5′ to the Inv-Y-Box with a 9 bp separation. In both genes, the two elements are arranged sequentially to allow orientation in the same direction. It is possible that this arrangement is required to enable interaction between the sterol regulatory

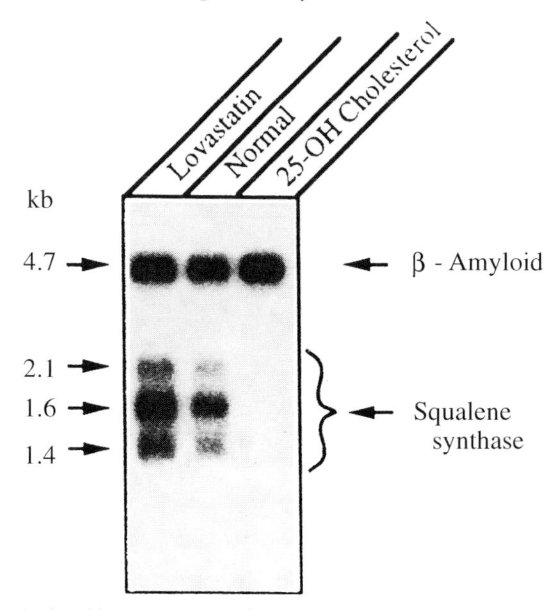

Figure 5 Northern blot analysis of human Poly(A^+) RNA from HepG2 cells grown under various conditions. Poly(A^+) RNA, obtained from HepG2 cells grown in the presence of lipid depleted serum (Normal) and under the same conditions but supplemented with either $5\,\mu g\,ml^{-1}$ lovastatin (Lovastatin) or $5\,\mu g\,ml^{-1}$ 25-hydroxycholesterol (25-OH Cholesterol), was twice purified on oligo(dT) cellulose and analyzed by Northern blot analysis.[35] HSS mRNA appears at three size transcripts of 1.4 kb, 1.6 kb and 2.1 kb. β-amyloid Poly(A^+) RNA is used as an unregulated reference (reproduced by permission of The American Society for Biochemistry and Molecular Biology from *J. Biol. Chem.*, 1993, **268**, 12 818).

element binding protein (SREBP) and the NF-Y transcription factors, which were shown to be involved in the interaction with these two sequence elements.[72]

Sequences within this 69 bP DNA, including the SRE-1 *cis*-acting element, show strong binding to the purified nuclear transcription factor ADD1/SREBP-1c[73] by mobility shift assay and footprinting analyses, further indicating the importance of this sequence element in the sterol-controlled transcriptional regulation of the HSS gene.[28]

2.09.7 CHROMOSOMAL MAPPING OF THE GENE AND SUBCELLULAR LOCALIZATION OF THE ENZYME

2.09.7.1 Chromosomal Mapping of the Human Gene

Employing a yeast artificial chromosome (YAC) DNA containing the human squalene synthase gene in a fluorescence *in situ* hybridization (FISH) analysis, the human squalene synthase gene (*FDFT1*) was mapped to chromosome 8. This assignment was confirmed by polymerase chain reaction analysis of a somatic cell hybrid containing human chromosome 8. A combination of somatic cell hybrid regional mapping panel and fractional length analysis of the FISH signal localized *FDFT1* to 8p22-p23.1.[74]

2.09.7.2 Chromosomal Mapping of the Mouse Gene

The mouse *Fdft1* gene was mapped to the proximal end of chromosome 14 by employing a rat cDNA for the hybridization of distinct restriction fragments length variants (RFLVs) followed by a linkage map construction.[75]

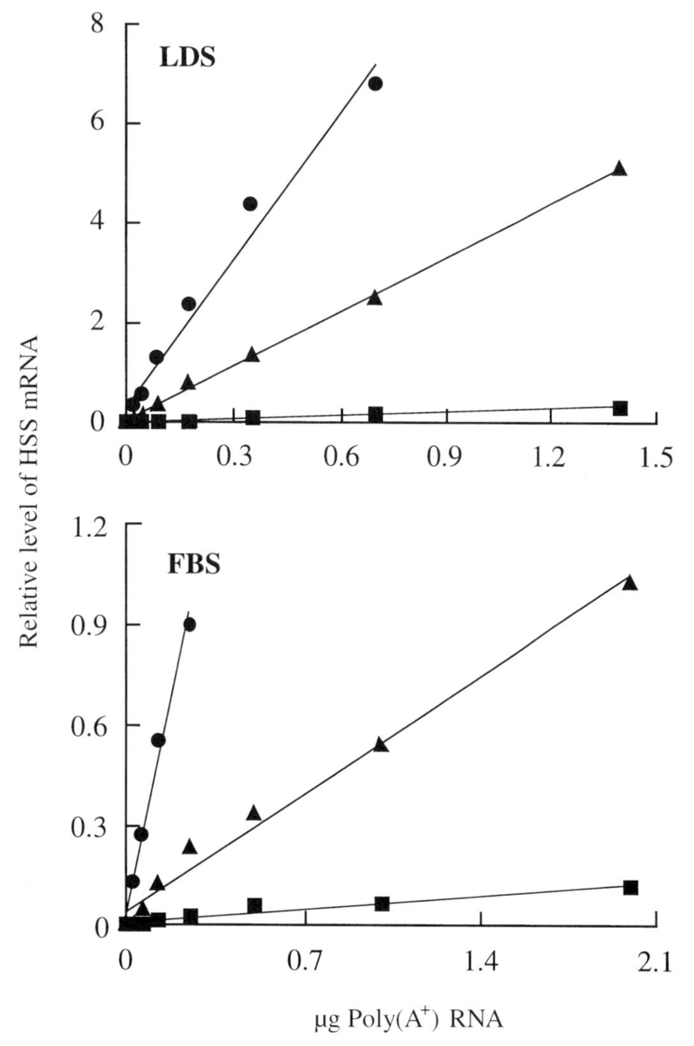

Figure 6 Quantitation of HSS mRNA from HepG2 cells grown under various conditions. P^{32}-Poly(A^+) RNA, obtained from HepG2 cells, was analyzed in slot blots, probed for either β-amyloid or HSS mRNAs and quantitated by PhosphoImager.[35] β-amyloid probe was used as a standard for nonregulated gene transcription. The plots depict HSS mRNA levels in cells grown in normal serum (fetal bovine serum, FBS, lower panel) or in the presence of lipid-depleted serum (LDS, upper panel) (▲) and under the same conditions but supplemented with either 5 µg ml^{-1} lovastatin (●) or 5 µg ml^{-1} 25-hydroxycholesterol (■) (reproduced by permission of The American Society for Biochemistry and Molecular Biology from *J. Biol. Chem.*, 1993, **268**, 12818).

2.09.7.3 Subcellular Localization of the Rat Enzyme

Various enzymes in the early part of the cholesterol biosynthetic pathway including HMG-CoA reductase,[76,77] mevalonate kinase,[78] and FPP synthase[79] were shown to be associated with the peroxisomes by a variety of methodologies including subcellular fractionation and immunoelectron microscopy localization. Biochemical activities of RSS[80] as well as different enzymes involved in the conversion of lanosterol to cholesterol have been reported to be present in peroxisomes.[81] This, together with a report that mevalonic acid could be converted to cholesterol *in vitro* by peroxisomes in the presence of a cytosolic fraction,[82] indicate the peroxisome as a possible site for cholesterol biosynthesis. However, the availability of anti-RSS monospecific antibodies enabled, using immunoelectron microscopy and subcellular fractionation studies, the localization of RSS exclusively to the endoplasmic reticulum (ER).[69] In addition, the localization of squalene epoxidase and squalene cyclase to the ER[83] exclude the peroxisomes from being an independent site for cholesterol production.

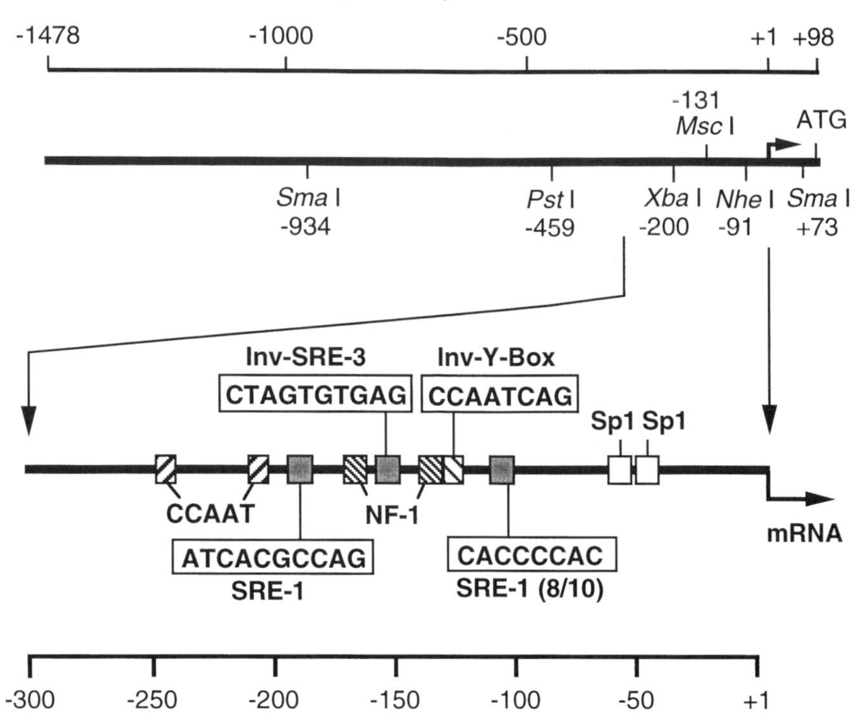

Figure 7 Schematic representation of the HSS promoter and potential sequence elements at the 5′ flanking region of the HSS gene. The relative positions of endonuclease restriction sites at the 5′ flanking region are shown.[28] The type and positions of various putative transcription and regulatory elements found within the 300 nt flanking the transcription start site are indicated in marked boxes. The sequences of the three SRE elements and the Inv-Y-Box are boxed (reproduced by permission of The American Society for Biochemistry and Molecular Biology from *J. Biol. Chem.*, 1995, **270**, 21 958).

2.09.8 ADDENDUM

In a later report, Guan *et al.* have extended their studies on the transcriptional regulation of the human squalene synthase gene. By using mutational analyses on a human squalene synthase promoter-luciferase reporter they have identified sequence elements in the promoter responsible for the regulation. They have demonstrated that mutation of an HSS-SRE-1 element present in the promoter significantly reduced, but did not abolish, the response to change in sterol concentration. Mutation scanning indicates that two additional DNA promoter sequences are involved in sterol-mediated regulation. The first sequence contains an inverted SRE-3 element (Inv-SRE-3) and the second contains an inverted Y-Box (Inv-Y-Box) sequence. A single mutation in any of these sequences reduced, but did not completely remove, the response to sterols. Combination mutation studies showed that the human squalene synthase promoter activity was abolished only when all three elements were mutated simultaneously. Co-expression of SRE-1 or SRE-2 binding proteins (SREBP-1 or SREBP-2) with the human squalene synthase promoter-luciferase reporter resulted in a dramatic increase of promoter activity. Gel mobility shift studies indicate differential binding of the SREBPs to regulatory sequences in the human squalene synthase promoter.[84] In addition, differential activation of mutant human squalene synthase promoter constructs by the two SREBPs was observed.[83] These later studies indicate that the transcription of the human squalene synthase promoter gene is regulated by multiple regulatory elements in the promoter.

In an attempt to explain studies that have shown that endotoxin (LPS) administration to Syrian hamsters markedly increased hepatic HMG-CoA reductase activity, protein mass, and mRNA levels, but only produced a modest increase in hepatic cholesterol synthesis, Memom *et al.* have examined the effect of LPS and cytokines on the activity, protein mass, and mRNA level of squalene synthase. They have demonstrated that LPS administration produces a marked decrease in the mRNA levels of squalene synthase. This decrease in squalene synthase mRNA occurred very rapidly (90 min after LPS) and required relatively small doses of LPS (1 μg per 100 g body weight). LPS also significantly decreased squalene synthase activity and protein mass. In addition, LPS produced a marked decrease in squalene synthase mRNA, activity, and protein levels when the basal levels of squalene synthase expression were increased fourfold by prior treatment with the bile acid binding resin, colestipol. Tumor necrosis factor and interleukin-1, which mediate many of the metabolic effects of LPS, also decreased hepatic squalene synthase activity and mRNA levels.[85] Taken together, these results suggest that the discordant regulation of HMG-CoA reductase and squalene synthase during the host response to infection and inflammation may have substantial effects on the regulation of substrate flux into the nonsterol pathways of mevalonate metabolism.

ACKNOWLEDGMENTS

We express our appreciation to Drs. Mary Ellen Digan, Faizu Kathawala, and Larry Perez for their help in the review of this manuscript. In addition, we thank Dr. Larry Perez for providing drawings of the reaction schemes.

2.09.9 REFERENCES

1. A. Endo and K. Hasumi, *Nat. Prod. Rep.*, 1993, **10**, 541.
2. J. A. Herd, C. M. Ballantyne, J. A. Farmer, J. J. Ferguson, K. L. Gould, P. H. Jones, M. S. West, and A. M. Gotto, Jr., *Circulation*, 1996, **94**(8), Suppl. I, I-597, #3496.
3. G. N. Levine, J. F. Keaney, Jr., and J. A. Vita, *N. Engl. J. Med.*, 1995, **332**, 512.
4. G. Popjak and W. S. Agnew, *Mol. Cell. Biochem.*, 1979, **27**, 97.
5. C. D. Poulter and H. C. Rilling, in "*Biosynthesis of Isoprenoid Compounds*," eds. J. W. Porter and S.L. Spurgeon, John Wiley and Sons, New York, 1981, vol. 1, p. 413.
6. H. C. Rilling, *Biochem. Soc. Trans.*, 1985, **13**, 997.
7. C. D. Poulter, *Acc. Chem. Res.*, 1990, **23**, 70.
8. S. A. Biller, K. Neuenschwander, M. M. Ponpipom, and C. D. Poulter, *Curr. Pharm. Des.*, 1996, **2**, 1.
9. V. Manne, C. S. Ricca, J. G. Brown, A. V. Tuomari, N. Yan, P. Dinesh, R. Schmidt, M. J. Lynch, C. P. Ciosek, Jr., J. M. Carboni, S. Robinson, E. M. Gordon, M. Barbacid, B. R. Seizinger, and S. A. Biller, *Drug Dev. Res.*, 1995, **34**, 121.
10. P. Alaei, E. E. MacNulty, and N. S. Ryder, *Biochem. Biophys. Res. Commun.*, 1996, **222**, 133.
11. J. L. Goldstein and M. S. Brown, *Nature*, 1990, **343**, 425.
12. J. A. Glomset, M. H. Gelb, and C. C. Farnsworth, *Trends Biochem. Sci.*, 1990, **15**, 139.
13. Y. Kamiya, A. Sakurai, S. Tamura, N. Takahashi, E. Tsuchiya, K. Abe, and S. Fukui, *Agric. Biol. Chem.*, 1979, **43**, 363.
14. A. R. Schmidt, C. J. Schneider, and J. A. Glomset, *J. Biol. Chem.*, 1984, **259**, 10 175.
15. J. R. Faust, J. L. Goldstein, and M. S. Brown, *Proc. Natl. Acad. Sci. USA*, 1979, **76**, 5018.
16. N. S. Ryder and M.-C. Dupont, *Biochem. J.*, 1985, **230**, 765.
17. P. Nussbaumer, G. Dorfstatter, I. Leitner, K. Mraz, H. Vyplel, and A. Stutz, *J. Med. Chem.*, 1993, **36**, 2810.
18. I. Abe, J. C. Tomesch, S. Wattanasin, and G. D. Prestwich, *Nat. Prod. Rep.*, 1994, **11**, 279.
19. A. Nadin and K. C. Nicolaou, *Angew. Chem.*, 1996, **35**, 1623.
20. N. S. Watson and P. A. Procopiou, *Prog. Med. Chem.*, 1996, **33**, 331.
21. I. Shechter and K. Bloch, *J. Biol. Chem.*, 1971, **246**, 7690.
22. A. A. Qureshi, E. Beytia, and J. W. Porter, *J. Biol. Chem.*, 1973, **248**, 1848.
23. W. S. Agnew and G. Popjak, *J. Biol. Chem.*, 1978, **253**, 4574.
24. G. Kuswik-Rabiega and H. C. Rilling, *J. Biol. Chem.*, 1987, **262**, 1505.
25. K. Sasiak and H. C. Rilling, *Arch. Biochem. Biophys.*, 1988, **260**, 622.
26. I. Shechter, E. Klinger, M. L. Rucker, R. G. Engstrom, J. A. Spirito, M. A. Islam, B. R. Boettcher, and D. B. Weinstein, *J. Biol. Chem.*, 1992, **267**, 8628.
27. T. L. McKenzie, G. Jiang, J. R. Straubhaar, D. G. Conrad, and I. Shechter, *J. Biol. Chem.*, 1992, **267**, 21 368.
28. G. Guan, G. Jiang, R. L. Koch, and I. Shechter, *J. Biol. Chem.*, 1995, **270**, 21 958.
29. B. Haendler, D. Soltis, R. Movva, H. P. Kocher, S. Ha, F. Kathawala, and G. Nemecek, "Book of Abstracts, 201st ACS National Meeting, 1991," American Chemical Society, Washington DC, BIOT 92.
30. S. M. Jennings, Y. H. Tsay, T. M. Fisch, and G. W. Robinson, *Proc. Natl. Acad. Sci. USA*, 1991, **88**, 6038.
31. M. Fegueur, L. Richard, A. D. Charles, and F. Karst, *Curr. Genet.*, 1991, **20**, 365.
32. G. W. Robinson, Y. H. Tsay, K. B. Kienzle, C. A. Smith-Monroy, and R. W. Bishop, *Mol. Cell. Biol.*, 1993, **13**, 2706.
33. P. V. LoGrasso, D. A. Soltis, and B. R. Boettcher, *Arch. Biochem. Biophys.*, 1993, **307**, 193.
34. D. Zhang, S. M. Jennings, G. W. Robinson, and C. D. Poulter, *Arch. Biochem. Biophys.*, 1993, **304**, 133.
35. G. Jiang, T. L. McKenzie, D. G. Conrad, and I. Shechter, *J. Biol. Chem.*, 1993, **268**, 12 818.
36. C. Summers, F. Karst, and A. D. Charles, *Gene*, 1993, **136**, 185.
37. D. A. Soltis, G. McMahon, S. L. Caplan, D. A. Dudas, H. A. Chamberlin, A. Vattay, D. Dottavio, M. L. Rucker, R. G. Engstrom, S. A. Cornell-Kennon, and B. R. Boettcher, *Arch. Biochem. Biohys.*, 1995, **316**, 713.
38. T. Inoue, T. Osumi, and S. Hata, *Biochim. Biophys. Acta*, 1995, **1260**, 49.
39. T. Nakashima, T. Inoue, A. Oka, T. Nishino, T. Osumi, and S. Hata, *Proc. Natl. Acad. Sci. USA*, 1995, **92**, 2328.
40. K. M. Hanley, O. Nicolas, T. B. Donaldson, C. Smith-Monroy, G. W. Robinson, and G. M. Hellmann, *Plant Mol. Biol.*, 1996, **30**, 1139.
41. K. Poralla, University of Tubingen, Germany, personal communication.
42. M. N. Ashby, S. Y. Kutsunai, S. Ackerman, A. Tzagoloff, and P. A. Edwards, *J. Biol. Chem.*, 1992, **267**, 4128.
43. M. R. Jackson, T. Nilsson, and P. A. Peterson, *EMBO J.*, 1990, **9**, 3153.
44. P. V. LoGrasso, S. Cornell-Kennon, and B. R. Boettcher, *Bioorg. Chem.*, 1994, **22**, 294.
45. D. C. Fulton, M. Tait, and D. R. Threlfall, *Phytochemistry*, 1995, **38**, 1137.

46. K. A. Mookhtiar, S. S. Kalinowski, D. Zhang, and C. D. Poulter, *J. Biol. Chem.*, 1994, **269**, 11 201.
47. B. R. Boettcher and S. A. Cornell-Kennon, Novartis Pharmaceuticals Corporation, East Hanover, NJ, USA, unpublished results.
48. K. Hanley and J. Chappell, *Plant Physiol.*, 1992, **98**, 215.
49. M. Prashad, F. G. Kathawala, and T. Scallen, *J. Med. Chem.*, 1993, **36**, 1501.
50. M. Prashad, *J. Med. Chem.*, 1993, **36**, 631.
51. D. Zhang and C. D. Poulter, *J. Am. Chem. Soc.*, 1995, **117**, 1641.
52. M. B. Jarstfer, B. S. J. Blagg, D. H. Rogers, and C. D. Poulter, *J. Am. Chem. Soc.*, 1996, **118**, 13 089.
53. J. D. Bergstrom, M. M. Kurtz, D. J. Rew, A. M. Amend, J. D. Karkas, R. G. Bostedor, V. S. Bansal, C. Dufresne, F. L. VanMiddlesworth, O. D. Hensens, J. M. Liesch, D. L. Zink, K. E. Wilson, J. Onishi, J. A. Milligan, G. Bills, L. Kaplan, M. Nallin-Omstead, R. G. Jenkins, L. Huang, L. M. S. Meinz, L. Quinn, R. W. Burg, Y. L. Kong, S. Mochales, M. Mojena, I. Martin, F. Palaez, M. T. Diez, and A. W. Alberts, *Proc. Natl. Acad. Sci. USA*, 1993, **90**, 80.
54. K. Hasumi, K. Tachikawa, K. Sakai, S. Murakawa, N. Yoshikawa, S. Kumazawa, and A. Endo, *J. Antibiot.*, 1993, **46**, 689.
55. S. Lindsey and H. J. Harwood, Jr., *J. Biol. Chem.*, 1995, **270**, 9083.
56. J. V. Schloss, *Acc. Chem. Res.*, 1988, **21**, 348.
57. T. Tanimoto, K. Onodera, T. Hosoya, Y. Takamatsu, T. Kinoshita, K. Tago, H. Kogen, T. Fujioka, K. Hamano, and Y. Tsujita, *J. Antibiot.*, 1996, **49**, 617.
58. H.-P. Fischer and D. Bellus, *Pestic. Sci.*, 1983, **14**, 334.
59. C. R. Johnson, B. R. Boettcher, R. E. Cherpeck, and M. G. Dolson, *Bioorg. Chem.*, 1990, **18**, 154.
60. E. J. Corey and R. P. Volante, *J. Am. Chem. Soc.*, 1976, **98**, 1291.
61. P. R. Ortiz de Montellano, J. S. Wei, W. A. Vinson, R. Castillo, and A. S. Boparai, *Biochemistry*, 1977, **16**, 2680.
62. T. R. Vedananda, R. E. Damon, J. B. Fell, T. E. Hughes, B. R. Boettcher, and T. J. Scallen, "Book of Abstracts, 212th ACS national Meeting, Orlando, FL," American Chemical Society, Washington, DC, 1996, MEDI 056.
63. I. Shechter, P. Gu, G. Jiang, T. J. Onofrey, R. O. Cann, A. Castro, and T. A. Spencer, *Bioorg. Med. Chem. Lett.*, 1996, **6**, 2585.
64. J. B. Fell, T. R. Vedananda, R. E. Damon, and T. Scallen, "Book of Abstracts, 207th ACS National Meeting, San Diego, CA", American Chemical Society, Washington, DC, 1994, MEDI 193.
65. D. Amin, R. Z. Rutledge, S. J. Needle, D. J. Hele, K. Neuenswander, R. C. Bush, G. E. Bilder, and M. H. Perrone, *Naunyn-Schmiedeberg's Arch. Pharmacol.*, 1996, **353**, 233.
66. G. J. Smith, R. G. Davidson, C. Dunkley, G. R. Brown, K. B. Mallion, and F. McTaggart, *Atherosclerosis*, 1994, **109**, 252.
67. S. A. Biller, M. J. Sofia, B. DeLange, C. Forster, E. M. Gordon, T. Harrity, L. C. Rich, and C. P. Ciosek, Jr., *J. Am. Chem. Soc.*, 1991, **113**, 8522.
68. T. W. Harrity, R. J. George, C. P. Ciosek, Jr., S. A. Biller, Y. Chen, J. K. Dickson, Jr., O. M. Fryszman, K. J. Jolibois, L. K. Kunselman, R. M. Lawrence, J. V. H. Logan, D. R. Magnin, L. C. Rich, D. A. Slusarchyk, R. B. Sulsky, and R. E. Gregg, *Circulation*, 1995, **92**(8), Suppl. I, I-103, #0484.
69. K. D. Stamellos, J. E. Shackelford, I. Shechter, G. Jiang, D. G. Conrad, G.-A. Keller, and S. K. Krisans, *J. Biol. Chem.*, 1993, **268**, 12 825.
70. G. Guan and I. Shechter, Uniformed Services University of the Health Sciences, Bethesda, MD, USA, unpublished results.
71. J. R. Smith, T. F. Osborne, M. S. Brown, J. L. Goldstein, and G. Gil, *J. Biol. Chem.*, 1988, **263**, 18 480.
72. J. Ericsson, S. M. Jackson, and P. A. Edwards, *J. Biol. Chem.*, 1996, **271**, 24 359.
73. P. Tonzonoz, J. B. Kim, R. A. Graves, and B. M. Spiegelman, *Mol. Cell. Biol.*, 1993, **13**, 4753.
74. I. Shechter, D. G. Conrad, I. Hart, R. C. Berger, T. L. McKenzie, J. Bleskan, and D. Patterson, *Genomics*, 1994, **20**, 116.
75. C. L. Welch, X. Yu-Rong, I. Shechter, R. Farese, M. Mehrabian, S. Mehdizadeh, C. H. Warden, and A. J. Lussis, *J. Lipid Res.*, 1996, **37**, 1406.
76. G. A. Keller, M. C. Barton, D. J. Shapiro, and S. J. Singer, *Proc. Natl. Acad. Sci. USA*, 1985, **82**, 770.
77. G. A. Keller, M. Pazirandeh, and S. K. Krisans, *J. Cell Biol.*, 1986, **103**, 875.
78. K. D. Stamellos, J. E. Shackelford, R. D. Tanaka, and S. K. Krisans, *J. Biol. Chem.*, 1992, **267**, 5560.
79. S. K. Krisans, J. Ericsson, P. A. Edwards, and G. A. Keller, *J. Biol. Chem.*, 1994, **269**, 14 165.
80. J. Ericsson, E. L. Appelkvist, A. Thelin, T. Chojnacki, and G. Dahlner, *J. Biol. Chem.*, 1992, **267**, 18 708.
81. E. L. Appelkvist, M. Reinhart, R. Fisher, J. Billheimer, and G. Dallner, *Arch. Biochem. Biophys.*, 1990, **282**, 318.
82. S. L. Thompson, R. Burrows, R. J. Laub, and S. K. Krisans, *J. Biol. Chem.*, 1987, **262**, 17 420.
83. I. Shechter, Uniformed Services University of the Health Sciences, Bethesda, MD, USA, unpublished results.
84. G. Guan, P.-H. Dai, T. F. Osborne, J. B. Kim, and I. Shechter, *J. Biol. Chem.*, 1997, **272**, 10 295.
85. R. A. Memon, I. Shechter, A. H. Moser, J. K. Shigenaga, C. Grunfeld, and K. R. Feingold, *J. Lipid Res.*, 1997, **38**, 1620.

2.10 Squalene Epoxidase and Oxidosqualene : Lanosterol Cyclase—Key Enzymes in Cholesterol Biosynthesis

IKURO ABE
University of Shizuoka, Japan

and

GLENN D. PRESTWICH
University of Utah, Salt Lake City, UT, USA

2.10.1 INTRODUCTION

Squalene epoxidase (SE) and oxidosqualene:lanosterol cyclase (OSLC) are two pivotal enzymes in cholesterol biosynthesis. The reaction steps catalyzed by these two enzymes provide fertile ground for basic studies in bioorganic chemistry. SE is the only known noncytochrome P-450 enzyme that epoxidizes an alkene, and selects only a single face of one of six trisubstituted alkenes. On the other hand, OSLC catalyzes a remarkable polyene cyclization reaction to form a total of six new carbon–carbon bonds and seven stereocenters in a single reaction.[1] In order to control the stereospecificity of these reactions, precise molecular interactions are required for the enzyme–substrate complexes. In addition, regulation of the levels of these enzymes *in vivo* has clinical importance for modulation of cholesterol biosynthesis. HMG-CoA reductase inhibitors, which are currently clinically used as cholesterol-lowering drugs, could in principle suppress all post-mevalonate biosynthetic pathways, including biologically important nonsteroidal isoprenoids (e.g., dolichol, ubiquinone, isopentenyl tRNA, and protein prenylation), which play important roles in regulation of normal cellular processes. In fact, levels of HMG-CoA reductase tend to increase due to upregulation of gene transcription and translation; clinically, this is counterbalanced by upregulation of LDL-receptor levels leading to a net lowering of serum cholesterol.[2] Nonetheless, selective inhibition of cholesterol biosynthesis is a desirable pharmaceutical goal, and enzyme inhibitors for SE and OSLC have been potential targets for the design of such therapeutic agents.[3]

The SE and OSLC enzymes have elicited intense chemical and biochemical interest for many years, and the unstable, membrane-bound enzymes have now been purified, cloned, and sequenced for the first time from both vertebrate and nonvertebrate sources. As a result, sequence information and recombinant proteins are now available for further characterization of the enzymes. Furthermore, development of several potent enzyme inhibitors and active-site probes have provided new mechanistic and stereochemical insights into the enzyme reactions. Comparative studies of the enzymes from fungal, plant, and several vertebrate sources offer tantalizing clues to species-specific differences in sensitivity to inhibitors. Studies of SE and OSLC enzymes are now progressing rapidly and promise to reveal the intimate three-dimensional structural details of the enzyme-catalyzed processes as well as the regulation mechanism of cholesterol biogenesis.

2.10.2 SQUALENE EPOXIDASE

2.10.2.1 General Considerations for the Enzyme Mechanism

Squalene epoxidase (SE) (EC 1.14.99.7) catalyzes the conversion of squalene to (3*S*)-2,3-oxidosqualene (Scheme 1). In addition to oxygen, SE requires FAD, NADPH (NADH in the case of fungal SE), NADPH-cytochrome P-450 reductase (EC 1.6.2.4),[4] and a supernatant protein factor (SPF) (not required for fungal enzyme).[5] The SPF, a 47 kDa protein, has been purified, cloned, and shown to be involved in intermembrane squalene transport.[6–9] The SPF can be replaced by selected detergents such as Triton X-100 in the assay system.

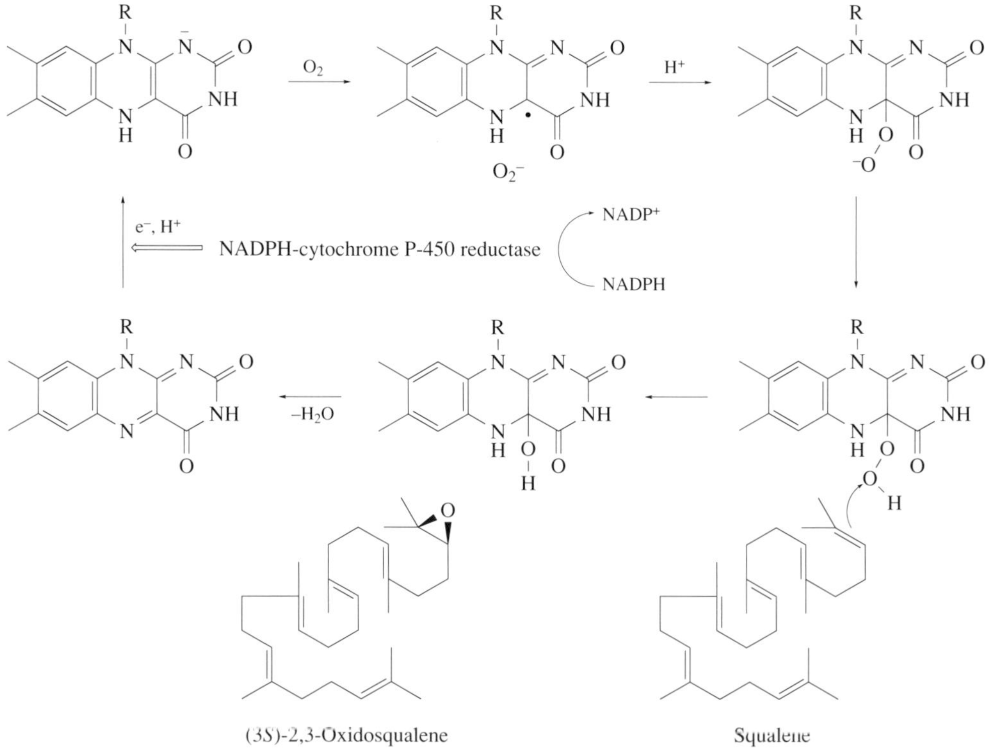

(3*S*)-2,3-Oxidosqualene

Squalene

Scheme 1

The NADPH-cytochrome P-450 reductase (P450R), a microsomal 77 kDa flavoprotein, has been well characterized[10–13] and is thought to be involved in catalyzing the transfer of reducing equivalents from the reduced pyridine nucleotide to the epoxidase. The role of the reductase is to shuttle electrons from NADPH via its FAD and FMN prosthetic groups to the microsomal cytochrome P-450 as a component of the mixed-function oxygenase system. The NADPH-, FAD-, and FMN-binding domains of P450R have been identified by sequence comparisons with other flavoproteins;[14] site-directed mutagenesis experiments have identified amino acid residues necessary for binding of FMN and NADPH.[15–17] Crystallization and preliminary X-ray studies of P450R have been reported.[18,19]

For optimum electron transfer to the active site of SE, a functional complex formation between P450R and SE would be required (Figure 1). A hydrophobic membrane anchor domain of P450R at the N-terminal may be involved in directing the P450R binding to SE as proposed for the interaction with cytochrome P-450.[20] Furthermore, certain charged amino acid residues may be important for the intermolecular electron transfer process by forming ionic bridges between the two proteins.[21–24] Site-directed mutagenesis of the acidic clusters of P450R (^{207}Asp-Asp-Asp^{209} and ^{213}Glu-Glu-Asp^{215}) have demonstrated that both cytochrome P-450 and cytochrome *c* interact with this region.[22] However, as yet there are no studies of the interactions between P450R and SE. Finally, it should be noted that P450R is a highly charged protein (97 Asp and Glu residues and 74 Arg and Lys residues out of a total of 677 amino acids of rat P450R); reconstitution of monooxygenase activities with P450R are all ionic-strength dependent, confirming that electrostatic interactions are significant.[13,25–29]

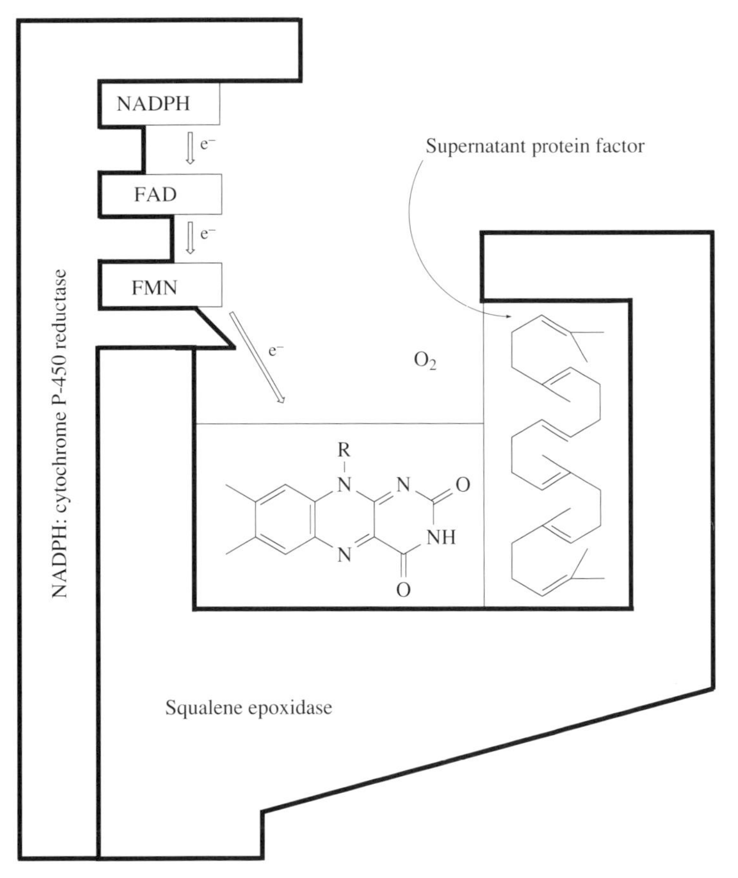

Figure 1 Squalene epoxidase is a mixed-function oxygenase.

Although the oxidation of alkenes is common for P-450 and other metal-containing enzymes, SE is the only known noncytochrome P-450 enzyme that epoxidizes an unactivated alkene.[30] The nonmetallic flavoprotein monooxygenase induces a splitting of the O—O bond, inserting one oxygen atom into the substrate (squalene) and reducing the other atom to H_2O. The mechanism of flavoprotein-mediated epoxidation is not well understood, but one of the possible mechanisms would proceed via formation of flavin C(4a)-hydroperoxide intermediate (Scheme 1).[31] The reduced form of the flavoprotein would initially react with molecular oxygen by a one-electron transfer to produce superoxide anion (O_2^{-}); after spin inversion of the initially formed radical pair, the C(4a)-hydroperoxide anion is formed. Uptake of a protein yields the neutral hydroperoxide that can now act as an electrophile, and then oxygen transfer to the terminal carbon–carbon double bond of the squalene molecule occurs to give the epoxide. The overall reaction of the epoxidation reaction can thus be described by Equation (1).

$$\text{squalene} + O_2 + \text{NADPH} + H^+ \rightarrow (3S)2,3\text{-oxidosqualene} + \text{NADP}^+ + H_2O \qquad (1)$$

It has been reported that the levels of SE *in vivo* may be regulated by the concentrations of both endogenous and exogenous sterols, perhaps through a common feedback mechanism.[32,33] Interestingly, SE has an extremely low specific activity compared to HMG-CoA reductase or squalene synthase in Hep G2 cells; immunoblotting and Northern blot analysis indicated that both expressed SE and mRNA for SE are rather low.[34] Accumulation of squalene was observed in human renal carcinoma cells supplemented with exogenous cholesterol.[35] Furthermore, an increased level of expression of SE mRNA and protein content was observed in human cell lines grown in 10% lipoprotein-deficient fetal bovine serum (FBS) for 48 h when compared with the culture in the presence of FBS.[36] From these studies, SE is thought to control the throughput from squalene to sterols in cholesterol biogenesis. As described below, both NB-598, a potent vertebrate SE inhibitor, and L-654,969, a potent HMG-CoA reductase inhibitor, increased SE activity in Hep G2 cells as a result of blocking the feedback regulation mechanism. Further, NB-598 dramatically increased LDL receptor levels, but the increase in HMG-CoA reductase activity was less than that induced by L-654,969; nonetheless, NB-598 inhibited cholesterol synthesis more potently than L-654,969.[37] These results suggest that the SE inhibitor did not affect the synthesis of nonsteroidal isoprenoids that are believed to regulate the HMG-CoA reductase activity post-transcriptionally.[37] Note that HMG-CoA reductase is considered to be controlled through multivalent regulation; both sterols and the nonsterol mevalonate-derived metabolites are involved.[38–41] SE inhibitors are thus expected to be highly potent therapeutic agents.

Finally, SE also catalyzes conversion of 2,3-oxidosqualene to 2,3:22,23-dioxidosqualene (DOS).[42,43] The DOS is further converted to oxysterols such as 24,25-epoxycholesterol, which are thought to regulate the coordinate expression of the genes for HMG-CoA reductase, the LDL-receptor, and other enzymes in the cholesterol biogenesis pathway.[44,45] How SE participates in such a regulation pathway offers challenges for both pharmaceutical and basic research.

2.10.2.2 Enzymology and Molecular Biology

As with squalene synthase and oxidosqualene cyclase (OSC), SE is a microsomal membrane-associated enzyme. SE from rat liver has been extensively studied by the Bloch group in the 1970s,[46–48] and was first purified to homogeneity by Ono and his co-workers.[49,50] The enzyme could be efficiently solubilized by Triton X-100, and was purified 143-fold to homogeneity by conventional column chromatography techniques in the presence of this detergent with a yield of 4%. Thiol-protecting reagents such as DTT and glycerol were useful in stabilizing activity. A final chromatofocusing step gave 90–95% pure enzyme with an apparent subunit size of 51 kDa. With standard assay conditions using 0.2 unit of NADPH-cytochrome P-450 reductase, apparent K_M values of 13 μM for squalene and 5 μM for FAD were reported. The purified enzyme showed no distinct absorption spectrum in the visible region, and was insensitive to SKF 525A, metyrapone, or other standard cytochrome P-450 inhibitors.[50] Further, during the conversion of [3-^{3}H]squalene into [^{3}H]2,3-oxidosqualene, no exchange of the labeled hydrogen was observed, supporting the nonmetal involved mechanism.[51] In contrast, alkene epoxidation reactions catalyzed by metalloenzymes often proceed with the exchange of a vinylic proton with the medium.[52] For both pig and rat liver SE,[53] efficient affinity purifications were developed by employing novel affinity matrices based on the use of two different trisnorsqualene amines that act as slow tight-binding inhibitors of each of the two SE enzymes (see Section 2.10.2.3).[54] The purified pig SE was shown to be a 55 kDa protein. In order to prevent aggregation of the enzyme, the presence of a weak, reversible, competitive inhibitor such as *N*,*N*-dimethyldodecylamine, which can be readily removed by DEAE chromatography, was effective.[53]

The cDNAs encoding several SEs have been cloned and sequenced from vertebrates (rat,[55] mouse[56]), the yeast *Saccharomyces cerevisiae*,[57] and *Candida albicans*.[58,59] Partial sequences of human SE[36,60] and pig SE[61] are also available (Figure 2). The yeast SE (*ERG1* gene) was cloned from a yeast mutant strain resistant for allylamine (Terbinafine), a potent inhibitor specific for fungal SE (see

Section 2.10.2.3). The 1488 bp open reading frame (ORF) encodes a predicted protein of 496 amino acids with molecular mass of 55 kDa.[57] *ERG1* is located on the right arm of chromosome 7 of *S. cerevisiae*.[62] Next, rat SE cDNA was cloned by the Ono group by screening yeast transformants expressing a rat cDNA library in the presence of Terbinafine. The cDNA for rat SE contained a 1719 bp ORF encoding a 64 kDa protein.[55] This vertebrate SE contained 77 additional N-terminal amino acids, and the nucleotide and deduced amino acid sequences showed 38.3% (659/1719) and 30.2% (173/573) identity, respectively, with those of allylamine-resistant yeast SE. Human SE has been isolated from a genomic library and was shown to consist of 11 exons and located in the neighborhood of chromosome 8q telomere.[63] According to the Kyte-Doolittle hydropathy plot analysis (Figure 3), the vertebrate SEs have one possible transmembrane domain at the N-terminal (Leu^{27}-Tyr^{43} in rat SE), which is absent in the yeast enzyme.[55] This region also contained a basic amino acid cluster (^{95}Lys-Arg-Arg-Arg-Lys^{99} in rat SE), and proteolytic cleavage at this site during

```
Homo sapiens              MWTFLGIATFTYFYKKFGDFITLANREVLLCVLVFLSLGLVLSYRCRHRNGGLLGRQRSGSQFALFSDIL  70
Rattus norvegicus         MWTFLGIATFTYFYKKCGD-VTLANKELLLCVLVFLSLGLVLSYRCRHRNGGLLGRHQSGSQFAAFSDIL  69
Mus musculus              MWTFLGIATFTYFYKKCGD-VTLANKELLLCVLVFLSLGLVLSYRCRHRHGGLLGRHQSGAQFAAFSDIL  69
Sus scrofa (partial)      MWTFLGIATFTYFYKKCGDFVSLANKELLLGVLVFLSLGLVLSYRCRYRNGALLGRQQSGSQFAVFSDIL  69
Saccharomyces cerevisiae  M---------------------------------------------------------------------   1
Candida albicans          M---------------------------------------------------------------------   1

                                                                     Dinucleotide-Binding Site
(H.s.)  SGLPFIGFFWAKSPPESENKEQLGARRRRKGTNISETSLIGTAACTSTSSQNDPEVIIVGAGVLGSALAAVLSRDGRKVTVIERDLKE  158
(R.n.)  SALPLIGFFWAKSPPESEKKEQLESKRRRKEVNLSETTLTGAATSVSTSSVTDPEVIIIGSGVLGSALATVLSRDGRTVTVIERDLKE  157
(M.m.)  SALPLIGFFWAKSP-ESEKKEQLESKKCRKEIGLSETTLTGAATSVSTSFVTDPEVIIVGSGVLGSALAAVLSRDGRKVTVIERDLKE  156
(S.s.)  SALP-IGFFWAKSPSGSEKKEQLGSRRGKKGSNISETTLVGAAASPLISSQNDPEIIIVGSGVLGSALAAVLSRDGRTVTVIERDLKE  157
(S.c.)  ------------------------------------SAVNVAPELINADNTITYDAIVIGAGVIGPCVATGLARKGKKVLIVERDWAM   53
(C.a.)  ------------------------------------------------SSV-KYDAIIIGAGVIGPTIATAFARQGRKVLIVERDWSK   40
                                                                ^  ^ ^^ ^   ^  ^ ^ ^  ^   ^^^

(H.s.)  PDRIVGEFLQPGGYHVLKDLGFGDTVEGLDAQVVNGYMIHDQESKSEVQIPYPLSE--------------------------------  214
(R.n.)  PDRILGECLQPGGYRVLRELGLGDTVESLNAHHIHGYVIHDCESRSEVQIPYPVSE--------------------------------  213
(M.m.)  PDRIVGELLQPGGYRVLQELGLGDTVEGLNAHHIHGYIVHDYESRSEVQIPYPLSE--------------------------------  212
(S.s.)  PDRILGEYLQPGGCHVLKDLGLEDTMEGIDAQVVDGYIIHDQESKSEVQIPFPLSE--------------------------------  213
(S.c.)  PDRIVGELMQPGGVRALRSLGMIQSINNIEAYPVTGYTVFFN--GEQVDIPYPYKAD--IPKVEKLKDLVKDG-NDKVLEDSTIHIKD  136
(C.a.)  PDRIVGELMQPAGIKALRELGMIKAINNIRAVDCTGYYIKYY--DETITIPYPLKKDACITNPVKPVPDAVDGVNDKLDSDSTLNVDD  126
        ^^^^ ^^  ^^ ^   ^  ^^         ^    ^^            ^^ ^

(H.s.)  --NNQVQSGRAFHHGRFIMSLRKAAMAEPNAKFIGGVVLQLLEEDDVVMGVQYKDKETGDIKEL-HAPLTVVADGLFSKFRKSLVSNK  299
(R.n.)  --NNQVQSGVAFHHGKFIMSLRKAAMAEPNVKFIEGVVLRLLEEDDAVIGVQYKDKETGDTKEL-HAPLTVVADGLFSKFRKNLISNK  298
(M.m.)  --TNQVQSGIAFHHGRFIMSLRKAAMAEPNVKFIEGVVLQLLEEDDAVIGVQYKDKETGDTKEL-HAPLTVVADGLFSKFRKSLISSK  297
(S.s.)  --NNHVQSGRAFRHGRFIMGLRKAAMAEPNAKFIEGTVLQLLEEEDVV........................................
(S.c.)  YEDDERERGVAFVHGRFLNNLRNITAQEPNVTRVQGNCIEILKDEKNEVVGAKVDIDGRGKVEFK-AHLTFICDGIFSRFRKELHPDH  223
(C.a.)  WDFDERVRGAAFHHGDFLMNLRQICRDEPNVTAVEATVTKILRDPLDPNTVIGVQTKQPSGTVDYHAKLTISCDGIYSKFRKELSPTN  214
                ^ ^^ ^^ ^   ^^     ^^^     ^     ^                        ^ ^^   ^^  ^ ^^^ ^

(H.s.)  V-SVSSHFVGFLMKNAPQFKANHAELIL-ANPSPVLIYQISSSETRVLVDIRG-EMPRNLREYMVEK----IYPQIPDHLKEPFLEAT  380
(R.n.)  V-SVSSHFVGFIMKDAPQFKANFAELVL-VDPSPVLIYQISPSETRVLVDIRG-ELPRNLREYMTEQ----IYPQIPDHLKESFLEAC  379
(M.m.)  V-SVSSHFVGFLMKDAPQFKPNFAELVL-VNPSPVLIYQISSSETRVLVDIRG-ELPRNLREYMAEQ----IYPQLPDHLKESFLEAS  378
(S.s.)  ........................................................................................
(S.c.)  VPTVGSSFVGMSLFNAKNPAPMHGHVILGSDHMPILVYQISPEETRILCAYNSPKVPADIKSWMIKD----VQPFIPKSLRPSFDEAV  307
(C.a.)  VPTIGSYFIGLYLKNAELPAKGKGHVLL-GGHAPALIYSVSPTETRVLCVYVSSKPPSAANDAVYKYLRDNILPAIPKETVPAFKEAL  301
        ^    ^ ^ ^     ^           ^     ^ ^ ^  ^  ^^^ ^        ^                ^  ^      ^ ^^

(H.s.)  DNSHLRSMPASFLPP---SSVKKRGVLLLGDAYNMRHPLTGGGMTVAFKDIKLWRKLLKGIPDLYDDAAIFEAKKSFYWA--RKTSHS  463
(R.n.)  QNARLRTMPASFLPP---SSVNKRGVLLLGDAYNLRHPLTGGGMTVALKDIKIWRQLLKDIPDLYDDAAIFQAKKSFFWS--RKRSHS  462
(M.m.)  QNGRLRTMPASFLPP---SSVNKRGVLILGDAYNLRHPLTGGGMTVALKDIKLWRQLLKDIPDLYDDAAIFQAKKSFFWS--RKRTHS  461
(S.s.)  ........................................................................................
(S.c.)  SQGKFRAMPNSYLPA---RQNDVTGMCVIGDALNMRHPLTGGGMTVGLHDVVLLIKKIGDL--DFSD-REKVLDELLDYHFERKSYDS  389
(C.a.)  EERKFRIMPNQYLSAMKQGSENHKGFILLGDSLNMRHPLTGGGMTVGLNDSVLLAKLLHPKFVEDFDDHQLIAKRLKTFHRKRKNLDA  389
             ^ ^^   ^      ^    ^    ^^  ^ ^^^^^^^^^^^   ^                ^               ^^

(H.s.)  FVVNILAQALYELFSATDDSLHQLRKACFLYFKLGGECVAGPVGLLSVLSPNPLALIGHFFAVAIYAVYFCFKSEPWITKPRALLSSS  551
(R.n.)  FVVNVLAQALYELFSATDDSLRQLRKACFLYFKLGGECLTGPVGLLSILSPDPLLLIRHFFSVAVYATYFCFKSEPWATKPRALFSSG  550
(M.m.)  FVVNVLAQALYELFSATDDSLHQLRKACFLYFKLGGECVTGPVGLLSILSPHPLVLIRHFFSVAIYATYFCFKSEPWATKPRALFSSG  549
(S.s.)  ........................................................................................
(S.c.)  -VINVLSVALYSLFAADSDNLKALQKGCFKYFQRGGDCVNKPVEFLSGVLPKPLQLTRVFFAVAFYTIYLNMEERGFLGLPMALLEGI  476
(C.a.)  -VINTLSISLYSLFAADKKPLRILRNGCFKYFQRGGECVNGPIGLLSGMLPFPMLLFNHFFSVAFYSVYLNFIERGLLGFPLALFEAF  476
         ^ ^ ^   ^^ ^^ ^    ^  ^   ^^ ^^  ^^ ^   ^   ^^   ^ ^  ^   ^^ ^^ ^  ^

(H.s.)  AVLYKACSVIFPLIYSEMKYMVH*  574
(R.n.)  AILYKACSIIFPLIYSEMKYLVH*  573
(M.m.)  AILYKACSILFPLIYSEMKYLVH*  572
(S.s.)  .......................
(S.c.)  MILITAIRVFTPFLFGELIG*  496
(C.a.)  EVLFTAIVIFTPYLWNEIVR*  496
```

Figure 2 Comparison of the deduced amino acid sequences of SEs from six species: human, *Homo sapiens* SE; rat, *Rattus norvegicus* SE; mouse, *Mus musculus* SE; pig, *Sus scrofa* SE (partial); yeast, *Saccharomyces cerevisiae* SE; candida, *Candida albicans* SE. Conserved residues in at least three sequences are boldfaced; hyphens indicate gaps introduced to maximize alignment. The putative FAD binding site (the dinucleotide binding site) is indicated. The N-terminal hydrophobic region (underlined) is predicted to be membrane associated.

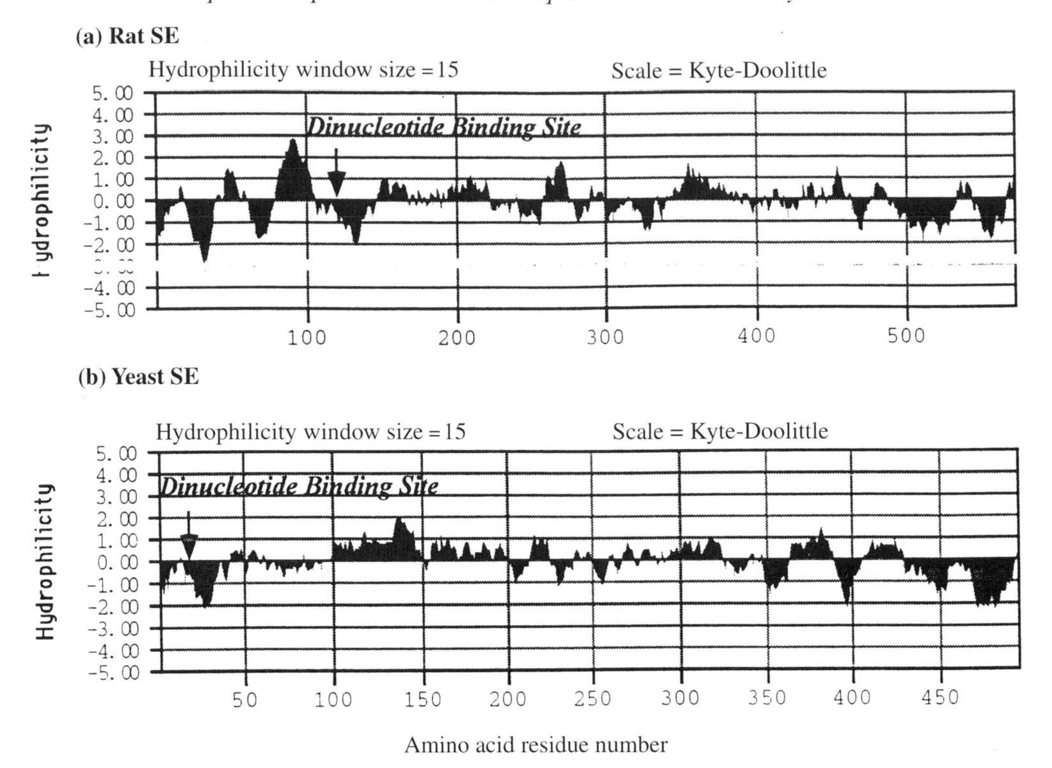

Figure 3 Hydropathy plots for (a) rat SE and (b) yeast *S. cerevisiae* SE. The deduced amino acid sequence was analyzed by MacVector 4.1 sequence analysis software (Kodak). The putative FAD binding site (dinucleotide binding site) is indicated.

protein purification could rationalize the 13 kDa difference between the cDNA encoding the 64 kDa protein and the isolated 51 kDa protein.[43] Furthermore, vertebrate SEs (human, rat, pig) have a possible N-glycosylation site (N102 in rat SE) adjacent to this cluster, but preliminary results obtained by the authors showed absence of protein glycosylation.[64]

A truncated recombinant rat SE (Glu^{100}-His^{573}) without the N-terminal putative transmembrane domain (and with an additional hexahistidine tag at the C-terminal for simple purification by Ni-chelate affinity chromatography) has been constructed and functionally expressed in *E. coli*.[43] This recombinant enzyme (Δ^{99}His) showed properties very similar to those of the native enzyme with regard to the requirements for NADPH, FAD, SPF, or Triton X-100, and P450R. Both native and recombinant rat SE had similar pH dependencies and sensitivity to most inhibitors. However, two differences were apparent. First, the recombinant enzyme (spec. act. 170 nmol mg^{-1} min^{-1})[43] was much more active than the native SE (6.2 nmol mg^{-1} min^{-1}),[50] suggesting altered conformation that conferred a higher V_{max}. Moreover, two photolabile micromolar inhibitors of native rat SE failed to inhibit the recombinant enzyme at concentrations up to 200 μM (see Section 2.10.2.3).

Vertebrates SEs require exogenous FAD for their activity. In flavoenzymes, FAD can be covalently bound to the apoenzymes, for example, between the 8α-methyl group of the isollaoxazine ring and nucleophilic groups of His, Cys, or Tyr.[65,66] It can also be found via high-affinity noncovalent interactions, for example hydrogen bonding between the 2′-hydroxyl group of adenosyl ribose and the carboxyl group of a glutamate of the dinucleotide-binding motif[67–69] (Figure 4). However, neither native nor recombinant rat SE (Δ^{99}His) contained any detectable FAD, and thus SE must be an apoenzyme with easily dissociable FAD.[43] Detergent solubilization might result in a conformation from which flavin is readily lost. For example, a similar situation has been reported for human neutrophil cytochrome b_{558}, a flavoprotein lacking bound FAD.[70] Although the amino acid sequences surrounding the FAD covalent attachment site in different flavoproteins bear little homology, a distinct noncovalent FAD-binding site shows high sequence identity (the dinucleotide-binding motif) in many FAD-containing enzymes of diverse function. This highly conserved domain is present in both vertebrate and yeast SEs at the N-terminus (Figure 5),[55] and shows significant similarity with the nucleotide-binding motif containing the β1-sheet-α-helix-β2-sheet beginning with the Gly-X-Gly-X-X-Gly sequence between the first β-sheet and the α-helix.[67,68] Thus, in rat SE, the

γ-carboxylate group of E152 is likely to form a hydrogen bond to 2′-OH group of the adenosyl ribose of FAD. Sakakibara and Ono used site-directed mutagenesis to demonstrate that alterations in this region of rat SE resulted in loss of enzyme activity.[34]

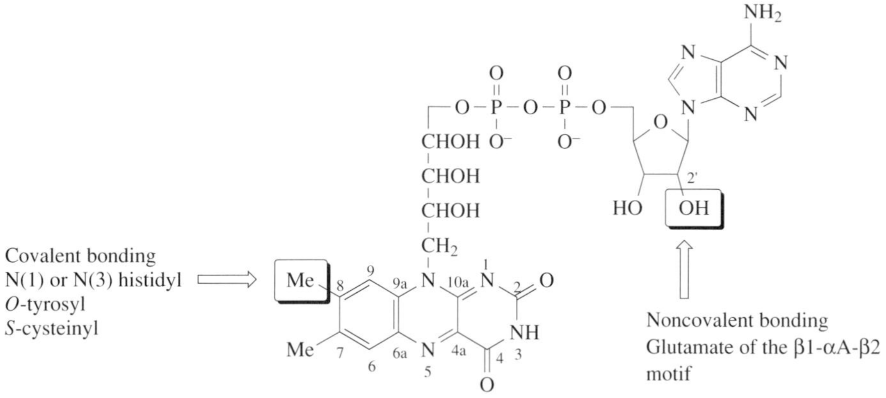

Figure 4 Covalent and noncovalent bonding of FAD in flavoproteins.

```
Squalene Epoxidase
Human                125 E V I I V G A G V L G S A L A A V L S R D G R K V T V I E R  154
Rat                  124 E V I I I G S G V L G S A L A T V L S R D G R T V T V I E R  153
Mouse                123 E V I I V G S G V L G S A L A A V L S R D G R K V T V I E R  152
Pig                  125 E I I I V G S G V L G S A L A A V L S R D G R K V T V I E R  154
Yeast                 20 D A I V I G A G V I G P C V A T G L A R K G K K V L I V E R   49
Candida                7 D A I I I G A G V I G P T I A T A F A R Q G R K V L I V E R   36
                                   ^   ^     ^       ^             ^             *

Flavoproteins in which FAD is not Covalently Bound
Pseudomonas PHBAH      4 Q V A I I G A G P S G L L L G Q L L H K A G I D N V I L E R   33
Human GSHR            22 D Y L V I G G G S G G L A S A R R A A E L G A R A A V V E S   51
Pig liver FMO          4 R V A I V G A G V S G L A S I K C C L E E G L E P T C F E R   33
                             ^   ^     ^       ^             ^             *

Flavoproteins in which FAD is Covalently Bound
Human MAO A           15 D V V V I G G G I S G L S A A K L L T E Y G V S V L V L E A   44
Streptomyces CO        7 P A V V I G T G Y G A A V S A L R L G E A G V Q T L M L E M   36
                             ^   ^     ^       ^             ^             *
```

Figure 5 Comparison of the putative FAD binding site (the dinucleotide binding site; β1-sheet–α-helix–β2-sheet) sequence of vertebrate and yeast SEs. The amino acid sequences for several flavoproteins are also presented.[68] The abbreviations used are: PHBAH, *p*-hydroxy benzoic acid hydroxylase; GSHR, glutathione reductase; FMO, flavin-containing monooxygenase; MAO, monoamine oxidase; CO, cholesterol oxygenase. Frequently occurring residues are boldfaced; conserved Gly or Ala residues (^) are indicated. The conserved glutamate residues (*) are thought to noncovalently bind to 2′-hydroxy group of adenosyl ribose by hydrogen bonding.

2.10.2.3 Substrate Specificity and Inhibition Studies

Trisnorsqualene alcohol (TNSA) was the first potent squalene-derived, noncompetitive inhibitor of vertebrate SE (IC_{50} = 4 μM, K_i = 4 μM for pig SE).[71] With the exception of the bisnorsqualene alcohol, analogues with extended or truncated carbon skeletons were poor inhibitors.[72] The primary alcohol functionality and the entire trisnorsqualenoid moiety were important for inhibition, and the absence of classical competitive inhibition suggested a more complex mode of action (Table 1).

Trisnorsqualene cyclopropylamine (TNS-CPA) was next determined to be a highly selective, slow tight-binding inhibitor of vertebrate SE (IC_{50} = 2 μM, K_i = 2.4 μM, and $k_{inact} = 0.055\ min^{-1}$ for pig SE).[73] Subsequently, an affinity resin based on the use of the TNS-CPA pharmacophore was successfully employed in the purification of SE from pig liver.[54] It is interesting to note that TNS-methylamine (IC_{50} = 1.0 μM for rat SE) was more effective for rat SE than TNS-CPA; consequently, the affinity purification of native rat SE was accomplished on a resin bearing this pharmacophore.[54]

Analogues or minor modifications of TNS-CPA, such as TNS *N*-methylcyclopropylamine (IC_{50} = 100 μM for pig SE), showed significantly reduced potency (Table 1).[74–76] In human Hep G2

Table 1 Inhibition of vertebrate SE by squalene analogues; IC_{50} values are reported for pig liver SE unless otherwise indicated

X	IC_{50} (μM)	*Ref.*	*X*	IC_{50} (μM)	*Ref.*
CH_2OH (TNSA)	4	70	$CH{=}CF_2$	4.5[b]	80
CH_2OCH_3	300	70	$CH{=}CHF$[c]	>100[b]	80
CHO	200	70	CH_2OH	400	70
COOH	>400	70	$C{\equiv}CH$	>400	79
$COOCH_3$	>400	70	$C{\equiv}CCH_3$	80[b]	79
CH_2OOH	4	70	(*E*)-$CH{=}C{=}CHCH_3$	50[b]	79
CH_2SH	30	70	$CH(OH)C{\equiv}CH$	>400	78
CH_2NH_2	200	70	$CH{=}$*c*-C_3H_5	NI	78
$CH_2N(CH_3)_2$	20	72	(*E*)-$CH{=}C(CH_3)CH{=}CH_2$	NI	78
CH_2NHEt	200	72	(*E*)-$CH{=}C(CH_3)$-*c*-C_3H_5	NI	78
CH_2NH(*i*-Pr)	NI[a]	72	$CH{=}C{=}CH_2$	NI	78
CH_2NH-*c*-C_3H_5 (TNS-CPA)	2	72	$CH{=}CBr_2$	NI	78
$CH_2N(CH_3)$-*c*-C_3H_5	100	72	$CH{=}CCl_2$	>400	78
$CH_2N(O)(CH_3)$-*c*-C_3H_5	200	72			
NH-*c*-C_3H_5	4	72			
CH_2O-*c*-C_3H_5	42[b]	73			
N(NO)-*c*-C_3H_5	>400	74			
N_3	>400	74			
$CH{=}CF_2$ (TNS-DFM)	5.4[b]	80			
$CH{=}CHF$[c]	100[b]	80			
$CH{=}C(CH_3)F$[d]	47[b]	80			
$CH_2CH{=}CF_2$	>100[b]	80			
$(CH_3)C{=}CF_2$	>100[b]	80			
$CH{=}C(CH_3)CF_2H$	nd[e]	84			
$CH{=}C(CH_3)CH_2F$	>150	86			
$CH{=}C(CH_2F)_2$	>150	86			
$CH{=}C(CH_3)CF_3$	>150	86			
$CH{=}C(CF_3)_2$	150[f]	84			
$CH{=}C(CH_3)CN$	nd	84			
$CH{=}C(CH_3)CH_2Si(CH_3)_3$	nd	84			
(*E*)-$CH{=}C(CH_3)$-*c*-C_3H_5	>400	78			
$CH{=}$*c*-C_3H_5	>400	78			
$CH{=}C{=}CH_2$	NI	78			
$CH{=}CCl_2$	NI	78			
$CH{=}CBr_2$	NI	78			
$CH(OH)C{\equiv}CH$	400	78			
$CH(O)C{\equiv}CH$	NI	78			
(*E*)-$CH{=}C(CH_3)CH{=}CH_2$	>400	78			
(*E*)-$CH{=}C{=}CHCH_3$	60[b]	79			
$C{\equiv}CH$	200[b]	79			
$C{\equiv}CCH_3$	100[b]	79			
$CH_2N(OH)CH_3$	13	75			
$CH(OH)CH_2NO_2$	270	75			

[a] No inhibition. [b] For rat liver SE. [c] Isomeric mixture consisted of 9:1 *Z*/*E* configurations. [d] Ratio of iosmers was 7:1 *Z*/*E*. [e] Not determined. [f] For yeast SE.

cells, TNS-CPA and TNS *N*-methylcyclopropylamine showed inhibition of cholesterol biosynthesis from [^{14}C]acetate with the IC_{50} values of 1.0 μM and 0.5 μM, respectively.[77] Cells incubated with TNS-CPA accumulated [^{14}C]squalene, while TNS *N*-methylcyclopropylamine accumulated [^{14}C]oxidosqualene and [^{14}C]dioxidosqualene as a result of the dual ability of TNS *N*-methylcyclopropylamine to inhibit both SE and OSC.[77] Similarly, 2-aza-2,3-dihydrosqualene, a known OSC inhibitor, which will be described later, was also shown to be a good inhibitor of SE (IC_{50} = 2.4 μM for rat SE).[77] Furthermore, a close analogue, TNS methyl hydroxylamine (IC_{50} = 13 μM and 5 μM for pig SE and OSC, respectively) was reported to be the first compound to show essentially equipotent inhibition of both SE and OSC.[76] Finally, a number of squalene analogues containing alkyne, allene, and diene functionalities showed only weak inhibition of SE (IC_{50} > 400 μM), with the exception of the allene (IC_{50} = 60 μM for rat SE), the methylacetylene (IC_{50} = 100 μM for rat SE), and the bis(methylacetylene) compound (IC_{50} = 80 μM for rat SE) (Table 1).[79,80] Molecular mechanics calculations indicated that a good inhibitor should possess hydrophobic substituents on an unpolarized, unsaturated system; additionally, the presence of a *pro*-C-3 hydroxyl group can enhance inhibitory potency.[79]

A terminal difluoroalkene analogue, TNS-difluoromethylidene (TNS-DFM) (IC_{50} = 5.4 μM, K_i = 4 μM, k_{inact} = 0.16 min^{-1} for rat SE) and its bisfunctional analogue (symmetrical tetrafluoro compound) (IC_{50} = 4.5 μM, K_i = 8 μM, k_{inact} = 0.12 min^{-1} for rat SE) showed time-dependent inhibition of rat liver SE.[81] The time-dependency of the inhibition, substrate protection against inactivation, and structure–activity data all indicated mechanism-based enzyme inactivation by a reactive functionality. We have envisaged two possible modes of inactivation of SE by TNS-DFM: (i) generation of a reactive acyl fluoride could lead to acylation of an active-site nucleophile, or (ii) a difluoromethyl ketone could be produced that could then act as a slow tight-binding inhibitor (Scheme 2). The preparation of [*vinyl*-^{3}H]-labeled TNS-DFM and the doubly labeled bisfunctional analogue from [^{3}H]TNS aldehyde and [^{3}H]-hexanorsqualene dialdehyde, respectively, gave high-specific-activity probes to test these possibilities.[82] Unfortunately, attempts to observe covalent attachment of tritiated probes to native rat SE or recombinant rat SE have been unsuccessful. These negative results offer modest support for the second mechanism. However, it is possible that an active-site flavin or another cofactor is modified instead of the active-site polypeptide. This has been observed for irreversible inhibitors of cytochrome P-450 monooxygenases, which covalently modify the heme rather than the protein.[83] TNS-DFM has been reported to be orally active in mice and accumulation of squalene and dioxidosqualene was observed.[84] Synthesis and SE inhibitory activity of other fluorinated squalene derivatives (Table 1) have not revealed any novel or significant levels of inhibition.[85–87] Presumably, only the vinylic fluorine significantly influences the electron density of the terminal double bond in terms of interference with transfer of activated oxygen to give the epoxide.[87]

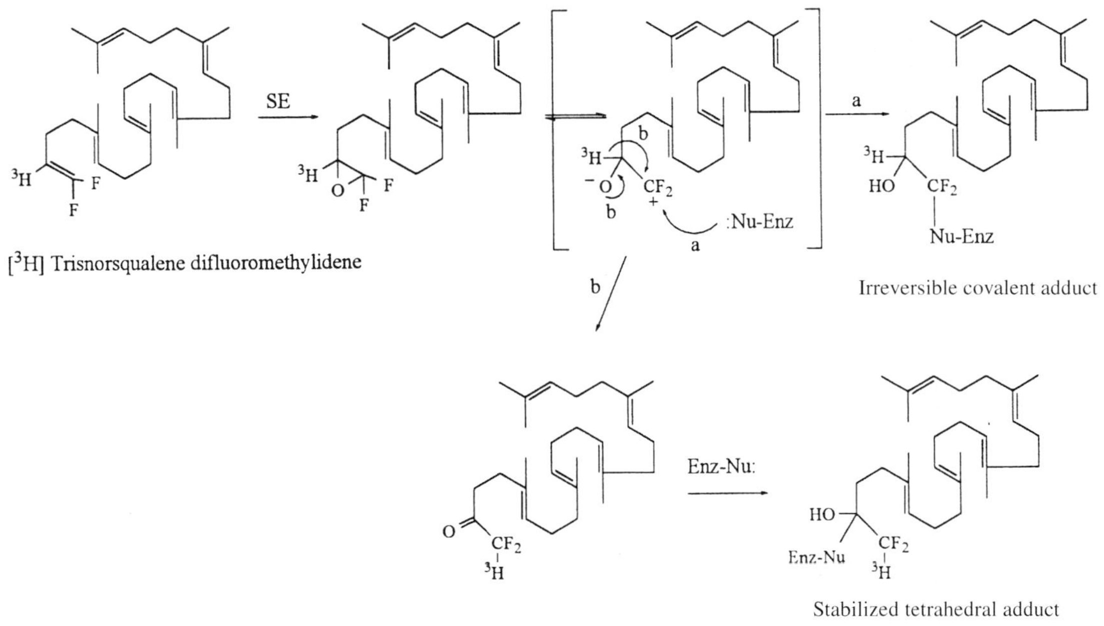

Scheme 2

SE can also catalyze the conversion of 2,3-oxidosqualene (OS) to (3*S*,22*S*)-2,3:22,23-dioxidosqualene (DOS), as demonstrated with both partially purified native pig SE[42] and with recombinant rat SE (Δ^{99}His).[43] Indeed, as described below, accumulation of both OS and DOS has been reported in liver homogenates treated with various OSC inhibitors.[88–90] The formation of DOS was reported to be approximately one-half the efficiency of OS.[42] A series of internal squalene epoxides, including (6*R*,7*R*)- and (6*S*,7*S*)-6,7-OS, (10*R*,11*R*)- and (10*S*,11*S*)-10,11-OS, were prepared and tested for SE inhibition with purified pig liver SE.[91,92] The IC_{50} values obtained were >28 μM, 6.7 μM, 25 μM, and >56 μM, respectively. Interestingly, (6*S*,7*S*)-6,7-OS was shown to be converted to (3*S*,6*S*,7*S*)-2,3:6,7-DOS as a single product. In contrast, its enantiomer, (6*R*,7*R*)-6,7-OS was converted to a mixture of regioisomeric (3*S*,6*R*,7*R*)-2,3:6,7-DOS and (3*S*,18*R*,19*R*)-2,3:18,19-DOS in a 5:12 molar ratio (Scheme 3). By way of comparison, 26-hydroxysqualene, a competitive inhibitor of SE (IC_{50} = 10 μM, K_i = 4 μM for pig SE), was converted to a 3:1 mixture of the proximal 2,3-epoxide and distal 22,23-epoxide (Scheme 3).[93] Finally, it should be noted that several internal oxidosqualenes, (6*S*,7*S*)-6,7-OS, (10*S*,11*S*)-10,11-OS, and (10*R*,11*R*)-10,11-OS, have been isolated from natural sources.[94–96]

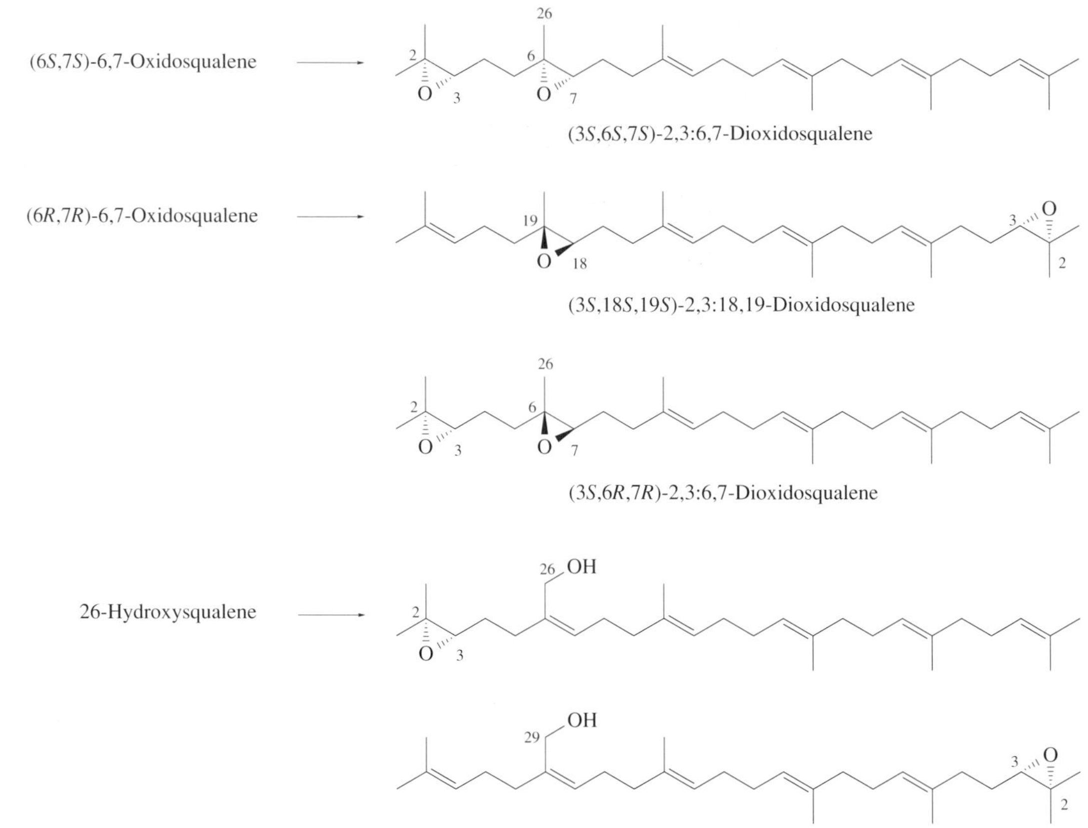

(6*S*,7*S*)-6,7-Oxidosqualene → (3*S*,6*S*,7*S*)-2,3:6,7-Dioxidosqualene

(6*R*,7*R*)-6,7-Oxidosqualene → (3*S*,18*S*,19*S*)-2,3:18,19-Dioxidosqualene

(3*S*,6*R*,7*R*)-2,3:6,7-Dioxidosqualene

26-Hydroxysqualene →

Scheme 3

It has been reported that tellurium, more specifically tellurite (TeO_3^{2-}) selectively affects SE activity *in vivo* and *in vitro*.[97,98] When rats are fed with a tellurium-containing diet, a peripheral neuropathy characterized by a transient demyelination of the sciatic nerves is observed. This neuropathy is correlated with the accumulation of squalene in rat tissues. It is not well understood why only cholesterol biosynthesis in the brain is specifically affected. In *in vitro* studies, the SE activity suffered 50% inhibition in the presence of 5 μM K_2TeO_3 in noncompetitive manner with respect to squalene and a mixed-type inhibition with respect to FAD, whereas the activities of cytochrome P-450 monoxygenases were relatively unaffected.[96] It was suggested that SE inhibition might involve the action of tellurite as a reversible sulfhydryl reagent, since SE is inhibited by a thiol-modifying reagent, *N*-ethylmaleimide.[97] As there are two conserved Cys residues (C490, C500 in rat SE) in all the known SEs, site-directed mutagenesis experiments of these residues would be interesting. Interestingly, although TeO_3^{2-} shows an IC_{50} value of about 10 μM with purified rat SE and with recombinant rat SE, neither SeO_3^{2-} nor SO_3^{2-} show any inhibition at concentrations up to 200 μM.[99]

Naftifine (IC_{50} = 1.1 μM for *C. albicans* SE; 144 μM for rat SE) and Terbinafine (IC_{50} = 0.03 μM for *C. albicans* SE; 77 μM for rat SE) are two prototypical examples of Sandoz Research Institute's allylamine antimycotics, known to selectively inhibit fungal SE (Figure 6).[100–105] Extensive structure–activity relationship (SAR) studies for replacement of the naphthalene moiety led to the discovery of SDZ 87-469 ((*E*)-3-chloro-*N*-(6,6-dimethyl-2-hepten-4-ynyl)-*N*-methylbenzo[*b*]thiophene-7-methanamine) (IC_{50} = 0.011 μM for *C. albicans* SE; 43 μM for rat SE).[106,107] Further, replacement of the (*E*)-1,3-enyne structural element of Terbinafine by a phenyl group resulted in the discovery of the antifungal benzylamines such as butenafine (IC_{50} = 0.045 μM for *C. albicans* SE; 23 μM for guinea pig SE).[106–109] These studies demonstrated that two lipophilic domains linked by a spacer of appropriate length and a polar center at a defined position in the spacer are the general requirements for high activity of allylamine antimycotics.[110–112] Kinetic analysis suggested that naftifine and Terbinafine were reversible, noncompetitive inhibitors of *C. albicans* SE with respect to squalene,

FAD, and NADPH.[103] Since the allylamines show no obvious structural resemblance to squalene, it seemed unlikely that they would function as substrate analogues. One possible mechanism of action postulates that the allylamine molecular binds weakly to two separate sites on the epoxidase, for example, the naphthalene ring of Terbinafine could bind to the substrate-binding site of the enzyme, while the allylamine side-chain could interact with a second site.[113] Another possibility would be an interaction with either the flavin cofactor binding site or to a specific lipid-binding domain of SE, leading to a change in the conformation of the enzyme.[113] The high degree of the species selectivity for inhibition of SE is correlated with the observation that rat SE showed only 30% identity to yeast enzyme at the amino acid sequence level. Recently, Sakakibara and Ono demonstrated that in the Terbinafine-resistant yeast SE,[57] interestingly, only single point mutation (C → T in the nucleotide sequence) took place and as a result, Leu^{251} was changed to Phe residue.[34] The Leu^{251} could be located proximal to the aryl ring binding site of yeast SE, and thus the binding of Terbinafine would be hampered by the replacement of this δ-branched residue with a β-branched aromatic residue in the mutant.

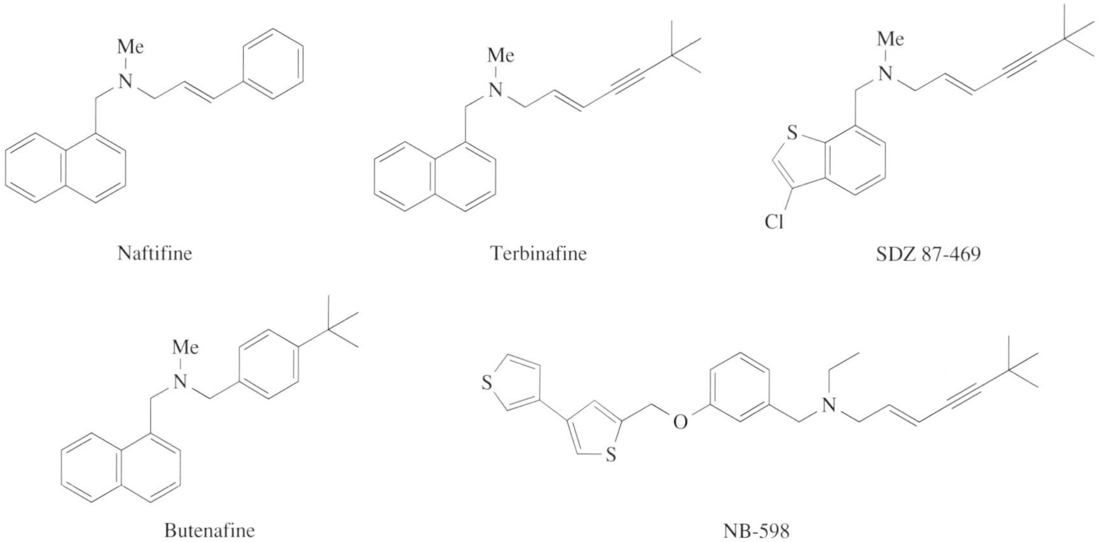

Naftifine Terbinafine SDZ 87-469

Butenafine NB-598

Figure 6 SE inhibitors: allylamines.

In order to develop potent SE inhibitors specific for the mammalian enzymes, an extensive medicinal chemistry program was undertaken at the Banyu Pharmaceutical Company, using the structure of Terbinafine as a lead for SAR optimization.[114,115] These studies demonstrated that extension of the carbon skeleton between tertiary nitrogen and a naphthyl group or substitution of a phenyl for the naphthyl group resulted in a dramatic increase in potency for the mammalian SEs.[114] It was further suggested that the allylamine side-chain binding site of both fungal and vertebrate enzymes had similar stereoelectronic requirements. In contrast, the binding site of vertebrate SE for the aromatic ring seems to be significantly different from that of fungal enzyme, thus providing the high degree of selectivity between the two enzymes (Figure 7).[116]

NB-598 ((*E*)-*N*-ethyl-*N*-(6,6-dimethyl-2-hepten-4-ynyl)-3-[(3,3′-bithiophen-5-yl)-methoxy]-benzene-methanamine) was thus identified as the new lead compound for development of a therapeutic drug. It is a highly potent, specific inhibitor of vertebrate SE ($IC_{50} = 0.75$ nM, $K_i = 0.68$ nM for human Hep G2 SE, and $IC_{50} = 4.4$ nM for rat liver SE),[117] and the inhibition was competitive with respect to squalene for these systems. Interestingly, NB-598 showed noncompetitive kinetics with partially purified pig liver SE.[42] NB-598 did not show significant antifungal activity against *C. albicans* and it had no effect on OSC activity. Moreover, *S. cerevisiae* SE was weakly inhibited in a partially noncompetitive manner.[118] In Hep G2 cells, NB-598 strongly inhibited cholesterol synthesis from [^{14}C]acetate in a dose-dependent manner and caused intracellular accumulation of [^{14}C]squalene. Multiple administration of oral doses of NB-598 decreased serum cholesterol levels in dogs, while the level of free fatty acids, phospholipids, and triacylglycerol were unaffected.[117,119,120] Moreover, NB-598 treatment dramatically increased the LDL receptor level in Hep G2 cells without a concomitant elevation of HMG-CoA reductase activity.[37,121] Thus, when compared with the HMG-CoA reductase inhibitor L-654,969, NB-598 may not inhibit the synthesis of nonsterol derivative(s) of mevalonate that regulate HMG-CoA reductase activity at the posttranslational level.[37] From

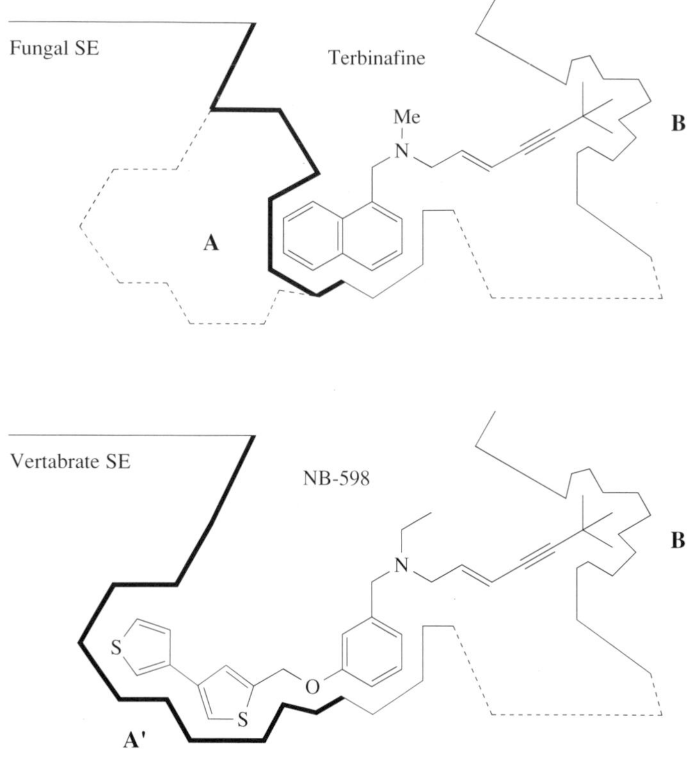

Figure 7 The two-site model for Terbinafine and NB-598 binding.

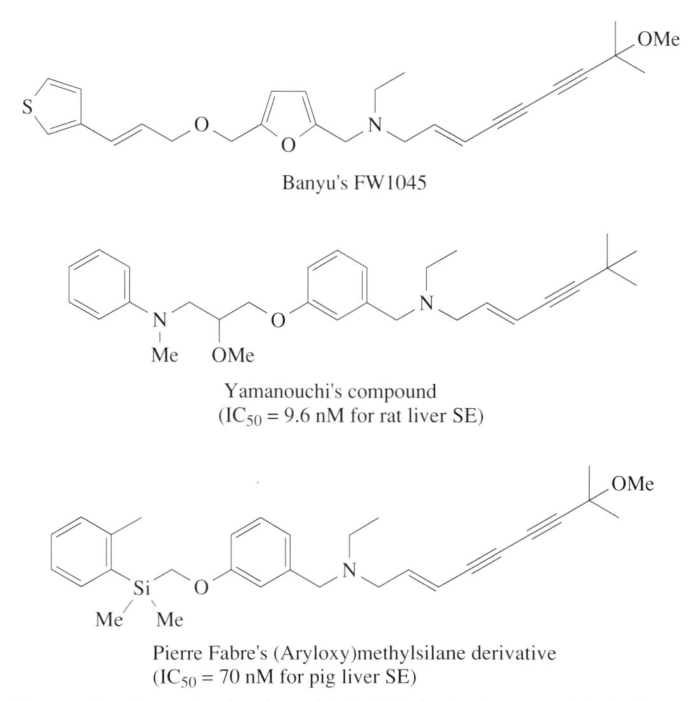

Figure 8 Recently developed NB-598 derivatives as SE inhibitors.

these observations, NB-598 is expected to be highly effective in the treatment of hypercholesterolemia. SAR studies in the 1990s have led to further development of several new inhibitors with equivalent or improved *in vitro* and *in vivo* biological profiles,[122,123] including Banyu's FW-1045,[124] a new amino alcohol derivative by the Yamanouchi Company,[125] and a series of (aryloxy)methylsilane derivatives developed at the Centre de Recherche Pierre Fabre[126–128] (Figure 8).

Several photoaffinity ligands for vertebrate SE have been synthesized in order to elucidate the active site structure of the enzyme: (i) squalene analogues having a diazoacetate photophore such as TNSA-DZA[129,130] and a variety of diazo and diazirine modifications;[75] (ii) the NB-598 analogues with diazirine[131] and with benzophenone (BP) photophores (PDA-I and PDA-II);[132] and (iii) the FAD analogue with a pendant BP photophore (BZDC-FAD)[133] for exploring the FAD binding site[134] (Figure 9). These compounds are now being used for the photoaffinity labeling of rat SE. In particular, the BP derivatives were found to be excellent photoaffinity probes,[135] since (a) BPs are chemically more stable than diazo esters, aryl azides, and diazirines, (b) BPs can be manipulated in ambient light and can be activated at 350–360 nm, avoiding protein-damaging wavelengths, and (c) BPs react preferentially with unreactive C—H bonds, even in the presence of solvent water and bulk nucleophiles, frequently with remarkable site specificity.[136] The authors have also succeeded in photoaffinity labeling the taxol binding sites on tubulin and P-glycoprotein by using a BZDC-taxol derivative.[137]

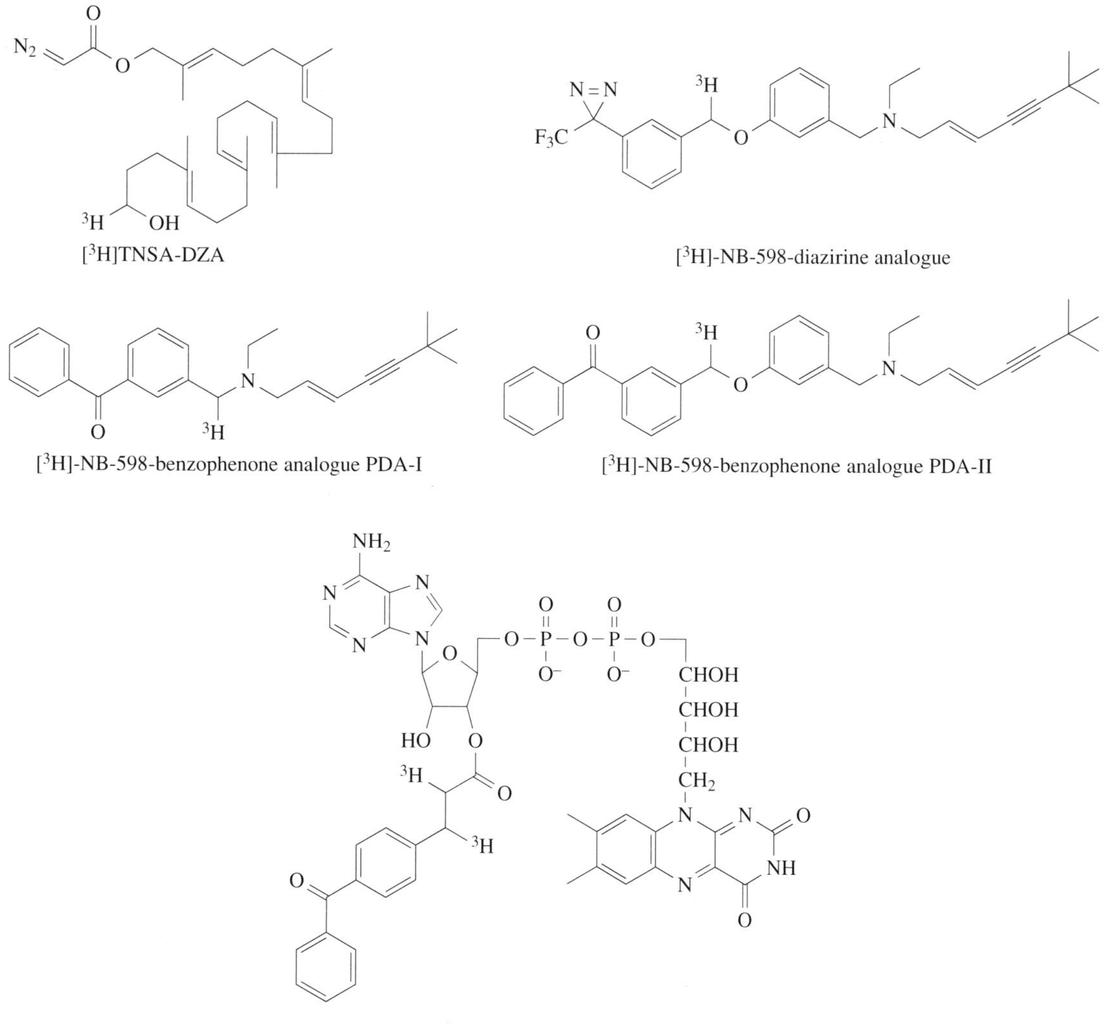

[^{3}H]TNSA-DZA

[^{3}H]-NB-598-diazirine analogue

[^{3}H]-NB-598-benzophenone analogue PDA-I

[^{3}H]-NB-598-benzophenone analogue PDA-II

[^{3}H]-FAD-BZDC analogue

Figure 9 Photoaffinity ligands for SE.

2.10.3 OXIDOSQUALENE : LANOSTEROL CYCLASE

2.10.3.1 General Considerations for Enzyme Mechanism

Oxidosqualene:lanosterol cyclase (OSLC) (EC 5.4.99.7) catalyzes the conversion of (3*S*)-2,3-oxidosqualene to lanosterol, forming a total of six new carbon–carbon bonds and seven stereocenters in a single reaction. Other oxidosqualene cyclases (OSCs) transform the same substrate to a wide variety of naturally occurring polycyclic triterpenes.[1] The relationships among enzyme structure, the cyclization mechanism, and the products(s) formed have provided intriguing problems for three generations of bioorganic chemists. In principle, subtle modifications of the structure of the active site of the enzyme are sufficient to produce dramatically different products. The enzyme provides a catalyst, a template, and stabilization of intermediates to guide the course of each cyclization reaction.

The formation of lanosterol is initiated by an acid-catalyzed opening of the oxirane ring of oxidosqualene folded in a chair–boat–chair conformation; participation by a neighboring π-bond gives an initial six-membered ring intermediate (not detected), which then undergoes a series of cation-induced polyalkenic cyclizations to a tetracyclic C-20 cation (Scheme 4). This intermediate C-20 protosterol cation, analogues of which have been trapped, then undergoes a series of 1,2-methyl and hydride shifts, and subsequent proton abstraction by the enzyme yields the lanosterol skeleton (Scheme 4).[138–140] For entropic reasons and based on experimental evidence, van Tamelen suggested that a single transition state for a complex polycyclization was unlikely, and that the cyclization should proceed through a series of discrete, conformationally rigid, partially cyclized carbocationic intermediates.[140,141] Indeed, monocyclic and bicyclic triterpenes, which are thought to be "trapped", partially cyclized intermediates, have been isolated from natural sources.[142–144]

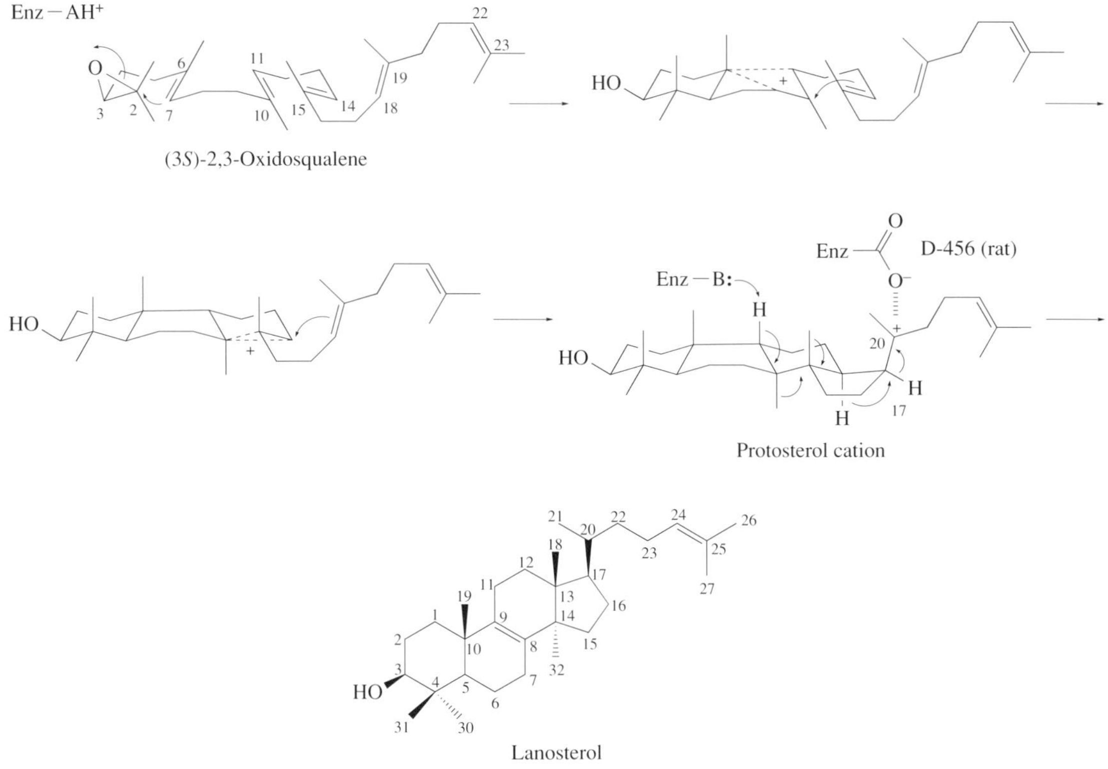

Enz—AH⁺

(3*S*)-2,3-Oxidosqualene

Enz—B:

D-456 (rat)

Protosterol cation

Lanosterol

Scheme 4

During the enzymatic cyclization of oxidosqualene, the six-membered C-ring is formed via a thermodynamically unfavorable secondary ("6.6.6-fused") tricyclic cation in a formally anti-Markovnikov addition. The Δ^{14} and the Δ^{18} double bonds in the substrate must be held in close proximity by the enzyme, thereby ensuring bond formation between these centers in an anti-Markovnikov sense.[140] van Tamelen also proposed that such an intermediate cation should be formulated as a nonclassical species to best rationalize the partitioning between five- and six-membered rings, to maintain the enzyme-enforced proximity of carbons, and to preserve stereochemistry during cyclization.[140] However, Corey recently proposed that, during cyclization of 2,3-oxidosqualene, the C-ring of the sterol nucleus is formed by ring closure to a Markovnikov intermediate with the five-membered C-ring first, then followed by ring expansion to a six-membered C-ring (Scheme 5).[145] This proposal is based on their observation that 20-oxa-2,3-oxidosqualene was cyclized to an unusual product with a five-membered C-ring, but this controversial interpretation fails to take into account the steric and electronic perturbations on the cyclization mechanism engendered by the CH_2 to O substitution.

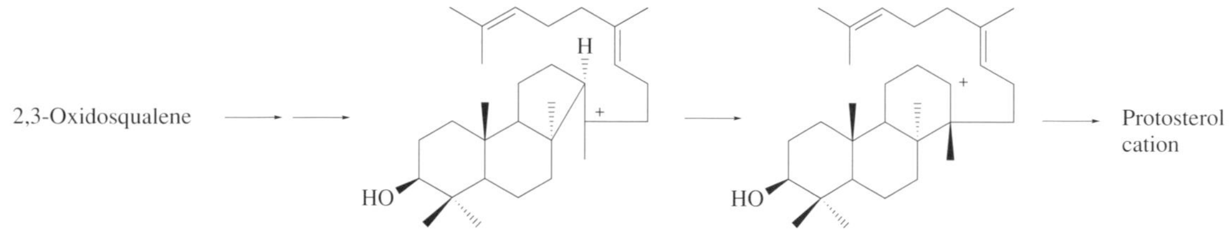

Scheme 5

Johnson proposed a mechanism for OSC in which "axial delivery of negative point charges" by the enzyme could stabilize the developing cationic centers as ion pairs (Figure 10).[146–149] In this model, no particular conformational control would be required by the enzyme, and the expected transition state stabilization should account for the boat ring-B (delivery of a point charge to the α-face at *pro*-C-8) as well as the anti-Markovinkov closure of ring-C (another point charge proximal to the α-face at *pro*-C-13 but not at *pro*-C-14). Further, such charge delivery to the β-face at *pro*-C-10 was proposed to be important in enhancing the rate and efficiency of the overall cyclization process.

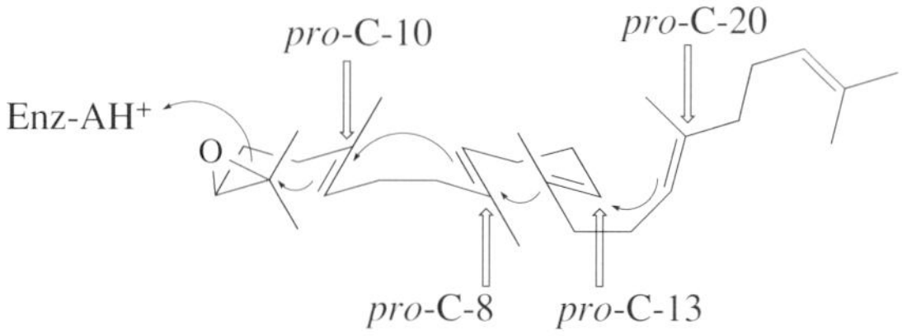

Figure 10 Axial delivery of negative point charges as proposed by Johnson.[146–149]

Significant progress in understanding the primary and secondary structure of the active sites of OSC enzymes has occurred since 1990. Indeed, there is evidence for the involvement of a negative point charge from a conserved Asp residue that apparently stabilizes the C-20 protosterol cation.[150–152] Other steric and electrostatic factors, such as cation-π interactions involving the incipient cation and the electron-rich faces of aromatic sidechains of Phe, Tyr, or Trp, could also shelter transient cationic cyclization species in analogy to known receptor–ligand and enzyme–substrate complexes.[153–155] Interestingly, the presence of a repetitive "QW motif", a highly conserved α-helix-turn motif rich in aromatic amino acids, has been identified as being unique to the cyclase enzymes.[156,157] The occurrence of six QW motifs in each eukaryotic OSC or in bacterial squalene cyclases (SCs) has led the authors to postulate that the aromatic residues of this motif may act as hydrophobic "negative point charges". Accordingly, the QW motifs would play a functional role in catalysis by stabilizing the developing cationic centers of the cyclizing substrate through cation-π interactions.[158] Crystallographic evidence[159] for *A. acidocaldarius* SC suggests, however, that the QW motifs are structural in nature, not catalytic.

Trapping of intermediate analogues of the protosterol cation was achieved by Corey and his co-workers.[160,161] Based on the observation that the 20-oxa-2,3-oxidosqualene[160] and (20*E*)-20,21-dehydro-2,3-oxidosqualene[161] were enzymatically transformed to protosterols having a 17β-side chain, the protosterol cation was proposed to have the 17β-oriented side chain. The formation of the natural C-20*R* stereochemistry of lanosterol was envisaged to occur via a least-motion pathway, involving only a small ($<60°$) rotation about the C-17–C-20 bond. The β-orientation of the side chain at C-17 and the strong steric interaction with the *cis*-14β-methyl substituent serves to hinder rotation about the C-17–C-20 bond, thereby controlling the configuration at C-20 in lanosterol formation.[161] Enzymatic formation of the protosterol 3β-hydroxyprotosta-17(20)[16,21-*cis*],24-diene was reported from microsomal preparations of the fungus *Cephalosporium caerulens*.[162] This diene is a biosynthetic precursor of fusidic acid, an antibiotic possessing the unique protosterol skeleton.[163] In this case, the cyclization reaction proceeded without the backbone rearrangement, and the initially formed C-20 protosterol cation was stabilized by proton elimination of H-17 to form a double-bond between C-17 and C-20.[163,164]

2.10.3.2 Enzymology and Molecular Biology

Purifications of several OSC enzymes have been reported from vertebrate,[165–168] plant,[169–172] and yeast[172–174] sources. As previously described by Bloch and co-workers,[175–177] the microsomal membrane-bound vertebrate OSCs required selected detergents and a narrow range of salt concentrations to maintain activity. As in the case of SE, the OSCs were efficiently solubilized by Triton X-100 and purified to homogeneity in the presence of the detergent. Thiol-protecting reagents such as DTT and glycerol were useful in stabilizing activity. Purification of detergent-solubilized vertebrate OSLCs was achieved using a simple combination of DEAE-cellulose and hydroxylapatite column chromatography, affording several milligrams of homogeneous enzyme from 100 g of liver.[165] The purified rat and pig liver enzymes showed single bands on SDS-PAGE with molecular masses of 78 kDa and 75 kDa, respectively; the N-terminal residues of the purified enzymes were found to be blocked.[165] As isoelectric point of 5.5, and an apparent K_M value for (3*S*)-oxidosqualene of 55 μM were reported for rat OSLC.[164] The enzyme showed a broad pH optimum within a pH range of 6.0–8.0.[178] Further, as described below, affinity labeling experiments using a high specific activity tritiated form of the first potent mechanism-based irreversible inactivator of the cyclase, 29-methylidene-2,3-oxidosqualene ([^{3}H]29-MOS),[179] demonstrated that the vertebrate OSLCs were specifically labeled and gave single bands on fluorogram with apparent molecular masses ranging from 70 kDa to 80 kDa (rat, 78 kDa; dog 73 kDa; pig, 75 kDa; and human, 73 kDa).[165]

Several lanosterol synthases have been cloned and sequenced from vertebrates (human[180–182] and rat[152,183]), *S. cerevisiae*,[184,185] *C. albicans*,[186,188] and *Schizosaccharomyces pombe*,[189] and an oxidosqualene-cycloartenol cyclase (OSCC) from the plant *Arabidopsis thaliana*[190] (Figure 11). The human OSC gene is located on chromosome 21q22.3, and Northern blot analysis revealed that a 4.4 kbp OSLC mRNA is prominent in all tissues including brain, heart, lung, kidney, pancreas, and liver.[182] These OSCs have been expressed in the OSLC-deficient yeast mutant strain (*erg7*). The *erg7* gene was recently cloned and sequenced, revealing two codon changes: an Asp-to-Ochre stop at position 359, and a Gly-to-Glu at position 359.[191] In contrast to the situation for SE, OSC could not be functionally expressed in *E. coli*. The predicted molecular masses of OSCs ranged from 80 kDa to 90 kDa, and the deduced amino acid sequences showed significant similarity to each other.

The authors have reported molecular cloning and the first functional expression of rat liver OSLC.[151] The cDNA contained a 2199 bp ORF encoding an 83 kDa protein and the rat liver OSLC showed 84.7% amino acid identity (621/733) with human OSLC, 40.2% identity (295/733) with *S. cerevisiae* OSLC, 39.0% identity (286/733) with *C. albicans* OSLC, 41.6% identity (305/733) with *S. pombe* OSLC, and 44.2% identity (324/733) with *A. thaliana* OSCC. It is interesting that rat lanosterol synthase showed higher similarity to plant cycloartenol synthase than yeast lanosterol synthase. The vertebrate OSLC also showed substantial identity to two prokaryotic SCs that directly cyclize squalene into the pentacyclic triterpene hopene: 26.3% identity (193/733) with *Zymomonas mobilis* SC,[192] and 16.6% identity (122/733) with thermoacidophilic *Alicyclobacillus acidocaldarius* SC,[193] suggesting a very ancient lineage of the vertebrate cyclase enzymes as proposed by Ourisson and co-workers.[194,195] In general, the amino acid sequence is more highly conserved in the C-terminal region than in the N-terminus. Except for these cyclases, no other significant sequence similarities were found in the GenBank or EMBL databases.

From sequence comparisons of eukaryotic OSCs and bacterial SCs, was identified the QW motif, a highly conserved repetitive motif rich in aromatic amino acids: ([K/R][G/A]XX[F/Y/W][L/I/V]XXXQXXXGXW).[156] There are six repeats of the QW motif in rat OSLC, four at the C-terminal and two at the N-terminal (Figure 12). The motif was well conserved both in eukaryotic OSCs and prokaryotic SCs. According to the PepPlot program,[196] a typical QW motif contains part of a β-strand at the amino terminus and a turn at the carboxyl terminus. It was proposed that the aromatic amino acids of the QW motif play a functional role in catalysis by stabilizing the carbocationic intermediates through cation-π interactions,[158] but crystallographic data suggests otherwise. The QW motif appears to stabilize the enzyme structure by connecting surface α-helices.[160]

Of 733 amino acids of rat OSLC, 188 residues (26%) are completely conserved in all five known OSCs (human, rat, yeast, *Candida*, fission yeast). Overall, rat OSLC contains a disproportionately higher number of aromatic amino acid residues (relative to this 26%) that are completely conserved: Phe (14 out of 28 residues), Trp (14 out of 24 residues), and Tyr (19 out of 35 residues). [152,185] Corey and co-workers have reported that when each of the conserved Trp residues in yeast OSLC was mutated to Phe, the mutant still retained ability to complement the OSLC deficiency in *erg7* yeast strain, suggesting that these Trp residues are not individually essential for enzyme activity.[189] The negatively charged Asp and Glu residues in rat OSC are less highly conserved; Asp (6 out of 36, D21, D92, D368, D456, D575, D627; D456 is labeled with 29-MOS) and Glu (9 out of 43, E127,

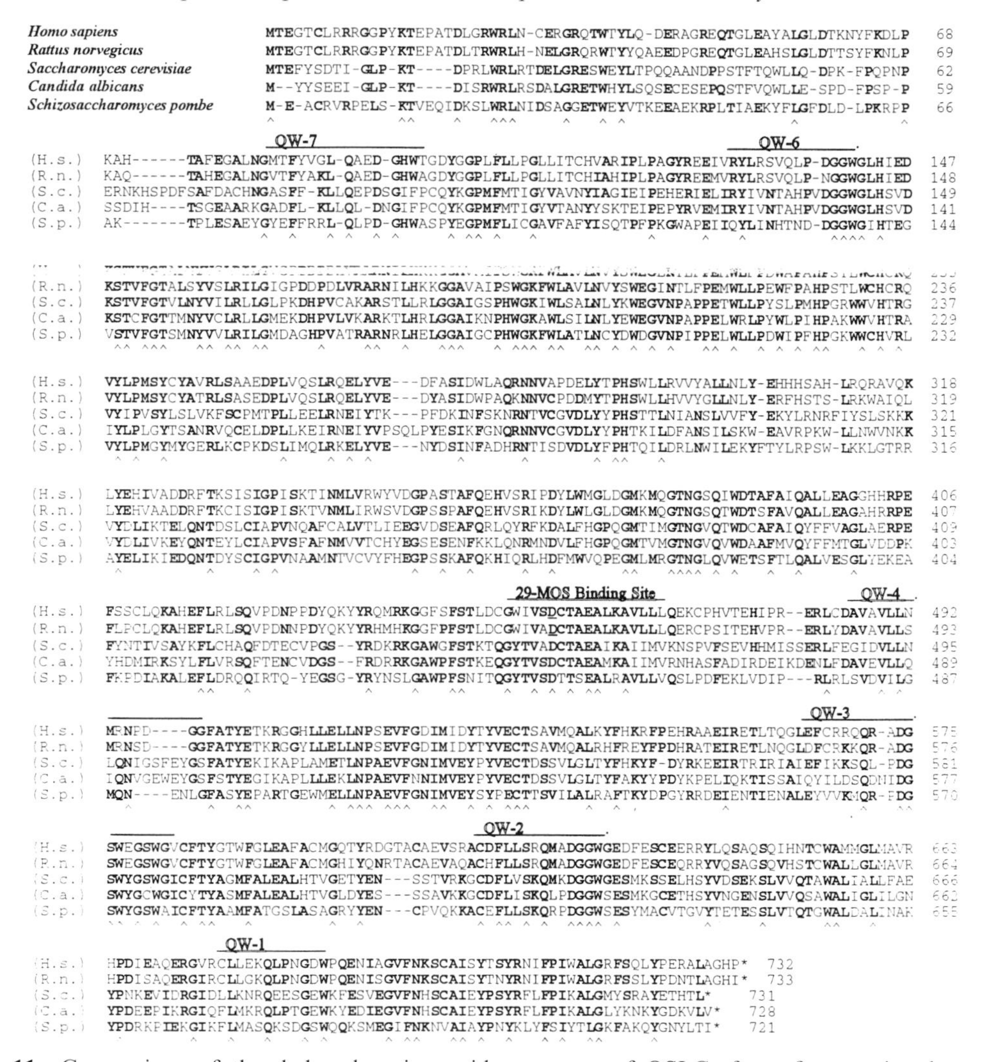

Figure 11 Comparison of the deduced amino acid sequences of OSLCs from five species: human, *H. sapiens*; rat, *R. norvegicus*; yeast, *S. cerevisiae*; candida, *C. albicans*; fission yeast, *S. pombe*. Conserved residues in at least three sequences are boldfaced; hyphens indicate gaps introduced to maximize alignment. The QW motifs and the 29-MOS binding site (DCTAEA) are indicated.

E215, E263, E460, E505, E514, E520, E533, E632). Cation-stabilizing "negative point charges" may arise from either anionic or from aromatic residues at the active site of the enzyme. The DCTAEA motif, the labeling site of the mechanism-based irreversible inhibitor, 29-MOS, is located just N-terminal to the QW-4 motif. This sequence is well conserved in all the known OSCs (Figure 13). Indeed, the Asp residue (D456) that is implicated in stabilization of the C-20 cation after tetracyclization but prior to hydride methyl migrations is conserved in all OSCs.[150–152] Furthermore, there are three conserved Cys (C534, C585, C617), six conserved Ser (S150, S356, S446, S577, S622, S648), eight conserved Thr (T16, T151, T155, T331, T382, T458, T535, T587), four conserved His (H145, H226, H233, H290), two conserved Lys (K15, K195), and seven conserved Arg (R26, R163, R178, R235, R261, R365, R436). The presence of an essential cysteinyl group in the active site of the enzyme has been previously suggested, since the OSLC activity can be enhanced by the addition of thiol-protecting reagent and inhibited by the sulfhydryl reagents such as *p*-chloromercuribenzenesulfonic acid and *N*-ethylmaleimide.[168,172,174,197] In contrast, diethyl pyrocarbonate, a histidyl-selective reagent, does not inhibit OSLC activity.[168,174] Finally, a disproportionately higher number of Gly (30 out of 59 residues) and Pro (14 out of 38 residues) are also conserved, suggesting important conserved elements of secondary structure.

According to the Kyte–Doolittle hydropathy plot analysis, the OSCs are moderately hydrophilic proteins (Figure 14). They have a hydrophilic region at the N-terminus; surprisingly for this membrane-associated protein, there were no significantly hydrophobic regions that could serve as

QW - 1

Human	672	RGVRCLLEKQLPNGDW
Rat	673	RGIRCLLGKQLPNGDW
Yeast	675	RGIDLLKNRQEESGEW
Candida	671	RGIQFLMKRQLPTGEW
Fission yeast	664	KGIKFLMASQKSDGSW
Plant	702	RAARYLINAQMENGDF
Bacteria-a	576	RGVQYLVETQRPDGGW
Bacteria-z	595	KGINWLAQNQDEEGLW

QW - 2

Human	614	RACDFLLSRQMADGGW
Rat	615	QACHFLLSRQMADGGW
Yeast	617	KGCDFLVSKQMKDGGW
Candida	613	KGCDFLISKQLPDGGW
Fission yeast	606	KACEFLLSKQRPDGGW
Plant	640	KACEFLLSKQQPSGGW
Bacteria-a	518	KALDWVEQHQNPDGGW
Bacteria-z	536	KAVAWLKTIQNEDGGW

QW - 3

Human	562	QGLEFCRRQQRADGSW
Rat	563	QGLDFCRKKQRADGSW
Yeast	568	IAIEFIKKSQLPDGSW
Candida	563	SAIQYILDSQDNIDGSW
Fission yeast	557	NALEYVVKMQRPDGSW
Plant	591	KAVKFIESIQAADGSW
Bacteria-a	470	RAVEYLKREQKPDGSW
Bacteria-z	488	AAVDYLLKEQEEDGSW

QW - 4

Human	485	DAVAVLLNMRNPDGGF
Rat	486	DAVAVLLSMRNSDGGF
Yeast	488	EGIDVLLNLQN-IGSF
Candida	482	DAVEVLLQIQN-VGEW
Fission yeast	480	LSVDVILGMQNENLGF
Plant	514	EAVNVIISLQNADGGL
Bacteria-a	402	KGFRWIVGMQSSNGGW
Bacteria-z	420	RAMEWTIGMQSDNGGW

QW - 5

Bacteria-a	335	KAGEWLLDRQI-TVPGDW
Bacteria-z	350	SALSWLKPQQILDVKGDW

QW - 6

Human	126	EIVRYLRSVQLP-DGGW
Rat	127	EMVRYLRSVQLP-NGGW
Yeast	127	ELIRYIVNTAHPVDGGW
Candida	119	EMIRYIVNTAHPVDGGW
Fission yeast	123	EIIQYLINHTND-DGGW
Plant	149	EMRRYLYNHQNE-DGGW

QW - 7

Human	79	NGMTFYVGLQAED-GHW
Rat	80	NGVTFYAKLQAED-GHW
Yeast	79	NGASFFKLLQEPDSGIF
Candida	72	KGADFLKLLQ-LDNGIF
Fission yeast	76	YGYEFFRRLQLPD-GHW
Plant	99	RGLDFYSTIQAHD-GHW
Bacteria-a	17	RAVEYLLSCQ-KDEGYW
Bacteria-z	24	KATRALLEKQQQD-GHW

Figure 12 Summary of the highly-conserved repetitive QW motifs ([K/R][G/A]XX[F/Y/W][L/I/V]XXXQXXXGXW) (α-helix-turn motif) in OSC from eight species: human *Homo sapiens*; rat, *R. norvegicus* OSLC; yeast, *S. cerevisiae* OSLC; candida, *C. albicans* OSLC; fission yeast, *S. pombe* OSLC; plant, *A. thaliana* OSCC (cycloartenol synthase); bacteria-a, *A. acidocaldarius* SC; bacteria-z. *Z. mobilis* SC. Frequently occurring residues are boldfaced; hyphens indicate gaps introduced to maximize alignment.

Lanosterol Synthase (OSLC)

Rat	439	H K G G F P F S T L D C G W I V A D C T A E A L K A V L L L	468
Human	438	R K G G F S F S T L D C G W I V S D C T A E A L K A V L L L	467
S. cerevisiae	439	R K G A W G F S T K T Q G Y T V A D C T A E A I K A I I M V	468
C. albicans	433	R K G A W P F S T K E Q G Y T V S D C T A E A M K A I I M V	462
Sc. prombe	434	S L G A W P F S N I T Q G Y T V S D T T S E A L R A V L L V	463
		^ * * * # ^ ^ ^	

Cycloartenol Synthase (OSCC) (Plant)

A. thaliana	466	S K G A W P F S T A D H G W P I S D C T A E G L K A A L L L	495
		^ * * * # ^ ^ ^	

Squalene-Hopene Synthase (Bacteria)

A. acidocaldarius	356	K P G G F A F Q F D N V Y Y P D V D D T A V V V W A L N T L	385
Z. mobilis	375	K P G G W A F Q Y R N D Y Y P D V D D T A V V T M A M D R A	404
		^ * * * # ^ ^ ^	

Figure 13 Comparison of the 29-MOS binding site sequence (just N-terminal of the QW-4 motif) with vertebrate, fungal, plant, and bacterial OSC/SCs. Only vertebrate OSLCs are labeled by [^{3}H]29-MOS at the first Asp residue (#) of the DCTAEA motif. Frequently occurring residues are boldfaced; conserved Gly or Ala or Thr residues (∧), and aromatic residues (*) are indicated.

membrane-spanning regions. Thus, microsomal OSCs appear to associate weakly to membranes; this may explain the ease of solubilization under mild conditions. Furthermore, in rat OSLC, there are five possible N-glycosylation sites (N383, N517, N606, N692, and N698); however, incubation with N-glycosidase did not affect OSLC activity.[178]

2.10.3.3 Substrate Specificity and Inhibition Studies

The substrate specificity of OSLC has been probed with a variety of substrate analogues.[1] In general, the enzyme accepts most modifications in the *pro*-side-chain region of 2,3-oxidosqualene, including 2,3:22,23-dioxidosqualene,[198,199] 22,23-dihydro-2,3-oxidosqualene,[198] 22,23-dihydro-22-methylene-2,3-oxidosqualene,[200] 30-acetyl-2,3-oxidosqualene,[201] and oxidosqualene analogues with truncated side chains.[202–204] As described, 20-oxa-2,3-oxidosqualene[160] and (20*E*)-20,21-dehydro-2,3-oxidosqualene[161] were enzymatically transformed to protosterol derivatives having 17β-oriented

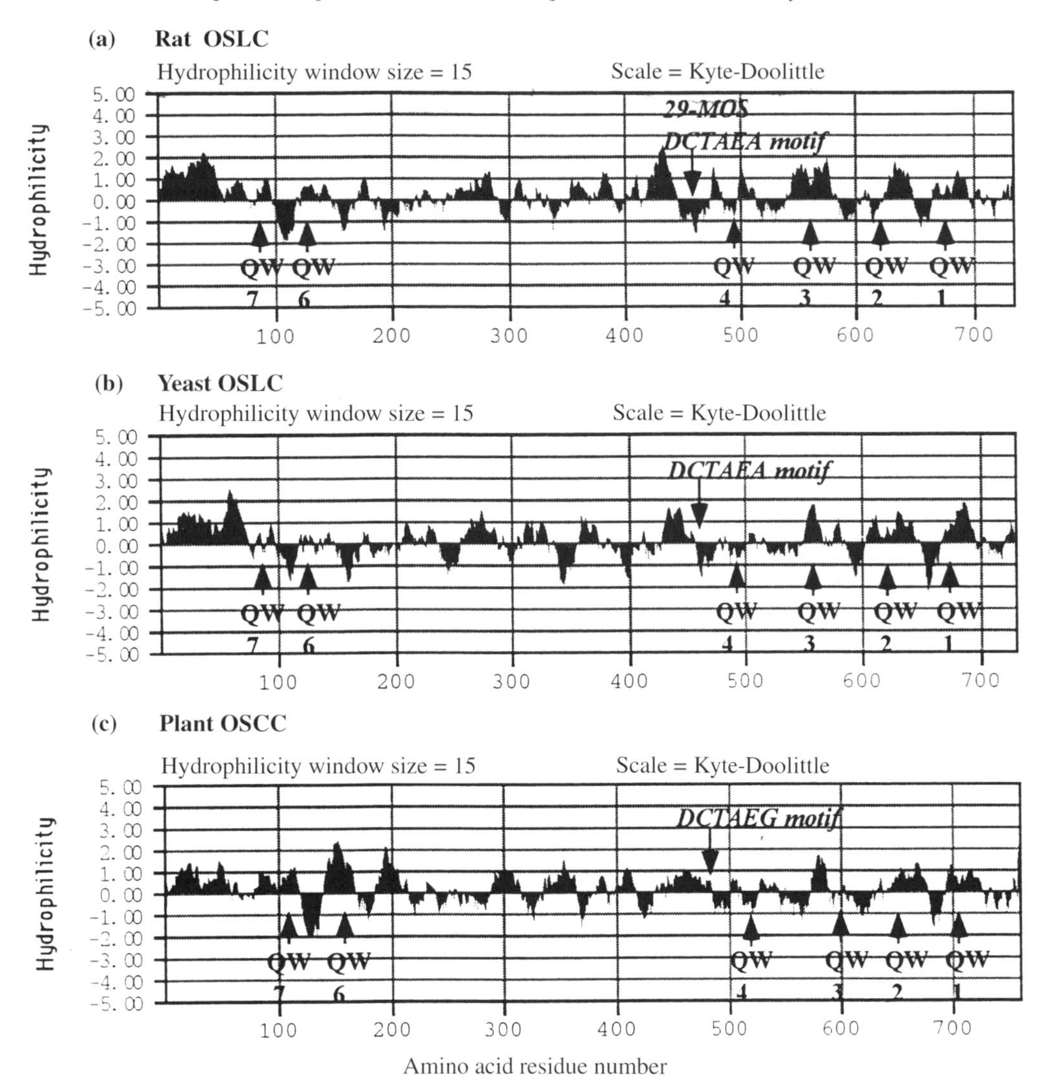

Figure 14 Hydropathy plots for (a) rat OSLC, (b) yeast *S. cerevisiae* OSLC, and (c) plant *A. thaliana* cycloartenol synthase OSCC. The deduced amino acid sequence was analyzed by MacVector 4.1 sequence analysis software (Kodak). The QW motifs and the 29-MOS binding site (DCTAEA) are indicated. Plant and yeast OSLCs are not labeled by [^{3}H]29-MOS.

side chains. A unique tetracyclic compound with a 6.6.5-fused A/B/C ring system and a pendant four-membered "D" ring has been also identified as a minor cyclization product of 20-oxa-2,3-oxidosqualene (Scheme 6).[145] Further, certain modifications at the C-29 methyl group were also acceptable; 29-hydroxy-2,3-oxidosqualene[205] and 29-trimethylsilyl-2,3-oxidosqualene[206] were all transformed to lanosterol analogues. 29-Methylidene-2,3-oxidosqualene, the mechanism-based irre-

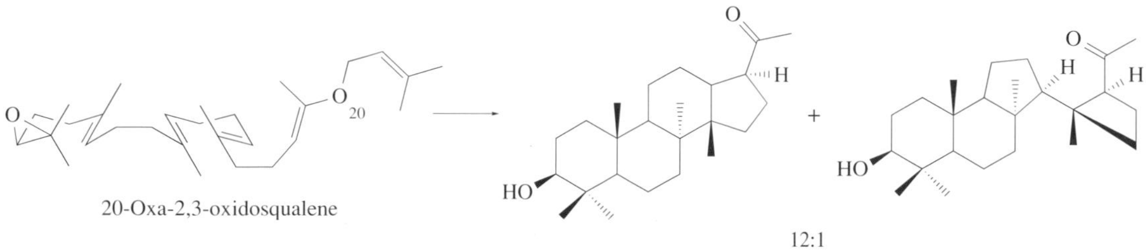

20-Oxa-2,3-oxidosqualene

12:1

Scheme 6

versible inhibitor, could also be cyclized if the concentration was less than the K_I value of 4.4 μM.[179] However, 29-cyclopropyl-2,3-oxidosqualene was not accepted as a substrate for cyclization.[206]

The importance of the Δ^{18} double bonds was also examined; for example, 18,19-dihydro-2,3-oxidosqualene, a substrate lacking the Δ^{18} double bond, was cyclized to 6.6.5-fused tricyclic product.[207] On the other hand, 15-desmethyl-18,19-dihydro-2,3-oxidosqualene was enzymatically transformed to a tricyclic product with a six-membered C-ring, a reaction that apparently involves hydrogen transfer from the side chain to the cationic C-ring intermediate (Scheme 7).[208] Further, the 2,3-oxidosqualene with the unnatural Z-stereochemistry at Δ^{18} was also cyclized to the mixture of 6.6.5-fused tricyclic products, both having *trans/syn/trans* A/B/C ring junctions (Scheme 8).[209] In contrast, truncated oxidosqualene analogues possessing Δ^{18} double bonds with the unnatural Z stereochemistry produced the norlanosterols having the unnatural 20*S* stereochemistry.[210,211] The presence of the Δ^{10} and Δ^{14} double bonds of 2,3-oxidosqualene were essential for the overall cyclization process. Neither 10,11-dihydro-2,3-oxidosqualene nor 14,15-dihydro-2,3-oxidosqualene afforded any cyclization products.[212]

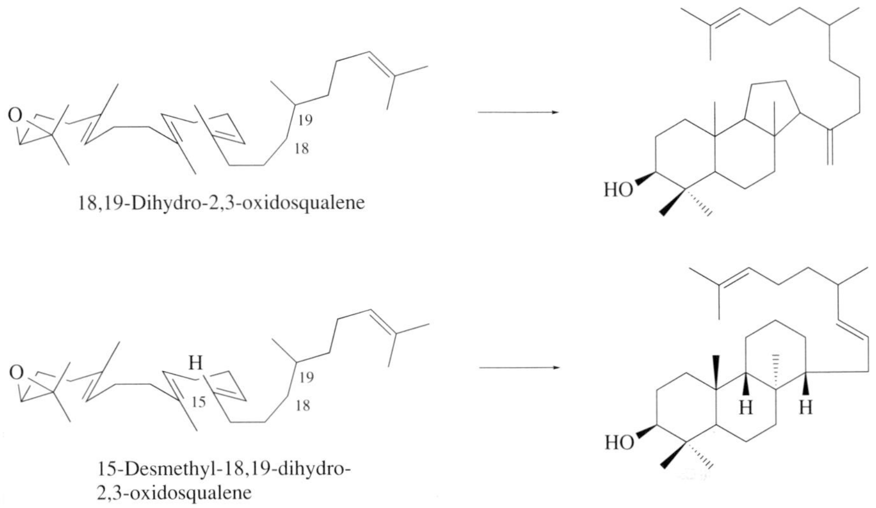

18,19-Dihydro-2,3-oxidosqualene

15-Desmethyl-18,19-dihydro-2,3-oxidosqualene

Scheme 7

18Z-2,3-Oxidosqualene

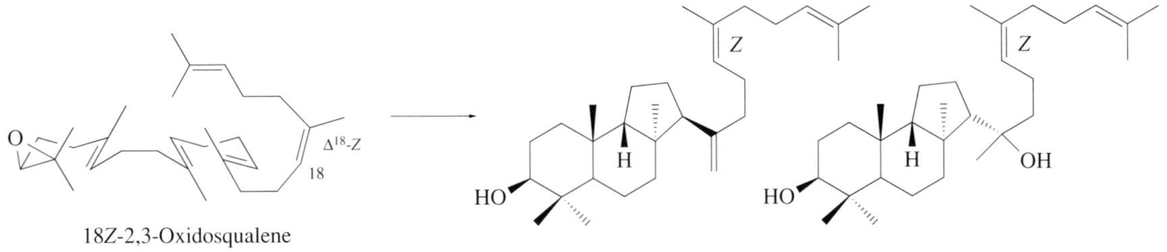

Scheme 8

The methyl group at C-10 is crucial to the correct folding of the substrate. 10,15-Didesmethyl-2,3-oxidosqualene was enzymatically converted to a unique tetracyclic compound with a 6.6.5-fused A/B/C ring system and a pendant four-membered "D" ring (Scheme 9).[213,214] Interestingly, 10,15-didesmethyl-2,3-oxidosqualene also functioned as an effective time-dependent irreversible inhibitor of the yeast OSLC.[211] Moreover, the liver enzyme cyclized 15-desmethyl-2,3-oxidosqualene to the 13-desmethyl lanosterol analogue almost as efficiently as it cyclized the natural substrate (Scheme 9).[215] Modifications of the methyl groups in 26-hydroxy-2,3-oxidosqualene,[205] 26-methylidene-2,3-oxidosqualene,[179] and 26-trimethylsilyl-2,3-oxidosqualene[206] were tolerated; in contrast, 26-cyclopropyl-2,3-oxidosqualene was not cyclized.[206] Substitution of the methyl groups with ethyl groups at C-10 and C-15 of oxidosqualene afforded only monocyclic products.[216] Furthermore, 27-methylidene-2,3-oxidosqualene was converted to a lanosterol analogue, which was the first demonstration that a vinyl substituent could also undergo an enzyme-mediated 1,2-shift instead of a hydrogen or methyl group (Scheme 10).[215] Finally, OSLC could also catalyze cyclization of partially cyclized mono- and bicyclic substrate analogues albeit in low yields.[218,219]

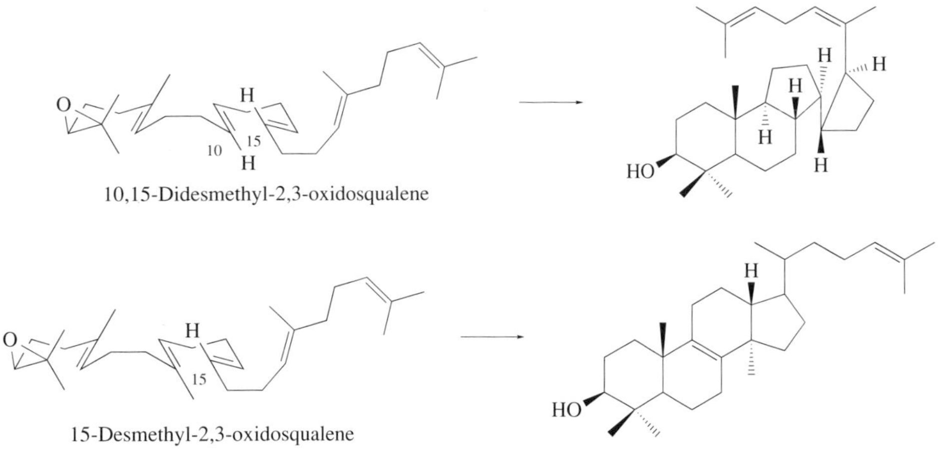

10,15-Didesmethyl-2,3-oxidosqualene

15-Desmethyl-2,3-oxidosqualene

Scheme 9

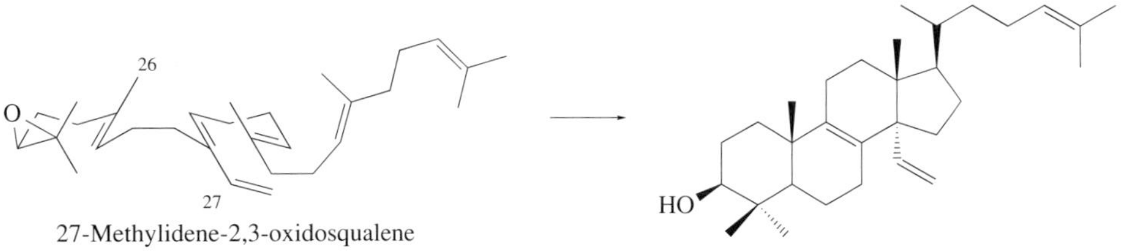

27-Methylidene-2,3-oxidosqualene

Scheme 10

With respect to the substituents on the oxirane ring of the substrate, OSLC tolerated modifications at the *pro*-α side of the incipient sterol "plane". However, stereoelectronic effects operating at the β-face of the conformationally folded substrate dramatically altered the course of the cyclization pathway.[220–222] For example, 2,3-*trans*-1-hydroxy-2,3-oxidosqualene[222] and 2,3-*trans*-1-methylidene-2,3-oxidosqualene,[223] were enzymatically converted to 4α-hydroxymethyl and 4α-vinyl analogues of lanosterol, respectively. In contrast, 2,3-*cis*-1-hydroxy-2,3-oxidosqualene afforded a partially cyclized bicyclic product (Scheme 11).[222] Interestingly, 1,1′-bisnor-2,3-oxidosqualene[145,224] and 2,3-*cis*-1′-nor-2,3-oxidosqualene were not cyclized, while 2,3-*trans*-1′-nor-2,3-oxidosqualene was cyclized.[220] Furthermore, neither 4-nor-2,3-oxidosqualene nor homooxidosqualene afforded any detectable cyclization products.[225] Previously, eucaryotic OSCs were thought to accept only (3*S*)-2,3-oxidosqualene, but not (3*R*)-enantiomer as a substrate.[226] However, it has since been reported that recombinant yeast OSLC could cyclize the (3*R*)-enantiomer to 3-*epi*-lanosterol at a rate of approximately 0.02 times that for (3*S*)-enantiomer.[145]

The first mechanism-based irreversible inhibitor of vertebrate OSLC, 29-methylidene-2,3-oxidosqualene (29-MOS) showed an IC_{50} value of 0.5 μM, an apparent K_i value of 4.4 μM, a k_{inact} value of 221 min^{-1}, and a partition ratio of 3.8 for pig liver OSLC.[176,225] Other 29-functionalized 2,3-oxidosqualene analogues such as 29-difluoromethylidene-2,3-oxidosqualene ($IC_{50} = 3.0$ μM, $K_i = 10.2$ μM for rat OSLC) were also irreversible inhibitors of OSLC, suggesting that initiation, not termination, of cyclization was rate-limiting for these suicide substrates.[228] In contrast, truncation of the side chain such as 29-difluoromethylidene-hexanor-2,3-oxidosqualene ($IC_{50} = 60$ μM for rat OSLC) suppressed inactivation and irreversibility.[228] Surprisingly, both 29-cyclopropyl-2,3-oxidosqualene and 26-cyclopropyl-2,3-oxidosqualene failed to act as irreversible inhibitors ($IC_{50} > 400$ μM); 29-hydroxy-2,3-oxidosqualene was quantitatively converted to the corresponding 21-hydroxy lanosterol analogue.[205] This compound was converted in five steps to 21-methylidene lanosterol, which was shown to be identical to the cyclization product of 29-MOS.[82] Therefore, the allylic 21-methylidene protosterol cation can undergo backbone rearrangement and release from the active site in competition with active site alkylation as described below.

Affinity labeling experiments using [^{3}H]29-MOS demonstrated that the vertebrate OSCs were specifically labeled even in crude microsomal preparations, giving single radioactive bands with apparent molecular masses ranging from 70 kDa to 80 kDa. In contrast, yeast and plant OSCs were not labeled, suggesting subtle species-specific differences in the active site structures of the enzymes.[165] (The authors found that bacterial SC from *A. acidocaldarius* was also specifically labeled with

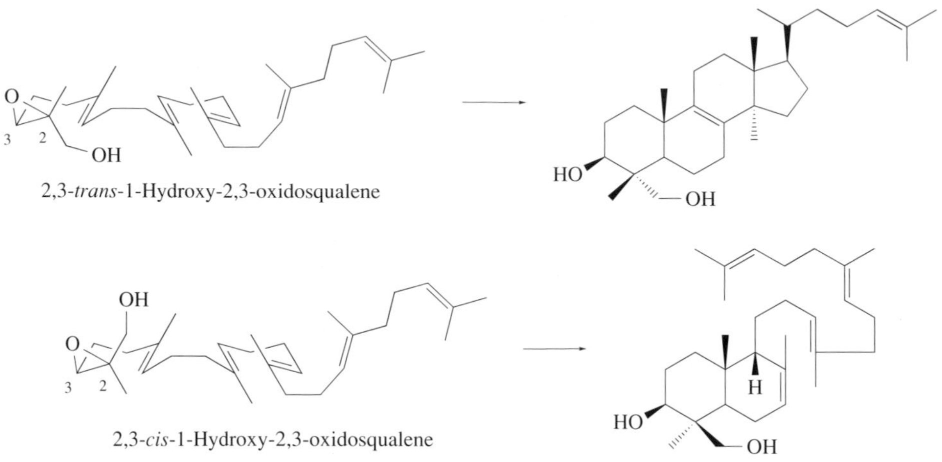

2,3-*trans*-1-Hydroxy-2,3-oxidosqualene

2,3-*cis*-1-Hydroxy-2,3-oxidosqualene

Scheme 11

[^{3}H]29-MOS.)[229] The proposed mechanism of inhibition involves initial cyclization to the 21-methylidene-protosterol cation as proposed for lanosterol formation. However, in addition to the backbone rearrangement that gives 21-methylidene lanosterol,[228] this allylic cation can be trapped by an active-site nucleophile, resulting in covalent bond formation and concomitant irreversible inactivation (Scheme 12).

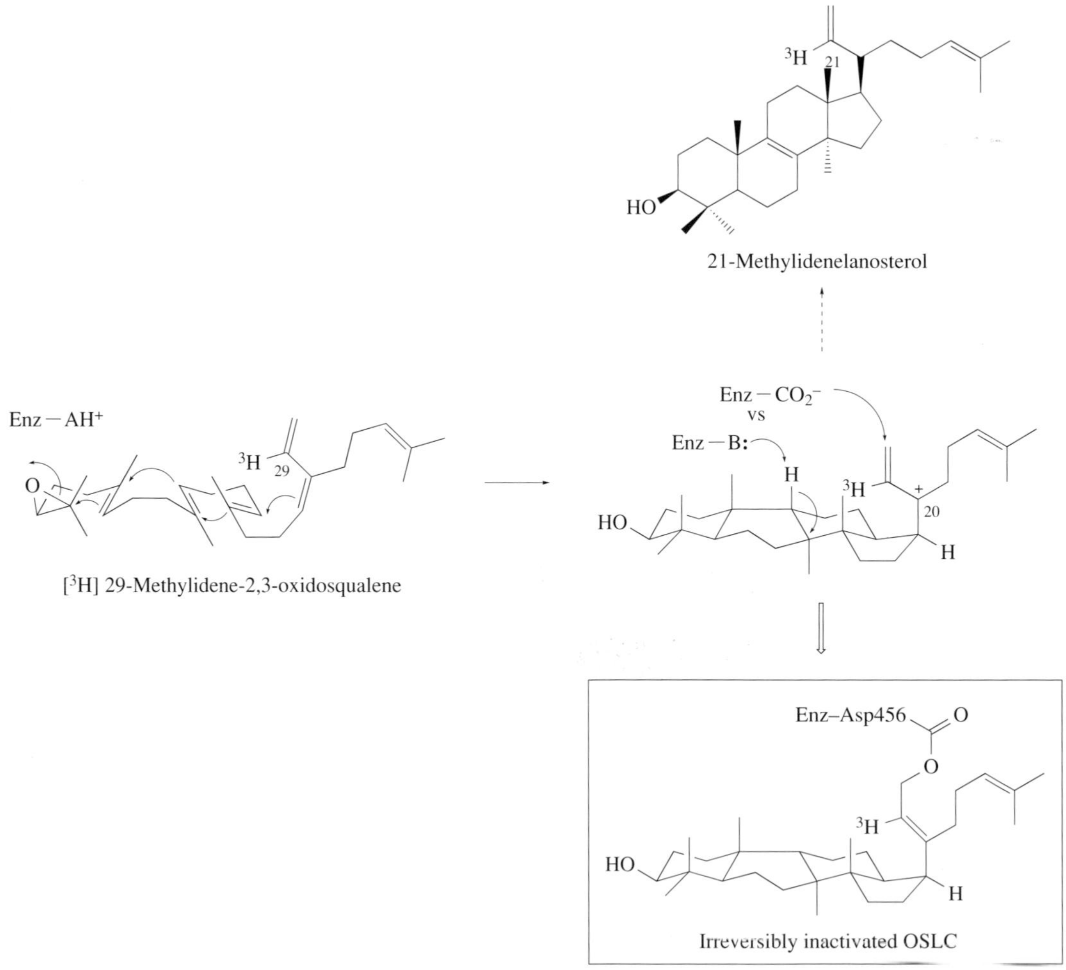

21-Methylidenelanosterol

[^{3}H] 29-Methylidene-2,3-oxidosqualene

Irreversibly inactivated OSLC

Scheme 12

Covalently modified peptide fragments were isolated and sequenced from purified rat liver OSLC.[150,151] The labeled peptide contained a DCTAE motif, just N-terminal of the QW-4 motif, which is well conserved in all the known OSCs despite the fact that neither yeast nor plant OSCs are labeled with [^{3}H]29-MOS. Radiosequencing revealed that the Asp residue of the DCTAEA motif was labeled with the suicide substrate.[150,152] The anionic residue, D-456, is thus implicated in stabilization of the C-20 cationic center of the protosterol cation prior to the backbone rearrangement. Site-directed mutagenesis experiments will provide a further test for this hypothesis. Griffin and co-workers have carried out the site-directed mutagenesis experiments in this region of yeast OSLC (V454A, D456A, C457A, T458A, and E460A) using a yeast mutant strain bearing a genetic disruption of the *ERG7* gene, and demonstrated that the V454A and D456A mutation failed to complement the *ERG7* disruption.[191] Furthermore, Poralla and co-workers have reported that a point mutation of the Asp residues of the corresponding DDTAVV motif of bacterial *A. acidocaldarius* SC caused almost complete loss of cyclase activity.[230] An earlier report of the ability of a carboxylic group modifying reagent, such as Belleau's reagent N-(ethoxycarbonyl)-2-ethoxy-1,2-dihydroquinoline, to inactivate the microsomal pig OSLC, may also relate to this critical Asp residue.[168]

As mentioned, the presence of a cysteinyl residue essential for catalysis in the active site of the OSC has been proposed. A thiol-modifying reagent, *N*-ethylmaleimide, showed a potent, time-dependent inhibition toward partially purified pig OSLC ($K_d = 2.9$ mM, $k_{inact} = 0.089$ min^{-1}).[168] However, the presence of a reactive SH group at the active site of the enzyme could not be confirmed since no protection against inactivation was observed in the presence of a competitive inhibitor as a protecting reagent. Recently, a series of squalene derivatives have been designed by Cattel and co-workers that could covalently modify an active site SH group (Figure 15).[197,231] Among them, squalene maleimide was a time-dependent irreversible inhibitor of vertebrate OSLC. On the other hand, the Ellman-type 3-carboxy-4-nitrophenyl-dithio-1,1′,2-trisnorsqualene (CNDT-TNS) was found to be an irreversible inhibitor of yeast OSLC ($K_i = 1.2$ mM, $k_{inact} = 0.220$ min^{-1}). In both cases, substrate protection was observed to a rather limited extent, providing partial support for the existence of an essential thiol group within the active site of the enzyme. However, the low k_{inact} values and bulky groups in place of the compact oxirane ring cast doubt on the specificity of this

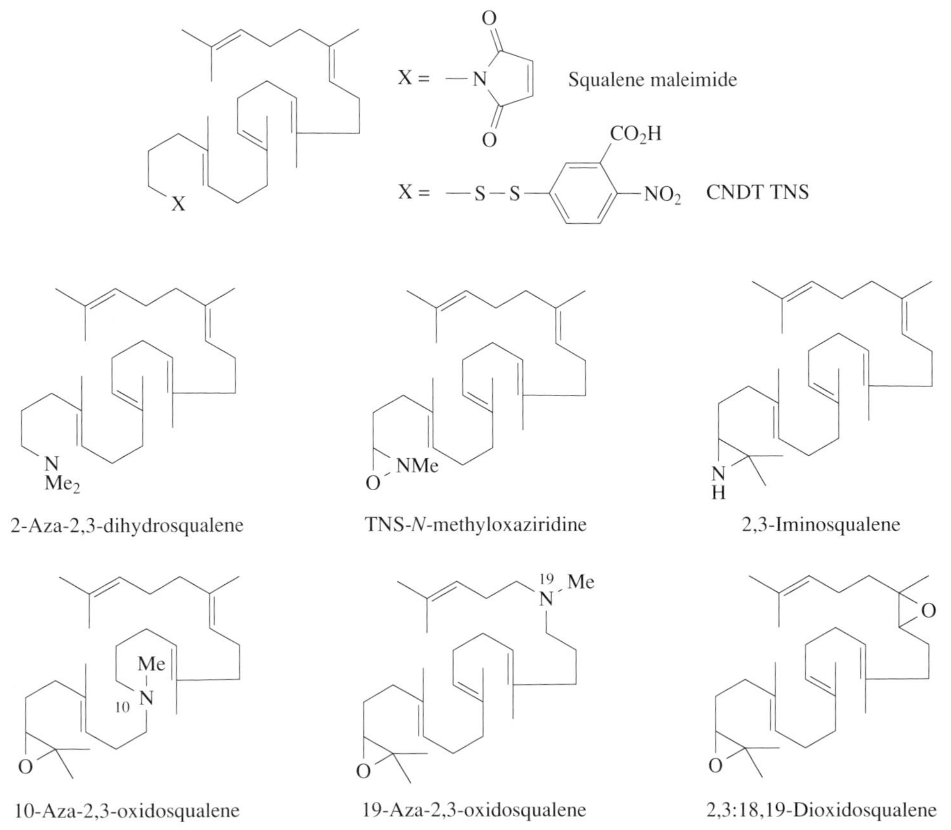

Figure 15 OSC inhibitors: acyclic OS analogues.

reagent for the substrate-binding site. Rat OSLC has three conserved Cys residues in the sequence, so that site-directed mutagenesis experiments may provide further information on the function of these residues.

Azasqualene analogues possessing nitrogen atoms situated at positions corresponding to the carbenium ion of high energy intermediates or transition states involved during cyclization of 2,3-oxidosqualene have been shown to inhibit cyclase activity efficiently (Figure 16).[197,232–235] At physiological pH, the secondary and tertiary amines would be protonated and would resemble the structure and charge of the incipient carbocationic intermediate. 2-Aza-2,3-dihydrosqualene (IC_{50} = 7.5 μM for rat OSLC), its N-oxide (IC_{50} = 3.7 μM for rat OSLC), and their derivatives are good inhibitors of vertebrate OSLC.[234–236] They also inhibit fungal and plant OSCs, and as the free amine forms, show modest inhibition of SE.[197] Trisnorsqualene *N*-methyloxaziridine showed highest inhibitory activity among the azasqualene derivatives (IC_{50} = 1.5 μM for rat OSLC; IC_{50} = 2.5 μM for yeast OSLC); however, the oxaziridine was unstable in aqueous media and proved to be a modest inhibitor of cholesterol production in 3T3 fibroblast cultures.[237] 10-Aza-2,3-oxidosqualene was designed as a cyclase-activated high energy intermediate that could mimic the bicyclic C-8 carbonium ion formed during OS cyclization[238–240] Only the (6*E*) isomer strongly inhibited OSLCs from vertebrates, yeast, and fungi (IC_{50} values varying from 3 to 5 μM). Similarly, 19-aza-2,3-oxidosqualene was designed as a cyclase-activated high energy intermediate analogue to mimic the tetracyclic C-20 carbocation intermediate during oxidosqualene cyclization, and showed potent, selective inhibition of vertebrate OSLC (IC_{50} = 1.5 μM for pig OSLC); it did not inhibit yeast or plant OSCs.[197,240,241] Finally, the classical, 2,3-iminosqualene remains one of the most potent inhibitors of OSLC (IC_{50} = 0.4 μM for rat OSLC);[90,242] nonetheless, it is epoxidized and subsequently cyclized to 24,25-iminolanosterol in fungi.[243]

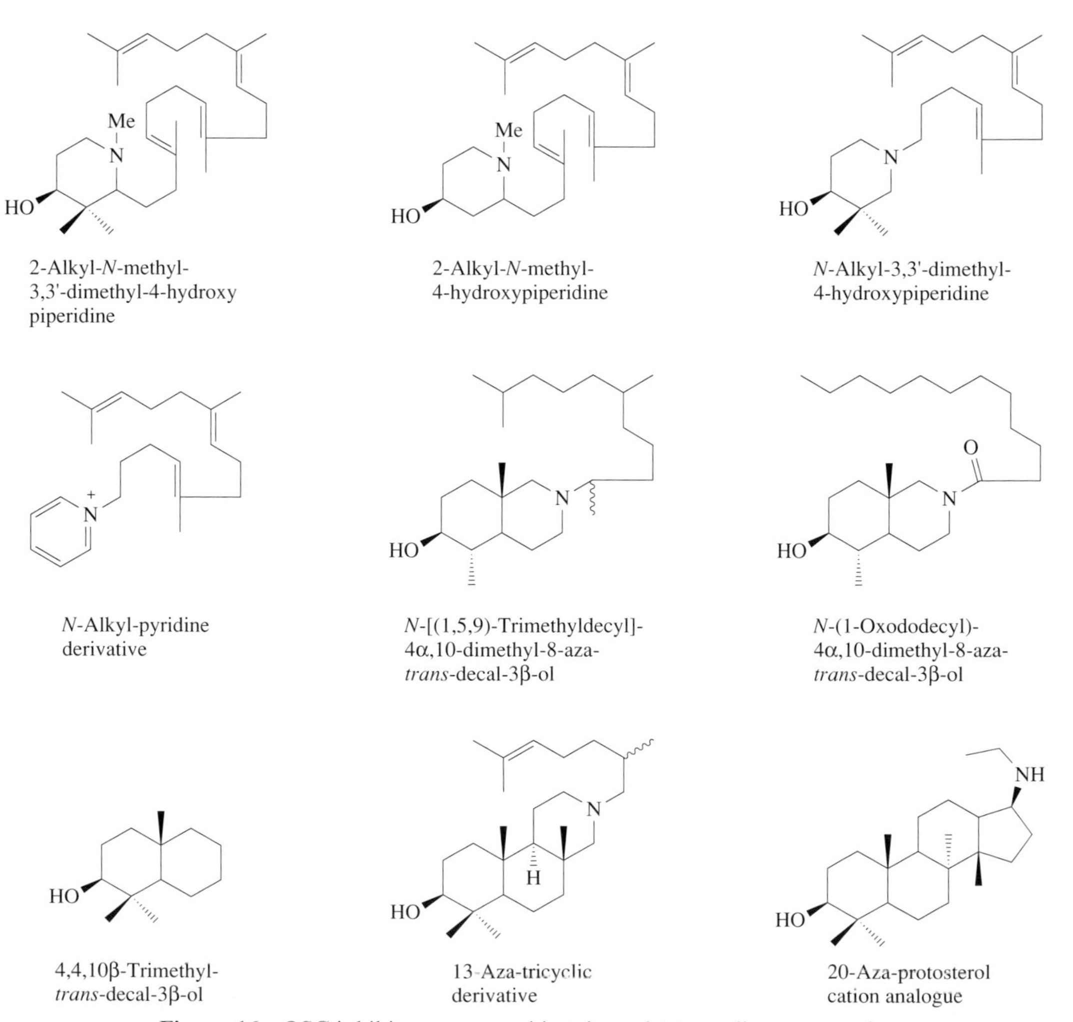

Figure 16 OSC inhibitors: mono-, bi-, tri-, and tetracyclic compounds.

Similarly, monocyclic, bicyclic, and tricyclic compounds, have been synthesized with a nitrogen atom situated at *pro*-C-8, *pro*-C-10, and *pro*-C-13, respectively, a position corresponding to the carbonium ion of high energy intermediates or transition states involved during cyclization of 2,3-oxidosqualene. Each was shown to be a potent inhibitor of OSC (Figure 16). Monocyclic 2-alkyl-*N*-methyl-3,3′-dimethyl-4-hydroxypiperidine (IC_{50} = 1.4 μM for pig OSLC),[244,245] and 2-alkyl-*N*-methyl-4-hydroxypiperidine (IC_{50} = 0.3 μM for pig OSLC)[245,246] showed strong inhibition. Another monocyclic *N*-alkyl-3,3′-dimethyl-4-hydroxypiperidine was shown to be a strong inhibitor of plant OSC, with an IC_{50} value of 1 μM for maize seedling cycloartenol cyclase.[247] A pyridine-containing OSC inhibitor [*N*-(4*E*,8*E*)-5,9,13-trimethyl-4,8,12-tetradecatrien-l-yl]pyridinium cation inhibits OSC (IC_{50} = 0.36 μM for *C. albicans* OSLC) at concentrations more than 100-fold lower than does the directly analogous piperidinium derivative.[248,249] A bicyclic 8-azadecalin derivative, *N*-(1,5,9)-trimethyldecyl-4α,10-dimethyl-8-aza-*trans*-decal-3β-ol (IC_{50} = 2 μM for rat OSLC) was designed to mimic a high-energy intermediate bearing a positive charge at C-8; this material was a strong inhibitor of plant cycloartenol cyclase, but did not inhibit β-amyrin synthase.[250] The 8-azadecalin was found to be a potent inhibitor of cholesterol biosynthesis in 3T3 fibroblasts.[251] In addition, treatment of 3T3 fibroblasts with 8-azadecalin also resulted in a marked reduction in HMG-CoA reductase activity through accumulation of dioxidosqualene and oxysterols.[251] Several azadecalin analogues have been synthesized and tested for their activities.[252–255] Among them, a bicyclic amide analogue, *N*-(1-oxododecyl)-4α,10-dimethyl-8-aza-*trans*-decal-3β-ol, has been reported to be a ten-fold more potent, competitive inhibitor of OSLC *in vitro* (IC_{50} = 0.11 μM for rat OSLC); it also inhibited cholesterol biosynthesis in HepG2 cells (IC_{50} = 0.70 μM).[252,256] The prototypical *trans*-decalin derivative, 4,4,10β-trimethyl-*trans*-decal-3β-ol (IC_{50} = 9μM for rat OSLC) showed lower potency than 8-azadecalin.[88,89,235] Finally, 13-aza-tricyclic derivatives displayed little inhibition of OSCs (IC_{50} > 300 μM for rat OSLC).[247] Relatively modest potency of inhibition by a 20-aza analogue of the protosterol cation (IC_{50} = 22 μM for yeast OSLC) suggested that the protosterol cation is not strongly bound by the enzyme.[257]

2,3:18,19-Dioxidosqualene (DOS) (Figure 16), which was synthesized and isolated as a mixture of four diastereoisomers, proved to be one of the best inhibitors of OSLC (IC_{50} = 0.11 μM for rat OSLC).[256] The regioisomers 2,3:6,7-DOS, 2,3:10,11-DOS, and 6,7:14,15-DOS also showed good inhibitory activity with IC_{50} values of 21, 13, and 9 μM, respectively, for rat OSLC.[258] The fact that these compounds could be potentially generated *in vivo* constitutes a remarkable difference relative to other OSLC inhibitors. Indeed, incubation of 6,7-oxidosqualene or 10,11-oxidosqualene with rat liver microsomes led to the formation of mixtures of the corresponding DOS, resulting from the epoxidation at their terminal double bonds.[91,92] Recently, synthesis of each of the four stereoisomers of 2,3:18,19-DOS has been reported.[259] Inhibition assay with purified pig liver OSC demonstrated that (3*S*,18*R*,19*R*)DOS was the best inhibitor (IC_{50} = 27 nM, K_i = 21 nM), while (3*S*,18*S*,19*S*)DOS also showed high activity (IC_{50} = 62 nM, K_i = 76 nM). In contrast, (18*E*)- and (18*Z*)-20-oxa-22,23-dihydro-2,3-oxidosqualene were found to be only weakly active competitive inhibitors (IC_{50} = 80 and 120 μM, and K_i = 40 and 60 μM, respectively, for rat OSLC).[260] A similar analogue, (18*E*)-20-oxa-2,3-oxidosqualene, was enzymatically converted to a protosterol derivative.[160] Recently, synthesis of (15*E*)-16-oxa-2,3-oxidosqualene and its (15*Z*)-isomer has been reported.[261] The 16-oxa analogue might be useful for trapping of the 6.6.5. tricyclic, Markovnikov-type intermediate cation during lanosterol formation.

A number of sulfur-containing 2,3-oxidosqualene analogues were synthesized by the Oehlschlager group, and determined to be extremely potent inhibitors of OSLCs (Figure 17).[262–265] In particular, oxidosqualene analogues in which sulfur has replaced carbons C-18 and C-19 were reported to have subnanomolar IC_{50} values for both vertebrate and fungal OSCs: S-18 (IC_{50} = 0.08 nM for rat OSLC; 0.22 nM for *C. albicans* OSLC) and S-19 (IC_{50} = 0.82 nM for rat OSLC; 3.2 nM for *C. albicans* OSLC).[263,265] Other thia oxidosqualenes with sulfur substitutions at C-6, C-8, C-10, C-11 and C-20 showed potent inhibition with the IC_{50} values at nanomolar level (IC_{50} = 69 nM, 680 nM, 69 nM, 540 nM, and 200 nM, respectively for *C. albicans* OSLC). Sulfoxides were generally much less active than corresponding thioethers. The reasons for the ca. 1000-fold higher potency relative to other potent inhibitors of OSLC such as 2,3:18,19-dioxidosqualene (IC_{50} = 0.11 μM for rat OSLC) are unclear. Indeed, these unusually low IC_{50} values have been revised in light of data obtained with purified rat OSC: S-18 (IC_{50} = 0.05 μM for rat OSLC) and S-19 (IC_{50} = 0.26 μM for rat OSLC).[265] Inhibition kinetics with purified vertebrate OSLCs demonstrated that the S-18 compound was a time-dependent, irreversible inhibitor of pig OSLC (Ki = 1.5 μM, k_{inact} = 0.06 min^{-1}, partition ratio = 16), while the inhibition by S-19 was reversible and not time-dependent.[265] Surprisingly, the inhibition of rat OSLC by the S-18 compound was reversible and more potent (K_i = 0.037 μM). It was recently demonstrated that pig OSLC could be covalently modified with [17-^{3}H]S-18 and

[22-^{3}H]S-18, two tritium-labeled radioisotopomers of S-18.[266] Covalent modification of OSC would require partial cyclization of S-18 at the active site of the enzyme with trapping of a cationic intermediate by an active-site nucleophile. Retention of the tritium label for both radioisotopomers excluded a possibility of an attack at C-20 with transfer of the side chain to the active site; alternatively, nucleophilic trapping could occur on a bicyclic or tricyclic intermediate (Scheme 13).

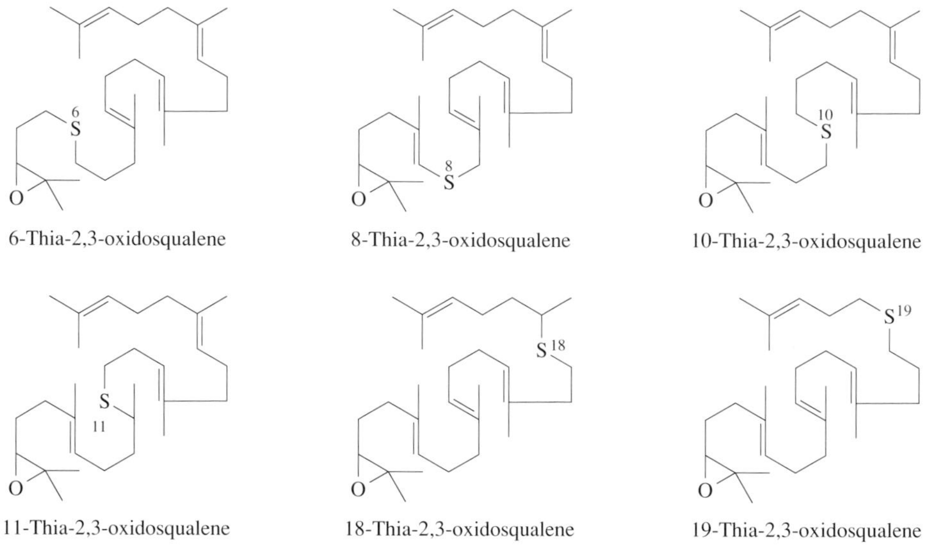

6-Thia-2,3-oxidosqualene 8-Thia-2,3-oxidosqualene 10-Thia-2,3-oxidosqualene

11-Thia-2,3-oxidosqualene 18-Thia-2,3-oxidosqualene 19-Thia-2,3-oxidosqualene

Figure 17 OSC inhibitors: sulfur-containing OS analogues.

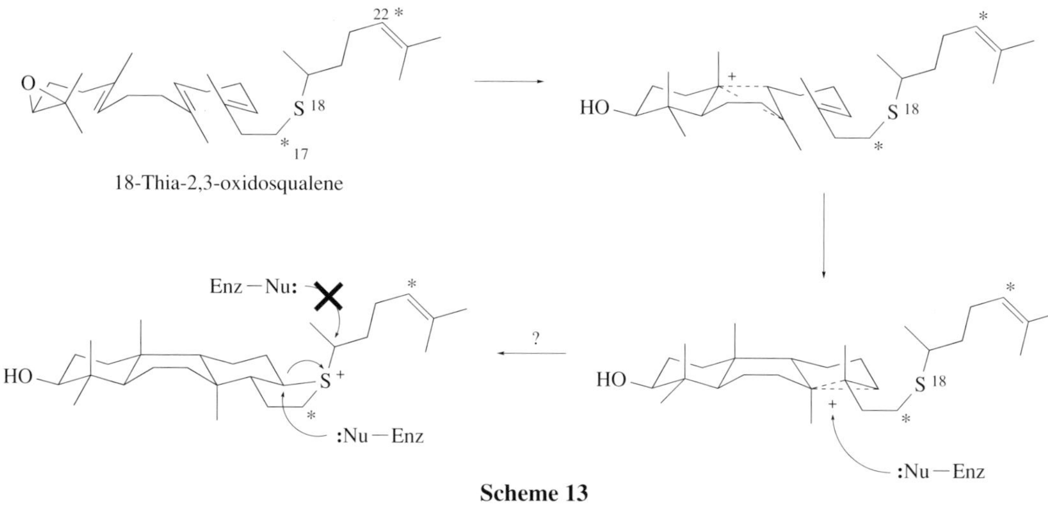

18-Thia-2,3-oxidosqualene

Enz—Nu: :Nu—Enz :Nu—Enz

Scheme 13

Scientists at Hoffman-La Roche reported that a series of benzophenone-containing compounds such as Ro-43-6300 (IC_{50} = 0.7 μM for *C. albicans* OSLC) (Figure 18) were potent inhibitors of fungal OSLCs (inhibition activities toward vertebrate OSLC were not reported).[270] The inhibitory activity was dependent on the distance between the nitrogen atom of the tertiary amine and the carbonyl group of the benzophenone moiety. The similarity in shape and size between the protosterol cation and the benzophenone inhibitor has been pointed out.[267] Another aromatic analogue that might adopt a conformation to allow a kind of high-energy intermediate mimicry, also showed strong inhibition of fungal OSLC (IC_{50} = 2.0 μM for *C. albicans* OSLC).[268] Other potent inhibitors, such as *N*-(1-*n*-dodecyl)-imidazole (IC_{50} = 3.9 μM for rat OSC) and its derivatives,[269] U18666A (3*β*-[2-(diethylamino)ethoxy]androst-5-en-17-one) (IC_{50} = 0.8 μM for rat OSLC),[235] a known inhibitor of sterol biogenesis in vertebrate and yeast cells, were also shown to be excellent inhibitors of vertebrate enzyme. Incubation of Hep G2 cells with U18666A resulted in a biphasic response of HMG-CoA reductase activity; reductase activity decreased at concentrations lower than 3 μM, but increased at higher concentrations.[270,271] AMO1618, a potent plant growth retardant, did not inhibit vertebrate OSLC.[233]

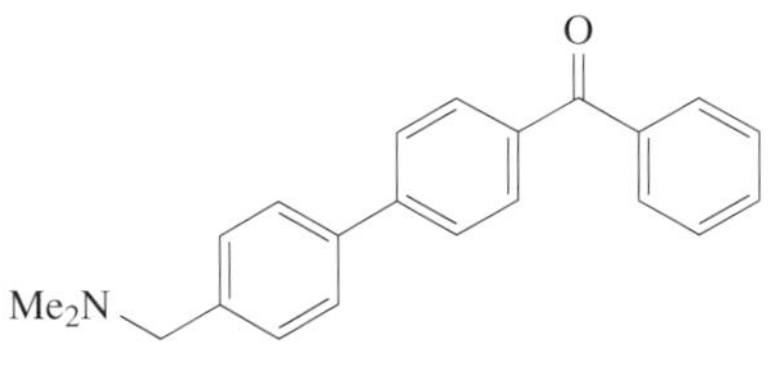

Hoffman-La Roche
Ro-43-6300

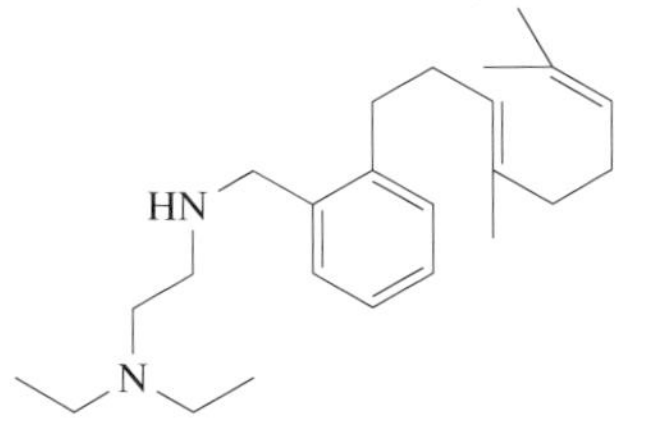

Hoffman-La Roche's
Aromatic ring-containing analogue

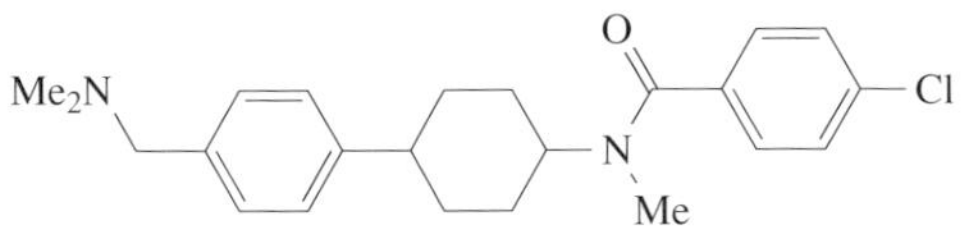

Thomae BIBX 79

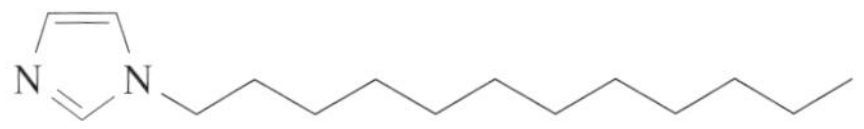

N-dodecylimidazole

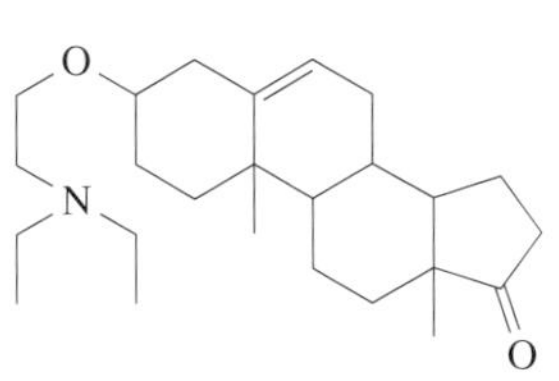

U18666A

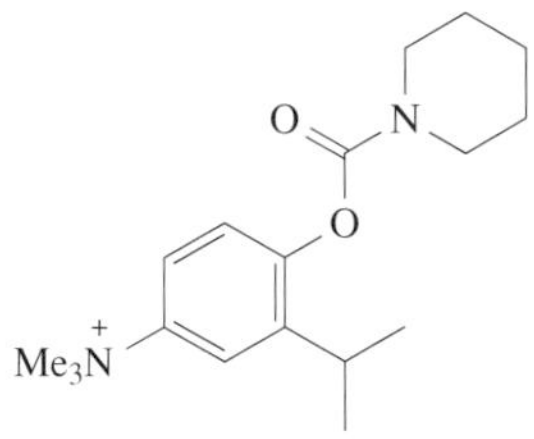

AMO1618

Figure 18 OSC inhibitors: aromatic-ring and miscellaneous compounds.

The recently disclosed compound BIBX 79 from K. Thomae GmbH (Figure 18) was also designed as an analogue of the protosterol cation, and was found to be an extremely potent, specific inhibitor of OSLC ($IC_{50} = 6$ nM for human Hep G2 OSLC).[270] In Hep G2 cells, BIBX 79 strongly inhibited cholesterol synthesis ($IC_{50} = 4$ nM). Within a concentration range of 10^{-7} to 10^{-9} M, in which a partial inhibition of OSLC was observed, HMG-CoA reductase activity was also decreased; at higher concentrations of BIBX 79 that totally blocked OSLC, an increase in HMG-CoA reductase activity was found.[272] This effect of BIBX 79 on HMG-CoA reductase is thought to be mediated by oxysterols that are formed by the cyclization of 2,3:22,23-DOS as described below. BIBX 79 is thus expected to be a highly promising therapeutic agent.

Inhibition of OSLC is known to result in the accumulation of 2,3-OS and (3*S*,22*S*)-2,3:22,23-DOS. As described above, SE catalyzes the conversion of 2,3-OS to 2,3:22,23-DOS.[42,43] Recent kinetic studies have shown cyclization of 2,3:22,23-DOS in preference to 2,3-OS by liver OSLC to give (24*S*)-24,25-epoxylanosterol.[199,272] The epoxylanosterol was further converted to (24*S*)-24,25-epoxycholesterol, which is proposed to regulate the coordinate expression of the genes for HMG-CoA reductase, the LDL-receptor, and other enzymes in the cholesterol biogenesis pathway.[44,45,273–275] Indeed, when Hep G2 cells were treated with 2,3:22,23-DOS or (24*S*)-24,25-epoxycholesterol, almost complete suppression of HMG-CoA reductase activity was observed.[51] This indicates that, at a cellular level, the OSLC inhibitors might act synergistically through a combination of two mechanisms: (i) by decreasing the amount of lanosterol formed, and (ii) by suppressing the HMG-CoA reductase through the formation of oxysterols such as (24*S*)-24,25-epoxycholesterol.[199] This could make the OSLC a particularly attractive target enzyme for the design of cholesterol-lowering drugs. In order to assess the cyclase as a potential pharmaceutical target, it will be important to determine whether inhibitors are able to act synergistically on cholesterol biosynthesis by repressing the regulatory HMG-CoA reductase.[51]

2.10.4 CONCLUSIONS

In the last several years, there have been significant advances in understanding the structures and mechanisms of SE and OSLC. However, many questions remain unanswered, particularly in the areas of structure, mechanism, and regulation. First of all, elucidation of the three-dimensional structures of these enzymes are prerequisite for further understanding of the molecular interactions involved in substrate binding and catalysis. For example, elucidation of the active site geometry of the SE–P450R–FAD–NADPH–substrate complex in squalene epoxidation and the role of the repeated QW motifs in cyclization of squalene and oxidosqualene invite experimental efforts. Structural information about the active site of the enzyme should facilitate rational design of new, more selective inhibitors of the enzymes.

Second, there are still many fundamental questions of enzyme mechanism that remain unresolved. For example, the actual mechanism of the flavoprotein-mediated epoxidation reaction and the electron transfer from P450R to SE enzyme are poorly understood at present. Further, in lanosterol formation, it will be of interest to know why the backbone rearrangement takes place after cyclization reaction, and why partially cyclized intermediates cannot be isolated during the reaction. How are the transition states and high-energy intermediates stabilized by the active site of the enzyme? The relationship between cyclization mechanism and active site structure of the various OSC and SC enzymes presents a formidable challenge.

Finally, the regulation of sterol biosynthesis in plants, fungi, and animals is a prime target for selective agricultural and pharmaceutical compounds. It will be necessary to study how these enzymes are expressed, how their enzyme activities are regulated *in vivo*, how they interact with other enzymes in the isoprenoid metabolism, and where they are localized within the cell.

ACKNOWLEDGMENTS

We express our appreciation to an excellent group of co-workers at Stony Brook and the University of Utah whose contributions are cited in the text, in particular Professor Stephanie E. Sen, Dr. Xiao-yi Xiao, Dr. Mei Bai, Dr. Pamela Denner-Ancona, Ms. Brenda A. Madden, Ms. Hee-Kyoung Lee, Dr. Dale M. Marecak, Dr. Yi Feng Zheng and Mr. Tongyun Dang. Collaborations with the research teams of Professors Teruo Ono, Karl Poralla, John H. Griffin, and Allan C. Oehlschlager have greatly expanded our horizons. Financial support of work at Stony Brook, now being continued at The University of Utah, has been provided by the Center for Biotechnology/New York State Science and Technology Foundation and the National Institutes of Health, with earlier support from Kirin Breweries, Inc. and Sandoz Research Institute. This contribution is dedicated to the memory of William S. Johnson, a friend and mentor who introduced one of us (G.D.P.) to the mystique of polyolefin cyclizations.

2.10.5 REFERENCES

1. I. Abe, M. Rohmer, and G. D. Prestwich, *Chem. Rev.*, 1993, **93**, 2189.
2. J. L. Goldstein and M. S. Brown, *Nature*, 1990, **343**, 425.
3. I. Abe, J. C. Tomesch, S. Wattanasin, and G. D. Prestwich, *Nat. Prod. Rep.*, 1994, **11**, 279.
4. T. Ono, S. Ozasa, F. Hasegawa, and Y. Imai, *Biochim. Biophys. Acta*, 1977, **486**, 401.
5. M. Nakamura and R. Sato, *Biochem. Biophys. Res. Commun.*, 1979, **89**, 900.
6. J. B. Ferguson and K. Bloch, *J. Biol. Chem.*, 1977, **252**, 5381.
7. Y. A. Saat and K. Bloch, *J. Biol. Chem.*, 1976, **251**, 5155.
8. I. W. Caras and K. Bloch, *J. Biol. Chem.*, 1979, **254**, 11 816.
9. E. J. Friedlander, I. W. Caras, L. Fen, H. Lin, and K. Bloch, *J. Biol. Chem.*, 1980, **255**, 8042.
10. J. L. Vermilion, D. P. Ballou, V. Massey, and M. J. Coon, *J. Biol. Chem.*, 1981, **256**, 266.
11. T. D. Porter and C. B. Kasper, *Proc. Natl. Acad. Sci. USA*, 1985, **82**, 973.
12. A. L. Shen and C. B. Kasper, in "Cytochrome P-450," eds. A. L. Shen and C. B. Kasper, Springer-Verlag, New York, 1993, p. 35.
13. G. C. M. Smith, D. G. Tew, and C. R. Wolf, *Proc. Natl. Acad. Sci. USA*, 1994, **91**, 8710.
14. T. D. Porter and C. B. Kasper, *Biochemistry*, 1986, **25**, 1682.
15. A. L. Shen, T. D. Porter, T. E. Wilson, and C. B. Kasper, *J. Biol. Chem.*, 1989, **264**, 7584.
16. A. L. Shen, M. J. Christensen, and C. B. Kasper, *J. Biol. Chem.*, 1991, **266**, 19 976.
17. D. S. Sem and C. B. Kasper, *Biochemistry*, 1993, **32**, 11 548.
18. S. Djordjevic, D. L. Roberts, M. Wang, T. Shea, M. G. W. Camitta, B. S. S. Masters, and J. J. P. Kim, *Proc. Natl. Acad. Sci. USA*, 1995, **92**, 3214.
19. Q. Zhao, G. Smith, S. Modi, M. Paine, R. C. Wolf, D. Tew, L.-Y. Lian, W. U. Primrose, G. C. Roberts, and H. P. C. Driessen, *J. Structural Biol.*, 1996, **116**, 320.
20. S. D. Black, J. S. French, C. H. J. Williams, Jr., and M. J. Coon, *Biochem. Biophys Res. Commun.*, 1979, **91**, 1528.
21. Y. Nisimoto, *J. Biol. Chem.*, 1986, **261**, 14 232.
22. A. L. Shen and C. B. Kasper, *J. Biol. Chem.*, 1995, **270**, 27 475.
23. S. G. Nadler and H. W. Strobel, *Arch. Biochem. Biophys.*, 1991, **290**, 277.

24. T. Shimizu, T. Tateishi, M. Hatano, and Y. Fujii-Kuriyama, *J. Biol. Chem.*, 1991, **266**, 3372.
25. A. I. Voznesensky and J. B. Schenkman, *J. Biol. Chem.*, 1992, **267**, 14 669.
26. A. I. Voznesensky and J. B. Schenkman, *Eur. J. Biochem.*, 1992, **210**, 741.
27. Y. Nisimoto and D. E. Edmondson, *Eur. J. Biochem.*, 1992, **204**, 1075.
28. A. I. Voznesensky and J. B. Schenkman, *J. Biol. Chem.*, 1994, **269**, 15 724.
29. D. S. Sem and C. B. Kasper, *Biochemistry*, 1995, **34**, 12 768.
30. K. Büch, H. Stransky, and A. Hager, *FEBS Lett.*, 1995, **376**, 45.
31. V. Massey, *J. Biol. Chem.*, 1994, **269**, 22 459.
32. Y. Hidaka, T. Satoh, and T. Kamei, *J. Lipid Res.*, 1990, **31**, 2087.
33. T. Satoh, Y. Hidaka, and T. Kamei, *J. Lipid Res.*, 1990, **31**, 2095.
34. J. Sakakibara, and T. Ono, *Tanpakushitu Kakusan Kouso* (*Protein Nucleic Acid Enzyme*), 1994, **39**, 1508.
35. R. Gonzalez, J. P. Carlson, and M. E. Dempsey, *Arch. Biochem. Biophys.*, 1979, **196**, 574.
36. P. Nakamura, J. Sakakibara, T. Izumi, A. Shibata, and T. Ono, *J. Biol. Chem.*, 1996, **271**, 8053.
37. Y. Hidaka, H. Hotta, Y. Nagata, Y. Iwasawa, M. Horie, and T. Kamei, *J. Biol. Chem.*, 1991, **266**, 13 171.
38. M. S. Brown and J. L. Goldstein, *J. Lipid Res.*, 1980, **21**, 505.
39. T. F. Osborne, G. Gil, J. L. Goldstein, and M. S. Brown, *J. Biol. Chem.*, 1988, **263**, 3380.
40. L. H. Cohen and M. Griffioen, *Biochem. J.*, 1988, **255**, 61.
41. M. Nakanishi, J. L. Goldstein, and M. S. Brown, *J. Biol. Chem.*, 1988, **263**, 8929.
42. M. Bai, X.-y. Xiao, and G. D. Prestwich, *Biochem. Biophys. Res. Commun.*, 1992, **185**, 323.
43. A. Nagumo, T. Kamei, J. Sakakibara, and T. Ono, *J. Lipid Res.*, 1995, **36**, 1489.
44. T. A. Spencer, *Acc. Chem. Res.*, 1994, **27**, 83.
45. L. L. Smith and B. J. Johnson, *Free Rad. Biol. Med.*, 1989, **7**, 285.
46. S. Yamamoto and K. Bloch, *J. Biol. Chem.*, 1970, **245**, 1670.
47. H.-H. Tai and K. Bloch, *J. Biol. Chem.*, 1972, **247**, 3767.
48. T. Ono and K. Bloch, *J. Biol. Chem.*, 1975, **250**, 1571.
49. T. Ono, K. Takahashi, S. Odani, H. Konno, and Y. Imai, *Biochem. Biophys. Res. Comm.*, 1980, **96**, 522.
50. T. Ono and Y. Imai, *Methods Enzymol.*, 1985, **110**, 375.
51. O. Boutaud, M. Ceruti, L. Cattel, and F. Schuber, *Biochem. Biophys. Res. Commun.*, 1995, **208**, 42.
52. F. P. Guengerich, *Crit. Rev. Biochem. Mol. Biol.*, 1990, **25**, 97.
53. M. Bai and G. D. Prestwich, *Arch. Biochem. Biophys.*, 1992, **293**, 305.
54. P. Denner-Ancona, M. Bai, H.-K. Lee, I. Abe, and G. D. Preswich, *BioMed. Chem. Lett.*, 1995, **5**, 481.
55. J. Sakakibara, R. Watanabe, Y. Kanai, and T. Ono, *J. Biol. Chem.*, 1995, **270**, 17.
56. K. Kosuga, S. Hata, T. Osumi, J. Sakakibara, and T. Ono, *Biochim. Biophys. Acta*, 1995, **1260**, 345.
57. A. Jandrositz, F. Turnowsky, and G. Högenauer, *Gene*, 1991, **107**, 155.
58. N. Ishii, M. Yamamoto, M. Arisawa, and Y. Aoki, *GenBank Data Base*, 1996, accession no. D88252.
59. B. Favre and N. S. Ryder, *Gene*, 1997, **189**, 119.
60. The WashU-Merck Project, *GenBank Data Base*, 1995, accession nos T66122 and R50895.
61. H.-K. Lee and G. D. Prestwich, unpublished results.
62. K. M. Landl, B. Klösch, and F. Turnowsky, *Yeast*, 1996, **12**, 609.
63. M. Nagai, J. Sakikibara, K. Wakui, Y. Fukushima, S. Igarashi, S. Tsuji, M. Arakawa, and T. Ono, *Genomics*, 1997, **44**, 141.
64. I. Abe, P. Denner-Ancona, H.-K. Lee, and G. D. Prestwich, unpublished results.
65. D. E. Edmonton and R. F. De Francesco, in "Chemistry and Biochemistry of Flavoenzymes," eds. D. E. Edmonton and R. F. De Francesco, CRC Press, Boca Raton, FL, 1991, Vol. I, p. 73.
66. T. P. Singer and W. S. McIntire, *Methods Enzymol.*, 1984, **106**, 369.
67. R. K. Wierenga, P. Terpstra, and W. G. J. Hol, *J. Mol. Biol.*, 1986, **187**, 101.
68. C. W. Abell, R. M. Stewart, P. J. Andrews, and S.-W. Kwan, *Heterocycles*, 1994, **39**, 933.
69. B. P. Zhou, D. A. Lewis, S.-W. Kwan, and C. W. Abel. *J. Biol. Chem.*, 1995, **270**, 23 653.
70. Y. Nisimoto, H. Otsuka-Murakami, and D. J. Lambeth, *J. Biol. Chem.*, 1995, **270**, 16 428.
71. S. E. Sen and G. D. Prestwich, *J. Am. Chem. Soc.*, 1989, **111**, 1508.
72. S. E. Sen, C. Wawrzeńczyk, and G. D. Prestwich, *J. Med. Chem.*, 1990, **33**, 1698.
73. S. E. Sen and G. D. Prestwich, *J. Am. Chem. Soc.*, 1989, **111**, 8761.
74. M. R. Angelastro, N. P. Peet, and P. Bey, US Pat. 5 051 534 (1991).
75. G. M. Anstead, S. E. Sen, and G. D. Prestwich, *Bioorg. Chem.*, 1991, **19**, 300.
76. G. M. Anstead, H.-K. Lin, and G. D. Prestwich, *Bioorg. Med. Chem. Lett.*, 1993, **3**, 1319.
77. W. A. Van Sickle, M. R. Angelastro, P. Wilson, J. R. Cooper, A. Marquart, and M. A. Flanagan, *Lipids*, 1992, **27**, 157.
78. N. S. Ryder, M. C. Dupont, and I. Frank, *FEBS Lett.*, 1986, **204**, 239.
79. S. E. Sen and G. D. Prestwich, *J. Med. Chem.*, 1989, **32**, 2152.
80. M. Ceruti, F. Viola, G. Grosa, G. Balliano, L. Delprino, and L. Cattel, *J. Chem. Res.* (*S*), 1988, 18.
81. W. R. Moore, G. L. Schatzman, E. T. Jarvi, R. S. Gross, and J. R. McCarthy, *J. Am. Chem. Soc.*, 1992, **114**, 360.
82. B. A. Madden and G. D. Prestwich, unpublished results.
83. R. B. Silverman, in "Mechanism-based Enzyme Inactivation: Chemistry and Enzymology," ed. R. B. Silverman, CRC press, Boca Raton, FL, 1988, Vol. II, p. 191.
84. E. T. Jarvi, J. R. McCarthy, and M. L. Edwards, *Eur. Pat. Appl.* 0 448 934 A2 (1991).
85. J. Mann and G. P. Smith, *J. Chem. Soc. Perkin Trans. 1*, 1991, 2884.
86. A. Duriati, M. Jung, J. Mann, and G. P. Smith, *J. Chem. Soc. Perkin Trans. 1*, 1993, 3073.
87. M. Ceruti, S. Amisano, P. Milla, F. Viola, F. Rocco, M. Jung, and L. Cattel, *J. Chem. Soc. Perkin Trans. 1*, 1995, 889.
88. J. A. Nelson, M. R. Czanny, T. A. Spencer, J. S. Limanek, K. R. McCrae, and T. Y. Chang, *J. Am. Chem. Soc.*, 1978, **100**, 4900.
89. T. Y. Chang, E. S. Schiavoni, Jr., K. R. McCrae, J. A. Nelson, and T. A. Spencer, *J. Biol. Chem.*, 1979, **254**, 11 258.
90. E. J. Corey, P. R. Ortiz de Montellano, K. Lin, and P. D. G. Dean, *J. Am. Chem. Soc.*, 1967, **89**, 2797.
91. J.-L. Abad, J. Casas, J. Abian, and A. Messeguer, *BioMed. Chem. Lett.*, 1993, **3**, 2581.

92. J.-L. Abad, J. Casas, F. Sanchezbaeza, and A. Messeguer, *J. Org. Chem.*, 1995, **60**, 3648.
93. M. Bai, X.-y. Xiao, and G. D. Prestwich, *BioMed. Chem. Lett.*, 1991, **1**, 227.
94. L. De Napoli, E. Fattorusso, S. Magno, and L. Mayol, *Phytochemistry*, 1982, **21**, 782.
95. H. Kigoshi, M. Ojika, Y. Shizuri, H. Niwa, and K. Yamada, *Tetrahedron Lett.*, 1982, **23**, 5413.
96. H. Kigoshi, M. Ojika, Y. Shizuri, H. Niwa, and K. Yamada, *Tetrahedron*, 1986, **42**, 3789.
97. M. Wagner, A. D. Toews, and P. Morell, *J. Neurochem.*, 1995, **64**, 2169.
98. M. Wagner-Recio, A. D. Toews, and P. Morell, *J. Neurochem.*, 1991, **57**, 1891.
99. P. Denner-Ancona and G. D. Prestwich, unpublished results.
100. G. Petranyi, N. S. Ryder, and A. Stütz, *Science*, 1984, **224**, 1239.
101. A. Stütz and G. Petranyi, *J. Med. Chem.*, 1984, **27**, 1539.
102. A. Stütz, *Angew. Chem. Int. Ed. Engl.*, 1987, **26**, 320.
103. N. S. Ryder and M.-C. Dupont, *Biochem. J.*, 1985, **230**, 765.
104. N. S. Ryder, I. Frank and M.-C. Dupont, *Antrimicrob. Agents Chemother.*, 1986, **29**, 858.
105. N. S. Ryder, A. Stütz, and P. Nussbaumer, in "Regulation of Isopentenoid Metabolism," eds. N. S. Ryder, A. Stütz, and P. Nussbaumer, American Chemical Society, Washington, DC, 1992, Vol. 497, p. 192.
106. P. Nussbaumer, G. Petranyi, and A. Stütz, *J. Med. Chem.*, 1991, **34**, 65.
107. P. Nussbaumer, N. S. Ryder, and A. Stütz, *Pestic. Sci.*, 1991, **31**, 437.
108. T. Maeda, M. Takase, A. Ishibashi, T. Yamamoto, K. Sasaki, T. Arika, M. Yokoo, and K. Amemiya, *Yakugaku Zasshi*, 1991, **111**, 126.
109. W. Iwatani, T. Arika, and H. Yamaguchi, *Antimicrob. Agents Chemother.*, 1993, **37**, 785.
110. P. Nussbaumer, G. Dorfstätter, M. A. Grassberger, I. Leitner, J. G. Meingassner, K. Thirring, and A. Stütz, *J. Med. Chem.*, 1993, **36**, 2115.
111. P. Nussbaumer, G. Dorfstätter, I. Leitner, K. Mraz, H. Vyplel, and A. Stütz, *J. Med. Chem.*, 1993, **36**, 2810.
112. P. Nussbaumer, I. Leitner, and A. Stütz, *J. Med. Chem.*, 1994, **37**, 610.
113. N. S. Ryder, *Biochem. Soc. Trans.*, 1990, **18**, 45.
114. H. Takezawa, M. Hayashi, Y. Iwasawa, M. Hosoi, Y. Idia, Y. Tsuchiya, M. Horie, and T. Kamei, *Eur. Pat. App.* 0 318 860 A2 (1989).
115. H. Takezawa, M. Hayashi, Y. Iwasawa, M. Hosoi, Y. Iida, Y. Tsuchiya, M. Horie, and T. Kamei, *Eur. Pat. Appl.* 0 395 768 A1 (1990).
116. Y. Iwasawa and M. Horie, *Drugs Future*, 1993, **18**, 911.
117. M. Horie, Y. Tsuchiya, M. Hayashi, Y. Iida, Y. Iwasawa, Y. Nagata, Y. Sawasaki, H. Fukuzumi, K. Kitani, and T. Kamei, *J. Biol. Chem.*, 1990, **265**, 18 075.
118. T. Satoh, M. Horie, H. Watanabe, Y. Tsuchiya, and T. Kamei, *Biol. Pharm. Bull.*, 1993, **16**, 349.
119. M. Horie, Y. Sawasaki, H. Fukuzumi, K. Watanabe, Y. Iizuka, Y. Tsuchiya, and T. Kamei, *Atherosclerosis*, 1991, **88**, 183.
120. Y. Nagata, M. Horie, Y. Hidaka, M. Yonemoto, M. Hayashi, H. Watanabe, F. Ishida, and T. Kamei, *Chem. Pharm. Bull.*, 1992, **40**, 436.
121. M. Horie, Y. Iwasawa, T. Satoh, A. Shimizu, Y. Nagata, and T. Kamei, *Biochem. Pharmacol.*, 1993, **46**, 297.
122. Y. Tsuchiya, T. Nomoto, M. Hayashi, Y. Iwasawa, H. Masaki, M. Ohkubo, Y. Sakuma, Y. Nagata, T. Satoh, and T. Kamei, *Eur. Pat. Appl.* 0 448 078 A2 (1991).
123. Y. Tsuchiya, T. Nomoto, M. Mitsuya, K. Nonoshita, M. Hayashi, T. Sato, Y. Sawasaki, and T. Kamei, *World. Pat.* 9 324 478 A1 (1993).
124. Y. Iwasawa, Y. Tsuchiya, Y. Iida, M. Hayashi, H. Masaki, T. Nomoto, Y. Sakuma, M. Kato, Y. Watanabe, H. Takezawa, M. Hosoi, Y. Sawasaki, K. Kitani, and T. Kamei, Presented at the XIIth International Symposium on Medicinal Chemistry, Basel, Switzerland, 1992.
125. K. Matsuda, H. Harada, R. Tsuzuki, K. Morihira, N. Ito, H. Kakuta, and Y. Iizumi, *World. Pat.* 9 312 069 A1 (1993).
126. J.-P. Gotteland, I. Brunel, F. Gendre, J. Désiré, A. Delhon, D. Junquéro, P. Oms, and S. Halazy, *J. Med. Chem.*, 1995, **38**, 3207.
127. J.-P. Gotteland, A. Delhon, D. Junquéro, P. Oms, and S. Halazy, *Biomed. Chem. Lett.*, 1996, **6**, 533.
128. J. P. Gotteland, D. Junquéro, P. Oms, A. Delhon, and S. Halazy, *Med. Chem. Res.*, 1996, **6**, 333.
129. M. Bai, Ph.D. Dissertation. University at Stony Brook, Stony Brook, New York USA, 1991.
130. H.-K. Lee and G. D. Prestwich, unpublished results.
131. M. Ceruso and G. D. Prestwich, *BioMed. Chem. Lett.*, 1994, **4**, 2179.
132. P. Denner-Ancona, Ph.D. Dissertation, The University at Stony Brook, Stony Brook, New York USA, 1995.
133. D. M. Marecak and G. D. Prestwich, unpublished results.
134. J. Doussière, G. Buzenet, and P. V. Vignais, *Biochemistry*, 1995, **34**, 1760.
135. G. D. Prestwich, G. Dormán, J. T. Elliott, D. M. Marecak, and A. Chaudhary, *Photochem. Photobiol.*, 1997, **65**, 222.
136. G. Dormán and G. D. Prestwich, *Biochemistry*, 1994, **33**, 5661.
137. I. Ojima, O. Duclos, G. Dormán, B. Simonot, G. D. Prestwich, S. Rao, K. A. Lerro, and S. B. Horwitz, *J. Med. Chem.*, 1995, **38**, 3891.
138. A. Eschenmoser, L. Ruzicka, O. Jeger, and D. Arigoni, *Helv. Chim. Acta.*, 1995, **38**, 1890.
139. G. Stork and A. W. Burgstahler, *J. Am. Chem. Soc.*, 1955, **77**, 5068.
140. E. E. van Tamelen, *J. Am. Chem. Soc.*, 1982, **104**, 6480.
141. E. E. van Tamelen and D. R. James, *J. Am. Chem. Soc.*, 1977, **99**, 950.
142. R. B. Boar, L. A. Couchman, A. J. Jaques and M. J. Perkins, *J. Am. Chem. Soc.*, 1984, **106**, 2476.
143. Y. Arai, M. Hirohara, H. Ageta, and H. Y. Hsü, *Tetrahedron Lett.*, 1992, **33**, 1325.
144. A. F. Barrero, E. A. Manzaneda R., and R. A. Manzaneda R., *Tetrahedron*, 1990, **46**, 8161.
145. E. J. Corey, S. C. Virgil, H. Cheng, C. H. Baker, S. P. T. Matsuda, V. Singh, and S. Sarshar, *J. Am. Chem. Soc.*, 1995, **117**, 11 819.
146. W. S. Johnson, S. J. Telfer, S. Cheng, and U. Schubert, *J. Am. Chem. Soc.*, 1987, **109**, 2517.
147. W. S. Johnson, S. D. Lindell, and J. Steele, *J. Am. Chem. Soc.*, 1987, **109**, 5852.
148. W. S. Johnson, *Tetrahedron*, 1991, **47**, xi.
149. W. S. Johnson, R. A. Buchanan, W. R. Bartlett, F. S. Tham, and R. K. Kullnig, *J. Am. Chem. Soc.*, 1993, **115**, 504.

150. I. Abe and G. D. Prestwich, *J. Biol. Chem.*, 1994, **269**, 802.
151. I. Abe and G. D. Prestwich, *Lipids*, 1995, **30**, 231.
152. I. Abe and G. D. Prestwich, *Proc. Natl. Acad. Sci. USA*, 1995, **92**, 9274.
153. R. A. Kumpf and D. A. Dougherty, *Science*, 1993, **261**, 1708.
154. A. McCurdy, L. Jimenez, D. A. Stauffer, and D. A. Dougherty, *J. Am. Chem. Soc.*, 1992, **114**, 10 314.
155. M. F. Hibert, S. Trumpp-Kallmeyer, A. Bruinvels, and J. Hoflack, *Mol. Pharmacol.*, 1991, **40**, 8.
156. K. Poralla, A. Hewelt, G. D. Prestwich, I. Abe, I. Reipen, and G. Sprenger, *Trends Biochem. Sci.*, 1994, **19**, 157.
157. K. Poralla, *Bioorg. Med. Chem. Lett.*, 1994, **4**, 285.
158. D. A. Dougherty, *Science*, 1996, **271**, 163.
159. K. U. Wendt, K. Poralla, and G. E. Schulz, *Science*, 1997, **277**, 1811.
160. E. J. Corey and S. C. Virgil, *J. Am. Chem. Soc.*, 1991, **113**, 4025.
161. E. J. Corey, S. C. Virgil, and S. Sarshar, *J. Am. Chem. Soc.*, 1991, **113**, 8171.
162. A. Kawaguchi, H. Kobayashi, and S. Okuda, *Chem. Pharm. Bull.*, 1973, **21**, 577.
163. W. O. Godtfredsen, H. Lorck, E. E. van Tamelen, J. D. Willett, and R. B. Clayton, *J. Am. Chem. Soc.*, 1968, **90**, 208.
164. L. J. Mulheirn and E. Caspi, *J. Biol. Chem.*, 1971, **246**, 2494.
165. I. Abe, M. Bai, X.-y. Xiao, and G. D. Prestwich, *Biochem. Biophys. Res. Commun.*, 1992, **187**, 32.
166. M. Kusano, I. Abe, U. Sankawa, and Y. Ebizuka, *Chem. Pharm. Bull.*, 1991, **39**, 239.
167. W. R. Moore and G. L. Schatzman, *J. Biol. Chem.*, 1992, **267**, 22 003.
168. A. Duriatii and F. Schuber, *Biochem. Biophys. Res. Commun.*, 1988, **151**, 1378.
169. I. Abe, Y. Ebizuka, and U. Sankawa, *Chem. Pharm. Bull.*, 1988, **36**, 5031.
170. I. Abe, U. Sankawa, and Y. Ebizuka, *Chem. Pharm. Bull.*, 1989, **37**, 536.
171. I. Abe, Y. Ebizuka, S. Seo, and U. Sankawa, *FEBS Lett.*, 1989, **249**, 100.
172. I. Abe, U. Sankawa, and Y. Ebizuka, *Chem. Pharm. Bull.*, 1992, **40**, 1755.
173. E. J. Corey and S. P. T. Matsuda, *J. Am. Chem. Soc.*, 1991, **113**, 8172.
174. T. Hoshino, H. J. Williams, Y. Chung, and A. I. Scott, *Tetrahedron*, 1991, **47**, 5925.
175. G. Balliano, F. Viola, M. Ceruti, and L. Cattel, *Arch. Biochem. Biophys.*, 1992, **293**, 122.
176. S. Yamamoto, K. Lin, and K. Bloch, *Proc. Natl. Acad. Sci. USa*, 1969, **63**, 110.
177. I. Shechter, F. W. Sweat, and K. Bloch, *Biochim. Biophys. Acta*, 1970, **220**, 463.
178. I. Abe and G. D. Prestwich, unpublished results.
179. X.-y. Xiao and G. D. Prestwich, *J. Am. Chem. Soc.*, 1991, **113**, 9673.
180. C. H. Baker, S. P. T. Matsuda, D. R. Liu, and E. J. Corey, *Biochem. Biophys. Res. Commun.*, 1995, **213**, 154.
181. C. K. Sung, M. Shibuya, U. Sankawa, and Y. Ebizuka, *Biol. Pharm. Bull.*, 1995, **18**, 1459.
182. M. Young, H. M. Chen, M. D. Lalioti, and S. E. Antonarakis, *Human Genetics*, 1996, **97**, 620.
183. M. Kusano, M. Shibuya, U. Sankawa, and Y. Ebizuka, *Biol. Pharm. Bull.*, 1995, **18**, 195.
184. E. J. Corey, S. P. T. Matsuda, and B. Bartel, *Proc. Natl. Acad. Sci. USA*, 1994, **91**, 2211.
185. Z. Shi, C. J. Buntel, and J. H. Griffin, *Proc. Natl. Acad. Sci. USA*, 1994, **91**, 7370.
186. R. Kelly, S. M. Miller, M. H. Lai, and D. R. Kirsch, *Gene*, 1990, **87**, 177.
187. C. J. Buntel and J. H. Griffin, *J. Am. Chem. Soc.*, 1992, **114**, 9711.
188. C. A. Roessner, C. Min, S. H. Hardin, L. W. Harris-Haller, J. C. McCollum, and A. I. Scott, *Gene*, 1993, **127**, 149.
189. E. J. Corey, S. P. T. Matsuda, C. H. Baker, A. Y. Ting, and H. Cheng, *Biochem. Biophys. Res. Commun.*, 1996, **219**, 327.
190. E. J. Corey, S. P. T. Matsuda, and B. Bartel, *Proc. Natl. Acad. Sci. USA*, 1993, **90**, 11 628.
191. J. H. Griffin, C. J. Buntel, and J. J. Siregar, *Proc. Natl. Acad. Sci. USA*, 1998, in press.
192. I. G. Reipen, K. Poralla, H. Sahm, and G. A. Sprenger, *Microbiology*, 1995, **141**, 155.
193. D. Ochs, C. Kaletta, K.-D. Entian, A. Beck-Sickinger, and K. Poralla, *J. Bacteriol.*, 1992, **174**, 298.
194. G. Ourisson, M. Rohmer, and K. Poralla, *Annu. Rev. Microbiol.*, 1987, **41**, 301.
195. G. Ourisson, *Pure Appl. Chem.*, 1989, **61**, 345.
196. J. Devereux, P. Haeberli, and O. Smithies, *Nucleic Acid Res.*, 1984, **12**, 387.
197. L. Cattel, M. Ceruti, G. Balliano, F. Viola, G. Grosa, F. Rocco, and P. Brusa, *Lipids*, 1995, **30**, 235.
198. E. J. Corey, and S. K. Gross, *J. Am. Chem. Soc.*, 1967, **89**, 4561.
199. O. Boutaud, D. Dolis, and F. Schuber, *Biochem. Biophys. Res. Comm.*, 1992, **188**, 898.
200. C. Anding, R. Heintz, and G. Ourisson, *C. R. Acad. Sci. Paris D*, 1973, **276**, 205.
201. J. Bujons, R. Guajardo, and K. S. Kyler, *J. Am. Chem. Soc.*, 1988, **110**, 604.
202. E. E. van Tamelen, K. B. Sharpless, J. D. Willett, R. B. Clayton, and A. L. Burlingame, *J. Am. Chem. Soc.*, 1967, **89**, 3920.
203. R. J. Anderson, R. P. Hanzlik, K. B. Sharpless, E. E. van Tamelen, and R. B. Clayton, *Chem. Comm.*, 1969, 53.
204. E. J. Corey and H. Cheng, *Tetrahedron Lett.*, 1996, **37**, 2709.
205. X.-y. Xiao and G. D. Prestwich, *Tetrahedron Lett.*, 1991, **32**, 6843.
206. X.-y. Xiao, Ph.D. Dissertation, State University of New York, Stony Brook, USA, 1991.
207. E. E. van Tamelen, K. B. Sharpless, R. Hanzlik, R. B. Clayton, A. L. Burlingame, and P. C. Wszolek, *J. Am. Chem. Soc.*, 1967, **89**, 7150.
208. E. E. van Tamelen, E. J. Leopold, S. A. Marson, and H. R. Waespe, *J. Am. Chem. Soc.*, 1982, **104**, 6480.
209. A. Krief, J.-R. Schauder, E. Guittet, C. Herve du Penhoat, and H.-Y. Lallemand, *J. Am. Chem. Soc.*, 1987, **109**, 7910.
210. M. Hérin, P. Sandra, and A. Krief, *Tetrahedron Lett.*, 1979, **33**, 3103.
211. A. Krief, P. Pasau, and L. Quéré, *Bioorg. Med. Chem. Lett.*, 1991, **1**, 365.
212. E. J. Corey and W. E. Russey, *J. Am. Chem. Soc.*, 1966, **88**, 4751.
213. E. J. Corey, S. C. Virgil, D. R. Liu, and S. Sarshar, *J. Am. Chem. Soc.*, 1992, **114**, 1524.
214. E. J. Corey, P. R. Ortiz de Montellano, and H. Yamamoto, *J. Am. Chem. Soc.*, 1968, **90**, 6254.
215. E. E. van Tamelen, R. P. Hanzlik, K. B. Sharpless, R. B. Clayton, W. J. Richter, and A. L. Burlingame, *J. Am. Chem. Soc.*, 1968, **90**, 3284.
216. T. Hoshino, E. Ishibashi, and K. Kaneko, *J. Chem. Soc. Chem. Commun.*, 1995, 2401.
217. J. C. Median, R. Guajardo, and K. S. Kyler, *J. Am. Chem. Soc.*, 1989, **111**, 2310.
218. E. E. van Tamelen and J. H. Freed, *J. Am. Chem. Soc.*, 1970, **92**, 7206.

219. E. E. van Tamelen and R. E. Hopla, *J. Am. Chem. Soc.*, 1979, **101**, 6112.
220. R. B. Clayton, E. E. van Tamelen, and R. G. Nadeau, *J. Am. Chem. Soc.*, 1968, **90**, 820.
221. L. O. Crosby, E. E. van Tamelen, and R. B. Clayton, *Chem. Comm.*, 1969, 532.
222. J. C. Medina and K. S. Kyler, *J. Am. Chem. Soc.*, 1988, **110**, 4818.
223. X.-y. Xiao, S. E. Sen, and G. D. Prestwich, *Tetrahedron Lett.*, 1990, **31**, 2097.
224. E. J. Corey, K. Lin, and M. Jautelat, *J. Am. Chem. Soc.*, 1968, **90**, 2724.
225. E. E. van Tamelen, A. D. Pedlar, E. Li, and D. R. James, *J. Am. Chem. Soc.*, 1977, **99**, 6778.
226. D. H. R. Barton, T. R. Jarman, K. C. Watson, D. A. Widdowson, R. B. Boar, and K. Damps, *J. Chem. Soc. Perkin Trans. 1*, 1975, 1134.
227. B. A. Madden and G. D. Prestwich, *J. Org. Chem.*, 1994, **59**, 5488.
228. B. A. Madden and G. D. Prestwich, *BioOrg. Med. Chem. Lett.*, 1997, *7*, 309.
229. I. Abe, T. Dang, Y. F. Zheng, B. A. Madden, C. Feil, K. Poralla, and G. D. Prestwich, *J. Am. Chem. Soc.*, 1997, **119**, 11 333.
230. C. Feil, R. Süssmuth, G. Jung, and K. Poralla, *Eur. J. Biochem.*, 1996, **242**, 51.
231. G. Ballinao, G. Grosa, P. Milla, F. Viola, and L. Cattel, *Lipids*, 1993, **28**, 903.
232. L. Cattel, M. Ceruti, G. Balliano, F. Viola, G. Grosa, and F. Schuber, *Steroids*, 1989, **53**, 363.
233. M. Cruti, G. Balliano, F. Viola, L. Cattel, N. Gerst, and F. Schuber, *Eur. J. Med. Chem.*, 1987, **22**, 199.
234. L. Cattel, M. Ceruti, F. Viola, L. Delprino, G. Balliano, A. Duriatti, and P. Bouvier-Navé, *Lipids*, 1986, **21**, 31.
235. A. Duriatti, P. Bouvier-Navé, P. Benveniste, F. Schuber, L. Delprino, G. Balliano, and L. Cattel, *Biochem. Pharmacol.*, 1985, **34**, 2765.
236. G. Balliano, P. Milla, M. Cerutti, L. Carrano, F. Viola, P. Brusa, and L. Cattel, *Antimicrobial Agents Chemother.*, 1994, **38**, 1904.
237. M. Ceruti, F. Viola, G. Balliano, G. Grosa, N. Gerst, F. Schuber, and L. Cattel, *Eur. J. Med. Chem.*, 1988, **23**, 533.
238. M. Ceruti, G. Balliano, F. Viola, G. Grosa, F. Rocco, and L. Cattel, *J. Med. Chem.*, 1992, **35**, 3050.
239. G. Balliano, P. Milla, M. Ceruti, F. Viola, L. Carrano, and L. Cattel, *FEBS Lett.*, 1993, **320**, 203.
240. F. Viola, P. Brusa, G. Balliano, M. Ceruti, O. Boutaud, F. Schuber, and F. Cattel, *Biochem. Pharm.*, 1995, **50**, 787.
241. M. Ceruti, F. Rocco, F. Viola, G. Balliano, G. Grosa, F. Dosio, and L. Cattel, *Eur. J. Med. Chem.*, 1993, **28**, 675.
242. G. Popjak, A. Meenan, and W. D. Nes, *Proc. R. Soc. Lond. B.*, 1987, **232**, 273.
243. W. D. Nes and E. J. Parish, *Lipids*, 1988, **23**, 375.
244. D. S. Dodd and A. C. Oehlschlager, *J. Org. Chem.*, 1992, **57**, 2794.
245. A. Perez, A. C. Oehlschlager, I. Abe, G. D. Prestwich, and D. S. Dodd, unpublished results.
246. D. S. Dodd, A. C. Oehlschlager, N. H. Georgopapadakou, A.-M. Polak, and P. G. Hartman, *J. Org. Chem.*, 1992, **57**, 7226.
247. M. Taton, P. Benveniste, A. Rahier, W. S. Johnson, H.-t. Liu, and A. R. Sudhakar, *Biochemistry*, 1992, **31**, 7892.
248. I. C. Rose, B. A. Sharpe, R. C. Lee, J. H. Griffin, J. O. Capobianco, D. Zakula, and R. C. Goldman, *Bioorg. Med. Chem.*, 1996, **4**, 97.
249. R. C. Goldman, D. Zakula, J. O. Capobianco, B. A. Sharpe, and J. H. Griffin, *Antimicrob. Agents Chemother.*, 1996, **40**, 1044.
250. M. Taton, P. Benveniste, and A. Rahier, *Biochem. Biophys. Res. Comm.*, 1986, **138**, 764.
251. N. Gerst, A. Duriatti, F. Schuber, M. Taton, P. Benveniste, and A. Rahier, *Biochem. Pharmacol.*, 1988, **37**, 1955.
252. M. W. Wannamaker, P. P. Waid, W. A. Van Sickle, J. R. McCarthy, P. K. Wilson, G. L. Schatzman, and W. R. Moore, *J. Med. Chem.*, 1992, **35**, 3581.
253. K. K. Ruhl, L. Anzalone, E. D. Arguropoulos, A. K. Gayen, and T. A. Spencer, *Bioorg. Chem.*, 1989, **17**, 108.
254. T. Hoshino, N. Kobayashi, E. Ishibashi, and S. Hashimoto, *Biosci. Biotech. Biochem.*, 1995, **59**, 602.
255. M. M. Barth, J. L. Binet, D. M. Thomas, D. C. de Fornel, S. Samreth, F. J. Schuber, and P. P. Renaut, *J. Med. Chem.*, 1996, **39**, 2302.
256. M. W. Wannamaker, P. P. Waid, W. R. Moore, G. L. Schatzman, W. A. Van Sickle, and P. K. Wilson, *Bioorg. Med. Chem. Lett.*, 1993, **3**, 1175.
257. E. J. Corey, D. C. Daley, and H. Cheng, *Tetrahedron Lett.*, 1996, **37**, 3287.
258. J.-L. Abad, J. Casas, F. Sánchez-Baeza, and A. Messeguer, *J. Org. Chem.*, 1993, **58**, 3991.
259. J.-L. Abad, M. Guardiola, J. Casas, F. Sánchez-Baeza, and A. Messeguer, *J. Org. Chem.*, 1996, **61**, 7603.
260. M. Ceruti, F. Viola, F. Dosio, L. Cattel, P. Bouvier-Navé, and P. Ugliengo, *J. Chem. Soc. Perkin Trans. 1*, 1988, 461.
261. J. Park, C. Min, H. Williams, and A. I. Scott, *Tetrahedron Lett.*, 1995, **36**, 5719.
262. Y. F. Zheng, A. C. Oehlschlager, and P. G. Hartman, *J. Org. Chem.*, 1994, **59**, 5803.
263. Y. F. Zheng, A. C. Oehlschlager, N. H. Georgopapadakou, P. G. Hartman, and P. Scheliga, *J. Am. Chem. Soc.*, 1995, **117**, 670.
264. Y. F. Zheng, D. S. Dodd, A. C. Oehlschlager, and P. G. Hartman, *Tetrahedron*, 1995, **51**, 5255.
265. D. Stach, Y. F. Zheng, A. L. Perez, A. C. Oehlschlager, I. Abe, G. D. Prestwich, and P. G. Hartman, *J. Med. Chem.*, 1997, **40**, 201.
266. I. Abe, W. Liu, A. C. Oehlschlager, and G. D. Prestwich, *J. Am. Chem. Soc.*, 1996, **118**, 9180.
267. S. Jolidon, A. Polak-Wyss, P. G. Hartman, and P. Guerry, in "Recent Advances in the Chemistry of Anti-infective Agents," eds. P. H. Bentley and R. Ponsfor, The Royal Society of Chemistry, London, 1993, p. 223.
268. S. Jolidon, A.-M. Polak, P. Guerry, and P. G. Hartman, *Biochem. Soc. Trans.*, 1990, **18**, 47.
269. E. I. Mercer, P. K. Morris, and B. C. Baldwin, *Comp. Biochem. Physiol. B*, 1985, **80**, 341.
270. A. Boogaard, M. Griffioen, and L. H. Cohen, *Biochem. J.*, 1987, **241**, 345.
271. S. R. Panini, R. C. Sexton, and H. Rudney, *J. Biol. Chem.*, 1984, **259**, 7767.
272. M. Mark, P. Müller, R. Maier, and B. Eisele, *J. Lipid Res.*, 1996, **37**, 148.
273. J. A. Nelson, S. R. Steckbeck, and T. A. Spencer, *J. Biol. Chem.*, 1981, **256**, 1067.
274. J. A. Nelson, S. R. Steckbeck, and T. A. Spencer, *J. Am. Chem. Soc.*, 1981, **103**, 6974.
275. T. A. Spencer, A. K. Gayen, S. Phirwa, J. A. Nelson, F. R. Taylor, A. A. Kandutsch, and S. K. Erickson, *J. Biol. Chem.*, 1985, **260**, 13 391.

2.11
Cycloartenol and Other Triterpene Cyclases

KARL PORALLA
Eberhard-Karls-Universität Tübingen, Germany

2.11.1 INTRODUCTION

2.11.1.1 General

Triterpenes comprise a group of isoprenoids typically containing 30 C-atoms which originate from six isoprenyl residues. About 40 cyclic triterpenes of the formula $C_{30}H_{50}$ and about 140 of the formula $C_{30}H_{50}O$ are known.[1] A significant number of these are believed to be cyclization products synthesized from squalene ($C_{30}H_{50}$) or from oxidosqualene ($C_{30}H_{50}O$). The majority of these triterpenes occur in higher plants as secondary metabolites, the functions of most of which are as yet unknown.

The isoprenoid field is one of the fascinating interfaces between chemistry, biochemistry and biology. This field is progressing in two dimensions: in width—ever more isoprenoids are being isolated and their structure elucidated; in depth—our knowledge of relationships between biochemistry, function and evolution is undergoing further refinement. Isoprenoids play an important role mediating plant–plant, plant–insect and plant–pathogen interactions. New facts concerning intraorganismic mediation have emerged in the study of brassinosteroids as plant hormones.[2] Even a new biosynthetic route for the isoprenyl unit has been detected during studies of hopanoid biosynthesis.[3]

The early development of the isoprenoid field was accompanied by the award of 19 Nobel prizes, as was pointed out in the interesting book by Nes and McKean.[4] It is evident from this that the field of triterpenes has contributed significantly to progress in isoprenoid research.

Plants contain a huge variety of cyclic triterpenes that do not occur at all in animals and fungi. The corresponding cyclases will be treated in this chapter. Furthermore, hopanoids occur not only in higher plants and ferns but also in a large variety of prokaryotes and will therefore also be included. The fungal and animal lanosterol cyclase is dealt with in Chapter 2.10.

2.11.1.2 The Catalytic Versatility of Triterpene Cyclases

Before focusing on specific triterpene cyclases, it is necessary to consider the catalytic variability of these enzymes. To understand this variability we will dissect the catalytic action of triterpene cyclases. In comparison to the ionization-dependent mono-, sesqui- and diterpene cyclases, triterpene cyclases never use a pyrophosphorylated compound as a substrate (see Chapters 2.05, 2.06, and 2.08). In contrast, the initial step of the triterpene cyclization reaction starting from squalene (**1**) or oxidosqualene (**2**) is protonation-dependent. In typical triterpene cyclizations, 12 to 20 covalent bonds are broken or formed, 7 to 9 chiral centers are established and 4 to 5 rings are built. The conclusion is therefore justified that triterpene cyclases catalyze the most complex one-step reaction known in biochemistry and chemistry. The complex triterpene cyclase reaction can be dissected into the following steps:

1. Binding of squalene (or oxidosqualene) in a distinct conformation to the active site of the cyclase. Due to the free rotation of five C—C bonds many conformers are possible.
2. Initiation of cyclization by protonation.
3. Propagation of cyclization.
4. Termination of cyclization at a precise stage and concomitant formation of a terminal cation.
5. Carbon skeleton rearrangement by concerted 1,2-shifts of hydride and methyl groups.
6. Elimination of a proton.

Step 5 does not necessarily occur. At steps 1, 4, and 5 variations are also possible. Hence, numerous cyclization products have been found in nature. They can be subdivided into distinct groups: variation in the number and stereochemistry of individual rings, different types of hydride and methyl shifts and different sites of final proton abstraction.

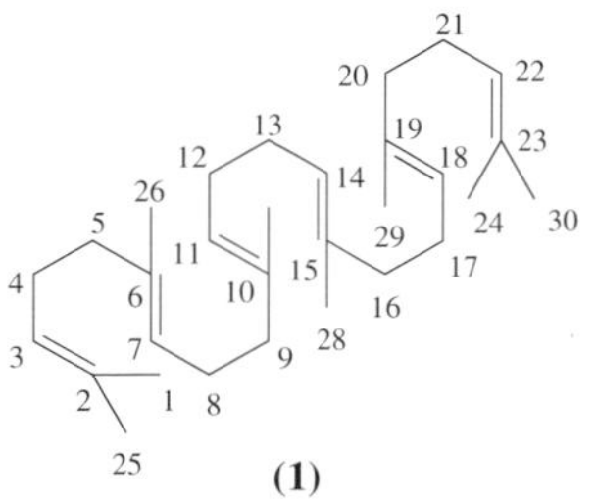

(1) 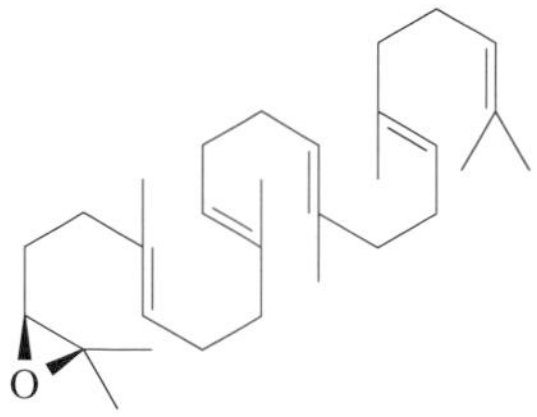 (2)

Physiological substrates for triterpene cyclases are all derivatives of squalene. They comprise: all-*trans*-squalene (**1**) and all-*trans* (3*S*)-2,3-oxidosqualene (**2**). In the case of the synthesis of α-onocerin (**5**) it seems only logical that dioxidosqualene (**3**) and the bicyclic structure (**4**) should be the substrate. Thus, a twofold cyclization reaction has to occur.[5]

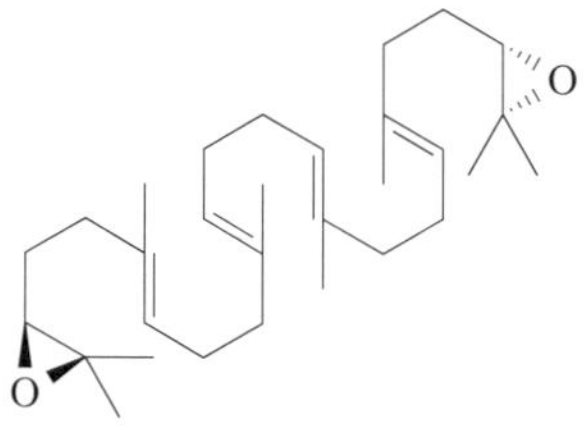

(3) 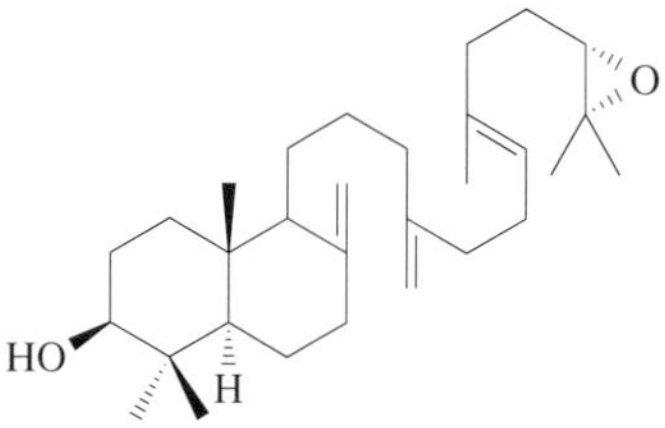 (4) 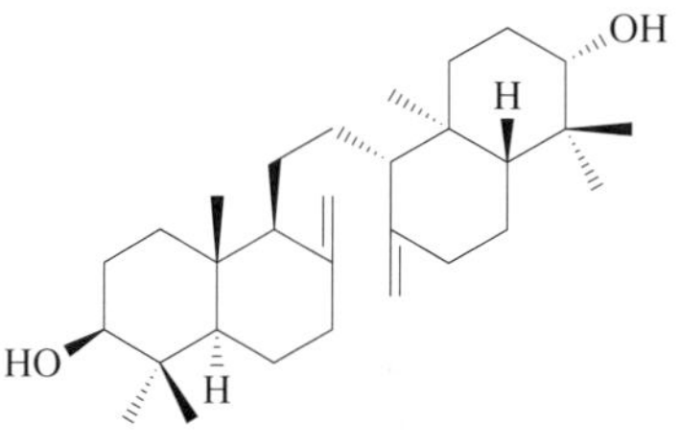 (5)

Examples of cyclic triterpenes with fewer than four rings are mono- (achilleol A (**6**)), di- (polypoda-8(26),13,17,21-tetraen-3-ol (**7**)), and tricyclic compounds (malabarica-14(26),17,21-trien-3-ol (**8**)).[1] Cyclases have not yet been characterized for these compounds, but these examples conform to an enzyme-mediated cyclization process in accordance with the Biogenetic Isoprene rule.[6,7] It should also be borne in mind that very irregular products such as ambrein (**9**) could also be synthesized by cyclases.[1]

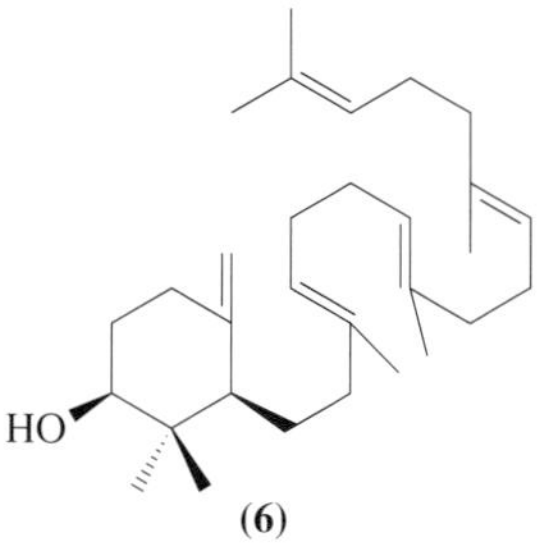

(6) 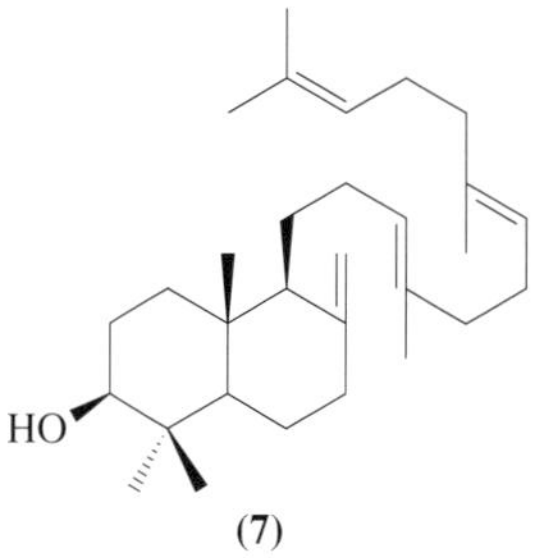 (7)

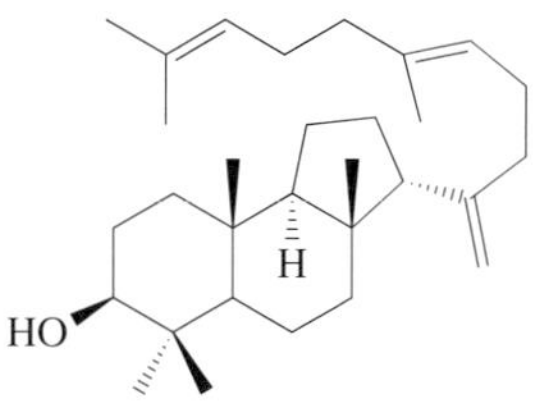

(8) 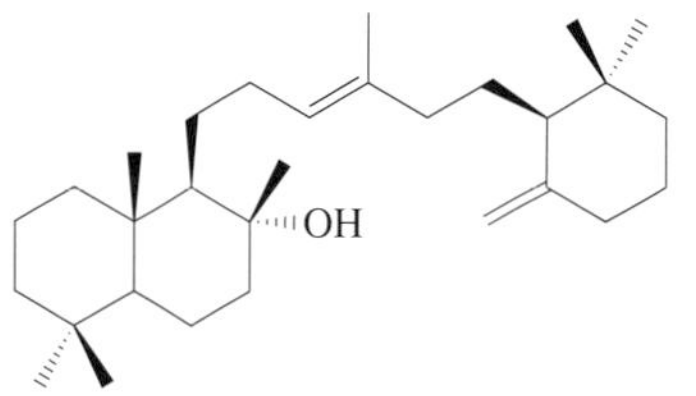 (9)

An abundant variety of at least 71 tetracyclic and pentacyclic triterpenes is known, very probably formed by different cyclases via the intermediate dammarenyl cation (**10**).[1] The dammarenyl cation is in an all-chair conformation. A selection of these compounds includes dammara-20,24-dienol (**11**), euphol (**12**), bacchara-12,21-dienol (**13**), and fernenol (3β-form) (**14**).[1] Tetracyclic compounds with different ring D structures are known; in dammara-20,24-dienol ring D is five-membered, while in bacchara-12,21-dienol it is six-membered. Sometimes extensive 1,2-shifts occur, as in the case of euphol and fernenol. Also ascribed to this group are triterpenes with a hopane and gammacerane skeleton. In the latter two cases, however, the involvement of the intermediary dammarenyl cation (**10**) is unlikely.

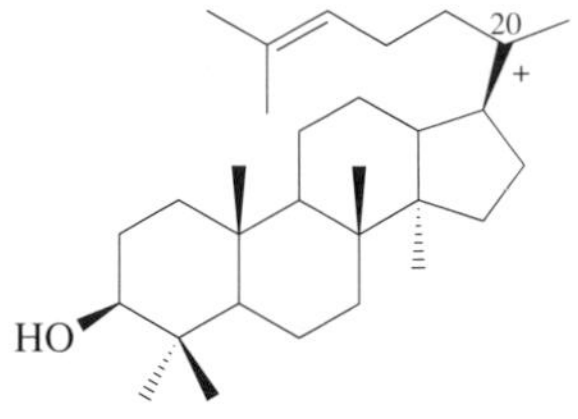

(**10**)

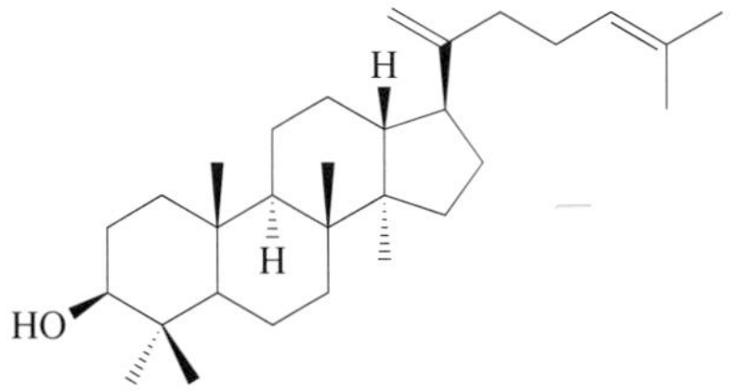

(**11**) 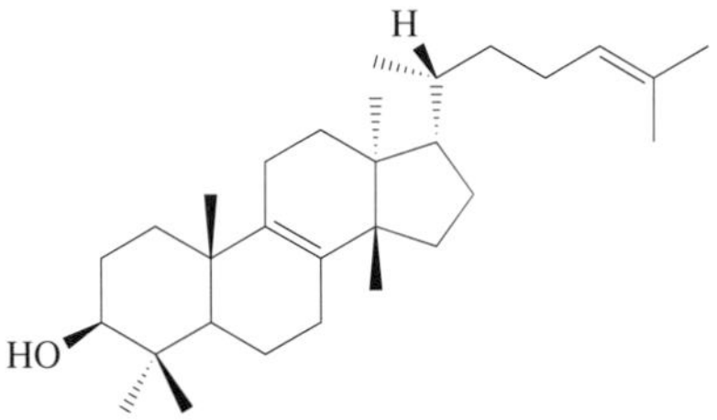 (**12**)

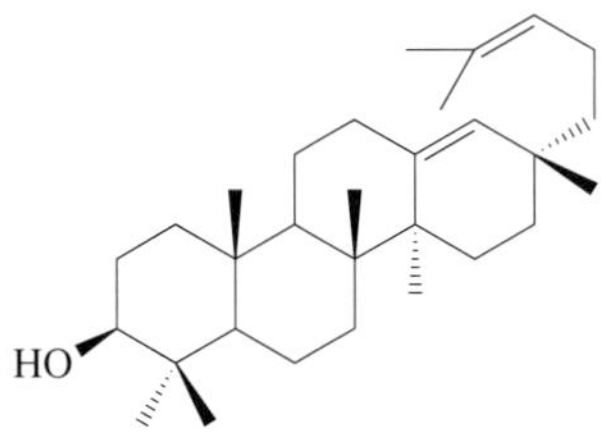

(**13**) 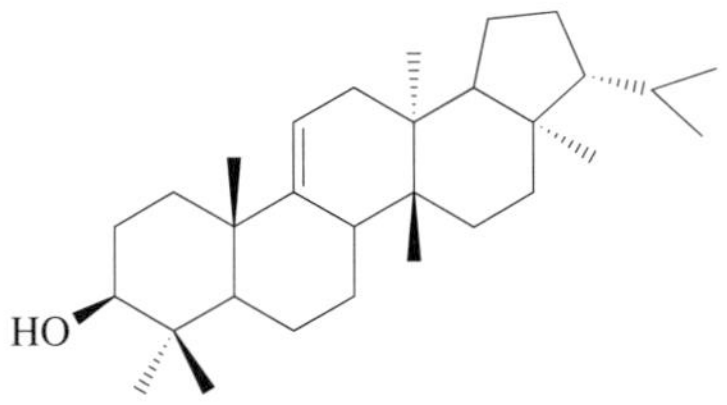 (**14**)

The most common triterpenes are formed via the intermediate protosteryl cation (**15**). This compound has a boat-conformation in ring B. About 11 cyclization products of this group are known, including lanosterol and cycloartenol (**16**). Further examples are parkeol (**17**), cucurbita-5,24-dienol (**18**) and isoarborinol (**19**).[1] Following carbon skeleton rearrangement, the methyl groups at C-13 and C-14 have a configuration opposite to that of dammarenyl derivatives (**12**) and (**14**).

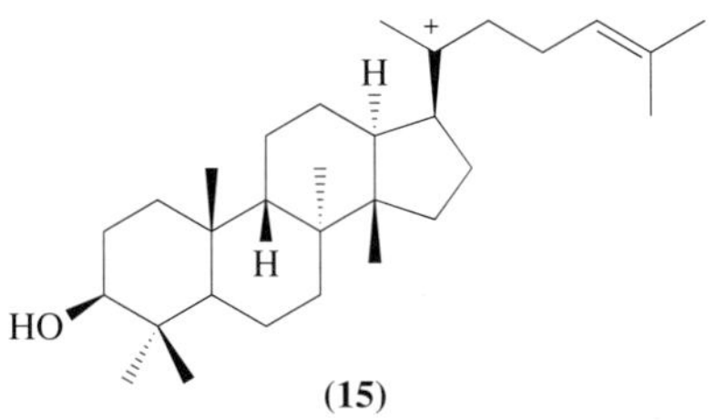

(**15**) 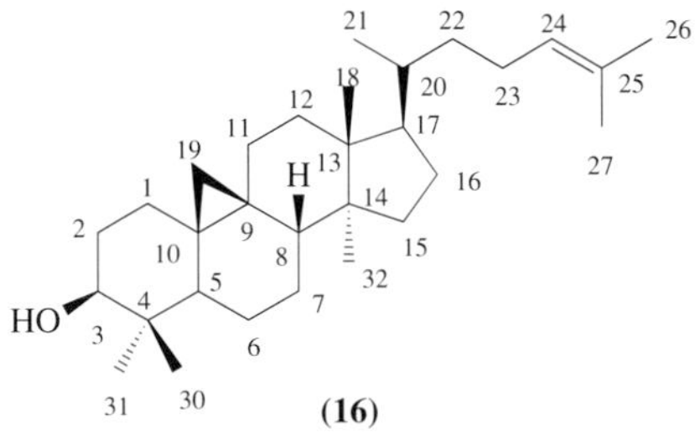 (**16**)

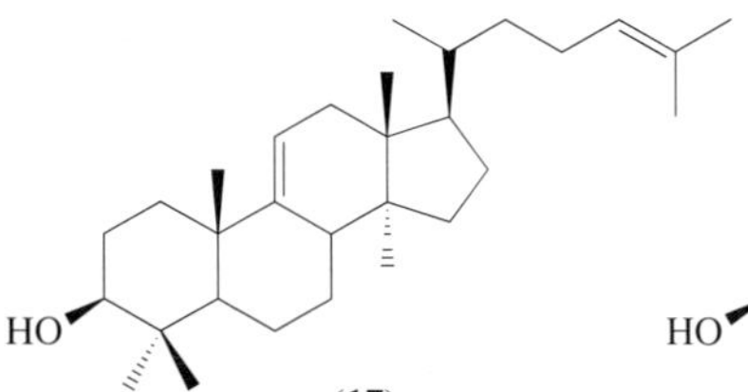

(**17**) 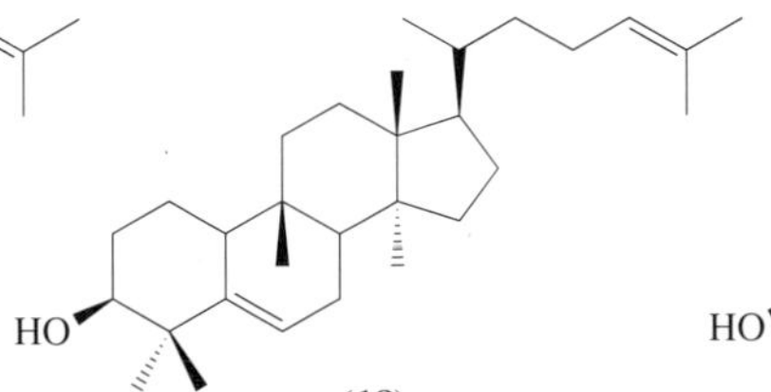 (**18**) 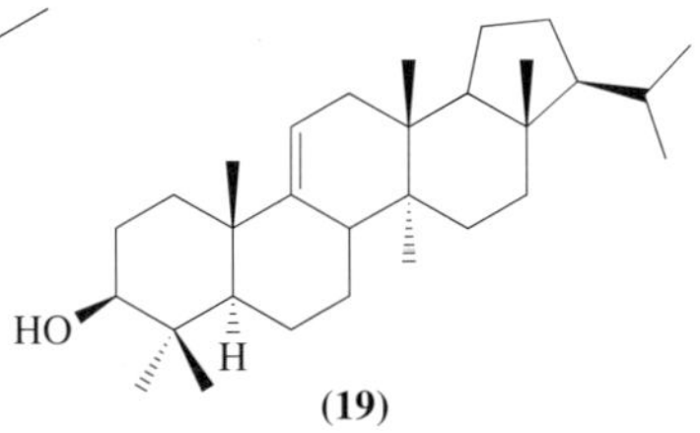 (**19**)

About 100 hypothetical cyclization products are known, all of them derived from squalene or oxidosqualene. Only sesquiterpene cyclases are more versatile and possess more variability around a common theme as compared with triterpene cyclases. All the triterpene cyclase products are in fact connected not only by structure but also by the cyclization mechanism of the triterpene cyclases and—as will be demonstrated later—by a common phylogenetic origin of the corresponding cyclases.

Several excellent reviews have been published which cover aspects of this review, such as that on triterpene cyclases by Abe *et al.*,[8] and those on inhibitors by Mercer[9] and Abe *et al.*[10]

2.11.2 CYCLOARTENOL CYCLASE

Cycloartenol (**16**) is an obligatory constituent in photosynthetic eukaryotes.[11] It also occurs in a high percentage in the amoebas *Acanthamoeba polyphaga*, *Naegleria lovaniensis* and *N. gruberi*.[12,13] It is likely that cycloartenol will also be detected in other nonphotosynthetic protozoa. Cycloartenol is synthesized from all-*trans* (*S*)-2,3-oxidosqualene (**2**) and is the precursor of the membrane sterols sitosterol, stigmasterol and campesterol or dihydrobrassicasterol in plants as well as of the plant hormones brassinosteroids.[2,11]

2.11.2.1 Biochemistry of Cycloartenol Cyclase

Cycloartenol cyclase (systematic name: (*S*)-2,3-oxidosqualene mutase (cyclizing, cycloartenol forming), EC 5.4.99.8) catalyzes the reaction shown in Scheme 1. An intermediate in this reaction is the protosteryl cation (**15**). This cation has a boat configuration in ring B and therefore differs from the dammarenyl cation (**10**) in the configuration of this ring. The reaction closely resembles the formation of lanosterol (see Chapter 2.10). At least 10 intermediate cations have to be formed during the formation of cycloartenol. At the end of the reaction, the proton is abstracted from C-19, and not from C-9, as in the case of lanosterol formation, thereby forming the additional cyclopropane ring.

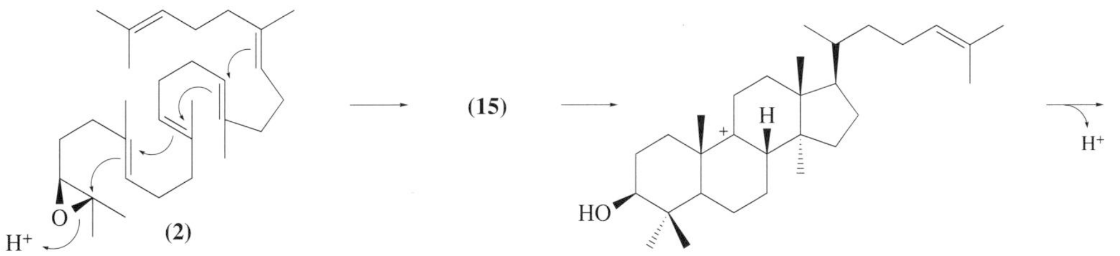

O H+ (2) (15) H + HO H+ (16)

Scheme 1

2.11.2.1.1 Purification

Progress in the purification of cycloartenol cyclase was slow due to its hydrophobicity and membrane location. Specific detergents of high purity were required to solubilize the enzyme in its native state from membrane preparations. In addition, specific purification procedures have had to be developed for each individual enzyme.

In 1988 and 1989 two cycloartenol cyclases were reported to be purified.[14,15] The purification procedure principally consisted in the preparation of microsomes, solubilization of the cyclase by Triton X-100, chromatography on hydroxyl apatite, isoelectric focusing, DEAE-cellulose chromatography and a tandem HPLC. Highly purified preparations were obtained from pea seedlings and from a cell suspension culture of *Rhabdosia japonica* which appeared on SDS-polyacrylamide gel electrophoresis as a single homogeneous band. The estimated molecular weights were 54 kDa and 55 kDa respectively. These values contrast with the 86 kDa found for the cloned cycloartenol cyclase of *Arabidopsis thaliana* (see Section 2.11.2.2).[16]

2.11.2.1.2 Reaction mechanism

Experiments on specific aspects of the reaction mechanism of cycloartenol cyclase were performed with unpurified enzyme preparations from *Ochromonas malhamensis* and *Zea mays* seedlings.[17,18] Abstraction of the hydrogen C-19 and formation of the cyclopropane ring are the distinctive steps as compared with lanosterol formation (see Chapter 2.10). The conclusions from the above experiments were: (1) cyclopropane ring closure takes place with retention of configuration of the hydrogens at C-19 and (2) abstraction of the hydrogen at C-19 is presumably performed by an enzymic subsite which is situated above the C-9 carbocation.

2.11.2.1.3 Substrate specificity

Similar to lanosterol cyclase, (*S*)-2,3-oxidosqualene is the physiological substrate of cycloartenol cyclase. The enantiomer (*R*)-2,3-oxidosqualene does not act as a substrate.[19] 22,23-Dihydro-22-methylene-2,3-oxidosqualene is a substrate analogue with an alteration in the *pro*-side chain region resulting in the synthesis of 24-methylenecycloartenol.[20] In contrast to lanosterol cyclase, for which numerous substrate analogues have been tested, only a few have been tested with cycloartenol cyclase.[8] It is to be expected that cycloartenol cyclase will exhibit a broad substrate specificity similar to lanosterol cyclase (see Chapter 2.10).

2.11.2.1.4 Inhibitors of cycloartenol cyclase

Inhibitors of sterol cyclases were reviewed in articles by Mercer[9] and by Abe *et al.*[10] Many inhibitors have been synthesized and tested against lanosterol cyclase due to their medical potential as hypocholesterolemic and antifungal drugs (see Chapter 2.10). Relatively few inhibitors have been tested against cycloartenol cyclase.

The most interesting group of inhibitors are mechanism-based inhibitors which are intermediate analogues of the oxidosqualene cyclization. They will be discussed in more detail owing to the pioneering work performed with these inhibitors.[21–24] The main features of these inhibitors are their analogy to oxidosqualene or to partially cyclized intermediates, together with the presence of a nitrogen atom at positions corresponding to a carbocation formed during cyclization. In this context the first compounds were structural variants of 2-aza-2,3-dihydrosqualene (**20**). All of these, with the exception of isopropyl derivatives, noncompetitively inhibited the cycloartenol cyclase of maize seedlings,[24] the *N*-oxide analogues (**21**) of these compounds being even more potent inhibitors. Mono-, bi-, and tricyclic compounds have also been synthesized.[24] The monocyclic *N*-alkyl-hydroxy-piperidine (**22**) and an 8-azadecalin derivative (**23**) were shown to be the strongest inhibitors of cycloartenol cyclase ($I_{50} = 1–2\ \mu M$). Tricyclic compounds were much less active. 2,3-Dihydro-2,3-iminosqualene (**24**) is also a nitrogen containing inhibitor though not a transition state analogue; it is mentioned here as another example of a nitrogen containing cycloartenol cyclase inhibitor.[25]

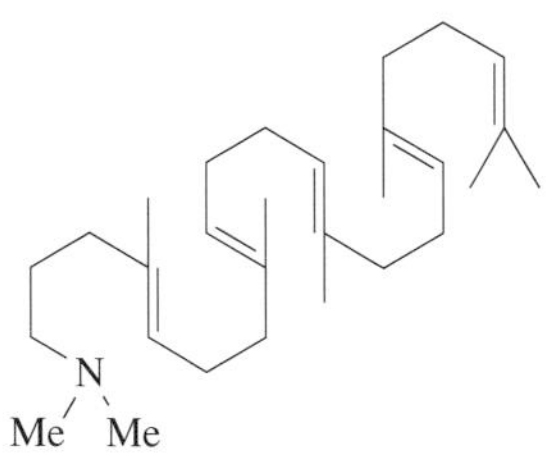

N Me Me

(20)

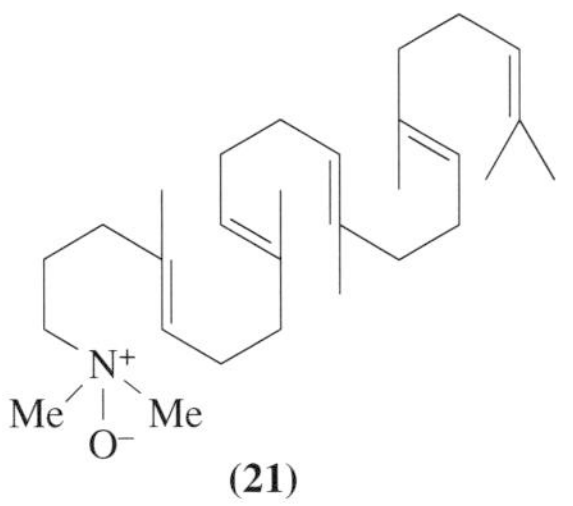

N+ Me O− Me

(21)

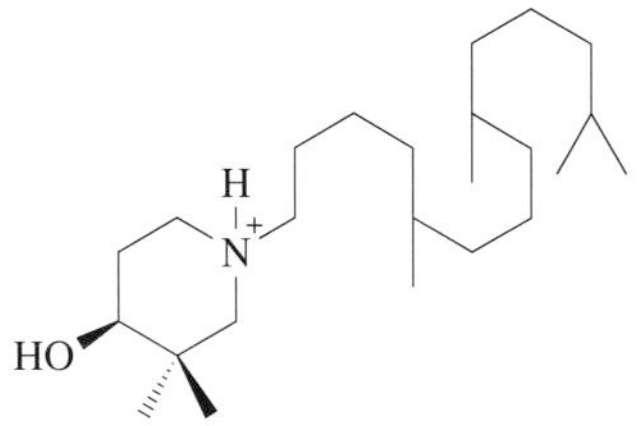

H N+ HO

(22)

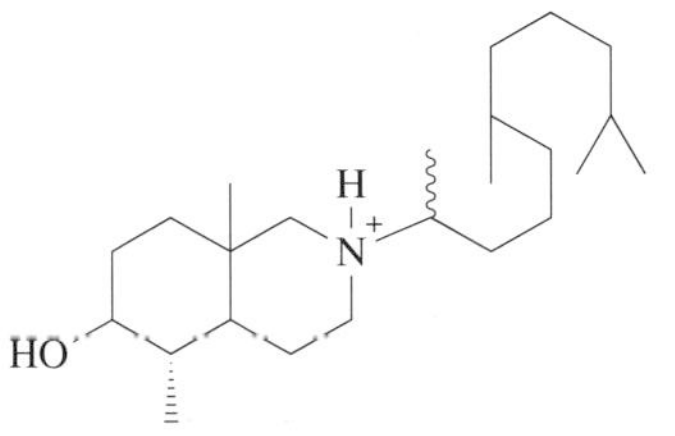

H N+ HO

(23)

HN

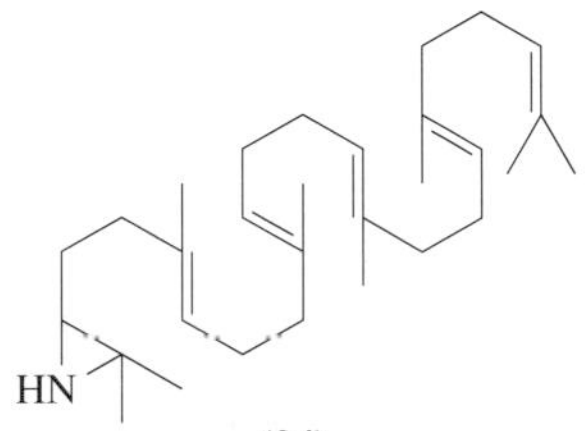

(24)

Other compounds which are not analogues of oxidosqualene—such as the herbicide AMO 1618 ((2-isopropyl-5-methyl-4-trimethylammonium chloride)phenyl-1-piperidine carboxylate) (**25**)—are inhibitors of cycloartenol cyclase.[26]

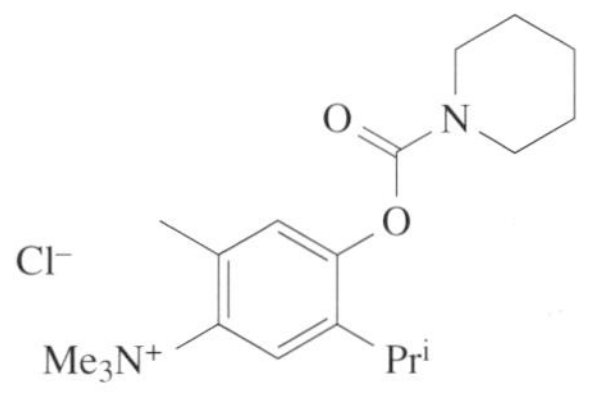

Cl⁻ Me$_3$N$^+$ Pr^i O N O

(25)

2.11.2.2 Genetics of Cycloartenol Cyclase

Cloning strategies for the isolation of triterpene cyclase genes diverge widely. In the case of the cycloartenol cyclase gene an interesting but laborious strategy was chosen by Matsuda and co-workers.[16] A yeast strain was transformed with an *Arabidopsis thaliana* cDNA expression library, the plasmids of which contained a constitutive yeast promoter. Permeabilized cells of pools of colonies were assayed for enzyme activity by thin-layer chromatography and using nonradiolabeled oxidosqualene. The homogenate from one out of about 10 000 transformants cyclized oxidosqualene to cycloartenol. The corresponding plasmid contained a coding region of 2277 basepairs encoding a protein with significant homology to known triterpene cyclases that exhibited cycloartenol cyclase activity. The gene encoding this cDNA was named CAS1 (*H*-**c**yclo**a**rtenol **s**ynthase).

2.11.3 AMYRIN CYCLASES AND OTHER TRITERPENE CYCLASES

α- (**29**) and β-Amyrin (**30**) occur in higher plants, but have never been found in bacteria.[1] They are synthesized from 2,3-oxidosqualene via the dammarenyl cation (**10**) having an all-chair conformation and further hypothetical cationic species as the baccharanyl (**26**), lupenyl (**27**) and oleanyl (**28**) cations depicted in Scheme 2.[8] The dammarenyl cation differs from the protosteryl cation (**15**) in the conformation of ring B. Hence, this type of cyclization shares common features with hopene and tetrahymanol cyclizations. Ring E is built by complex rearrangement reactions, according to the current view. This mechanism is quite different from the E-ring formation in hopene and tetrahymanol.[27]

2.11.3.1 Biochemistry of Amyrin Cyclases

β-Amyrin cyclase was purified from pea seedlings in eight steps as a soluble and homogeneous fraction by SDS-PAGE.[28] Preparation of microsomes and solubilization of active cyclase by Triton X-100 was followed by steps very similar to those described for the purification of cycloartenol cyclase. The molecular weight of the most abundant protein visible by SDS-PAGE had a molecular weight of 35 kDa, in contrast to the molecular weights of other triterpene cyclases (see Table 1). A similar purification procedure has been described for the β-amyrin cyclase of *Rabdosia japonica*.[15]

Truncated substrates have been cyclized by β-amyrin cyclase, e.g., 24,30-bisnor-2,3-oxidosqualene to 29,30-bisnoramyrin.[37] In addition, a bicyclic analogue of 2,3-oxidosqualene (**31**) was cyclized to β-amyrin.[38] This processing of a partially cyclized substrate shows clearly that the polycyclization reaction need not be a concerted reaction in the strict sense. The reaction can be initiated not only at C-2 of oxidosqualene but also at a later, partially cyclized, stage.

The same transition state analog inhibitors used for cycloartenol cyclase were tested against α- and β-amyrin cyclase. The monocyclic *N*-alkyl-hydroxypiperidine was much less active against the

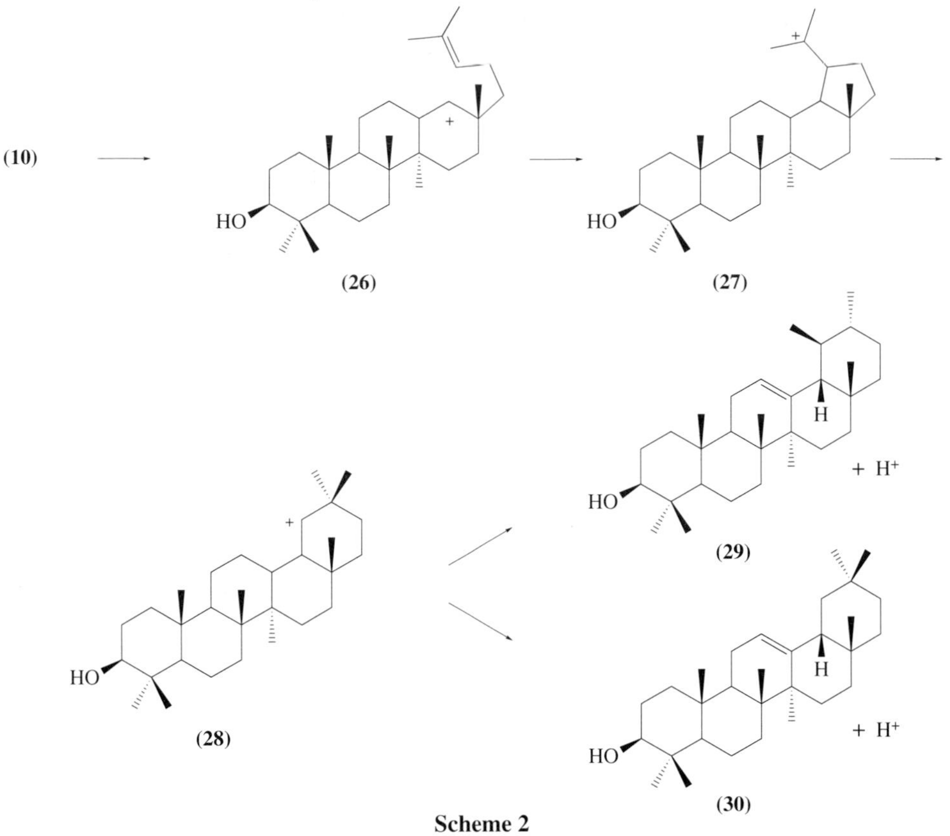

(10) → (26) → (27) →

(28) → (29) + H+

(28) → (30) + H+

Scheme 2

Table 1 Size and molecular weight of bacterial squalene hopene cyclases and plant triterpene cyclases. OCC, oxidosqualene cycloartenol cyclase; OLC, oxidosqualene lupeol cyclase.

Source	*Amino acids*	*Molecular weight* (kDa)	*Ref.*
Alicyclobacillus acidocaldarius	631	71.5	29
Alicyclobacillus acidoterrestris	634	71.2	30
Zymomonas mobilis	658	74.1	31
Bradyrhizobium japonicum	658	73.4	32
Methylococcus capsulatus	654	74.0	33
Rhodopseudomonas palustris	654	72.3	34
Synechocystis sp. PCC6803	646	72.0	35
Arabidopsis thaliana OCC	759	86.4	16
Arabidopsis thaliana OLC	757	87.3	36

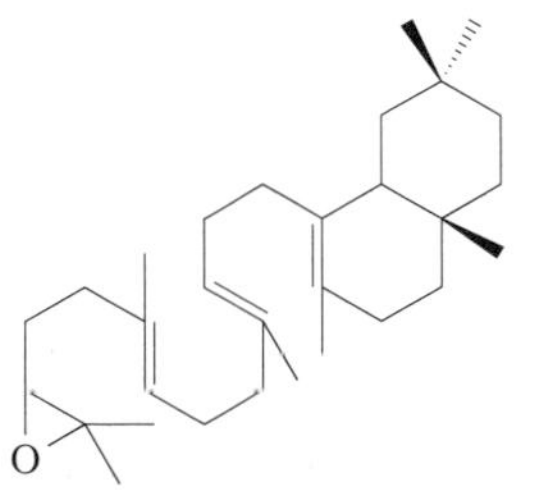

(31)

α(β)-amyrin cyclases of *Rubus fruticosus* or pea seedlings as compared with cycloartenol cyclase.[24] *Cis/trans* isomeric 20-oxa vinyl ether analogues, with oxygen at a position corresponding to C-20 of oxidosqualene, are modest inhibitors of β-amyrin cyclase but efficient against lanosterol cyclase.[39]

2.11.3.2 Genetics of Lupeol Cyclase

Searching the GenBank database for sequences similar to *Arabidopsis thaliana* cycloartenol cyclase, a similar sequence was found.[36,40] This sequence was cloned, labeled and used for screening an *A. thaliana* cDNA plasmid library by colony hybridization. A positive clone was sequenced and expressed in yeast deficient in lanosterol synthesis. A cell-free extract of the transformed yeast synthesized lupeol (**32**) from oxidosqualene. Lupeol cyclase is 54% identical to cycloartenol cyclase (for further data see Table 2). This strategy of cloning and characterization is applicable to further *A. thaliana* triterpene cyclase genes once significant parts of the genome are sequenced.

(32)

Table 2 Similarity (upper part) and identity values (lower part) for triterpene cyclases from prokaryotes and eukaryotes. Similarity was calculated with the software ALIGN Plus version 2.0 and identity with CLUSTAL/PALIGN. SHC, squalene hopene cyclase; OSC, oxidosqualene lanosterol cyclase; OCC, oxidosqualene cycloartenol cyclase; OLC, oxidosqualene lupeol cyclase.

Triterpene cyclase	*M.c.*	*B.j.*	*R.p.*	*Z.m.*	*Syn.*	*A.ac.*	*A.at.*	*H.s.*	*S.c.*	*A.t.*	*A.t. OLC*
SHC *M. capsulatus*		61	60	60	47	49	48	34	33	35	37
SHC *B. japonicum*	49		86	71	49	47	47	38	34	38	39
SHC *R. palustris*	51	79		70	49	49	47	38	34	39	38
SHC *Z. mobilis*	50	59	60		50	48	45	37	32	36	37
SHC *Synechocystis* sp.	37	39	38	39		51	51	37	33	38	38
SHC *A. acidocaldarius*	41	40	40	40	43		77	38	34	37	37
SHC *A. acidoterrestris*	39	38	38	38	41	66		38	35	39	38
OSC *H. sapiens*	28	29	30	28	28	30	28		51	55	55
OSC *S. cerevisiae*	28	28	25	24	26	28	29	41		48	45
OCC *A. thaliana*	30	28	30	29	31	30	31	45	39		68
OLC *A. thaliana*	30	28	29	28	29	30	30	40	35	56	

2.11.4 HOPENE AND TETRAHYMANOL CYCLASES

Hopene (**33**) and tetrahymanol (**34**) cyclases are included in a single chapter, as these compounds are very similar; the positions of the methyl groups are the same, though in the case of the gammacerane derivative tetrahymanol (**34**), ring E is expanded. Moreover, it is known that tetrahymanol cyclase produces diplopterol (= hopan-22-ol) (**35**) as a by-product[41] and that some bacteria produce tetrahymanol in addition to hopanoids.[42,43] The former is in cells probably as a by-product of certain hopene cyclases.

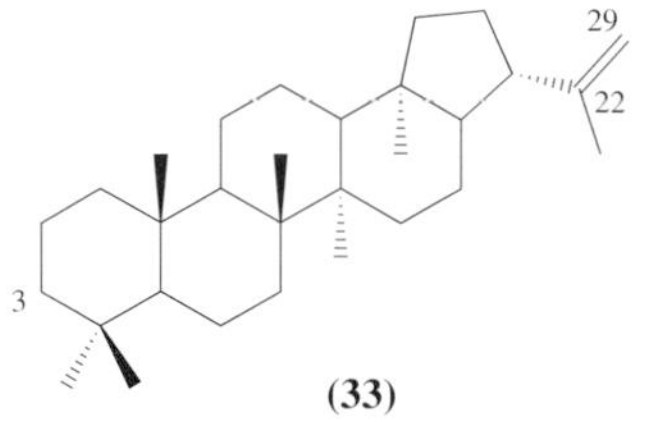

(33) 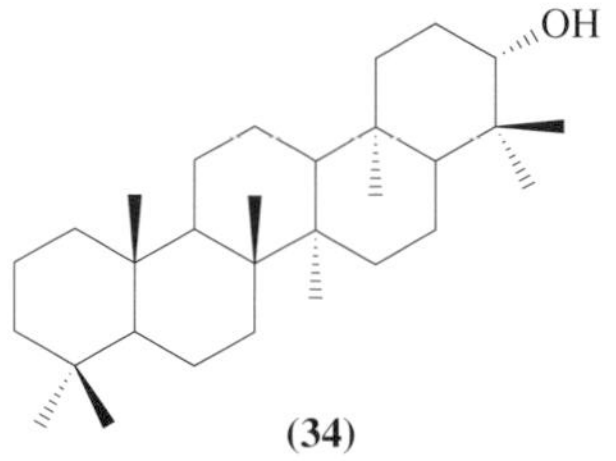 (34) 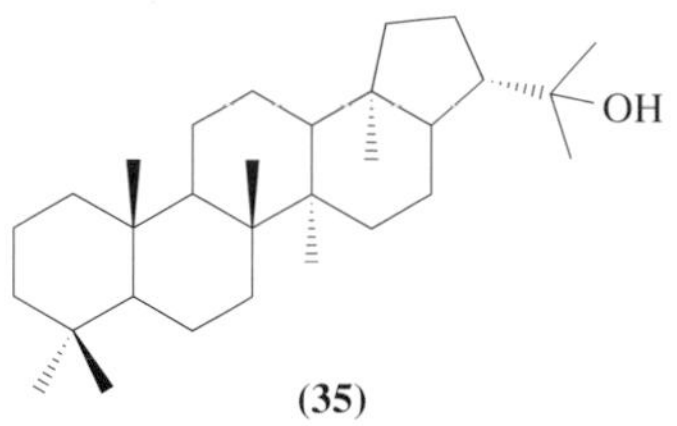 (35)

2.11.4.1 Occurrence of Hopanoids and Tetrahymanol

Hopanoids were first detected in the dammar resin of the tropical tree *Hopea* (*Dipterocarpaceae*), named in honor of the eighteenth-century British botanist John Hope. 3-Hydroxy- and 3-keto-hopanoids have been found in scattered taxa of higher plants.[1] Deoxyhopanoids, on the other hand, occur mainly in bacteria, in a few fungi, and frequently in ferns.[1] A very informative review treats the trove of isoprenoids found in ferns.[44]

The most prominent occurrence of hopanoids is in *Bacteria*. In the second prokaryotic domain, that of the *Archaea*, they have never been detected. Hopanoids have been isolated from about 100 bacterial strains.[45] Their occurrence is mainly concentrated in the groups of cyanobacteria, *Rhodospirillaceae* and methylotrophic bacteria. They also frequently occur in taxonomically unrelated dinitrogen-fixing bacteria.[46–48] Bacterial hopanoids mainly differ from plant hopanoids by the absence of an oxygen function at C-3 and the fact that the main hopanoids carry an elongated side chain.[27] As shown by several types of experiments, hopanoids have the same lipid-condensing function as cholesterol has in higher organisms.[49]

Tetrahymanol occurs in some lower eukaryotes, e.g., the ciliate *Tetrahymena*, the anaerobic ciliate *Trimyema compressum* and the low fungus *Piromonas communis*[50–52] as well as other lower eukaryotes, especially anaerobic ones. It was very surprising to find tetrahymanol occurring in bacteria as well, e.g., in *Rhodopseudomonas palustris* and the phylogenetically related *Bradyrhizobium japonicum*.[43,44]

2.11.4.2 Biochemistry of Hopene Cyclase

Two types of hopene cyclases must be postulated. One type uses oxidosqualene as substrate, resulting in the synthesis of hopan-3β-ol. This is very probably the case in higher plants where one finds mainly the oxygenated type of hopane derivatives. A second type of hopene cyclase using squalene as substrate (Scheme 3) probably occurs in ferns and definitely does in bacteria. This cyclase uses all-*trans*-squalene as substrate and transforms it via the hopanyl cation **(36)** to hop-22(29)-ene **(33)** and diplopterol (= hopan-22-ol) **(35)**.[53] This reaction is highly exergonic (see Section 2.11.6.2).

2.11.4.2.1 Purification

Three bacterial hopene cyclases have been purified to homogeneity. The first example was the hopene cyclase from the Gram-positive, thermoacidophilic bacteria *Alicyclobacillus* (formerly *Bacillus*) *acidocaldarius*.[53–55] This bacterium grows optimally at 60 °C, so hopene cyclase was conveniently purified at room temperature. The enzyme was solubilized by Triton X-100 from cytoplasmic membranes; followed by chromatography on DEAE-cellulose, phenyl Sepharose and two gel-filtrations on Sephacryl-S500 and Sephacryl-S300. These steps resulted in a 900-fold purification in comparison to the cell-free extract. The purified hopene cyclase of *A. acidocaldarius* produces about 80% hopene and 20% diplopterol from squalene.

A very different procedure was adopted for the purification of hopene cyclase from *Rhodopseudomonas palustris*.[56] This bacterium was selected because it is Gram-negative, photosynthetic and, besides hopene, diplopterol and elongated hopanoids, it also contains tetrahymanol. After isolation of the cytoplasmic membranes, the enzyme was solubilized by CHAPS and solubilized proteins precipitated at 60 °C, whereby a significant enrichment of hopene cyclase resulted. Further purification was achieved by successive chromatography on DEAE Sephacel, octyl Sepharose and

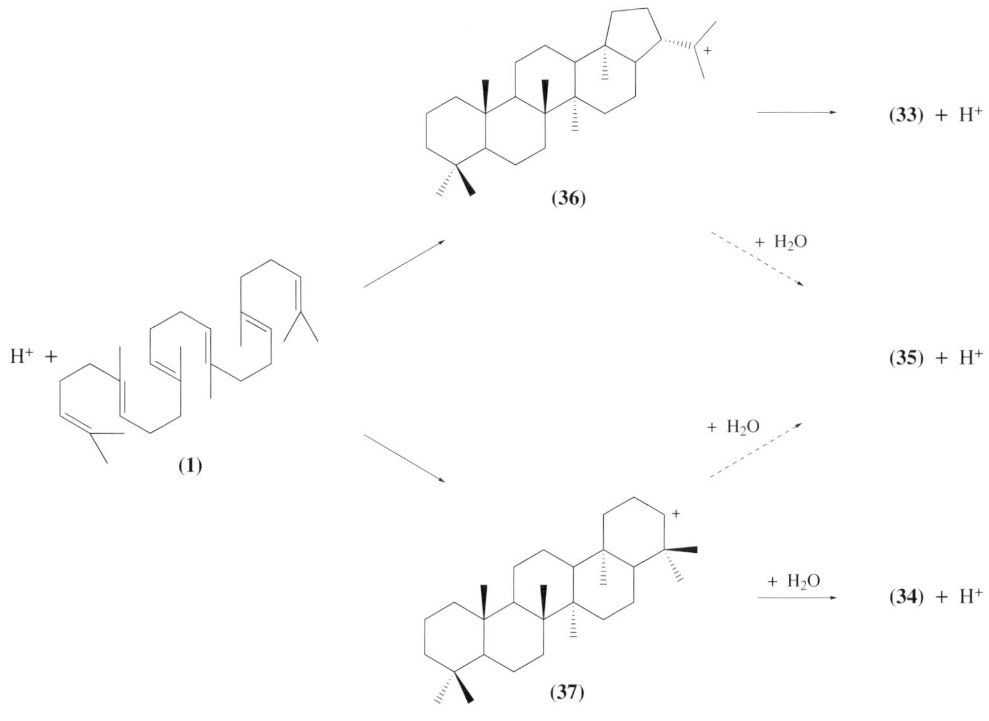

Scheme 3

Blue Sepharose. The fact that Blue Sepharose gave the best results for obtaining a pure enzyme preparation is surprising since this material is normally used for adsorption of NAD(P)-dependent enzymes. During all purification steps, though hopene and diplopterol were observed in the enzyme assay, production of tetrahymanol never was. The possibility therefore exists that the enzyme produces tetrahymanol as an unspecific side-product only in the native lipid environment; another possibility is that a second cyclase activity was inactivated from the outset of the purification procedure. The *R. palustris* cyclase shows a significant difference from the *A. acidocaldarius* cyclase in that it produces nearly the same amount of diplopterol as hopene.

Hopene cyclase from the Gram-negative, ethanol-producing *Zymomonas mobilis* was also purified to homogeneity. Here the purification was hampered by the considerable instability of the enzyme. Solubilization of the cyclase from the membrane fraction by Triton X-100 was followed by chromatography on octyl-Sepharose, isoelectric focusing, gel filtration on Sephacryl S300 HR, DEAE-cellulose, and preparative native PAGE in the presence of taurodeoxycholate.[57] Properties of the purified hopene cyclases are listed in Table 3.

Table 3 Biochemical properties of different squalene hopene cyclases and squalene tetrahymanol cyclase. n.d., not determined.

Source	K_m *for squalene* (μM)	*Optimum pH*	*Optimum temperature* (°C)
Alicyclobacillus acidocaldarius	9	6.0	60
Rhodopseudomonas palustris	n.d.	6.5	30
Zymomonas mobilis	12	6.0	25.5
Tetrahymena thermophila	18	7.0	30

2.11.4.2.2 Substrate specificity

Enantiomers of oxidosqualene have to be considered as unusual substrates for hopene cyclases. A cell-free system of *Acetobacter pasteurianum* converts (3*R*)- and (3*S*)-2,3-oxidosqualene to

3α-hydroxy and 3β-hydroxyhopanoids.[58] Usually, oxidosqualene is not expected to occur in hopanoid-synthesizing bacteria, with the exception of *Methylococcus capsulatus* which, besides hopanoids, also synthesizes sterols.[59]

2,3-Dihydrosqualene (**38**) is a fairly good substrate for hopene cyclase in a cell-free extract of *A. acidocaldarius*.[60] This compound is converted to a 1:1 mixture of (20*R*)-dammar-13(17)-ene (**39**) and (20*R*)-dammar-12-ene (**40**) (see Scheme 4). Both products have the 20*R* configuration (opposite to lanosterol), suggesting that the stereochemistry of the cyclization reaction is controlled by the cyclase. In these compounds the D-ring is five-membered, in comparison to hopene. Thus the formation of a six-membered ring-D (an anti-Markovnikov cyclization) is dependent on assistance from the terminal double bond. This result also shows that pentacyclic triterpene-forming enzymes already have the hidden potential to form tetracyclic triterpenes. Also intriguing is the fact that hopene cyclase contains a proton-eliminating property which is not regiospecific. This fact begs the question whether in this case the same amino acid residue abstracts the proton, as in the case of hopene cyclization. The terminal tetracyclic cation would then have to move to bring the corresponding region (C-12/C-13) into the correct position. Alternatively, another amino acid residue may adopt the proton-abstracting function.

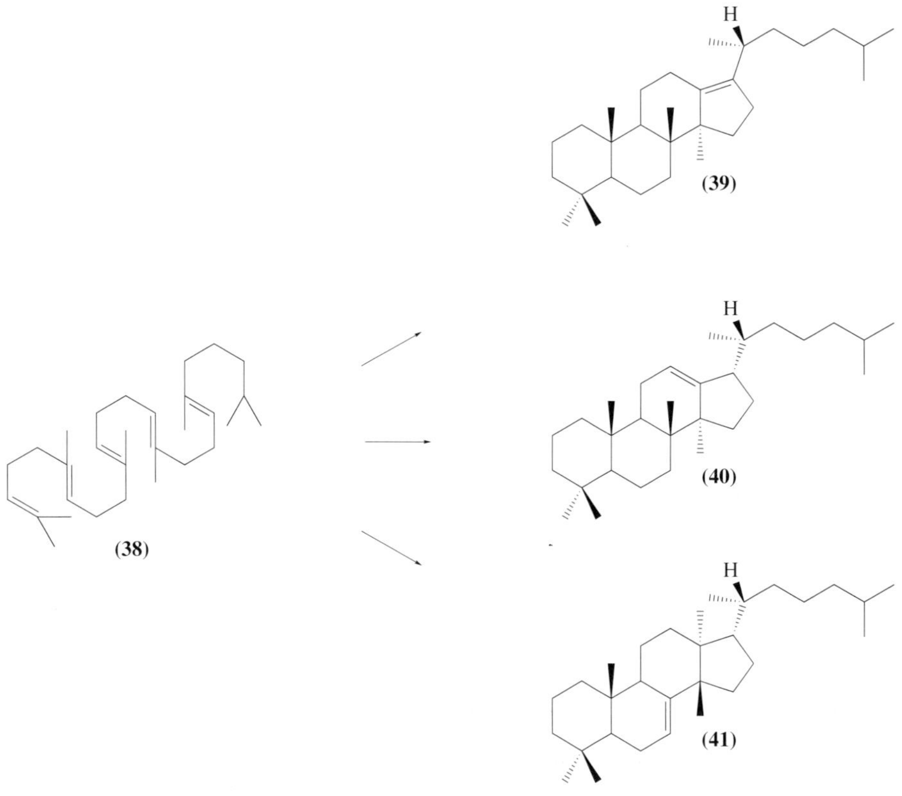

H

(39)

H

(38)

(40)

H

(41)

Scheme 4

Highly enriched hopene cyclase preparations from *A. acidocaldarius* convert, in very low yield, homogeraniol and *E,E*-homofarnesol to bicyclic and tricyclic compounds containing a furan ring. Bishomofarnesyl ether is cleaved and also cyclized to the same tricyclic compound as above. Homofarnesylcarboxylate, homofarnesyl methyl ether, farnesol, geraniol and nerol were not substrates for hopene cyclase.[54]

2.11.4.2.3 *Product specificity*

In all preparations of the *A. acidocaldarius* cyclase, including highly purified ones, the enzyme produces, besides hop-22(29)-ene (**33**), up to 20% diplopterol (hopan-22-ol) (**35**).[53–55] The oxygen in diplopterol is presumed to be derived from water, as was proved for tetrahymanol cyclase.[61] The ratio of the products hopene:diplopterol was independent of pH in the assay mixture.[53] The reason why diplopterol is produced may be that for the cyclase it is inherently difficult to exclude water during the proton-elimination process which can compete for quenching with the hopenyl cation.

In very pure cyclase preparations involving replacement of Triton X-100 by other detergents, additional conversion products in amounts of 0.5–1.5%, compared with hopene and diplopterol, have been detected. Such products were not previously detected because of the multiple peaks of Triton X-100 in the gas chromatographic assay. These minor components, which all have a molecular mass of 410 (identical to the mass of squalene and hopene) and which have gas chromatographic retention times between squalene and hopene, include eupha-7,24-diene (**42**), compound (**11**), and two further dammaradienes.[62] This result is very interesting because it shows that hopene cyclase possesses some limited potential for synthesis of tetracyclic triterpenes and that methyl shifts are feasible. Furthermore, this result demonstrates that the catalytic process may be inherently difficult to keep on track.

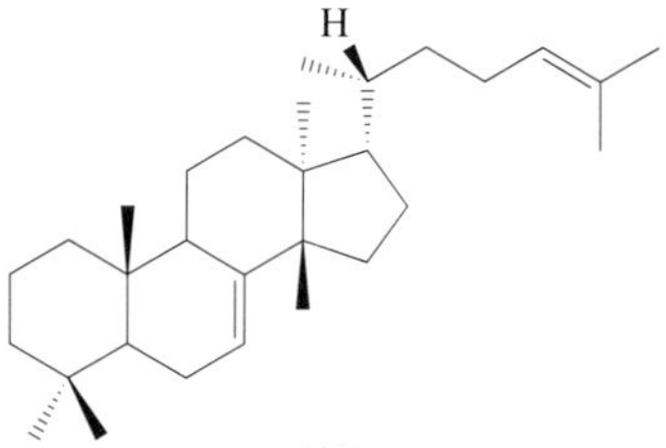

(**42**)

2.11.4.2.4 *Poisons and inhibitors*

Several sulfhydryl reagents and the histidine reagent diethyl pyrocarbonate proved to be effective poisons for hopene cyclase.[53] These results demonstrate that cysteine and histidine residues participate in catalysis or are otherwise essential in function and structure. Hopene cyclase, like eukaryotic oxidosqualene sterol cyclases, is strongly and competitively inhibited by the herbicide AMO 1618 (**25**), which bears no simple similarity to the substrate.[53]

Besides the above-mentioned examples, compounds more or less similar to squalene were also tested and proved to function as inhibitors, e.g., 2,3-dihydro-2-azasqualene (**20**), its *N*-oxide (**21**) and its *N,N*-diethyl analogue, as well as 2,3-dihydro-2,3-iminosqualene (**24**). These compounds also selectively inhibit bacteria containing hopanoids.[63]

An efficient competitive inhibition was observed with dodecyltrimethylammonium bromide and dodecyldimethylamine-*N*-oxide (**43**) and their homologues.[55] The concentrations for 50% inhibition are about 0.5 μM. Mutants of *A. acidocaldarius* that are resistant against 60 μM dodecyltrimethylammonium bromide possess a wild-type hopene cyclase, as was demonstrated by inhibition experiments.[55]

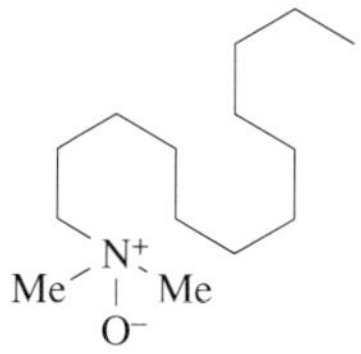

(**43**)

2.11.4.2.5 *Circular dichroism measurements*

Circular dichroism (CD) is suitable for obtaining information about the secondary structure of hopene cyclase of *A. acidocaldarius*. The secondary structure estimated by this method were α-helicity, 38%; β-sheets, 0%; turns, 32%; and random structures, 30%. These results are in accord with structure-prediction programs. Interestingly, no β-sheets were found. Measurements were also performed with several mutant hopene cyclases and gave essentially the same results.[64] These results are in accordance with the x-ray structure of hopene cyclase (see Section 2.11.6.2).

2.11.4.3 Biochemistry of Tetrahymanol Cyclase

Squalene-tetrahymanol cyclases of *Tetrahymena pyriformis* and *T. thermophila* catalyze the formation of tetrahymanol from all-*trans*-squalene (Scheme 3).[41,65] The postulated intermediate is the gammaceryl cation (**37**). Diplopterol (**35**) was detected as a minor conversion product.[41] Caspi and co-workers unambiguously demonstrated that the hydroxy group of tetrahymanol originates from water and not from molecular oxygen, as in the case of sterols.[61] In the formation of both tetrahymanol and diplopterol, the hydroxyl ion is added from the α-face. It is also worth stressing that in the cyclization process leading to tetrahymanol, three energetically unfavorable anti-Markovnikov ring formations occur. By comparison, in sterol cyclization only a single anti-Markovnikov cyclization occurs, and for hopene cyclization two are involved.

2.11.4.3.1 Purification

Tetrahymanol cyclase was highly enriched from the ciliate *Tetrahymena thermophila*.[41] After homogenization and centrifugation the sedimented crude membrane fraction was solubilized with octylthioglucoside. The solute was subjected to chromatography on DEAE-trisacryl, hydroxyapatite, and FPLC on Mono Q. K_m values for the optimal temperature and optimal pH are shown in Table 3. The molecular weight of the enriched main band in SDS-PAGE was 72 kDa. It is also very interesting to note that this cyclase is inactivated by Tween 80 and digitonin. Its activity can be recovered, however, by supplementing the assay buffer with octylthioglucoside above its critical micellar concentration. This procedure might be applicable to other membrane-bound proteins.[41]

2.11.4.3.2 Substrate specificity

A cell-free extract from *T. pyriformis* cyclizes all-*trans* pentaprenyl ether to a tetracyclic scalarane-type sesterterpene. This compound has a six-membered ring D. All-*trans*-hexaprenyl ether is converted to bicyclo-, tricyclo-, tetracyclo-, and pentacycloprenyl methyl ethers lacking a hydroxyl group.[66] Normally, tetrahymanol cyclase synthesizes an alcohol as product. Also intriguing is the fact that proton abstraction in these cyclization products occurs at different sites in the carbon skeleton. Hence, similar to hopene cyclase experiments with dihydrosqualene as substrate, one has to assume several potential proton-abstracting sites or else the terminal cation moves to a single abstraction site. Furthermore, the observed cyclization products argue for a stepwise and not a concerted cyclization process.

Racemic (*RS*)-2,3-oxidosqualene is transformed by a cell-free system from *Tetrahymena pyriformis* to gammacerane-3α,21α-diol and gammacerane-3β,21α-diol.[65] In addition, a third component with a 2,3,4-trimethylcyclohexanone moiety was detected. There are strong suggestions that this compound is formed by an incomplete cyclization of the (3*R*)-enantiomer only.[67]

The catalytic plasticity of tetrahymanol cyclase was also demonstrated by the transformation of 2,3-dihydrosqualene (**38**) to euph-7-ene (**41**) in good yield (Scheme 4).[29] This compound corresponds to eupha-7,24-diene (**42**) which is a minor product of the hopene cyclase reaction (see Section 2.11.4.2.3). It is surprising that in the formation of eupha-7-ene numerous 1,2-shifts are observed and the cyclization reaction is not terminated by the addition of water. Both events do not occur in the formation of tetrahymanol. A further interesting observation is the very different cyclization product obtained in comparison to the products of dihydrosqualene cyclization by the *A. acidocaldarius* cyclase (see Section 2.11.4.2.3).

2.11.4.3.3 Poisons and inhibitors

p-Chlormercuribenzoate and diethyl pyrocarbonate are poisons for tetrahymanol cyclase, indicating that cysteine and histidine residues are essential for catalysis or structure of the cyclase. 2,3-Iminosqualene (**24**) and *N*,*N*-dimethyldodecylamine-*N*-oxide (**43**) are effective inhibitors with I_{50} values of 50 nM and 30 nM, respectively.[41]

2.11.4.4 Genetics of Hopene Cyclases

2.11.4.4.1 Cloning strategies

The hopene cyclase of the thermophilic bacterium *A. acidocaldarius* was the first triterpene cyclase to be cloned, sequenced and expressed.[29] The strategy of reverse genetics was applied. After purification, 24 N-terminal amino acids were sequenced and several DNA probes were constructed based on this sequence, taking into account the high GC-content of *A. acidocaldarius* DNA. By hybridization of these DNA probes with genomic DNA, a fragment was identified and ligated into plasmid pUC19. Transformed *Escherichia coli* showed hopene cyclase activity. The plasmid insert was sequenced and the N-terminus of the translated open reading frame proved identical to the N-terminus of the purified cyclase.

Based on this first hopene cyclase gene, DNA probes were constructed for the isolation of further cyclase genes. In this manner, the hopene cyclase gene of *Zymomonas mobilis* was cloned and expressed.[31] Based on the knowledge of the first two hopene cyclase genes and of the first lanosterol cyclase genes, PCR methods using degenerate primers were deployed to isolate the hopene cyclase genes of *A. acidoterrestris*, *Methylococcus capsulatus* and *Rhodopseudomonas palustris*.[30,33,34] *M. capsulatus* is the only bacterium containing steroids as well as hopanoids and consequently must possess a sterol cyclase.[59] Hybridization and PCR experiments failed using conserved stretches of the hopene cyclase gene as probes and primers, respectively, in an attempt to isolate the first bacterial lanosterol cyclase gene.[68]

A very different method was applied to isolate the hopene cyclase gene from *Bradyrhizobium japonicum*.[32] PCR, using genomic DNA as a template and degenerate primers homologous to conserved motifs of sterol and hopene cyclase genes, resulted in a 150-bp product. This product proved similar to the corresponding region of different hopene cyclase genes. Based on the 150-bp product, homologous primers were synthesized and PCR was applied to a pool of *E. coli* strains containing a cosmid gene library of *B. japonicum* genomic DNA. The large pools were narrowed down until two clones were identified that yielded a PCR-product of the expected length. From one of these clones the hopene cyclase gene was sequenced, permitting a hopanoid biosynthesis operon to be detected upstream.

In a genome-sequencing project, a hopene cyclase gene in the cyanobacterium *Synechocystis* has been detected.[35] Data and references on cloned triterpene cyclases are given in Table 1.

2.11.4.4.2 Hopanoid biosynthesis operons

An open reading frame for which the deduced amino acid sequence has a low but, nevertheless, distinct similarity to different eukaryotic squalene synthases was detected upstream of the hopene cyclase gene of *M. capsulatus*. Significant similarity in signature motifs was found by comparison to phytoene, dehydrosqualene and *Bradyrhizobium* synthases. Synthesis of squalene was demonstrated in transformed *E. coli* in which the wild-type does not form squalene at all but synthesizes its precursor farnesyl diphosphate.[69]

Upstream of the hopene cyclase gene in both, *Zymomonas mobilis* and *Bradyrhizobium japonicum*, three open reading frames were identified with pairwise similar sequences.[31,70] Two open reading frames have a low but distinct similarity in the specific amino acid motifs to genes of phythoene, dehydrosqualene and squalene synthases. A third open reading frame is related to phytoene desaturases. Neither organism contains carotenoids. Therefore, the genes are obviously related to hopanoid biosynthesis. The similarities suggest that in these bacteria dehydrosqualene is first biosynthesized by either one of the synthases and that, in a second step, squalene is formed from dehydrosqualene by an enzyme which is similar to phytoene desaturase. A second synthase may be essential for synthesis of an as yet unknown but polar product from farnesyl diphosphate.[71]

2.11.5 CONCLUSIONS FROM GENETIC STUDIES

2.11.5.1 Similarity of Triterpene Cyclases

That all hopene cyclases are similar is quite natural (see Table 2). The degree of similarity and identity roughly parallels the phylogenetic similarity of the bacteria (see Figure 1). Cyclases from

Gram-positive and Gram-negative bacteria are clustered, with one exception: cyanobacterial hopene cyclase is included in the cluster of Gram-positives. This is also interesting, in that these bacteria have an additional QW motif in common (see Section 2.11.5.2.2). Hopene cyclases are distantly related to sterol and other triterpene cyclases. It is very interesting to observe that lupeol cyclase forming a pentacyclic product is more closely related to cycloartenol cyclase than to hopene cyclases. On the basis of this result, one may speculate that both lupeol and the other plant triterpene cyclases originally evolved from a cycloartenol cyclase or a precursor thereof. It can be anticipated that plant hopene cyclases are also more similar to cycloartenol cyclases than to bacterial hopene cyclases. Only when quite detailed phylogenetic trees are available will a decision be possible on the viability of these speculations. Lanosterol cyclases are also included in the phylogenetic tree (Figure 2). Thus we observe that all sterol cyclases (and derivatives thereof) occur in a single cluster. They have, therefore, not taken multiple routes off from a precursor cyclase. Moreover, an early branching of cycloartenol cyclase is evident.

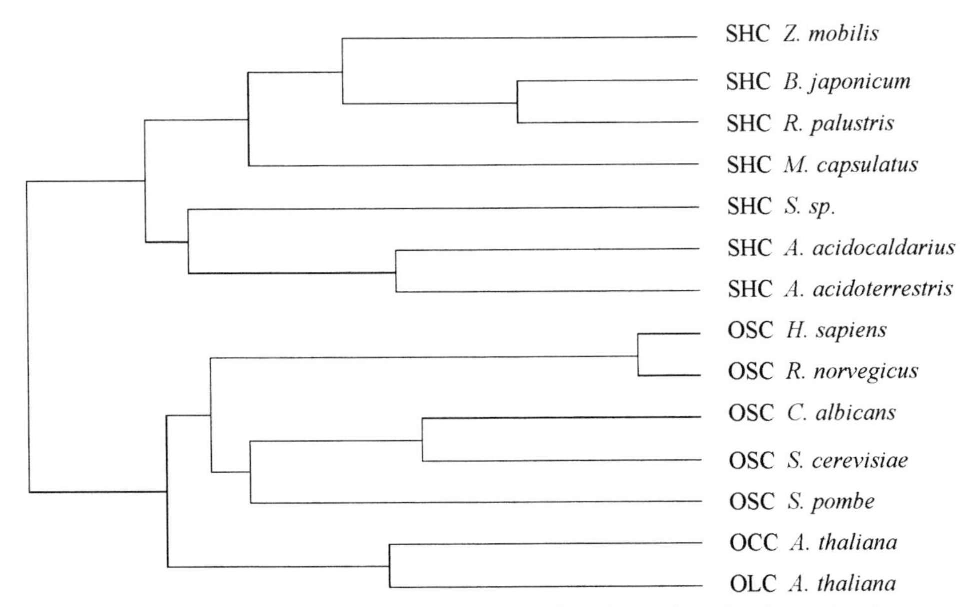

Figure 1 Dendrogram according to CLUSTAL program of prokaryotic and eukaryotic triterpene cyclases. SHC, squalene hopene cyclase; OSC, oxidosqualene lanosterol cyclase; OCC, oxidosqualene cycloartenol cyclase; OLC, oxidosqualene lupeol cyclase.

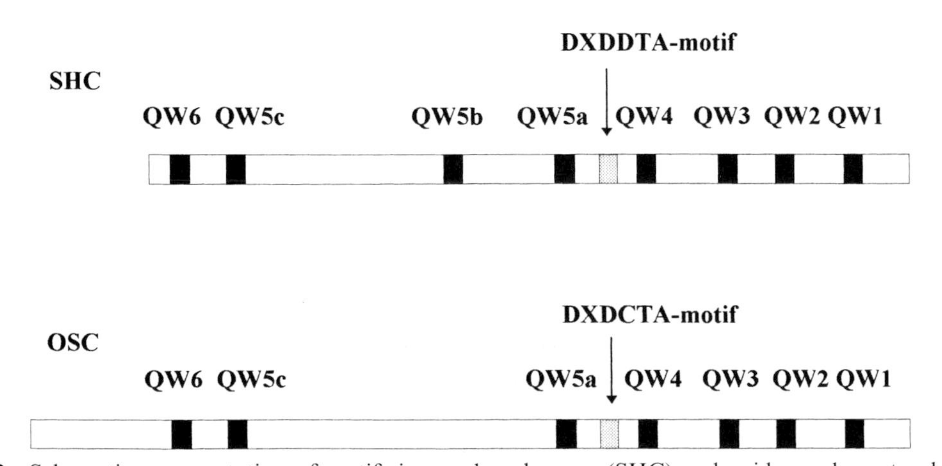

Figure 2 Schematic representation of motifs in squalene hopene (SHC) and oxidosqualene sterol cyclases (OSC). The black boxes represent the QW motifs. For historical reasons numbering is from the C-terminus. The shaded box represents the DXDDTA and DXDCTA motif respectively. QW5b occurs in Gram-positive bacteria only and QW5c is truncated at the N-terminus.

Triterpene cyclases have no similarity to any other terpene cyclases (see Chapters 2.05, 2.06, 2.07 and 2.08) with the exception of two common motifs occurring in kaurene and abietadiene synthases also (see Section 2.11.5.2).[72,73]

Eukaryotic, as compared with prokaryotic terpene cyclases, contain about 100 additional amino acids, mostly at the N-terminus. When aligning the 14 known triterpene cyclases it is not possible to spot large conserved regions. Small conserved areas are scattered over the total sequence. The C-terminus appears to be more conserved than the N-terminus. Of the 37 strictly conserved amino acids, 13 occur in the QW motifs (see Section 2.11.5.2.2) and 11 are aromatic. The high number of conserved aromatic acids points to their important function in triterpene cyclases.

Ourisson and co-workers have cited chemical and functional considerations as indicating a common phylogenetic origin for triterpene cyclases.[74] Molecular oxygen independent biosynthesis suggests that hopanoids or similar molecules are the phylogenetic precursors of membrane sterols. Though the core of the hypothesis is now confirmed, many details remain to be clarified. In any event, the hypothesis of Ourisson was of inspirational value to workers in this field.[74]

2.11.5.2 Common Amino Acid Motifs

During studies of rat lanosterol cyclase, peptides were characterized which were bound covalently to 29-methylene-2,3-oxidosqualene (see Chapter 2.10).[75] One peptide contained the DD/CTA motif. The analogue reacted covalently at the first aspartyl residue of this motif (see Chapter 2.10). This motif can also be aligned with the DD/CTA motif of hopene cyclases. The above studies also led to detection of a second specific motif, namely the QW motif.

2.11.5.2.1 The DD/CTA motif

The possible function of the DCTA motif of lanosterol cyclase is explained in detail in Chapter 2.10. The same motif also occurs in cycloartenol and lupeol cyclases. The homologous motif DDTA occurs in all hopene cyclases. They evidently seem to fulfill an essential function which may be impaired by mutation of these residues.

The Asp residues (Asp376 and Asp377) of the DDTA motif in hopene cyclase of *Alicyclobacillus acidocaldarius* have been systematically mutated to Glu, Asn, Gly, and Arg respectively.[64] With the exception of the exchange Asp376Glu, all other substitutions resulted in complete or near-complete loss of enzyme activity. Compared with the wild-type enzyme, the specific activity of the Asp376Glu mutant enzyme was reduced to 10%, accompanied by a significant decrease in the apparent V_{max}, while the apparent K_m remained unchanged. CD measurements indicated that the mutations did not affect the secondary structure (see Section 2.11.4.2.5).[64] In this respect, acidic amino acids may function as negative point charges to stabilize intermediate carbocations.[76] Other Asp residues (Asp374, Asp530 and Asp536) may not be essential for catalysis because they have been substituted by Glu without loss of activity. An exchange to Asn significantly alters activity. In all mutant cyclases, an alteration of the product specificity was never detected.[64]

It is interesting to note that a DDTA motif also occurs in three kaurene synthase As and in abietadiene synthase, again in a more or less N-terminal region.[72,73] By analogy to the triterpene cyclases, this motif may serve a similar function in diterpene and triterpene cyclases. This motif is diagnostic for terpene cyclases exhibiting a protonation-initiated cyclization reaction.[73]

2.11.5.2.2 The QW motif

The knowledge of the sequence of certain peptides from rat lanosterol cyclase,[75] together with visual inspection of *Alicyclobacillus acidocaldarius* hopene cyclase, permitted the recognition of a characteristic repeat.[77] This nontandem repeat was subsequently found in all triterpene cyclases at homologous sequence positions (Figure 2). This motif is highly specific for this group of proteins. Both a loose (a) and a restrictive version (b) of the consensus sequence is feasible:

(a) R/K G/A $(X)_2$ Y/F/W *L* $(X)_3$ Q $(X)_{2-5}$ G X W
(b) Y/F/W L $(X)_3$ Q $(X)_3$ G X W

The distance between the first aromatic amino acid and the Gln is strictly conserved. The most conserved portion is between Q and the C-terminal W. Hence the motif was called the QW motif. In Gram-positive bacteria and cyanobacteria there exists an additional motif at the N-terminus. Secondary structure prediction programs, predict a regular secondary structure for the QW motifs. At the N-terminus there is always an α-helix and starting with Q there is a loop structure. Forthcoming X-ray structural experiments should settle the question of secondary structure.

It is worth mentioning that triterpene cyclases contain 11–13% aromatic amino acids, more than double the percentage of aromatic amino acids found in proteins of *E. coli* or *S. cerevisiae*. Besides occurring in the QW motif, a significant percentage of the aromatic amino acids are found in other stretches. It is therefore likely that aromatic amino acids outside the QW motif will have essential functions, presumably in membrane insertion or in catalysis.

Prior to discussing the possible function of the QW motif, a model by Johnson for the active site of triterpene cyclases will be introduced.[76] From general mechanistic considerations which were confirmed by experiments with N-containing inhibitors mimicking high energy intermediates (see Section 2.11.2.1.4) it is accepted that transient carbocationic intermediates are formed during the polycyclization process. Depending on the two pathways for cyclization (dammarenyl or protosteryl cation intermediate), the cations are specifically located in the active site of the cyclase. Johnson developed the idea that the intermediate cations might be stabilized and positioned by interaction with negative point charges. These point charges would be nucleophilic carboxylic groups or aromatic residues.

It is known that aromatic residues form complexes with cations and cationic groups.[78] It is therefore realistic to assign to one or both of the aromatic amino acid residues in the QW motif a function in complexing the carbocations which are generated during the polycyclization of the substrate.[77,79] Furthermore, a key role for a repeat failure in the cyclase may be suspected, because the substrates squalene and oxidosqualene are composed of repeating building blocks. These ideas accord with the above model proposed by Johnson.

If one tries to develop a possible pathway for the evolution of triterpene cyclases, it seems readily conceivable that a triterpene cyclase synthesizing a monocyclic triterpene may evolve in a stepwise manner, by repeated addition (or evolution) of certain similar elements, to a polycycle-forming enzyme. Alternatively, mutation of certain subsites of the active site of a pentacyclic triterpene-forming cyclase might have led to evolution of tetra-, tri- and bicyclic triterpene-forming cyclases.[79] The phylogenetic tree (Figure 1) indicates an evolution from a cycloartenol cyclase to a lupeol cyclase forming a pentacyclic triterpene. It is not possible as yet to assign a proven role to specific amino acid residues.

Mutation experiments in the QW motif 2 (see Figure 2) of hopene cyclase of *Alicyclobacillus acidocaldarius*, including mutations at positions Lys518, Trp522, Gln527, Gly531, and Trp533, did not result in inactivation or alteration of product specificity.[78]

2.11.6 FUTURE LINES OF RESEARCH

2.11.6.1 Biological Problems

Studying the product specificity of hopene cyclase has raised one question in particular: is a specific cyclase necessarily required for every known triterpene, or does a nonspecific cyclase, at least in some cases, produce two or more triterpenes with more or less specific functions in secondary metabolism? Furthermore, nonspecificity in catalysis may be introduced by post-translational modification (e.g., proteolytic alteration of a cyclase).

For understanding specific problems of evolution it is always better to have a phylogenetic tree with numerous branches. As yet, a tetrahymanol cyclase sequence is missing from the tree. Some cycloartenol cyclases from protozoa, lower plants, and ferns should be sequenced to be sure that all cycloartenol cyclases have the same progenitor. A second possibility exists that cycloartenol cyclases were invented several times from a common precursor cyclase. It will also be necessary to sequence hopene cyclases from ferns and higher plants to elucidate whether these cyclases branched off from cycloartenol cyclases (as lupeol cyclase) or from bacterial hopene cyclases. Tackling this area efficiently should be possible on the basis of the known sequences using PCR with degenerate primers and with genomic DNA as template. It should also be possible to make cDNA libraries from tissues which produce certain triterpenes in large amounts and to use PCR on these cDNA libraries.

2.11.6.2 Cyclase Structures

Crystallization and x-ray structure elucidation of hopene cyclase of *A. acidocaldarius* at 2.9 Å resolution has been achieved.[79,80] The structure reveals a dimeric monotopic membrane protein which partly dips by a nonpolar plateau into the cytoplasmic membrane. From this plateau the active site is accessible by a channel structure. The active site of the hopene cyclase is located in a large central cavity that is of suitable size to bind squalene in its required conformation and is lined by aromatic amino acid residues. The structure supports a mechanism in which the acid starting the reaction by protonating a carbon–carbon double bond at C-2 of squalene is Asp376 that is coupled to His451. For the proton abstraction a water molecule in the hydrogen-bonding network of Gln262:Glu45:Glu93:Arg127 is envisioned. Surface α-helices in both α-helix barrels of the cyclase are often connected by the described QW motif. These QW motifs form hydrogen bonds between the outer helices that tighten the cyclase structure. This additional tightening is essential because the cyclization reaction is highly exergonic. Considerations relating the QW motif to catalysis are wrong.[81]

2.11.6.3 Catalytic Antibodies

Cationic cyclization is one of the most important carbon–carbon bond forming reactions in chemistry and biochemistry. Numerous examples of these reactions have been presented in this review article. It is therefore significant that a research group has undertaken the task to isolate monoclonal catalytic antibodies for a cationic cyclization.[81] The challenge was to synthesize an appropriate hapten possessing similarity to a high-energy intermediate of the proposed reaction. By this method, a way was found to achieve the specific synthesis of decalin systems starting from a linear diene substrate.[82] It is to be expected that more complex reactions involving catalytic antibodies will be accomplished in the near future. N-containing inhibitors (**22**) and (**23**), which are analogues of high-energy intermediates, may play an important role in hapten design. This field will perhaps develop more rapidly if it takes into account somatic hypermutation during the evolution of the catalytic antibodies.[83]

ACKNOWLEDGMENTS

The work of the author was supported by the Deutsche Forschungsgemeinschaft (SFB 323; Po 117/16-1) and by the European Community (Biotechnology of Extremophiles). The author wishes to thank Guy Ourisson for inspiration and motivation, Michel Rohmer for cooperation and assistance with the chemistry, and Seiichi Matsuda for discussions. Help in the preparation of figures and tables by Anette Tippelt and Corinna Feil is acknowledged.

2.11.7 REFERENCES

1. J. Buckingham (ed.), Dictionary of Natural Products on CD-ROM, Version 5 : 1, Chapman and Hall, London, 1996.
2. J. Li, P. Nagpal, V. Vitart, T. C. McMorris, and J. Chory, *Science*, 1996, **272**, 398.
3. M. Rohmer, M. Seemann, S. Horbach, S. Bringer-Meyer, and H. Sahm, *J. Am. Chem. Soc.*, 1996, **118**, 2564.
4. W. R. Nes and M. L. McKean, "Biochemistry of Steroids and Other Isopentenoids" University Park Press, Baltimore, MD, 1977.
5. M. G. Rowan, P. D. G. Dean, and T. W. Goodwin, *FEBS Lett.*, 1971, **12**, 229.
6. A. Eschenmoser, L. Ruzicka, O. Jeger, and D. Arigoni, *Helv. Chim. Acta*, 1955, **38**, 1890.
7. G. Stork and A. W. Burgstahler, *J. Am. Chem. Soc.*, 1955, **77**, 5068.
8. I. Abe, M. Rohmer, and G. D. Prestwich, *Chem. Rev.*, 1993, **93**, 2189.
9. E. I. Mercer, *Lipids*, 1991, **26**, 584.

10. I. Abe, J. C. Tomesch, S. Wattanasin, and G. D. Prestwich, *Nat. Prod. Rep.*, 1994, **11**, 279.
11. P. Benveniste, *Annu. Rev. Plant Physiol.*, 1986, **37**, 275.
12. D. Raederstorff and M. Rohmer, *Biochem. J.*, 1985, **231**, 609.
13. D. Raederstorff and M. Rohmer, *Eur. J. Biochem.*, 1987, **164**, 427.
14. I. Abe, Y. Ebizuka, and U. Sankawa, *Chem. Pharm. Bull.*, 1988, **36**, 5031.
15. I. Abe, Y. Ebizuka, S. Seo, and U. Sankawa, *FEBS Lett.*, 1989, **249**, 100.
16. E. J. Corey, S. P. T. Matsuda, and B. Bartel, *Proc. Natl. Acad. Sci. USA*, 1993, **90**, 11 628.
17. L. J. Altman, C. Y. Han, A. Bertolino, G. Handy, D. Laungani, W. Muller, S. Schwartz, D. Shanker, W. H. de Wolf, and F. J. Yang, *J. Am. Chem. Soc.*, 1978, **100**, 3235.
18. S. Seo, A. Uomori, Y. Yoshimura, K. Takeda, H. Seto, Y. Ebizuka, H. Noguchi, and U. Sankawa, *J. Chem. Soc., Perkin Trans. 1*, 1989, 261.
19. P. Heintz and P. Benveniste, *Phytochemistry*, 1970, **9**, 1499.
20. C. Anding, R. Heintz, and G. Ourisson, *Comp. Rend. Acad. Sci. D*, 1973, **276**, 205.
21. M. Ceruti, L. Delprino, L. Cattel, P. Bouvier-Navé, A. Duriatti, F. Schuber, and P. Benveniste, *J. Chem. Soc., Chem. Commun.*, 1985, 1054.
22. M. Taton, P. Benveniste, and A. Rahier, *Pure Appl. Chem.*, 1987, **59**, 287.
23. M. Taton, M. Ceruti, L. Cattel, and A. Rahier, *Phytochemistry*, 1996, **43**, 75.
24. M. Taton, P. Benveniste, A. Rahier, W. S. Johnson, H.-t. Liu, and A. Sudhakar, *Biochemistry*, 1992, **31**, 7892.
25. A. Duriatti, P. Bouvier-Navé, P. Benveniste, F. Schuber, L. Delprino, G. Balliano, and L. Cattel, *Biochem. Pharmacol.*, 1985, **34**, 2765.
26. W. D. Nes, T. J. Douglas, J.-T. Lin, E. Heftmann, and L. G. Paleg, *Phytochemistry*, 1982, **21**, 575.
27. G. Ourisson and M. Rohmer, *Acc. Chem. Res.*, 1992, **25**, 403.
28. I. Abe, U. Sankawa, and Y. Ebizuka, *Chem. Pharm. Bull.*, 1989, **37**, 536.
29. D. Ochs, C. Kaletta, K.-D. Entian, A. Beck-Sickinger, and K. Poralla, *J. Bacteriol.*, 1992, **174**, 298.
30. EMBL accession number X80766.
31. I. G. Reipen, K. Poralla, H. Sahm, and G. A. Sprenger, *Microbiology*, 1995, **141**, 155.
32. M. Perzl, P. Müller, K. Poralla, and E. L. Kannenberg, *Microbiology*, 1997, **143**, 1235.
33. EMBL accession number Y09978.
34. EMBL accession number Y09979.
35. T. Kaneko and further 23 authors, *DNA Res.*, 1996, **3**, 109.
36. GenBank accession number U49919.
37. E. J. Corey and S. K. Gross, *J. Am. Chem. Soc.*, 1968, **90**, 5045.
38. H. Horan, J. P. McCormick, and D. Arigoni, *J. Chem. Soc., Chem. Commun.*, 1973, 73.
39. M. Ceruti, F. Viola, F. Dosio, L. Cattel, P. Bouvier-Navé, and P. Ugliengo, *J. Chem. Soc., Perkin Trans.* 1, 1988, 461.
40. S. P. T. Matsuda and co-workers, unpublished.
41. J. Saar, J.-C. Kader, K. Poralla, and G. Ourisson, *Biochim. Biophys. Acta*, 1991, **1075**, 93.
42. G. Kleemann, K. Poralla, G. Englert, H. Kjosen, S. Liaasen-Jensen, S. Neunlist, and M. Rohmer, *J. Gen. Microbiol.*, 1990, **136**, 2551.
43. E. L. Kannenberg, T. Härtner, J.-M. Bravo, M. Perzl, K. Poralla, and M. Rohmer, Abstract H64 for "8. International Conference on Molecular Plant–Microbe Interactions", Knoxville, TN, 1996.
44. T. Murakami and N. Tanaka, *Progr. Chem. Org. Nat. Products*, 1988, **54**, 1.
45. H. Sahm, M. Rohmer, S. Bringer-Meyer, G. A. Sprenger, and R. Welle, *Adv. Microb. Physiol.*, 1993, **35**, 247.
46. A. M. Berry, O. T. Harriott, R. A. Moreau, S. F. Osman, D. R. Benson, and A. D. Jones, *Proc. Natl. Acad. Sci. USA*, 1993, **90**, 6091.
47. C. Vilchèze, P. Llopiz, S. Neunlist, K. Poralla, and M. Rohmer, *Microbiology*, 1994, **140**, 2749.
48. E. L. Kannenberg, M. Perzl, and T. Härtner, *FEMS Microbiol. Lett.*, 1995, **127**, 255.
49. G. Ourisson, M. Rohmer, and K. Poralla, *Annu. Rev. Microbiol.*, 1987, **41**, 301.
50. Y. Tsuda, A. Morimoto, T. Sano, Y. Inubushi, F. B. Mallory, and J. T. Gordon, *Tetrahedron Lett.*, 1965, **19**, 1427.
51. S. Holler, N. Pfennig, S. Neunlist, and M. Rohmer, *Eur. J. Protistol.*, 1993, **29**, 42.
52. P. Kemp, D. G. Lander, and C. G. Orpin, *J. Gen. Microbiol.*, 1984, **130**, 27.
53. B. Seckler and K. Poralla, *Biochim. Biophys. Acta*, 1986, **881**, 356.
54. S. Neumann and H. Simon, *Biol. Chem. Hoppe-Seyler*, 1986, **367**, 723.
55. D. Ochs, C. H. Tappe, P. Gärtner, R. Kellner, and K. Poralla, *Eur. J. Biochem.*, 1990, **194**, 75.
56. G. Kleemann, R. Kellner, and K. Poralla, *Biochim. Biophys. Acta*, 1994, **1210**, 317.
57. C. Tappe and K. Poralla, unpublished.
58. M. Rohmer, C. Anding, and G. Ourisson, *Eur. J. Biochem.*, 1980, **112**, 541.
59. M. Rohmer, P. Bouvier, and G. Ourisson, *Eur. J. Biochem.*, 1980, **112**, 557.
60. I. Abe and M. Rohmer, *J. Chem. Soc., Perkin Trans. 1*, 1994, 783.
61. J. M. Zander, J. B. Greig, and E. Caspi, *J. Biol. Chem.*, 1970, **245**, 1247.
62. C. Pale, M. Rohmer, C. Feil, and K. Poralla, unpublished.
63. G. Flesch and M. Rohmer, *Arch. Microbiol.*, 1987, **147**, 100.
64. C. Feil, R. Süssmuth, G. Jung, and K. Poralla, *Eur. J. Biochem.*, 1996, **242**, 51.
65. P. Bouvier, Y. Berger, M. Rohmer, and G. Ourisson, *Eur. J. Biochem.*, 1980, **112**, 549.
66. J.-M. Renoux and M. Rohmer, *Eur. J. Biochem.*, 1986, **155**, 125.
67. I. Abe and M. Rohmer, *J. Chem. Soc., Chem. Commun.*, 1991, 902.
68. A. Tippelt and K. Poralla, unpublished.
69. A. Tippelt, L. Jahnke, and K. Poralla, *Biochim. Biophys. Acta*, in press.
70. E. L. Kannenberg, M. Perzl, P. Müller, and K. Poralla, *Plant Soil*, 1996, **186**, 107.
71. A. I. Koukkou, C. Drainas, and M. Rohmer, *FEMS Microbiol. Lett.*, 1996, **140**, 277.
72. T.-P. Sun and Y. Kamiya, *Plant Cell*, 1994, **6**, 1509.
73. B. Stofer-Vogel, M. R. Wildung, G. Vogel, and R. Croteau, *J. Biol. Chem.*, 1996, **271**, 23 262.
74. M. Rohmer, P. Bouvier, and G. Ourisson, *Proc. Natl. Acad. Sci. USA*, 1979, **76**, 847.
75. I. Abe and G. D. Prestwich, *J. Biol. Chem.*, 1994, **269**, 802.

76. W. S. Johnson, *Tetrahedron*, 1991, **47**, XI.
77. K. Poralla, A. Hewelt, G. D. Prestwich, I. Abe, I. Reipen, and G. Sprenger, *Trends Biochem. Sci.*, 1994, **19**, 157.
78. C. Feil and K. Poralla, unpublished.
79. K. U. Wendt, C. Feil, A Lenhart, K. Poralla, and G. E. Schulz, *Protein Sci.*, 1997, **6**, 722.
80. K. U. Wendt, K. Poralla, and G. E. Schulz, *Science*, 1997, **277**, 1811.
81. K. Poralla, *Bioorg. Med. Chem. Lett.*, 1994, **4**, 285.
82. J. Hasserodt, K. D. Janda, and R. A. Lerner, *J. Am. Chem. Soc.*, 1997, **119**, 5993.
83. P. A. Patten, N. S. Gray, P. L. Yang, C. B. Marks, G. J. Wedemayer, J. J. Boniface, R. C. Stevens, and P. G. Schultz, *Science*, 1996, **271**, 1086.

2.12
Carotenoid Genetics and Biochemistry

GREGORY ARMSTRONG
Eidgenössiche Technische Hochschule, Zürich, Switzerland

2.12.1 GENERAL INTRODUCTION

Carotenoids represent one of the most widely distributed and structurally diverse classes of natural pigments, with crucial functions in photoprotection, photosynthesis, and nutrition. Among prokaryotes, red, orange, and yellow carotenoids are accumulated by all anoxygenic photosynthetic bacteria and cyanobacteria, and by many species of nonphotosynthetic bacteria. Eukaryotes including all algae, plants and some fungi synthesize these pigments. Dietary carotenoids and metabolites thereof also play important biological roles in noncarotenogenic organisms, such as mammals, birds, amphibians, fish, crustaceans, and insects.

Major advances have been made during the last 10 years in our understanding of the molecular genetics, biochemistry, and regulation of prokaryotic carotenoid biosynthesis. These developments have important implications for eukaryotes and make increasingly attractive the prospects for genetic engineering of carotenoid content in microbes and plants for humanitarian and commercial purposes. This chapter will focus on the most recent results in the field of prokaryotic carotenoid biosynthesis, with particular emphasis on the molecular–genetic dissection of pigment biosynthetic pathways. For additional background on this topic, interested readers are referred to any of a number of thorough reviews of carotenoid biosynthesis in specific groups of prokaryotes,[1,2] such as anoxygenic photosynthetic bacteria,[3] nonphotosynthetic myxobacteria[4] and cyanobacteria,[5] or eukaryotes, such as fungi[6] and higher plants.[7,8] More general overviews of carotenoid biosynthesis that integrate data obtained from both prokaryotes and eukaryotes[9–11] or focus primarily on biochemical aspects are also available.[12,13]

2.12.2 OVERVIEW OF CAROTENOID BIOSYNTHESIS

2.12.2.1 Natural Distribution

Carotenoids are a major class of lipophilic isoprenoids that range in color from deep red to light yellow. They are perhaps most familiar to us in everyday life as the dominant pigments in many storage roots, fruits, and flowers. The carrot root, tomato, and red pepper fruits, and daffodil and marigold petals represent a few of the most obvious examples of carotenoid-containing tissues. Carotenoids are produced by all chlorophyll (Chl)-containing photosynthetic eukaryotes from algae to higher plants, as well as by many species of fungi.[14,15] An examination of the ability to synthesize carotenoids among prokaryotes reveals their ancient evolutionary origin. These pigments are ubiquitous among both bacteriochlorophyll (Bchl)-containing anoxygenic photosynthetic and Chl-containing oxygenic photosynthetic prokaryotes (i.e., cyanobacteria), not to mention their wide distribution among nonphotosynthetic bacteria.

2.12.2.2 Chemical Structures

The number of chemically distinct carotenoids described in the literature has increased dramatically since the 1950s, and now includes some 600 structurally unique pigments.[16,17] Following the established tradition in reviews of carotenoid biosynthesis, the carotenoids mentioned in this chapter will be referred to throughout by their trivial names, with the corresponding semisystematic names (Table 1) given in parentheses at their first mention. Complete IUPAC–IUB rules for the nomenclature of carotenoids are given elsewhere.[18] General aspects of the chemical structures, physical properties, proposed biosynthetic schemes, species-specific distributions, cellular localization, and functions of carotenoids not covered in the reviews listed in Section 2.12.1 have been discussed extensively elsewhere.[14,15,19–22] The highlights of these reviews relevant to the situation in prokaryotes are summarized below in the remainder of Section 2.12.2 and in Section 2.12.3.

Table 1 Semisystematic nomenclature for carotenoids mentioned in the text or figures.

Trivial name	*Semisystematic name*[a]
Adonixanthin	(3*S*,3′*R*)-3,3′-dihydroxy-*β*,*β*-caroten-4-one
Astaxanthin	3,3′-dihydroxy-*β*,*β*-carotene-4,4′-dione
Canthaxanthin	*β*,*β*-carotene-4,4′-dione
α-Carotene	*β*,*ε*-carotene
β-Carotene	*β*,*β*-carotene
γ-Carotene	*β*,*ψ*-carotene
ζ-Carotene	7,8,7′,8′-tetrahydro-*ψ*,*ψ*-carotene
β-Cryptoxanthin	(3*R*)-*β*,*β*-caroten-3-ol
β-Cryptoxanthin monoglucoside	(3*R*)-3-(*β*-D-glucosyloxy)-*β*,*β*-carotene
Demethylspheroidene	3,4-didehydro-1,2,7′,8′-tetrahydro-*ψ*,*ψ*-caroten-1-ol
Demethylspheroidenone	1-hydroxy-3,4-didehydro-1,2,7′,8′-tetrahydro-*ψ*,*ψ*-caroten-2-one
Dihydro-*β*-carotene	7,8-dihydro-*β*,*ψ*-carotene
4,4′-Diaponeurosporene	7,8-dihydro-4,4′-diapocarotene
4,4′-Diapophytoene	7,8,11,12,7′,8′,11′,12′-octahydro-4,4′-diapocarotene
Echinenone	*β*,*β*-caroten-4-one
3-Hydroxyechinenone	(3*S*)-3-hydroxy-*β*,*β*-caroten-4-one
3′-Hydroxyechinenone	3′-hydroxy-*β*,*β*-caroten-4-one
Hydroxyneurosporene	1,2,7′,8′-tetrahydro-*ψ*,*ψ*-caroten-1-ol
Hydroxyspheroidene	1′-methoxy-3′,4′-didehydro-1,2,7,8,1′,2′-hexahydro-*ψ*,*ψ*-caroten-1-ol
Isorenieratene	*φ*,*φ*-carotene
β-Isorenieratene	*β*,*φ*-carotene
Lutein	*β*,*ε*-carotene-3,3′-diol
Lycopene	*ψ*,*ψ*-carotene
Methoxyneurosporene	1-methoxy-1,2,7′,8′-tetrahydro-*ψ*,*ψ*-carotene
Myxobactin	1′-glucosyloxy-3,4,3′,4′-tetrahydro-1′,2′-dihydro-*β*,*ψ*-carotene
Myxobactone	1′-glucosyloxy-3′,4′-didehydro-1′,2′-dihydro-*β*,*ψ*-caroten-4-one
Myxoxanthophyll	2′-(*β*-L-rhamnopyranosyloxy)-3,4′-didehydro,1′,2′,dihydro-*β*,*ψ*-carotene-3,1′-diol
Neurosporene	7,8-dihydro-*ψ*,*ψ*-carotene
Phoenicoxanthin	(3*S*)-3-hydroxy-*β*,*β*-carotene-4,4′-dione
Phytoene	7,8,11,12,7′,8′,11′,12′-octahydro-*ψ*,*ψ*-carotene
Phytofluene	7,8,11,12,7′,8′-hexahydro-*ψ*,*ψ*-carotene
Rhodoxanthin	4′,5′-didehydro-4,5′-retro-*β*,*β*-carotene-3,3′-dione
Spheroidene	1-methoxy-3,4-didehydro-1,2,7′,8′-tetrahydro-*ψ*,*ψ*-carotene
Spheroidenone	1-methoxy-3,4-didehydro-1,2,7′,8′-tetrahydro-*ψ*,*ψ*-caroten-2-one
7,8,11,12-Tetrahydrolycopene	7,8,11,12-tetrahydro-*ψ*,*ψ*-carotene
β-Zeacarotene	7′,8′-dihydro-*β*,*ψ*-carotene
Zeaxanthin	(3*R*,3′*R*)-*β*,*β*-carotene-3,3′-diol
Zeaxanthin monoglucoside	(3*R*,3′*R*)-3′-(*β*-D-glucosyloxy)-*β*,*β*-caroten-3-ol
Zeaxanthin diglucoside	(3*R*,3′*R*)-3,3′-bis(*β*-D-glucosyloxy)-*β*,*β*-carotene

[a] Stereochemistry is given if applicable and when known. Several different naturally occurring stereoisomers of astaxanthin and lutein have been reported.[10,16]

Most naturally occurring carotenoids are hydrophobic tetraterpenoids containing a C_{40} methyl-branched hydrocarbon backbone derived from the successive 1′–4 homoallylic-allylic condensations of eight C_5 isoprene units (Figure 1). In addition, novel carotenoids with longer or shorter backbones occur in some species of nonphotosynthetic bacteria. For example, members of the genera *Bacillus*, *Corynebacterium*, *Flavobacterium*, and *Sarcina* produce carotenoids with C_{45} or C_{50} backbones through additional C_5 isoprene condensations. Triterpenoid carotenoids containing C_{30} backbones resulting from six consecutive C_5 isoprene condensations, as opposed to partial degradation of a C_{40} backbone, are found in species of *Pseudomonas*, *Staphylococcus*, and *Streptococcus*.

The term "carotenoid" actually encompasses both carotenes and xanthophylls. A "carotene" refers to a hydrocarbon carotenoid. Carotene derivatives that contain one or more oxygen atoms, in the form of hydroxy-, methoxy-, oxo-, epoxy-, carboxy-, or aldehydic functional groups, or within glycosides, glycoside esters, or sulfates, are collectively known as "xanthophylls." Xanthophylls are far more structurally diverse and abundant among carotenogenic organisms than are their carotene precursors. Carotenoids are furthermore described as being acyclic, monocyclic, or bicyclic depending on whether the ends of the hydrocarbon backbone have been cyclized to yield aliphatic or aromatic ring structures.

An additional layer of structural complexity is added by the staggering number of potential geometrical isomers for any given carotenoid given that each double bond in the polyene backbone

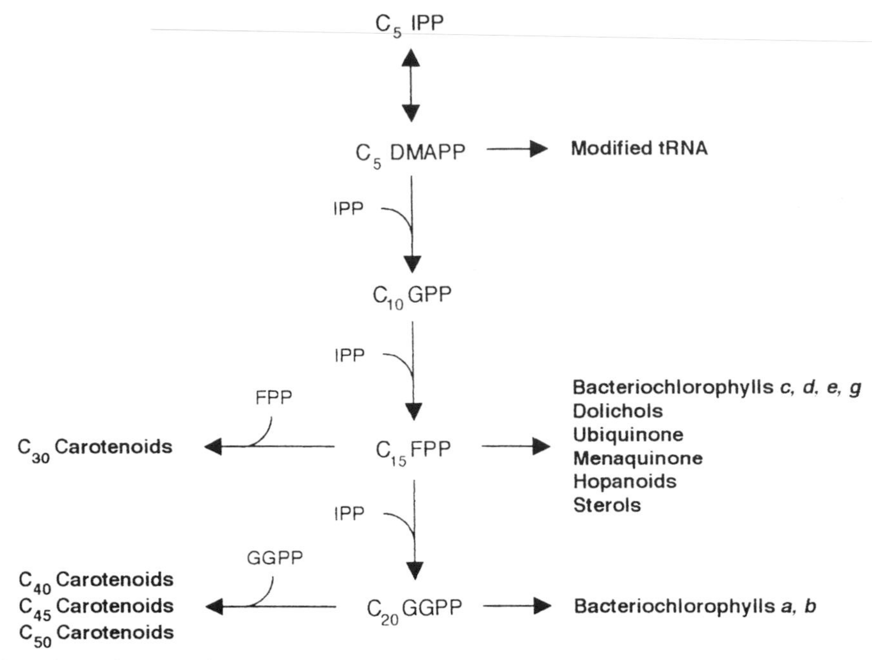

Figure 1 Overview of general isoprenoid biosynthesis. Products of this pathway found in some or all prokaryotes are given. DMAPP, dimethylallyl pyrophosphate; FPP, farnesyl pyrophosphate; GPP, geranyl pyrophosphate; GGPP, geranylgeranyl pyrophosphate; IPP, isopentenyl pyrophosphate.

can assume either a *cis* or a *trans* conformation. In fact, however, the vast majority of naturally occurring carotenoids exist primarily or exclusively as all-*trans* isomers. Phytoene (7,8,11,12,7′,8′,11′,12′-octahydro-ψ,ψ-carotene) represents the most widely distributed and notable exception, typically occurring as the 15,15′-*cis*-isomer in both prokaryotes and eukaryotes (Figure 2). Certain nonphotosynthetic bacteria, such as species of *Flavobacterium* and *Mycobacterium*, do, however, accumulate all-*trans*-phytoene, and some eukaryotes possess *cis* isomers of phytoene other than the 15,15′-form.

2.12.2.3 Organization of the Biochemical Pathway

C_{40} phytoene, formed by a 1′-2-3 condensation of two molecules of the C_{20} precursor geranylgeranyl pyrophosphate (GGPP) and a subsequent 1′-1 rearrangement, and C_{30} 4,4′-diapophytoene (7,8,11,12,7′,8′,11′,12′-octahydro-4,4′-diapocarotene) formed by the analogous condensation of two molecules of C_{15} farnesyl pyrophosphate (FPP), are the progenitors of all other carotenoids (Figure 2). The reactions leading to the synthesis of phytoene and 4,4′-diapophytoene, also known as dehydrosqualene, are carried out by soluble or membrane-associated enzymes. Later steps in carotenoid biosynthesis, however, are conducted by membrane-bound enzymes. In anoxygenic photosynthetic bacteria, nonphotosynthetic bacteria, and fungi, three or four consecutive desaturations of phytoene lengthen the polyene chromophore and yield neurosporene (7,8-dihydro-ψ,ψ-carotene) or lycopene (ψ,ψ-carotene), respectively (Figure 3). None of the desaturation intermediates normally accumulate in more than trace amounts. Cyanobacteria, algae, and plants, in contrast, desaturate phytoene four times in two discrete biosynthetic segments, first producing ζ-carotene (7,8,7′,8′-tetrahydro-ψ,ψ-carotene) and thereafter converting it to lycopene. The absorption properties and hence color of a given carotenoid are determined by the number of conjugated double bonds within the hydrocarbon backbone. Phytoene, for example, contains only three conjugated double bonds and absorbs UV light, whereas ζ-carotene and more highly unsaturated carotenoids possess seven or more conjugated double bonds and thus absorb visible wavelengths. An increasing degree of apparently nonenzymatic isomerization of the carotene products to the all-*trans* conformation usually accompanies the multiple desaturations of 15,15′-*cis*-phytoene. Symmetric ζ-carotene has been firmly established as a desaturation product in cyanobacteria, whereas uncertainty exists as to whether the nonaccumulating desaturation intermediate in other bacteria is the asymmetric ζ-carotene isomer 7,8,11,12-tetrahydrolycopene (7,8,11,12-tetrahydro-ψ,ψ-carotene) or sym-

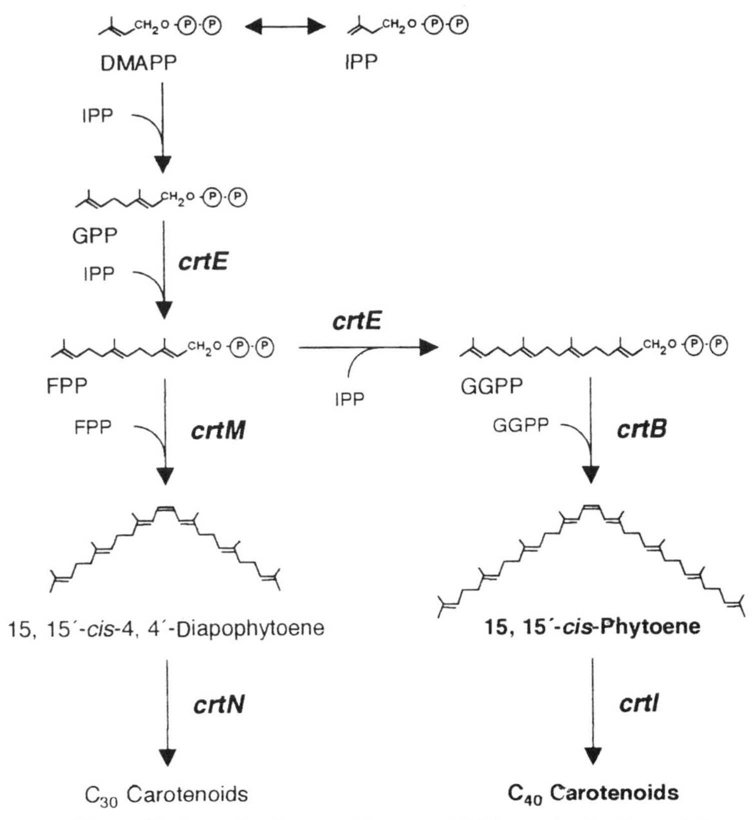

Figure 2 The carotenoid-specific branch of general isoprenoid biosynthesis. Bacterial *crt* genes required for specific biosynthetic conversions are given here and the corresponding enzymes are listed in Table 2. Bold typeface indicates the widespread natural occurrence of C_{40} carotenoids synthesized from phytoene.

metric ζ-carotene. Lycopene cyclization frequently follows carotene desaturation and precedes xanthophyll formation in the biosynthetic sequence (Figure 3). Carotenoid desaturation and cyclization are targets for many known carotenoid biosynthesis inhibitors. During xanthophyll formation the carotenoid structures are modified with oxygen-containing functional groups such that the end product pigments are often species-specific (Figures 4–6). Introduction of hydroxy groups often precedes glycosylation and/or fatty acid ester formation during the late stages of carotenoid biosynthesis in bacteria.

2.12.3 LOCALIZATION AND FUNCTIONS OF CAROTENOIDS

2.12.3.1 Localization

Carotenoids are relatively hydrophobic molecules and are therefore typically associated with biological membranes and/or are noncovalently bound to specific proteins. In pigment–protein complexes, carotenoids sometimes display substantial bathochromic shifts in their absorption maxima. Most carotenoids accumulated in anoxygenic photosynthetic bacteria are associated with the membrane-bound light-harvesting and reaction center Bchl-binding polypeptides of the bacterial photosynthetic apparatus. Similarly, the bulk of the carotenoids in cyanobacteria are noncovalently bound to the Chl-binding light-harvesting and reaction center polypeptides of photosystems I and II. In addition, some cyanobacterial carotenoids can be found as constituents of the outer membrane of the cell wall. Unlike anoxygenic photosynthetic bacteria, cyanobacteria possess novel carotenoid-binding proteins that lack Chl. Lipophilic carotenoid-binding proteins have been found both in the outer membrane of the cell envelope and in the cytoplasmic membrane. Soluble zeaxanthin ((3*R*,3′*R*)-β,β-carotene-3,3′-diol)-binding proteins have also been described in cyanobacteria.

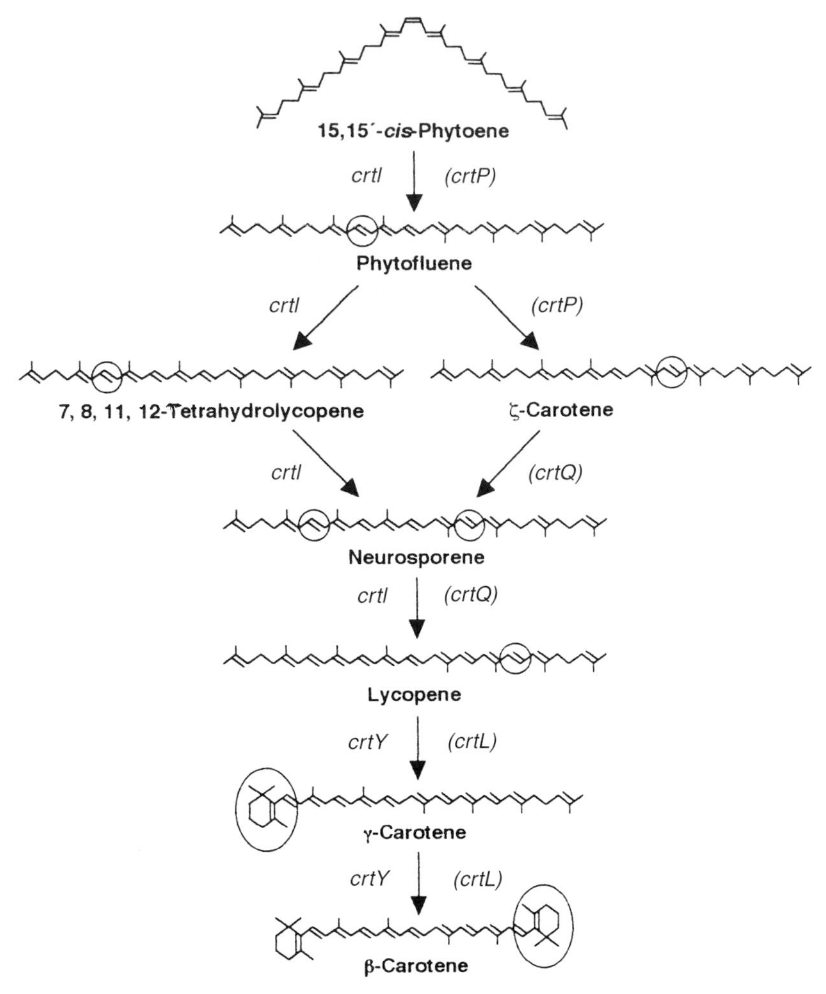

Figure 3 Two genetically distinct pathways exist for the conversion of phytoene to β-carotene (β,β-carotene). *crt* genes required for specific biosynthetic conversions are given here either on the left for anoxygenic photosynthetic bacteria and nonphotosynthetic bacteria, or on the right in parentheses for cyanobacteria. The corresponding enzymes are listed in Table 2. Structural alternations to the carotenoids during biosynthesis are circled for emphasis.

Among nonphotosynthetic bacteria, carotenoids and their glycosides have been found in cytoplasmic and cell wall membranes. In many species, however, the exact cellular location of carotenoids and whether or not they are bound to specific proteins are not known.

2.12.3.2 Functions

The paramount function of carotenoids in all photosynthetic organisms, including anoxygenic photosynthetic bacteria and cyanobacteria, is to prevent or minimize photooxidative damage. Carotenoids protect against the potentially lethal combination of oxygen, light, and photosensitizing Bchl or Chl molecules by quenching both triplet excited states of the photosensitizers and excited state singlet oxygen. In addition to their role in photooxidative protection, carotenoids located in the light-harvesting protein antenna complexes of anoxygenic photosynthetic bacteria and cyanobacteria increase the cross-section for the absorption of radiant energy, which is ultimately transferred via Bchl or Chl molecules to the photosynthetic reaction center, the site of primary charge separation. A photoprotective function has also been postulated for cyanobacterial carotenoid-binding proteins that do not contain Chl, which in some cases have been demonstrated to increase in abundance in response to light stress and carbon limitation.

Among nonphotosynthetic bacteria, carotenoids also offer photooxidative protection against photosensitizing porphyrin molecules such as protoporphyrin IX and heme. In addition, carotenoids have been proposed to regulate membrane fluidity. This putative function may provide an explanation for the evolutionary maintenance of carotenoid biosynthesis in some species of nonphotosynthetic bacteria that are not obviously exposed to photooxidative stress in their normal habitats.

In addition to these functions of carotenoid pigments in prokaryotes, certain plant and algal carotenoids participate in an epoxidation–deepoxidation reaction called the xanthophyll cycle to help dissipate excess radiant energy, and provide the biosynthetic precursors of abscisic acid, an essential plant hormone, and retinal, the photoreceptor pigment for phototaxis in algae. Among humans and a number of animals, dietary carotenoids which contain the half-structure of β-carotene (Figure 3) serve as provitamin A, retinal, and retinoic acid sources, and are hence critical for nutrition, vision, and development, respectively. Finally, carotenoids provide natural coloration to birds, reptiles, amphibians, fish, and various invertebrates.

2.12.4 CAROTENOID BIOSYNTHESIS GENES

2.12.4.1 Gene Nomenclature

The *crt* designation for genes that encode carotenoid biosynthesis enzymes (Table 2) was first introduced in studies of *Rhodobacter capsulatus*[48] and has now been adopted by most researchers in the field.[1,5,27] The parallel *car* nomenclature, which was originally used to denote genetic loci involved in carotenoid synthesis in the fungus *Phycomyces blakesleeanus*[6,49] and subsequently in the nonphotosynthetic bacterium *Myxococcus xanthus*,[50] continues to be applied as a designation for carotenoid biosynthesis regulatory genes identified in *M. xanthus*[4] (Table 3).

2.12.4.2 Anoxygenic Photosynthetic Bacteria

The distribution of specific carotenoids in anoxygenic purple and green photosynthetic bacteria has been reviewed.[14,57] Biosynthetic pathways for a number of species have been proposed on the basis of considerations of pigment structure, chemical inhibition of carotenoid biosynthesis, precursor labeling studies, and in some cases, mutant analyses. Indeed, the isolation of pigmentation mutants of *Rhodobacter sphaeroides* over 40 years ago led to discovery of the photoprotective role of carotenoids.[58–60]

2.12.4.2.1 **Rhodobacter** *species*

The molecular genetics of carotenoid biosynthesis in anoxygenic photosynthetic bacteria have thus far been analyzed in detail in two closely related species, *R. capsulatus* and *R. sphaeroides*,[3] both of which represent genetically well-characterized systems with respect to anoxygenic photosynthesis in general.[54] These metabolically versatile purple nonsulfur bacteria normally accumulate a mixture of two acyclic xanthophylls, spheroidene (1-methoxy-3,4-didehydro-1,2,7′,8′-tetrahydro-ψ,ψ-carotene) and spheroidenone (1-methoxy-3,4-didehydro-1,2,7′,8′-tetrahydro-ψ,ψ-caroten-2-one; Figure 4), whose abundance and distribution depend on the oxygen tension in the growth medium.[57,61,62] In the genus *Rhodobacter* carotenoids are absolutely required for survival only in the simultaneous presence of both light and oxygen. Such a situation arises during the transition from dark aerobic chemoheterotrophic growth to illumination under anoxygenic conditions, which initiate photosynthetic membrane development and Bchl *a* accumulation. Thus, it has proven possible to isolate *Rhodobacter* species mutants lacking carotenoids or containing altered pigment compositions.[3] Such mutants can be stably maintained under several different growth conditions that are not photooxidative.

The application of both classical and molecular genetic approaches since the 1970s has ultimately led to the conclusion that both *R. capsulatus* and *R. sphaeroides* contain 50 kilobase pair chromosomal regions, roughly 1% of their total genomes, that encode most of the functions required for photosynthesis, including all known Bchl *a* and carotenoid biosynthesis enzymes.[54] Isolation of these photosynthesis-related genes by *in vivo* functional complementation of point mutations allowed the generation of gene-disrupted, pigment-deficient mutants, and the alignment of physical and genetic maps of the chromosomal regions required for carotenoid biosynthesis.[48,63–70]

Table 2 Bacterial carotenoid biosynthesis genes and enzymes.

Gene	*Enzyme*	*Species*	*Ref.*
Phytoene/4,4′-diapophytoene formation			
crtE	GGPP synthase	*Erwinia herbicola* strain Eho10	23
		Erwinia herbicola strain Eho13	24
		Erwinia uredovora	25
		Flavobacterium sp. strain R1534	26
		Myxococcus xanthus	27
		Rhodobacter capsulatus	28
		Rhodobacter sphaeroides	29
		Streptomyces griseus	30
crtB	Phytoene synthase	*Agrobacterium aurantiacum*	31
		Erwinia herbicola strain Eho10	23
		Erwinia herbicola strain Eho13	24
		Erwinia uredovora	25
		Flavobacterium sp. strain R1534	26
		Myxococcus xanthus	27
		Rhodobacter capsulatus	28
		Rhodobacter sphaeroides	32
		Streptomyces griseus	30
		Synechococcus sp. strain PCC7942	33
		Synechocystis sp. strain PCC6803	34
		Thermus thermophilus	35
crtM	Diapophytoene synthase	*Staphylococcus aureus*	36
Phytoene/4,4′-diapophytoene desaturation			
crtI	Phytoene desaturase (CrtI-type)	*Agrobacterium aurantiacum*	31
		Erwinia herbicola strain Eho10	23
		Erwinia herbicola strain Eho13	24
		Erwinia uredovora	25
		Flavobacterium sp. strain R1534	26
		Myxococcus xanthus	37
		Rhodobacter capsulatus	28, 38
		Rhodobacter sphaeroides	32
		Streptomyces griseus	30
crtP	Phytoene desaturase (CrtP-type)	*Synechococcus* sp. strain PCC7942	39
		Synechocystis sp. strain PCC6803	40
crtN	Diapophytoene desaturase	*Staphylococcus aureus*	36
crtQ	ζ-Carotene desaturase	*Anabaena* sp. strain PCC7120[a]	41
Lycopene cyclization			
crtY	Lycopene cyclase (CrtY-type)	*Agrobacterium aurantiacum*	31
		Erwinia herbicola strain Eho10	42
		Erwinia herbicola strain Eho13	24
		Erwinia uredovora	25
		Flavobacterium sp. strain R1534	26
		Streptomyces griseus	30
crtL	Lycopene cyclase (CrtL-type)	*Synechococcus* sp. strain PCC7942	41
ORF6	Lycopene cyclization	*Myxococcus xanthus*[b]	27
Acyclic xanthophyll formation			
crtC	Hydroxyneurosporene synthase	*Myxococcus xanthus*	27
		Rhodobacter capsulatus	28
		Rhodobacter sphaeroides	29
crtD	Methoxyneurosporene desaturase	*Myxococcus xanthus*[c]	27
		Rhodobacter capsulatus	28
		Rhodobacter sphaeroides	29, 44
crtF	Hydroxyneurosporene *O*-methyltransferase	*Rhodobacter capsulatus*	28
		Rhodobacter sphaeroides	29
crtA	Spheroidene monooxygenase	*Rhodobacter capsulatus*	28, 45, 46
		Rhodobacter sphaeroides	29

Table 2 (continued)

Gene	Enzyme	Species	Ref.
Cyclic xanthophyll formation			
crtZ	β-Carotene hydroxylase	*Agrobacterium aurantiacum*	31
		Alcaligenes PC-1	31
		Erwinia herbicola	42
		Erwinia uredovora	25
		Flavobacterium sp. strain R1534	26
crtX	Zeaxanthin glucosyltransferase	*Erwinia herbicola* strain Eho10	42
		Erwinia herbicola strain Eho13	24
		Erwinia uredovora	25
crtW	β-C-4-oxygenase	*Agrobacterium aurantiacum*	31
		Alcaligenes PC-1	47
Postulated biosynthetic function			
ORF2	Carotenoid desaturase	*Myxococcus xanthus*[c]	27
crtT	Carotenoid methyltransferase	*Streptomyces griseus*[c]	30
crtU	Carotenoid desaturase	*Streptomyces griseus*[c]	30
crtV	Carotenoid methylesterase	*Streptomyces griseus*[c]	30

[a] The *crtQ* gene does not seem to be present in other species of prokaryotes, including cyanobacteria. [b] Predicted *M. xanthus* gene product is not homologous to CrtY or CrtL. [c] Functions predicted on the basis of protein sequence similarities to other known proteins. The *crtT*, *crtU*, and *crtV* gene designations should be considered provisional until the carotenoid substrates of the corresponding gene products have been identified.

Table 3 Bacterial carotenoid biosynthesis regulatory genes and gene products.

Gene	Gene product function	Mutant phenotype	Species	Ref.
carA[a]	DNA-binding repressor	Light-independent *carBA*[b] expression and carotenoid accumulation; decreased light-inducible *carC* expression	*Myxococcus xanthus*	27
carD[c]	Unknown	Same as *carQ* with additional pleiotropic developmental defects	*Myxococcus xanthus*	51
carQ	RNA polymerase σ factor	Loss of light-inducible *carBA*, *carC*, and *carQRS* expression, and carotenoid accumulation	*Myxococcus xanthus*	52
carR	Anti-σ factor	Light-independent *carBA*, *carC*, and *carQRS* expression, and carotenoid accumulation	*Myxococcus xanthus*	52
carS	Transcription factor?	Loss of light-inducible *carBA* expression and carotenoid accumulation	*Myxococcus xanthus*	52
crtS	RNA polymerase σ factor	No carotenoid accumulation	*Streptomyces setonii*	53
ppsR	DNA-binding repressor	Derepression of *crt*, *bch*, and *puc* expression, and carotenoid accumulation under aerobic conditions	*Rhodobacter capsulatus* *Rhodobacter sphaeroides*	54 55
tspO	Environmental sensor?	Derepression of *crt*, *bch*, and *puc* expression, and carotenoid accumulation under aerobic conditions	*Rhodobacter capsulatus* *Rhodobacter sphaeroides*	28 29, 56

[a] The *carA* locus may correspond to ORF10 and/or ORF11, which encode putative DNA-binding proteins. [b] The *carBA* locus encodes two putative regulatory genes (ORF10, ORF11) and all known *crt* biosynthetic genes, except for the unlinked *carC* (*crtI*) gene. [c] Gene sequence has not been reported.

The nucleotide sequences of seven clustered carotenoid biosynthesis genes, *crtA*, *crtB*, *crtC*, *crtD*, *crtE*, *crtF*, and *crtI*, were subsequently determined in *R. capsulatus*[28,38,45,46] and *R. sphaeroides*.[29,32,44] The *crt* gene organization in both species suggests the existence of a minimum of four transcriptional operons: *crtA*, *crtIB*, *crtDC*, and *crtEF*.[28,29] However, both the phenotypes resulting from polar mutations and transcript mapping in *R. capsulatus* and *R. sphaeroides* suggest that the actual operon structure may be more complex.[71] The available data imply the existence of separate promoters for

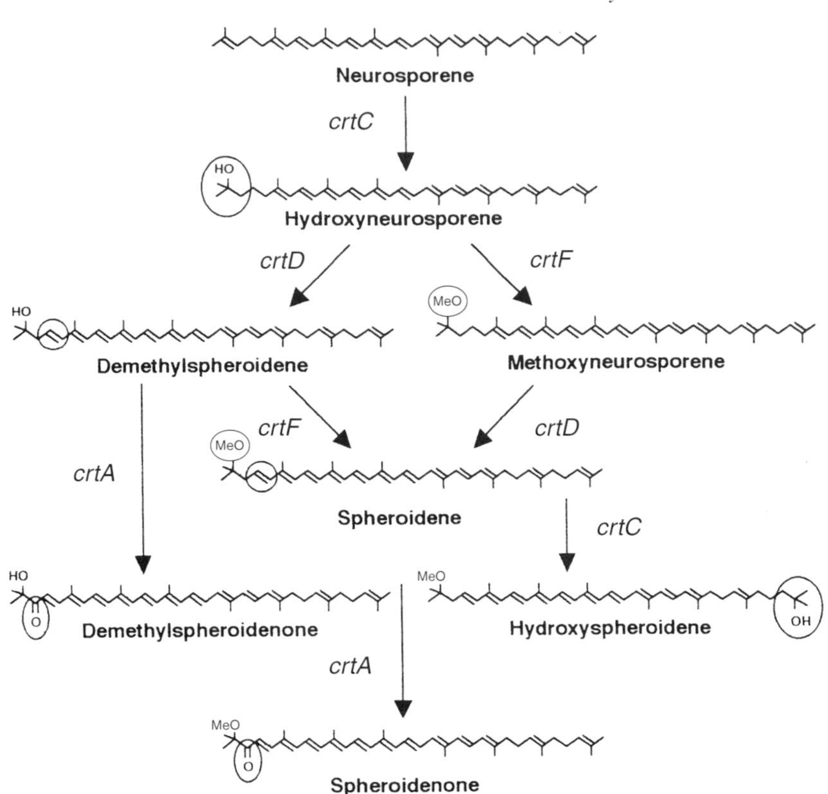

Figure 4 Acyclic xanthophyll biosynthetic pathway in the genus *Rhodobacter*. *crt* genes required for specific carotenoid biosynthetic reactions are given here and the corresponding enzymes are listed in Table 2. Structural alternations to the carotenoids during biosynthesis are circled for emphasis.

crtI, *crtB*, *crtD*, and *crtC*, two promoters for *crtE*, and multiple transcripts overlapping several genes.[29,32,69,72,73] Indeed, the operons containing *crtA*, *crtE*, and *crtF* belong to superoperons that allow their cotranscription with Bchl *a* biosynthesis (*bch*) genes and the B870 light-harvesting antenna and reaction center polypeptide (*puf*) genes, which encode Bchl *a*- and carotenoid-binding pigment–protein complexes of the photosynthetic membrane.[29,45,72] The chromosomally unlinked *puc* genes encoding the Bchl *a*- and carotenoid-binding B800–850 peripheral light-harvesting antenna polypeptides are coregulated with *crt* and *bch* genes by separate mechanisms described in Section 2.12.6.1.1.

The seven *crt* genes common to *R. capsulatus* and *R. sphaeroides* encode all of the enzymes necessary to convert the isoprenoid pyrophosphate carotenoid precursor geranyl pyrophosphate (GPP) into spheroidene and spheroidenone (Table 2). The gene products involved in the conversion of neurosporene to these end products have been functionally assigned to specific biosynthetic steps primarily on the basis of the carotenoids accumulated in *R. capsulatus* pigmentation mutants.[3,63] Such mutants containing abnormal complements of colored carotenoids have generally proven to be viable. Mutations in *crtF*, which cause the accumulation of demethylspheroidene (3,4-didehydro-1,2,7′,8′-tetrahydro-ψ,ψ-caroten-1-ol and demethylspheroidenone (1-hydroxy-3,4-didehydro-1,2,7′,8′-tetrahydro-ψ,ψ-caroten-2-one; Figure 4) are, however, deleterious to cells for unknown reasons and lead to the appearance of second site mutations that either abolish the synthesis of these carotenoids or prevent accumulation of the carotenoid-binding light-harvesting polypeptides.[65] In contrast, *crtD* mutations, which cause the accumulation of hydroxyneurosporene (1,2,7′,8′-tetrahydro-ψ,ψ-caroten-1-ol) and methoxyneurosporene (1-methoxy-1,2,7′,8′-tetrahydro-ψ,ψ-carotene; Figure 4), arise naturally at a relatively high frequency.[61,64,74]

Assignment of functions to the *crtB*, *crtE*, and *crtI* genes required for phytoene formation and desaturation proved technically more difficult than for genes involved in colored carotenoid biosynthesis.[1,3] All mutants blocked in this part of the biochemical pathway display a blue-green phenotype resulting from the loss of colored carotenoids and the continued production of Bchl. Blue-green mutants are capable of surviving by anoxygenic photosynthesis but are photo-oxidatively killed when simultaneously exposed to light and oxygen, such as during the transition from aerobic to photosynthetic growth.[48,58–60] Pigment analysis of different blue-green strains of *R. capsulatus* revealed that *crtI* mutants accumulate the colorless carotenoid phytoene (Figure 2), in contrast to *crtB* and *crtE* mutants.[68,69,73] As discussed in detail elsewhere,[1,3] the functions of *crtB* and *crtE* were ultimately resolved by *in vivo* functional analysis of homologous genes from nonphotosynthetic bacteria of the genus *Erwinia*[75,76] (Section 2.12.4.3.1) and from the cyanobacterium *Synechococcus* sp. strain PCC7942.[33] The functional assignments of the *crtB* and *crtI* genes of *R. capsulatus* have also been confirmed using the corresponding mutants of this bacterium as hosts for *in vivo* analyses of several plant carotenoid biosynthesis genes.[77–80]

Carotenoid biosynthesis in *R. capsulatus* and *R. sphaeroides* is expected to begin with two successive 1′–4 additions of isopentyl pyrophosphate (IPP) to C_{10} GPP and C_{15} FPP, yielding C_{20} GGPP (Figures 1 and 2), as discussed in Section 2.12.5.2.1. Interestingly, *R. capsulatus* and *R. sphaeroides crtE* mutants lack colored carotenoids but continue to produce relatively large amounts of certain Bchl-containing pigment–protein complexes[29,66] even though Bchl synthesis also utilizes GGPP. It has recently been suggested that some but not all *R. sphaeroides crtE* mutants actually contain an unidentified Bchl derivative rather than the normal Bchl *a*, but this report awaits confirmation.[29] In any case, genus *Rhodobacter crtE* mutants are viable, indicating that they must synthesize essential quinones. This suggests that the generally accepted biosynthetic scheme for the formation of GGPP is probably incomplete, as has recently been proposed.[1,3,29,81] One possible explanation for the viability of *crtE* mutants could be the existence of at least two separate pools of GGPP formed by distinct GGPP synthases, including a carotenoid-specific enzyme. This model is consistent with the observation that normally noncarotenogenic *Escherichia coli* cells genetically transformed with *crt* genes from species of *Erwinia* require the *crtE* gene in order to produce more than traces of carotenoids.[25,42] Thus, the low endogenous activity of the *E. coli* GGPP synthase cannot replace the activity of the presumably carotenoid-specific GGPP synthase from *Erwinia* species.[75,76]

Because GGPP is used by bacteria for several purposes (Figure 1),[82] the first carotenoid-specific biochemical reaction in *Rhodobacter* species, and indeed in most carotenogenic prokaryotes and eukaryotes, is the formation of C_{40} phytoene (Figure 2). In this reaction two molecules of GGPP undergo a tail-to-tail condensation catalyzed by CrtB, the phytoene synthase. Thereafter, the conversion of phytoene to the colored carotenoid end products requires five *crt* genes that are predicted to encode enzymes which perform several types of reactions, including the desaturations mediated by the structurally related CrtI and CrtD proteins, hydration carried out by CrtC, methylation conducted by CrtF and the addition of a keto group catalyzed by CrtA (Figures 3 and 4). Structural and functional homologues of several of these enzymes have since been found in other organisms (Table 2), as discussed in Section 2.12.5. An unusual feature of phytoene desaturation in *Rhodobacter* species in comparison to other bacteria and fungi is that neurosporene, rather than lycopene is the final pigment produced.[63,68,69,73,77] 7,8,11,12-Tetrahydrolycopene (Figure 3) has been proposed in the literature as one of the desaturation intermediates in the genus *Rhodobacter* on the basis of studies with other species of anoxygenic photosynthetic bacteria.[14,57] However, absorption spectra and chromatographic analysis of pigments produced by *in vitro* carotenoid-synthesizing systems derived from crude *R. capsulatus* cell extracts[68] or containing purified *R. capsulatus* CrtI isolated from overexpressing *E. coli* strains[83] raise the possibility, however, that symmetric ζ-carotene may be the true intermediate. Further details of the biochemical characterization of CrtI are given in Section 2.12.5.2.3.

2.12.4.3 Nonphotosynthetic Bacteria

Carotenoids are produced by many phylogenetically distinct groups of nonphotosynthetic bacteria, although the ability to synthesize these pigments often differs from strain to strain within a single genus.[10,14,19] This observation suggests that carotenoids are not required for viability under all growth conditions in some of the species studied to date. The C_{40} carotenoids accumulated by nonphotosynthetic bacteria vary widely in structure. In addition, some species instead synthesize novel C_{30}, C_{45}, or C_{50} carotenoids. The genetics of C_{40} carotenoid biosynthesis in nonphotosynthetic bacteria have been most thoroughly described in two species of *Erwinia*, represented by *E. herbicola* strains Eho10 and Eho13 and *E. uredovora*, and in the myxobacterium *Myxococcus xanthus*, in which carotenogenesis is light-induced. Molecular genetic information about C_{40} carotenoid biosynthesis in *Agrobacterium aurantiacum*, *Alcaligenes* PC-1, *Thermus thermophilus*, *Streptomyces griseus*, *Streptomyces setonii*, *Flavobacterium* sp. strain R1534, *Mycobacterium aurum*, and the C_{30} carotenoid biosynthesis pathway of *Staphylococcus aureus* has been obtained.

2.12.4.3.1 **Erwinia** *species*

The study of carotenoid biosynthesis in the genus *Erwinia*, the members of which are phytopathogenic soil bacteria, has thus far been performed almost exclusively using an indirect approach in *Escherichia coli*, a genetically amenable host. The latter bacterium normally lacks carotenoids but is capable of expressing the fortuitously clustered genus *Erwinia crt* genes, which were originally cloned by screening for yellow pigmented *E. coli* colonies after genetic transformation with cosmid libraries of DNA from *Erwinia* species.[25,84,85] The *Erwinia uredovora crt* genes have since been shown to be expressed in bacterial hosts other than *E. coli*, including *Agrobacterium tumefaciens*, *R. sphaeroides*, and *Zymomonas mobilis*.[86–88] In contrast, transfer of the *R. capsulatus crt* gene cluster to *E. coli* did not result in carotenoid accumulation[64] although other bacteria, including *Agrobacterium radiobacter*, *A. tumefaciens*, *Azotomonas insolita*, *Paracoccus denitrificans*, and *Pseudomonas stutzeri*, proved capable of synthesizing these pigments after transformation with the *R. sphaeroides crt* genes.[55,89]

Determination of the structures and compositions of carotenoids from *Erwinia herbicola*, *Erwinia uredovora*, and the corresponding transformed *E. coli* strains revealed the end products to be primarily β-carotene-derived cyclic xanthophylls (Figure 5): β-cryptoxanthin monoglucoside ((3*R*)-3-(β-D-glucosyloxy)-β,β-carotene), zeaxanthin, zeaxanthin monoglucoside ((3*R*,3′*R*)-3′-(β-D-glucosyloxy)-β,β-caroten-3-ol), and zeaxanthin diglucoside ((3*R*,3′*R*)-3,3′-bis(β-D-glucosyloxy)-β,β-carotene).[25,90] Interestingly, *E. coli* strains carrying genus *Erwinia crt* genes were found to have similar pigment distributions but several-fold higher specific carotenoid contents than the corresponding *Erwinia* species. Because *E. coli* cells do not normally make carotenoids, it is reasonable to assume that they also lack specific carotenoid-binding proteins. Therefore, the question of the cellular location of carotenoids produced by *E. coli* transformed with genus *Erwinia crt* genes was of interest. The fact that such transformed *E. coli* were found to be resistant to photooxidative killing when exposed to the combination of long wavelength UV light and lipophilic photosensitizing molecules suggested that the carotenoids accumulated in cell membranes.[91]

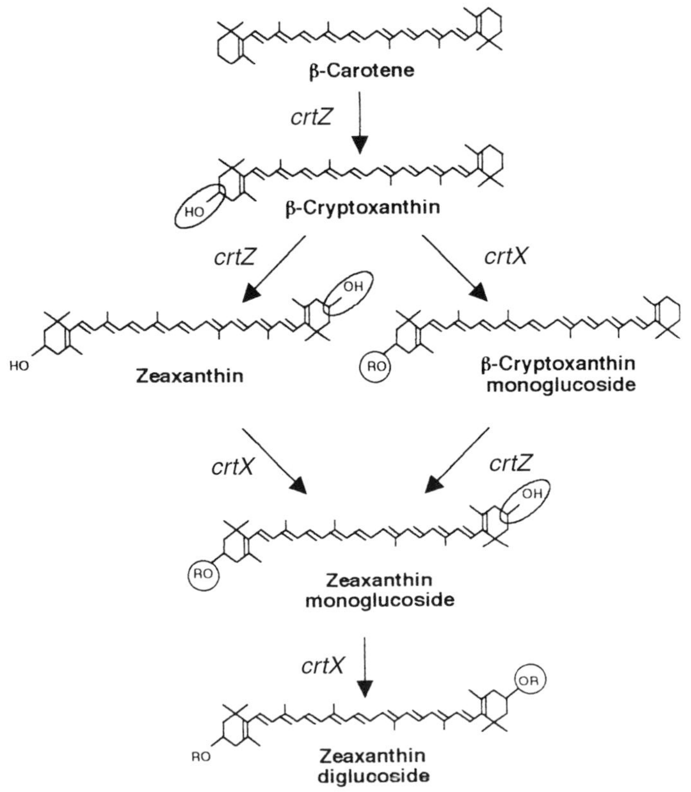

Figure 5 Cyclic xanthophyll glycoside biosynthesis pathway in the genus *Erwinia*. *crt* genes required for specific carotenoid biosynthetic reactions are given here and the corresponding enzymes are listed in Table 2. R = β-D-glucoside. Structural alternations to the carotenoids during biosynthesis are circled for emphasis.

The cloning of *crt* genes from *Erwinia* species stimulated great interest in the field of bacterial carotenoid biosynthesis. First, the structural nature of the pigments accumulated in the genus *Erwinia* and in transformed *E. coli* strains made it clear that a biochemical pathway similar in part to that used by cyanobacteria, algae, and plants, all of which contain zeaxanthin as a major pigment, must be active.[1,14] In contrast, the specialized end product xanthophylls of the genetically characterized genus *Rhodobacter* carotenoid biosynthesis pathway (Figure 4) are found in only a few species of bacteria.[57] Second, the use of *E. coli* strains that express one or more of the genus *Erwinia crt* genes has proved invaluable for isolating and/or performing *in vivo* functional analyses on the basis of restored or altered pigment accumulation of novel *crt* genes from cyanobacteria,[33,34,41,43,92–94] anoxygenic photosynthetic bacteria,[92] and other species of nonphotosynthetic bacteria.[30,31,47] This strategy has also allowed the identification and study of eukaryotic carotenoid biosynthesis genes from fungi,[95] algae,[96–98] and plants.[99–107]

Nucleotide sequence comparisons and mutational analyses of the cloned *crt* gene clusters from *E. herbicola* strains Eho10 and Eho13 and from *E. uredovora* have defined six genes, *crtB*, *crtE*, *crtI*, *crtX*, *crtY*, and *crtZ*, that encode enzymes (Table 2) required for the synthesis of glucosides of zeaxanthin and cryptoxanthin (Figures 2, 3, and 5) from isoprenoid pyrophosphate carotenoid precursors.[23–25,42,90] The *E. herbicola* strain Eho10 *crt* gene cluster was also found to be uniquely interrupted by an additional open reading frame (ORF), ORF6, of unknown function.[42,90] In *Erwinia* species the organization of the *crt* genes mandates the existence of at least two transcriptional operons, *crtE*(ORF6)*XYIB* and *crtZ*. The existence of the former is supported by primer extension analysis of the 5′ end of mRNA transcripts encoding *Erwinia herbicola* strain Eho13 *crtE*, as well as the expression in *E. coli* of the *crtE* promoter region fused to a reporter gene.[24]

Sequence comparison of the *E. herbicola* Eho10 and *E. uredovora crt* gene clusters with that of *R. capsulatus* rapidly allowed the identification of structural homologues of *crtB*, *crtE*, and *crtI*, which encode the three earliest enzymes specifically required for carotenoid biosynthesis in the latter bacterium.[23,25,108] The functions of *crtX*, *crtY*, and *crtZ* were determined (Figure 5) and the function of *crtI* confirmed (Figure 3) by analyzing the C_{40} carotenoids accumulated in *E. coli* strains containing partially deleted or mutated genus *Erwinia crt* gene clusters,[25,42] and by restoring the wildtype phenotype to an *R. sphaeroides crtI* mutant.[88] Mutations in *crtE* and *crtB* were found, to prevent the accumulation of phytoene or later C_{40} carotenoid biosynthesis intermediates. Therefore, the functions of *crtE* and *crtB* (Figure 2) were determined by adding radiolabeled isoprenoid pyrophosphates to extracts of *E. coli* or *A. tumefaciens* transformed with one or both of these genes and examining the products formed.[75,76] Genus *Erwinia* carotenoid biosynthetic reactions that had not been previously described in studies of *Rhodobacter* species were cyclization, performed by CrtY, hydroxylation, mediated by CrtZ, and glucosylation, catalyzed by CrtX (Table 2). CrtE,[109] CrtI[110], CrtX,[111] CrtY,[112,113] and CrtZ[112] were subsequently overexpressed in *E. coli* and their functions confirmed by biochemical analyses, as described in Sections 2.12.5.2.1 and 2.12.5.2.3–2.12.5.2.5.

2.12.4.3.2 Myxococcus xanthus

The 1990s have seen the emergence of the myxobacterium *M. xanthus* as the preeminent model system for classical and molecular genetic studies of blue light-induced carotenogenesis, a phenomenon found in certain nonphotosynthetic bacteria and fungi.[4] The molecular details of the regulatory circuits involved in light regulation will be discussed in Section 2.12.6.2.2, whereas this section will focus on the carotenoid biosynthesis pathway itself.

Early work with *M. xanthus* demonstrated that carotenoid accumulation was strictly light-dependent and occurred when cells entered the stationary phase, thereby preventing cellular photolysis associated with a 16-fold increase in accumulation of the photosensitizer protoporphyrin IX in cell membranes.[114] Furthermore, protoporphyrin IX was identified as the blue light photoreceptor for both carotenogenesis and photolysis, thus establishing a direct link between these two processes.[115,116] The cytoplasmic membrane was identified as a major site of carotenoid accumulation in the genus *Myxococcus*.[117]

Proposals for carotenoid biosynthetic pathways in *Myxococcus* species[4,118] have relied heavily on the structural determination of pigments from bacterial cultures grown in the presence or absence of classical prokaryotic carotenoid biosynthesis inhibitors.[119] Many of these compounds which prevent either phytoene desaturation or lycopene cyclization.[13] Typical C_{40} end product carotenoids of the genus *Myxococcus* include fatty acid esters of myxobactone (1′-glucosyloxy-3′,4′-didehydro-

1′,2′-dihydro-β,ψ-caroten-4-one) and myxobactin (1′-glucosyloxy-3,4,3′,4′-tetrahydro-1′,2′-dihydro-β,ψ-carotene), both monocyclic carotenoid glucosides thought to arise from γ-carotene (β,ψ-carotene; Figure 3). Similar carotenoids are also present in the genera *Cystobacter* and *Stigmatella*. In addition to fatty acid esters of myxobactone and myxobactin, surprisingly large amounts of phytoene have been observed in wildtype strains of some *Myxococcus* species, including *M. xanthus*.[118,119] The significance of this observation is unclear because phytoene has too few conjugated double bonds to be effective in the photoprotective quenching of excited state protoporphyrin IX molecules.

Genetic studies of carotenoid biosynthesis in *M. xanthus* using point and transposon Tn5 pigmentation mutants have revealed two phenotypic classes.[4] One class of putative regulatory mutants, those denoted Car^{c}, display constitutive (i.e., light-independent) synthesis of carotenoids. A second class of putative regulatory/biosynthetic mutants, those denoted Car^{-}, do not accumulate carotenoids under any conditions. All of the genetic loci found to be associated with either of these phenotypes were originally designated *car*. This nomenclature has been retained for loci strictly involved in the regulation of carotenogenesis, whereas molecularly cloned and characterized loci that encode carotenoid biosynthesis enzymes have since been renamed using the *crt* system.[27]

Analysis of Car^{-} mutants initially revealed the existence of two unlinked genetic loci, *carB* and *carC*, representing putative carotenoid biosynthetic functions.[50] The light-induced *carB* locus was also found to be linked to *carA*, a regulatory locus associated with the Car^{c} phenotype. A 16 kilobase region surrounding the *carA* locus was subsequently cloned in a shuttle vector.[121] Although mutations in both *carB* and *carC* resulted in the absence of colored carotenoids under all conditions, a *carC* mutant accumulated phytoene in the light.[120] Therefore, *carB* and *carC* were proposed to be responsible for the synthesis and desaturation of phytoene, respectively (Figures 2 and 3). The light-induced *carC* gene was subsequently cloned and found to represent the *M. xanthus* homologue of *crtI* (Table 2).[37] Transposon mutagenesis and transcriptional analysis of the cloned *carA–carB* region provided evidence for the existence of a light-induced *carBA* operon that coupled several different carotenoid biosynthesis functions to the *carA* regulatory function.[122]

The nucleotide sequence of the *carBA* region has recently been determined, allowing the alignment of individual gene sequences with the genetic map generated through mutant studies.[27] The analysis of gene function has been complicated by the operon structure and the potential for polar inactivation by Tn5 insertions not only of the disrupted gene but also of downstream, transcriptionally coupled genes. Nevertheless, the available data indicate that the *carBA* region encodes 11 genes, including four of known function, *crtB*, *crtE*, *crtC*, and *crtD* (Figures 2 and 4), two putative carotenoid biosynthesis genes, ORF2 and ORF6, two putative regulatory genes, ORF10 and ORF11, and three genes of unknown function, ORF7, ORF8, and ORF9. These genes seem to be organized in a *crtE*ORF2*crtBDC*ORF6–ORF7–ORF8–ORF9–ORF10–ORF11 operon, designated here as the *M. xanthus crt* gene cluster for the sake of simplicity. Functions were assigned to the *crtE*, *crtB*, *crtC*, and *crtD* genes on the basis of mutant phenotypes and the structural similarities of the gene products to other known carotenoid biosynthesis enzymes (Table 2).[27] For example, *carB* transposon mutations resulting in a Car^{-} phenotype were found to interrupt the *crtE* gene. An insertional mutation within ORF7 seemed to have no effect on the esterification of the end product carotenoids, suggesting that ORF7 and the downstream ORF8 and ORF9 are not directly involved in carotenoid biosynthesis. One unusual feature of *M. xanthus* is the physical separation of *crtI* from the other clustered *crt* genes.[37] Whether this unique genomic arrangement is related to the high degree of regulation of *crtI* expression, as discussed in Section 2.12.6.2.2, is unknown.

The clustered *crtE*, *crtB*, *crtC*, and *crtD* genes, together with the unlinked *crtI* gene, should be sufficient to direct the synthesis of neurosporene, lycopene, and their hydroxy xanthophyll derivatives (Figures 3 and 4). ORF6 encodes a novel protein required for lycopene cyclization (Table 2). The product of ORF2 is predicted to be structurally related to bacterial carotenoid desaturases, such as CrtI and CrtD, and may be required late in the biochemical pathway. Therefore, the biosynthetic functions provided by the five known *crt* genes, and those postulated for ORF6 and ORF2, could be sufficient to produce myxobactin. In order to synthesize the end product carotenoids found in the genus *Myxococcus*, enzymes that catalyze glucosylation, addition of a keto group and fatty acid sugar esterification would have to be encoded by as yet unidentified genes.[4,27,118] Experimental approaches such as the introduction of site-directed nonpolar mutations into the *crt* gene cluster, *in vivo* complementation of mutants, and biochemical analyses of purified enzymes may be necessary to complete the molecular genetic description of carotenoid biosynthesis in *M. xanthus*.

2.12.4.3.3 **Agrobacterium aurantiacum *and* Alcaligenes *PC-1***

crt genes from marine bacteria, including a *crt* gene cluster from *A. aurantiacum* and two of the corresponding genes from *Alcaligenes* PC-1, have been cloned.[31,47] *A. aurantiacum* grown under several different conditions accumulates primarily adonixanthin ((3*S*,3′*R*)-3,3′-dihydroxy-*β*,*β*-caroten-4-one) and astaxanthin (3,3′-dihydroxy-*β*,*β*-carotene-4,4′-dione; Figure 6),[123] the latter a pigment of major commercial interest (Section 2.12.7).[10,14] The *A. aurantiacum crt* gene cluster was cloned by screening for altered pigmentation in colonies of *E. coli* containing genus *Erwinia crt* genes that allow *β*-carotene accumulation in the former organism.[31]

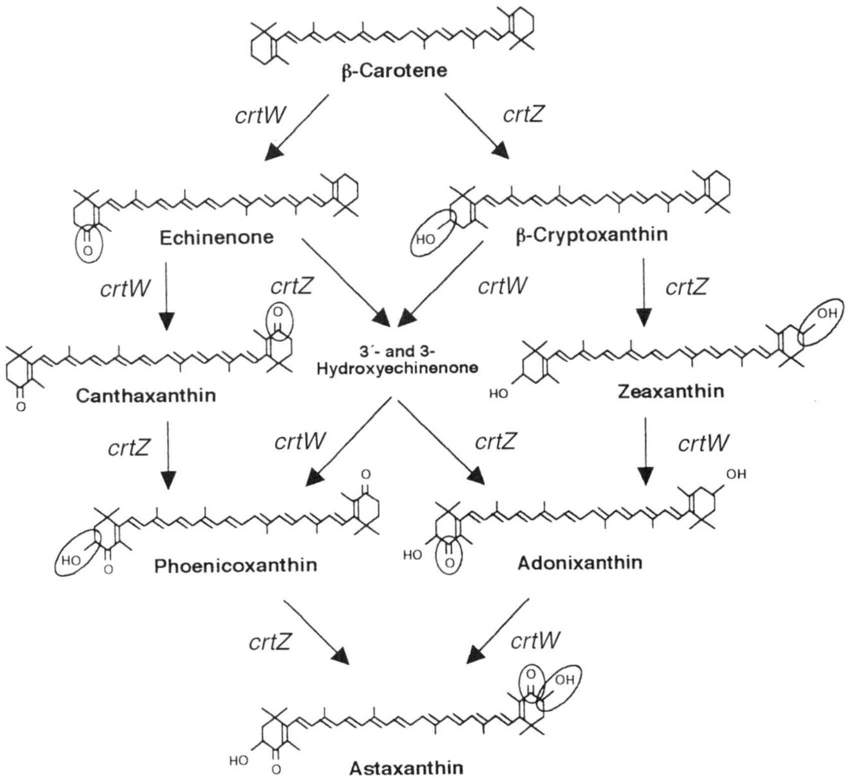

Figure 6 Cyclic xanthophyll biosynthesis pathway in *A. aurantiacum*. *crt* genes required for specific carotenoid biosynthetic reactions are given here and the corresponding enzymes are listed in Table 2. The structures of 3-hydroxyechinenone ((3*S*)-3-hydroxy-*β*,*β*-caroten-4-one) and 3′-hydroxyechinenone (3′-hydroxy-*β*,*β*-caroten-4-one)[15,30] have been omitted here for ease of presentation. Structural alternations to the carotenoids during biosynthesis are circled for emphasis.

The cloned gene cluster includes five genes, *crtB*, *crtI*, *crtW*, *crtY*, and *crtZ*, that may be organized in a *crtWZYIB* operon. In contrast to other known bacterial *crt* gene clusters, no *crtE* gene was found in the sequenced region, although the cloning strategy and subsequent functional characterization of the *A. aurantiacum crt* genes do not exclude the possibility that a flanking region of the genome might encode *crtE*. Homologies between the predicted *crtB*, *crtI*, *crtY*, and *crtZ* gene products and cognate enzymes of the genus *Erwinia*, combined with analyses of pigment content in *E. coli* strains containing different combinations of *Erwinia uredovora* and *A. aurantiacum crt* genes, allowed assignments of gene function (Table 2). The novel *crtW* gene identified in *A. aurantiacum* and *Alcaligenes* PC-1 was found to be required for the addition of a keto group to cyclic carotenes and xanthophylls.[31,47] The resulting biosynthetic pathway for the synthesis of astaxanthin from GGPP suggests that CrtZ, the *β*-carotene hydroxylase, and CrtW, the *β*-C-4-oxygenase, can each accept several different substrates (Figure 6).

2.12.4.3.4 **Thermus thermophilus**

Thermus thermophilus, an aerobic thermophilic bacterium normally found in hot springs, has also recently served as a model for classical and molecular genetic studies of carotenoid biosynthesis.

The genus *Thermus* had long been known to produce carotenes, but the structures of the polar carotenoid end products reported to comprise up to 60% of the outer membrane lipids were not initially determined.[10,14] Attempts to clone *T. thermophilus crt* genes in *E. coli* by screening for carotenoid production at temperatures that allow growth of the latter bacterium were unsuccessful. Therefore, the novel strategy of gene cloning by screening for carotenoid overproduction in the homologous host was adopted.[35] This strategy resulted in the isolation of orange rather than the normal yellow bacterial colonies and ultimately led to the identification of a *T. thermophilus* gene that displayed homology to *crtB* from other bacteria (Table 2). Because the amounts of all carotenoids normally found in *T. thermophilus* were higher in the strain containing an additional plasmid-borne copy of *crtB*, it was suggested that phytoene synthase was a rate-limiting biosynthetic enzyme in this organism (Figure 2). Carotenoid underproducing and overproducing mutants of *T. thermophilus* have also been identified.[124,125] Carotenoid overproduction was observed to severely inhibit bacterial growth at 80 °C but not at 70 °C. On the other hand, bacterial survival rates after UV irradiation were slightly lowered by carotenoid underproduction, but were substantially raised by carotenoid overproduction.[125] Genetic transformation with the *T. thermus crtB* gene of a mutant bacterial strain that accumulates 13-fold more carotenoids than the wildtype and is thought to be mutated at a locus other than *crtB* resulted in a further increase in total carotenoid content to 20-fold that of the wildtype.[124]

The *T. thermophilus crt* genes appeared to be clustered but, unlike other bacterial *crt* gene clusters characterized thus far, seem to be encoded on a large endogenous plasmid rather than on the chromosome.[126] It is, however, interesting to note that an early study of the genus *Erwinia* suggested that the ability to produce yellow pigments might be plasmid-encoded in at least some species,[84] though not in those that have since been examined in detail (Section 2.12.4.3.1). The unique end product carotenoids of *T. thermophilus* have been identified as novel branched fatty acid esters derived from glucosides of zeaxanthin (Figure 5).[127] Based on their unusual structures, these carotenoids have been proposed to regulate membrane stability at high temperatures. The structural nature of these pigments indicates that carotenoid biosynthesis in *T. thermophilus* likely follows a pathway similar to that used by the genus *Erwinia* (Figures 2, 3, and 5).

2.12.4.3.5 **Streptomyces griseus**

In the genus *Streptomyces*, by analogy to the situation described in the genus *Myxococcus* (Section 2.12.4.3.2), carotenogenesis can be light-induced, constitutive, or completely absent.[53] A cryptic *crt* gene cluster has recently been cloned from one Car$^-$ strain of *S. griseus* by genetic transformation of another Car$^-$ bacterial strain in which at least some of the cryptic genes were found to be expressed.[30] Certain strains of the genus *Streptomyces* accumulate aromatic β-carotene-derived cyclic carotenes such as isorenieratene (φ,φ-carotene), β-isorenieratene (β,φ-carotene), and hydroxy xanthophyll derivatives thereof. These aromatic carotenoids also occur, for example, in certain green sulfur anoxygenic photosynthetic bacteria of the genus *Chlorobium*.[10,14,57] The cloned cryptic *S. griseus crt* genes were initially revealed to direct the synthesis of lycopene (Figure 3), which could be converted to isorenieratene by the products of the endogenous *crt* genes of the colorless mutant host.[30]

Sequence analysis of the *S. griseus crt* gene cluster has revealed the presence of seven genes, *crtB*, *crtE*, *crtI*, *crtU*, *crtV*, *crtT*, and *crtY*, which may be organized in *crtEIBV* and *crtYTU* transcriptional operons. Sequence relationships between the predicted *crtB*, *crtE*, *crtI*, and *crtY* gene products and homologous enzymes from other bacteria, as well as functional analyses of *S. griseus crtE* in an *E. coli* strain transformed with *Erwinia uredovora crt* genes and of *S. griseus crtB* and *crtE* transferred to the colorless mutant of *S. griseus* allowed functional assignments for these genes (Table 2). The three novel *crt* genes, *crtT*, *crtU*, and *crtV*, were designated as such because their products displayed structural homologies to the genus *Rhodobacter* CrtF and CrtI/CrtD enzymes and to CheB, a protein–glutamate methyltransferase, respectively.[30] Although *crtT*, *crtU*, and *crtV* have been postulated to be involved in the conversion of β-carotene to isorenieratene via β-isorenieratene, which is thought to involve the aromatization of each of the aliphatic β-rings accompanied by the migration of a methyl group, no direct evidence exists to support this proposal. Primer extension analysis suggests the presence of a promoter 5′ to *crtE* that likely regulates expression of the proposed *crtEIBV* operon.

2.12.4.3.6 **Staphylococcus aureus**

Two *Staphylococcus aureus crt* genes representing part of a larger, and as yet uncharacterized gene cluster involved in carotenoid biosynthesis have been identified by their expression in *E. coli* and in an unpigmented strain of *Staphylococcus carnosus*.[36] Members of the genus *Staphylococcus* have been demonstrated to accumulate a wide variety of C_{30} carotenoids originating from FPP rather than GGPP (Figures 1 and 2). The early reactions of C_{30} carotenoid biosynthesis involving 4,4′-diapophytoene formation and its desaturation to 4,4′-diaponeurosporene (7,8-dihydro-4,4′-diapocarotene)[10,14,19] parallel the reaction sequence well established for the C_{40} pathway in anoxygenic photosynthetic bacteria and nonphotosynthetic bacteria. Some of the varied pigment end products accumulated during the late stages of cell growth in the genus *Staphylococcus* ultimately arise from 4,4′-diaponeurosporene, and require several oxidation, glucosylation, and acylation reactions. The *S. aureus* strain from which the *crt* genes were cloned accumulates 4,4′-diaponeurosporene during the early stages of growth.[19,36]

Sequence analysis of the two thus far identified *S. aureus crt* genes, *crtM* and *crtN*, indicates that their products are the respective C_{30} carotenoid biosynthesis homologues of CrtB and CrtI in the C_{40} pathway (Figure 2; Table 2) and that they probably form a *crtNM* operon. Confirming these observations, both *E. coli* and an *S. carnosus* pigmentation mutant carrying the two *S. aureus crt* genes accumulate 4,4′-diaponeurosporene.[36] Furthermore, cell extracts prepared from *E. coli* transformed with either *S. aureus crtM* alone or with both *crtM* and *crtN* are capable of incorporating radiolabeled FPP into 4,4′-diapophytoene and 4,4′-diaponeurosporene, respectively.

2.12.4.3.7 ***Other nonphotosynthetic bacteria***

crt gene clusters involved in C_{40} carotenoid biosynthesis in two other species of bacteria have also been cloned. A cluster of five *crt* genes from *Flavobacterium* R1534 that direct the synthesis of zeaxanthin in *E. coli* has been characterized.[26] Some members of the genus *Flavobacterium* accumulate zeaxanthin (Figure 5), whereas others contain C_{45} and C_{50} carotenoids.[10,14] The five genes have been reported to correspond to *crtB*, *crtE*, *crtI*, *crtY*, and *crtZ* from other bacteria (Table 2), and are organized into at least two transcriptional operons. This combination of genes would be consistent with the known genetic requirements for the synthesis of zeaxanthin in the genus *Erwinia* (Figures 2, 3, and 5).

An as yet largely uncharacterized *crt* gene cluster has also been obtained from *Mycobacterium aurum*.[128] Isorenieratene, also known as leprotene, is a major end product in *M. aurum*,[10,14] suggesting that the carotenoid biosynthesis pathway which operates in this bacterium may resemble that active in *S. griseus*.[30]

2.12.4.4 Cyanobacteria

Carotenoids are found universally among the Chl *a*- and Chl *b*-containing oxygenic photosynthetic cyanobacteria.[5,14] Typical pigments accumulated include β-carotene, the cyclic xanthophylls echinenone (β,β-caroten-4-one), canthaxanthin (β,β-carotene-4,4′-dione), zeaxanthin, and the cyclic xanthophyll glycoside myxoxanthophyll (2′-(β-L-rhamnopyranosyloxy)-3,4′-didehydro,1′,2′-dihydro-β,ψ-carotene-3,1′-diol), which contains a rhamnose sugar moiety. As mentioned previously, β-carotene and zeaxanthin are major carotenoids in plants and in some other bacteria, such as species of *Erwinia* and *Flavobacterium* (Figure 5). Echinenone and canthaxanthin are two intermediates of astaxanthin synthesis in *A. aurantiacum* (Figure 6).

2.12.4.4.1 **Synechococcus, Synechocystis,** ***and*** **Anabaena** ***species***

Studies of *Synechococcus* sp. strain PCC7942,[33,39,43,94,129,130] *Synechocystis* sp. strain PCC6803,[34,40,131] and *Anabaena* sp. strain PCC7120[41,93] have provided substantial information about the genes required for the biosynthesis of cyclic carotenes in cyanobacteria. *Synechococcus* PCC7942 and *Synechocystis* PCC6803 have been demonstrated to accumulate the normal spectrum of cyanobacterial carotenoids.[130–132] Nevertheless, the genetics of an entire cyanobacterial carotenoid biosynthetic pathway have not yet been established, in part because the *crt* genes of cyanobacteria have been found to be dispersed rather than clustered in the genome. Four molecularly characterized *crt* genes, denoted *crtB*, *crtP*, *crtQ*, and *crtL* in the current nomenclature,[1,5] are required for the biosynthetic reactions that convert GGPP to β-carotene (Figures 2 and 3).

Very few cyanobacterial mutants with novel pigmentation resulting from altered carotenoid compositions have been described in the literature.[5] Colored carotenoids are strictly required to protect against photooxidative damage during photosynthetic growth and even mutants blocked late in the biosynthetic pathway would be difficult to detect visually due to the masking effects of Chl *a* and Chl *b*. Therefore, two alternative approaches involving selection of mutants and screening for pigmentation changes in a heterologous host have been taken to isolate cyanobacterial *crt* genes.

The first strategy followed was to select for resistance to inhibitors of carotenoid biosynthesis.[40,94,130] This approach relied on the observations that so-called bleaching (i.e., chlorosis-inducing) herbicides, such as 4-chloro-5-methylamino-2-(3-trifluromethylphenyl)-pyridazin-3(2H)one (SAN 9879; norflurazon), inhibit the desaturation of phytoene in Chl *a*- and Chl *b*-containing organisms, and that substituted triethylamines such as 2-(4-methylphenoxy)triethylamine (MPTA) and 2-(4-chlorophenylthio)triethylamine (CPTA) inhibit lycopene cyclization.[13] Genetic selections to identify norflurazon-resistant cyanobacterial strains led to the cloning and sequencing of herbicide resistance genes from *Synechococcus* PCC7942[39,129] and *Synechocystis* PCC6813.[40] A series of norflurazon-resistant mutants were generated in both species of cyanobacteria and analyzed with respect to their carotenoid contents, *in vitro* phytoene desaturase activities, and molecular sites of mutation.[131,132] These experiments led to the conclusion that phytoene desaturation was the rate-limiting step in cyanobacterial carotenoid biosynthesis, in contrast to the findings in *T. thermophilus* that implicated phytoene synthase as the rate-limiting enzyme.[35] The final and most direct proof that the *Synechococcus* PCC7942 *crtP* (formerly *pds*) gene encoded phytoene desaturase was obtained by *in vivo* restoration of carotenoid accumulation to *E. coli* strains that contained some *Erwinia uredovora crt* genes but lacked *crtI*.[92] This and several similar experiments performed with plant homologues of *crtP*[78,99,133] demonstrated conclusively that the CrtP-type phytoene desaturase from Chl *a*- and Chl *b*-containing prokaryotes and eukaryotes catalyzes, in contrast to the CrtI enzyme, only two successive desaturations to yield ζ-carotene (Figure 3). A selection strategy was also employed to clone and characterize the *crtL* (formerly *lcy*) gene that confers MPTA resistance and which has been shown to encode the cyanobacterial lycopene cyclase.[43,94]

The second approach that has been used to clone a cyanobacterial *crt* gene was to proceed directly to a library-based screen based on pigment accumulation.[93] To this end, a cDNA library from *Anabaena* PCC7120 was transformed into carotene-accumulating *E. coli* strains that carried combinations of *crt* genes from other bacteria. Unusually pigmented transformants were characterized, resulting in the isolation and sequencing of *crtQ* (formerly *zds*), a gene whose product allowed the *in vivo* synthesis of lycopene from ζ-carotene and neurosporene, but not from phytoene (Figure 3).[41,93]

The *crtB* (formerly *psy* or *pys*) homologues of *Synechococcus* PCC7942[33] and *Synechocystis* PCC6803[34] (Table 2) were identified because they are physically adjacent on the chromosome to the respective *crtP* genes of these cyanobacteria.[39,40] Despite this bit of serendipity, a feature that generally distinguishes cyanobacteria from other carotenogenic prokaryotes is the chromosomal dispersal of the *crt* genes. Only *crtP* and *crtB* are physically linked to one another, and mutational data suggest that even they do not form an obligate transcriptional operon. The function of *crtB* in cyanobacteria was confirmed by *in vivo* functional complementation of *E. coli* strains transformed with the *Erwinia uredovora crt* genes and by following the incorporation of radiolabeled GGPP added to cell extracts of the transformed *E. coli*.[33,34] The two known cyanobacterial carotenoid desaturases, CrtP and CrtQ, have been overexpressed in *E. coli* and isolated in an active state for biochemical studies,[134,135] as described in Section 2.12.5.2.3.

2.12.5 CAROTENOID BIOSYNTHESIS ENZYMES

Classical biochemical analyses of carotenoid biosynthesis have been reviewed elsewhere.[12] The membrane-bound enzymes that convert phytoene to the colored end product carotenoids, have, however, historically proven recalcitrant to this type of approach. Many structural features of carotenoid biosynthesis enzymes have therefore been deduced from the primary amino acid sequences encoded by the corresponding genes.[9] Such an analysis based on molecular genetics, as well as several biochemical studies of purified prokaryotic enzymes, have led to the following general picture of the evolution of carotenoid biosynthesis.

2.12.5.1 Evolutionary Conservation

CrtE (GGPP synthase), CrtB (phytoene synthase), and CrtI (phytoene desaturase) (Figures 2 and 3) have been found to be structurally conserved not only in anoxygenic photosynthetic bacteria such as *Rhodobacter* species and nonphotosynthetic bacteria represented by *Erwinia* species, but also to possess structural and functional homologues in eukaryotes.[23,108] In particular, CrtE- and CrtB-related enzymes are present among all carotenogenic organisms that have been examined to date.[8,9] In the early 1990s, the observations were made that bacterial CrtE displayed structural similarities with GGPP synthases from the fungus *Neurospora crassa*[136] and red pepper fruit,[137] bacterial CrtB[23] with a tomato protein preferentially expressed during fruit ripening,[138] which was later shown to functionally represent phytoene synthase,[79,139,140] and bacterial CrtI with the *N. crassa* phytoene desaturase.[77,141] In the mid-1990s, eukaryotic homologues of the xanthophyll biosynthesis enzymes CrtZ and CrtW (Figures 5 and 6) from nonphotosynthetic bacteria were identified in the higher plant *Arabidopsis thaliana*[106] and in the green alga *Haematococcus pluvialis*,[96–98] respectively.

Not all carotenoid biosynthesis enzymes have, however, been evolutionarily conserved among all organisms. In fact, both the membrane-bound carotenoid desaturases and lycopene cyclases can be subdivided into two structurally distinct groups[1,5] depending on whether they occur in anoxygenic photosynthetic bacteria, nonphotosynthetic bacteria and fungi (CrtI-type desaturases[23,77,141] and CrtY-type cyclases[25,31]), or in cyanobacteria, algae, and higher plants (CrtP-type desaturases[39,78,99,133] and CrtL-type cyclases[43,103]). Thus, the structural conservation of certain carotenoid biosynthesis enzymes depends upon lifestyle: does the organism in question survive using oxygenic Chl-based photosynthesis or not? The reason for the existence of two separate structural classes of carotenoid desaturases and lycopene cyclases is not known, although several conceivable explanations exist (Section 2.12.8.1).

2.12.5.2 Structural and Biochemical Features

2.12.5.2.1 GGPP synthase

As shown in Figure 1, GGPP is used for several different biosynthetic pathways in bacteria.[82] However, the experimental results that (i) carotenoid-deficient mutants of bacteria that lack the GGPP synthase activity encoded by the *crtE* gene can be isolated, (ii) that most bacterial *crt* gene clusters encode *crtE*, and (iii) that carotenoid biosynthesis can be engineered in *E. coli* strains only in the presence of a *crtE* gene (Sections 2.12.4.2.1 and 2.12.4.3.1, 2.12.4.3.2, and 2.12.4.3.5), suggest that CrtE may represent a carotenoid-specific GGPP synthase. Thus, although the production of GGPP itself cannot be said to belong exclusively to the carotenoid branch of isoprenoid metabolism, this reaction is probably critical to pigment biosynthesis in most, if not all, carotenogenic bacteria. CrtE and its eukaryotic homologues have been identified in a number of organisms, including *Rhodobacter* species,[28,29] nonphotosynthetic bacteria,[23,25,27,30] the fungus *N. crassa*,[136] and higher plants.[8,137,142]

Bacterial CrtE shares several conserved structural motifs with various isoprenyl pyrophosphate synthases, which constitute a superfamily of structurally and functionally related proteins that includes both prokaryotic and eukaryotic GGPP, FPP, and hexaprenyl pyrophosphate synthases[143,144] (see Chapter 2.04). These enzymes function as homodimers that require either Mg^{2+} or Mn^{2+} for activity, and catalyze 1′–4 isoprenoid pyrophosphate condensations between C4 of the homoallylic C_5 building block IPP and C1 of various allylic substrates. The biochemical specificity of a given isoprenyl pyrophosphate synthase is manifested at three levels: (i) the chain lengths of the allylic isoprenoid substrate(s) accepted and the product(s) synthesized; (ii) the stereochemistry of the double bonds in the allylic substrate(s); and (iii) the stereochemistry of the new double bond, generally all-*trans*, introduced by the condensation with IPP.

FPP and hexaprenyl pyrophosphate synthases were originally observed to contain three short conserved structural motifs, termed domains I, II, and II, that were rich in charged amino acids, such as Arg, Asp, and Gln.[145] Subsequent protein sequence comparisons have revealed that the isoprenyl pyrophosphate synthase superfamily members in fact share five conserved stretches of amino acids.[143] Domains I and II, which have been proposed to be involved directly or indirectly through Mg^{2+} bridges in binding the pyrophosphate moiety of the substrate,[145] were subsequently identified in bacterial and eukaryotic GGPP synthases.[75,108,136,137,146] Mutagenesis studies of eukaryotic FPP synthases suggest that the Arg and Asp residues within these highly charged regions may indeed be involved in substrate and product binding and/or catalysis itself.[147–149] It is also worth noting that a mutated *N. crassa* homologue of bacterial CrtE that carries the amino acid substitution Ser-336 to Asn-336, a change which alters a conserved residue located immediately C-terminal to domain II, is almost inactive *in vitro*.[95] Thus, these data provide compelling albeit indirect evidence that domains I and II of bacterial CrtE enzymes may be crucial for activity.

Bacterial carotenoid biosynthesis can, in general, be said to start with the 1′–4 condensation of IPP with GPP (Figure 2), based on the *in vivo* and *in vitro* substrate preferences of the genus *Erwinia* CrtE enzyme.[75,76,109] This condensation yields first FPP, which is then extended by the addition of another molecule of IPP to produce GGPP. In contrast, the fungal and higher plant structural homologues of CrtE have been found to accept dimethylallyl pyrophosphate (DMAPP) as the initial allylic substrate *in vitro*.[95,137,150,151] Both eubacterial and eukaryotic GGPP synthases produce GGPP exclusively.

2.12.5.2.2 *Phytoene synthase*

CrtB, the phytoene synthase, is the most widely distributed and functionally conserved C_{40} carotenoid biosynthesis enzyme described thus far, and seems to be universally present among both carotenogenic prokaryotes and eukaryotes (Table 2).[9] This enzyme uses an isoprenyl pyrophosphate substrate, but differs from GGPP synthase and other members of the superfamily of isoprenoid pyrophosphate synthases described in Section 2.12.5.2.1 in that it does not catalyze a 1′–4 condensation between homoallylic and allylic substrates, but rather a 1′–2–3 condensation of two molecules of GGPP followed by a 1′–1 rearrangement.

CrtB and eukaryotic homologues thereof have been identified thus far in *Rhodobacter* species,[28,32] numerous nonphotosynthetic bacteria,[23,25,27,30,31,35] several cyanobacteria,[33,34] the fungus *N. crassa*,[152] and in higher plants.[8,79,80,138,140,153] CrtM, the diapophytoene synthase of the C_{30} carotenoid biosynthesis pathway of *S. aureus*, is also a CrtB homologue.[36] Bacterial CrtB in addition displays significant structural conservation with eukaryotic squalene synthase.[75,153] This enzyme carries out a reaction in some respects similar to those catalyzed by CrtB and CrtM (Figure 2), namely the condensation of two molecules of FPP to produce the C_{30} sterol hydrocarbon precursor squalene, a structural relative of diapophytoene in which the 15,15′-*cis*-double bond of the C_{30} carotenoid is reduced using NADPH as a cofactor (see Chapter 2.09).

Bacterial CrtB enzymes contain several conserved sequence motifs including one particularly highly conserved stretch of about 35 amino acids also found in tomato phytoene synthase and eukaryotic squalene synthase that includes a number of both positively and negatively charged residues.[33,75,154] A portion of this motif resembles domain II,[3,108] a conserved sequence feature originally identified in other types of isoprenyl pyrophosphate synthases (Section 2.12.5.2.1).[136,137,145] As in the case of bacterial CrtE proteins, the charged residues in the domain II-like motif of CrtB are likely to be involved in binding the pyrophosphate moiety of the substrate GGPP and/or catalysis itself.

Although bacterial CrtB has not yet been isolated and studied in a pure state, *in vivo* functional assays and *in vitro* tests with cell extracts were critical for demonstrating the phytoene synthase activity of this enzyme. *Erwinia uredovora* CrtB[76] and *Synechococcus* PCC7942 CrtB[33] were expressed in strains of *E. coli* that contained a genus *Erwinia crtE* gene to enable them to produce large amounts of GGPP (Section 2.12.4.2.1), in addition to which *E. uredovora* CrtB was expressed in *A. tumefaciens*.[76] Conversion of GGPP into phytoene could be demonstrated both *in vitro* and *in vivo* only when bacteria were transformed with the *crtB* gene. An earlier study of purified red pepper phytoene synthase, the first enzyme specific to carotenoid biosynthesis to be purified from any organism, had indicated that this monomeric Mn^{2+}-dependent protein is capable of condensing two molecules of GGPP to form phytoene via an unstable cyclopropylcarbinyl pyrophosphate intermediate.[155] It seems plausible that bacterial CrtB is also a monomeric Mn^{2+}-dependent enzyme because of its structural similarity to higher plant phytoene synthase.[23]

2.12.5.2.3 *CrtI- and CrtP-type carotenoid desaturases*

Two structurally, functionally, and mechanistically distinct classes of desaturases that introduce double bonds into the polyene chromophores of carotenoids have thus far been defined.[1,5,9] CrtI-type phytoene desaturases include those of *Rhodobacter* species,[28,32,38] various nonphotosynthetic bacteria,[23,25,30,31,37] and fungi,[77,141,156] as well as the diapophytoene desaturase (CrtN) of *S. aureus* (Table 2).[36] Furthermore, other structurally related CrtI-type enzymes with known or probable functions other than phytoene (diapophytoene) desaturation include the ζ-carotene desaturase (CrtQ) of the cyanobacterium *Anabaena* PCC7120,[41] the methoxyneurosporene desaturases (CrtD) of *Rhodobacter* species[28,44] and *M. xanthus*,[27] and two putative carotenoid desaturases of *M. xanthus* (ORF2)[27] and *S. griseus* (CrtU).[30]

All CrtI-type enzymes, exemplified by *R. capsulatus* CrtI and CrtD, possess two highly conserved regions located at their respective N- and C-termini.[28] The function of the C terminal region is unknown, although it has been postulated to be involved in protein dimer interactions by analogy with other known disulfide oxidoreductases not involved in carotenoid biosynthesis.[77] The conserved N-terminal region clearly represents a putative β-sheet–α-helix–β-sheet motif of an ADP-binding fold,[23,36,41,77,108] a well-known structural feature of oxidoreductase enzymes that is typically involved in the interaction with cofactors such as FAD, NAD, or NADP.[157] Evidence that the putative ADP-binding motif is indeed important for CrtI-type enzyme activity comes from comparative gene sequence analysis of the wildtype *R. capsulatus crtD* allele[3,54] and a mutated version that produces an inactive CrtD enzyme and is predicted to contain the amino acid exchange Gly-13 to Arg-13.[23,28] Such an exchange would significantly alter the structure of a highly conserved portion of the ADP-binding motif.

CrtP-type enzymes are represented by the phytoene desaturases of cyanobacteria,[39,40] an alga,[158] and various higher plants,[8,78,99,133] as well as a recently discovered higher plant ζ-carotene desaturase[102] that is structurally distinct from CrtQ.[41] Because the *crtQ* gene seems thus far to be unique to *Anabaena* PCC7120, one might speculate that other species of cyanobacteria may possess as yet undiscovered homologues of the CrtP-type higher plant ζ-carotene desaturase.

The CrtI- and CrtP-type enzymes are structurally unrelated, with the exception that both classes of proteins share N-terminal putative ADP-binding folds.[39,78,99,133] Both CrtI- and CrtP-type bacterial carotenoid desaturases also have, apparently coincidentally, molecular masses of approximately 55 kDa. Despite these minor similarities, however, bacterial CrtI- and CrtP-type carotenoid desaturases also display several important functional and mechanistic differences. The overexpression in *E. coli* and reconstitution in an active state of CrtI from *Erwinia uredovora*[110] and *Rhodobacter* species,[83,88] CrtQ from *Anabaena* PCC7120[135] and CrtP from *Synechococcus* PCC7942[134] have allowed detailed *in vitro* studies of the biochemical properties of these enzymes. As summarized below, the data obtained thus far with respect to substrate and product specificities and chemical inhibition of enzyme activities confirm observations previously made on the basis of intermediates accumulated *in vivo* in functional complementation experiments.[32,77,88,92,93]

CrtI-type phytoene desaturases usually carry out four consecutive desaturations accompanied by an apparently nonenzymatic isomerization to convert 15,15′-*cis*-phytoene to predominantly all-*trans*-lycopene (Figure 3).[77,92,110] Exceptionally, CrtI in *Rhodobacter* species[77,83,88,92] and CrtN in *S. aureus*[36] perform only three desaturations, yielding mostly all-*trans*-neurosporene and all-*trans*-4,4′-diaponeurosporene, respectively. In contrast to the generally accepted biosynthetic scheme shown in Figure 3, purified *R. capsulatus* CrtI seems to be able to utilize symmetric ζ-carotene as an *in vitro* desaturation substrate (Section 2.12.4.2.1).[83] 4,4′-diapophytoene, the C_{30} equivalent of phytoene (Figure 2), could not, however, be desaturated by this enzyme. *R. capsulatus* and *E. uredovora* CrtI have both been reported to require FAD, probably in a freely exchangeable form, rather than NAD or NADP as an *in vitro* desaturation cofactor, in addition to which the former enzyme has been found to be active in the absence of ATP.[83,110] Apparently contradictory *in vitro* data obtained using crude extracts from *E. coli* cells that overexpress *R. sphaeroides* CrtI suggest, however, that ATP stimulates phytoene desaturation and that FAD is not required for the reaction.[32] No NAD, NADP, or FAD requirement for ζ-carotene desaturation could be demonstrated *in vitro* using overexpressed *Anabaena* PCC7120 CrtQ, another CrtI-type desaturase. This enzyme activity was, however, found to be oxygen-dependent.[135] Despite its structural similarity to CrtI-type phytoene desaturases, purified CrtQ was found to be incapable of desaturating phytoene, but accepted different geometric isomers of ζ-carotene as well as β-zeacarotene (7′,8′-dihydro-β,ψ-carotene) as substrates. CrtI-type enzymes have been shown to be sensitive to the classical microbial phytoene desaturation inhibitor diphenylamine, but not to bleaching herbicides such as norflurazon.[13,83,92,135,159] Interestingly, large variations in the cellular levels of *R. capsulatus* CrtI produced through its overexpression in the homologous host have little effect on the total amount of colored carotenoids accumulated, suggesting that phytoene desaturation does not limit carotenoid biosynthesis in this bacterium.[38]

CrtP-type phytoene desaturases, unlike CrtI enzymes, introduce two double bonds to convert 15,15′-*cis*-phytoene into predominantly *cis*-isomers of ζ-carotene,[92,134,137,160] and are strongly inhibited by norflurazon[39,40,137] but not by diphenylamine.[159] Analysis of a number of cyanobacterial strains selected for norflurazon resistance has led to the identification of a collection of mutations in the respective *crtP* genes of *Synechococcus* PCC7942[132] and *Synechocystis* PCC6803[131] that cause this phenotype. Norflurazon resistance can be conferred by amino acid substitutions of Arg-195 to Pro-195, Leu-320 to Pro-320, Val-403 to Gly-403, and Leu-436 to Arg-436 in *Synechococcus* PCC7942 CrtP. Mutations of Arg-195 to Pro-195, Cys-195, or Ser-195 yield the same result in *Synechocystis* PCC6803 CrtP, emphasizing the critical role of Arg-195 in the sensitivity of CrtP-type phytoene desaturases to norflurazon. Arg-195 lies within a small region of cyanobacterial CrtP that seems to be structurally conserved in other oxidoreductases that process noncarotenoid substrates and which contains a leucine-repeat that is thought to represent a protein dimerization motif.[132] As a rule, norflurazon-resistant cyanobacterial mutants display strongly reduced phytoene desaturase activities *in vitro*, as well as increased phytoene accumulation and a concomitant decrease in colored carotenoid content *in vivo*. This has led to the proposal that CrtP limits carotenoid biosynthesis in cyanobacteria,[132] in contrast to the situation observed for the CrtI-type phytoene desaturase of *R. capsulatus*.[38] Another type of *Synechococcus* PCC7942 norflurazon-resistant mutant has also been identified in which *crtP* gene promoter activity seems to have been increased, leading to an overexpression of CrtP.[132] The differential sensitivities of CrtI- and CrtP-type phytoene desaturases to chemical inhibition have been recently exploited for the genetic engineering of resistance to bleaching herbicides in cyanobacteria and plants (Section 2.12.7). In contrast to the FAD requirement found for purified bacterial CrtI enzymes, the activity of *Synechococcus* PCC7942 CrtP could be stimulated by exogenous NAD or NADP but not FAD.[134] With respect to the lack of an FAD requirement, it is worth mentioning that a higher plant homologue of CrtP from red pepper has been isolated with tightly bound FAD[133] and that the daffodil homologue has been shown to be activated by incorporation of FAD as a nonexchangeable and possibly covalently-bound cofactor.[161] Were such an activation process also to exist in cyanobacteria, it might explain the inability to demonstrate an FAD requirement for phytoene desaturation using exogenous FAD.[134]

2.12.5.2.4 CrtY- and CrtL-type lycopene cyclases

Many species of bacteria produce carotenoids that contain two β-ionone ring end groups such as β-carotene and its xanthophyll derivatives (Figures 3, 5, and 6).[5,10,14] The proposed mechanism for the cyclization of lycopene involves proton attack at C-2 and C-2′, which generates an unstable carbonium ion intermediate. The division of lycopene β-cyclases into two structurally distinct species-specific classes is rather analogous to that described for carotenoid desaturases (Section 2.12.5.2.3), although as discussed below and in contrast to the desaturases, both types of lycopene cyclase are functionally equivalent.

CrtY-type lycopene cyclases, which have thus far been found in several species of nonphotosynthetic bacteria (Table 2),[25,30,31,42] contain conserved N-terminal putative $\beta\alpha\beta$-ADP-binding folds,[108] by analogy to the carotenoid desaturases (Section 2.12.5.2.3). CrtL-type lycopene cyclases have been described in cyanobacteria[43] and higher plants,[8,103–105] and a structurally related CrtL-type enzyme of cyclic xanthophyll biosynthesis, capsanthin/capsorubin synthase, has also been identified in red pepper.[100,103] In contrast to higher plants and algae, prokaryotes have only rarely been reported to synthesize carotenoids that contain one β-ionone and one ε-ionone ring each, such as α-carotene (β,ε-carotene) and lutein (β,ε-carotene-3,3′-diol).[10,14] Results with the higher plant *Arabidopsis thaliana* demonstrate conclusively that the ε-cyclization and β-cyclization reactions are performed by distinct but structurally related CrtL-type ε- and β-lycopene cyclases, respectively.[104]

The deduced amino acid sequences of the bacterial CrtY- and CrtL-type lycopene cyclases display very little structural conservation outside of conserved N-terminal putative $\beta\alpha\beta$-ADP-binding folds,[43] and two other short stretches of amino acids of unknown function that have been identified after comparison with higher plant CrtL-type lycopene cyclases.[103,105] Despite these limited structural similarities, both CrtY- and CrtL-type bacterial carotenoid cyclases have predicted molecular masses of about 45 kDa and are functionally similar. This has been shown by *in vitro* studies with crude or purified preparations of genus *Erwinia* CrtY isolated from overexpressing *E. coli* strains[112,113] and by functional *in vivo* complementation with *Erwinia uredovora* CrtY[107] and *Synechococcus* PCC7942 CrtL[43] of genetically transformed strains of *E. coli* that carry other bacterial *crt* genes and therefore accumulate lycopene cyclase substrates.

The enzymatic activity of *E. herbicola* CrtY was originally studied in crude bacterial lysates after its overexpression.[112] These experiments demonstrated the *in vitro* conversion of all-*trans*-lycopene to all-*trans*-β-carotene (Figure 3), although the assay was not highly reproducible. To circumvent this problem, *Erwinia uredovora* CrtY was purified from *E. coli* after its overexpression and reconstituted as an active NADH- or NADPH-dependent enzyme.[113] This cofactor dependency, while in agreement with the presence of the putative ADP-binding motif in the enzyme, is puzzling because the biochemical mechanism postulated for lycopene cyclization does not require a net transfer of electrons. The exact role of the cofactor therefore remains to be determined. In contrast to CrtY, no bacterial CrtL-type lycopene cyclase has yet been purified and studied *in vitro*. However, functional analysis by expression of *Synechococcus* PCC7942 CrtL[43] or *Erwinia uredovora* CrtY[107] in *E. coli* strains genetically engineered to accumulate different carotenoids has allowed an *in vivo* comparative analysis of the substrates and products of the cyclization reaction catalyzed by both enzymes. Taken together with the *in vitro* biochemical data available for CrtY,[112,113] the picture emerges that CrtY and CrtL are functionally very similar if not equivalent. Both enzymes are capable of performing two successive β-cyclization reactions to convert acyclic lycopene to bicyclic β-carotene via monocyclic γ-carotene, to convert acyclic or neurosporene to bicyclic dihydro-β-carotene (7,8-dihydro-β,ψ-carotene) via monocyclic β-zeacarotene.[43,107] Genus *Erwinia* CrtY cannot, however, cyclize *cis* isomers of lycopene.[112,113] It will be of interest to examine the specificity of CrtL-type lycopene cyclases for *cis* vs. all-*trans* carotenoid isomers when purified CrtL becomes available. Because monocyclic carotenoids such as γ-carotene are almost undetectable in *in vivo* functional assays, it has been postulated that CrtL-type lycopene cyclases may be active as homodimers, thus ensuring that both ends of the initial substrate molecule are cyclized.[94] ζ-Carotene can reportedly serve as an additional substrate for CrtY-mediated cyclization *in vivo*,[107] but not *in vitro*.[113] *Synechococcus* PCC7942 CrtL, however, proved incapable of cyclizing ζ-carotene *in vivo*.[43] Both prokaryotic and eukaryotic CrtL-type carotenoid cyclases have been demonstrated to be sensitive to chemical inhibitors of lycopene cyclization such as CPTA and MPTA.[43,103,105] An MPTA-resistant *Synechococcus* PCC7942 mutant was shown to have acquired inhibitor tolerance by mutating the *crtL* gene promoter, such that overexpression of CrtL is likely to occur.[43]

In *Myxococcus xanthus*, a novel gene required for lycopene cyclization, ORF6, encodes a protein with no predicted structural similarities to either CrtY- or CrtL-type carotenoid cyclases.[27] Whether the gene product represents a new class of lycopene β-cyclase or rather encodes an accessory factor directly or indirectly required for the cyclization reaction in this bacterium remains to be established through *in vivo* or *in vitro* functional tests.

2.12.5.2.5 Xanthophyll biosynthesis enzymes

Biochemical interconversions in xanthophyll biosynthesis (Figures 4–6) are more species-specific among prokaryotes and eukaryotes than are the reactions leading to phytoene formation (Figure 2) or carotene biosynthesis (Figure 3). Perhaps for this reason, relatively little is known about the properties of individual xanthophyll biosynthesis enzymes either on a structural basis, or from *in vitro* reconstitution experiments or *in vivo* studies in which they have been expressed in genetically modified *E. coli* strains that produce carotenoids.

Several *Rhodobacter* species xanthophyll biosynthesis (Figure 4) enzymes whose amino acid sequences have been deduced from the corresponding gene sequences[28,29] have structural homologues in other bacteria (Table 2). Both CrtC, the hydroxyneurosporene synthase, and CrtD, the CrtI-type methoxyneurosporene desaturase of *Rhodobacter* species (Section 2.12.5.2.3), are predicted to be about 30% identical to their respective homologues in *M. xanthus*.[27] No structural homologue of genus *Rhodobacter* CrtA, the molecular oxygen-dependent spheroidene monooxygenase, has yet been described in other bacteria or eukaryotes.

CrtF, the genus *Rhodobacter* hydroxyneurosporene-*O*-methyltransferase, which catalyzes the *S*-adenosyl-L-methionine (SAM)-dependent methylation of a terminal hydroxy group,[63] shares significant structural similarity with a range of SAM-dependent enzymes that methylate substrates other than carotenoids.[3,29] This observation has led to a proposal for the existence of a 30 amino acid long SAM-binding motif within *Rhodobacter* species CrtF.[3] A 30% identical putative homologue of CrtF, denoted CrtT, has been predicted to be encoded within the *crt* gene cluster of *S. griseus*.[30]

Some structural and biochemical information is available about CrtZ[25,42,112] and CrtX,[25,42,111] the two enzymes involved in cyclic xanthophyll biosynthesis in the genus *Erwinia* (Figure 5). CrtZ, the β-carotene hydroxylase, is predicted to have homologues that are about 55% identical in several other species of nonphotosynthetic marine bacteria (Table 2)[31] and more recently has been demonstrated to be the structural homologue of a functionally equivalent higher plant enzyme that has been postulated to be active as a homodimer.[106] The latter study also identified within the central portion of the predicted polypeptide sequences of the bacterial and plant CrtZ enzymes a highly conserved stretch of 10 amino acids, termed motif 1, of unknown function. Earlier *in vitro* studies using cyanobacterial membrane preparations had established that the hydroxylation of β-carotene in this system was catalyzed by a molecular oxygen-dependent monooxygenase whose activity was stimulated by NADPH.[162] Overexpression of *Erwinia herbicola* CrtZ in *E. coli* and preparation of bacteria lysates allowed an *in vitro* examination of the requirements for β-carotene hydroxylation.[112] Enzyme activity for the conversion of β-carotene to zeaxanthin via β-cryptoxanthin ((3*R*)-β,β-caroten-3-ol; Figure 5) was obtained only with a cytoplasmic but not with a membrane fraction and could be stimulated by the addition of NADPH, NADH, or ascorbate. *In vivo* functional assays of *Erwinia uredovora*[92,107] and *A. aurantiacum* CrtZ[31] expressed in *E. coli* strains modified to accumulate various carotenoids have provided additional information about the substrate specificity of this enzyme. When one neglects monohydroxylated carotenoids, *E. uredovora* CrtZ can recognize not only β-carotene, but also β-zeacarotene and dihydro-β-carotene as the initial hydroxylation substrates,[92,107] whereas *A. aurantiacum* CrtZ can additionally utilize echinenone and canthaxanthin in the branched pathway for the synthesis of astaxanthin (Figure 6).[31]

CrtX, the zeaxanthin glucosyltransferase, has thus far been characterized only from *E. herbicola*.[111] This enzyme is, however, likely to be structurally related to other bacterial xanthophyll biosynthesis enzymes because carotenoid glycosylation, and specifically glucosylation, is a relatively common reaction among different groups of bacteria.[5,10,14] Overexpression of *Erwinia herbicola* CrtX in *E. coli* and examination of its enzymatic activity *in vitro* in bacterial extracts demonstrated that the weakly membraneassociated enzyme utilized zeaxanthin and UDP-glucose as cosubstrates in the synthesis of zeaxanthin diglucoside via zeaxanthin monoglucoside.[111] Furthermore, the C-terminal half of CrtX was found to contain a region that was structurally conserved in several other UDP-glucosyltransferases or UDP-glucuronosyltransferases that are not involved in carotenoid biosynthesis but do bind UDP-activated substrates. This finding led to a proposal for a 46 amino acid long structural motif involved in binding the UDP moiety.[111]

In two species of astaxanthin-producing marine bacteria, the novel CrtW enzyme, β-C-4-oxygenase, involved in the late stages of cyclic xanthophyll biosynthesis (Figure 6), has been identified as a *crt* gene product.[31,47] *A. aurantiacum* and *Alcaligenes* PC-1 CrtW, which are 75% identical, are predicted to be structurally homologous to a functionally similar, though perhaps not identical, enzyme that is present in the astaxanthin-producing green alga *H. pluvialis*.[96–98] The algal and bacterial CrtW enzymes shared four highly conserved regions at the amino acid sequence level, but no function has yet been associated with these structural features.[97] Neglecting the carotenoid intermediates that contain a single keto group (Figure 6), *A. aurantiacum* CrtW can recognize β-carotene, β-cryptoxanthin, and zeaxanthin as substrates for oxygenation *in vivo* in carotenogenic *E. coli* cells.[31]

2.12.6 REGULATION OF CAROTENOID BIOSYNTHESIS

2.12.6.1 Anoxygenic Photosynthetic Bacteria

Carotenoid accumulation in anoxygenic photosynthetic bacteria has long been known to be regulated by two key environmental factors, oxygen and light.[3,61] Carotenoids are usually associated with specific Bchl-containing pigment–protein complexes and thus accumulate preferentially under conditions which favor the formation of the intracytoplasmic photosynthetic membrane. This process is induced by low oxygen tensions and the extent of the photosynthetic membrane increases with decreasing light intensities. No regulatory genes which influence wildtype carotenoid levels and *crt* gene expression independently of Bchl levels and *bch* gene expression have yet been identified in anoxygenic photosynthetic bacteria. This observation is not, however, surprising given the observed correlation between carotenoid and Bchl accumulation, with the latter taken as a measure of photosynthetic membrane development.[163]

2.12.6.1.1 Rhodobacter *species*

The expression of most genus *Rhodobacter crt* genes increases several-fold under anaerobic photosynthetic vs. semiaerobic or aerobic conditions, and is reflected in changes in gene promoter activities[46,56,72,164] and mRNA levels.[32,46,69] Two genus *Rhodobacter* regulatory genes, *ppsR* and *tspO* (Table 3), have recently been demonstrated to encode products that normally repress not only carotenoid and Bchl *a* pigment levels but also *crt*, *bch*, and *puc* gene expression several-fold under aerobic growth conditions that reduce or eliminate the need for photosynthetic pigments.[55,56,164–166] *ppsR* (formerly *crtJ* or ORF469) and *tspO* (formerly *crtK* or ORF160) were initially thought to be specifically required for carotenoid biosynthesis on the basis of mutant studies. The altered carotenoid content phenotypes originally reported for the respective transposon and interposon gene disruption mutations that defined *crtJ*[66,73] and *crtK*[28,69] seem, in retrospect, to have been caused by second site mutations in other *crt* genes.[56,167] The mechanism by which TspO functions is unknown, although this hydrophobic outer membrane protein has been proposed to serve as an environmental sensor. Both TspO[56] and its mRNA[46] are more abundant in anaerobic photosynthetic vs. aerobically grown cells. Interestingly, TspO displays about 35% deduced amino acid sequence identity with mammalian peripheral-type benzodiazepine receptors localized in the mitochondria.[3,56,168] PpsR, in contrast, is predicted to be a DNA-binding protein containing a conserved helix–turn–helix motif found in other known bacterial transcriptional repressors.[3,55] PPSR apparently recognizes a conserved palindromic sequence found 5′ to several operons containing *crt*, *bch*, and *puc* genes,[28,54,55,164] and probably functions by repressing transcription of these operons when bacterial cultures contain sufficient levels of oxygen for respiration.

In addition to the repression of pigment levels and gene expression exerted by TspO and PpsR under aerobic conditions, a mechanism that links the ratio of carotenoid end products to the ratio between the two types of Bchl *a*-binding light-harvesting protein antenna complexes has recently been identified in *R. sphaeroides*.[169] It had long been known that the conversion of spheroidene to spheroidenone catalyzed by CrtA (Figure 4) requires molecular oxygen as a source for the keto group.[57] Reexamination of this reaction under various growth conditions has revealed that the spheroidenone to spheroidene ratio increases at high light intensities.[169] Intriguingly, at a given oxygen tension and light intensity, this ratio correlates with the ratio of the peripheral B800–850 light-harvesting complex to the inner-core B875 light-harvesting complex. Furthermore, spheroidene was found to be preferentially associated with the B800–850 antenna complex, while spheroidenone predominated in the B875 antenna complex. These data have led to the model that the cellular redox poise either detects or determines the ratio of these carotenoids, and hence couples carotenoid content to the ratio of the two types of light-harvesting protein antenna complexes. This type of regulation represents one of the first indications in bacteria that the structural diversity among carotenoid end products may, at least in some organisms, have a biological function.

2.12.6.2 Nonphotosynthetic Bacteria

2.12.6.2.1 Erwinia *species*

Little information is available about the regulation of carotenoid biosynthesis in the genus *Erwinia* and the regulation of the *crt* genes has not been studied directly. Pigment accumulation during growth in the presence or absence of glucose was determined for a number of *E. herbicola* strains.[84] Carotenoid synthesis was found to be repressed by glucose in both *Erwinia herbicola* Eho10 and transformed *E. coli* carrying the *E. herbicola crt* genes, and to be dependent on cyclic adenosine monophosphate. By analogy with other bacteria, these results suggest transcriptional regulation of the *crt* genes in both *E. herbicola* and *E. coli*, and a possible upregulation of expression under conditions of carbon limitation, such as in older, stationary phase cultures. This type regulation was not, however, observed in a number of other pigmented *E. herbicola* strains.

2.12.6.2.2 Myxococcus xanthus

In *M. xanthus* a number of elegant experiments have led to the most complete genetic description of the regulation of light-induced carotenoid biosynthesis thus far available from any organism.[4,170] The genetically unlinked putative regulatory loci *carA* and *carR*, mutations in which lead to Car^c constitutive carotenoid production phenotypes, were identified during transposon mutagenesis of *M. xanthus*.[50,171] Shortly thereafter, a further regulatory locus closely linked to *carR* and later termed *carQ* was discovered because of its associated Car$^-$-deficient carotenoid phenotype.[120,121] Further genetic and molecular characterization of the *carR* region revealed the presence of not two but three translationally coupled genes, *carQRS*, organized in a light-inducible operon.[52,172] Mutations in the newly identified regulatory gene, *carS*, were also observed to result in a Car$^-$ phenotype. Recently identified mutants defective in another unlinked regulatory locus, *carD*, display a Car$^-$ phenotype in addition to being deficient in developmental processes, such as starvation-induced formation of fruiting bodies and sporulation.[51]

A series of genetic studies have defined the interplay between the *carA*, *carD*, *carQ*, *carR*, and *carS* regulatory loci (Table 3), and the light-inducible promoters of the *M. xanthus crt* gene cluster, the *crtI* operon and the *carQRS* operon.[27,37,50–52,121,122,170,172] Although transcription of all three promoters is induced by light, the mechanisms regulating them are nevertheless distinct. For example, the *crtI* promoter is the most highly regulated, displaying a 400-fold induction by light, but only under conditions of carbon starvation or upon cessation of cell growth.[4,37] This regulation explains previous observations that colored carotenoid accumulation in *M. xanthus* occurs only in stationary phase cells[114] and pinpoints CrtI-mediated phytoene desaturation as a key regulatory step. The *crt* gene cluster promoter, in contrast, is only 20-fold inducible by light and is insensitive to the carbon supply.[4,50] In addition to these qualitative and quantitative differences in promoter regulation by environmental stimuli, the *crt* gene cluster, and the *crtI* and *carQRS* operons also differ in their direct and indirect endogenous genetic requirements for light induction.[4,170]

Some of the structural features of the deduced *car* gene products have proved helpful in understanding the genetic data with respect to light regulation. CarQ was recently reported to be structurally related to a group of proteins known as extracytoplasmic function σ factors of RNA polymerase, while CarR seems to be an inner membrane protein and CarS displays no similarity to other known proteins.[52,170] The identity of CarA has proved more elusive, although characterization of the *M. xanthus crt* gene cluster operons from the wildtype and a *carA* mutant have revealed the nature of the nucleotide changes that lead to the loss of CarA function.[27] Two closely linked point mutations at the 3′ end of the operon apparently inactivate ORF10 and/or ORF11, both of which encode putative regulatory proteins that contain helix–turn–helix DNA-binding motifs. These findings are consistent with the model that CarA is a transcriptional regulator indirectly involved in upregulation of *crtI* expression in response to light and directly responsible for repression of expression of the *crt* gene cluster in the dark.[170] The characteristic Car^c phenotype of a *carA* mutant apparently results from residual CrtI-mediated desaturation of the phytoene synthesized by the products of the derepressed *crt* gene cluster operon.[120]

One recent and comprehensive model incorporating CarA, CarQ, CarR, and CarS (Table 3) as regulatory factors that control blue light-induced carotenogenesis in *M. xanthus* can be summarized as follows.[170] Basal transcription of the *carQRS* operon in dark-grown bacterial cultures allows small amounts of the inner membrane protein CarR, a putative anti-σ factor, and associated σ factor CarQ to be sequestered at the cytoplasmic membrane. As discussed in Section 2.12.4.3.2, protoporphyrin IX accumulated in the cytoplasmic membrane acts as the blue light receptor for carotenogenesis and for photooxidative damage leading to photolysis.[114–116] When dark-grown *M. xanthus* cells are exposed to light, protoporphyrin IX generates an as yet unidentified signal leading to the degradation of CarR and the release of CarQ from the membrane. The latter then interacts with the core RNA polymerase, allowing transcription of the light-induced *crtI* and *carQRS* operons. Newly synthesized CarR is destroyed by the protoporphyrin IX-generated signal, but increasing amounts of CarQ, CarS, and CrtI accumulate. CarS activates transcription of the *crt* gene cluster operon and thereby allows the accumulation of carotenoid biosynthesis enzymes and CarA. The latter eventually competes with CarS to repress transcription of the *crt* gene cluster operon. Before this happens, however, all of the enzymes needed to produce the fatty acid esters of myxobactone and myxobactin have been made. Carotenoids then accumulate in the cytoplasmic membrane and quench the protoporphyrin IX-generated signal for the destruction of CarR, thus completing the regulatory cycle. Although the above model does not explicitly discuss the role of CarD, this protein has been reported along with CarQ to be necessary for the light-induced transcription of the *crtI* and *carQRS* operons, and may help to integrate developmental and environmental signals that activate carotenogenesis.[51]

2.12.6.2.3 Streptomyces setonii

As discussed in Section 2.12.4.3.5, carotenogenesis is also light induced in some strains of the genus *Streptomyces*. In *S. setonii*, the light-inducibility is subject to genetic variation and is lost at a relatively high frequency.[53] The isolation of a revertant of a Car⁻ mutant led to the identification of *crtS*, a regulatory gene that restored colored carotenoid accumulation. Based on its deduced amino acid sequence, CrtS seems not to be a biosynthetic enzyme but rather to represent an alternative σ factor of RNA polymerase that activates carotenoid accumulation by permitting the expression of cryptic *crt* genes in *S. setonii* (Table 3). Indeed, *S. setonii crtS* was also found to be effective in activating carotenoid accumulation in a Car⁻ strain of *S. griseus* bearing cryptic *crt* genes.[30] By analogy with light-induced carotenogenesis in *M. xanthus*, one might speculate that CrtS fulfills a role similar to that of the proposed extracytoplasmic function σ factor CarQ.[170]

2.12.6.3 Cyanobacteria

Cyanobacterial carotenoid levels and composition are regulated by various environmental factors, in particular light intensity and quality.[5] No genes involved in the specific regulation of carotenoid biosynthesis have, however, been identified, nor has the regulation of cyanobacterial *crt* genes been studied. One might anticipate, however, that endogenous factors that simultaneously control the accumulation of carotenoids, Chl *a* and Chl *b*, and pigment-binding proteins of the photosynthetic apparatus and outer membrane will be present.

2.12.7 GENETIC ENGINEERING OF CAROTENOID BIOSYNTHESIS

One of the most exciting prospects on the horizon in the field of carotenoid biosynthesis is the development of alternative strategies for the production of carotenoids of nutritional, medical, and esthetic interest.[10] Attempts are in progress to supplement classical genetics and breeding techniques with genetic engineering to produce interesting carotenoids in noncarotenogenic hosts or tissues, and to mix and match biosynthetic pathways from different organisms in carotenogenic hosts.

As one example, dietary vitamin A deficiency, a severe worldwide nutritional problem in underdeveloped countries and one that is particularly acute among children,[173] is currently being approached with the goal of using genetic engineering to achieve β-carotene (i.e., provitamin A) accumulation in staple foods that normally lack carotenoids. The main target of this effort is the carotenoid-deficient endosperm, or seed storage tissue, of rice, the staple food of much of the world's population. Building on efforts that have led to the production of transgenic rice plants that accumulate phytoene in their seed storage tissues,[174] one future strategy could be to express bacterial *crtI* and *crtY* genes (Figure 3) to allow this phytoene to be converted into β-carotene in the plastids of rice seeds.

That a bacterial carotenoid biosynthesis gene can function in a higher plant has previously been demonstrated by the engineering of herbicide resistance based on the known differences between the CrtI- and CrtP-type phytoene desaturases in their sensitivities to norflurazon,[159] which inhibits the latter enzyme in cyanobacteria and plants.[39,132] Norflurazon-resistant tobacco plants were produced by genetic transformation with a modified *E. uredovora crtI* gene fused at the 5′ end to a gene sequence encoding a higher plant plastid transit peptide to direct the recombinant enzyme to the proper cellular compartment.[175,176] The *E. uredovora crtI* gene has also been used to introduce the trait of norflurazon resistance into the cyanobacterium *Synechococcus* PCC7942.[177] A by-product of the experiments with tobacco has been the observation that coexpression of the bacterial CrtI-type and the endogenous eukaryotic CrtP-type phytoene desaturases alters the distribution of end product xanthophylls in transgenic plants.[176]

Naturally occurring carotenoids such as lycopene, β-carotene, lutein, zeaxanthin, canthaxanthin, astaxanthin, and rhodoxanthin (4′,5′-didehydro-4,5′-retro-β,β-carotene-3,3′-dione) are of considerable commercial interest as coloring agents for food, pharmaceuticals, cosmetics, and animal feed.[10,178] Several of these pigments, including β-carotene, canthaxanthin, and astaxanthin, are currently commercially produced by total chemical synthesis. The recent genetic elucidation of bacterial carotenoid biosynthetic pathways leading to the accumulation of zeaxanthin, canthaxanthin, and astaxanthin (Figures 5 and 6) may offer interesting alternatives for the *in vivo* production of these pigments.[25,26,31,42,47] Introduction of portions of or intact *crt* gene clusters of *Erwinia* species into other bacteria, such as *A. tumefaciens* and *Z. mobilis*,[86] or eukaryotes, such as the yeast *Saccharomyces cerevisiae*,[179] has already been demonstrated to allow production of substantial amounts of lycopene and β-carotene in these organisms.

2.12.8 CONCLUSIONS

2.12.8.1 The Lessons of Prokaryotic Carotenoid Biosynthesis

In exploring the general principles of prokaryotic carotenoid biosynthesis as elucidated in anoxygenic photosynthetic bacteria, nonphotosynthetic bacteria, and cyanobacteria, several recurring themes can be discerned. First, with the notable exception of the cyanobacteria, *crt* genes encoding biosynthetic enzymes are generally clustered.[23–34,36,39–43,90,126,128] In particular, potential *crtIB* (*crtNM*) transcriptional operons have been found in all *crt* gene clusters thus far except for that of *M. xanthus*. *crt* gene clusters and multigene operons may provide a selective advantage by allowing bacteria to rapidly and simultaneously adjust the amounts of several enzymes to activate or shut down different steps in pigment biosynthesis. Evidence for the coordination of *crt* gene regulation by species-specific mechanisms has grown as more and more species of bacteria are examined.[27,30,32,46,53,69,72,122,164] Cloned *crt* genes have thus far been found to be chromosomally encoded in all bacteria other than *T. thermophilus*.[126]

Second, all molecularly characterized bacterial *crt* gene clusters other than that of *A. aurantiacum* contain a *crtE* gene.[23–31,42] This observation and the pigment-deficient phenotypes observed for *crtE* mutants support the hypothesis that carotenogenic bacteria contain multiple GGPP synthases, at least one of which supplies FPP and GGPP for essential functions such as quinone production (Figure 1) and another of which, encoded by *crtE* (Figure 2), that is required for carotenoid pigment accumulation.[1,3,29,81] It is worth noting that higher plants, including red pepper and *Arabidopsis thaliana*, apparently possess multiple genes encoding putative GGPP synthases.[8,142]

Third, as evident from the genetically characterized prokaryotic carotenoid biosynthetic pathways (Figures 3–6) and biochemical studies with purified proteins,[83,109,113,135] many of the carotenoid biosynthesis enzymes are promiscuous in terms of their substrate requirements. Thus, starting from a relatively small pool of unique types of enzymatic reactions, it is possible to generate the tremendous structural diversity found among naturally occurring carotenoids by using different combinations of *crt* genes, and hence biosynthetic steps, in different organisms. How much of this diversity has been actively selected for during evolution remains to be determined. A recent report of the *in vivo* formation of *R. sphaeroides* light-harvesting pigment–protein complexes containing novel carotenoids produced by a chimeric *R. sphaeroides*–*E. herbicola* biosynthetic pathway is one example of the type of study that may help to answer this question.[88]

Finally, the structural conservation of soluble CrtE-homologous GGPP synthases and membrane-associated CrtB-homologous phytoene synthases among all bacteria (Table 2) and, indeed, among all carotenogenic organisms, is perhaps to be expected, given that GGPP and phytoene are universal early intermediates in the C_{40} pigment biosynthetic pathway (Figure 2).[8,9] More surprising, however, are the structural differences between the functionally related membrane-bound carotene biosynthetic enzymes of anoxygenic photosynthetic bacteria, nonphotosynthetic bacteria, and fungi on one hand, and those of Chl-containing oxygenic photosynthetic cyanobacteria, algae, and plants on the other. The CrtI-type carotenoid desaturases (CrtI, CrtN, CrtD) and the CrtY-type lycopene cyclases of organisms that lack Chl appear almost totally unrelated in their primary sequences to the functionally related CrtP-type phytoene desaturases and CrtL-type lycopene cyclases, respectively, of Chl *a*- and Chl *b*-containing organisms.[9] The *Anabaena* PCC7120 ζ-carotene desaturase, CrtQ, seems to represent an evolutionary oddity in that it is a CrtI-type desaturase that has no identified counterpart in other cyanobacteria.[41,102] Less is known about the evolutionary conservation of xanthophyll biosynthesis because so few gene sequences of functionally related enzymes are available in cyanobacteria and eukaryotes. Recent data suggest, however, that at least two enzymes of cyclic xanthophyll biosynthesis, CrtZ and CrtW, are indeed structurally conserved between nonphotosynthetic bacteria on the one hand and algae or higher plants on the other.[97,106] Further research may help to establish whether the observed structural dichotomy among carotenoid desaturases and lycopene cyclases results from convergent evolution and reflects the appearance of these particular carotenoid biosynthetic reactions as two separate events in Chl *a*- and Chl *b*-containing organisms, and in all other carotenogenic organisms.[43,99] Alternatively, these particular enzyme groups may have been placed under strong selective constraints to diverge structurally and mechanistically in response to the changes in the global environment during evolution, such as the advent of oxygenic photosynthesis.

ACKNOWLEDGMENTS

I thank Catharina, Gordon, and Christina for their patience and understanding during the preparation of this chapter. I would also like to acknowledge the contributions of numerous researchers in the field of carotenoid biosynthesis who were kind enough to supply reprints and preprints of their work for this project.

2.12.9 REFERENCES

1. G. A. Armstrong, *J. Bacteriol.*, 1994, **176**, 4795.
2. G. A. Armstrong, *Ann. Rev. Microbiol.*, 1997, **51**, 629.
3. G. A. Armstrong, in "Advances in Photosynthesis: Anoxygenic Photosynthetic Bacteria," eds. R. E. Blankenship, M. T. Madigan, and C. E. Bauer, Kluwer, Dordrecht, 1995, vol. 2, p. 1135.
4. D. A. Hodgson and F. J. Murillo, in "Myxobacteria II," eds. M. Dworkin and D. Kaiser, American Society for Microbiology, Washington DC, 1993, p. 157.
5. J. Hirschberg and D. Chamovitz, in "Advances in Photosynthesis: The Molecular Biology of Cyanobacteria," ed. D. Bryant, Kluwer, Dordrecht, 1994, vol. 1, p. 559.
6. E. Cerdá-Olmedo, in "Phycomyces," eds. E. Cerdá-Olmedo and E. D. Lipson, Cold Spring Harbor Laboratories, New York, 1987, p. 199.
7. G. E. Bartley, P. A. Scolnik, and G. Giuliano, *Annu. Rev. Plant Physiol. Plant Mol. Biol.*, 1994, **45**, 287.
8. G. E. Bartley and P. A. Scolnik, *Plant Cell*, 1995, **7**, 1027.
9. G. A. Armstrong and J. E. Hearst, *FASEB J.*, 1996, **10**, 228.
10. E. A. Johnson and W. A. Schroeder, *Adv. Biochem. Eng./Biotech.*, 1995, **53**, 119.
11. G. Sandmann, *Eur. J. Biochem.*, 1994, **223**, 7.
12. P. M. Bramley, in "Methods in Plant Biochemistry: Enzymes of Secondary Metabolism," ed. P. J. Lea, Academic Press, London, 1993, vol. 9, p. 281.
13. P. M. Bramley, in "Carotenoids in Photosynthesis," eds. A. Young and G. Britton, Chapman & Hall, London, 1993, p. 127.
14. T. W. Goodwin, "The Biochemistry of Carotenoids: Plants," Chapman & Hall, London, 1980, vol. 1, p. 1.
15. G. Britton, "The Biochemistry of Natural Pigments," Cambridge University Press, Cambridge, 1983, p. 23.
16. O. Straub, in "Key to Carotenoids," eds. H. Pfander, M. Gerspacher, M. Rychener, and R. Schwabe, Birkhäuser Verlag, Basel, 1987, p. 1.
17. D. Kull and H. Pfander, in "Carotenoids: Isolation and Analysis," eds. G. Britton, S. Liaaen-Jensen, and H. Pfander, Birkhäuser Verlag, Basel, 1995, vol. 1A, p. 295.
18. B. C. L. Weedon and G. P. Moss, in "Carotenoids: Isolation and Analysis," eds. G. Britton, S. Liaaen-Jensen, and H. Pfander, Birkhäuser Verlag, Basel, 1995, vol. 1A, p. 27.
19. R. F. Taylor, *Microbiol. Rev.*, 1984, **48**, 181.
20. G. Britton, in "Carotenoids: Spectroscopy," eds. G. Britton, S. Liaaen-Jensen, and H. Pfander, Birkhäuser Verlag, Basel, 1995, vol. 1B, p. 13.
21. B. Demmig-Adams, A. M. Gilmore, and W. W. Adams, III, *FASEB J.*, 1996, **10**, 403.
22. H. A. Frank and R. J. Cogdell, *Photochem. Photobiol.*, 1996, **63**, 257.
23. G. A. Armstrong, M. Alberti, and J. E. Hearst, *Proc. Natl. Acad. Sci. USA*, 1990, **87**, 9975.
24. K.-Y. To, E.-M. Lai, L.-Y. Lee, T.-P. Lin, C.-H. Hung, C.-L. Chen, Y.-S. Chang, and S.-T. Liu, *Microbiology*, 1994, **140**, 331.
25. N. Misawa, M. Nakagawa, K. Kobayashi, S. Yamano, Y. Izawa, K. Nakamura, and K. Harashima, *J. Bacteriol.*, 1990, **172**, 6704.
26. L. Pasamontes, D. Hug, M. Tessier, H.-P. Hohmann, J. Schierle, and A. P. G. M. van Loon, *Gene*, 1997, **185**, 35.
27. J. A. Botella, F. J. Murillo, and R. Ruiz-Vazquez, *Eur. J. Biochem.*, 1995, **233**, 238.
28. G. A. Armstrong, M. Alberti, F. Leach, and J. E. Hearst, *Mol. Gen. Genet.*, 1989, **216**, 254.
29. H. P. Lang, R. J. Cogdell, S. Takaichi, and C. N. Hunter, *J. Bacteriol.*, 1995, **177**, 2064.
30. G. Schumann, H. Nürnberger, G. Sandmann, and H. Krügel, *Mol. Gen. Genet.*, 1996, **252**, 658.
31. N. Misawa, Y. Satomi, K. Kondo, A. Yokoyama, S. Kajiwara, T. Saito, T. Ohtani, and W. Miki, *J. Bacteriol.*, 1995, **177**, 6575.
32. H. P. Lang, R. J. Cogdell, A. T. Gardiner, and C. N. Hunter, *J. Bacteriol.*, 1994, **176**, 3859.
33. D. Chamovitz, N. Misawa, G. Sandmann, and J. Hirschberg, *FEBS Lett.*, 1992, **296**, 305.
34. I. Martínez-Férez, B. Fernández-González, G. Sandmann, and A. Vioque, *Biochim. Biophys. Acta*, 1994, **1218**, 145.
35. T. Hoshino, R. Fujii, and T. Nakahara, *Appl. Environ. Microbiol.*, 1993, **59**, 3150.
36. B. Wieland, C. Feil, E. Gloria-Maercker, G. Thumm, M. Lechner, J.-M. Bravo, K. Poralla, and F. Götz, *J. Bacteriol.*, 1994, **176**, 7719.
37. M. Fontes, R. Ruiz-Vázquez, and F. J. Murillo, *EMBO J.*, 1993, **12**, 1265.
38. G. E. Bartley and P. A. Scolnik, *J. Biol. Chem.*, 1989, **264**, 13 109.
39. D. Chamovitz, I. Pecker, and J. Hirschberg, *Plant Mol. Biol.*, 1991, **16**, 967.
40. I. M. Martínez-Férez and A. Vioque, *Plant Mol. Biol.*, 1992, **18**, 981.

41. H. Linden, N. Misawa, T. Saito, and G. Sandmann, *Plant Mol. Biol.*, 1994, **24**, 369.
42. B. S. Hundle, M. Alberti, V. Nievelstein, P. Beyer, H. Kleinig, G. A. Armstrong, D. H. Burke, and J. E. Hearst, *Mol. Gen. Genet.*, 1994, **245**, 406.
43. F. X. Cunningham, Jr., Z. Sun, D. Chamovitz, J. Hirschberg, and E. Gannt, *Plant Cell*, 1994, **6**, 1107.
44. E. Garí, J. C. Toledo, I. Gibert, and J. Barbé, *FEMS Microbiol. Lett.*, 1992, **72**, 103.
45. D. A. Young, M. B. Rudzik, and B. L. Marrs, *FEMS Microbiol. Lett.*, 1992, **74**, 213.
46. G. A. Armstrong, D. N. Cook, D. Ma, M. Alberti, D. H. Burke, and J. E. Hearst, *J. Gen. Microbiol.*, 1993, **139**, 897.
47. N. Misawa, S. Kajiwara, K. Kondo, A. Yokoyama, Y. Satomi, T. Saito, W. Miki, and T. Ohtani, *Biochem. Biophys. Res. Commun.*, 1995, **209**, 867.
48. H. C. Yen and B. Marrs, *J. Bacteriol.*, 1976, **126**, 619.
49. E. Cerdá-Olmedo and P. Reau, *Mutat. Res.*, 1970, **9**, 369.
50. J. M. Balsalobre, R. M. Ruiz-Vásquez, and F. J. Murillo, *Proc. Natl. Acad. Sci. USA*, 1987, **84**, 2359.
51. F. J. Nicolás, R. Ruiz-Vázquez, and F. J. Murillo, *Genes Dev.*, 1994, **8**, 2375.
52. S. J. McGowan, H. C. Gorham, and D. A. Hodgson, *Mol. Microbiol.*, 1993, **10**, 713.
53. F. Kato, T. Hino, A. Nakaji, M. Tanaka, and Y. Koyama, *Mol. Gen. Genet.*, 1995, **247**, 387.
54. M. Alberti, D. H. Burke, and J. E. Hearst, in "Advances in Photosynthesis: Anoxygenic Photosynthetic Bacteria," eds. R. E. Blankenship, M. T. Madigan, and C. E. Bauer, Kluwer, Dordrecht, 1995, vol. 2, p. 1083.
55. R. J. Penfold and J. M. Pemberton, *J. Bacteriol.*, 1994, **176**, 2869.
56. A. A. Yeliseev and S. Kaplan, *J. Biol. Chem.*, 1995, **270**, 21 167.
57. K. Schmidt, in "The Photosynthetic Bacteria," eds. R. K. Clayton and W. R. Sistrom, Plenum, New York, 1978, p. 729.
58. M. Griffiths, W. R. Sistrom, G. Cohen-Bazire, and R. Y. Stanier, *Nature*, 1955, **176**, 1211.
59. M. Griffiths and R. Y. Stanier, *J. Gen. Microbiol.*, 1956, **14**, 698.
60. W. R. Sistrom, M. Griffiths, and R. Y. Stanier, *J. Cell. Comp. Physiol.*, 1956, **48**, 473.
61. G. Cohen-Bazire, W. R. Sistrom, and R. Y. Stanier, *J. Cell. Comp. Physiol.*, 1957, **49**, 25.
62. S. Liaaen-Jensen, G. Cohen-Bazire, and R. Y. Stanier, *Nature*, 1961, **192**, 1168.
63. P. A. Scolnik, M. A. Walker, and B. L. Marrs, *J. Biol. Chem.*, 1980, **255**, 2427.
64. B. L. Marrs, *J. Bacteriol.*, 1981, **146**, 1003.
65. D. P. Taylor, S. N. Cohen, W. G. Clark, and B. L. Marrs, *J. Bacteriol.*, 1983, **154**, 580.
66. K. M. Zsebo and J. E. Hearst, *Cell*, 1984, **37**, 937.
67. J. M. Pemberton and C. M. Harding, *Curr. Microbiol.*, 1986, **14**, 25.
68. G. Giuliano, D. Pollock, and P. A. Scolnik, *J. Biol. Chem.*, 1986, **261**, 12 925.
69. G. Giuliano, D. Pollock, H. Stapp, and P. A. Scolnik, *Mol. Gen. Genet.*, 1988, **213**, 78.
70. S. A. Coomber, M. Chaudhri, A. Connor, G. Britton, and C. N. Hunter, *Mol. Microbiol.*, 1990, **4**, 977.
71. J. T. Beatty, in "Advances in Photosynthesis: Anoxygenic Photosynthetic Bacteria," eds. R. E. Blankenship, M. T. Madigan, and C. E. Bauer, Kluwer, Dordrecht, 1995, vol. 2, p. 1209.
72. D. A. Young, C. E. Bauer, J. C. Williams, and B. L. Marrs, *Mol. Gen. Genet.*, 1989, **218**, 1.
73. G. A. Armstrong, A. Schmidt, G. Sandmann, and J. E. Hearst, *J. Biol. Chem.*, 1990, **265**, 8329.
74. E. Garí, I. Gibert, and J. Barbé, *Mol. Gen. Genet.*, 1992, **232**, 74.
75. S. K. Math, J. E. Hearst, and C. D. Poulter, *Proc. Natl. Acad. Sci. USA*, 1992, **89**, 6761.
76. G. Sandmann and N. Misawa, *FEMS Microbiol. Lett.*, 1992, **69**, 253.
77. G. E. Bartley, T. J. Schmidhauser, C. Yanofsky, and P. A. Scolnik, *J. Biol. Chem.*, 1990, **265**, 16 020.
78. G. E. Bartley, P. V. Viitanen, I. Pecker, D. Chamovitz, J. Hirschberg, and P. A. Scolnik, *Proc. Natl. Acad. Sci. USA*, 1991, **88**, 6532.
79. G. E. Bartley, P. V. Viitanen, K. O. Bacot, and P. A. Scolnik, *J. Biol. Chem.*, 1992, **267**, 5036.
80. G. E. Bartley and P. A. Scolnik, *J. Biol. Chem.*, 1993, **268**, 25 718.
81. F. M. Hahn, J. A. Baker, and C. D. Poulter, *J. Bacteriol.*, 1996, **178**, 619.
82. M. M. Sherman, L. A. Petersen, and C. D. Poulter, *J. Bacteriol.*, 1989, **171**, 3619.
83. A. Raisig, G. Bartley, P. Scolnik, and G. Sandmann, *J. Biochem.*, 1996, **119**, 559.
84. K. L. Perry, T. A. Simonitch, K. J. Harrison-LaVoie, and S.-T. Liu, *J. Bacteriol.*, 1986, **168**, 607.
85. L.-Y. Lee and S.-T. Liu, *Mol. Microbiol.*, 1991, **5**, 217.
86. N. Misawa, S. Yamano, and H. Ikenaga, *Appl. Environ. Microbiol.*, 1991, **57**, 1847.
87. M. Nakagawa and N. Misawa, *Agric. Biol. Chem.*, 1991, **55**, 2147.
88. C. N. Hunter, B. S. Hundle, J. E. Hearst, H. P. Lang, A. T. Gardiner, S. Takaichi, and R. J. Cogdell, *J. Bacteriol.*, 1994, **176**, 3692.
89. J. M. Pemberton and C. M. Harding, *Curr. Microbiol.*, 1987, **15**, 67.
90. B. S. Hundle, P. Beyer, H. Kleinig, G. Englert, and J. E. Hearst, *Photochem. Photobiol.*, 1991, **54**, 89.
91. R. W. Tuveson, R. A. Larson, and J. Kagan, *J. Bacteriol.*, 1988, **170**, 4675.
92. H. Linden, N. Misawa, D. Chamovitz, I. Pecker, J. Hirschberg, and G. Sandmann, *Z. Naturforsch., Teil C*, 1991, **46**, 1045.
93. H. Linden, A. Vioque, and G. Sandmann, *FEMS Microbiol. Lett.*, 1993, **106**, 99.
94. F. X. Cunningham, Jr., D. Chamovitz, N. Misawa, E. Gannt, and J. Hirschberg, *FEBS Lett.*, 1993, **328**, 130.
95. G. Sandmann, N. Misawa, M. Wiedemann, P. Vittorioso, A. Carattoli, G. Morelli, and G. Macino, *J. Photochem. Photobiol. B Biol.*, 1993, **18**, 245.
96. T. Lotan and J. Hirschberg, *FEBS Lett.*, 1995, **364**, 125.
97. S. Kajiwara, T. Kakizono, T. Saito, K. Kondo, T. Ohtani, N. Nishio, S. Nagai, and N. Misawa, *Plant Mol. Biol.*, 1995, **29**, 343.
98. J. Breitenbach, N. Misawa, S. Kajiwara, and G. Sandmann, *FEMS Microbiol. Lett.*, 1996, **140**, 241.
99. I. Pecker, D. Chamovitz, H. Linden, G. Sandmann, and J. Hirschberg, *Proc. Natl. Acad. Sci. USA*, 1992, **89**, 4962.
100. F. Bouvier, P. Hugueney, A. d'Harlingue, M. Kuntz, and B. Camara, *Plant J.*, 1994, **6**, 45.
101. N. Misawa, M. R. Truesdale, G. Sandmann, P. D. Fraser, C. Bird, W. Schuch, and P. M. Bramley, *J. Biochem.*, 1994, **116**, 980.
102. M. Albrecht, A. Klein, P. Hugueney, G. Sandmann, and M. Kuntz, *FEBS Lett.*, 1995, **372**, 199.

103. P. Hugueney, A. Badillo, H.-C. Chen, A. Klein, J. Hirschberg, B. Camara, and M. Kuntz, *Plant J.*, 1995, **8**, 417.
104. F. X. Cunningham, Jr., B. Pogson, Z. Sun, K. A. McDonald, D. DellaPenna, and E. Gannt, *Plant Cell*, 1996, **8**, 1613.
105. I. Pecker, R. Gabbay, F. X. Cunningham, Jr., and J. Hirschberg, *Plant Mol. Biol.*, 1996, **30**, 807.
106. Z. Sun, E. Gantt, and F. X. Cunningham, Jr., *J. Biol. Chem.*, 1996, **271**, 24 349.
107. S. Takaichi, G. Sandmann, G. Schnurr, Y. Satomi, A. Suzuki, and N. Misawa, *Eur. J. Biochem.*, 1996, **241**, 291.
108. G. A. Armstrong, B. S. Hundle, and J. E. Hearst, *Methods Enzymol.*, 1993, **214**, 297.
109. M. Wiedemann, N. Misawa, and G. Sandmann, *Arch. Biochem. Biophys.*, 1993, **306**, 152.
110. P. D. Fraser, N. Misawa, H. Linden, S. Yamano, K. Kobayashi, and G. Sandmann, *J. Biol. Chem.*, 1992, **267**, 19 891.
111. B. S. Hundle, D. A. O'Brien, M. Alberti, P. Beyer, and J. E. Hearst, *Proc. Natl. Acad. Sci. USA*, 1992, **89**, 9321.
112. B. S. Hundle, D. A. O'Brien, P. Beyer, H. Kleinig, and J. E. Hearst, *FEBS Lett.*, 1993, **315**, 329.
113. G. Schnurr, N. Misawa, and G. Sandmann, *Biochem. J.*, 1996, **315**, 869.
114. R. P. Burchard and M. Dworkin, *J. Bacteriol.*, 1966, **91**, 535.
115. R. P. Burchard, S. A. Gordon, and M. Dworkin, *J. Bacteriol.*, 1966, **91**, 896.
116. R. P. Burchard and S. B. Hendricks, *J. Bacteriol.*, 1969, **97**, 1165.
117. H. Kleinig, *Biochim. Biophys. Acta*, 1972, **274**, 489.
118. S. Takaichi, H. Yazawa, and Y. Yamamoto, *Biosci. Biotech. Biochem.*, 1995, **59**, 464.
119. H. Reichenbach and H. Kleinig, in "Myxobacteria: Development and Cell Interactions," ed. E. Rosenberg, Springer, New York, 1984, p. 128.
120. A. Martínez-Laborda, J. M. Balsalobre, M. Fontes, and F. J. Murillo, *Mol. Gen. Genet.*, 1990, **223**, 205.
121. A. Martínez-Laborda and F. J. Murillo, *Genetics*, 1989, **122**, 481.
122. R. Ruiz-Vázquez, M. Fontes, and F. J. Murillo, *Mol. Microbiol.*, 1993, **10**, 25.
123. A. Yokoyama and W. Miki, *FEMS Microbiol. Lett.*, 1995, **128**, 139.
124. T. Hoshino, R. Fujii, and T. Nakahara, *J. Ferm. Bioeng.*, 1994, **77**, 423.
125. T. Hoshino, Y. Yoshino, E. D. Guevarra, S. Ishida, T. Hiruta, R. Fujii, and T. Nakahara, *J. Ferm. Bioeng.*, 1994, **77**, 131.
126. K. Tabata, S. Ishida, T. Nakahara, and T. Hoshino, *FEBS Lett.*, 1994, **341**, 251.
127. A. Yokoyama, G. Sandmann, T. Hoshino, K. Adachi, M. Sakai, and Y. Shizuri, *Tetrahedron Lett.*, 1995, **36**, 4901.
128. M. Houssaini-Iraqui, H. L. David, S. Clavel-Sérès, F. Hilali, and N. Rastogi, *Curr. Microbiol.*, 1993, **27**, 317.
129. D. Chamovitz, I. Pecker, G. Sandmann, P. Böger, and J. Hirschberg, *Z. Naturforsch., Teil C*, 1990, **45**, 482.
130. H. Linden, G. Sandmann, D. Chamovitz, J. Hirschberg, and P. Böger, *Pestic. Biochem. Physiol.*, 1990, **36**, 46.
131. I. Martínez-Férez, A. Vioque, and G. Sandmann, *Pestic. Biochim. Physiol.*, 1994, **48**, 185.
132. D. Chamovitz, G. Sandmann, and J. Hirschberg, *J. Biol. Chem.*, 1993, **268**, 17 348.
133. P. Hugueney, S. Römer, M. Kuntz, and B. Camara, *Eur. J. Biochem.*, 1992, **209**, 399.
134. P. D. Fraser, H. Linden, and G. Sandmann, *Biochem. J.*, 1993, **291**, 687.
135. M. Albrecht, H. Linden, and G. Sandmann, *Eur. J. Biochem.*, 1996, **236**, 115.
136. A. Carattoli, N. Romano, P. Ballario, G. Morelli, and G. Macino, *J. Biol. Chem.*, 1991, **266**, 5854.
137. M. Kuntz, S. Römer, C. Suire, P. Hugueney, J. H. Weil, R. Schantz, and B. Camara, *Plant J.*, 1992, **2**, 25.
138. J. Ray, C. Bird, M. Maunders, D. Grierson, and W. Schuch, *Nucleic Acids Res.*, 1987, **15**, 10 587.
139. C. R. Bird, J. A. Ray, J. D. Fletcher, J. M. Boniwell, A. S. Bird, C. Teulieres, I. Blain, P. M. Bramley, and W. Schuch, *BioTechnology*, 1991, **9**, 635.
140. P. M. Bramley, C. Teulieres, I. Blain, C. Bird, and W. Schuch, *Plant J.*, 1992, **2**, 343.
141. T. J. Schmidhauser, F. R. Lauter, V. E. A. Russo, and C. Yanofsky, *Mol. Cell. Biol.*, 1990, **10**, 5064.
142. A. Badillo, J. Steppuhn, J. Deruère, B. Camara, and M. Kuntz, *Plant Mol. Biol.*, 1995, **27**, 425.
143. A. Chen, P. A. Kroon, and C. D. Poulter, *Protein Sci.*, 1994, **3**, 600.
144. D. J. McGarvey and R. Croteau, *Plant Cell*, 1995, **7**, 1015.
145. M. N. Ashby and P. A. Edwards, *J. Biol. Chem.*, 1990, **265**, 13 157.
146. M. N. Ashby, S. Y. Kutsunai, S. Ackerman, A. Tzagoloff, and P. A. Edwards, *J. Biol. Chem.*, 1992, **267**, 4128.
147. P. F. Marrero, C. D. Poulter, and P. A. Edwards, *J. Biol. Chem.*, 1992, **267**, 21 873.
148. A. Joly and P. A. Edwards, *J. Biol. Chem.*, 1993, **268**, 26 983.
149. L. Song and C. D. Poulter, *Proc. Natl. Acad. Sci. USA*, 1994, **91**, 3044.
150. O. Dogbo and B. Camara, *Biochim. Biophys. Acta*, 1987, **920**, 140.
151. A. Laferrière and P. Beyer, *Biochim. Biophys. Acta*, 1991, **1077**, 167.
152. T. J. Schmidhauser, F. R. Lauter, M. Schumacher, W. Zhou, V. E. A. Russo, and C. Yanofsky, *J. Biol. Chem.*, 1994, **269**, 12 060.
153. S. Römer, P. Hugueney, F. Bouvier, B. Camara, and M. Kuntz, *Biochem. Biophys. Res. Commun.*, 1993, **196**, 1414.
154. G. W. Robinson, Y. H. Tsay, B. K. Kienzle, C. A. Smith-Monroy, and R. W. Bishop, *Mol. Cell. Biol.*, 1993, **13**, 2706.
155. O. Dogbo, A. Laferrière, A. d'Harlingue, and B. Camara, *Proc. Natl. Acad. Sci. USA*, 1988, **85**, 7054.
156. M. Ehrenshaft and M. Daub, *Appl. Environ. Microbiol.*, 1994, **60**, 2766.
157. N. S. Scrutton, A. Berry, and R. N. Perham, *Nature*, 1990, **343**, 38.
158. I. Pecker, D. Chamovitz, V. Mann, G. Sandmann, P. Böger, and J. Hirschberg, in "Research in Photosynthesis," ed. N. Murata, Kluwer, Dordrecht, 1993, vol. III, p. 11.
159. G. Sandmann and P. D. Fraser, *Z. Naturforsch., Teil C*, 1993, **48**, 307.
160. P. Beyer, M. Mayer, and H. Kleinig, *Eur. J. Biochem.*, 1989, **184**, 141.
161. S. Al-Babili, J. von Lintig, H. Haubruck, and P. Beyer, *Plant J.*, 1996, **9**, 601.
162. G. Sandmann and P. M. Bramley, *Biochim. Biophys. Acta*, 1985, **843**, 73.
163. A. J. Biel and B. L. Marrs, *J. Bacteriol.*, 1985, **162**, 1320.
164. S. N. Ponnampalam, J. J. Buggy, and C. E. Bauer, *J. Bacteriol.*, 1995, **177**, 2990.
165. R. J. Penfold and J. M. Pemberton, *Curr. Microbiol.*, 1991, **23**, 259.
166. M. Gomelsky and S. Kaplan, *J. Bacteriol.*, 1995, **177**, 1634.
167. D. W. Bollivar, J. Y. Suzuki, J. T. Beatty, J. M. Dobrowolski, and C. E. Bauer, *J. Mol. Biol.*, 1994, **237**, 622.
168. M. E. Baker and D. D. Fanestil, *Cell*, 1991, **65**, 721.
169. A. A. Yeliseev, J. M. Eraso, and S. Kaplan, *J. Bacteriol.*, 1996, **178**, 5877.
170. H. C. Gorham, S. J. McGowan, P. R. H. Robson, and D. A. Hodgson, *Mol. Microbiol.*, 1996, **19**, 171.

171. A. Martínez-Laborda, E. Montserrat, R. Ruiz-Vázquez, and F. J. Murillo, *Mol. Gen. Genet.*, 1986, **205**, 107.
172. D. A. Hodgson, *Mol. Microbiol.*, 1993, **7**, 471.
173. G. H. Toenniessen, in "Rice Biotechnology," eds. G. S. Khush and G. H. Toenniessen, CAB International, Wallingford, UK, 1991, p. 253.
174. P. K. Burkhardt, P. Beyer, J. Wünn, A. Klöti, G. A. Armstrong, M. Schledz, J. von Lintig, and I. Potrykus, *Plant J.*, 1997, **11**, 1071.
175. N. Misawa, S. Yamano, H. Linden, M. R. de Felipe, M. Lucas, H. Ikenaga, and G. Sandmann, *Plant J.*, 1993, **4**, 833.
176. N. Misawa, K. Masamoto, T. Hori, T. Ohtani, P. Böger, and G. Sandmann, *Plant J.*, 1994, **6**, 481.
177. U. Windhövel, B. Geiges, G. Sandmann, and P. Böger, *Plant Physiol.*, 1994, **104**, 119.
178. H. J. Nelis and A. P. De Leenheer, *J. Appl. Bacteriol.*, 1991, **70**, 181.
179. S. Yamano, T. Ishii, M. Nakagawa, H. Ikenaga, and N. Misawa, *Biosci. Biotech. Biochem.*, 1994, **58**, 1112.

2.13
Protein Prenylation

MICHAEL H. GELB, PAUL McGEADY, KOHEI YOKOYAMA,
and GEENG-FU JANG

University of Washington, Seattle, WA, USA

2.13.1 INTRODUCTION

Among the numerous post-translational modifications that proteins and peptides undergo in cells, the attachment of prenyl groups (prenylation) is a recently discovered event. The first prenylated polypeptide to be discovered is the mating factor from the fungus *Rhodospiridium toruloides*, which is an undecapeptide containing a C-terminal *S*-farnesyl-cysteine methyl ester (Figure 1).[1] The structure of this peptide, called rhodotorucine A, was established by direct structural techniques (Section 2.13.2). Other species of fungi including common baker's yeast (*Saccharomyces cerevisiae*) were found to produce farnesylated and methylated peptide mating factors, and one of them, tremerogen A-10, bears a hydroxylated farnesyl group (Figure 1).[2–4]

In the early 1980s, while the farnesylated fungal mating factors were being discovered, an independent series of studies with eukaryotic cells led to the discovery that animal cells contain prenylated proteins. High concentrations of compactin, an inhibitor of mevalonic acid biosynthesis, blocked the cell cycle progression of cultured fibroblasts, an effect which could be reversed not by the addition of cholesterol to the medium but by adding small amounts of mevalonic acid.[5,6] Since mevalonic acid is the first committed precursor of the isoprenoids, these results suggested that mevalonate or some isoprenoid derived from mevalonate other than cholesterol was required for cell proliferation. Studies with radiolabeled mevalonic acid added to campactin-treated cells revealed that a specific set of cellular proteins were radiolabeled.[7] When mevalonate radiolabeled at carbon-

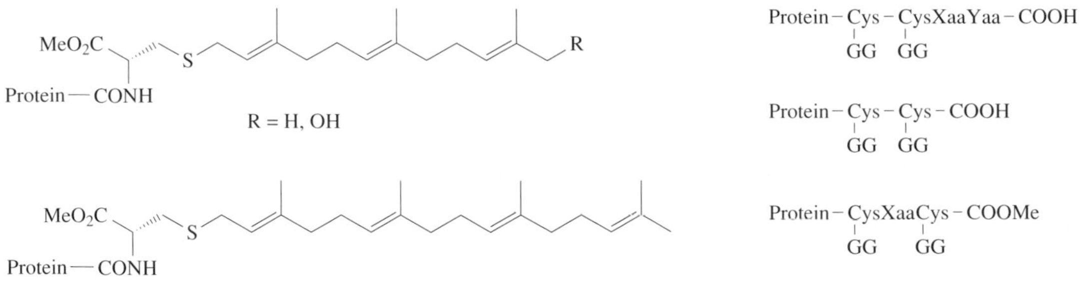

MeO₂C S R
Protein—CONH
R = H, OH

MeO₂C S
Protein—CONH

Protein–Cys–CysXaaYaa–COOH
GG GG

Protein–Cys–Cys–COOH
GG GG

Protein–CysXaaCys–COOMe
GG GG

Figure 1 Structure of protein prenyl groups: farnesylated and hydroxy-farnesylated proteins (upper left), singly geranylgeranylated proteins (lower left), doubly geranylgeranylated proteins (right). In all cases, the prenyl group is thioether linked to a cysteine. Note that some doubly geranylgeranylated proteins are C-terminally methylated, and some are not. All singly prenylated proteins are C-terminally methylated except for the subunits of phosphorylase kinase (see text).

1 was fed to cells, radiolabeled proteins were not formed. Since conversion of mevalonate into isoprenoids occurs with loss of carbon-1 as CO_2, this result suggested that some sort of isoprenoid was incorporated into proteins. Subsequent similar findings were reported with other types of animal cells.[8–10]

The first animal cell protein shown to become radiolabeled by feeding radiolabeled mevalonate to cells is the nuclear membrane-associated protein lamin B.[11,12] Rigorous structural analysis (Section 2.13.2) of the mevalonate-derived modification showed that it is a farnesyl group thioether-linked to the C-terminal cysteine of lamin B.[13] Thus, in the late 1980s it became clear that mammalian cells, like fungi, contain prenylated proteins. While structural studies were being carried out on lamin B, it was apparent that mammalian cells contain an additional type of prenyl group. Proteolysis of total protein from cells radiolabeled with mevalonic acid followed by gel filtration of the derived fragments revealed major and minor peaks of MWs of about 1 kDa and 0.5 kDa, respectively. The minor peak was found to contain the farnesyl group, while the major peak contained the geranylgeranyl group, which has 20 carbons, and thus is one isoprene unit longer than the farnesyl group[14,15] (Figure 1). At about the same time, it was found that Ras proteins were radiolabeled in mammalian cells grown in the presence of radiolabeled mevalonic acid, and the attached prenyl group was tentatively assigned as farnesyl based on the chromatographic properties of the radiolabeled fragment that was chemically released from the protein.[16–18]

Additional work has led to the identification of many prenylated proteins, including enzymes, signal transduction components, and structural proteins. This chapter gives a listing of currently known prenylated proteins and their structures and a brief description of techniques used to establish the structure of the prenyl group. Also given is a summary of enzymes that attach prenyl groups to proteins and those that further modify prenylated proteins. Finally, the functions of protein prenylation are discussed in the context of specific examples that are best understood. Several other reviews of protein prenylation are available.[19–28]

2.13.2 CHEMICAL STRUCTURES OF PROTEIN PRENYL GROUPS

2.13.2.1 Farnesylated Proteins

Most farnesylated proteins contain the lipid chain thioether linked to a cysteine residue at the C-terminus of the protein. The only known exceptions are the α and β subunits of rabbit skeletal muscle phosphorylase kinase which contain an *S*-farnesyl-cysteine that is the fourth residue from the C-terminus.[29] All known farnesylated proteins are synthesized as precursor proteins with the C-terminal sequence Cys-Ali-Ali-Xaa (Ali is usually, but not always, an aliphatic amino acid, and Xaa is a variety of different amino acids) (Table 1). After farnesylation of the SH group of the cysteine, the Ali-Ali-Xaa sequence is removed by endoproteolytic cleavage (Section 2.13.4), and the newly exposed α-carboxyl group of the farnesylated cysteine is usually, if not always methylated, although methylation appears to be reversible (Section 2.13.5). In the case of the phosphorylase kinase subunits, proteolysis and methylation do not occur. The fungal mating pheromones tremerogen

A-10 and A-9291-I are the only known examples in which the farnesyl group is hypermodified by hydroxylation of a terminal methyl group (Figure 1, Table 1).

The fungal mating pheromones were the first farnesylated polypeptides to be structurally characterized. In some cases, sufficient quantities of these materials were available so that their structures were established using NMR. The sequences of these peptides were determined by Edman degradation. Methyl iodide-promoted cleavage of the farnesyl group from these peptides was used as well. This reagent reacts with the farnesylated cysteine to give the *S*-methyl sulfonium salt which undergoes heterologous C—S bond cleavage to form a mixture of products derived from the farnesyl cation (Figure 2). These were detected by gas chromatography.

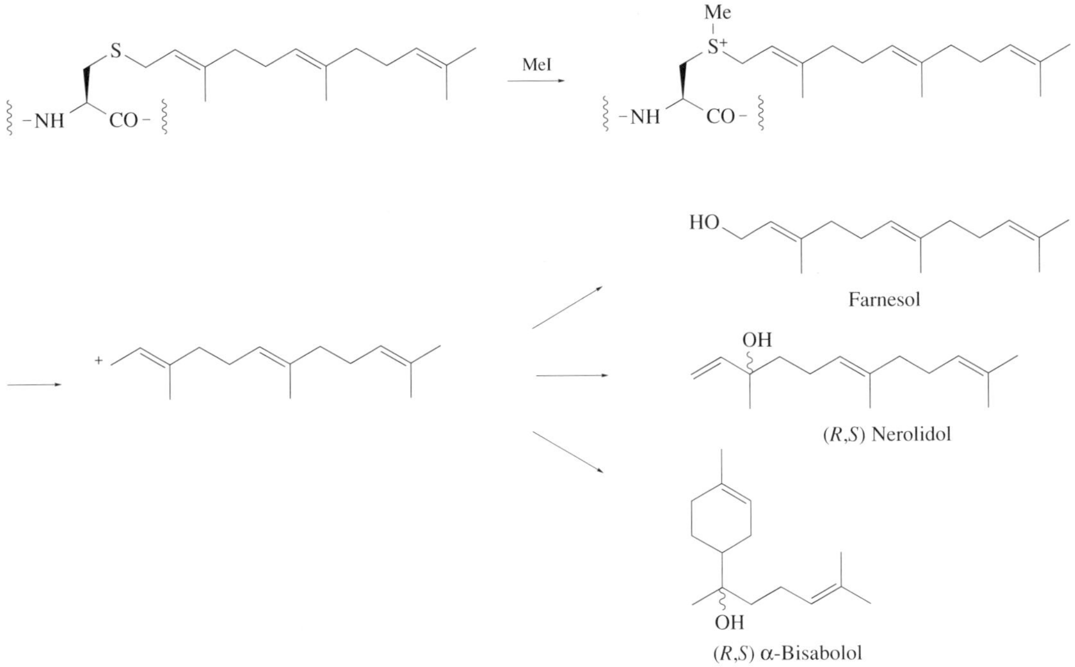

Figure 2 MeI-promoted cleavage of a thioether-linked farnesyl group from a protein to give farnesol as the major product and other minor product prenols resulting from rearrangement of the primary carbocation.

Lamin B was the first animal cell protein to be shown by direct structural techniques to contain a thioether-linked farnesyl group.[13] Since only microgram amounts of protein were available, the sensitive technique of gas chromatography/mass spectrometry was used to identify all-*trans*-2,6,10-trimethyl-dodecatriene, which was released from the protein by cleavage of the C—S bond with Raney nickel[56] (Equation (1)). This reaction is carried out in water with an overlay of pentane so that the released triene becomes trapped in the upper organic layer where it resists further reduction by Raney nickel. This technique can be used when at least 10 nmol of protein is available.

The most sensitive technique for analyzing protein farnesyl groups is to treat tissue culture cells[57] or whole animals[31,58] with ^{3}H- or ^{14}C-mevalonic acid or mevalonolactone to radiolabel the prenyl group. This is often done in the presence of hydroxymethylglutaryl-CoA reductase inhibitors (mevinolin or compactin) to suppress endogenous mevalonate production. After purification of the desired protein, often by immunoprecipitation, the protein is treated with MeI (Figure 2), and the radiolabeled prenols are analyzed by HPLC versus authentic standards.[57] The disadvantage of this method is that it provides only tentative structural data.

Methylation of farnesylated proteins is often demonstrated by monitoring the incorporation of a radiolabel into the investigated protein after feeding radiolabeled methionine to cell culture; this amino acid is taken up by cells and converted to *S*-adenosyl-methionine, the methyl donor. In a few cases, the detection of methylation of the C-terminal *S*-farnesyl-cysteine residue was carried out by

Table 1 Prenylated proteins.[a]

Protein	Function of protein	Prenyl group	C-terminal sequence prior to prenylation	C-terminal motif	Function of the prenyl group	Ref.
Enzymes						
Inositol triphosphate phosphatase	modulation of signal transduction	F[b] (HPLC)[c]	CVVQ	C(F)COOMe[d] (RL)[e]	membrane binding	30
Cyclic GMP phosphodiesterase	phototransduction	F (α-subunit) (HPLC) GG (β-subunit) (HPLC)	CCVQ (α-subunit) CCIL (β-subunit)	C(F)COOMe (α-subunit) C(GG)COOMe (β-subunit) (RL)	?	31
Phosphorylase kinase	signal transduction	F (MS)	CAMQ (α-subunit)	C(F)AMQ (MS)	?	29
			CLVS (β-subunit)	C(F)LVS (MS)		
Fungal mating pheromones	sexual discrimination					
A factor (*S. cerevisiae*)		F (MS)	CVIA	C(F)COOMe (MS)	required for inducing a receptor-based mating response	4
Rodotorucine A (*R. toruloidees*)		F (MS)	CTVA	C(F)COOH (MS)		32, 33
A-10 (*T. mesenterica*)		hydroxy-F (MS, NMR)	?	C(F)COOMe (MS)		3, 33, 34
A-9291-I (*T. brasiliensis*)		hydroxy-F (MS)	?	C(F)COOMe (MS)		35
M-factor (*S. pombe*)		F (RL)	CVIA	C(F)COOMe (RL)	?	36
G proteins (heterotrimeric, γ-subunit)	GTPase involved in receptor-based signal transduction					
Bovine brain G protein		GG (MS)	CAIL	C(GG)COOMe (RL)	membrane binding, binding to α-subunit	37
Transducin		F (MS)	CVLS	C(F)COOMe (RL)	membrane binding, binding to α-subunit	38
Kinases						
Rhodopsin kinase	inactivation of photoactivated rhodopsin	F (HPLC)	CVLS	C(F)COOMe (RL)	membrane binding	39
Casesin kinase-I (yeast)	?	?	LGCC	?	membrane binding	40

Miscellaneous proteins						
ANJI	plant chaperone	F (HPLC)	CAQQ	?	?	41
Hepatitis delta virus large antigen	viral structural protein	F (HPLC)	CRPQ	?	required for virion assembly	42
2′,3′-cyclic nucleotide 3′-phosphodiesterase	myelinogenesis	F and GG (HPLC)	CTII	?	membrane binding	43
PxF	?, localized to the outer surface of peroxisomes	F (HPLC)	CLIM	?	?	44
YDJ1 (yeast)	necessary for phosphorylation of cyclin Cln3	F (HPLC)	CASQ	?	membrane binding, necessary for growth at elevated temperature	45
Nuclear lamins	nuclear envelope structural proteins					
Pre-lamin A		F (HPLC)	CSIM	C(F)COOMe (RL), farnesylated C-terminus is removed by proteolysis	binding to the nuclear envelope	36
Lamin B		F (MS)	CLVM	C(F)COOMe (RL)	binding to the nuclear envelope	13
Small GTP-binding proteins[f]						
Rab	intracellular vesicle trafficking	GG (MS, HPLC)	CC, CXC, CCXX, CVLL (Rab 8)	C(GG)C(GG)COOH, C(GG)XC(GG)COOMe, C(GG)C(GG)XXCOOH (MS)	required for binding to Rab GDI	46, 47
Rac	actin polymerization	GG (HPLC)	CLLL (rac1) CSLL (rac2)	?	?	48
Ral	signal transduction	GG (HPLC)	CSLL	?	?	48
Rap	may modulate Ras activity	F (Rap 2a) (HPLC) GG (Rap 2b) (HPLC)	CAIQ (rap 2a) CVIL (rap 2b)	?	?	49
Ras	signaling of cell proliferation and other events	F (HPLC)	CVLS (H-Ras) CVVM (N-Ras) CVIM (K-Ras) CIIC (RAS1, yeast) CIIS (RAS2, yeast)	C(F)COOMe (RL)	membrane bindings and possibly other functions	27, 50
Rho	organization of actin cytoskeleton	F and GG (Rho B) (HPLC) F (Rho E) (HPLC) GG (Rho A) (MS)	CKVL (rho B) CTVM (rho E) CLVL (rho A)	C(F)COOMe (RL, MS for Rho A)	interaction with GDI and GDS, membrane binding	51–55

[a] This table contains only those proteins that have been shown to contain covalently attached prenyl groups. Only proteins that have a known biological function are included. This table contains most of the known prenylated proteins, but there is no assurance that it is absolutely comprehensive. All proteins are from mammalian sources unless otherwise indicated. [b] F and GG designate farnesyl and geranylgeranyl, respectively. [c] HPLC indicates that the structure of prenyl group is based on high-pressure liquid chromatographic analysis of the cleaved prenyl group versus authentic standards. MS indicates that the structure of the prenyl was established by mass spectrometry. [d] C(F) and C(GG) designate farnesyl or geranylgeranyl groups, respectively, thioether-linked to cysteine. COOMe and COOH designate the α-carboxyl methyl ester or α-carboxyl, respectively, of the C-terminal residue. [e] RL indicates that methylation was established by radiolabeling cells with radioactive methionine, submitting the protein to saponification, and analyzing the vapor phase for radiolabeled methanol. [f] Each family of GTPases contains several members. Rather than listing all family members, one or two examples are given for each unique prenylated C-terminal structure type.

mass spectrometric analysis of C-terminal proteolytic fragments[59] or by detection of radiomethylated cysteic acid methyl ester formed from oxidation and exhaustive proteolysis of the protein.[37]

2.13.2.2 Geranylgeranylated Proteins

Geranylgeranylated proteins are structurally more diverse than farnesylated proteins (Figure 1). All proteins modified by a single geranylgeranyl group contain a C-terminal *S*-geranylgeranyl-cysteine α-carboxylmethyl ester. Such proteins have been analyzed by the methods described above for farnesylated proteins. Some geranylgeranylated proteins contain two geranylgeranyl groups that are thioether linked to adjacent cysteines at or near the C-terminus of the protein (Figure 1). The only method reported for the complete determination of these structures is the use of mass spectrometry to detect the doubly lipidated and sometimes methylated C-terminal peptides derived by proteolysis of prenylated proteins.[46] Interestingly, proteins containing the structure motif *S*-geranylgeranyl-cysteinyl-*S*-geranylgeranyl-cysteinyl either at or near the C-terminus of the protein are not C-terminally methylated, whereas those containing a nonprenylated amino acid between the two lipidated cysteines are C-terminally methylated (Figure 1).[60–62]

2.13.3 PROTEIN PRENYLTRANSFERASES

Three distinct protein prenyltransferases that catalyze transfer of farnesyl or geranylgeranyl from farnesyl pyrophosphate (FPP) or geranylgeranyl pyrophosphate (GGPP) to proteins have been identified and purified from cytosol fractions of mammalian and yeast cells. Protein farnesyl-transferase (PFT) and protein geranylgeranyltransferase-I (PGGT-I) are closely related enzymes that transfer prenyl groups from FPP and GGPP, respectively, to proteins containing the C-terminal CysAliAliXaa motif.[63–67] The Xaa residue in the CysAliAliXaa motif plays a major role in recognition by these two enzymes.[68] PFT preferentially farnesylates CysAliAliXaa-containing proteins in which Xaa is serine, methionine, glutamine, cysteine, or possibly other residues. PGGT-I preferentially geranylgeranylates proteins having a C-terminal leucine or phenylalanine.

PFT and PGGT-I are heterodimers consisting of a common α-subunit and distinct β-subunits.[69] Apparent molecular weights of the subunits observed on SDS–PAGE are 48 kDa for the α-subunit, and 46 kDa for the PFT β-subunit, and 43 kDa for the PGGT-I β-subunit. cDNAs encoding the PFT subunits have been cloned from rat, bovine, and human sources.[70–73] The PFT α-subunit has 377 amino acids with a calculated molecular weight of 44 kDa, and the β-subunit has 437 amino acids with a calculated molecular weight of 48.6 kDa. cDNAs encoding the PGGT-I β-subunit have been cloned from rat and human libraries, and they encode polypeptides of 377 amino acid residues.[74] In *S. cerevisiae*, RAM2 and RAM1 (also called DPR1) genes encode PFT α- and β-subunits, respectively[17,75,76] (13–15), and they have 30% and 37% identity with mammalian PFT. The CDC43 (also called CAL1) gene has been shown to encode the β-subunit of yeast PGGT-I.[77]

CysAliAliXaa tetrapeptides are sufficient to be recognized and prenylated by PFT and PGGT-I, but there may be additional recognition determinants in certain protein substrates. The polylysine domain near the C-terminus of K-Ras 4B and regions of G protein γ-subunits upstream of the CysAliAliXaa motif enhance the interaction of these proteins with PFT and PGGT-I, respectively.[78,79]

Steady-state kinetic analysis shows that both PFT and PGGT-I can operate by a random sequential mechanism, but they show a strong tendency to bind prenyl pyrophosphate prior to prenyl acceptor;[80–82] yeast PFT seems to be strictly ordered.[83] Both PFT and PGGT-I have been shown to form stable binary complexes with FPP and GGPP, respectively, that can be isolated by gel filtration in a catalytically competent form.[66,84] Although PFT and PGGT-I are selective for their respective prenyl donors and acceptors, this specificity is not absolute.[80,85] PFT and PGGT-I are metallo-enzymes that contain one equivalent of tightly bound Zn^{2+} that is required for enzymatic activity[64,84,86] but not for tight binding of FPP and GGPP.

Protein geranylgeranyltransferase-II (PGGT-II), also known as Rab geranylgeranyltransferase, transfers geranylgeranyl groups to both cysteine residues of Rab proteins containing the C-terminal motifs CysXaaCys, CysCys, and CysCyxXaaXaa.[47] This enzymes has been purified from rat brain cytosol.[87,88] PGGT-II consists of 50 kDa and 38 kDa subunits, which are tightly bound together, and a weakly associated 95 kDa protein termed Rab Escort Protein 1 (REP 1). cDNAs encoding the three chains of PGGT-II have been cloned.[89,90] The 50 kDa and 38 kDa subunits show about

30% identity to their counterparts of PFT and PGGT-I, and thus the heterodimer probably constitutes the catalytic component.

Unlike PFT and PGGT-I, PGGT-II does not prenylate short peptides.[91,92] REP1 binds to Rab proteins and presents it to the other two subunits of PGGT-II leading to prenylation. A second REP protein, REP2, has been identified.[93] A defect in REP1 function has been shown to be responsible for the human disease choroideremia.[94]

Yeast homologues of the mammalian PGGT-II components of α, β, and REP have been identified as the BET4, BET2, and MS14 (also called MSR6) genes, respectively.[95–97] Like the mammalian PGGT-II, the products of BET4 and BET2 form a complex, and the addition of the MS14/MSR6 product supports the geranylgeranylation of the yeast Rab homologue Ypt1.[98]

2.13.4 PRENYL PROTEIN-SPECIFIC ENDOPROTEINASE

As described above, many prenylated proteins undergo selective proteolysis in which the three C-terminal amino acids are removed to leave a prenylated-cysteine as the new C-terminus of the protein. This was first established in the case of the fungal mating pheromones, where it was found that the gene coding for the phermone coded for three C-terminal residues that were not present in the mature peptide. Gutierrez *et al.* expressed in animal cells a mutant form of Ras that contains a unique tryptophan on the C-terminal side of Cys-186 and found that radiolabeled tryptophan incorporated into a Ras precursor. This was followed by loss of the radiolabel as the mature form of Ras accumulates.[99] Later, Hancock *et al.* used a reticulocyte lysate translation system to show that proteolysis and subsequent methylation of Ras occurs only if canine pancreatic microsomal membranes are included in the lysate.[100] Proteolysis of synthetic farnesylated and geranylgeranylated peptides has also been detected in homogenates from yeast[101] and bovine and rat liver microsomes.[102,103] In all cases, a single endoproteolytic event occurs to release an intact tripeptide (AliAliXaa) from the structure Cys(*S*-prenyl)AliAliXaa, and proteolysis only occurs if the peptide is prenylated. Cell fractionation studies by Jang *et al.* demonstrate that the protease is mainly localized in the endoplasmic reticulum.[103]

The prenyl protein-specific protease can cleave short prenylated peptides with the structure *N*-acetyl-Cys(*S*-prenyl)-AliAli or *N*-acetyl-Cys(*S*-prenyl)-AliAliXaa.[102,103] Peptides with the above structure but containing D-amino acids in the sequence are not protease substrates.[104] The prenyl protein-specific protease is not inactivated by a variety of standard protease inhibitors, but is sensitive to the thiol reagent *p*-chloromercuribenzoate.[102,103] Analogues of short prenylated peptides containing tetrahedral functional groups in place of the protease-susceptible amide were prepared, and some were found to be potent inhibitors of prenyl protein-specific protease.[105]

Prenyl protein-specific protease has been solubilized from membranes and partially purified.[106,107] The enzyme appears to be unstable, and its purification to homogeneity has not been reported.

2.13.5 PRENYL PROTEIN-SPECIFIC METHYLTRANSFERASE

The *S*-adenosyl-L-methionine-dependent methylation of the α-carboxyl group of *S*-prenyl-cysteine at the C-termini of prenylated proteins has been detected in membranes from mammalian cells and yeast.[108–111] The enzyme seems to methylate proteins ending in either *S*-farnesyl- or *S*-geranylgeranyl-cysteine,[112] and even simple compounds such as *N*-acetyl-*S*-farnesyl-L-cysteine, *N*-acetyl-*S*-geranylgeranyl-L-cysteine, and *S*-farnesylthiopropionic acid, but not *S*-farnesylthioacetic acid, can serve as substrates.[111,113,114] Steady-state kinetic analysis indicates that the enzyme displays an ordered Bi Bi mechanism with *S*-adenosyl-L-methionine binding first and *S*-adenosyl-L-homocysteine (one of the products) leaving last.[115] Methylation may be a reversible modification and thus may function in the regulation of prenylated proteins.[116] A methylesterase activity that acts on proteins with a C-terminal *S*-prenyl-cysteine methyl ester has been identified in bovine rod outer segment membranes,[117] but further work is needed to determine if this enzyme demethylates prenylated protein in intact cells.

Most Rab proteins contain two geranylgeranyl groups (Table 1), and it has been demonstrated that they are C-terminally methylated only if the C-terminal *S*-geranylgeranyl-cysteine is not immediately preceded by the second *S*-geranylgeranyl-cysteine (i.e., proteins containing Cys(*S*-

geranylgeranyl)XaaCys(*S*-geranylgeranyl) are methylated but those with Cys(*S*-geranylgeranyl)Cys (*S*-geranylgeranyl) are not).[60–62] Competitive substrate studies suggest that methylation of Rab proteins and proteins containing single prenyl groups are methylated by distinct transferases.[62]

2.13.6 SPECIFIC EXAMPLES OF PRENYLATED PROTEINS

2.13.6.1 Ras Proteins

The Ras GTP-binding proteins play a pivotal role in several signal transduction and differentiation processes.[118–120] It is currently thought that a significant fraction of human tumors are caused by mutations to Ras that lock it in its active GTP-bound state.[121,122] Farnesylation of all three types of human Ras (H-, N-, and K-Ras) and yeast Ras is required for Ras function.[27] Additionally, H-Ras and N-Ras, but not K-Ras4B, undergo palmitoylation at one or more cysteines just to the N-terminal side of the C-terminal *S*-farnesyl-cysteine methyl ester.[123] Unlike yeast and mammalian Ras, the Ras proteins from *Drosophila melanogaster*[124] and from the slime molds *Dictyostelium discoideum*[125,126] and *Physarum polycephalum*[127,128] might be geranylgeranylated because their C-terminal residue is Leu (CysAliAliLeu, see Section 2.13.3). Although necessary for the normal and oncogenic functions of Ras,[129] the specific functions of these post-translational modifications are, to a large extent, unclear. There appears to be general consensus that farnesylation together with palmitoylation (H- and N-Ras) or the polybasic region (K-Ras) allow Ras to bind to the intracellular surface of the plasma membrane.[130]

Blocking prenylation of Ras *in vivo* by mutating the CysAliAliXaa motif leads to cytosolic, biologically inactive protein.[27] However, the biochemical basis for this is not clear. At least two different but not mutually exclusive models have been postulated for the function of Ras prenylation: (i) membrane anchoring and (ii) direct protein–protein interaction mediated by the prenyl group.[27,131,132] Oncogenic Ras genes have been engineered to prevent farnesylation (mutation of CysAliAliXaa) and to allow N-terminal myristoylation (addition of an N-myristoyltranferase recognition sequence). Expression of these constructs in mammalian cells leads to transformation. This result suggests that, in the absence of farnesylation, myristoylation can restore the membrane localization of Ras, and this enables Ras function. This result argues against a role of the farnesyl group as a molecular handle for other signaling components to bind to. However, the story is more complex in that corresponding nononcogenic forms of these Ras constructs did not behave normally, suggesting that membrane binding is required for Ras-induced transformation but not for all of the cellular functions of these signaling proteins.[133,134] Mutation of the Xaa residue of CysAliAliXaa to Leu leads to a Ras protein that is geranylgeranylated rather than farnesylated, and such a Ras protein inhibits normal cell growth.[135] At present, it is difficult to interpret all of these findings.

Ras proteins stimulate the mitogen-activated protein kinase (MAP-kinase) cascade. It has been found that the protein kinase Raf, a target of Ras, when tagged with a CysAliAliXaa sequence and expressed in mammalian cells, can stimulate the MAP-kinase pathway in a Ras-independent manner.[136,137] This would indicate that in the Raf/MAP-kinase pathway, the function of Ras is to recruit Raf to the membrane, and this requires that the Ras be lipidated. However, in a partially purified, cell-free system from *Xenopus laevis* oocytes, it was found that Ras activates the MAP-kinase pathway in the absence of membranes, and this requires that Ras be farnesylated. Furthermore, this activation is, to a slight degree, dependent on the structure of the prenyl group.[138,139] These latter results indicate that there is some prenyl-mediated interaction of Ras with another protein or proteins in this system. Two possibilities are that the farnesyl group binds to one or more components of the MAP-kinase pathway or that farnesylation of Ras causes a structural change in this GTPase that renders it functional. It has also been shown that the efficiency of Ras GTP/GDP exchange catalyzed by the Ras nucleotide exchanger SOS is much higher if the Ras is farnesylated.[140] Taken together these studies indicate that, while some of the functions of Ras farnesylation can be explained by a simple membrane anchor model, this is probably not the only function of Ras farnesylation.

2.13.6.2 Transducin

In the visual transducin cycle, the heterotrimeric G protein transducin becomes activated upon interaction with photoactivated rhodopsin, and then it activates cGMP phosphodiesterase. The γ-

subunit of transducin, like many proteins of the visual transduction machinery, is farnesylated (Table 1). The γ-subunit is tightly bound to the β-subunit, and this heterodimer can reversibly bind the α-subunit. Interaction of the heterotrimeric G protein with photoactivated rhodopsin leads to the replacement of GDP bound to the α with GTP, and this causes α to dissociate from $\beta\gamma$. Two forms of transducin $\beta\gamma$ have been purified from bovine retina. The full-length form T$\beta\gamma$-2 has a C-terminal *S*-farnesyl-cysteine, and the α-carboxyl group of this residue is partially methylated. The truncated form T$\beta\gamma$-1 is not farnesylated because it lacks the C-terminal cysteine.[141,142] *In vitro* data suggest that only the lipidated form T$\beta\gamma$-2 forms a complex with the α-subunit[142] and that the farnesyl group is not required for $\beta\gamma$ dimer formation. This is reminiscent of the results described above for Ras proteins where farnesylation seems to affect the interaction of the farnesylated protein with another protein in the absence of membranes. Additional studies have shown that farnesylation of a variety of $\beta\gamma$ dimers is required for high-affinity binding to α-subunits,[143] although one exception has been reported.[144]

Additional studies show that transducin $\beta\gamma$ interacts most tightly with rhodopsin in membranes only if the latter protein is photobleached, if the transducin γ-subunit is farnesylated and methylated, and if the $\beta\gamma$ dimer is also bound to the α-subunit.[38] Furthermore, in the presence of photobleached rhodopsin, T$\beta\gamma$-2 but not T$\beta\gamma$-1 enhances the affinity of the α-subunit for GTP.[145] The high-affinity interaction of T$\beta\gamma$-2 with photobleached rhodopsin in the presence of α-subunit is not seen if a GTP analogue is present. This is presumably because binding of the GTP analogue to the α-subunit causes it to dissociate from $\beta\gamma$, and T$\beta\gamma$-2 alone binds weakly to rhodopsin. Thus it appears that both farnesylation and methylation of transducin are required for efficient coupling to rhodopsin, although a more detailed study of the role of methylation shows that methylation of T$\beta\gamma$-2 only modestly increases its coupling with photobleached rhodopsin.[146] Further work is needed to determine if the farnesyl and/or methyl group interacts directly with the membrane, with the α-subunit, with rhodopsin, or with a combination of these elements. Studies have shown that a farnesylated dodecapeptide peptide derived from the C-terminus of T$\beta\gamma$-2 blocks coupling of transducin with rhodopsin and the nonfarnesylated peptide is inactive.[147] No data was reported in this study for the methylated and farnesylated peptide.

Detailed studies by Bigay *et al.* cast doubt on the suggestion that farnesylation of transducin $\beta\gamma$ confers on this dimer the ability to bind tightly to α.[148] Transducin can be eluted from retinal membranes by washing with detergent-free buffer, and eluted $\beta\gamma$ is not associated with the α-subunit. These authors also report that the presence of detergent seems to increase the association of $\beta\gamma$ with α. These is synergism in the binding of lipidated $\beta\gamma$ and α to phospholipid vesicles, which suggests that these proteins are favored to associate at the membrane interface. Studies have shown that the N-terminus of α is fatty acylated by a mixture of saturated and unsaturated 14-carbon fatty acids.[149] The crystal structure of transducin heterotrimer is shown in Figure 3.[150] Although the protein that was crystalized is not lipidated (terminal amino acids at the lipidation sites have been deleted), one can easily see that the N-terminus of α and the C-terminus of γ are very close to each other. Thus, there is structural support for the notion that both the fatty acyl group on α and the farnesyl group on $\beta\gamma$ can be bound simultaneously to membranes. This would explain the synergism described by Bigay *et al.* and together these studies suggest that the farnesyl group of $\beta\gamma$ does not engage in high-affinity interaction with a site on the α-subunit.

2.13.7 METHYLATION

Very little is known about the functions of carboxymethylation of prenylated proteins. Studies with prenylated proteins and peptides indicate that methylation tends to increase the fraction of material bound to membranes;[151–155] this is presumably due to an increase in hydrophobicity when the negatively charged carboxylate is neutralized. As noted above, methylation may be a reversible process and thus serve to regulate the function of prenylated proteins, but strong evidence for this is lacking.

Prenylated cysteine derivatives such as *N*-acetyl-*S*-farnesyl-cysteine can block the methylation of prenylated proteins by serving as competitive substrates for the methyltransferase (Section 2.13.5). Thus, these compounds have been used in intact cells to probe the role of methylation of prenylated proteins in specific cellular events. These agents interfere with platelet aggregation,[156] f-Met-Leu-Phe-induced chemotaxic response of mouse peritoneal macrophage,[157] superoxide formation in neutrophils,[158,159] and release of insulin from pancreatic cells.[160] However, the physiological effects

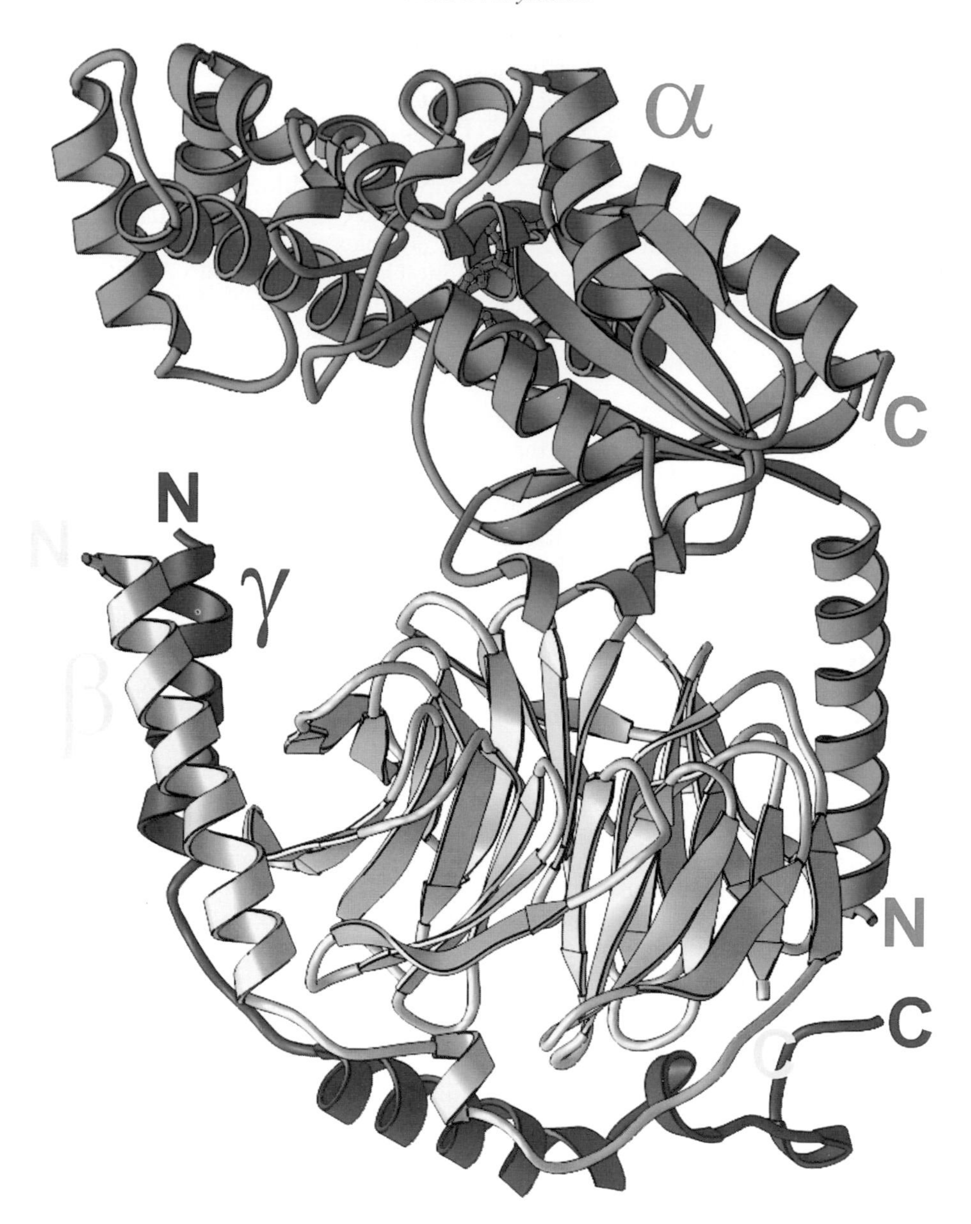

Figure 3 Structure of the transducin heterotrimer129. The α, β, and γ subunits are shown in green, yellow, and blue, respectively. Bound guanine nucleotide is shown in pink. The fatty acyl chain attached to the N-terminus of α and the farnesyl group attached to the C-terminus of γ are not present in the protein that was crystallized, but note that these lipids are predicted to lie close to each other and could very likely bind simultaneously to the same membrane plane. (Courtesy of R. Gaudet, Yale University.)

of these prenyl-cysteine analogues may be more complex. Studies with platelets show that these analogues inhibit platelet aggregation without inhibiting protein methylation.[161]

In *S. cerevisiae*, a mutant defective in the STE14 gene that encodes the methyltransferase is sterile[162] due in part to the failure of a-factor to become methylated; methylation is required for the a-factor response.[163] Interestingly, the mating pheromone of the yeast *R. toruloides* lacks a methyl ester (Table 1) and appears to be fully functional in the mating response.[164] The yeast Ras proteins RAS1 and RAS2 in STE14 mutant yeast are not methylated, and this does not seem to adversely affect the function of these proteins in yeast nor their membrane binding.[165]

2.13.8 PRENYLATION IN PLANTS AND NONFUNGAL MICROORGANISMS

Protein prenylation in plants has been discovered,[166,167] and only one protein of known function has been shown to be prenylated *in vivo*[168] (Table 1). The β-subunit of a plant PFT has been sequenced,[169] as have a number of small GTP-binding proteins, which contain typical C-terminal motifs that may specify prenylation (Section 2.13.3).[169–172]

Studies with radiolabeled mevalonate have demonstrated protein prenylation in tobacco suspension cultures,[167] and in both spinach leaves and seedlings.[166] This last observation is interesting because it allows the examination of prenylation in different organs under various conditions which affect the organism rather than the cell. These labeling studies have also suggested that plant proteins contain, in addition to the farnesyl and geranylgeranyl moieties, a number of other largely unidentified, covalently bound isoprenoids not found in other organisms.[166]

Prenylation also occurs in the pathogens *Giardia lamblia* and *Schistosoma mansoni*.[173,174] The growth of *Giardia* is blocked by inhibitors of mevalonate biosynthesis such as compactin. [^{3}H]mevalonate is metabolically incorporated into cellular proteins of approximate molecular weights 21–26 kDa and 50 kDa. Chromatographic analysis of the radioactivity cleaved from the protein by treatment with MeI (Section 2.13.2) reveals the existence of farnesyl and geranylgeranyl groups. Some of the 21 kDa radiolabeled proteins are immunoprecipitated by antihuman Ras antibodies, suggesting the existence of prenylated Ras and/or Ras-related proteins in *Giardia*.

2.13.9 FUTURE PROSPECTS

Elucidation of the structure of prenyl groups attached to proteins is well established, and several analytical methods have been developed for determining the prenylation status of proteins in cells. Protein prenyltransferases have been purified and cloned from mammalian sources and from yeast, and large quantities of enzymes are available for further studies. Since the farnesylation of Ras is critical for oncogenic transformation induced by these proteins, there has been a massive effort to obtain inhibitors of protein farnesyltransferase. Many inhibitors have been reported, and many cause Ras-transformed cells in culture to revert to a normal phenotype.[175] Studies with farnesyltransferase inhibitors and animal tumor models are showing promising results.[176] It may be noted that protein farnesyltransferase is the first well defined molecular target for the development of agents that block Ras-dependent oncogenesis.

There is general consensus in the field that the prenylation of proteins increases their affinity for membranes, and C-terminal methylation causes additional enhancement in membrane binding. Despite this knowledge, it is far from clear whether membrane anchoring is the function of protein prenyl groups in cells. A significant question that remains unanswered is whether the interaction of a prenylated protein with a nonprenylated protein, when both are present at the membrane–water interface, requires direct prenyl group–protein interactions. There is strong circumstantial evidence that protein–prenyl group contact occurs between proteins in the aqueous phase. This is based on the observation that Rab proteins, which contain two geranylgeranyl groups, are dissociated from membranes by forming a stoichiometric complex with a protein termed Rab-GDI, and formation of this heterodimer requires that the Rab be prenylated.[177] It seems incomprehensible that a protein with two 20-carbon hydrocarbon chains can desorb from membranes without these chains in intimate contact with a hydrophobic pocket on Rab or Rab-GDI.

2.13.10 REFERENCES

1. Y. Kamiya, A. Sakurai, S. Tamura, and N. Takahashi, *Biochem. Biophys. Res. Commun.*, 1978, **83**, 1077.
2. Y. Sakagami, A. Isogai, A. Suzuki, S. Tamura, C. Kitada, and M. Fujino, *Agric. Biol. Chem.*, 1978, **42**, 1093.
3. Y. Sakagami, A. Isogai, A. Suzuki, S. Tamura, C. Kitada, and M. Fujino, *Agric. Biol. Chem.*, 1979, **43**, 2643.
4. R. J. Anderegg, R. Betz, S. A. Carr, J. W. Crabb, and W. Duntze, *J. Biol. Chem.*, 1988, **263**, 18 236.
5. A. J. R. Habenicht, J. A. Glomset, and R. Ross, *J. Biol. Chem.*, 1980, **255**, 5134.
6. V. Quesney-Huneeus, M. H. Wiley, and M. D. Siperstein, *Proc. Natl. Acad. Sci. USA*, 1979, **76**, 5056.
7. R. A. Schmidt, C. J. Schneider, and J. A. Glomset, *J. Biol. Chem.*, 1984, **259**, 10 175.
8. M. Sinensky and J. Logel, *Proc. Natl. Acad. Sci. USA*, 1985, **82**, 3257.
9. W. A. Maltese and K. M. Sheridan, *J. Cell. Physiol.*, 1987, **133**, 471.
10. L. Sepp-Lorenzino, N. Azrolan, and P. S. Coleman, *FEBS Lett.*, 1989, **245**, 110.
11. S. L. Wolda and J. A. Glomset, *J. Biol. Chem.*, 1988, **263**, 5997.
12. L. A. Beck, T. J. Hosick, and M. Sinensky, *J. Cell Biol.*, 1988, **107**, 1307.
13. C. C. Farnsworth, S. L. Wolda, M. H. Gelb, and J. A. Glomset, *J. Biol. Chem.*, 1989, **264**, 20 422.

14. C. C. Farnsworth, M. H. Gelb, and J. A. Glomset, *Science*, 1990, **247**, 320.
15. H. C. Rilling, F. Breunger, W. W. Epstein, and P. F. Crain, *Science*, 1990, **247**, 318.
16. P. J. Casey, P. A. Solski, C. J. Der, and J. E. Buss, *Proc. Natl. Acad. Sci. USA*, 1989, **89**, 8323.
17. W. R. Scharfer, C. E. Trueblood, C.-C. Yang, M. P. Mayer, S. Rosenberg, C. D. Poulter, S.-H. Kim, and J. Rine, *Science*, 1990, **249**, 1133.
18. J. F. Hancock, A. I. Magee, J. E. Childs, and C. J. Marshall, *Cell*, 1989, **57**, 1167.
19. J. A. Glomset, M. H. Gelb, and C. C. Farnsworth, *Trends Biochem. Sci.*, 1990, **15**, 139.
20. J. A. Glomset and C. C. Farnsworth, *Annu. Rev. Cell Biol.*, 1994, **10**, 181.
21. W. A. Maltese, *FASEB J.*, 1990, **4**, 3319.
22. S. Clarke, *Annu. Rev. Biochem.*, 1992, **61**, 355.
23. P. J. Casey, *J. Lipid Res.*, 1993, **33**, 1731.
24. M. Sinensky and R. J. Lutz, *BioEssays*, 1992, **14**, 25.
25. J. Rine and S.-H. Kim, *New Biol.*, 1990, **2**, 219.
26. C. J. Der and A. D. Cox, *Cancer Cells*, 1991, **3**, 331.
27. A. D. Cox and C. J. Der, *Crit. Rev. Oncog.*, 1992, **3**, 365.
28. F. L. Zhang and P. J. Casey, *Annu. Rev. Biochem.*, 1996, **65**, 241.
29. L. M. Heilmeyer Jr., M. Serwe, C. Weber, J. Metzger, E. Hoffmann-Posorske, and H. E. Meyer, *Proc. Natl. Acad. Sci. USA*, 1992, **89**, 9554.
30. F. De-Smedt, X. Pesesse, A Boom, S. N. Schiffmann, and C. Erneux, *J. Biol. Chem.*, 1996, **271**, 10 419.
31. J. S. Anant, O. C. Ong, H. Y. Xie, S. Clarke, P. J. O'Brien, and B. K.-K. Fung, *J. Biol. Chem.*, 1992, **267**, 687.
32. Y. Kamiya, A. Sakurai, S. Tamura, and N. Takahashi, *Agric. Biol. Chem.*, 1979, **43**, 1049.
33. E. Tsuchiya, S. Fukui, Y. Kamiya, Y. Sakagami, and M. Fujino, *Biochem. Biophys. Res. Commun.*, 1978, **85**, 459.
34. M. Fujino, C. Kitada, Y. Sakagami, A. Isogai, S. Tamura, and A. Suzuki, *Naturwissenschaften*, 1980, **67**, 406.
35. Y. Ishibashi, Y. Sakagami, A. Isogai, and A. Suzuki, *Biochemistry*, 1984, **23**, 1399.
36. J. Davey, *EMBO J.*, 1992, **11**, 951.
37. H. K. Yamane, C. C. Farnsworth, H. Xie, W. Howald, B. K.-K. Fung, S. Clarke, M. H. Gelb, and J. A. Glomset, *Proc. Natl. Acad. Sci. USA*, 1990, **87**, 5866.
38. H. Ohguro, Y. Fukada, T. Takao, Y. Shimonishi, T. Yoshizawa, and T. Akino, *EMBO J.*, 1991, **10**, 3669.
39. J. Inglese, J. F. Glickman, W. Lorenz, M. G. Caron, and R. J. Lefkowitz, *J. Biol. Chem.*, 1992, **267**, 1422.
40. A. Vancura, A. Sessler, B. Leichus, and J. Kuret, *J. Biol. Chem.*, 1994, **269**, 19 271.
41. J.-K. Zhu, R. A. Bressan, and P. M. Hasegawa, *Proc. Natl. Acad. Sci. USA*, 1993, **90**, 8557.
42. J. C. Otto and P. J. Casey, *J. Biol. Chem.*, 1996, **271**, 4569.
43. D. A. De-Angelis and P. E. Braun, *J. Neurosci. Res.*, 1994, **39**, 386.
44. G. James, J. L. Goldstein, R. K. Pathak, R. G. Anderson, and M. S. Brown, *J. Biol. Chem.*, 1994, **269**, 14 182.
45. A. J. Caplan, J. Tsai, P. J. Casey, and M. G. Douglas, *J. Biol. Chem.*, 1992, **267**, 18 890.
46. C. C. Farnsworth, M. Kawata, Y. Yoshida, Y. Takai, M. H. Gelb, and J. A. Glomset, *Proc. Natl. Acad. Sci. USA*, 1991, **88**, 6196.
47. C. C. Farnsworth, M. C. Seabra, L. H. Ericsson, M. H. Gelb, and J. A. Glomset, *Proc. Natl. Acad. Sci. USA*, 1994, **91**, 11 963.
48. B. T. Kinsella, R. A. Erdman, and W. A. Maltese, *J. Biol. Chem.*, 1991, **266**, 9786.
49. F. X. Farrell, K. Yamamoto, and E. G. Lapetina, *Biochem. J.*, 1993, **289**, 349.
50. A. Fujiyama, S. Tsunasawa, F. Tamanoi, and F. Sakiyama, *J. Biol. Chem.*, 1991, **266**, 17 926.
51. M. Katayama, M. Kawata, Y. Yoshida, H. Horiuchi, T. Yamamoto, Y. Matsuura, and Y. Takai, *J. Biol. Chem.*, 1991, **266**, 12 639.
52. P. Adamson, C. J. Marshall, A. Hall, and P. A. Tilbrook, *J. Biol. Chem.*, 1992, **267**, 20 033.
53. R. Foster, K. Q. Hu, Y. Lu, K. M. Nolan, J. Thissen, and J. Settleman, *Mol. Cell. Biol.*, 1996, **16**, 2689.
54. S. A. Armstrong, V. C. Hannah, J. L. Goldstein, and M. S. Brown, *J. Biol. Chem.*, 1995, **270**, 7864.
55. Y. Hori, A. Kikuchi, M. Isomura, M. Katayama, Y. Miura, H. Fujioka, K. Kaibuchi, and Y. Takai, *Oncogene*, 1991, **6**, 515.
56. M. H. Gelb, C. C. Farnsworth, and J. A. Glomset in "Lipidation of Proteins: A Practical Approach," ed. A. J. Turner, IRL Press, Oxford, 1992.
57. C. C. Farnsworth, P. J. Casey, W. N. Howald, J. A. Glomset, and M. H. Gelb, *Methods*, 1990, **1**, 231.
58. J. S. Anant and B. K.-K. Fung, *Biochem. Biophys. Res. Commun.*, 1992, **183**, 468.
59. M. Kawata, C. C. Farnsworth, Y. Yoshida, M. H. Gelb, J. A. Glomset, and Y. Takai, *Proc. Natl. Acad. Sci. USA*, 1990, **87**, 8960.
60. T. E. Smeland, M. C. Seabra, J. L. Goldstein, and M. S. Brown, *Proc. Natl. Acad. Sci. USA*, 1994, **91**, 10 712.
61. M. H. Gelb, Y. Reiss, F. Ghomashchi, and C. C. Farnsworth, *Bioorg. Med. Chem. Lett.*, 1995, **8**, 881.
62. J. L. Giner and R. R. Rando, *Biochemistry*, 1994, **33**, 15 116.
63. Y. Reiss, J. L. Goldstein, M. C. Seabra, P. J. Casey, and M. S. Brown, *Cell*, 1990, **62**, 81.
64. J. F. Moomaw and P. J. Casey, *J. Biol. Chem.*, 1992, **267**, 17 438.
65. K. Yokoyama, G. W. Goodwin, F. Ghomashchi, J. Glomset, and M. H. Gelb, *Biochem. Soc. Trans.*, 1992, **20**, 489.
66. K. Yokoyama and M. H. Gelb, *J. Biol. Chem.*, 1993, **268**, 4055.
67. P. J. Casey and M. C. Seabra, *J. Biol. Chem.*, 1996, **271**, 5289.
68. K. Yokoyama, G. W. Goodwin, F. Ghomashchi, J. A. Glomset, and M. H. Gelb, *Proc. Natl. Acad. Sci. USA*, 1991, **88**, 5302.
69. M. C. Seabra, Y. Reiss, P. J. Casey, M. S. Brown, and J. L. Goldstein, *Cell*, 1991, **65**, 429.
70. W.-J. Chen, D. A. Andres, J. L. Goldstein, D. W. Russell, and M. S. Brown, *Cell*, 1991, **66**, 327.
71. W. J. Chen, D. A. Andres, J. L. Goldstein, and M. S. Brown, *Proc. Natl. Acad. Sci. USA*, 1991, **88**, 11 368.
72. N. E. Kohl, R. E. Diehl, M. D. Schaber, E. Rands, D. D. Soderman, B. He, S. L. Moores, D. L. Pompliano, S. Ferro-Novick, S. Powers, *et al.*, *J. Biol. Chem.*, 1991, **266**, 18 884.
73. C. A. Omer, A. M. Kral, R. E. Diehl, G. C. Prendergast, S. Powers, C. M. Allen, J. B. Gibbs, and N. E. Kohl, *Biochemistry*, 1993, **32**, 5167.
74. F. L. Zhang, R. E. Diehl, N. E. Kohl, J. B. Gibbs, B. Giros, P. J. Casey, and C. A. Omer, *J. Biol. Chem.*, 1994, **269**, 3175.

75. B. He, P. Chen, S. Y. Chen, K. L. Vancura, S. Michaelis, and S. Powers, *Proc. Natl. Acad. Sci. USA*, 1991, **88**, 11 373.
76. L. E. Goodman, S. R. Judd, C. C. Farnsworth, S. Powers, M. H. Gelb, J. A. Glomset, and F. Tamanoi, *Proc. Natl. Acad. Sci. USA*, 1990, **87**, 9665.
77. Y. Ohya, M. Goebl, L. E. Goodman, B. S. Petersen, J. D. Friesen, F. Tamanoi, and Y. Anraku, *J. Biol. Chem.*, 1991, **266**, 12 356.
78. G. L. James, J. L. Goldstein, and M. S. Brown, *J. Biol. Chem.*, 1995, **270**, 6221.
79. V. K. Kalman, R. A. Erdman, W. A. Maltese, and J. D. Robishaw, *J. Biol. Chem.*, 1995, **270**, 14 835.
80. K. Yokoyama, P. McGeady, and M. H. Gelb, *Biochemistry*, 1995, **34**, 1344.
81. D. L. Pompliano, E. Rands, M. D. Schaber, S. D. Mosser, N. J. Anthony, and J. B. Gibbs, *Biochemistry*, 1992, **31**, 3800.
82. F. L. Zhang, J. F. Moomaw, and P. J. Casey, *J. Biol. Chem.*, 1994, **269**, 23 465.
83. J. M. Dolence, P. B. Cassidy, J. R. Mathis, and C. D. Poulter, *Biochemistry*, 1995, **34**, 16 687.
84. Y. Reiss, M. S. Brown, and J. L. Goldstein, *J. Biol. Chem.*, 1992, **267**, 6403.
85. D. L. Pompliano, M. D. Schaber, S. D. Mosser, C. A. Omer, J. A. Shafer, and J. B. Gibbs, *Biochemistry*, 1993, **32**, 8341.
86. W. J. Chen, J. F. Moomaw, L. Overton, T. A. Kost, and P. J. Casey, *J. Biol. Chem.*, 1993, **268**, 9675.
87. M. C. Seabra, M. S. Brown, C. A. Slaughter, T. C. Sudhof, and J. L. Goldstein, *Cell*, 1992, **70**, 1049.
88. M. C. Seabra, J. L. Goldstein, T. C. Sudhof, and M. S. Brown, *J. Biol. Chem.*, 1992, **267**, 14 497.
89. D. A. Andres, M. C. Seabra, M. S. Brown, S. A. Armstrong, T. E. Smeland, F. P. Cremers, and J. L. Goldstein, *Cell*, 1993, **73**, 1091.
90. S. A. Armstrong, M. C. Seabra, T. C. Sudhof, J. L. Goldstein, and M. S. Brown, *J. Biol. Chem.*, 1993, **268**, 12 221.
91. R. Khosravi-Far, G. J. Clark, K. Abe, A. D. Cox, T. McLain, R. J. Lutz, M. Sinensky, and C. J. Der, *J. Biol. Chem.*, 1992, **267**, 24 363.
92. B. T. Kinsella and W. A. Maltese, *J. Biol. Chem.*, 1992, **267**, 3940.
93. F. P. Cremers, S. A. Armstrong, M. C. Seabra, M. S. Brown, and J. L. Goldstein, *J. Biol. Chem.*, 1994, **269**, 2111.
94. M. C. Seabra, M. S. Brown, and J. L. Goldstein, *Science*, 1993, **259**, 377.
95. R. Li, C. Havel, J. A. Watson, and A. W. Murray, *Nature*, 1993, **366**, 82.
96. Y. Jiang, G. Rossi, and S. Ferro-Novick, *Nature*, 1993, **366**, 84.
97. K. Fujimura, K. Tanaka, A. Nakano, and A. Toh-e, *J. Biol. Chem.*, 1994, **269**, 9205.
98. Y. Jiang and S. Ferro-Novick, *Proc. Natl. Acad. Sci. USA*, 1994, **91**, 4377.
99. L. Gutierrez, A. I. Magee, C. J. Marshall, and J. F. Hancock, *EMBO J.*, 1989, **8**, 1093.
100. J. F. Hancock, K. Cadwallader, and C. J. Marshall, *EMBO J.*, 1991, **10**, 641.
101. M. N. Ashby, D. S. King, and J. Rine, *Proc. Natl. Acad. Sci. USA*, 1992, **89**, 4613.
102. Y.-T. Ma and R. R. Rando, *Proc. Natl. Acad. Sci. USA*, 1992, **89**, 6275.
103. G. F. Jang, K. Yokoyama, and M. H. Gelb, *Biochemistry*, 1993, **32**, 9500.
104. Y.-T. Ma, A. Chaudhuri, and R. R. Rando, *Biochemistry*, 1992, **31**, 11 772.
105. Y.-T. Ma, B. A. Gilbert, and R. R. Rando, *Biochemistry*, 1993, **32**, 2386.
106. Y.-L. Chen, Y.-T. Ma, and R. R. Rando, *Biochemistry*, 1996, **35**, 3227.
107. G.-F. Jang and M. H. Gelb, *Biochemistry*, 1998, in press.
108. D. Perez-Sala, E. W. Tan, F. J. Canada, and R. R. Rando, *Proc. Natl. Acad. Sci. USA*, 1991, **88**, 3043.
109. R. C. Stephenson and S. Clarke, *J. Biol. Chem.*, 1992, **267**, 13 314.
110. R. C. Stephenson and S. Clarke, *J. Biol. Chem.*, 1990, **265**, 16 248.
111. M. H. Pillinger, C. Volker, J. B. Stock, G. Weissmann, and M. R. Philips, *J. Biol. Chem.*, 1994, **269**, 1486.
112. D. Perez-Sala, B. A. Gilbert, E. W. Tan, and R. R. Rando, *Biochem. J.*, 1992, **284**, 835.
113. E. W. Tan, D. Perez-Sala, F. J. Canada, and R. R. Rando, *J. Biol. Chem.*, 1991, **266**, 10 719.
114. C. Volker, P. Lane, C. Kwee, M. Johnson, and J. Stock, *FEBS Lett.*, 1991, **295**, 189.
115. Y. Q. Shi and R. R. Rando, *J. Biol. Chem.*, 1992, **267**, 9547.
116. R. R. Rando, *Biochem. Soc. Trans.*, 1996, **24**, 682.
117. E. W. Tan and R. R. Rando, *Biochemistry*, 1992, **31**, 5572.
118. L. Weismuller and F. Wittinghofer, *Cell. Signalling*, 1994, **6**, 247.
119. F. McCormick, *Curr. Opin. Genet. Dev.*, 1994, **4**, 71.
120. M. S. Marshall, *FASEB J.*, 1995, **9**, 1311.
121. T. Pawson and T. Hunter, *Curr. Opin. Genet. Dev.*, 1994, **4**, 1.
122. S. A. Moodie and A. Wolfman, *Trends Genet.*, 1994, **10**, 44.
123. C. M. H. Newman and A. I. Magee, *Biochim. Biophys. Acta*, 1993, **1155**, 79.
124. H. W. Brock, *Gene*, 1987, **51**, 129.
125. C. D. Reymond, R. H. Gomer, M. C. Mehdy, and R. A. Firtel, *Cell*, 1984, **39**, 141.
126. S. M. Robbins, J. G. Williams, K. A. Jermyn, G. B. Spiegelman, and G. Weeks, *Proc. Natl. Acad. Sci. USA*, 1989, **86**, 938.
127. P. Kozlowski, J. Fronk, and K. Toczko, *Biochim. Biophys. Acta*, 1993, **1173**, 357.
128. P. Kozlowski, Z. Tymowska, and K. Toczko, *Biochim. Biophys. Acta*, 1993, **1174**, 299.
129. J. B. Gibbs, A. Oliff, and N. E. Kohl, *Cell*, 1994, **77**, 175.
130. J. F. Hancock, H. Paterson, and C. J. Marshall, *Cell*, 1990, **63**, 133.
131. P. J. Casey, J. F. Moomaw, F. L. Zhang, Y. B. Higgins, and J. A. Thissen, *Recent Prog. Horm. Res.*, 1994, **49**, 215.
132. W. R. Schafer and J. Rine, *Annu. Rev. Genet.*, 1992, **30**, 209.
133. P. M. Lacal, C. Y. Pennington, and J. C. Lacal, *Oncogene*, 1988, **2**, 533.
134. J. E. Buss, P. A. Solski, J. P. Schaeffer, M. J. MacDonald, and C. J. Der, *Science*, 1989, **243**, 1600.
135. A. D. Cox, M. M. Hisaka, J. E. Buss, and C. J. Der, *Mol. Cell. Biol.*, 1992, **12**, 2606.
136. S. J. Leevers, H. F. Paterson, and C. J. Marshall, *Nature*, 1994, **369**, 411.
137. D. Stokoe, S. G. Macdonald, K. Cadwallader, M. Symons, and J. F. Hancock, *Science*, 1994, **264**, 1463.
138. T. Itoh, K. Kaibuchi, T. Masuda, T. Yamamoto, Y. Matsuura, A. Maeda, K. Shimizu, and Y. Takai, *J. Biol. Chem.*, 1993, **268**, 3025.
139. P. McGeady, S. Kuroda, K. Shimizu, Y. Takai, and M. H. Gelb, *J. Biol. Chem.*, 1995, **270**, 26 347.

140. E. Porfiri, T. Evans, P. Chardin, and J. F. Hancock, *J. Biol. Chem.*, 1994, **269**, 22 672.
141. Y. Fukada, T. Takao, H. Ohguro, T. Yoshizawa, T. Akino, and Y. Shimonishi, *Nature*, 1990, **346**, 658.
142. H. Ohguro, Y. Fukada, T. Yoshizawa, T. Saito, and T. Akino, *Biochem. Biophys. Res. Commun.*, 1990, **167**, 1235.
143. J. A. Iniguez-Lluhi, M. I. Simon, J. D. Robishaw, and A. G. Gilman, *J. Biol. Chem.*, 1992, **267**, 23 409.
144. D. E. Wildman, H. Tamir, E. Leberer, J. K. Northup, and M. Dennis, *Proc. Natl. Acad. Sci. USA*, 1993, **90**, 794.
145. Y. Fukada, H. Ohguro, T. Saito, T. Yoshizawa, and T. Akino, *J. Biol. Chem.*, 1989, **264**, 5937.
146. Y. Fukada, T. Matsuda, K. Kokame, T. Takao, Y. Shimonishi, T. Akino, and T. Yoshizawa, *J. Biol. Chem.*, 1994, **269**, 5163.
147. O. G. Kisselev, M. V. Ermolaeva, and N. Gautam, *J. Biol. Chem.*, 1994, **269**, 21 399.
148. J. Bigay, E. Faurobert, M. Franco, and M. Chabre, *Biochemistry*, 1994, **33**, 14 081.
149. T. A. Neubert, R. S. Johnson, J. B. Hurley, and K. A. Walsh, *J. Biol. Chem.*, 1992, **267**, 18 274.
150. D. G. Lambright, J. Sondek, A. Bohm, N. P. Skiba, H. E. Hamm, and P. B. Sigler, *Nature*, 1996, **379**, 311.
151. J. F. Hancock, K. Cadwallader, and C. J. Marshall, *EMBO J.*, 1991, **10**, 641.
152. K. Kato, A. D. Cox, M. M. Hisaka, S. M. Graham, J. E. Buss, and C. J. Der, *Proc. Natl. Acad. Sci. USA*, 1992, **89**, 6403.
153. Y. Fukada, T. Matsuda, K. Kokame, T. Takao, Y. Shimonishi, T. Akino, and T. Yoshizawa, *J. Biol. Chem.*, 1994, **269**, 5163.
154. F. Ghomashchi, X. Zhang, L. Liu, and M. H. Gelb, *Biochemistry*, 1995, **34**, 11 910.
155. J. R. Silvius and F. l'Heureux, *Biochemistry*, 1994, **33**, 3014.
156. H. Huzoor-Akbar, W. Wang, R. Kornhauser, C. Volker, and J. B. Stock, *Proc. Natl. Acad. Sci. USA*, 1993, **90**, 868.
157. C. Volker, R. A. Miller, W. R. McCleary, A. Rao, M. Poenie, J. M. Backer, and J. B. Stock, *J. Biol. Chem.*, 1991, **266**, 21 515.
158. M. R. Philips, M. H. Pillinger, R. Staud, C. Volker, M. G. Rosenfeld, G. Weissmann, and J. B. Stock, *Science*, 1993, **259**, 977.
159. J. Ding, D. J. Lu, D. Perez-Sala, Y.-T. Ma, J. F. Maddox, B. A. Gilbert, J. A. Badwey, and R. R. Rando, *J. Biol. Chem.*, 1994, **269**, 16 837.
160. G. M. Bokoch, *Curr. Opinion Cell. Biol.*, 1994, **6**, 212.
161. Y.-T. Ma, Y.-Q. Shi, Y.-H. Lim, S. H. McGrail, J. A. Ware, and R. R. Rando, *Biochemistry*, 1994, **33**, 5414.
162. S. Sapperstein, C. Berkower, and S. Michaelis, *Mol. Cell. Biol.*, 1994, **14**, 1438.
163. K. Kuchler, R. E. Sterne, and J. Thorner, *EMBO J.*, 1989, **8**, 3973.
164. T. Miyakawa, T. Tachikawa, Y. K. Jeong, E. Tsuchiya, and S. Fukui, *Biochem. Biophys. Res. Commun.*, 1987, **143**, 893.
165. C. Hrycyna, S. Sapperstein, S. Clarke, and S. Michaelis, *EMBO J.*, 1991, **10**, 1699.
166. E. Swiezewska, A. Thelin, G. Dallner, B. Andersson, and L. Ernster, *Biochem. Biophys. Res. Commun.*, 1993, **192**, 161.
167. S. K. Randall, M. S. Marshall, and D. N. Crowell, *Plant Cell*, 1993, **5**, 433.
168. J.-K. Zhu, R. A. Bressan, and P. M. Hasegawa, *Proc. Natl. Acad. Sci. USA*, 1993, **90**, 8557.
169. Z. Yang, C. L. Cramer, and J. C. Watson, *Plant Physiol.*, 1993, **101**, 667.
170. M. Matsui, S. Sasamoto, T. Kunieda, N. Nomura, and R. Ishizaki, *Gene*, 1989, **76**, 313.
171. N. Terryn, S. Anuntalabhochai, M. Van Montagu, and D. Inze, *FEBS Lett.*, 1992, **299**, 287.
172. K. Palme, T. Diefenthal, M. Vingron, C. Sander, and J. Schell, *Proc. Natl. Acad. Sci. USA*, 1992, **89**, 787.
173. H. D. Lujan, M. R. Mowatt, G.-Z. Chen, and T. E. Nash, *Molec. Biochem. Parasitol.*, 1995, **72**, 121.
174. G.-Z. Chen and J. L. Bennett, *Molec. Biochem. Parasitol.*, 1993, **59**, 287.
175. N. E. Kohl, S. D. Mosser, S. J. deSolms, E. A. Giuliani, D. L. Pompliano, S. L. Graham, R. L. Smith, E. M. Scolnick, A. Oliff, and J. B. Gibbs, *Science*, 1993, **260**, 1934.
176. K. S. Koblan, N. E. Kohl, C. A. Omer, N. J. Anthony, M. W. Conner, S. J. deSolms, T. M. Williams, S. L. Graham, G. D. Hartman, A. Oliff, and J. B. Gibbs, *Biochem. Soc. Trans.*, 1996, **24**, 688.
177. T. Musha, M. Kawata, and Y. Takai, *J. Biol. Chem.*, 1992, **267**, 9821.

2.14
Ginkgolide Biosynthesis

MATTHIAS SCHWARZ and DUILIO ARIGONI
Eidgenössische Technische Hochschule, Zürich, Switzerland

2.14.1 INTRODUCTION

2.14.1.1 *Ginkgo biloba*: Myth and Object of Biosynthetic Interest

The "maidenhair-tree" *Ginkgo biloba* has exerted a strong fascination upon both artists and scientists.[1,2] Goethe was inspired to compose a love-poem bearing its name, in which he developed profound philosophical thoughts based on the unique bipartite shape of the *Ginkgo* leaves. Charles Darwin gave it the nickname "living fossil," thus pointing out the remarkable fact that *G. biloba*, unlike most of its contemporaries, has subsisted on the earth for more than 150 million years since it first appeared in the Cretaceous Period, defying all major global changes with stoic calm. Surprisingly enough, the "living fossil" seems to be equally well prepared for the challenges of the modern age: a beautiful specimen situated near the center of Hiroshima is reported to have regrown from its charred stump only one year after the catastrophe of 1945. Further endowed with a very high tolerance toward modern day air pollution and an astonishing, yet largely unexplained resistance to almost all serious plant pests, *G. biloba* has become an ornamental tree frequently encountered along the streets of Occidental cities. It was in its original habitat of Asia, however, where man first discovered the therapeutic use of *Ginkgo* extracts more than two millennia before Christ. In the twentieth century, highly standardized *Ginkgo* leaf extracts are employed to treat capillary circulation disorders, often proving to be particularly effective in cases of age-related loss of cerebral performance. This therapeutic effect of the leaf extract has been ascribed to a class of unusual C_{20}-compounds, called the ginkgolides, which have been shown *in vitro* to be potent antagonists of the platelet activating factor (PAF).[3,4] Although first isolated in 1932 by Furukawa,[70] it was not until 1966 that the ginkgolides were assigned structures ((**1**)–(**4**)), based on the results of chemical and NMR-spectroscopic studies by Nakanishi[5] and on the simultaneously disclosed X-ray analysis data by Sakabe.[6] The proposed structures, which have repeatedly been confirmed ever since,[7,8] display a highly oxygenated system of six fused five-membered rings forming a compact cage and the rare structural element of a *t*-butyl group. The four original ginkgolides ((**1**)–(**4**)), as well as a fifth (**5**), discovered much later by Weinges,[9] differ from each other only in their hydroxy substitution pattern at three specific positions R^{1-3}. Following the 1971 publication of Nakanishi and Habaguchi[10] describing incorporation experiments using radioactive precursors (see Section 2.14.1.2), there were no further reports of biosynthetic studies dealing with the gingkolides prior to 1991, when the authors' studies, described in Sections 2.14.2 and 2.14.3, were initiated.[11] The success of these latter studies depended upon the development of a biological system that allowed the use of precurosrs labeled with stable isotopes (see Section 2.14.1.3). Moreover, the *Ginkgo* extracts yielding the ginkgolides were found to contain in addition phytosterols, such as sitosterol (**6**), which are formed by well-known biosynthetic pathways elucidated in numerous other biological systems. Sitosterol could thus be used as an internal standard monitoring the metabolic fate of the precursors administered to *G. biloba*.

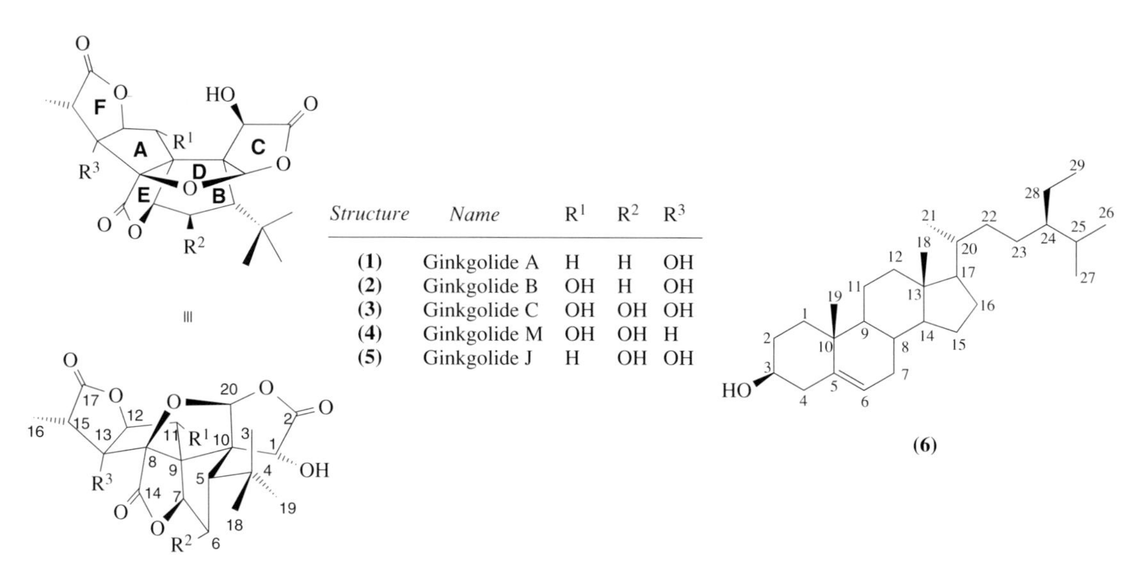

Structure	*Name*	R^1	R^2	R^3
(**1**)	Ginkgolide A	H	H	OH
(**2**)	Ginkgolide B	OH	H	OH
(**3**)	Ginkgolide C	OH	OH	OH
(**4**)	Ginkgolide M	OH	OH	H
(**5**)	Ginkgolide J	H	OH	OH

(**6**)

2.14.1.2 Previous Work on Ginkgolide Biosynthesis

In the first and only preliminary study on ginkgolide biosynthesis recorded in the literature, Nakanishi and Habaguchi described the results of a series of incorporation experiments with radioactive precursors.[10] Using the cotton-wick method[12] these investigators administered samples of [2-^{14}C]acetate (**7**), [2-^{14}C]mevalonate (**8**), and [Me-^{14}C]-L-methionine (**9**) to five month old *Ginkgo* plants. After suitable incubation periods, the plants were harvested and extracted, yielding radioactively labeled specimens of ginkgolide B (**2**), with total incorporation rates ranging from 0.0027% to 0.0048% (Scheme 1). The resulting labeled ginkgolide samples were separately submitted to Kuhn–Roth-degradation yielding, in each case, acetic acid (**10**), expected to correspond to C-15 and C-16 of (**2**), and pivalic acid (**11**), corresponding to C-3, C-4, C-5, C-18, and C-19. The degradation products (**10**)-I and (**11**)-I obtained from the ginkgolide sample through incorporation of (**7**) were further broken down by Schmidt-degradation to give $BaCO_3$ (**14**)-I and the corresponding noramines (**12**)-I and (**13**)-I, respectively, that were analyzed as their 2,4-dinitrophenyl benzoates. Based upon the distribution of the label within the degradation products, which is summarized in Scheme 1, Nakanishi and Habaguchi proposed the biogenetic hypothesis shown in Scheme 2, which posits the ginkgolides as diterpenes derived from tricyclic hydrocarbon intermediates of the pimarane and abietane type. Thus these authors invoked the *ent*-pimarenone cation (**15**) as an intermediate, that would be transformed into a spirocyclic triacid (**18**) by three major skeletal transformations, involving (i) a double contraction of rings B and C to form the central spiro[4.4]nonane moiety, (ii) a 1,2-migration of the C-13 methyl group to C-15, and (iii) a cleavage of ring A between C-3 and C-4 to generate an isopropylidene group at C-5. Subsequent S-adenosyl-methionine (SAM)-dependent methylation of the exocyclic double bond, followed (or preceded) by loss of C-3, would then produce an intermediate (**19**), with the characteristic ginkgolide carbon skeleton, that would require only a few additional oxidation steps to be converted into (**2**). This Scheme could explain the incorporation of radioactivity from [Me-^{14}C]-L-methionine (**9**) into the *t*-butyl group of (**2**)-III as well as the distribution of the radioactivity from [2-^{14}C]acetate (**7**) within the *t*-butyl and the secondary methyl group of (**2**)-I, as established by the results of the Schmidt-degradation of (**10**)-I and (**11**)-I. The observed 1.2:1-ratio between the molar specific activities of (**12**)-I and (**14**)-Ia, pointing to an even distribution of the label between C-15 and C-16 of (**2**)-I, further implies that the secondary methyl group of the ginkgolides must be identical with the methyl group that undergoes the above-mentioned 1,2-migration, as shown in Scheme 3. This, in turn, might suggest that the methyl migration would be initiated by epoxidation rather than by protonation of the exocyclic double bond of (**16**). For lack of other specific degradation reactions involving the ginkgolide carbon skeleton, Nakanishi and Habaguchi were unable to verify and further refine their biogenetic hypothesis. Any subsequent study on ginkgolide biosynthesis, however, would have to confront the challenge either of developing a new set of specific degradation reactions or of finding a biological system that would allow the use of precursors labeled with stable isotopes.

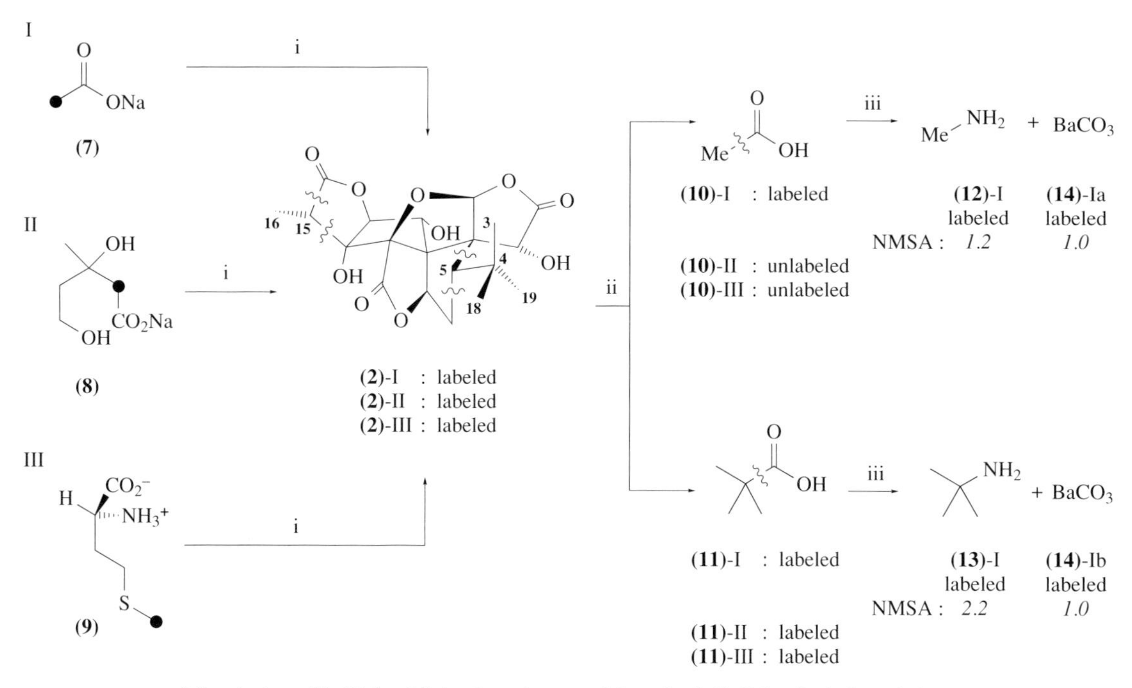

i, Incubation with *Ginkgo biloba* plants (cotton-wick method); ii, Kuhn–Roth-degradation; iii, Schmidt-degradation; NMSA = normalized molar specific activities

Scheme 1

2.14.1.3 The Biological System

Biosynthetic studies with plant organisms traditionally battle against the low ($\ll 1\%$) specific incorporation rates that are obtained when the putative biosynthetic precursors are administered to whole plants or plant parts, either by feeding them from solutions through the intact or chopped root system, by directly injecting them into the stem, or by spraying them onto leaves.[13] The reasons for the low specific incorporation (i.e., the high internal dilution) of the exogenous precursors in intact plant systems, as opposed, for example, to bacterial or fungal cultures, lie on one hand in the large quantities of endogenous, unlabeled material that are inherently present at the beginning of

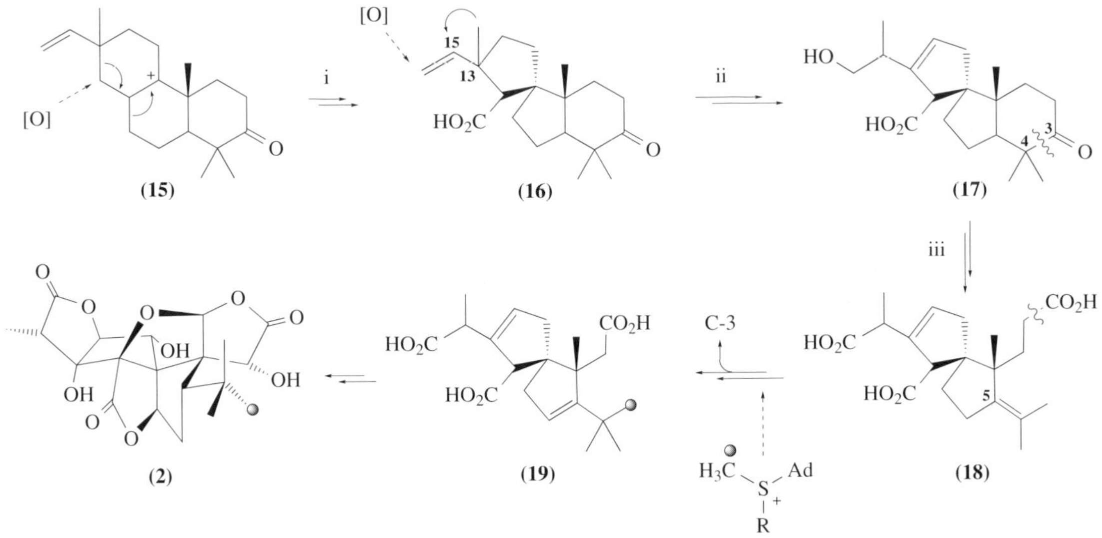

Scheme 2

Scheme 3

the incubation period, and on the other hand in the plant's unique ability to utilize CO_2 from the air as the sole carbon source for its biosynthetic activity. The low specific incorporation rates usually observed with plants therefore present a severe logistical barrier to successful stable isotope incorporation experiments, since isotopic enrichments of at least 0.5 atom% ^{13}C above natural abundance is generally required for successful NMR analysis of the labeled biosynthetic products. In order to overcome this problem, scientists have often resorted to the use of plant cell cultures, despite the difficulties that can be associated with raising stable plant cell lines. Although successful in numerous cases reported in the literature,[14] the latter strategy is hampered by the fact that many plant cells, including those of *G. biloba*, unfortunately lose most of their ability to synthesize secondary metabolites when propagated in artificial culture.

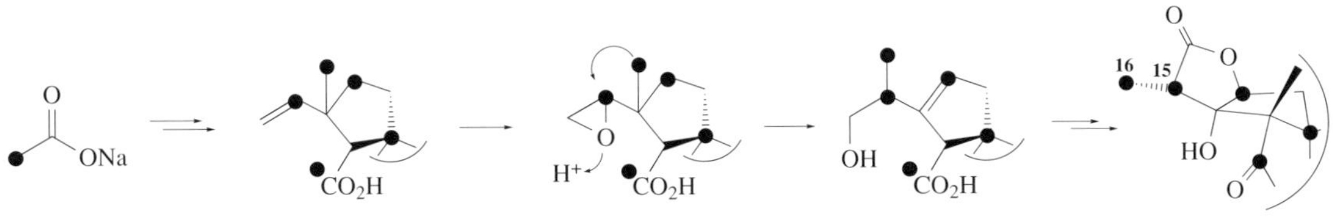

In light of these potential problems, the authors' attention was particularly attracted by a 1959 report in the *American Journal of Botany*.[15] In the course of morphological studies on plant embryo development, Ball had cultivated *Ginkgo* embryos, freshly excised from *Ginkgo* kernels, on sterile agar media containing standardized nutrient salts, different types of sugars in varying concentrations, as well as some optional additives, such as L-glutamine and auxin. Since the embryos were kept in the dark, the sugars in the nutrient medium served as their exclusive carbon source. After incubation times of several weeks, the etiolated seedlings had developed morphologically impeccable roots and, occasionally, even shoots. In the light of the above considerations, the *Ginkgo* embryo system seemed highly promising with regard to biosynthetic experiments, since it could be expected to combine the advantages of cell culturing methods (biomass multiplication during the incubation period, absence of light, standardized, controllable culture conditions) with the strengths of whole-plant cultivation techniques (biosynthetic competence, straightforward system setup).

After successful reproduction of Ball's data using a nutrient medium composed of 1% agar, 2% D-glucose, 0.03% L-glutamine, and Heller's salts, two questions remained to be answered: (i) were the *Ginkgo* embryos capable of synthesizing sufficient quantities of ginkgolides under the conditions chosen, and, if so, (ii) could specific incorporation rates be obtained that would permit the use of stable isotopes. Figure 1 displays a graphical representation of the ginkgolide yields that were obtained from 100 embryos incubated for 15, 30, or 45 days on 1 L of nutrient medium, as compared to 100 ungrown, freshly excised embryos. As can be seen, *Ginkgo* embryos raised under these conditions did indeed produce ginkgolides in sufficient amounts for biosynthetic experiments. More importantly, the bulk of the isolated ginkgolides had been newly formed during the 45-day incubation period, almost exclusively therefore from the carbon sources added to the nutrient medium.

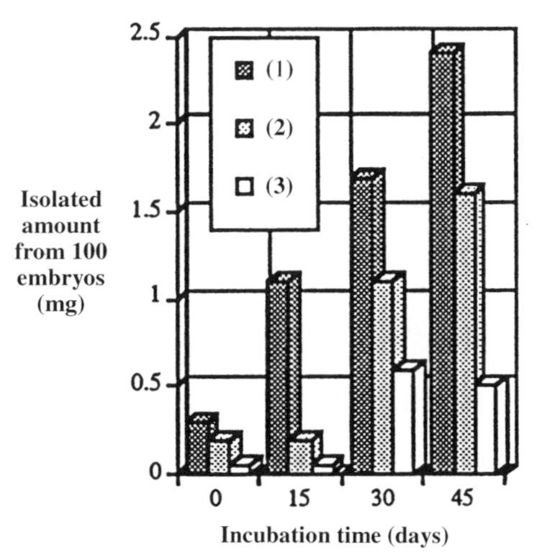

Figure 1 Gingkolide production by embryo cultures of *Gingko biloba*.

Given these preliminary observations, it was gratifying but not totally unexpected when application of [^{14}C]-labeled samples of D-glucose as biosynthetic precursors gave rise to staggering specific incorporation rates of around 50% for both ginkgolides and sitosterol (Table 1). By contrast, with [^{14}C]-labeled ($\pm$)-mevalolactones (MVL), there was a marked discrepancy between the specific incorporation rates detected for the ginkgolides and for sitosterol, the significance of which is discussed in Section 2.14.2. At this stage, it is sufficient to note that, in both cases, the observed levels of incorporation were still high enough to enable MVLs labeled with stable isotopes to be used as biosynthetic precursors. With a biological system in hand that paved the way for biosynthetic experiments based on the use of stable isotopes and thus on NMR-spectroscopic analysis of the derived labeled metabolites, it became crucial to have a rigorous assignment of the ^{1}H- and ^{13}C-NMR spectra of the two major ginkgolides, (**1**) and (**2**), as well as of sitosterol (**6**). As for the ginkgolides, literature data, based primarily on spectra obtained in DMSO,[7] were rigorously supplemented by the authors' studies relying on ^{1}H-, ^{13}C-, DEPT-, NOE-, HSQC-, and HMBC-experiments in acetone, methanol, and pyridine.[11] In the case of sitosterol, the literature assignments[16,17] were verified independently by DEPT-, DQFCOSY-, and HMQC-experiments.[11] Thus equipped with an appropriate biological system and the necessary analytical tools, the first experiment could be tackled—the incorporation of [U-^{13}C]-D-glucose.

Table 1 Average specific incorporation rates observed for different radioactive potential biosynthetic precursors of ginkgolides (**1**), (**2**), and (**3**), and of sitosterol (**6**).

Precursor [*% (w/w) of glucose*]	*Specific incorporation rates*	
	Ginkgolides	*Sitosterol*[a]
[U-^{14}C]-D-Glucose (100)	50–60%	40–50%
($\pm$)-[2-^{14}C]-MVL (2.5)	0.5%	40%
Na-[2-^{14}C]-acetate (5.0)	0.1%	n.d.
[methyl-^{14}C]-L-methionine ($\leqslant$0.001)	4×10^{-5}%	n.d.

[a]n.d., not determined.

2.14.2 ISOPENTENYL DIPHOSPHATE BIOSYNTHESIS

2.14.2.1 Unexpected Results from Incorporation of Labeled Glucose Samples

2.14.2.1.1 Incorporation of [U-^{13}C]-D-glucose

In the first incorporation experiment, 50 *Ginkgo* embryos were grown on 0.5 L of nutrient medium containing 1 g of a D-glucose sample labeled uniformly—i.e., equally at all six positions—with ^{13}C, [U-^{13}C]-D-glucose (**21a**), which had been blended with 9 parts of unlabeled D-glucose (**21**) prior to being added to the nutrient medium. Since the formalism used throughout this chapter to illustrate labeling patterns is to some extent misleading, yet at the same time crucial for a correct understanding of the complex matter, a brief review using the incorporation of (**21a**) as an instructive example is worthwhile (Scheme 4). Through glycolysis, the *Ginkgo* embryos metabolize the above blend to an acetyl-CoA pool which consists of 90% unlabeled molecules (**22**) and at most 10% of doubly labeled species (**22a**), ignoring any additional dilution by endogeneous material. Three acetate units from this pool are then statistically combined to give mevalonate and, with loss of one carbon atom, finally isopentenyl diphosphate (IPP) (**23**), possessing the complex labeling distribution shown in Scheme 4. The majority of the IPP-species thus generated, (D), are completely unlabeled and therefore irrelevant for the purpose of the experiment. The same applies to species (A), which is fully labeled, but not abundant enough to be of spectroscopic importance. Experience further shows that species (B1–B3), in which two out of three acetate units are labeled at a time, usually remain hidden as well, occurring as they do at least eight times less frequently than the major labeled species, (C1–C3), each with one labeled acetate unit per molecule of IPP. In biosynthetic schemes, which usually neglect the presence of species (A), (B1–B3) and (D) for the sake of lucidity, formulas (C1–C3) can be further compressed to a convenient, but potentially misleading shorthand form (**23a**). There is a fundamental difference in meaning between the shorthand formula (**23a**) and the species formula (A), even though in both cases all five carbon atoms are graphically highlighted. (A) implies that all three acetate units, that is all five carbons, are labeled at the same time and in the same molecule, whereas (**23a**) means that either the first, the second, or the third acetate-derived unit is labeled. By condensing four such IPP-molecules (**23a**) to form the aliphatic diterpene precursor geranylgeranyl diphosphate (GGPP) and by applying Nakanishi's biogenetic hypothesis[10] from there, one could predict the labeling pattern of the ginkgolides produced from [U-^{13}C]-D-glucose to be that designated as (**X**) in Figure 2.

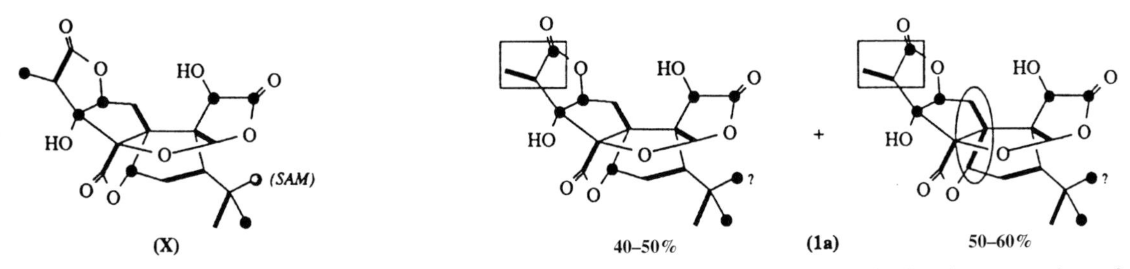

Figure 2 Expected (**X**) and observed labeling patterns for gingkolide A (**1a**) and after incorporation of [U-^{13}C]glucose.

In fact, careful analysis of the ^{13}C- and INADEQUATE-spectra of two ginkgolide samples, (**1a**) and (**2a**), isolated from the incorporation experiment, revealed a labeling pattern that was basically congruent with (**X**), while diverging in two important respects. The first, concerning the framed moiety in lactone ring F, will be discussed in detail in Section 2.14.3.1. As for the second, confined to the circled portion of the structure around the spiro-center, the spectroscopic data left no doubt that in at least half of the labeled molecules the three carbons C-7, C-9, and C-11 simultaneously carried a ^{13}C-label, thus forming a triad, i.e., a contiguous set of three ^{13}C-atoms. By contrast, the expected and the observed labeling patterns of sitosterol (**6a**) were shown to be identical, thus proving that the precursor (**21a**) had been metabolized correctly according to Scheme 4. The question therefore arose as to why the anomalous couplings were restricted only to the ginkgolides, or more precisely, to only three out of all the carbon atoms of the ginkgolide molecule. Close inspection of the critical substructure, which corresponds to the third IPP-unit of GGPP, revealed in fact a unique feature distinguishing it from the remaining three IPP units (Scheme 5).

The biogenetic hypothesis, as well as simple connectivity considerations, require that in the course of the transition from the putative tricyclic intermediates to the ginkgolides, e.g., (**15**)→(**1**), a skeletal

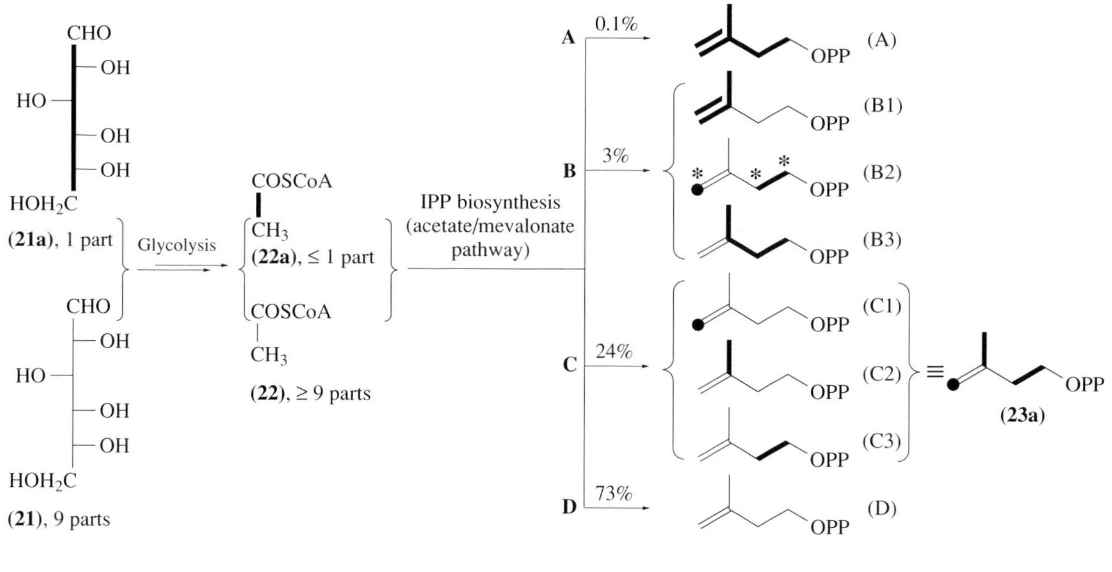

Scheme 4

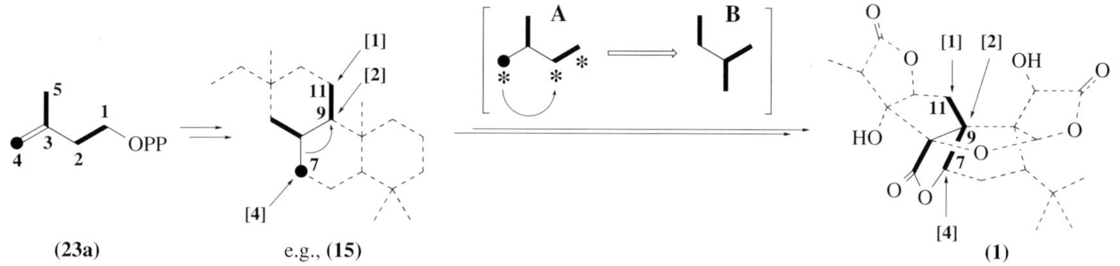

Scheme 5

rearrangement involving the third IPP unit take place, so as to create a new bond between the previously separated carbons C-7 and C-9. Taken to the level of the isopentane-skeleton, this rearrangement corresponds to the transition **A**→**B**, linking together C-4 and C-2 of IPP. Obviously, the *Ginkgo* embryos grown on tenfold diluted [U-^{13}C]-D-glucose had synthesized IPP-molecules carrying a triple ^{13}C-label at C-1, C-2, and C-4 in at least half of all species labeled at either one of these positions. Notably, if it had not been for the above-mentioned rearrangement, this multiple label would have remained undetected, since the three affected carbon atoms are normally detached from each other. (In principle, one should be able to detect β-couplings between C-2 and C-4, but since the corresponding coupling constants (typically 0–5 Hz) are normally in the range of the observed line-widths of the ^{13}C NMR signals, they frequently remain spectroscopically invisible.) From Scheme 4, it can be seen that IPP-molecules bearing a triple label of this type are indeed formed from [U-^{13}C]-D-glucose by the acetate-mevalonate-pathway (species B2), yet not in an abundance that would allow them to compete with the major species (C1) or (C3). Furthermore, no evidence had been found in the NMR spectra of (**1a**) and (**2a**) indicating the presence of the two concomitant B-subspecies (B1) and (B3). Interestingly, a very similar observation had been reported in 1981 in connection with studies on the biosynthesis of the sesquiterpene pentalenolactone (**26**) in *Streptomyces* UC5319 (Scheme 6).[18]

Incorporation of [U-^{13}C]-D-glucose in diluted form had afforded a pentalenolactone sample carrying, apart from the expected label, an unusual triple ^{13}C-label, which was also confined to a portion of the structure known to be formed by a rearrangement linking C-2 with C-4 derived from a common IPP-unit. Taken together, the above results strongly suggested that both *G. biloba* and *Streptomyces* UC5319 are equipped with an enzymatic machinery capable of synthesizing IPP by a pathway different than the classical acetate-mevalonate route. As a characteristic feature, the new pathway must involve a step in which a two-carbon unit **B** is formally inserted into a three-carbon unit **A**, as shown in Scheme 7. In order to verify and to further corroborate this hypothesis, four additional incorporation experiments with different singly-labeled glucose samples were undertaken.

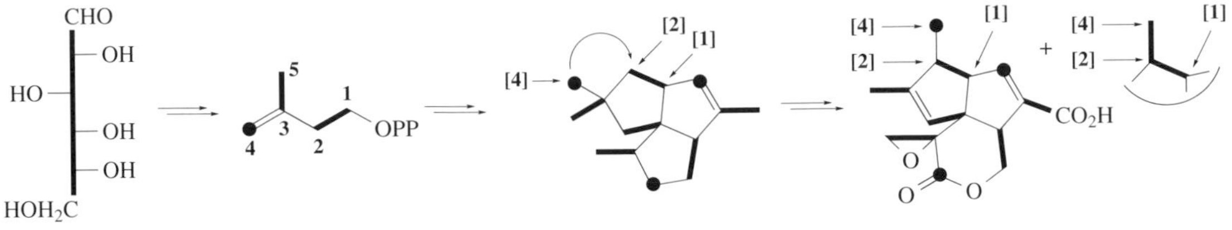

(**21a**), diluted (**23a**) (**25**) (**26**)

Scheme 6

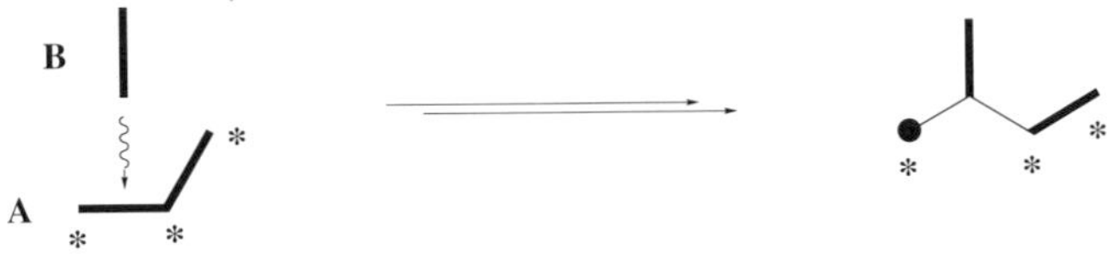

Scheme 7

2.14.2.1.2 Incorporation of [1-^{13}C]-, [2-^{13}C]-, [3-^{13}C]-, and [6-^{13}C]-D-glucose

Using the same conditions as for the first experiment, *Ginkgo* embryos were incubated with fivefold diluted samples of [1-^{13}C]-, [2-^{13}C]-, [3-^{13}C]-, and [6-^{13}C]-D-glucose, providing, after a 45-day incubation period, labeled samples of (**1**), (**2**), and (**6**), which were analyzed by ^{13}C-NMR spectroscopy. In order to identify the labeled carbon atoms, the signal intensities found in the ^{13}C NMR spectra were compared to the respective standard signal intensities as determined from spectra of the corresponding unlabeled compounds. For the sake of lucidity, the resulting labeling patterns of the ginkgolides and of sitosterol are summarized in Table 2 as reduced to the level of the individual IPP-units and compared to the expected labeling distribution based on the acetate-mevalonate-pathway. Two conclusions could be drawn from the comparison of the labeling patterns depicted in Table 2. First, the hypothesis that had emerged after incorporation of [U-^{13}C]glucose could indeed be confirmed, in that the labeling patterns detected for the IPP-units of the ginkgolides were clearly different from those generated by the operation of the classical acetate–mevalonate pathway, thus showing that the IPP-units of the ginkgolides are formed via a new, distinct pathway. Section 2.14.2.2 will summarize the currently available information on the new pathway for IPP-biosynthesis in *G. biloba* and other organisms. Even more exciting, however, was the fact that the label distributions within the IPP-units of the ginkgolides were clearly different from those characterizing the IPP units of sitosterol. *G. biloba* must thus be equipped with two distinct and well-segregated enzymatic machineries for the biosynthesis of the key intermediate of terpene biosynthesis, IPP. Based upon the labeling patterns summarized in Table 2 and the corresponding biogenetic correlation between the carbon atoms of IPP and glucose, shown in Scheme 8, it can be concluded that the IPP-units of sitosterol are formed by the operation of the classical acetate-mevalonate pathway (**I**), whereas the IPP-units building up the ginkgolides are assembled via a so-far undefined new pathway (**II**). Section 2.14.2.3 will cast additional light on the background and on the exciting implications of this discovery.

2.14.2.2 The Deoxyxylulose (Triose–Phosphate/Pyruvate) Pathway

A striking feature of the biogenetic correlations between glucose and IPP carbon atoms shown in Scheme 8 is the symmetry between the upper and the lower half of the glucose molecule, (1/6, 2/5, 3/4), in terms of the labeling patterns produced on the level of the derived IPP-units (**23a-I**) and

Table 2 Expected and observed labelling patterns of the IPP-units of ginkgolides A, B, and sitosterol after incorporation of different point-labelled glucose samples (○ 1/4 ●).

Precursor	*Expected labelling pattern based on the operation of the acetate–mevalonate pathway*	*Observed labelling patterns*	
		Ginkgolides (**1**) *and* (**2**)	*Sitosterol* (**6**)
[1-^{13}C]-D-Glucose	 OPP	 OPP	n.d.
[2-^{13}C]-D-Glucose	 OPP	 OPP	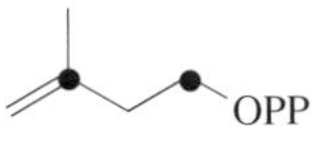 OPP
[3-^{13}C]-D-Glucose[a]	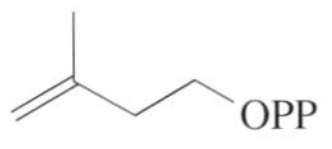 OPP	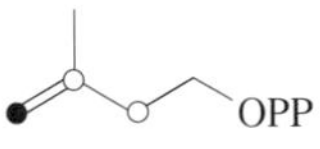 OPP	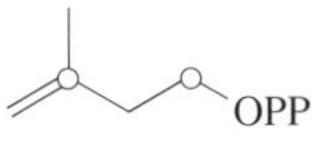 OPP
[6-^{13}C]-D-Glucose	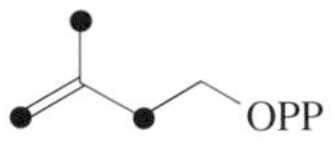 OPP	 OPP	 OPP

[a]A weak additional label was detected after incorporation of [3-^{13}C]-D-glucose within the IPP units of both ginkgolides and sitosterol, which, in both cases, corresponded to the label produced by the incorporation of [2-^{13}C]-D-glucose, thus indicating that the label had been transferred before the IPP biosynthesis had set in. It is not immediately evident to what process this label transfer could be attributed.

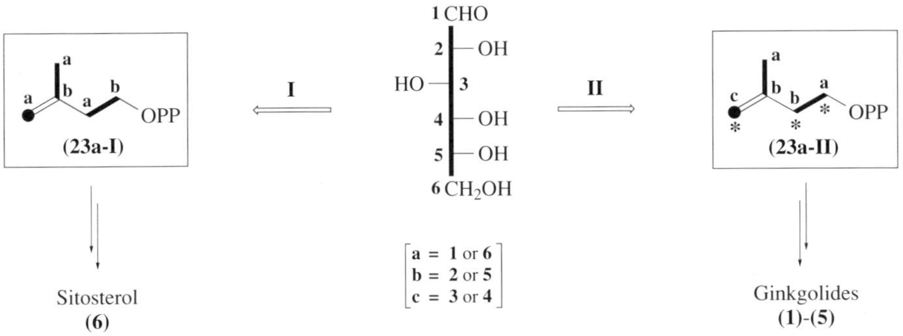

1 CHO
2 OH
HO 3
4 OH
5 OH
6 CH_2OH
I
II
a b a b OPP
(23a-I)
c b b a * OPP
(23a-II)
Sitosterol (6)
Ginkgolides (1)-(5)
a = 1 or 6
b = 2 or 5
c = 3 or 4

Scheme 8

(**23a-II**) (a, b, c). In the case of the acetate–mevalonate pathway (**I**), this symmetry is known to reflect the fact that during glycolysis the hexose skeleton is split into two trioses, dihydroxyacetone phosphate (DHAP) (**27**) and glyceraldehyde-3-phosphate (GAP) (**28**), or specifically (**27a**) and (**28a**) after incorporation of (**21a**) (Scheme 9).

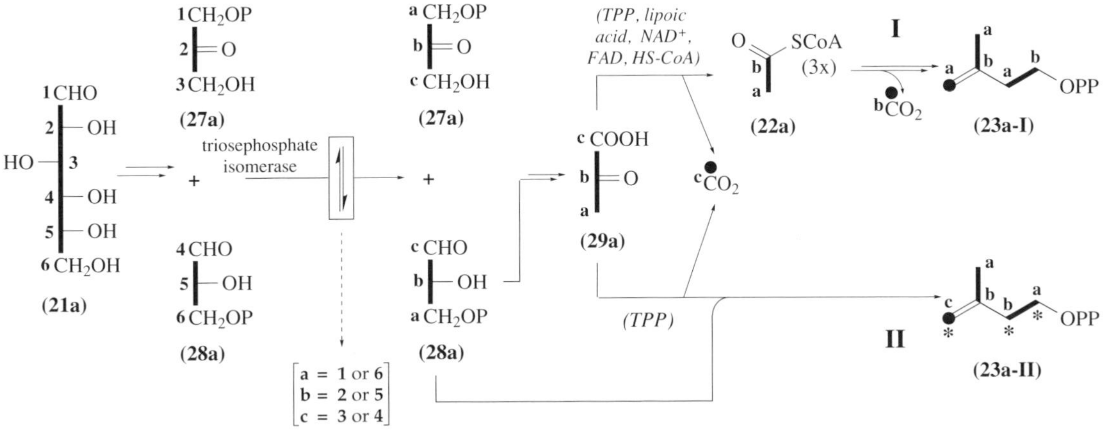

(21a)
(27a)
triosephosphate isomerase
(28a)
(29a)
(TPP, lipoic acid, NAD^+, FAD, HS-CoA)
O SCoA (3x)
(22a)
$^{b}CO_2$
$^{c}CO_2$
(TPP)
(23a-I)
(23a-II)
a = 1 or 6
b = 2 or 5
c = 3 or 4

Scheme 9

The trioses are interconverted by a further enzyme, triosephosphate isomerase (TIM), leading to a complete equilibration between the former glucose carbons 1/6 (a), 2/5 (b), and 3/4 (c) (see Scheme 9). Subsequently, (**28a**) is transformed into pyruvate (**29a**), and, with concomitant loss of carbon (c) as CO_2, into acetyl-CoA (**22a**). Three acetyl-CoA-units are required to build up IPP showing the characteristic biogenetic correlation pattern for the acetate–mevalonate pathway (**23a-I**). As mentioned above, the same symmetry between the two halves of the glucose molecule was also observed for the IPP molecules (**23a-II**), produced by the new pathway (**II**), yet with the difference that one (c)-carbon was conserved, as part of the starred C_3-unit designated as **A** in Scheme 7. This, in turn, raised the idea that this C_3-unit **A** could in fact be identified as a triose, such as (**27a**) or (**28a**). As for the C_2-unit **B**, which is inserted into the triose-unit according to Scheme 7, it displayed the typical labeling pattern, (a)–(b), of an acetyl-CoA-unit (**22a**), suggesting that it could also be derived from thiamine pyrophosphate (TPP)-catalyzed decarboxylation of pyruvate. However, as shown in Scheme 10, in contrast to the formation of (**22a**), the initial decarboxylation product, a formal equivalent of an acetyl anion, would not in this case be transferred to an oxidant (lipoic acid) but rather to the carbonyl group of the triose-unit (**27a**) or (**28a**), thus forming α-hydroxy-ketone (**30a**) or (**32a**). Both (**30a**) and (**32a**) are prone to undergo a ketol rearrangement affording the isomeric α-hydroxy-carbonyl compounds (**31a**) and (**33a**), respectively. Indeed, the first two steps of the biosynthesis of valine and leucine from two pyruvate units constitute a neat precedent for a TPP-catalyzed, decarboxylative addition of pyruvate onto a carbonyl function, followed by a ketol-rearrangement of the resulting α-hydroxy-ketone. It is this very ketol rearrangement that accomplishes the repeatedly mentioned insertion of the C_2- into the C_3-unit, thereby generating the characteristic isopentane skeleton of IPP. From (**31a**) or (**33a**), only a few reduction steps are required to afford finally IPP (**23a-II**) with the correct biogenetic correlation pattern. Since both routes shown in Scheme 10 produce the same correlation pattern relating the carbon atoms of glucose to those of IPP, (**23a-II**), it is intrinsically impossible to decide, based upon glucose incorporation experiments, whether the C_2-unit derived from pyruvate is added to C-2 of DHAP (**27**) or to C-1 of GAP (**28**).

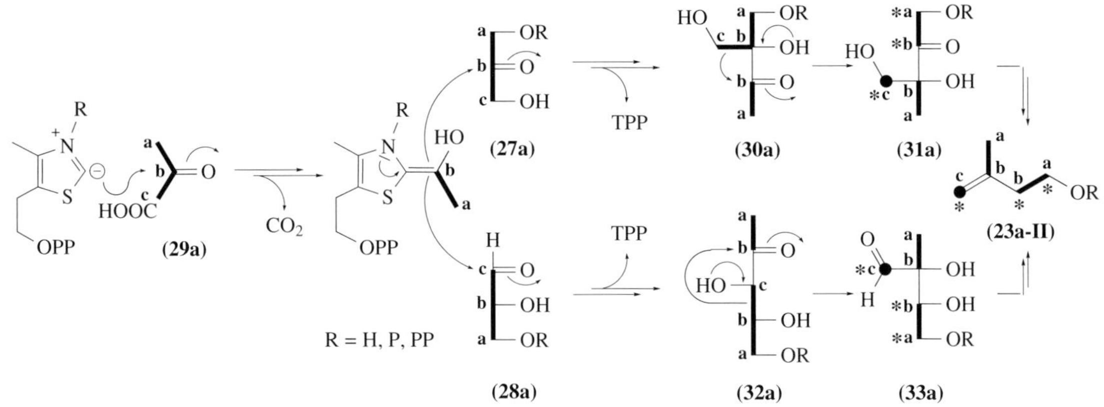

Scheme 10

In order to answer this question and, at the same time, to verify the mechanistic hypothesis developed so far (see Schemes 7, 9, and 10), it was necessary to synthesize one of the postulated intermediates in labeled form and to test it for incorporation into the IPP-units of the ginkgolides. This goal was eventually accomplished by Broers,[19] who was investigating in parallel studies the biosynthesis of the IPP units of ubiquinone and vitamin K in *Escherichia coli*. Since *E. coli* cultures offer many advantages over the *Ginkgo* embryos as a biological system, including at least 50-fold shorter incubation times and a less tedious culture setup, it was decided that the putative intermediates would be administered first to *E. coli*. Having already shown by preliminary experiments that *E. coli* also uses the newly detected pathway for IPP biosynthesis, Broers synthesized two deuterated samples of 1-deoxy-D-xylulose (**32b**) and (**32c**), the nonphosphorylated form of the postulated product of the TPP-catalyzed addition of pyruvate (**29**) onto C-1 of GAP (**28**). Interestingly, 1-deoxy-D-xylulose was by no means an unknown compound in *E. coli*, in which it had been shown to act as the precursor of both pyridoxine[20] (vitamin B_6) and of the thiazole moiety of thiamine[21] (vitamin B_1). Broers then administered (**32b**) and (**32c**) to *E. coli* cultures (see Scheme 11). Analysis of the ^2H NMR and mass spectra of the isolated ubiquinone samples showed that the labels from (**32b**) and (**32c**) had in fact been incorporated specifically into positions C-5 and C-1, respectively, of the resulting IPP units, with overwhelming specific incorporation rates of more than 90%, i.e., almost without internal dilution. Shortly thereafter, Cartayrade[22] was able to demonstrate, using the *Ginkgo*-embryo system, that both (**32b**) and (**32c**) were also specifically incorporated into the IPP units of ginkgolide A, with less dramatic, yet significant, specific incorporation rates of 1–2%. As further indicated by Scheme 11, the new pathway for IPP biosynthesis, which will be referred to as the deoxyxylulose pathway in the following text, seems to be widespread among isoprenoid biosynthesizing organisms. Cartayrade's further results[20] showing that the sage species *Salvia miltiorrhiza* is also equipped with the enzymatic machinery of the deoxyxylulose pathway will play an important role for the considerations presented in the following section. Furthermore, in cooperation with Cane's group the authors were able to show that IPP biosynthesis in the pentalenolactone-producing organism *Streptomyces*-UC5319 (see Scheme 6) relies in fact on the deoxyxylulose pathway as well.[23]

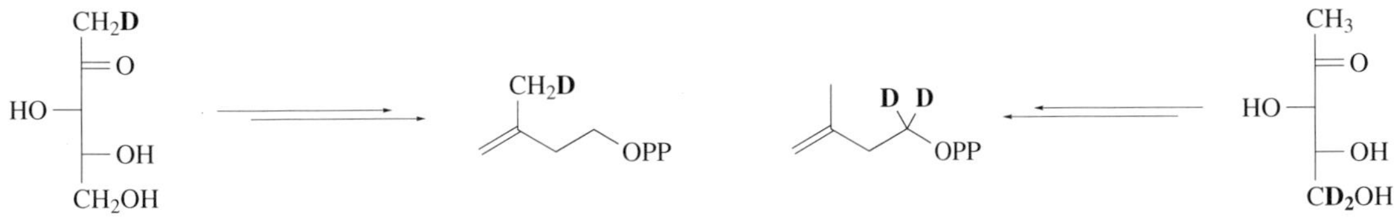

(**32b**) (**23b**) (**23c**) (**32c**)

Scheme 11

In the meantime, extensive independent investigations by Rohmer and his collaborators[24,25] on the early steps of isoprenoid biosynthesis in numerous microorganisms, including *E. coli*, *Alicyclobacillus acidoterrestris*, *Methylbacterium fujisawaense*, or *Zymomonas mobilis*,[26,27] have thoroughly documented the widespread occurrence and provided many crucial details of the triose phosphate–pyruvate/deoxyxylulose pathway (see Chapter 2.03). Moreover, the gene encoding deoxyxylulose phosphate synthase has now been cloned and expressed from both *E. coli* and peppermint (*Mentha x piperita*), and the corresponding gene has been identified in several other species.[28–30]

With regard to *Ginkgo*, the anomalous labeling patterns found in ginkgolide A (**1**) and B (**2**) after incorporation of different glucose samples could now be rationalized by the operation of the deoxyxylulose pathway. However, further questions remained to be answered. Where are the two pathways for IPP biosynthesis located and how is their strict segregation structurally realized? How can the slight—but significant—incorporation of the label from mevalolactone into the ginkgolides be explained, if the former is not an intermediate of the deoxyxylulose pathway?

2.14.2.3 Coexistence of Two Different Pathways for IPP Biosynthesis

Summarizing the results presented in the two previous Sections 2.14.2.1 and 2.14.2.2, it can be said that isoprenoid biosynthesis in *G. biloba* is characterized by a mechanistic and structural dichotomy whereby the IPP units of sitosterol (**6**) are assembled via the classical acetate pathway, whereas the IPP units of the ginkgolides (**1**)–(**5**) are formed by the deoxyxylulose pathway. This dichotomy, in turn, raises the question as to where the corresponding enzymatic machineries are located and how the strict segregation between them is structurally attained. Paradoxically, as will be shown in this section, these questions find their answers in an old controversy, which has divided plant physiologists since the 1970s. The latter emanates from the results of the classical studies by Goodwin and Mercer[31] who administered to corn sprouts radioactive samples of either mevalonate or CO_2 (as bicarbonate). To their surprise, they noticed that the label from mevalonate had exclusively been incorporated into sterols, whereas the activity related to CO_2 was primarily found in carotenes. In the following years, these original results were confirmed and further refined by numerous studies using different plant systems, as well as isolated organelles such as mitochondria and chloroplasts.[32–35] A synopsis of all results would state that plant cells are characterized by two distinguishable compartments in which isoprenoid biosynthesis can take place—the cytoplasm, which is responsible for sesquiterpene and sterol biosynthesis, and the organelles, which are in charge of monoterpene, diterpene, and carotene formation (plastids), and of ubiquinone biosynthesis (mitochondria). As opposed to cytosolic isoprenoid biosynthesis, the plastidial and mitochondrial counterpart does not accept mevalonate or acetate as precursors, whereas CO_2, glucose, or triose phosphates are well incorporated.[36,37]

The controversy was fueled by contradictory sets of experimental results culminating in the establishment of two contrary organisational models of plant isoprenoid biosynthesis, (a) and (b), shown in Figure 3.[33] The recurring motif of the first type of studies was the finding that neither isolated chloroplasts nor mitochondria incorporated mevalonic acid (MVA) or any MVA-phosphates as precursors for their isoprenoids, but they did accept IPP.[38–40] These results still held true in cases where the chloroplast membranes had been made permeable by hypotonic treatment.[41,42] Activity from MVA and MVA-phosphates was only incorporated, if a cytoplasmic fraction was added to the organelle preparations. The conclusions drawn from these results are summarized in Figure 3 (a). According to this model, IPP is exclusively formed *de novo* in the cytoplasm; neither chloroplasts (nor plastids in general) nor mitochondria are provided with the enzymes of the acetate pathway that would enable them to synthesize IPP from acetyl-CoA. Consequently, isoprenoid biosynthesis in the organelles would rely entirely on IPP imported from the cytoplasm.

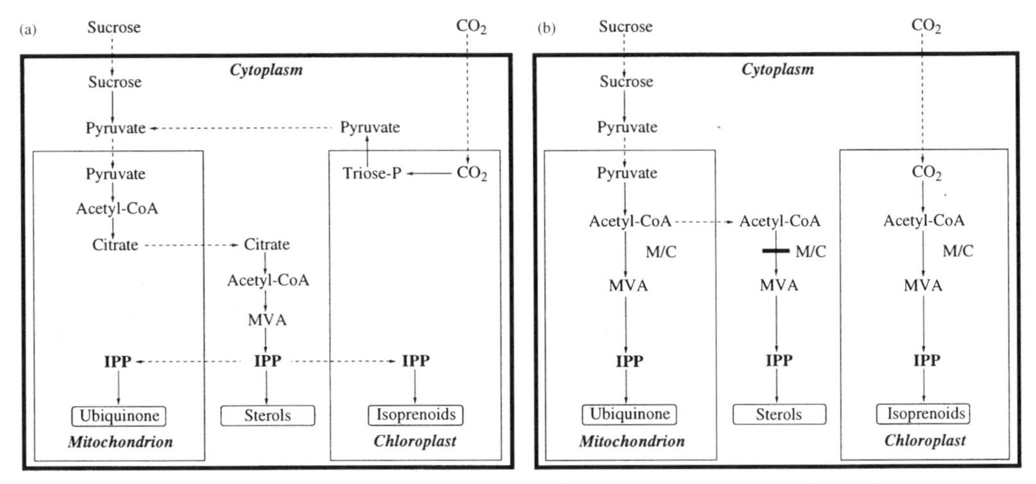

Figure 3 Two opposite organizational models of plant isoprenoid biosynthesis.

The second type of studies, including experiments with inhibitors of the rate-limiting enzyme of the acetate pathway, HMG-CoA reductase, produced results that were strictly incompatible with model (**a**), in showing that administration of these inhibitors (M = mevinolin, C = compactin), typically suppressed cytosolic sterol biosynthesis to a large extent, while leaving plastidial isoprenoid biosynthesis untouched.[43] Since it was difficult to see how plastidial isoprenoid biosynthesis could have remained unaffected by the shutdown of cytosolic IPP-production were it to rely exclusively on the import of IPP from the cytoplasm, a second organisational model (**b**) was put forward. According to model (**b**), each subcellular compartment capable of isoprenoid biosynthesis is equipped with its own enzymatic machinery for IPP formation based on the acetate pathway, with a *caveat* that the mitochondrial and plastidial HMG-CoA reductases constitute isoforms of the cytosolic enzymes displaying a very low mevinolin and compactin susceptibility. Model (**b**), however, while explaining the results from the inhibitor studies, could not in turn rationalize the differential incorporation of IPP vs. MVA and MVA phosphates into isolated organelles. By matching these literature data against the results obtained with *G. biloba*, all pending questions can be rationalized within a new scheme (Figure 4). As far as *Gingko* is concerned, it can be concluded that the IPP-units of sitosterol (**6**) are formed in the cytoplasm via the acetate–mevalonate pathway, whereas the IPP units of the diterpenoid gingkolides are assembled in specific organelles, most probably the plastids via the deoxyxylulose pathway. More importantly, if the dichotomy observed for IPP biosynthesis in *G. biloba* were generally applicable to higher plant cells, all the contradictory results presented above could now easily be explained. It would be easy to understand why isolated organelles had never accepted MVA or MVA phosphates as precursors, as neither of them is an intermediate of the deoxyxylulose pathway. Furthermore, it would be plain to see why HMG-CoA reductase inhibitors had not affected plastidial isoprenoid formation, since HMG-CoA reductase is not an enzyme of the triose phosphate–pyruvate/deoxyxylulose pathway.

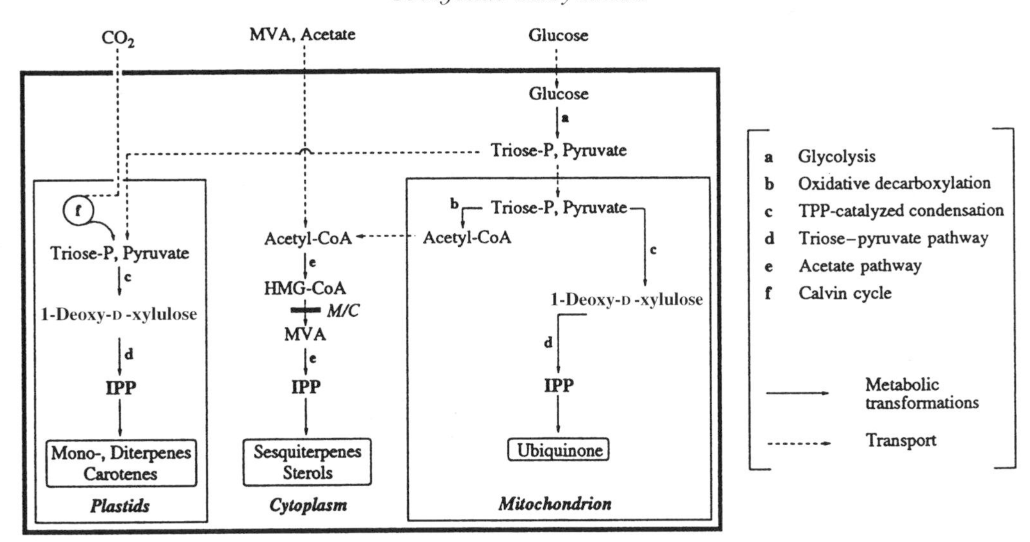

Figure 4 New model for the structural organization of plant isoprenoid biosynthesis.

In light of these considerations, little doubt remained that the new structural organization model of IPP-biosynthesis shown in Figure 4 would prove to be of general scope. Nonetheless, it seemed worthwhile to strive for additional evidence by examining another plant system capable of diterpene and sterol formation and known to allow the use of stable isotopes. Cartayrade[22] chose cell cultures of the sage species *S. miltiorrhiza*, which were known to produce the diterpene ferruginol,[44] and carried out incubations with $[1\text{-}^{13}C]$glucose, as well as with $[1\text{-}^{2}H]$- and $[5,5\text{-}^{2}H_2]$-1-deoxy-D-xylulose (**32b**) and (**32c**). From the labeling patterns revealed by the ferruginol samples as compared to those observed in the corresponding sitosterol samples, it could be concluded that the *Salvia* cells are characterized by exactly the same dichotomy as observed in *Ginkgo*. Thus the IPP units of ferruginol are synthesized via the deoxyxylulose pathway, whereas the IPP units of sitosterol are formed by the acetate pathway. In the meantime, the simultaneous operation of both pathways in higher plants[45] and in a species of *Streptomyces*[46] has been reported.

However, one result obtained with *Ginkgo* still seemed to contradict the organizational scheme presented in Figure 4. The addition of different MVL samples to the nutrient medium had not only labeled sitosterol (**6**), but also ginkgolides such as (**1**), although with a difference in specific incorporation rate of almost two orders of magnitude (40% vs. 0.5%). Since the NMR spectra of the isolated ginkgolide samples showed that the label had been incorporated in a specific manner into distinct and predictable positions, all arguments based on partial degradation of the precursor or on accompanying labeled impurities were invalid. However, as demonstrated by some of the above-mentioned experiments with isolated organelles, a compound immediately derived from mevalonic acid, IPP, is potentially able to cross the membrane between the cytoplasm and the ginkgolide-producing organelle. From the ratio between the specific incorporation rates of MVA into (**6**) and (**1**), it could be inferred that only about 1/80 (1–2%) of the IPP molecules synthesized in the cytoplasm would have to pass the membrane barrier in order to cause the observed labeling patterns resulting from MVA incorporation into the ginkgolides. Two puzzling, yet at the same time very suggestive, characteristics of all ginkgolide samples resulting from mevalolactone incorporations indicated that the proposed migration of IPP from the cytosol to the organelle is correct, but not exhaustive. Thus in all eleven incorporation experiments involving ^{13}C- and/or ^{2}H-labeled mevalolactone specimens, a pronounced asymmetry in the label intensities of the four IPP-units was observed. Typically, the carbon and hydrogen atoms of the fourth IPP-unit of the ginkgolide molecule, i.e., HC-12, C-13, HC-15, H_3C-16, and C-17, displayed only 20%, at best, of the ^{13}C- and ^{2}H-enrichments observed for the carbons and hydrogens corresponding to the first three IPP units. Incorporation of labeled glucose samples, on the other hand, afforded ginkgolides whose four IPP units displayed comparable isotopic enrichments.

Scheme 12 shows a comparison of the labeling patterns observed in ginkgolide A after incorporation of $[3\text{-}^{13}C]$glucose (**21e**) on one hand and of $[2\text{-}^{13}C]$mevalolactone (**24d**) on the other, two precursors which each label C-4 of IPP. In (**1d**), derived from (**24d**), a marked difference in the ^{13}C-enrichments of the fourth as compared to the first three IPP units is observed, whereas in the case of (**1e**), obtained after incorporation of (**21e**), the four IPP units are indistinguishable. In order to explain this asymmetry, it was postulated that not only IPP, but also the IPP-derived farnesyl diphosphate (FPP), which is abundant in the cytoplasm as a precursor of both sesquiterpenes and sterols, had penetrated into the ginkgolide-producing organelle where it had been extended by a fourth IPP unit to give GGPP.

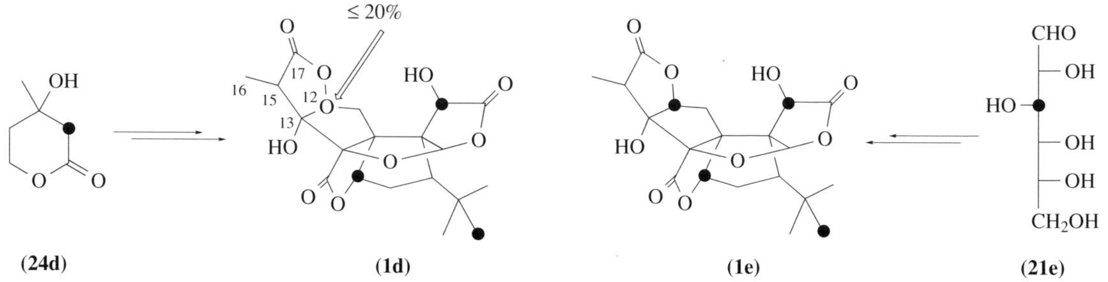

(24d) **(1d)** **(1e)** **(21e)**

Scheme 12

Indeed, studies with both bacteria[47] and plants[48] have established that GGPP synthase will accept IPP, dimethylallyl diphosphate (DMAPP), geranyl diphosphate (GPP), and FPP as substrates and extend them by the requisite number of IPP units to form GGPP. The relative extent of the FPP- vs. the IPP-passage would be reflected by the enrichments of the first three IPP-units (label due to FPP and IPP) vs. the enrichment of the fourth unit (label due to IPP only), and would thus amount to about four FPP molecules for each IPP molecule. The ultimate proof for the validity of this hypothesis was provided by the results of the incorporation of [3,4-$^{13}C_2$]mevalolactone (**24f**) (see Figure 5). The labeling patterns obtained in both sitosterol (**6f**) and ginkgolide A (**1f**) derived from (**24f**) are characterized by the fact that in numerous structural domains, labeled carbon atoms of distinct IPP units can potentially meet. The frequency of such encounters between labeled IPP units, which are spectroscopically detected by additional ^{13}C—^{13}C couplings, can normally be estimated based on the value of the specific incorporation rate. For example, in the case of (**6f**), the probability for a ^{13}C-couple to be adjacent to a second $^{13}C_2$-pair was calculated to be greater than 90%, based upon the coupling patterns observed in the ^{13}C NMR spectrum, which means that (**6f**) had been synthesized from an IPP pool consisting almost exclusively of labeled IPP molecules.

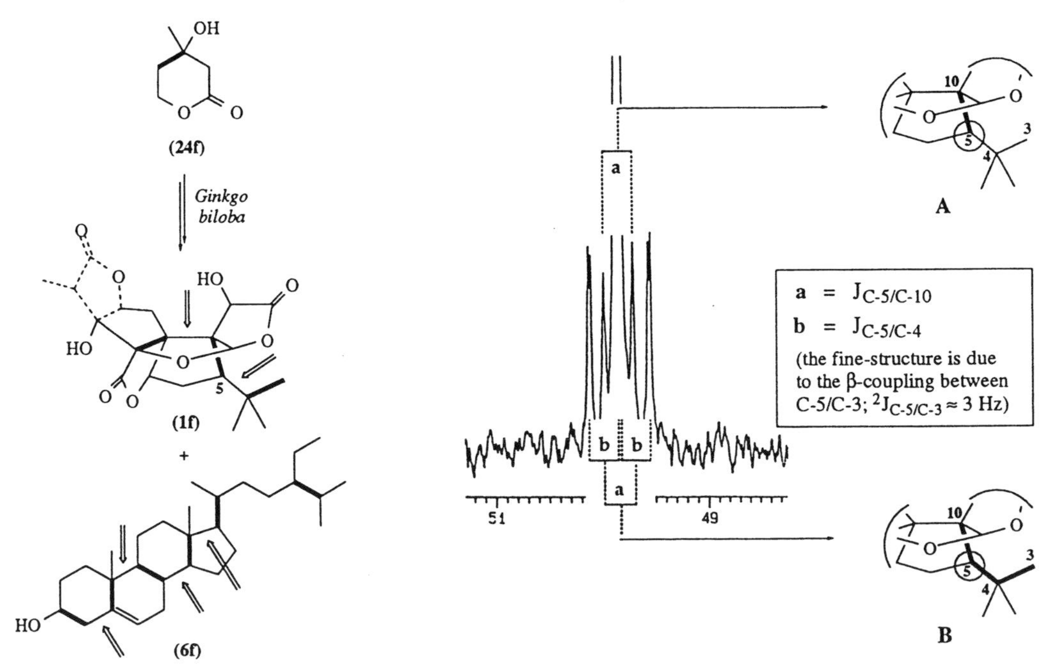

Figure 5 Labeling patterns observed for sitosterol (**6f**) and gingkolide (**1f**) after incorporation of [3,4-$^{13}C_2$]mevalolactone (**24f**). Coupling pattern displayed by the (C-5) signal of (**1f**).

This NMR spectroscopic result matched the data obtained from the mass spectrum, which showed that the isolated sample of (**6f**) was a mix of 40% unlabeled $^{13}C_0$-species (**6**), due to endogenous material present at the beginning of the feeding period and 40% fully labeled $^{13}C_{12}$-species, synthesized during the incubation period with no dilution. Minor amounts of all statistical label combination in between the two extremes made up for the remaining 20%. The value of the specific incorporation rate (40%), obtained from incorporation of radioactive MVL samples was calculated based upon the overall specific activity measured for the sitosterol sample isolated at the end of the incubation time, corresponding to the mean value of all specific activities of the above species. In the case of sitosterol (**6f**), the spectroscopic data were thus in good agreement with the value of the specific incorporation rate. The same was not true, however, for ginkgolide A (**1f**). Based upon the specific incorporation rate detected after incorporation of [2-^{14}C]mevalolactone, it was predicted that the probability of encounters between labeled IPP units would range around 0.5%. The fine structure of the signal corresponding to C-5 of (**1f**), depicted in Figure 5, revealed the presence of two species: **A**, with only the second IPP unit being labeled, and **B**, with the first and the second IPP units simultaneously carrying a label. From the 1:2 ratio between the areas under the signals corresponding to species **A** and **B**, it could be inferred that the frequency of encounters between labeled IPP units ranged between 60% and 70%. In other words, the first three IPP units of the aliphatic C_{20} precursor of these labeled ginkgolide molecules must have been assembled in a place where at least 60–70% of these IPP molecules were labeled, *eo ipso* certainly not in the ginkgolide-producing organelle itself. If, however, 80% of the label detected in the ginkgolides were in fact due to an FPP leakage in the organelle membrane, as postulated above, the spectroscopic data could easily be rationalized given the fact that the probability for encounters between labeled IPP units in the cytosolic FPP was shown to be 90% by the spectra of (**6f**). The decrease in the encounter frequency between labeled IPP units from 90%, as observed in (**6f**), to 70%, as seen in (**1f**), would be explained by the co-import of IPP molecules from the cytosol, which, considering the high dilution by plastid-produced IPP, give rise almost exclusively to species such as that designated as **A** in Figure 5.

At the end of this section on IPP biosynthesis in *G. biloba* and instead of a summary, it may be appropriate to consider a new, provocative line of thought. In the light of the considerations presented above, a series of results obtained in 1972[49] assume new significance. In the course of studies on ubiquinone biosynthesis, isolated rat liver mitochondria had been incubated with mevalonic acid and ATP on one hand and with IPP on the other hand. Interestingly, polyprenylation of *p*-hydroxybenzoic acid was shown to take place only in the presence of IPP, but not in the presence of MVA/ATP. The compelling similarity between these results and those obtained with isolated organelles of plant origin, as described above, raises the question as to whether the mechanistic and structural dichotomy of IPP biosynthesis detected in *G. biloba* and *S. miltiorrhiza* could equally apply to mammalian (and thus to human) cells. Could this perhaps explain the fortunate circumstance that hypercholesterolemia therapies based on the use of HMG-CoA reductase inhibitors do not seem to interfere with the vital mitochondrial ubiquinone production?

2.14.3 GINKGOLIDE BIOSYNTHESIS

2.14.3.1 Early Steps: Formation of the Tricyclic Intermediates from GGPP

2.14.3.1.1 Mechanism of the cyclization of GGPP to abietadiene

The previous section has described the unravelling of the sophisticated ways, by which *G. biloba* synthesizes the fundamental building block of isoprenoid biosynthesis, IPP (**21**), and has also cast some light on the subsequent transformation of IPP to the aliphatic diterpene precursor GGPP (**33**). In this section, it will be described how GGPP is converted first into tricyclic hydrocarbon intermediates of the abietane-type (Section 2.14.3.1) and, from there, eventually into the ginkgolides (**1**)–(**5**) (Section 2.14.3.2). The reader may be assured that the "living fossil" has not yet finished astounding him. The starting point for these considerations is a result already mentioned in passing in the previous section (see Figure 2). Incorporation of [U-^{13}C]-D-glucose had afforded samples of ginkgolide A and B, (**1a**) and (**2a**), whose labeling patterns in lactone ring F did not coincide with the pattern expected from the earlier biosynthetic model of Nakanishi and Habaguchi[10] based on apparent incorporation of [2-^{14}C]acetate (**7**). According to this model, the secondary methyl group of the ginkgolides would be identical to the methyl group undergoing a 1,2-migration in the transition from pimarane- to abietane-type intermediates. Applied to the incorporation of [U-^{13}C]glucose, this migration would have produced the labeling pattern designated as (**E3**) in Scheme

13. The label distribution actually observed in ginkgolides A and B, (**1a**) and (**2a**), unambiguously established instead that the methyl carbon that undergoes the 1,2-migration must subsequently have been oxidized to a carboxyl function. This finding suggests that it is a protonation of the double bond that triggers the migration of the methyl group. Given this inference, an odd result obtained after incorporation of [6,6,6-$^2H_3$]MVL (**24g**) assumes a deeper significance (Scheme 14).

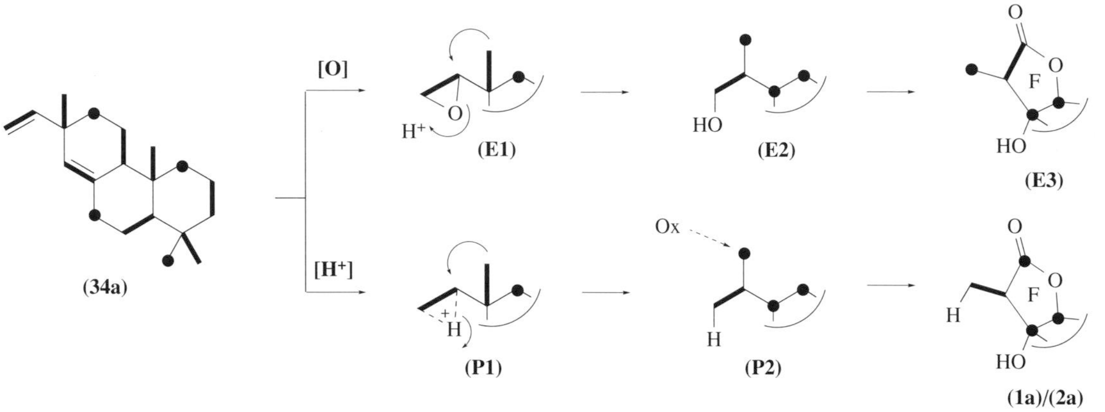

Scheme 13

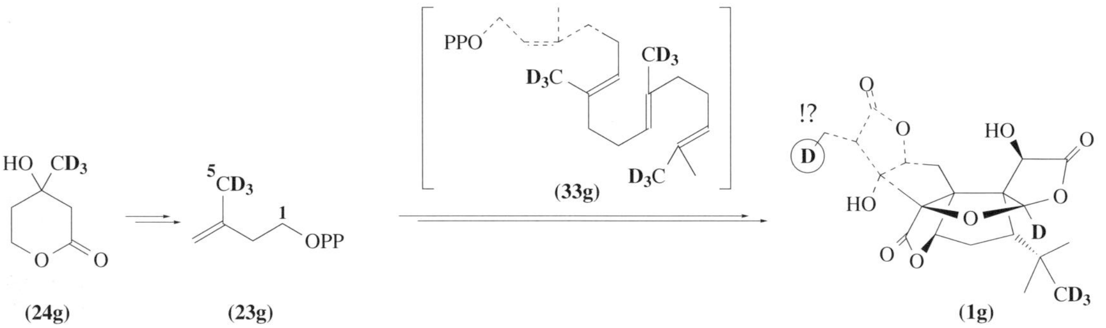

Scheme 14

As mentioned in the previous section (Scheme 12), all ginkgolide samples isolated from incorporation experiments with labeled MVL samples displayed an idiosyncratic asymmetry regarding the label intensities of the first three vs. the fourth IPP unit. Typically, the ^{13}C- or ^2H-enrichments detected for the fourth IPP unit ranged from 5% to 20%, at best, of the average values calculated for the first three units. It was therefore particularly surprising that the ^2H NMR spectra of the ginkgolide A sample (**1g**), obtained from incorporation of (**24g**), revealed the presence of a full equivalent of deuterium in the secondary methyl group, that is precisely at a position corresponding to the fourth, "silent" IPP unit of the aliphatic precursor (**33g**), and, moreover, derived from C-1 of IPP instead of the originally labeled C-5. Subsequently this result was further confirmed by the data obtained from incorporation of [6,6-$^2H_2$]glucose. Obviously, at a biosynthetic stage to be specified, a deuterium transfer from one of the first three IPP-units into the fourth IPP-unit had taken place. The possibility of a transfer from the first IPP-unit could be eliminated, since the ^2H spectra of (**1g**) established the presence of a trideuteromethyl group within the *t*-butyl group. Distance arguments further ruled out a transfer from the second unit, and by a process of elimination, attention was focused on the third IPP unit. The above result showing that the methyl migration is initiated by protonation of the exocyclic double bond suggested an intriguing hypothesis, according to which a deuterium transfer from the third to the fourth IPP unit would take place during the cyclization of GGPP (**33**) to a tricyclic intermediate such as 8,12-abietadiene (**37**). According to the classical notion of this process, adapted to the case of incorporation of [6,6,6-$^2H_3$]MVL (**24g**) in

Scheme 15-(**A**), GGPP (**33g**) first cyclizes to a bicyclic intermediate, either (+)-copalyl-PP (**35g**) or its (−)-enantiomer. Further cyclization of (**35g**) would afford the tricyclic pimarenyl cation or its C-13 enantiomer, the sandaracopimarenyl cation. Removal of a deuteron from C-14 by an external base can generate the first stable tricyclic intermediate, pimaradiene (**34g**), which is prone to undergo the aforementioned acid-catalyzed 1,2-methyl migration to give finally 8,12-abietadiene (or levopimaradiene) (**37g**).

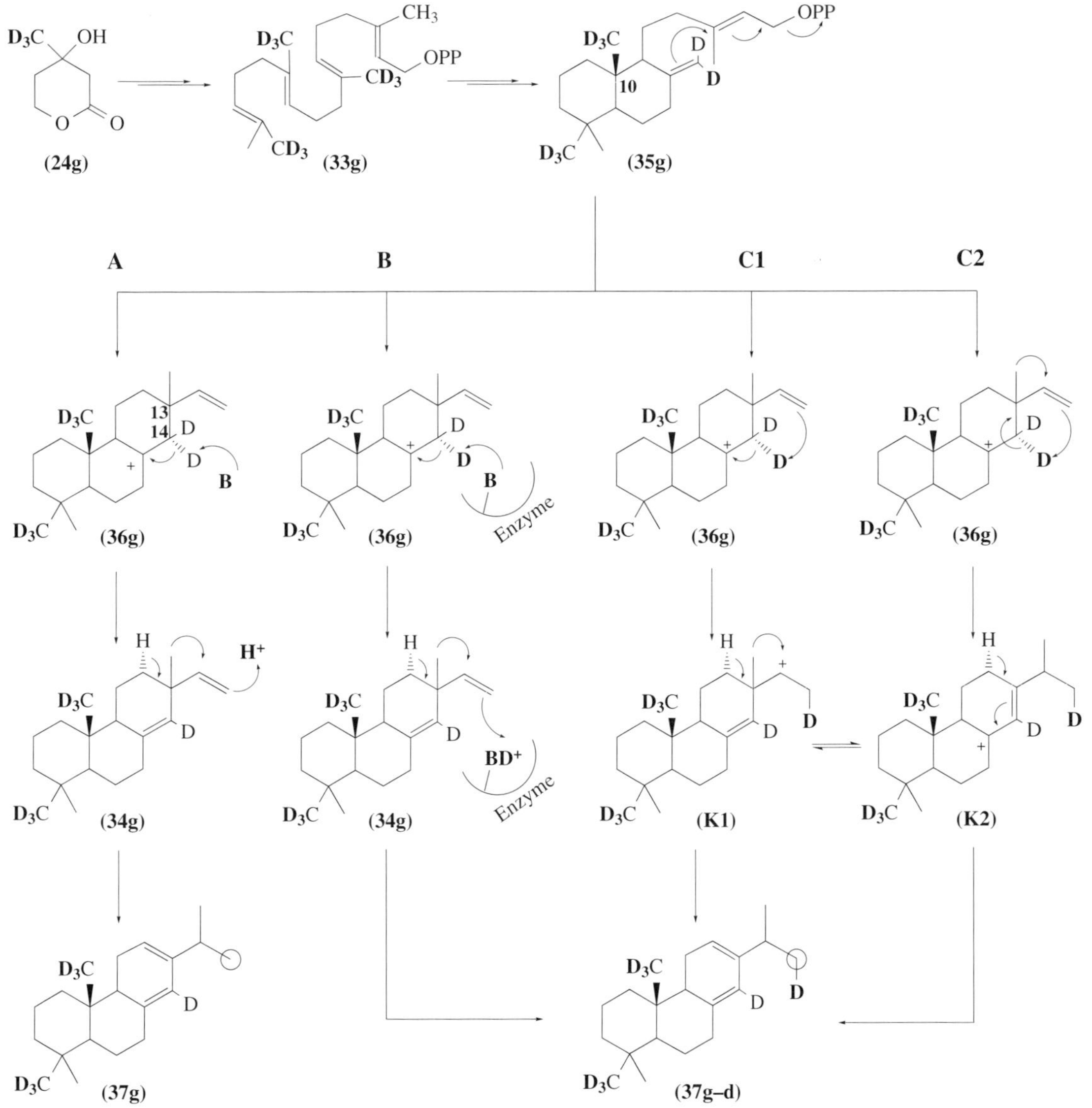

Scheme 15

This classical view of the cyclization process does not, however, explain the presence of a deuterium atom in the secondary methyl group of the ginkgolides. If, instead, the deuterium atom abstracted from C-14 of (**36g**) tarried at the catalytic site of the enzyme long enough to be used for the subsequent protonation of the exocyclic double bond of (**34g**), as shown in (**B**), an abietadiene specimen (**37g–d**) would be generated with a deuterium atom in the circled methyl group that ultimately becomes the secondary methyl group of the ginkgolides. An interesting shortcut variant of this process is (**C1**), which avoids the intervention of an external basic function by using the exocyclic double bond itself as an internal base. Deuterium abstraction from C-14 by the double bond could generate a second cation (**K1**), prone to undergo the methyl migration so as to afford (**37g–d**). The same process can also be formulated as a special case of a six-electron electrocyclic rearrangement involving five centers (**C2**), which would lead to (**37g–d**) via an isomeric cation (**K2**).

From an energetic point of view, both shortcut variants (**C**) would be preferred over the "sticky enzyme" postulate (**B**), since they each bypass an energy minimum, represented by (**34g**), by linking a highly exergonic reaction, namely a deprotonation α- to a cationic center, with the reverse, highly endergonic, process, the protonation of an isolated double bond. A report[50] on the diterpene cyclase of the pine species *Pinus pinaster* is of interest in this context, as it demonstrates that this enzyme is capable of catalyzing the direct cyclization of GGPP to a double bond isomer of (**37**), 7,13-abietadiene, without producing isolable intermediates such as pimaradiene (**34**). Moreover, the cyclization of GGPP to copalyl-PP and thence to (−)-abietadiene catalyzed by recombinant abietadiene synthase from grand fir (*Abies grandis*) has subsequently been shown to take place by a mechanism involving the net transfer of a proton originating at the C-19 methyl of GGPP, by way of the C-17 exomethylene of copalyl-PP and residing ultimately at the C-16 *pro-S* methyl of (−)-abietadiene.[51] The homoannular abietadiene (**37**) is in fact only one dehydrogenation step away from an authentic compound isolated from *Ginkgo* roots by Cartayrade.

2.14.3.1.2 The role of (+)-dehydroabietane

In the course of his search for intermediates of ginkgolide biosynthesis, Cartayrade[22] managed to extract from 2 kg of dried *G. biloba* roots ~50 μg of a compound that could be identified as dehydroabietane (**38**) by GC/MS-analysis. By comparing the c.d. curves and the retention times on chiral GC-columns,[48] he could further show that the absolute configuration of the sample isolated from *Ginkgo* was identical with that of a pine specimen, commonly referred to as the "normal" configuration ((10*S*, 5*S*), (+)-rotation). This finding conflicted, however, with one of the central postulates of the original Nakanishi biogenetic hypothesis which required that the tricyclic intermediates possess the opposite "*ent*"-configuration, seemingly incompatible with a biogenetic correlation between (+)-(**38**) and the ginkgolides. On the other hand, a result obtained from incorporation of [6,6-$^2H_2$]-D-glucose (**21h**) furnished evidence in support of a participation of (+)-(**38**) in ginkgolide biosynthesis (see Scheme 16(a)). The ^{2}H NMR spectrum of the ginkgolide A sample (**1h**) isolated from incorporation of (**21h**) displayed the deuterium distribution expected from the operation of glycolysis and the deoxyxylulose pathway in which nonquaternary carbon atoms corresponding to C-1 and C-5 of IPP carry a deuterium label. However, one of the two ginkgolide methylene carbons corresponding to C-1 of IPP, C-11, carried only one deuterium atom, whereas the other methylene center, C-6, had retained both of its original two deuterium atoms. In addition the pyridine-d_6 ^{2}H NMR spectrum, in which the signals for H-11α and H-11β are well separated, proved that one full deuterium equivalent was present in the α-position and no deuterium in the β-position, thus excluding the possibility of a nonspecific deuterium loss. The lack of one deuterium at C-11 (**1h**) can easily be rationalized by postulating that ring C of the tricyclic intermediate temporarily gains aromatic character during the course of the biosynthesis, for example by the conversion of abietadiene to dehydroabietane (**37h**)→(+)-(**38h**), as indicated in Scheme 16(a). Subsequently a proton would be introduced from the aqueous medium in such a way as to occupy the 11β-position in the biosynthetic products.

In order to establish which one of the two deuterium atoms attached to C-11 of (**37h**) is lost during this series of transformations, the *Ginkgo* embryos were incubated with [1-^{2}H]glucose (**21i**), as shown in Scheme 16(b). On the basis of what is known from other systems about the stereochemical course of the glycolytic enzyme reactions and of what has been established about the deoxyxylulase pathway in the previous section, one would expect IPP molecules formed from (**21i**) to have the C-1 configuration specified in (**23i**) (Scheme 16(b)). In the course of GGPP formation this configuration would be inverted in the first three and retained in the fourth IPP unit, giving rise to abietadiene species (**37i**) with a stereochemically well-defined labeling pattern. The configuration at C-6 of the ginkgolide A sample (**1i**) isolated from incorporation of (**21i**) served as an internal check on the validity of the stereochemical assumptions. Thus analysis of the ^{2}H NMR spectrum of (**1i**) revealed the presence of a full deuterium equivalent in position 6β and at the same time no trace of deuterium in position 6α, thereby proving that the deuterium distribution assigned to (**37i**) was indeed correct. Since, on the other hand, no deuterium was detected in positions 11α and 11β of (**1i**), it could be concluded that it is the circled deuterium atom, or, generally speaking, the axial of the two C-11 hydrogen atoms of (**37i**) that is lost during ginkgolide biosynthesis.

Further support for the alleged participation of (+)-dehydroabietane (+)-(**38**) in ginkgolide biosynthesis was provided by the results of the cytochrome-P450 inhibitor studies carried out by Neau and Walter[53] on young *Ginkgo* plants. Therein, it was demonstrated that above a certain

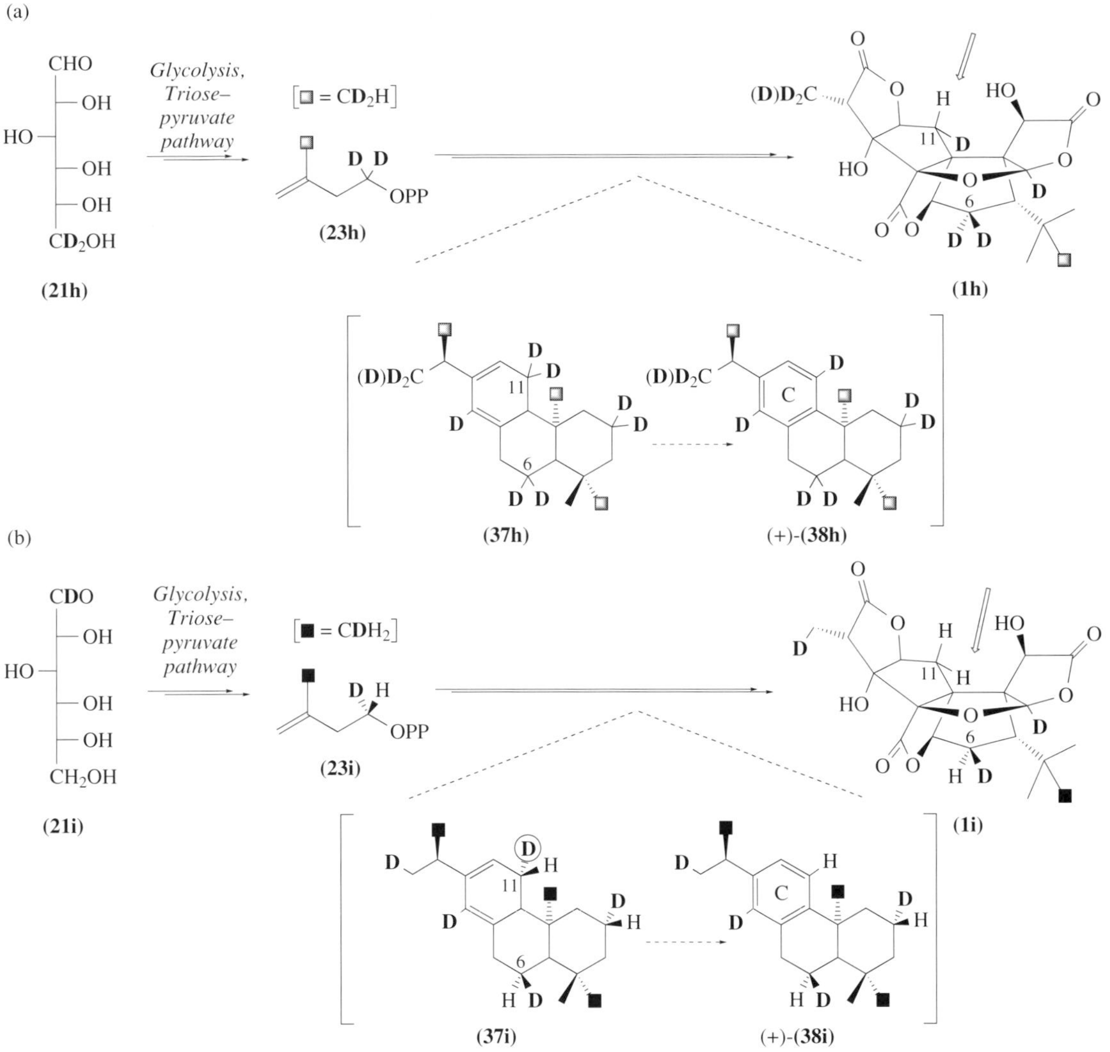

(a)

CHO, OH, HO, OH, OH, CD_2OH **(21h)** — *Glycolysis, Triose–pyruvate pathway* → [□ = CD_2H] D D OPP **(23h)** → **(1h)**

[**(37h)** ⇢ (+)-**(38h)**]

(b)

CDO, OH, HO, OH, OH, CH_2OH **(21i)** — *Glycolysis, Triose–pyruvate pathway* → [■ = CDH_2] D H OPP **(23i)** → **(1i)**

[**(37i)** ⇢ (+)-**(38i)**]

Scheme 16

inhibitor dose, the concentration of (+)-(**38**) increased by a factor of four, while the ginkgolide contents simultaneously decreased by a factor of three, suggesting that dehydroabietane could be the last hydrocarbon, i.e., oxygen-free, intermediate of the biogenetic ladder leading to the ginkgolides. Taken together with the conclusions of the previous section, this would imply that the early biogenetic steps up to the stage of dehydroabietane (+)-(**38**) are confined to the ginkgolide-producing organelle (probably a plastid) as illustrated by Scheme 17.

The dehydroabietane thus generated would then have to be transported into the cytoplasm in order to be further processed by cytochrome-P450-dependent monoxygenases located in the endoplasmatic reticulum (ER). There are many analogous examples in the literature of mono- or diterpene biosynthesis in which cyclization is initiated in specialized organelles such as plastids and the resultant products are oxidatively modified at the ER. A prominent case is the biosynthesis of the gibberellins, which involves formation of the hydrocarbon intermediate *ent*-kaurene in the chloroplasts, transport of the latter into the cytoplasm, and oxidative processing to the gibberellins at the ER.[54] In conclusion, all experimental results and literature data now fit smoothly into a general scheme covering the early steps of gingkolide biosynthesis that is at variance with the original biogenetic hypothesis with respect to the absolute configuration of the proposed tricyclic intermediates.

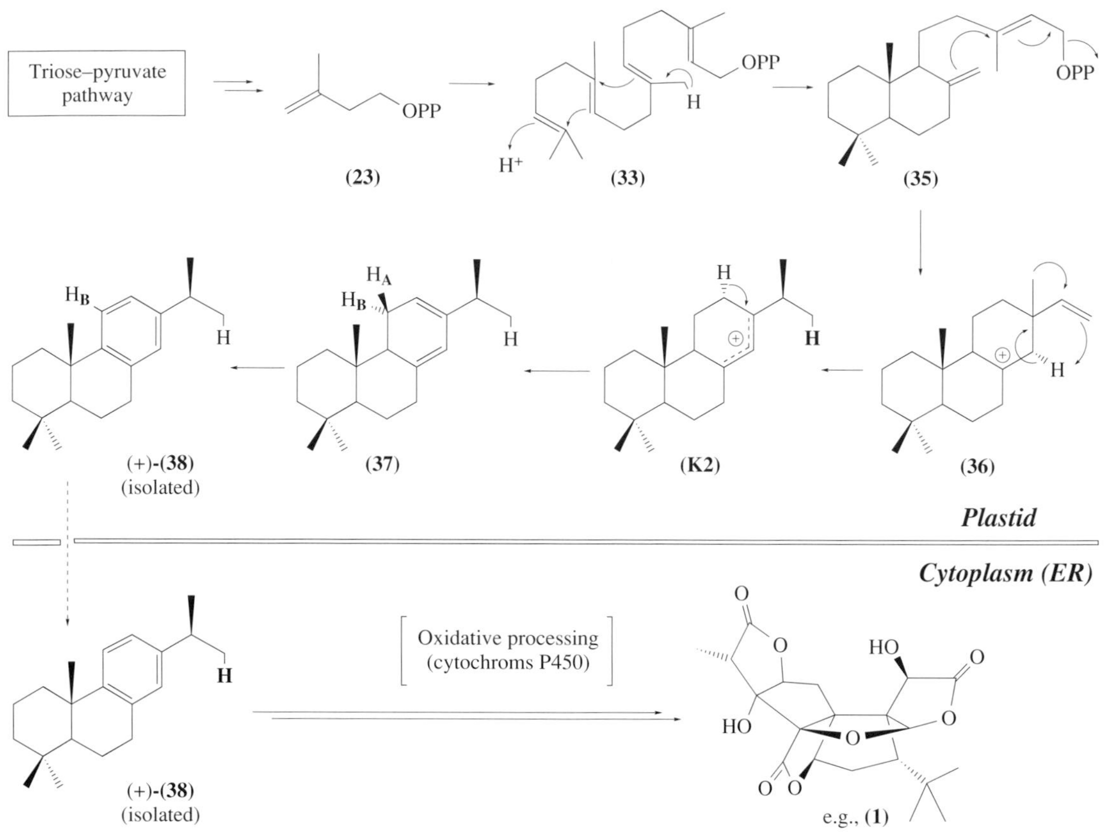

Scheme 17

2.14.3.2 Late Steps: Formation of the Ginkgolides from Tricyclic Hydrocarbon Intermediates

2.14.3.2.1 Formation of the t*-butyl group*

As described in Section 2.14.1.2, the *t*-butyl group of the ginkgolides was originally asserted[10] to be formed via SAM-dependent methylation of an isopropylidene group thought to be generated by cleavage of ring A of the tricyclic intermediates (Scheme 2). This postulate rested on the apparent specific incorporation of [Me-^{14}C]-L-methionine (**9**) into the *t*-butyl group of ginkgolide B and had as its corollary the proposed *ent*-configuration of the intervening tricyclic intermediates. In a preliminary attempt to reproduce the incorporation of (**9**) using the *Ginkgo* embryo system, however, the authors observed random distribution of the label over the entire ginkgolide carbon skeleton. Further doubts arose from the results of two incorporation experiments with ^{13}C-labeled glucose samples (Scheme 18).

Incubation of 50 *Ginkgo* embryos at a time with fivefold diluted samples of [6-^{13}C]-D-glucose (**21k**) and [2-^{13}C]-D-glucose (**21l**) provided samples of sitosterol, (**6k**) and (**6l**), as well as of two ginkgolides A/B, (**1k**)/(**2k**) and (**1l**)/(**2l**), respectively. The labeling patterns detected for the two sitosterol specimens (Scheme 18(a)), in particular for the sidechain moiety, displayed a well-known characteristic reflecting the processes involved in the biosynthesis of SAM from glucose. The two sitosterol carbon atoms known to be derived from SAM, C-28 and C-29, showed ^{13}C-enrichments whose relative intensities compared to the nonSAM-derived carbon atoms were higher after incorporation of (**21k**) than after incorporation of (**21l**). A detailed discussion of this phenomenon is beyond the scope of this chapter; suffice it to say that the methyl carbon of SAM is ultimately derived from either C-3 of serine (corresponding to C-1 and/or C-6 of glucose) or from C-2 of glycine (corresponding to C-2 and/or C-5 of glucose), either of which are transferred onto homocysteine via the cofactor tetrahydrofolic acid (THF). In this pathway, formation of glycine from serine is contingent upon the transfer of C-3 of serine onto THF, meaning that for every glycine molecule produced, one carbon corresponding to C-1 and/or C-6 of glucose is mandatorily incorporated into

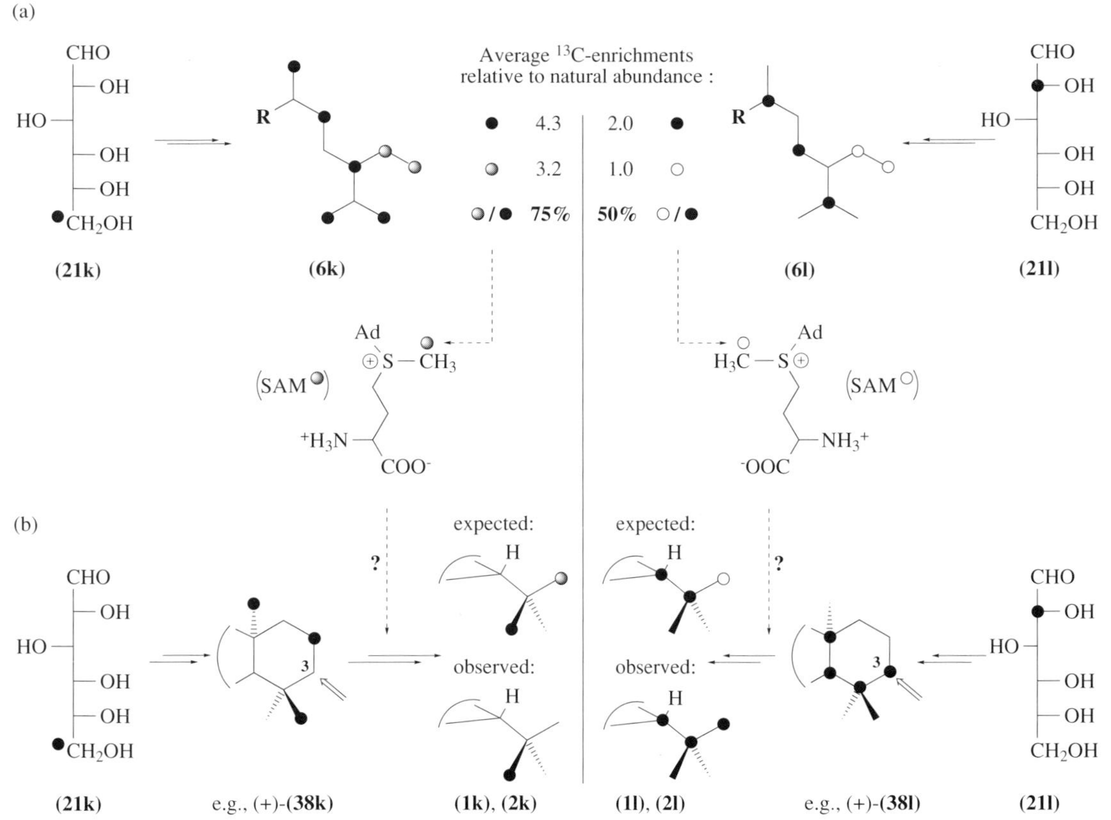

Scheme 18

SAM. In other words, the number of carbons incorporated into SAM from glycine, i.e. from C-2 and/or C-5 of glucose, can only equal, but can never exceed the number of carbons incorporated from serine, corresponding to C-1 and/or C-6 of glucose. Since glycine is involved in numerous other metabolic processes, the actual contribution of C-2 and/or C-5 of glucose towards SAM biosynthesis will always be less than that of carbons C-1 and/or C-6. Accordingly, in the case of labeled glucoses, the ^{13}C-enrichments of SAM-derived carbons (derived from C-1- and/or C-6 labeled glucose) always equal or exceed those resulting from C-2 and/or C-5-labeled glucose precursors. This phenomenon is clearly evident in the labeling patterns of the two sitosterol samples **(6k)** and **(6l)**. By contrast, this correlation was completely reversed in the case of the *t*-butyl group of the ginkgolide samples (Scheme 18(b)). In marked contrast to the sitosterol labeling patterns, the ginkgolides resulting from incorporation of **(21k)** did not show any additional label in the *tert*-butyl group other than that corresponding to C-5 of the first IPP-unit. On the other hand, the samples derived from **(21l)** carried an excess of one full ^{13}C-equivalent in addition to the label attributed to C-3 of the first IPP-unit. Within the framework of the SAM-hypothesis, this would require that *G. biloba* be equipped with a biosynthetic pathway enabling it to procure the methyl group of SAM exclusively from C-2 and/or C-5 of glucose, but not from C-1 and/or C-6.

Scheme 18(b) suggests a much simpler explanation for the labeling patterns observed for **(1k)**/**(2k)** and **(1l)**/**(2l)**. The distribution of isotopic label within the *t*-butyl group can easily be rationalized were the carbon atom corresponding to C-3 of the tricyclic precursor not to be lost during the course of gingkolide biosynthesis, as postulated, but rather preserved throughout the entire biosynthetic pathway. Considering the implications of this suggestion for the entire biogenetic hypothesis, it was now imperative to gather more stringent experimental evidence to confirm the apparent conservation of C-3.

A series of MVL incorporation experiments was therefore initiated aiming specifically at the center in question using an array of different isotopes (Scheme 19). Since C-3 of dehydroabietane (+)-**(38)** is derived from C-2 of IPP and thus from C-4 of MVL, the *Ginkgo* embryos were incubated with samples of [4-^{14}C]MVL (**24m**), [4,4-$^{2}H_2$]MVL (**24n**), [4,5-$^{13}C_2$]MVL (**24o**), and [3,4-$^{13}C_2$]MVL (**24f**). Incorporation of (**24m**) (Scheme 19(a)) furnished radioactive specimens of ginkgolides A, B, and C, (**1m**)–(**3m**), which were each recrystallized to constant specific radioactivity. Subsequent Kuhn–Roth degradation afforded pivalic acid samples with molar specific activities ranging from 60% to 62% of the molar specific activities determined for the respective ginkgolide specimens, thus suggesting that two out of three labeled carbon atoms had been excised from the ginkgolide molecule by this degradative procedure. The deviation from the theoretical value of 67% could be explained by a low-level labeling of the fourth IPP unit as discussed in the previous section. If C-3 of (+)-(**38m**) had been lost during gingkolide biosynthesis, the expected value for the ratio between the molar specific activities, corrected for the contribution of the fourth unit, would have been 46%, since only one out of two labeled carbons within (**1m**)–(**3m**) would have been isolated by the Kuhn–Roth degradation reaction. In confirmation of these observations, incorporation of (**24n**) (Scheme 19(b)) afforded a sample of ginkgolide A (**1n**), whose ^{2}H NMR spectrum displayed two signals of equal intensity that matched the ^{1}H NMR signals assigned to the *tert*-butyl group and to H-5. Both findings are very indicative, considering that neither H-3 nor H-5 of the tricyclic precursor (+)-(**38n**) were expected to be conserved based on the SAM-hypothesis for the formation of the *t*-butyl group (see transition (**17**)→→(**19**) in Scheme 2).

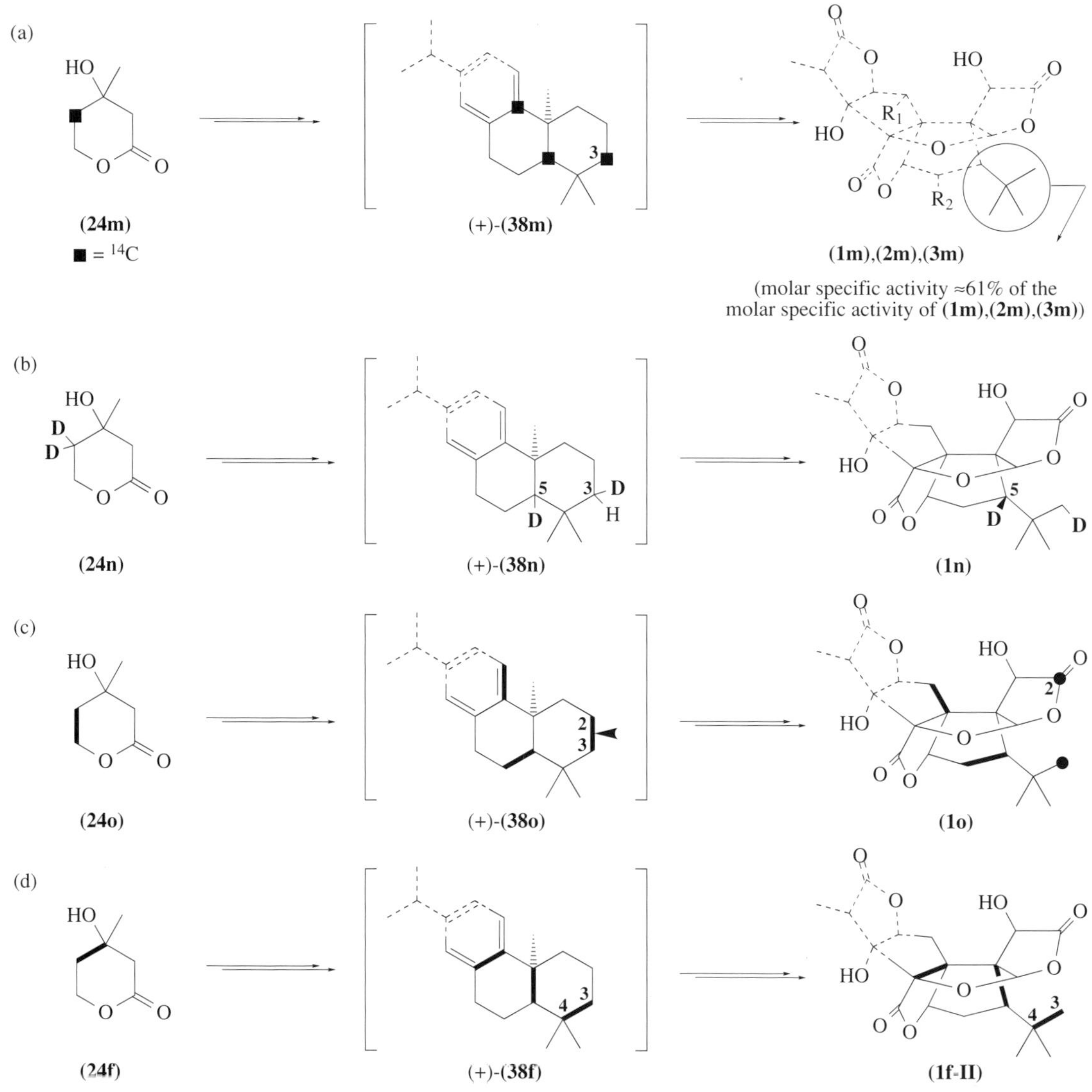

(a) (**24m**) ■ = ^{14}C (+)-(**38m**) (**1m**),(**2m**),(**3m**)

(molar specific activity ≈61% of the molar specific activity of (**1m**),(**2m**),(**3m**))

(b) (**24n**) (+)-(**38n**) (**1n**)

(c) (**24o**) (+)-(**38o**) (**1o**)

(d) (**24f**) (+)-(**38f**) (**1f-II**)

Scheme 19

Ginkgo embryos were also incubated with a doubly ^{13}C-labeled sample of MVL ((**24o**), [4,5-$^{13}C_2$]MVL) to provide samples of ginkgolides A and B, (**1o**) and (**2o**), with the labeling pattern shown in Scheme 19(c). From the observation of bond labels between C-5/6 and C-9/11 of the gingkolides, as revealed by the presence of the relevant doublets in the corresponding ^{13}C NMR spectra, it could be inferred that the respective two ^{13}C atoms had remained linked to each other for the duration of the entire biosynthetic sequence leading from (**24o**) via (+)-(**38o**) to (**1o**). More importantly, however, single labels were detected in the *t*-butyl group on one hand and at C-2 of (**1o**) and (**2o**) on the other hand, indicating that only the bond between C-2 and C-3 of the tricyclic precursor had been broken, but not the one between C-3 and C-4, as had been postulated earlier. To gain independent evidence for this crucial aspect of gingkolide biosynthesis, it was decided to carry out an experiment in which the bond in question, between C-3 and C-4 of the tricyclic intermediates, would be labeled (Scheme 19(d)). Incorporation of an undiluted synthetic sample of [3,4-$^{13}C_2$]MVL (**24f**) (99% ^{13}C per position) yielded a specimen of ginkgolide A (**1f-I**) with an interesting ^{13}C NMR spectrum, whose relevant region, comprising the signals for the quaternary and the methyl carbons of the *t*-butyl group, is depicted in Figure 6(a).

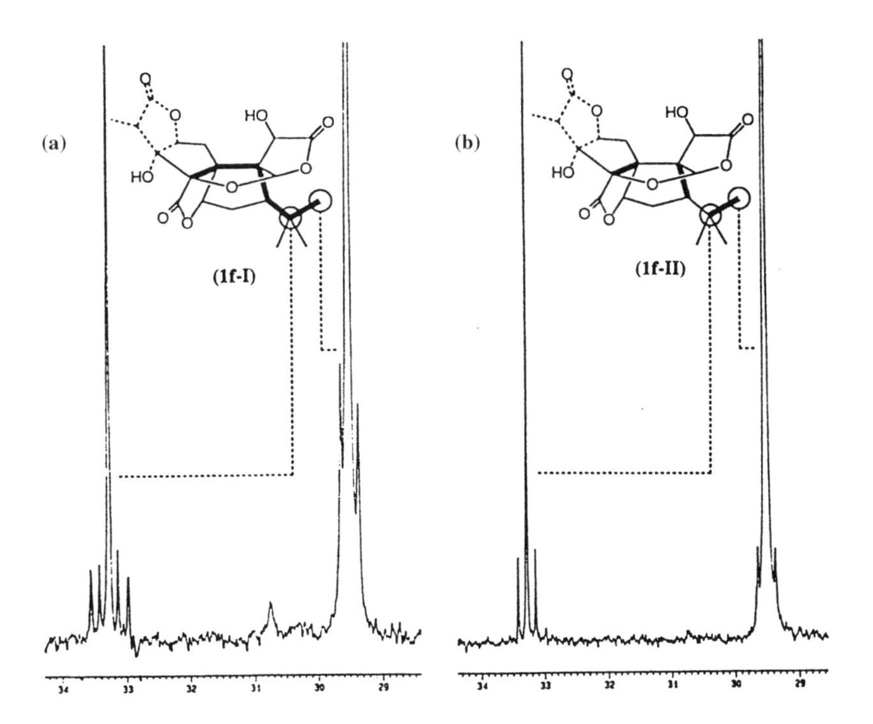

Figure 6 ^{13}C NMR spectra of gingkolide A samples (**1f-I**) (a) and (**1f-II**) (b) in MeOH-d_4. Signals for quaternary and methyl carbons of the *t*-butyl group.

A distinct coupling between the quaternary carbon and one methyl carbon can be seen from the signal patterns, thus establishing the presence of a bond label between C-3 and C-4 and proving the conservation of C-3 from the tricyclic precursors up to the ginkgolides. However, the signal for the quaternary carbon revealed it to be engaged in an additional coupling with the adjacent carbon C-5, which forms part of the second IPP unit. A detailed discussion of this observation has already been provided in Section 2.14.3.2 (see Figure 5). It is caused by the high cytosolic concentration of labeled IPP molecules used for the assembly of farnesyl diphosphate. In order to eliminate the extraneous couplings between the IPP-units while maintaining a sufficient label intensity to detect the intra-unit couplings, the precursor (**24f**) was now diluted with two parts of unlabeled MVL prior to addition to the nutrient medium. Isolation and NMR-spectroscopic analysis of the derived ginkgolide A sample (**1f-II**) disclosed the desired trenchant labeling pattern, shown in Figure 6(b). Not only had all inter-unit couplings disappeared, but at the same time the intra-unit coupling between C-3 and C-4 had persisted, being less intense than before due to the dilution of the precursor, yet still significant. In the light of the results summarized in Schemes 18 and 19, the proposed loss of C-3 of the tricyclic intermediates and subsequent recovery of a carbon atom from SAM, summarized in Scheme 20(a), could now be firmly excluded.

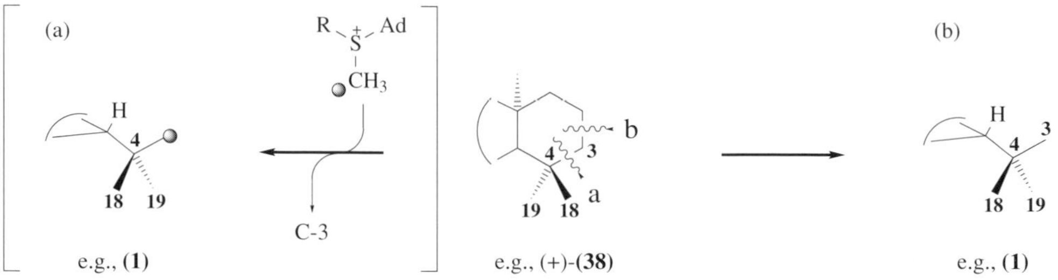

(a) e.g., (1) C-3 e.g., (+)-(38) (b) e.g., (1)

Scheme 20

It is certain that C-3 is conserved in the ginkgolides and that the bond between C-3 and C-4 remains intact during the entire biosynthetic pathway. Moreover, since C-2 is also preserved, albeit detached from C-3 (Scheme 19(c)), it is an inescapable conclusion that the cleavage of ring A of the tricyclic intermediate is brought about by breaking the bond between C-2 and C-3, as shown in Scheme 20(b). This inference, in turn, inevitably raised the question as to mechanism of this ring cleavage. It is particularly remarkable that the cleavage of the bond between C-2 and C-3 does not leave behind a visible scar in the form of a functional group on the C-3 side of the former linkage, while on the side of C-2 the carbonyl group of the lactone function bears witness of the bond breaking process. Apparently, the primary scar at C-3 is removed immediately after the bond cleavage by transferring a hydrogen atom from a neighbouring structural element, thereby forming the C-3 methyl group of the *t*-butyl group and a secondary scar at another, nearby location. A promising candidate for this secondary scar is the C-1 hydroxyl group, which is the only hydroxyl group common to all five ginkgolides. Indeed, the spatial proximity of C-1 and the *t*-butyl methyl groups is chemically documented by the behavior towards light of the 1-dehydro derivative of ginkgolide A (**39**)[55] (Scheme 21).

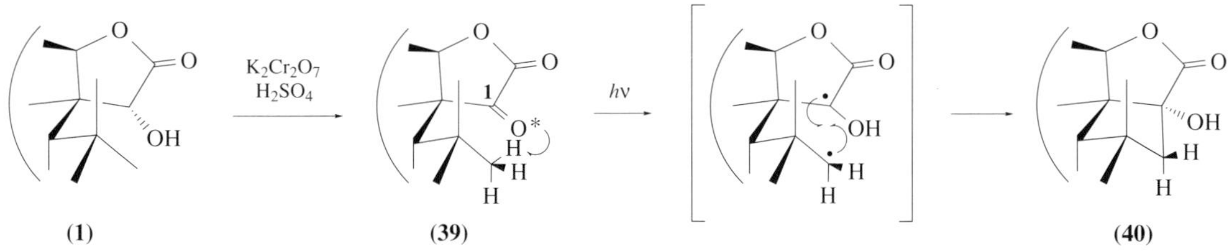

(1) $K_2Cr_2O_7$ H_2SO_4 (39) $h\nu$ (40)

Scheme 21

Obtained by moderately harsh oxidation of ginkgolide A (**1**), 1-dehydro-ginkgolide A (**39**) is readily converted to photodehydroginkgolide A (**40**) by mere exposure to daylight. The extreme ease of this reaction, which is an example of a Norrish type II photocyclization, reflects the steric proximity between the C-1 carbonyl group and the hydrogen atoms of the *t*-butyl group. Taking all these considerations into account, the mechanistic hypothesis presented in Scheme 22 can be advanced, beginning with an intermediate (**C1**) with an intact ring A. Several oxidation steps involving C-2 and C-20 would yield the cyclic hemiacetal (**C2**), which could be further oxidized to the corresponding alkoxy radical (**C4**), presumably by a cytochrome P450, or a mechanistically equivalent cofactor represented in Scheme 22 as an active Fe(IV)-oxo species.

(**C4**) is prone to undergo a radical β-cleavage to afford a first C-radical (**C5**), which, in turn, would readily rearrange to the more stable C-radical (**C6**) by a 1,5-shift of a hydrogen atom H_t from C-1 to C-3, thereby closing the primary scar of the bond cleavage at the expense of a secondary scar at C-1. The C-radical (**C6**) could now easily be hydroxylated by the Fe(III)-hydroxy species of the cofactor generated in the ring cleavage. The hypothesis of a radical cleavage involving the α,β-carbon–carbon bond of an alkoxy radical, such as (**C4**), relies upon a number of nonenzymatic literature precedents.[56] Notably, in one case, the radical cleavage is even reported to be

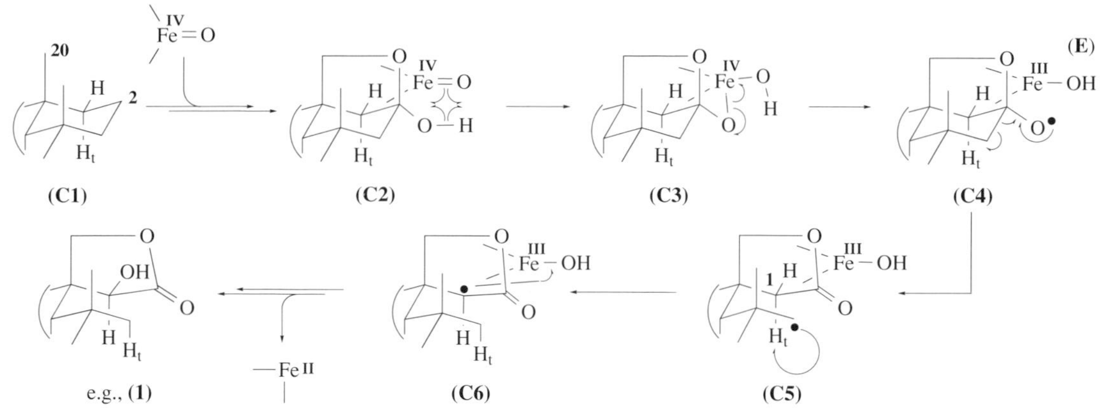

Scheme 22

followed by a 1,5-H shift.[57] The validity of the central postulate in the mechanistic hypothesis illustrated in Scheme 22 was tested by two MVL-incorporation experiments summarized in Scheme 23.

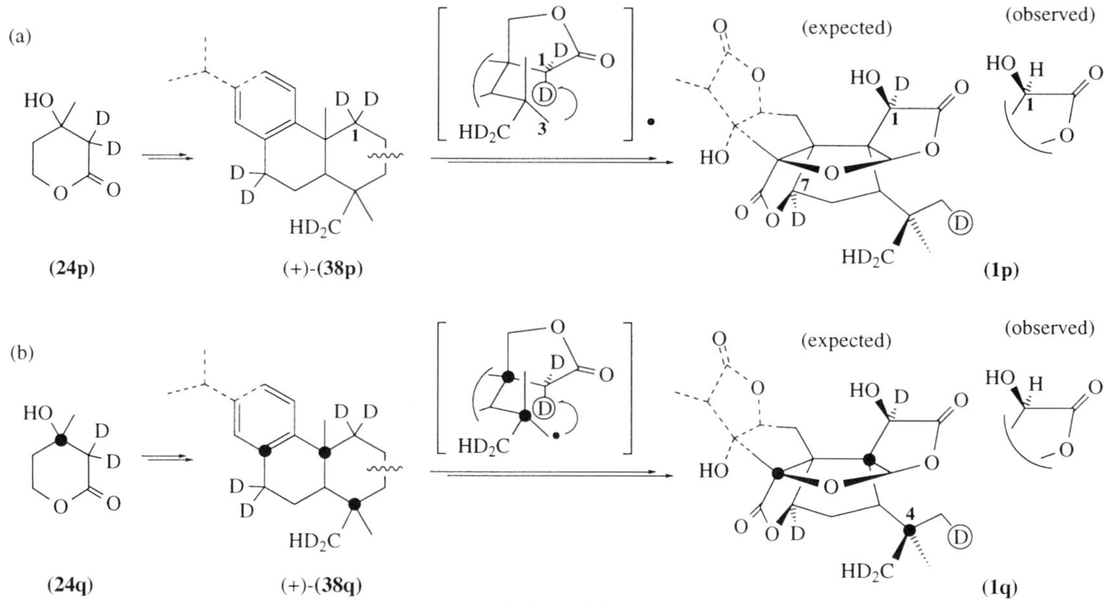

Scheme 23

In the first experiment (a), *Ginkgo* embryos were grown on 8 L of nutrient medium, containing 160 g of D-glucose and 4 g of [2,2-$^2H_2$]MVL (**24p**). Incorporation of (**24p**) was expected to afford dehydroabietane species (+)-(**38p**) carrying a double deuterium label at C-1, corresponding to the designated origin of the postulated 1,5-hydrogen migration. If the mechanistic hypothesis were correct, one of these two deuterium atoms should be transferred from C-1 to C-3 to generate a *t*-butyl group containing three deuterium equivalents as compared to the reference positions, C-1 and C-7, with one deuterium equivalent each. In the event, after the 45-day incubation period, 10 mg of ginkgolide A (**1p**) was isolated, whose ^{2}H NMR spectrum showed two signals, one corresponding to D-7 and the other to the *t*-butyl group, accounting for 3.6 deuterium equivalents with respect to the D-7 signal. Surprisingly, no trace of a signal could be detected at the chemical shift corresponding to D-1, indicating that, at some stage, one of the two hydrogen (deuterium) atoms at C-1 is replaced by a hydrogen atom from the medium. Among the numerous mechanistic scenarios that could account for this finding, the favored model is presented in Section 2.14.3.2.3. For the moment, it is important to note that, judging by the integral ratio of 3.6 between the *t*-butyl and the H-7 signal, the mechanistic hypothesis for the ring cleavage was confirmed. The over-integration of the *t*-butyl signal with respect to the theoretical value of 3.0 was attributed to a small shoulder, which could not be eliminated despite repeated efforts to repurify (**1p**). Although an inconspicuous detail, this observation nonetheless bluntly laid bare the limitations of the ^{2}H NMR approach as far as the stringency of quantitative arguments are concerned. It cannot be decided *a priori* to which extent the observed integrals are due to deuterium-labeled by-products in the sample. Especially in cases with low (<1%) deuterium enrichments, the smallest amounts of strongly labeled impurities are sufficient to confound the results.

A follow-up experiment was therefore carried out in which the deuterium transfer into the *t*-butyl group could be monitored by means of ^{13}C NMR spectroscopy (Scheme 23(b)). Again, the alleged starting point for the 1,5-hydrogen-shift, C-1, was to carry a deuterium label, but an additional "^{13}C-sensor" was now to be embedded into the target area, bound to report a deuterium transfer by the resulting shift of the corresponding ^{13}C NMR signal towards high field. This experimental approach was based on the well-known phenomenon that deuterium substitutions in α-, β-, and γ-positions to a ^{13}C atom shift the corresponding ^{13}C NMR signal towards higher field, typically by 0.5–1.0 ppm, 0.1 ppm, and 0.02–0.04 ppm, respectively. It is noteworthy that this strategy could only be envisaged because of the remarkably high (>60%) probability of encounters between labeled IPP units within the ginkgolide molecule, thereby guaranteeing that in at least 60% of the cases the deuterium atom originating from the second IPP unit would be transferred into an equally labeled, ^{13}C-containing, first IPP unit. A MVL sample that complies with the above conditions is [3-^{13}C-2,2-$^{2}H_2$]MVL (**24q**) that was synthesized in five steps (>98% ^{13}C, ^{2}H), based on literature procedures and administered to the *Ginkgo* embryos in a batch of 4 L of nutrient medium to afford about 4 mg of ginkgolide A (**1q**).

Figure 7 shows the ^{13}C NMR signal of the quaternary carbon atom of the *t*-butyl group, C-4. As can be seen, the intense signal corresponding to the natural ^{13}C-abundance is accompanied by three minor signals **A**, **B**, and **C**, which are shifted towards high field, relative to the major signal, by 0.24, 0.16, and 0.08 ppm, respectively, thus indicating the presence of three, two, and one deuterium atoms in the β-position with respect to the reporter ^{13}C-atom. The most important signal is **A**, since it unambiguously discloses the presence of species (**a**), which can only have arisen from the postulated deuterium transfer from C-1 to C-3. The appearance of signal **B** had been expected as well, since the ^{13}C-atom of the first IPP unit (C-4) has only a 70% chance, at best, to be paired with an equally labeled second IPP unit, and thus to receive a third deuterium atom at the β-position. In the remaining 30% of the cases, a hydrogen atom is transferred from C-1 to C-3, thereby giving rise to species (**b1**). While signals **A** and **B** had been predicted, it was not immediately obvious to what process the appearance of signal **C** could be attributed, until the ^{13}C NMR spectrum of the sitosterol sample (**6q**) isolated from the same incorporation experiment became available. The latter spectrum revealed that apart from the expected [^{13}C-$^{2}H_2$]IPP-species, a considerable amount of [^{13}C-$^{2}H_1$]- and [^{13}C-$^{2}H_0$]IPP-species had been formed from (**24q**) and used to assemble farnesyl diphosphate. Frequently observed in biosynthetic studies using MVA samples labeled at C-2,[58] deuterium (or tritium) loss from C-4 of IPP is explained by the rapid interconversion between IPP (**23**) and its double bond isomer dimethylallyl diphosphate. In the light of this result, the interpretation of the signal structure observed for C-4 of (**1q**) could now be completed: signal **B** is not only due to species (**b1**), but also to species (**b2**), produced by a deuterium transfer from C-1 into a first IPP-unit of the [^{13}C-$^{2}H_1$]-type. And finally, in analogy to **B**, signal **C** is rationalized either by hydrogen transfer from a non- (or partially) labeled second IPP unit into a first unit lacking one deuterium atom (**c1**), or by deuterium transfer into a first IPP unit devoid of deuterium (**c2**).

In conclusion, starting from the newly established biogenetic origin of the three *t*-butyl methyl groups, a novel mechanistic hypothesis for the cleavage of ring A of the tricyclic precursors of the gingkolides has been developed and experimentally verified. The new hypothesis explains the presence of two characteristic features of the ginkgolide molecule (*t*-butyl group and α-hydroxy-lactone ring C) as the consequence of a single radical fragmentation process with a concomitant radical 1,5-hydrogen shift. Endowed with this "radically" new insight into the processes that lead to the formation of the right-hand portion of the ginkgolide molecule, a fundamental stereochemical issue of ginkgolide biosynthesis could now be addressed afresh (Scheme 24).

Comparison of the bold-faced structure moieties **H-5/C-5/C-10/C-20** of the tricyclic intermediates, (+)-(**38**) and (−)-(**38**), with the corresponding structural feature in the ginkgolide (**1**) reveals that H-5 and C-20 are *trans*-oriented on the level of the precursors (+)-(**38**) and (−)-(**38**), but *cis*-oriented in the biosynthetic products such as (**1**). This realization prompts the trivial conclusion that at some point in the biosynthetic sequence the configuration at either C-5 or C-10 must be inverted. More precisely, C-5 is involved in the inversion of configuration if the tricyclic intermediates belong to the "*ent*"-series, while C-10 is affected if they form part of the "*normal*" series. An

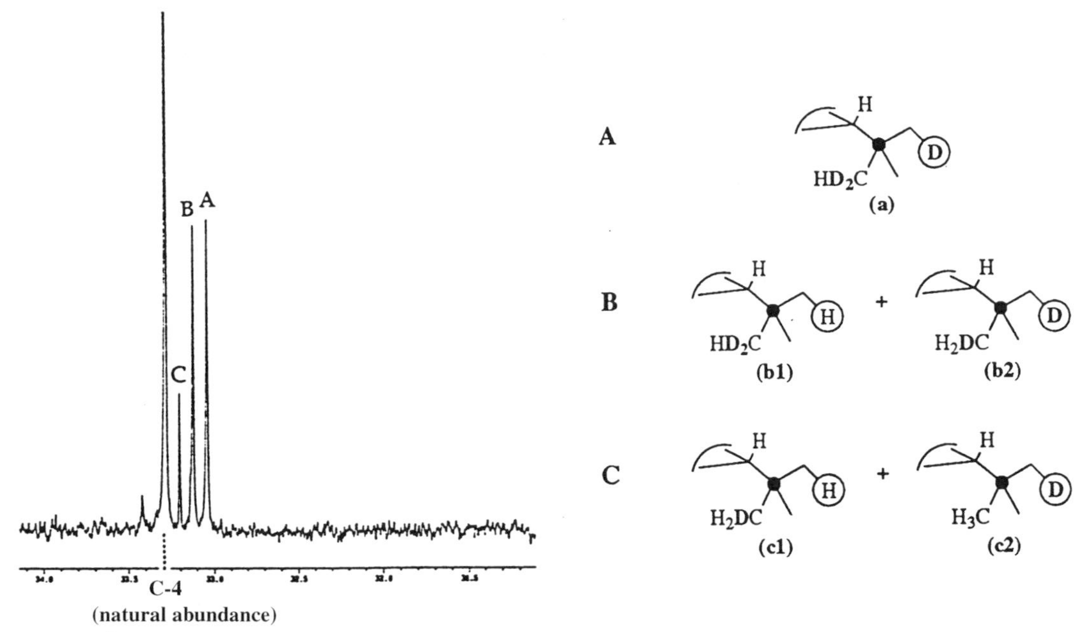

Figure 7 ^{13}C NMR spectrum of gingkolide A (**1q**) in MeOH-d_4. Signal for the quaternary carbon of the *t*-butyl group (C-4).

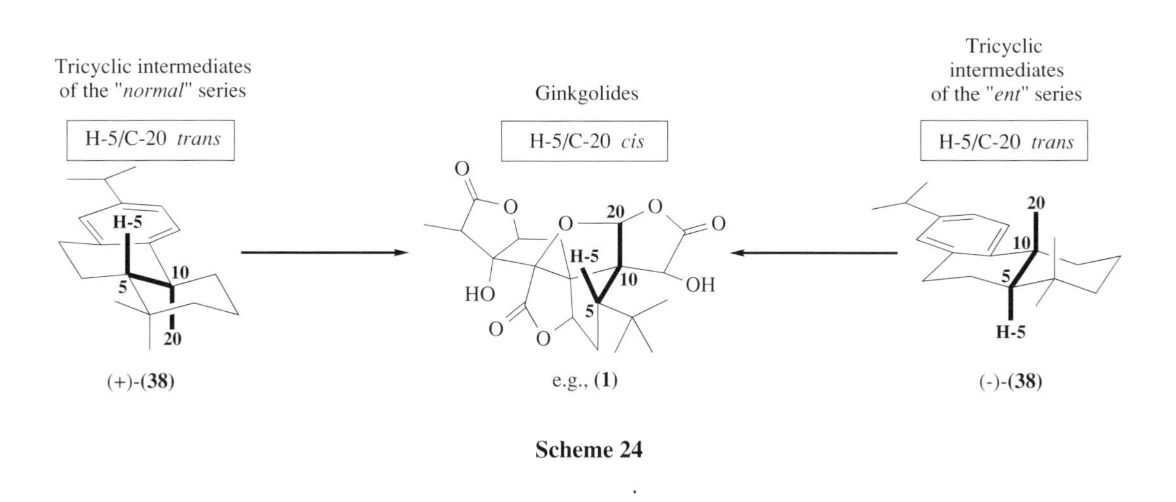

Scheme 24

overwhelming amount of evidence has already been adumbrated within this section establishing that: (i) the bond between C-3 and C-4 remains intact during the entire biosynthesis, thus ruling out the participation of SAM in the formation of the *t*-butyl group, (ii) the new ring cleavage mechanism involving the bond between C-2 and C-3 and a hydrogen shift from C-1 to C-3 does not affect C-5 and/or H-5, and (iii) incorporation of [4,4-$^2H_2$]MVL shows that H-5 is conserved. These results point to the conclusion that C-10 rather than C-5 must undergo an inversion of configuration, and, consequently, that the tricyclic intermediates of ginkgolide biosynthesis must have the "normal" configuration.

2.14.3.2.2 Absolute configuration of the tricyclic hydrocarbon intermediates

With the knowledge that the *t*-butyl methyl groups are derived from C-3, C-18, and C-19 of the tricyclic hydrocarbon intermediates, corresponding to C-2, C-4, and C-5 of IPP, respectively, it becomes theoretically possible to generate ginkgolides with chiral *t*-butyl groups from MVL samples such as (**24r**) and (**24s**) carrying different labels at C-4, C-2, and C-6 (Scheme 25). It is of crucial importance that the configuration of these *t*-butyl groups is determined exclusively by the stereochemical course of the cyclization reaction converting GGPP into "*ent*" or "*normal*" tricyclic intermediates, more precisely, by the conformation in which the aliphatic precursor is coiled up by the diterpene cyclase. Based on what is known about the biosynthesis[59] and the cyclization mode[60] of aliphatic terpene precursors, the configuration of the *t*-butyl group produced from a given MVL specimen can be predicted unambiguously for both the "*ent*" and the "*normal*" configuration of the tricyclic precursors, as shown in Scheme 25. Conversely, the configuration of a *t*-butyl group generated from a given MVL specimen can be used to decide whether the tricyclic intermediates are formed in the "*ent*" or in the "normal" series.

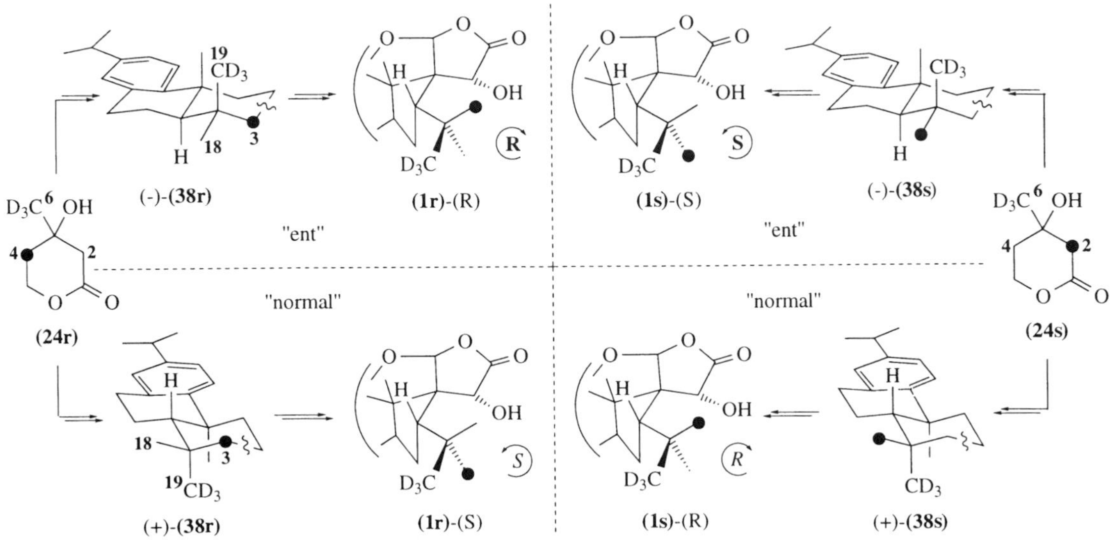

Scheme 25

A promising approach to the nontrivial problem of identifying the configuration of the chiral *t*-butyl groups generated biosynthetically from either (**24r**) or (**24s**) is offered by a reaction introduced in the previous section (see Scheme 21), the Norrish photocyclization of 1-dehydro-ginkgolide A (**39**) to photodehydro-ginkgolide A (**40**) (Scheme 26). If a sample of (**39**) with a chiral *t*-butyl group were to be submitted to the photocyclization conditions, a set of the three isotopomeric photoproducts (**40**)-x, (**40**)-y, and (**40**)-z would be formed in the case of an (*R*)-configuration, (**39**)-(*R*), and a complementary set (**40**)-x′, (**40**)-y′, and (**40**)-z′ would be generated in the case of an (*S*)-configuration, (**39**)-(*S*), depending on which one of the three homotopic methyl groups is engaged in the H˙-(or D˙-) abstraction. Due to the kinetic isotope effect favouring a H˙- over a D˙-abstraction, the formation of species (**40**)-x is expected to be three to six times slower than the formation of (**40**)-y, and (**40**)-z, according to literature data on Norrish type II photoreactions.[61–64] The same considerations apply to species (**40**)-z′, which should be discriminated against relative to species (**40**)-x′ and (**40**)-y′. Spectroscopically, this has an interesting consequence for the two signals corresponding to the geminal methyl groups of (**40**). In the case of an (*R*)-configurated *t*-butyl group, (**39**)-(*R*), the ^{13}C-enrichment will be found predominantly in the pseudoaxial methyl group (species (**40**)-z), while in the opposite case, (**39**)-(*S*), it will prevail in the pseudoequatorial methyl group (species (**40**)-x′). Provided an unambiguous assignment of the ^{13}C-NMR-signals for the two methyl groups of (**40**) were feasible, this would then constitute a method for determining the configuration of the *t*-butyl group of (**39**), and thus of (**1**). Notably, even if the photocyclization were not subject to a kinetic isotope effect, it would still be possible to distinguish between (*R*)- and (*S*)-configuration of the *t*-butyl group, by observing the differential deuterium γ-shifts to be expected in (**40**)-x (two deuterium atoms) vs. (**40**)-x′ (three deuterium atoms), or, analogously, in (**40**)-z vs. (**40**)-z′. In the event, the ^{1}H NMR signals of the geminal methyl groups of (**40**) were assigned on the basis of distinct NOEs with H-20 and H-5 for the pseudoaxial position and with H-6α for the pseudo-equatorial position.

After an HSQC-experiment had furnished the assignment of the correlated ^{13}C-NMR-signals, the ground was prepared for the decisive experiment. *Ginkgo* embryos were grown on 20 L of nutrient medium, split in two equal batches containing either (**24r**) or (**24s**), which had been synthesized in 5 g quantities by eight-step procedures. After 45 days, samples of ~10 mg of ginkgolide A (**1r**)

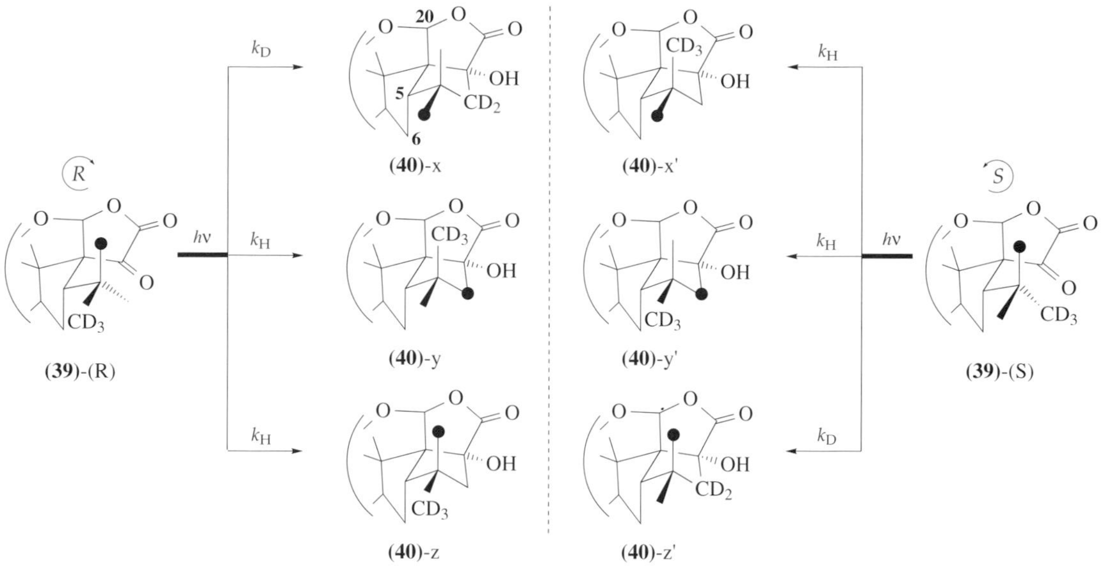

Scheme 26

and (**1s**) were isolated from the cultures, purified by HPLC, and subjected to oxidation and photocyclization. The relevant region of the ^{13}C NMR spectra of the resulting products (**40s**) and (**40r**) are depicted in Figure 8. In the product from incorporation of (**24s**) the ^{13}C-enrichment is located predominantly in the pseudoaxial methyl group, whereas the pseudoequatorial methyl group is only weakly labelled (Figure 8(a)). This implies that in the photoreaction species (**40**)-z is formed preferentially with respect to (**40**)-x; a preponderance of ca. 6:1 reflecting the value of k_H/k_D for the H-transfer step can be estimated from the relative intensities of the satellite signals. In addition, the chemical shift differences between the satellite signals due to (**40**)-z/(**40**)-x and the respective natural abundance signals display a ratio of 3:1, in accordance with the fact that in (**40**)-z there are three deuterium atoms contributing to the γ-effect on the ^{13}C-enriched carbon, whereas in (**40**)-x there are only two such γ-neighboring deuterium atoms. Clearly the *t*-butyl group of dehydro-ginkgolide A, and thus of ginkgolide A, derived from incorporation of (**24s**) must possess the (*R*)-configuration. By applying the same analysis to the photodehydro-ginkgolide A sample derived from incorporation of (**24r**), one obtains the complementary result (Figure 8(b)). In the latter case, the major isotopomeric species is (**40**)-x′, whose prevailing over (**40**)-z′ denotes an (*S*)-configurated *t*-butyl group in dehydro-ginkgolide A (**39**)-(*S*) and ginkgolide A (**1**)-(*S*). Again, the ratio of the shift differences of (**40**)-x′ vs. (**40**)-z′ is approximately 3:1. The observed specific formation of (**1s**)-(*R*) from (**24s**) and of (**1r**)-(*S*) from (**24r**) is possible only, as evidenced in Scheme 25, if the tricyclic hydrocarbon intermediates are biosynthesized in the "*normal*" rather than in the "*ent*" series. This vindicates the validity of the postulate put forward on the basis of the collective evidence discussed in previous sections. The newly gained knowledge that an inversion at the quarternary center C-10 must take place in going from the tricyclic hydrocarbon precursor to the ginkgolides paves the way for a fundamental revision of the original biogenetic scheme.

2.14.3.2.3 A revised biogenetic sequence

In the light of the results presented so far, there is little doubt that the dehydroabietane isolated by Cartayrade from *Ginkgo* roots and shown to have the normal configuration is in fact a central intermediate of ginkgolide biosynthesis. As mentioned at the end of Section 2.14.3.1, there is some evidence pointing to a scenario whereby (+)-(**38**) would be the final product of the plastidial portion of the biosynthetic sequence and would then be transported into the cytoplasm and further processed to the ginkgolides in a complex set of reactions involving several oxidation steps. For instance, one methyl group (C-17) of the isopropyl chain has to be oxidized to a carboxyl group at some stage in

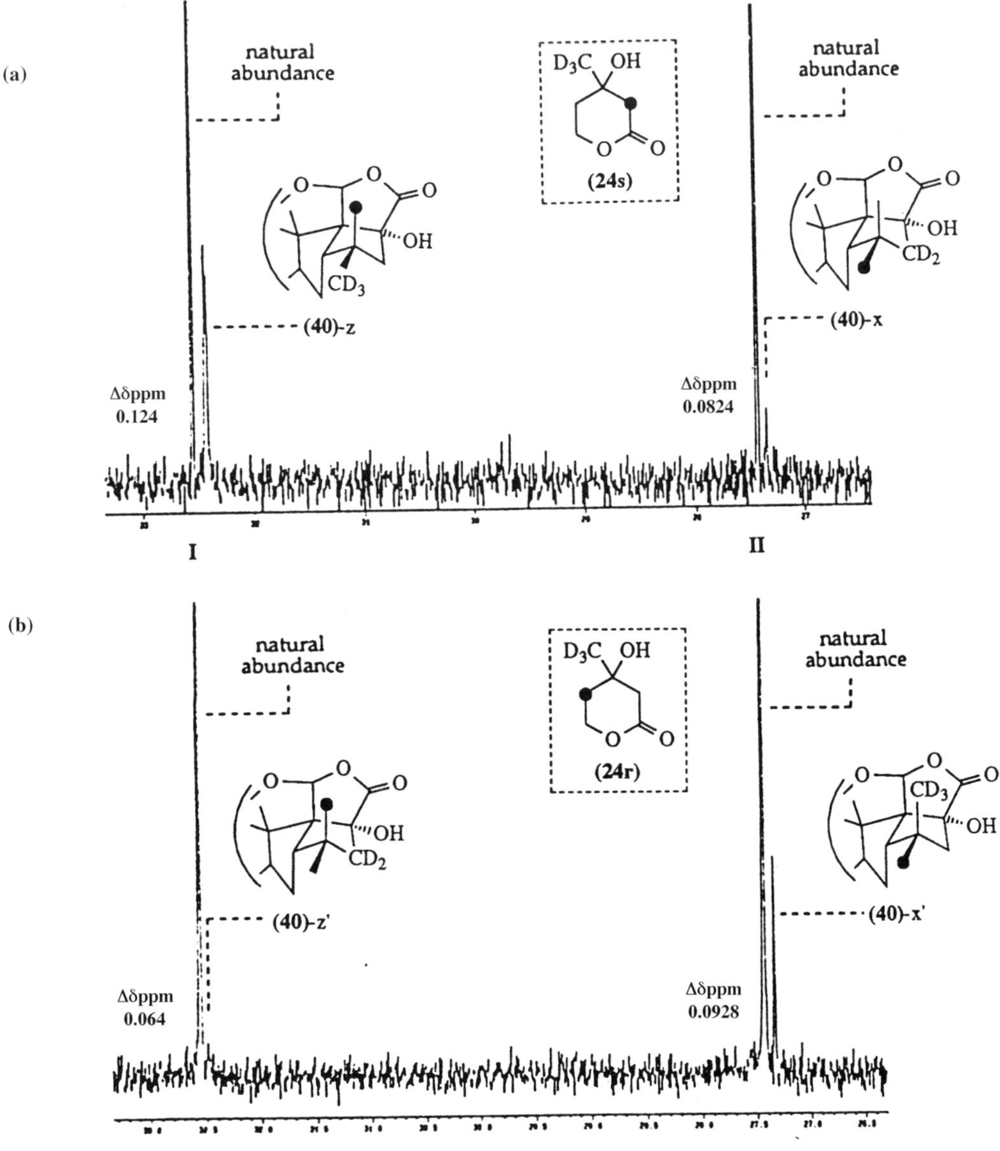

Figure 8 ^{13}C NMR spectra of photodehydro-gingkolide A samples (**40s**) (a) and (**40r**) (b) in pyridine-d$_5$. Signals for pseudoaxial (**I**) and pseudoequatorial (**II**) methyl group.

the biosynthesis. Since the available experimental data do not allow to pinpoint the stage at which this oxidation takes place, C-17 will be denoted as "R" in the sequel, where R can equal Me, CH_2OH, CHO, or COOH. Another necessary oxidative transformation is the cleavage of ring A, which has been extensively discussed in Section 2.14.3.2.1. Moreover, rings B and C need to be contracted to five-membered rings, probably in connection with oxygenations at different positions of the aromatic ring of (+)-(**38**). Finally, the original configuration at C-10 must be inverted, which can only be accomplished, given the quaternary nature of C-10, by temporarily breaking one of the bonds to either C-1, C-5, or C-9 and subsequently reforming it so that the required *cis*-relationship between C-20 and H-5 is established. The labelling pattern obtained after incorporation of [U-^{13}C]glucose (bond label between C-10 and C-20) rules out any scenarios in which C-20 is removed and another carbon reintroduced from the opposite face. Scheme 27 presents a plausible mechanistic hypothesis, which associates the inversion at C-10 with the contraction of ring B.

According to this model, (+)-(**38**) is first oxidized at C-12 to give ferruginol (**41**) a compound known to occur in many other plants. Further oxidation of (**41**) at the para position affords the

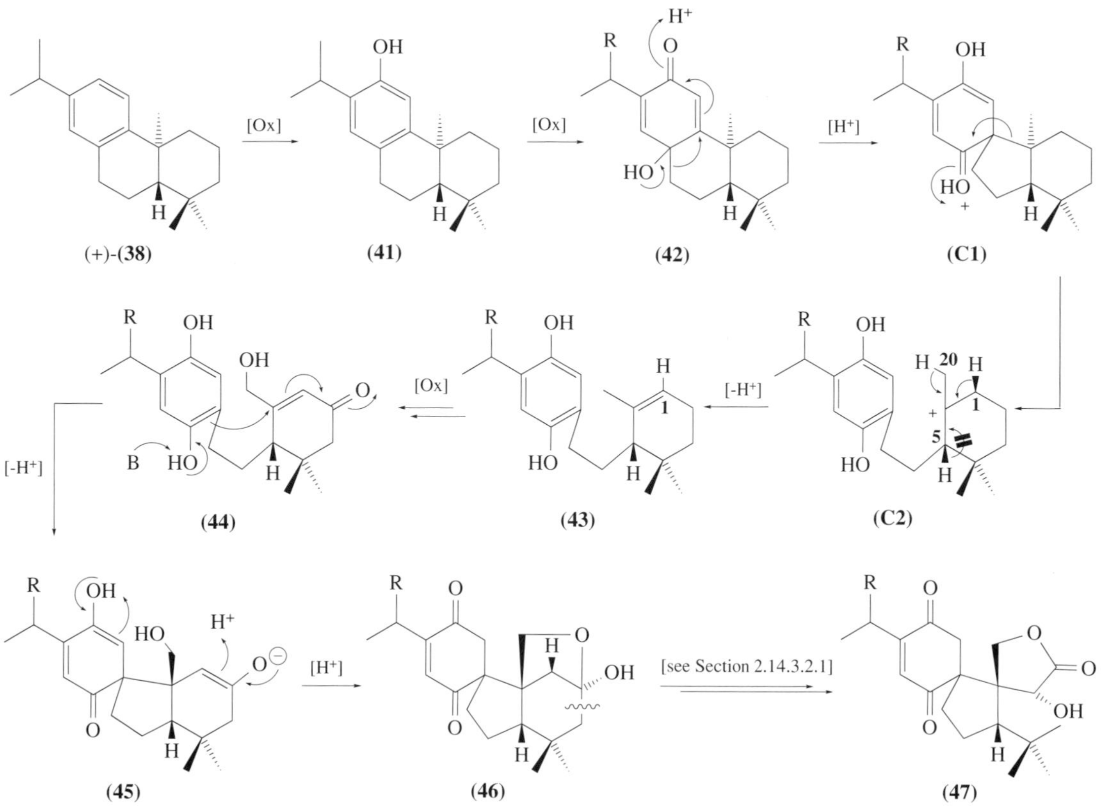

[R = Me, CH_2OH, CHO, CO_2H]

Scheme 27

ketol (**42**), which, under acid catalysis, is prone to undergo a dienone-phenol rearrangement affording the cation (**C1**). Among all possible follow-up reactions, that involving cleavage of the bond C-9/C-10 is the most interesting with regard to the required inversion at C-10 since it generates a carbocation (**C2**) with planar geometry at C-10. Nonenzymatic precedents for dienone–phenol rearrangements followed by cleavage of a carbon–carbon bond to regenerate aromaticity are reported in the literature for related systems.[65,66] The positive charge of (**C2**) can be saturated by elimination of either H-20 or H-1, but not H-5, as the latter has been shown to be conserved during ginkgolide biosynthesis. In the sequel, elimination of H-1 will be favored over elimination of H-20 because it could seamlessly explain the results from incorporation of (**24p**) and (**24q**), described in Section 2.14.3.2.1, that had revealed that one of the two C-1 hydrogens is replaced by a hydrogen atom from the medium during ginkgolide formation (see Scheme 23).

The ring-B-seco compound thus formed (**43**) could be oxidized in the two allylic positions of the double bond to give the enone (**44**), which is well suited for an intramolecular Michael addition affording again a tricyclic intermediate (**45**). Provided that the carbon nucleophile attacks the double bond from the opposite side with respect to the position of H-5, the overall inversion of the configuration at C-10, as well as the formation of a contracted ring B can be rationalized. Stereospecific incorporation of two protons from the medium (at C-1 and C-11, respectively) followed by (spontaneous?) formation of the cyclic hemiacetal moiety would then lead to the intermediate (**46**), which is eminently well suited for undergoing the cleavage of ring A illustrated in Scheme 22 (see Section 2.14.3.2.1). Resting in part on enzymatic and nonenzymatic precedents reported in the literature, the mechanistic hypothesis presented in Scheme 27 is strongly corroborated by several pieces of experimental evidence. For example, biomimetic studies with (**41**) established its chemical potential for transformation into ring-B-seco products such as (**43**)[53] (see Scheme 28).

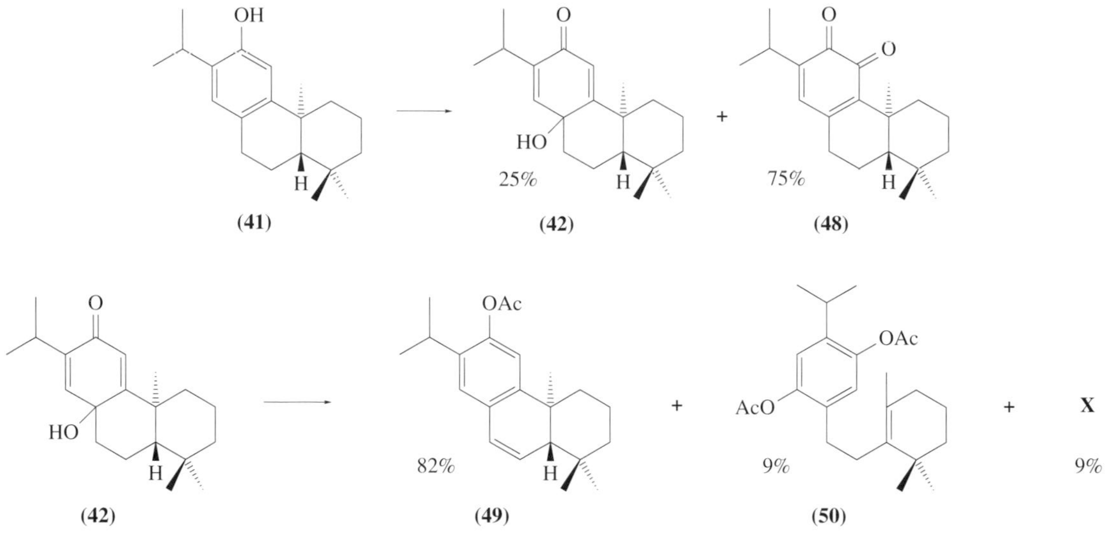

OH (41) → O HO 25% (42) + O O 75% (48)

O HO (42) → OAc 82% (49) + OAc AcO 9% (50) + X 9%

Scheme 28

Treatment of ferruginol (**41**) with benzeneseleninic anhydride[68] afforded the para-quinol (**42**) along with major quantities of the ortho-quinone (**48**). Subjecting the para-quinol (**42**) to the Thiele-conditions, (Ac_2O, trace H_2SO_4) furnished a mixture of one major product, 6,7-dehydroferruginol acetate (**49**), and two minor products, one of which was identified as the ring-B-seco diacetate (**50**) and the other as an isomeric tricyclic diacetate (**X**). The synthetic ring-B-seco product (**50**) differs from the postulated ring-B-seco intermediate (**43**) only in terms of the position of the isolated double bond. Although it is expected that the chemical route will afford the thermodynamically more stable product carrying a tetrasubstituted double bond (**50**), there is *a priori* no need to assume that the same has to apply to the enzymatic reaction as well.

Further substantiation of the hypothetical sequence shown in Scheme 27 was provided by the isolation of a new substance from the mother liquors of *Ginkgo* extracts, subsequently shown by X-ray analysis to have the structure (**51**).[69] A comparison of the structures shown in Scheme 29 reveals that (**51**), which can be interpreted as the product of a double elimination of water from (**52**), displays a striking similarity to the proposed end product (**47**) of the hypothetical reaction sequence of Scheme 27, differing from the latter only by the presence of two additional hydroxyl groups at C-20 and C-7 and in the oxidation state of C-12. The fact that the steric position of the oxygen function at C-7 of (**51**) is opposite to that observed in the ginkgolides is both puzzling and indicative. While ruled out as a direct intermediate of ginkgolide biosynthesis, (**51**) can be viewed as a shunt product due to an anomalous hydroxylation at C-7. This would imply that an immediate precursor of (**52**) with no hydroxyl group at C-7 (**53**) could constitute the missing link between the postulated product intermediate (**47**) and the corresponding hydroxylated product (**54**) with the correct configuration at C-7 (Scheme 30). Only a few redox steps and a well precedented rearrangement are required to convert (**54**) into ginkgolide A (**1**). Most other ginkgolides are accessible from (**1**) by hydroxylation at the appropriate positions.

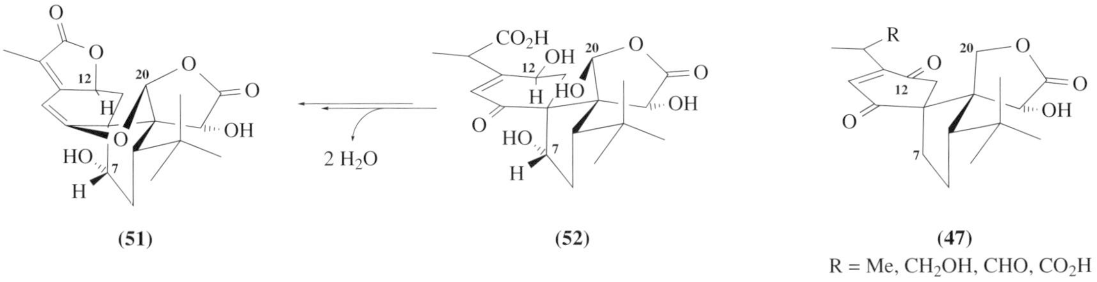

(51) ← $2\ H_2O$ (52) (47)

R = Me, CH_2OH, CHO, CO_2H

Scheme 29

(47) (53) (54)

(1) (55)

Scheme 30

2.14.4 SUMMARY

A few general thoughts may be allowed at the end of this chapter on ginkgolide biosynthesis. *G. biloba* has indeed lived up to its reputation of being an extraordinary creature deviating from the norm in many respects. Over and over again the authors found themselves confronted with unexpected and puzzling results while analyzing the data from a given incorporation experiment, yet never to the point of complete perplexity, as if the "living fossil" was condescendingly unveiling its age-old secrets bit by bit, while at the same time teaching the authors to be humble and unprejudiced. Undoubtedly, the *Ginkgo* embryo system (see Section 2.14.1.3) constituted the foundation for the success of the biosynthetic investigations, by rendering possible stable isotope tracer experiments. It remains to be seen whether the embryo methodology will be applicable to other plant systems as well and prove to have a similar impact on the progress of biosynthetic studies as in *G. biloba*. The results obtained have been integrated into a new scheme of ginkgolide biosynthesis (see Section 2.14.3) which is at variance with the major postulates of a former biogenetic hypothesis, such as the absolute configuration of the tricyclic intermediates or the events involved in the cleavage of ring A and the formation of the *t*-butyl group. Beyond the specific biosynthesis of the ginkgolides, the *Ginkgo* also afforded enlightening insights into the early steps of isoprenoid biosynthesis (Section 2.14.2).

2.14.5 REFERENCES

1. P. F. Michel, "Ginkgo biloba—l'arbre qui a vaincu le temps," L'Art du Vivant, Editions du Félin.
2. P. F. Michel and D. Hosford, in "Ginkgolides—Chemistry, Biology, Pharmacology, and Clinical Perspectives," ed. P. Braquet, J. R. Prous Science, Barcelona, 1988, vol. 1, p. 1.
3. P. Braquet, *Med. Res. Rev.*, 1991, **111**, 295.
4. W. Schwabe, "Tebonin—aus der W. Schwabe Ginkgo-Forschung," Dr. Willmar Schwabe GmbH, Karlsruhe, 1992, p. 7.
5. M. Maruyama, A. Terahara, Y. Nakadeira, M. C. Woods, Y. Takagi, and K. Nakanishi, *Tetrahedr. Lett.*, 1967, 299, 303, 309, 315.
6. N. Sakabe, S. Takada, and K. Okabe, *Chem. Commun.*, 1967, 259.
7. C. Roumestand, B. Perly, and P. Braquet, in "Ginkgolides—Chemistry, Biology, Pharmacology, and Clinical Perspectives," ed. P. Braquet, J. R. Prous Science, Barcelona, 1988, vol. 1, p. 49.
8. L. Dupont, in "Ginkgolides—Chemistry, Biology, Pharmacology, and Clinical Perspectives," ed. P. Braquet, J. R. Prous Science, Barcelona, 1988, vol. 1, p. 69.
9. K. Weinges, M. Hepp, and H. Jaggy, *Liebigs Ann. Chem.*, 1987, 521.
10. K. Nakanishi and K. Habaguchi, *J. Am. Chem. Soc.*, 1971, **93**, 3546.
11. M. Schwarz, Ph.D. Thesis, Eidgenössische Technische Hochschule, Zürich, 1994.
12. K. Saito and Z. Kasai, *Phytochem.*, 1969, **8**, 2177.
13. S. A. Brown and L. R. Wetter, *Progr. Phytochem.*, 1972, **3**, 1.

14. Y. Tomita and Y. Ikeshiro, *J. Chem. Soc. Chem. Commun.*, 1987, 1311.
15. E. Ball, *Amer. Journ. Bot.*, 1959, **46**, 130.
16. I. Rubinstein, L. J. Goad, A. D. H. Clague, and L. H. Mulheirn, *Phytochem.*, 1976, **15**, 195.
17. S. Seo, A. Uomori, Y. Yoshimura, K. Takeda, H. Seto, Y. Ebizuka, H. Noguchi, and U. Sankawa, *J. Chem. Soc. Perkin Trans. 1*, 1988, 2407.
18. D. E. Cane, T. Rossi, A. M. Tillmann, and J. P. Pachlatko, *J. Am. Chem. Soc.*, 1981, **103**, 1838.
19. S. T. J. Broers, Ph.D. Thesis, Eidgenössische Technische Hochschule, Zürich, 1994.
20. R. E. Hill, B. G. Sayer, and I. D. Spenser, *J. Am. Chem. Soc.*, 1989, **111**, 1916.
21. S. David, B. Estramareix, J.-C. Fischer, and M. Thérisod, *J. Am. Chem. Soc.*, 1981, **103**, 7341.
22. A. Cartayrade, Post-Doctoral Report, Eidgenössische Technische Hochschule, Zürich, 1994.
23. S. Eppacher, B.S. Thesis, Eidgenössische Technische Hochschule, Zürich, 1996.
24. G. Flesch and M. Rohmer, *Eur. J. Biochem.*, 1988, **175**, 405.
25. M. Rohmer, B. Sutter, and H. Sahm, *J. Chem. Soc. Chem. Commun.*, 1989, 1471.
26. M. Rohmer, M. Knani, P. Simonin, B. Sutter, and H. Sahm, *Biochem. J.*, 1993, **295**, 517.
27. M. Rohmer, M. Seemann, S. Horbach, S. Bringer-Meyer, and H. Sahm, *J. Am. Chem. Soc.* 1996, **118**, 2564.
28. G. Sprenger, U. Schörken, T. Weigert, S. Grolle, A. A. de Graaf, S. V. Taylor, T. P. Begley, S. Bringer-Meyer, and H. Sahm, *Proc. Natl. Acad. Sci. USA*, 1997, **94**, 12 857.
29. L. M. Lois, N. Campos, S. R. Putra, K. Danielsen, M. Rohmer, and A. Boronat, *Proc. Natl. Acad. Sci. USA*, 1998, **95**, 2105.
30. B. M. Lange, M. R. Wildung, D. McCaskill, and R. Croteau, *Proc. Natl. Acad. Sci. USA*, 1998, **95**, 2100.
31. T. W. Goodwin and E. I. Mercer, "The Control of Lipid Metabolism," Academic Press, London, 1963, p. 37.
32. B. Liedvogel, *J. Plant Physiol.*, 1986, **124**, 211.
33. J. C. Gray, *Adv. Bot. Res.*, 1987, **14**, 25.
34. T. J. Bach, *Plant Physiol. Biochem.*, 1987, **25**, 163.
35. H. Kleinig, *Annu. Rev. Plant Physiol. Plant Mol. Biol.*, 1989, **40**, 39.
36. D. Schultze-Siebert and G. Schultz, *Plant Physiol. Biochem.*, 1987, **25**, 145.
37. D. Schultze-Siebert and G. Schultz, *Plant Physiol.*, 1987, **84**, 1233.
38. K. Kreuz and H. Kleinig, *Planta*, 1981, **153**, 578.
39. K. Kreuz and H. Kleinig, *Eur. J. Biochem.*, 1984, **141**, 531.
40. F. Lütke-Brinkhaus, B. Liedvogel, and H. Kleinig, *Eur. J. Biochem.*, 1984, **141**, 537.
41. F. D. Moore and D. C. Shephard, *Protoplasma*, 1977, **92**, 167.
42. R. Bäuerle, F. Lütke-Brinkhaus, B. Ortmann, S. Berger, and H. Kleinig, *Planta*, 1990, **181**, 229.
43. T. J. Bach, *Rec. Adv. Phytochem.*, 1990, **24**, 1.
44. Y. Tomita, M. Annaka, and Y. Ikeshiro, *J. Chem. Soc. Chem. Commun.*, 1989, 108.
45. D. Arigoni, S. Sanger, C. Latzel, W. Eisenreich, A. Bacher, and M. H. Zenk, *Proc. Natl. Acad. Sci. USA*, 1997, **94**, 10 600.
46. H. Seto, H. Watanabe, and K. Furihata, *Tetrahedron Lett.* 1996, **37**, 7979.
47. I. Takahashi and K. Ogura, *J. Biochem.*, 1982, **92**, 1527.
48. O. Dogbo and B. Camara, *Biochim. Biophys. Acta*, 1987, **920**, 140.
49. K. Momose and H. Rudney, *J. Biol. Chem.*, 1972, **247**, 3930.
50. B. Laprebrand, A. Laferrière, A. Saint-Guily, and J. Walter, "2nd Symposium of the European Network on Plant Terpenoids", Strasbourg, 1994, Poster P6.
51. M. M. Ravn, R. M. Coates, R. Jetter, and R. B. Croteau, *Chem. Commun.* 1998, 21.
52. W. A. König, *J. High Res. Chrom.*, 1990, **13**, 328.
53. E. Neau and J. Walter, Internal report (Nestlé, Institut H. Beaufour, W. A. Schwabe Arzneimittel), 1994.
54. J. E. Graebe, *Annu. Rev. Plant. Physiol.*, 1987, **38**, 419.
55. K. Nakanishi, in "Ginkgolides—Chemistry, Biology, Pharmacology, and Clinical Perspectives," ed. P. Braquet, J. R. Prous Science, Barcelona, 1988, vol. 1, p. 27.
56. P. Dowd and W. Zhang, *Chem. Rev.*, 1993, **93**, 2091.
57. T. I. Macdonald and D. E. O'Dell, *J. Org. Chem.*, 1981, **46**, 1501.
58. L. J. Goad and T. W. Goodwin, *Progr. Phytochem.*, 1972, **3**, 113.
59. G. Popják and J. W. Cornforth, *Biochem. J.*, 1966, **101**, 553.
60. A. Eschenmoser, L. Ruzicka, O. Jeger, and D. Arigoni, *Helv. Chim. Acta*, 1955, **38**, 1890.
61. D. R. Coulson and N. C. Yang, *J. Am. Chem. Soc.*, 1966, **88**, 4511.
62. P. J. Wagner, *Acc. Chem. Res.*, 1971, **4**, 168.
63. F. D. Lewis, *J. Am. Chem. Soc.*, 1970, **92**, 5602.
64. A. Padwa and W. Bergmark, *Tetrahedron Lett.*, 1968, **55**, 5795.
65. P. J. Kropp, *J. Am. Chem. Soc.*, 1963, **85**, 3280.
66. A. K. Banerjee, M. C. Carrasco, and C. A. Peña, *Tetrahedron*, 1990, **46**, 4133.
67. M. Schwarz, Post-Doctoral Report, Eidgenössische Technische Hochschule, Zürich, 1996.
68. R. H. Burnell, M. Jean, and D. Poirier, *Can. J. Chem.*, 1987, **65**, 775.
69. H. Jaggy, Dr. Willmar Schwabe GmbH, Karlsruhe, personal communication, 1994.
70. S. Furukawa, *Sci. Papers Inst., Phys. Chem. Res. (Japan)*, 1932, **19**, 27.

Author Index

This Author Index comprises an alphabetical listing of the names of the authors cited in the text and the references listed at the end of each chapter in this volume.

Each entry consists of the author's name, followed by a list of numbers, for example

Templeton, J. L., 366, 385[233] (350, 366), 387[370] (363)

For each name, the page numbers for the citation in the reference list are given, followed by the reference number in superscript and the page number(s) in parentheses of where that reference is cited in the text. Where a name is referred to in text only, the page number of the citation appears with no superscript number. References cited in both the text and in the tables are included.

Although much effort has gone into eliminating inaccuracies resulting from the use of different combinations of initials by the same author, the use by some journals of only one initial, and different spellings of the same name as a result of the transliteration processes, the accuracy of some entries may have been affected by these factors.

Subject Index

PHILIP AND LESLEY ASLETT
Marlborough, Wiltshire, UK

Every effort has been made to index as comprehensively as possible, and to standardize the terms used in the index in line with the IUPAC Recommendations. In view of the diverse nature of the terminology employed by the different authors, the reader is advised to search for related entries under the appropriate headings.

The index entries are presented in letter-by-letter alphabetical sequence. Compounds are normally indexed under the parent compound name, with the substituent component separated by a comma of inversion. An entry with a prefix/locant is filed after the same entry without any attachments, and in alphanumerical sequence. For example, 'diazepines', '1,4-diazepines', and '2,3-dihydro-1,4-diazepines' will be filed as:-

diazepines
1,4-diazepines
1,4-diazepines, 2,3-dihydro-

The Index is arranged in set-out style, with a maximum of three levels of heading. Location references refer to volume number (in bold) and page number (separated by a comma); major coverage of a subject is indicated by bold, elided page numbers; for example;

triterpene cyclases, **299–320**
  amino acids, 315

See cross-references direct the user to the preferred term; for example,

olefins *see* alkenes

See also cross-references provide the user with guideposts to terms of related interest, from the broader term to the narrower term, and appear at the end of the main heading to which they refer, for example,

thiones
  see also thioketones

WITHDRAWAL